Foreword

Fungi represent the fifth kingdom of organisms, which is characterized—second only to prokaryotes—by a huge number of diverse species. Even more, fungi have developed a tremendous variety in lifestyles, biochemical properties, and morphological characters, the latter having been a permanent challenge for defining species and their identification. They have conquered practically all habitats, from deep sea water to desert soil, and from prokaryotes to mammals, leading to an array of positive but also negative impacts on mankind. On the negative side, fungi are known as pathogens of plants—a situation which seriously affects crop plantations all around the earth—but also of higher fungi, of lower eukaryotes, and of all animals up to mammals and men. Also, their versatile metabolism provided them with efficient abilities to colonize almost all material, leading to biodeterioration of various organic materials including paintings and covers, which allowed them to settle in buildings and flats resulting in indoor contamination as a major problem of today. Yet there are also numerous benefits: many fungi are known as beneficial symbionts of plants, such as plant tissue endosymbionts and mycorrhizas. In fact, the earth would be devoid of plants in the absence of the latter. Finally several fungi have been domesticated by humans, either for their use in agriculture (such as for biocontrol of plant or invertebrate pathogens or in plant growth protection and stimulation), for the preparation of feed- and foodstuff, and as efficient producers of biotechnological products such as primary metabolites, numerous enzymes, and antibiotics. In the area of modern molecular biotechnology, fungi such as *Pichia pastoris* have become important high-throughput hosts for the production of recombinant proteins of bacterial to human origin. Last but not least, fungi like *Saccharomyces cerevisiae*, *Neurospora crassa*, and *Aspergillus nidulans* have become model systems for basic biochemical and genetic research, and an impressing amount of our textbook knowledge would not be available without them. In the current genomic age, elucidation of the genome inventory of about 50 multicellular asco- and basidiomycetes and the same number of yeasts has been completed and opened new avenues for their investigation.

In view of this steadily increased interest in fungi, also the methods needed for their isolation and identification, as well as their genetic manipulation and monitoring of gene expression and protein production, have become refined and complemented. This book aims at presenting an inventory of techniques and methods that are currently in use for studying fungi: it contains 57 chapters dedicated to description of these techniques, starting from concepts of cultivation, enumeration, and visualization of fungi; molecular approaches for detection and quantification; measurement of relevant enzymes and methods for their application; and the use of bioinformatic tools to investigate fungal genomes.

As a professional reference, this book is aimed at all people who work with fungi and should be useful both to academic institutions and research teams, as well as to teachers, graduate and postgraduate students.

Vienna, Austria   Prof. Christian P. Kubicek

Foreword

It gives me immense pleasure to write a foreword for *Laboratory Protocols in Fungal Biology* of Springer, USA edited by Dr. Vijai Kumar Gupta and Dr. Maria G. Tuohy. After going through the content of this laboratory protocol, I feel that it is a wonderful attempt done by Dr. Gupta to compile together all the information about the subject that will be highly useful to all mycologists around the globe. I am sure that this volume will be highly useful to all those concerned with fungi and their biology, including environmental and public health officers and professionals in the field of interest. The volume is really exhaustive covering almost all the aspects of fungal biology. It will also be of interest to postgraduate students in this field and also for one and all interested in Fungi. Additionally it will be of great market value. This effort of Dr. Gupta's is admirable.

Varanasi, India   Prof. R.S. Upadhyay

Preface

The interaction between fungi and their environment is central to many natural processes that occur in the biosphere. The hosts and habitats of these eukaryotic microorganisms are very diverse; fungi are present in every ecosystem on Earth. The fungal kingdom is equally diverse, consisting of seven different known phyla. Yet detailed knowledge is limited to relatively few species. The relationship between fungi and humans has been characterized by the juxtaposed viewpoints of fungi as infectious agents of much dread and their exploitation as highly versatile systems for a range of economically important biotechnological applications. Understanding the biology of different fungi in diverse ecosystems as well as their interactions with living and nonliving is essential to underpin effective and innovative technological developments.

The tools and techniques of molecular biology, once reserved for mammalian and bacterial systems, have been adapted and optimized for the analysis of fungal species at the molecular level. Rapid screening techniques based on screening specific regions in the DNA of fungi have been used in species comparison and identification and are now being extended across fungal phyla with the ultimate goal being the assembly of the "Fungal Tree of Life" by the US National Science Foundation. Within a decade after the Human Genome Sequence was published, genome sequencing technology has been adapted to yield the complete genome sequences of not only fungi of commerce and medical relevance, but other more isoteric species. Post-genomics approaches and systems biology are now also being applied to understanding the details of fungal biology and the interactions between fungi, their hosts, and their environment. The majority of fungi are multicellular eukaryotic systems and therefore may be excellent model systems by which to answer fundamental biological questions. A greater understanding of the cell biology of these versatile eukaryotes will underpin efforts to engineer (e.g., "humanize") certain fungal species to provide novel cell factories for production of proteins for pharmaceutical applications. Finally, renewed interest in all aspects of the biology and biotechnology of fungi may also enable the development of "one pot" microbial cell factories to meet consumer energy needs into the twenty first century. To realize this potential and to truly understand the diversity and biology of these eukaryotes, continued development of scientific tools and techniques is essential.

This publication aims to provide a detailed compendium of analytical methods used to investigate different aspects of mycology, including fungal biology and biochemistry, genetics, phylogenetics, genomics, proteomics, molecular enzymology, and biotechnological applications, in a manner that reflects the many recent developments of relevance to scientists investigating the Kingdom of Fungi.

Galway, Ireland

Vijai Kumar Gupta
Maria G. Tuohy
Manimaran Ayyachamy
Anthonia O'Donovan
Kevin M. Turner

Contents

Contributors

M.Z. Abdin Department of Biotechnology, Jamia Hamdard University, New Delhi, Delhi, India

Malik M. Ahmad Department of Biotechnology, Jamia Hamdard University, New Delhi, Delhi, India

Pravej Alam Department of Biotechnology, Jamia Hamdard University, New Delhi, Delhi, India

Eduardo Alves Department of Phytopathology, Federal University of Lavras, Lavras, Minas Gerais, Brazil

David L. Andrews Department of Plant Pathology, University of Georgia, Athens, GA, USA

Manimaran Ayyachamy Department of Biochemistry, School of Natural Sciences, National University of Ireland, Galway, Ireland

Lourdes Baeza-Montañez Instituto de Hortofruticultura Subtropical y Mediterránea "La Mayora", Consejo Superior de Investigaciones Científicas (IHSM-UMA-CSIC), Estación Experimental "La Mayora", Algarrobo-Costa, Málaga, Spain

R. Bagyalakshmi Sankara Nethralaya, Larsen and Toubro Microbiology Research Centre, Chennai, Tamil Nadu, India

Paramjit K. Bajwa School of Environmental Sciences, University of Guelph, Guelph, ON, Canada

Eugenia Barros Department of Biosciences, Council for Scientific and Industrial Research (CSIR), Brummeria, Pretoria, South Africa

Gaurav Bhavsar Department of Oncology, University College London Cancer Institute, London, UK

Jonas Blomberg Department of Medical Sciences, Uppsala Academic Hospital, Uppsala University, Uppsala, Sweden

Priscilla Braglia Sir William Dunn School of Pathology, University of Oxford, Oxford, UK

Dieter Buchheidt Third Department of Internal Medicine, Mannheim University Hospital, Mannheim, Germany

Virginia Casado Department of Microbiologia y Genetica—CIALE, Universidad de Salamanca, Salamanca, Spain

Kerry Chester Department of Oncology, University College London Cancer Institute, London, UK

Nadezhda I. Chigineva All-Russian Collection of Microorganisms (VKM IBPM RAS, Pushchino, Russia), G. K. Skryabin Institute of Biochemistry and Physiology of Microorganisms, Russian Academy of Science, Pushchino, Moscow Region, Russia

Yangrae Cho Department of Plant and Environmental Protection Sciences, University of Hawaii at Manoa, Honolulu, HI, USA

Finola E. Cliffe Department of Biochemistry, School of Natural Sciences, National University of Ireland Galway, Galway, Ireland

Rebecca Creamer Department of Entomology, Plant Pathology, and Weed Science, New Mexico State University, Las Cruces, NM, USA

Tanya E.S. Dahms Department of Chemistry and Biochemistry, University of Regina, Regina, SK, Canada

Marcelo de Carvalho Alves Department of Soil and Rural Engineering, Campus of the Federal University of Mato Grosso, Federal University of Mato Grosso, Cuiaba, Mato Grosso, Brazil

José J. de Vega-Bartol Department of Microbiologia y Genetica—CIALE, Universidad de Salamanca, Salamanca, Spain

José M. Díaz-Mínguez Department of Microbiologia y Genetica—CIALE, Centro Hispano Luso de Investigaciones Agrarias, Universidad de Salamanca, Salamanca, Spain

Svetlana S. Eremina All-Russian Collection of Microorganisms (VKM IBPM RAS, Pushchino, Russia), G. K. Skryabin Institute of Biochemistry and Physiology of Microorganisms, Russian Academy of Science, Pushchino, Moscow Region, Russia

Ronnie Eriksson Livsmedelsverket, Uppsala, Sweden

Vitaly Erukhimovitch Analytical Equipment Unit, Ben-Gurion University of the Negev, Beer-Sheva, Israel

Raquel González Fernández Department of Biochemistry and Molecular Biology, University of Córdoba, Córdoba, Spain

Bride Foster Department of Oncology, University College London Cancer Institute, London, UK

María D. García-Pedrajas Instituto de Hortofruticultura Subtropical y Mediterránea "La Mayora", Consejo Superior de Investigaciones Científicas (IHSM-UMA-CSIC), Estación Experimental "La Mayora", Málaga, Spain

Gagan Garg Department of Chemistry and Biomolecular Sciences, Macquarie University, Sydney, NSW, Australia

Rajeeva Gaur Department of Microbiology, Dr. R.M.L. Avadh University, Faizabad, Uttar Pradesh, India

Christos Georgiou Department of Biology, University of Patras, Patras, Achaia, Greece

Roberto A. Geremia Laboratoire d'Ecologie Alpine, CNRS/UJF, Université Joseph Fourier, Grenoble, France

Mélanie Gerphagnon Université Blaise Pascal, Aubière, France

Bianca Gielesen DSM Biotechnology Center, Delft, Zuid Holland, The Netherlands

Annie Juliet Gnanam College of Natural Science, Institute for Cellular and Molecular Biology, University of Texas at Austin, Austin, TX, USA

Scott E. Gold United States Department of Agriculture—Agricultural Research Unit (USDA–ARS), Toxicology and Mycotoxin Research Unit, Athens Georgia, USA

Konstantinos Grintzalis Department of Biology, University of Patras, Patras, Achaia, Greece

Vijai Kumar Gupta Molecular Glycobiotechnology Group, Department of Biochemistry, School of Natural Sciences, National University of Ireland Galway, Galway, Ireland

Assistant Professor of Biotechnology, Department of Science, Faculty of Arts, Science & Commerce, MITS University, Rajasthan, India

Marc B. Habash School of Environmental Sciences, University of Guelph, Guelph, ON, Canada

Richard C. Hamelin Department of Forest Sciences, The University of British Columbia, Vancouver, BC, Canada

Laurentian Forestry Centre, Natural Resources Canada, Quebec, QC, Canada

Janelle M. Hare Department of Biology and Chemistry, Morehead State University, KY, USA

Nicole K. Harner School of Environmental Sciences, University of Guelph, Guelph, ON, Canada

Ladislav Homolka Department of Ecology of Microorganisms, Institute of Microbiology, Academy of Sciences of the Czech Republic, Prague, Czech Republic

Mahmoud Huleihel Department of Virology and Developmental Genetics, Ben-Gurion University of the Negev, Beer-Sheva, Israel

Natalya E. Ivanushkina All-Russian Collection of Microorganisms (VKM IBPM RAS, Pushchino, Russia), G. K. Skryabin Institute of Biochemistry and Physiology of Microorganisms, Russian Academy of Science, Pushchino, Moscow Region, Russia

Saleem Javed Department of Biochemistry, Jamia Hamdard University, New Delhi, India

V.K. Jayaraman Scientific and Engineering Computing Group (SECG), Centre for Development of Advanced Computing (C-DAC), University of Pune, Pune, Maharashtra, India

Marlène Jobard LMGE UMR CNRS, U.F.R. Sciences et Technologies, Aubière Cedex, France

Magnus Jobs School of Health and Social Studies, Högskolan Dalarna, Uppsala University, Falun, Sweden

Bernhard Kluger Department for Agrobiotechnology (IFA-Tulln), University of Natural Resources and Life Sciences Vienna, Tulln, Austria

Galina A. Kochkina All-Russian Collection of Microorganisms (VKM IBPM RAS, Pushchino, Russia), G. K. Skryabin Institute of Biochemistry and Physiology of Microorganisms, Russian Academy of Science, Pushchino, Moscow Region, Russia

Christian P. Kubicek Department of Chemical Engineering, Vienna University of Technology, Vienna, Austria

Ellen L. Lagendijk Department of Molecular Microbiology and Biotechnology, Leiden University, Leiden, The Netherlands

Hung Lee University of Guelph, School of Environmental Sciences, Guelph, ON, Canada

De-Wei Li Valley Laboratory, The Connecticut Agricultural Experiment Station, Windsor, CT, USA

Jose L. Lopez-Ribot Department of Biology, South Texas Center for Emerging Infectious Diseases, The University of Texas at San Antonio, San Antonio, TX, USA

Gilvaine Ciavareli Lucas Department of Phytopathology, Federal University of Lavras, Lavras, Minas Gerais, Brazil

Departamento de Fitopatologia, Universidade Federal de Lavras, Caixa postal, Lavras, Minas Gerais, Brazil

Gengkon Lum Department of Computer Science and Information Systems, Youngstown State University, Youngstown, OH, USA

Hui Ma Department of Chemistry, National University of Singapore, Singapore

Alexandra McAleenan Clinical Sciences Centre, Imperial College London, London, UK

H.N. Madhavan Sankara Nethralaya, Larsen and Toubro Microbiology Research Centre, Chennai, Tamil Nadu, India

Cathal S. Mahon Department of Pharmaceutical Chemistry, University of California—San Francisco, San Francisco, CA, USA

Minna Mäki Program Leader, NAT, Orion Diagnostica Oy, Espoo, Finland

P.T. Manoharan Department of Botany, Vivekananda College, Madurai, Tamil Nadu, India

Segula Masaphy Department of Applied Microbiology and Mycology, MIGAL, Kiryat Shmona, Israel

Maria D. Mayan Fundación CHUAC, Biomedical Research Center—INIBIC, A Coruña, Galicia, Spain

Vera Meyer Department of Applied and Molecular Microbiology, Berlin University of Technology, Berlin, Germany

Xiang Jia Min Department of Biological Sciences, Center for Applied Chemical Biology, Youngstown State University, Youngstown, OH, USA

Sonal Modak Bioinformatics Centre, University of Pune, Pune, Maharashtra, India

C.N. Mortensen Department of Agriculture and Ecology, University of Copenhagen, Copenhagan, Taastrup, Denmark

Dirk Mueller-Hagen Department of Applied and Molecular Microbiology, Technische Universität Berlin, Berlin, Germany

Suman Mukherjee Laboratory of Biochemistry and Genetics, NIDDK, National Institutes of Health, Bethesda, MD, USA

Susann Müller Department of Environmental Microbiology, Helmholtz Centre for Environmental Research—UFZ, Leipzig, Saxonia, Germany

S. Chandra Nayaka Department of Studies in Biotechnology, Asian Seed Health Centre, University of Mysore, Mysore, Karnataka, India

Christopher Nguyen Department of Plant and Environmental Protection Sciences, University of Hawaii at Manoa, Honolulu, HI, USA

Jonathan Niño Department of Microbiologia y Genetica—CIALE, Universidad de Salamanca, Villamayor, Salamanca, Spain

S.R. Niranjana Department of Studies in Biotechnology, University of Mysore, Mysore, Karnataka, India

Jesús V. Jorrín Novo Department of Biochemistry and Molecular Biology, University of Córdoba, Córdoba, Spain

Anthony J. O'Donoghue Department of Pharmaceutical Chemistry, University of California—San Francisco, San Francisco, CA, USA

Anthonia O'Donovan Discipline of Biochemistry, School of Natural Sciences, National University of Ireland, Galway, Ireland

Mary C. O'Loughlin Department of Life Sciences, University of Limerick, Castletroy, Limerick, Ireland

Miruna Oros-Sichler Institute for Epidemiology and Pathogen Diagnostics, Julius Kühn Institut, Braunschweig, Lower Saxony, Germany

Jean-Paul Ouedraogo Department Applied and Molecular Microbiology, Institute of Biotechnology, Berlin University of Technology, Berlin, Germany

Svetlana M. Ozerskaya All-Russian Collection of Microorganisms (VKM IBPM RAS, Pushchino, Russia), G. K. Skryabin Institute of Biochemistry and Physiology of Microorganisms, Russian Academy of Science, Pushchino, Moscow Region, Russia

Brejesh Kumar Pandey Molecular Plant Pathology Laboratory, Central Institute for Subtropical Horticulture, Indian Council of Agricultural Research, Lucknow, Uttar Pradesh, India

M. Pandi Department of Molecular Microbiology, School of Biotechnology, Madurai Kamaraj University, Madurai, Tamil Nadu, India

Ioannis Papapostolou Department of Biology, University of Patras, Patras, Achaia, Greece

Biplab C. Paul Department of Chemistry and Biochemistry, University of Regina, Regina, SK, Canada

Zahi Paz Department of Plant Pathology, University of Georgia, Athens, GA, USA

Pilar Pérez Departamento de Microbiología CSIC/Universidad de Salamanca, Instituto de Biología Funcional y Genómica (IBFG), Salamanca, Spain

Kugen Permaul Department of Biotechnology and Food Technology, Durban University of Technology, Durban, Kwa-Zulu-Natal, South Africa

Christopher G. Pierce Department of Biology, South Texas Center for Emerging Infectious Diseases, The University of Texas at San Antonio, San Antonio, TX, USA

Edson Ampélio Pozza Departamento de Fitopatologia, Universidade Federal de Lavras, Caixa postal, Lavras, Minas Gerais, Brazil

Prashant Prabhakar Department of Biotechnology, Dr. D.Y. Patil University, Pune, Maharashtra, India

H.S. Prakash Department of Studies in Biotechnology, Asian Seed Health Centre, University of Mysore, Mysore, Karnataka, India

P. Rajapriya Department of Microbiology, Srinivasan College of Arts and Science, Perambalur, Tamil Nadu, India

Arthur F.J. Ram Department of Molecular Microbiology and Biotechnology, Leiden University, Leiden, BE, The Netherlands

M. Venkata Ramana Department of Studies in Microbiology, University of Mysore, Mysore, Karnataka, India

Shoba Ranganathan Department of Chemistry and Biomolecular Sciences, Macquarie University, Sydney, NSW, Australia

Department of Biochemistry, Yong Loo Lin School of Medicine, National University of Singapore, Singapore, Singapore

Serena Rasconi Department of Biology, University of Oslo, Oslo, Norway

Juan C. Ribas Departamento de Microbiología CSIC/Universidad de Salamanca, Senior Scientist from the Spanish Research Council (Consejo Superior de Investigaciones Científicas, CSIC), Instituto de Biología Funcional y Genómica (IBFG), Salamanca, Spain

Terri L. Richardson School of Environmental Sciences, University of Guelph, Guelph, ON, Canada

Jeyabalan Sangeetha Department of Zoology , Karnataka University, 580003, Dharwad, Karnataka, India

Thomas Scheper Chip Technology Institute for Technical Chemistry, University of Hannover, Hannover, Lower Saxony, Germany

Jochen Schmid Department of Chemistry of Biogenic Resources, Technische Universität München, Straubing, Bavaria, Germany

Cor D. Schoen Department of Bio-Interactions and Plant Health, Plant Research International B. V, Wageningen, The Netherlands

Denise Schöfbeck Department for Agrobiotechnology (IFA-Tulln), University of Natural Resources and Life Sciences Vienna, Tulln, Austria

Rainer Schuhmacher Department for Agrobiotechnology (IFA-Tulln), University of Natural Resources and Life Sciences Vienna, Tulln, Austria

Shimantika Sharma Department of Biotechnology, Dr. D.Y. Patil University, Pune, Maharashtra, India

Mary Shier Department of Biochemistry, National University of Ireland, Galway, Ireland

Sukhdeep Sidhu School of Environmental Sciences, University of Guelph, Guelph, ON, Canada

Volker Sieber Chemistry of Biogenic Resources, Technische Universität München, Straubing, Bavaria, Germany

Télesphore Sime-Ngando UMR CNRS 6023, Université Blaise Pascal, Clermont II, Aubière, Cedex, France

Suren Singh Department of Biotechnology and Food Technology, Durban University of Technology, Durban, Kwa-Zulu-Natal, South Africa

Kornelia Smalla Julius Kühn Institut, Federal Research Centre for Cultivated Plants, Institute for Epidemiology and Pathogen Diagnostics, Braunschweig, Lower Saxony, Germany

Laelie A. Snook Department of Human Health and Nutritional Sciences, Guelph, Ontario, Canada

Birgit Spiess Third Department of Internal Medicine, Mannheim University Hospital, Mannheim, Germany

Akhil Srivastava Department of Plant and Environmental Protection Sciences, University of Hawaii at Manoa, Honolulu, HI, USA

Frank Stahl Chip Technology Institute for Technical Chemistry, University of Hannover, Hannover, Germany

Dawn Elizabeth Stephens Department of Biotechnology and Food Technology, Durban University of Technology, Durban, Kwa-Zulu-Natal, South Africa

Vega Tello Department of Microbiologia y Genetica—CIALE, Universidad de Salamanca, Salamanca, Spain

Devarajan Thangadurai Department of Botany, Karnataka University, Dharwad, Karnataka, India

K. Lily Therese Sankara Nethralaya, Larsen and Toubro Microbiology Research Centre, Vision Research Foundation, Chennai, Tamil Nadu, India

Berend Tolner Department of Oncology, University College London Cancer Institute, London, UK

Jack T. Trevors School of Environmental Sciences, University of Guelph, Guelph, ON, Canada

Clement K.M. Tsui Department of Forest Sciences, The University of British Columbia, Vancouver, BC, Canada

Maria G. Tuohy Department of Biochemistry, School of Natural Sciences, National University of Ireland, Galway, Ireland

Katherine D. Turner School of Natural Sciences, Centre for Chromosome Biology, National University of Ireland Galway, Galway, Ireland

Kevin M. Turner Manufacturing Sciences and Technology, Pfizer Ireland Pharmaceuticals, The Pfizer Biotech Campus at Grange Castle, Dublin, Ireland

A.C. Udayashankar Department of Studies in Biotechnology, Asian Seed Health Centre, University of Mysore, Mysore, Karnataka, India

R.S. Upadhyay Department of Botany, Centre of Advanced Study, Banaras Hindu University, Varanasi, Uttar Pradesh, India

Priya Uppuluri Department of Biology, South Texas Center for Emerging Infectious Diseases, The University of Texas at San Antonio, San Antonio, TX, USA

Marco van den Berg Applied Biochemistry and Screening, DSM Biotechnology Center, Delft, Zuid-Holland, The Netherlands

Cees A.M.J.J. van den Hondel Department of Molecular Microbiology and Biotechnology, Leiden University, Leiden, BE, The Netherlands

Alexander N. Vasilenko All-Russian Collection of Microorganisms (VKM IBPM RAS, Pushchino, Russia), G. K. Skryabin Institute of Biochemistry and Physiology of Microorganisms, Russian Academy of Science, Pushchino, Moscow Region, Russia

Kim Vigor Department of Oncology, University College London Cancer Institute, London, UK

Jeannette Vogt Department of Environmental Microbiology, Helmholtz Centre for Environmental Research—UFZ, Leipzig, Saxonia, Germany

Bin Wang Westmead Hospital, Centre of Virus Research, Westmead Millennium Institute, University of Sydney, Westmead, NSW, Australia

Gerlinde Wiesenberger Institute of Applied Genetics and Cell Biology, University of Natural Resources and Life Sciences Vienna, Tulln, Austria

Akshay Yadav Scientific and Engineering Computing Group (SECG), Centre for Development of Advanced Computing (C-DAC), University of Pune, Pune, Maharashtra, India

Naomichi Yamamoto Department of Environmental Health, Graduate School of Public Health, Seoul National University, 1 Gwanak-ro, Gwanak-gu, Seoul, Korea

Susanne Zeilinger Research Area Gene Technology and Applied Biochemistry, Institute for Chemical Engineering, Vienna University of Technology, Vienna, Austria

Shaobin Zhong Department of Plant Pathology, North Dakota State University, Fargo, ND, USA

Lucie Zinger Laboratoire d'Ecologie Alpine, CNRS/UJF, Université Joseph Fourier, Grenoble, France

Safety Norms and Regulations in Handling Fungal Specimens

1

Finola E. Cliffe

Abstract

This chapter provides basic safety information required when handling fungal cultures and when performing the procedures outlined in this manual. Several topics are discussed including routine precautions when working with fungal organisms.

Keywords

Fungi • Mycology • Health and safety • Biosafety • Biosafety levels

Introduction

Biosafety measures designed to ensure the safety of laboratory workers include the use of various primary and secondary barriers, many of which are due to the advent of new technologies in the fields of materials science and engineering. Personnel undertaking the protocols in this manual may come across potentially hazardous materials such as pathogenic and infectious biological fungal agents, in addition to toxic chemicals and carcinogenic, mutagenic, or teratogenic reagents. In the case of fungal specimens, it has long been acknowledged that laboratory workers can attain infections from the agents they work with.

There have been many reported cases of laboratory-acquired infection, with countless more cases undoubtedly left unreported. Inhalation appears to be the most prominent route of exposure. Fungal hyphae in nature develop structures such as conidia on fruiting bodies or hyphal elements that develop into transmissible subsegments, which are ultimately designed for optimum dispersal in air. These elements are designed to be readily discharged, resistant to desiccation, and to remain aloft for long periods of time. Once inhaled by a host, the conidia develop into the yeast phase and can be found in the tissue of infected hosts [1]. Even with the advances in biosafety training and education, laboratory-acquired

F.E. Cliffe (✉)
Department of Biochemistry,
Molecular Glycobiotechnology Group, School of Natural Sciences, National University of Ireland Galway,
University Road, Galway, Ireland
e-mail: fcliffe@gmail.com

V.K. Gupta et al. (eds.), *Laboratory Protocols in Fungal Biology: Current Methods in Fungal Biology*,
Fungal Biology, DOI 10.1007/978-1-4614-2356-0_1, © Springer Science+Business Media, LLC 2013

fungal infections continue to occur. The dimorphic fungi *Blastomyces dermatitidis*, *Coccidioides immitis*, and *Histoplamsa caspsulatum*, for example, were found to be responsible for the majority of laboratory-acquired fungal infections in the United States [2–4]. Laboratory-associated pulmonary infections have occurred following the inhalation of conidia from mold-form cultures of *B. dermatitidis* [5, 6] and local infections from the accidental parenteral inoculation with infected tissues or cultures containing yeast forms of the fungus [7, 8] have been documented. Various reports of laboratory-associated *C. immitis* are reported in the literature prior to 1980 [9–11] including a case recorded by Nabarro [12] where a biochemist developed an intense acute infection after working with a colonial growth. Laboratory-associated histoplasmosis occurs mainly through inhalation of conidia produced by the mold form of the fungus [4, 13]; however, cutaneous infections have occurred due to accidental inoculation [14, 15].

These incidences indicate the ongoing occurrences of laboratory-acquired infections as a result of simple and preventable laboratory errors. As mentioned, the bulk of laboratory-acquired fungal infections are caused by inhalation of infectious conidia from the mold form, resulting in pulmonary infections; for example, the simple processes of opening of a culture plate lid can result in the release of large numbers of conidia [16]. To reduce the risk of infection it is practical to handle all fungal cultures under the conditions of biosafety laboratory containment BSL-2 or BSL-3 [17].

New biosafety technologies and associated guidelines have been developed to considerably improve ways to safely use fungal material. An enhanced understanding of the risks associated with various manipulations of many agents transmissible by different routes has enabled laboratory workers to apply appropriate biosafety practices to specific laboratory areas. These safety guidelines include engineering controls, management policies, work practices and procedures, as well as medical interventions. However, users must always progress with the caution

associated with good laboratory practice, under the supervision of personnel responsible for implementing laboratory safety programs at their institutions.

Biosafety Levels

Several biosafety levels (BSL) have been developed for laboratories to provide increasing levels of staff and environmental protection. BSLs are guidelines that describe appropriate containment equipment, facilities, and procedures for use by laboratory workers. The BSLs range from biosafety level 1 (BSL 1) to biosafety level 4 (BSL 4), and each BSL is based on the increased risk associated with the pathogenicity of the microorganisms encountered. Most clinical microbiology laboratories follow BSL 2 practices. When working with highly infectious agents for which the risk of aerosol transmission is greater, laboratories should follow BSL 3 practices.

BSL-1 is suitable for working with fungal agents that are not known to cause disease in healthy humans. BSL-1 practices, safety equipment, and facility design and construction are appropriate for undergraduate and secondary educational training and teaching laboratories, and for other laboratories in which work is done with defined and characterized strains of viable microorganisms not known to consistently cause disease in healthy adult humans. It is important to remember, however, that many agents not ordinarily associated with disease processes in humans are opportunistic pathogens and may cause infection in the young, the aged, and immunodeficient or immunosuppressed individuals. BSL-1 represents a basic level of containment that relies on standard microbiological practices with no special primary or secondary barriers recommended, other than a sink for hand washing.

BSL-2 should be used for work involving fungal agents that pose a moderate potential hazard to laboratory workers. These agents include the large group of opportunistic fungal pathogens such as *Aspergillus* spp. and *Fusarium* spp. Some protocols can be carried out on an open bench providing the potential for aerosol production is low [17]. Although organisms regularly employed at Biosafety Level 2 are not known to be transmissible by the aerosol route, procedures with aerosol or high splash potential that may increase the risk of such personnel exposure must be conducted in primary containment equipment, or in devices such as a biological safety cabinet (BSC) or safety centrifuge cups. Personal protective equipment (PPE), such as splash shields, face protection, gowns, and gloves should be used as appropriate. In addition, secondary barriers such as hand-washing sinks and waste decontamination facilities must be accessible to decrease the chance of environmental contamination [18].

BSL-3 is appropriate for work with infectious agents, which may cause serious or potentially lethal diseases as a result of inhalation. The fungal pathogens *C. immitis* and *H. capsulatum* fall into this group. Autoinoculation, ingestion, and exposure to infectious aerosols are the main hazards to personnel working with these organisms. All laboratory operations should be performed in a BSC or other enclosed apparatus, such as a gastight aerosol generation chamber. Secondary barriers for this level include controlled access to the laboratory and ventilation requirements that minimize the release of infectious aerosols from the laboratory. Within this level, primary and secondary barriers to protect personnel in contiguous areas, the community, and the environment from exposure to potentially infectious aerosols have been highlighted [18].

At present, no fungal agents have been classified for use at BSL-4. BSL-4 practices, safety equipment, and facility design and construction are applicable for work with hazardous and exotic agents that pose a high individual risk of life-threatening disease, which may be transmitted via the aerosol route, and for which there is no available vaccine or therapy. The primary hazards to personnel working with Biosafety Level 4 agents are respiratory exposure to infectious aerosols, mucous membrane or broken skin exposure to infectious droplets, and autoinoculation. All manipulations of potentially infectious diagnostic materials, isolates, and naturally or experimentally infected animals pose a high risk of exposure and infection to laboratory personnel, the community, and the environment. The laboratory worker's complete isolation from aerosolized infectious materials is accomplished primarily by working in a Class III BSC or in a full-body, air-supplied, positive-pressure personnel suit. The BSL-4 facility itself is generally a separate building or completely isolated zone with complex, specialized ventilation requirements and waste management systems to prevent release of viable agents to the environment [18].

The safety plan of a laboratory should address general considerations, chemical safety, and section-specific safety. In the case of mycology laboratories, as with all laboratories, each section requires a site-specific risk assessment to address biohazard considerations and to outline measures for staff protection. Table 1.1 outlines an example of the type of assessment that should be performed.

Materials (See Note 1)

1. Sterile distilled water
2. PPE such as coats, gowns, gloves, masks, face shields, safety glasses
3. Ethanol (70%)
4. Biosafety cabinet
5. Eyewash station
6. Hand washing sinks
7. HEPA filtered respirators or masks
8. Plasticwear (substitute for glass)
9. Centrifuge safety cups
10. Containers for transport of specimens, waste, and sharps.
11. Biohazard bags
12. Biohazard labels
13. Automatic or mechanical pipetting devices

Table 1.1 Example of a risk assessment for mycology (abridged list of tasks) [19]

Task	Exposure risk	Biosafety level	Personal protective equipment
Plating/inoculation of specimens			
Place specimens on media	Low	BSL-1	Lab coat, gloves, face shield
Inoculate primary culture media with specimens	Moderate	BSL-2	Lab coat, gloves, BSC
Storage/disposal/retrieval of specimens	Low	BSL-1	Lab coat, gloves
Culture reading			
Prepare fungal wet mounts (KOH-calcofluor) and/or India ink preps on isolates	Moderate	BSL-2	Lab coat, gloves, biosafety cabinet (BSC)
Read fungal wet mounts (KOH-calcofluor) and/or India ink preps on isolates	Low	BSL-2	Lab coat/gloves
Examine fixed smears	Low	BSL-1	Lab coat
Examine sealed cultures	Low	BSL-1	Lab coat
Manipulation of yeast isolated in culture, subculture of colonies, preparation of wet mounts/smears	Low	BSL-1	Lab coat
Manipulation of yeast, preparation of suspension	Low to moderate	BSL-1	Lab coat, gloves, face shield, or BSC
Manipulation of molds isolated in culture, subculture of colonies or broth, preparation of wet mounts/smears/fixed slides	Moderate	BSL-2	Lab coat, gloves, BSC
Transfer of molds between BSC and incubators and/or storage areas	Moderate	BSL-1	Lab coat
Nucleic acid probe or antigen detection performed on mold isolates	Moderate	BSL-2	Lab coat, gloves, BSC
Transport of fungal blood cultures between bench/incubator	Moderate	BSL-1	Lab coat/gloves
Subculture/inoculate fungal blood culture	High	BSL-2	Lab coat, gloves, BSC

Methods

Routine Precautions When Working with Fungal Cultures

The following practices are recommended for all laboratories handling potentially dangerous fungal agents:

1. Limit access to work areas. Close doors during work with research materials and lock when staff are not present in the laboratory.
2. Decontaminate all work surfaces after each working day using an appropriate disinfectant and fungicide such as 2% amphyl solution. Decontaminate all spills of viable material immediately and all liquid or solid wastes that have come in contact with viable material.
3. Aerosol-containment safety carriers much be used in the centrifuge for use with all infectious and potentially infectious materials.
4. Do not pipette by mouth.
5. Do not allow eating, drinking, smoking, or application of cosmetics in the work area. Do not store food in refrigerators that contain laboratory supplies.
6. Wash hands with soap or detergent after handling viable materials or removing gloves, and before leaving the laboratory. Do not handle telephones, doorknobs, or other common utensils without washing hands.
7. When handling viable materials, minimize creation of aerosols. For example, aerosol-containment safety carriers must be used in the centrifuge for use with all infectious and potentially infectious materials.
8. Wear laboratory coats (preferably disposable) when in work area, but do not wear them away from the work area. Observers who are not handling infectious and potentially infectious material, but are present in the laboratory where such an activity is in

progress, will wear a protective gown and mask if necessary.

9. Wear disposable gloves when handling viable materials. These should be disposed of as biohazardous waste. Change gloves if they are directly contaminated. Do not wear gloves away from the work area.

10. Any potentially contaminated material should be autoclaved for 30 min at 121 °C prior to reuse.

11. Use sharps only when no alternatives (e.g., safety devices or nonsharps) exist.

12. Wear eye/face protection if splashes or sprays are anticipated.

13. Transport materials outside of the laboratory using secondary containment.

14. Transfer materials to other facilities according to federal and international regulations.

15. Be acquainted with written instructions for laboratory procedures and proper responses to emergencies.

16. Report spills, exposures, illnesses, and injuries immediately.

17. Control pest populations. Windows in the laboratory that can be opened must be equipped with screens to exclude insects.

18. Use furniture that is easy to clean (i.e., with smooth, waterproof surfaces) and as few seams as possible.

19. Keep biohazard waste in covered containers free from leaks. Use biohazard bags as required by institutional procedure. Dispose of according to institutional procedure.

20. When *C. immitis* or *H. capsulatum* is used, all isolates should be maintained on slants to prevent aerosolization when cultures are manipulated.

21. Seal all plated media with a gas-permeable tape before removing them from the biosafety cabinet.

22. All excess materials are to be discarded in biohazard bags secured with tape and placed in containers designated for disposal of biohazardous waste.

23. All laboratory surfaces that come in contact with fungal specimens or cultures should be decontaminated with 10% bleach that is freshly prepared each day [19].

Experimentation

Administrative controls established prior to commencing experimentations should be used. These include

1. Substitution of hazardous materials with less hazardous materials in experiments.

2. Documentation of laboratory procedures that state safe work practices and the use of PPE.

Engineering controls that are mechanical in nature must also be used. Such controls employ a mechanical way to isolate the worker from the hazard. In fungal culturing such controls include:

1. The availability of needles or blades that are self-sheathing or automatically retracted and are known as "safer sharps."

2. The use of plastic instead of glass for vials, flasks, beakers, etc.

3. Containers for waste collection should be sturdy, compatible with the waste to be collected, labeled suitably, and kept closed when not in use.

4. Biological safety cabinets should be employed as often as possible. These cabinets protect the worker by scrubbing the particulate-laden air currents using high efficient particulate air (HEPA) filters.

5. PPE is employed to firstly provide a barrier to possible routes of entry on a workers body and secondly to protect clothing and shoes so that laboratory contaminants are not transferred from the laboratory to external areas. PPE must be selected as appropriate to the work being undertaken.

6. The building ventilation system must be so constructed that air from the laboratory is not recirculated to other areas within the building. Air may be high-efficiency particulate air (HEPA) filtered, reconditioned, and recirculated within that laboratory. When exhaust air from the laboratory is discharged to the outside of the building, it must be dispersed away from occupied buildings and air intakes. Depending on the agents in use, this air may be discharged through HEPA filters. A heating, ventilation, and air-conditioning (HVAC) control system may be installed to

prevent sustained positive pressurization of the laboratory.

7. An autoclave for the decontamination of contaminated waste material should be available in the containment laboratory. If waste has to be transferred from the containment laboratory for decontamination and disposal, it must be transported in sealed, unbreakable, and leakproof vessels according to national or international regulations [20].

Accidents

1. Workers must vacate the work area in the event of contamination (e.g., breakage of culture tubes, etc.) until the whole area has been decontaminated.
2. If a culture tube/plate is broken, several steps must be taken
   (a) All staff must exit the room and close door as they leave.
   (b) To decontaminate the room, the appropriate PPE must be worn (mask, gown, gloves, shoes, and hair cover).
   (c) Enter the room and cover spill with paper towels soaked thoroughly with disinfectant such as 2% amphyl solution. Leave the towels to stand for 1 h before cleaning up the spill and keep the area wet with amphyl to prevent dried particles from becoming airborne. If required, disinfect all contaminated equipment, i.e., specimen containers, culture tubes, pens, etc.
   (d) Exit area and do not re-enter for 1 h.
   (e) Contaminated clothing must be discarded or autoclaved for 1 h at 121 °C.
3. The laboratory supervisor must be informed of the accident and an incident report must be completed.
4. For accidents involving the eye,
   (a) Go to the eye wash station and call for help.
   (b) Eyes must be washed carefully and thoroughly for at least 15 min to remove chemicals or particles.
   (c) Seek medical attention.
   (d) The laboratory supervisor must be informed and an incident report must be completed.

Emergency Preparedness and Response

Emergencies and disruptions to the normal working environment, such as hurricanes and other disasters, can occur. Possible disruptions can include spills, exposures, injuries, power or water loss, equipment failure, fire, or flooding. The response to each of these emergencies will depend on the individual and institutional circumstances, which are too varied to discuss in detail. However, a written emergency and evacuation plan should be put in place and communicated to all personnel for such situations to circumvent employee injury or contamination via fungal agents.

To prepare the mycology laboratory for an emergency the following steps should be taken:

1. Cover all external windows with wood.
2. Ensure all refrigerators and freezers are connected to emergency electrical outlets.
3. Incubators should be locked to prevent breakage and subsequent dispersion of cultures containing potentially pathogenic fungi.
4. Computer terminals and electronic equipment should be disconnected and moved to one room and covered with plastic.
5. Work and log books and other essential paperwork should also be moved to one room and covered with plastic.

Information and Training

Findings from several studies have suggested that formal training in mycology laboratories is inadequate and must be supplemented. All of the information discussed in this chapter must be communicated to laboratory personnel by management. Lack of communication of safety guidelines can result in health and safety incidences.

1. Every laboratory worker should know the location and proper use of PPE and first aid equipment (e.g., eye wash stations, etc.).
2. Emergency telephone numbers for emergency services and personnel should be clearly posted and known to all workers.
3. At least one employee should be trained in first aid procedures.
4. Specific training for use of hazardous fungal agents should be available for all personnel.

5. A written training plan complemented with completed employee training records is strongly recommended for every laboratory to ensure awareness of routine laboratory practice as well as safe mycology laboratory practice [18, 21].

Notes

Work surfaces must be disinfected prior to commencing work, after work, and in the instance of spillages. Various types of disinfectants and in the case of mycology, fungicides, are available for use and can be determined as appropriate to the work being undertaken. In particular, aldehydes such as formaldehyde and glutaraldehyde, halogens such as iodine and chlorine compounds, and phenol derivative proprietary products such as Amphyl, Lysol, and Vesphene are effective fungicidal agents [22]. Many other commercial disinfectants are also available; however, their product information sheet and material safety data sheet (MSDS) must be consulted to ensure appropriate use.

References

1. Gilchrist MJR, Fleming DO (2000) Biosafety precautions for *Mycobacterium tuberculosis* and other airborne pathogens. In: Fleming DO, Hunt DL (eds) Biological safety: principles and practices. ASM Press, Washington, DC, pp 209–221
2. Pike RM (1976) Laboratory-associated infections: summary and analysis of 3921 cases. Health Lab Sci 13:105–114
3. Pike RM (1978) Past and present hazards of working with infectious agents. Arch Pathol Lab Med 102:333–336
4. Sewell D (1995) Laboratory-associated infections and biosafety. Clin Microbiol Rev 8(3):389–405
5. Baum GL, Lerner PI (1970) Primary pulmonary blastomycosis: a laboratory-acquired infection. Ann Intern Med 73(2):263–265
6. Denton JF, Di Salvo AF, Hirsch ML (1967) Laboratory-acquired North American blastomycosis. JAMA 199(12):935–936
7. Graham WR Jr, Callaway JL (1982) Primary inoculation blastomycosis in a veterinarian. J Am Acad Dermatol 7(6):785–786
8. Larsh HW, Schwarz J (1977) Accidental inoculation blasmycosis. Cutis 19:334–336
9. Smith DT, Harrell ER (1948) Fatal coccidioidomycosis—a case of a laboratory infection. Am Rev Tuberc 57(4):368–374
10. Wilson JW, Smith CE, Plunkett OA (1953) Primary cutaneous coccidioidomycosis; the criteria for diagnosis and a report of a case. Calif Med 79(3):233–239
11. Fiese MJ (1958) Coccidioidomycosis. CC Thomas, Illinois
12. Nabarro JD (1948) Primary pulmonary coccidioidomycosis; case of laboratory infection in England. Lancet I:982–984
13. Murray JF, Howard D (1964) Laboratory-acquired histoplasmosis. Am Rev Respir Dis 89(5):631–640
14. Tesh RB, Schneida JD Jr (1966) Primary cutaneous histoplasmosis. N Engl J Med 275(11):597–599
15. Tosh FE, Yates JL, Brasher CA, Balhuize J (1964) Primary cutaneous histoplasmosis—report of case. Ann Intern Med 114(1):118–119
16. Singh K (2009) Laboratory-acquired infections. Clin Infect Dis 49(1):142–147
17. Warnock DW (2000) Mycotic agents of human disease. In: Fleming DO, Hunt DL (eds) Biological safety: principles and practices. ASM Press, Washington, pp 111–120
18. Burnett LC, Lunn G, Coico R (2005) Biosafety: guidelines for working with pathogenic and infectious microorganisms. Current protocols in microbiology. Wiley, Hoboken, NJ, pp 1A..–A..14
19. Atlas RM (1993) Handbook of microbiological media. CRC Press, Boca Raton
20. World Health Organisation (2004) Laboratory Biosafety Manual. WHO, Geneva. [cited 2011 6th July]; 3rd. Available from: http://www.who.int/csr/resources/publications/biosafety/Biosafety7.pdf
21. Nye MB, Beard MA, Body BA (2006) Diagnostic mycology: controversies and consensus—what should laboratories do? Part II. Clin Microbiol Newsl 28(17):129–134
22. Carter GR, Wise DJ (2004) Essentials of veterinary bacteriology and mycology, 6th edn. Wiley-Blackwell, Ames, Iowa

Methods of Cryopreservation in Fungi

2

Ladislav Homolka

Abstract

Traditional method of the routine subculturing by transfer of fungal cultures from staled to fresh media is not a very practical means of storing large numbers of fungal cultures. It is time-consuming, prone to contamination, and does not prevent genetic and physiological changes. At present, besides freeze-drying (lyophilization), cryopreservation seems to be the best preservation technique available for fungi.

Keywords

Cryopreservation • Fungi collections • Liquid nitrogen • Perlite • Subculturing • Fungi storage

Introduction

Serious mycological (and generally biological) work requires a reliable source of cultures (i.e., well-defined and taxonomically determined starting material), which is ensured by its safe long-term storage. This implies the fundamental and growing importance of culture collections not only for preservation of the endangered genofond (and consequently the biodiversity), but also as a principal source of material for biotechnological processes, research, and teaching. The first and most important problem to be solved is the long-term maintenance of this material.

Collections of fungi were originally kept by serial transfers from staled to fresh media. This routine subculturing is not a very practical method for storing large numbers of fungal cultures. It is time-consuming, prone to contamination, and does not prevent genetic and physiological changes (degeneration, aging) during long-term and frequent subculturing [1]. Over the years, various storage methods have been developed in order to eliminate these disadvantages. Their common feature is at least partial suppression of growth and metabolism of the cultures. Among them, keeping fungal cultures in sterile water [2–8] was surprisingly efficient (especially with lower fungi, but also with some basidiomycetes) and experiences its revival. In some fungi, preservation under a layer of mineral oil, in silica gel, soil, or sand [9–13] was successful. These

L. Homolka (✉)
Department of Ecology of Microorganisms,
Institute of Microbiology, Academy
of Sciences of the Czech Republic, Vídeňská 1083,
Prague 4 142 20, Czech Republic
e-mail: homolka@biomed.cas.cz

V.K. Gupta et al. (eds.), *Laboratory Protocols in Fungal Biology: Current Methods in Fungal Biology*,
Fungal Biology, DOI 10.1007/978-1-4614-2356-0_2, © Springer Science+Business Media, LLC 2013

methods enabled a reduction in fungal growth and extended the time intervals between transfers to fresh media. Nevertheless, the method of serial subculturing is still used in collections with limited financial support or in the majority of other collections as a backup preservation method.

Searching for improved or new methods resulted through different intermediate steps in the introduction of lyophilization and cryopreservation of fungal cultures [14]. Application of the most extended method of culture preservation—freeze-drying (lyophilization), tested for sporulating fungi as early as 1945 [15]—is rather limited in the case of basidiomycetes and other fungi nonsporulating in vitro [16, 17]. Most attempts to revitalize dehydrated hyphae of fungi have failed, except for some successes [18–20]; nevertheless, the real absence of spores must be always carefully checked. Despite this, several attempts have recently been made using modified protocols [21–23] and the growing interest in this technique can be seen at present. An important role played in the whole process, besides the freeze-drying, is also the freezing rate and the lyoprotectant used [24, 25]. Freeze-drying of fungi has several important advantages over all other maintenance methods. Cultures can be stored easily in dense packing without any special requirements and need not be revived on agar slants prior to dispatch. The product is light, inactive, and dry, enabling easy distribution by mail.

The new or modified methods have been frequently used and evaluated [5, 6, 10, 26–31], but they are not generally applicable. Often a specific preservation protocol is necessary even for individual strains of the same species. At present, besides freeze-drying (lyophilization), cryopreservation seems to be the best preservation technique available for filamentous fungi [14, 26, 32]. A very comprehensive and detailed overview of the methods and results of the cryopreservation in microorganisms was published by Hubálek [33].

Cryogenic technique for long-term storage of large numbers of fungal species was introduced to ATCC in 1960 and the results have been very satisfactory [34, 35]. The technique was consecutively introduced to many other prominent collections, e.g., CAB International Mycological Institute [1], etc. In certain collections—e.g., IFO (Institute for Fermentation Osaka)—nonsporulating cultures of basidiomycetes are stored by cryopreservation at −80°C in electric freezers [36].

Mycelium and/or spore suspensions with or without a cryoprotectant in sealed glass ampoules were originally used for cryopreservation of filamentous fungi. Later, glass ampoules were replaced with safer polypropylene cryovials and/or straws. Agar blocks immersed in an appropriate cryoprotectant were originally used as carriers of fungal mycelium for the cryopreservation process [37]. A useful straw technique with agar miniblocks for the preservation of fungi in liquid nitrogen was developed by Elliott in 1976 [38] and improved by Stalpers et al. [39]. Another technique using straws in cryotubes without a cryoprotectant solution was described by Hoffmann [40]. A modified Hoffmann's technique was compared with the original agar block one in our paper [29]. Commercial preservation systems with polystyrene beads as carriers were used for cyopreservation of conidia of enthomopathogenic fungi [41] and of sporulating *Aspergillus fumigatus* cultures at −80°C [42]. Porous ceramic beads were employed for cryopreservation of several sporulating fungal cultures and for a *Saccharomyces cerevisiae* culture at −70°C by Palágyi et al. [43]. It is symptomatic that these techniques have not been used for nonsporulating filamentous fungi. In this context it should be mentioned that as early as 1978 Feltham et al. described a method of preservation of bacteria on glass beads at −76°C [44]. Some reports [45–47] indicate that cryopreservation at −80°C is suitable for many fungal cultures, including basidiomycetes. Nevertheless, Leeson et al. [48] state that to completely stabilize frozen cultures, the temperature must be sufficiently reduced to both minimize metabolism and prevent ice crystal formation, which can cause physical damage during storage. The temperature limit securing prevention of formation of such ice crystals is −139°C. This is why at present many culture collections start to keep their cultures at −150°C in ultralow-temperature electric freezers, which are sometimes equipped with liquid nitrogen supply.

The cryopreservation process includes freezing and thawing and the protocol of these procedures plays an important role [49, 50]. In principle, there are two kinds of freezing protocols: a slow (controlled) one and a fast (uncontrolled) one, which both have been used for cryopreservation of fungi [51, 52]. Generally, too low freezing rates cause excessive dehydration and concentration of the solution leading to cell damage; on the contrary, too fast freezing leads to insufficient dehydration and formation of abundant ice crystals with lethal consequences. Nevertheless, different fungal cultures exhibit different sensitivities to freezing conditions and to the presence and concentration of cryoprotectants. A freezing rate of 1°C per minute is usually used for cryopreservation of fungi; in the author's experiments with sensitive mutant strains of *Agaricus bisporus*, the freezing rate 0.5°C per minute was successfully used.[1] Lately, cryomicroscopic methods have been used to study the process of freezing and thawing of fungal cultures [14, 24, 53]. Successful cryopreservation depends on the cryoprotectant used [54]. At present, dimethylsulfoxid and glycerol are the most widespread [55]. The method of cryogenic culture maintenance seems to be mostly successful also in nonsporulating cultures [14, 32, 56, 57]. An overview study on the influence of the cryopreservation process on survival of taxonomically very broad spectrum of fungi published by Smith [14] showed that there was no obvious link between taxonomic grouping and the response of the fungi to freezing and thawing. This was confirmed in our study [31]. Similar studies were carried out also in edible mushrooms *Lentinus edodes* [58] and the genus *Pleurotus* [59]. Davell and Flygh [60] showed that even an ectomycorrhizal fungus *Cantharellus cibarius* can be successfully cryopreserved when a sufficient number of cryoprotocols is tested. Cryopreservation of spores of vesicular–arbuscular mycorrhizal fungi was described by Douds and Schenck [61].

Beyond survival, another principal requirement for the successful preservation of fungal strains is maintenance of their genetic and physiological features, such as growth, morphology, and metabolite production. In our experiments with some white-rot basidiomycetes, no negative effect of cryopreservation or the used cryoprotective on production of ligninolytic enzymes was found [62]. The complete revival of cryopreserved cultures (evaluated mostly by measuring the colony diameter) is generally still uncertain. The survival rate varies between 60 and 100% [21, 58, 59, 63, 64]. Only a few studies of the genetic stability of cryopreserved fungi have been performed. Singh et al. [47] confirmed the genetic stability of 11 cryogenically preserved edible mushroom strains by comparing random amplified polymorphic DNA (RAPD) and internal transcribed spacer (ITS) profiles. Using polymerase chain reaction (PCR) fingerprinting, Ryan et al. [65] checked the genetic stability of several isolates of *Fusarium oxysporum* and *Metarhizium anisopliae*. Other studies include confirmation of the genetic stability of *Uncinula necator* conidia after storage at −80°C [45] and investigation of the influence of mid-term cryopreservation at −80°C on 15 isolates of 10 basidiomycete species, for which the DNA fingerprint patterns were unchanged [66]. All of these reports were solely based on fingerprinting methods, which are not suitable for the detection of minute yet important changes in the genome, such as point or short indel mutations. Rather, sequencing approaches are required to successfully detect these mutations. This approach was used in our recent study [67]. Considering the above data, there is a continuous need for developing, improving, optimizing, and combining of preservation procedures, because the present methods are not applicable to all fungal cultures. Although many of these fungi can be grown in pure cultures on solid media, their growth is often attenuated and their morphology and other characteristics changed, which can result in their complete loss. The number of characteristics evaluating the success of preservation should be increased.

---

[1] Homolka, unpublished results.

As mentioned previously, cryopreservation, namely in liquid nitrogen, seems to be the most reliable, safe, and prospective method of a long-term maintenance of most fungal species, especially those not amenable to freeze-drying. It is probably the only storage technique that can ensure genomic and phenotypic stability. But not even the aforementioned cryopreservation method is applicable to preservation of all fungal cultures in the present form. According to the literature as well as the author's personal experience, especially the maintenance of basidiomycetes is challenging. Many of these fungi do not form asexual spores, their dominant life form, the vegetative mycelium, is sensitive to environmental conditions and therefore not amenable to freeze-drying.

To address these issues, a method of cryopreservation using perlite as a carrier for fungal mycelia was developed in the author's laboratory (perlite protocol or PP) [28] and then successfully verified for 442 basidiomycete strains [30]. Perlite is a unique aluminosilicate volcanic mineral that retains substantial amounts of water that can be released when needed—a feature that seems to have a dominant effect on cryopreservation success. The PP can be used for cryopreservation of taxonomically different groups of fungi, including yeasts [31], and works relatively well for fungi that cannot survive other routine preservation procedures. Expanded perlite was used as a solid support in solid-state fermentations [68]; otherwise it is used in many applications, particularly in the construction, horticulture, and other various industrial fields. It is recommended as an efficient purifying agent and as a carrier for pesticides, feed concentrates, herbicides, and other similar applications.

Perlite Protocol (PP)

The protocol is suitable for maintaining a broad spectrum of fungal cultures of different origin. It was verified in several culture collections (e.g., in Finland, the Netherlands, USA, Czech Republic, etc.) with great success and now it is routinely used there.

Materials

1. Distilled water.
2. Agar Difco.
3. Glycerol p.a. Sigma.
4. Isopropyl alcohol 100% Sigma (alternatively).
5. Agricultural-grade perlite (Agroperlit, GrowMarket s.r.o., Prague, Czech Republic, http://www.growmarket.cz/produkt/agroperlit-8l)—particles 1–2 mm.
6. Dried wort extract Sladovit, Malthouse Bruntál, Bruntál, Czech Republic, diluted to a density of 4° Balling scale with distilled water (further wort in the text); or preferably MYA medium: Malt-extract Difco 25 g, Yeast-extract Difco 2 g, Glycerol p.a., Serva 50 g, distilled water ad 1,000 mL. pH adjusted to 6.5 with 1 M KOH solution.
7. Ethanol (70%).
8. Liquid nitrogen (LN).
9. Nunc CryoTube Vials 1.8 mL screw-capped (Nalgene/Nunc, Rochester, USA).
10. Cork borer.
11. Lancet and small spoon.
12. Rule for measuring of colony diameter.
13. Water bath.
14. Balling hydrometer (saccharometer) (alternatively).
15. Hot-air sterilizer.
16. Autoclave.
17. Refrigerator (about 4°C).
18. Deep-freezer (−80°C).
19. Microscope equipped with phase contrast.
20. Laminar flow box.
21. Thermostat (incubator) for 24°C.
22. Container for storing samples in liquid nitrogen (e.g., HARSCO TW-5K container, Harsco, Camp Hill, USA).
23. Programmable freezer for controlled freezing of cryovials with mycelia (e.g., IceCube 1800 freezer, SY-LAB Geraete GmbH., Neupurkersdorf, Austria); or alternatively Cryo 1°C Freezing Container "Mr. Frosty" (Nalgene Labware)[2].

---

[2] All chemicals and devices named can be replaced with other ones produced by other renowned companies.

Methods

The methods given below describe general procedures for cryopreservation of fungi using the perlite protocol (PP).

Strains (Cultures)

The starting cultures are kept on wort agar slants or dishes (wort 4° Balling, 1.5% agar Difco) at 4°C or other media suitable for the growth of the strains destined for the procedure (e.g., MYA medium with 1.5% agar, etc.) and transferred to the fresh medium every 6 months.

1. Prepare an agar medium, sterilize it in an autoclave (121°C, 20 min.), pour it into sterile plastic Petri dishes (diameter 100 mm, 30 mL per dish), and let it cool down in a laminar flow box.
2. In a laminar flow box cut out an agar plug (6 mm diameter) from the actively growing part of a colony on a Petri dish with a cork borer, place it on a Petri dish with fresh medium using a sterile lancet and then let the dish incubate for 14 days at 24°C. Then put the dish(es) into a refrigerator and keep it at about 4°C.

Culture Preparation and Freezing–Thawing Protocols

Fungal cultures are grown directly in firmly closed sterile plastic cryovials (1.8 mL) with 200 mg of perlite (Agroperlit, agricultural grade) moistened with 1 mL of wort (4° Balling) or other medium (e.g., MYA) enriched with 5% glycerol as a cryoprotectant. For sterilization of the cork borer, lancet, and small spoon use a hot-air sterilizer (150°C, 30 min.).

1. Distribute perlite into cryovials (200 mg per vial), flood it with 1.8 mL of the medium enriched with 5% of glycerol, and sterilize vials in autoclave (121°C, 20 min.). In a laminar flow box cut out an agar plug (6 mm diameter) from the actively growing part of a colony on a Petri dish using a cork borer, place it using a sterile lancet on the surface of perlite in the cryovial, close the vial firmly and let it incubate for 14 days at 24°C.
2. Freeze the cryovials with perlite overgrown by mycelium in a programmable freezer (or alternatively in a "Mr. Frosty" container in a deep-freezer) to −70°C at a freezing rate of 1°C per minute. Then place them in LN in a container.
3. Take the stored frozen cultures in cryovials out of the LN container, transfer them to a water bath (37°C), and leave them there until the ice is completely thawed (thawing—reactivation of cultures). Prior to opening, disinfect the surface of cryovials with 70% ethanol.

Viability Test

1. After thawing, separate at least partially the perlite particles overgrown with mycelium by shaking, the content of the cryovials (two parallels of each strain) divide into three approximately equal aliquots each and these plate onto wort (or an other) agar medium in Petri dishes (diameter 100 mm) using a small sterile spoon.
2. Incubate the cultures in Petri dishes at 24°C for 14 days. Strains exhibiting survival of at least four out of six separate aliquots are considered viable.

Growth Estimation and Morphological Analysis

1. Growth of cultures measure as a mean diameter increase of a growth-covered zone (in mm) during a 14-day incubation at 24°C on the respective agar medium in Petri dishes (diameter 100 mm) inoculated with perlite aliquots from cryovials before freezing and after reactivation. Measure six zones (three aliquots from two cryovials) for each strain. The first occurrence of growth varies between frozen cultures, with some strains showing signs of re-growth within 2 days but most strains reactivating within 7 days after plating.

2. Use the same procedure except for freezing and thawing for growth measurement of the control.

3. Carry out the morphological analysis on control cultures and on those arising from the viability tests. Check the selected macroscopic features (colony color, reverse color, texture of the mycelium) and microscopic features (hyphal branching, presence/absence of clamp connections, presence/absence of hyphal vacuolization, etc.) using a microscope.

4. If possible, estimate also other characteristics of the resulting cultures (e.g., enzyme or metabolite production, etc.) according to your consideration.

References

1. Onions AHS (1971) Preservation of fungi. In: Booth C (ed) Methods in microbiology, vol 4. Academic, New York, London, pp 113–115

2. Marx DH, Daniel WJ (1976) Maintaining cultures of ectomycorrhizal and plant pathogenic fungi in sterile water cold storage. Can J Microbiol 22:338–341

3. Ellis JJ (1979) Preserving fungus strains in sterile water. Mycologia 71:1072–1075

4. Richter DL, Bruhn JN (1989) Revival of saprotrophic and mycorrhizal basidiomycete cultures from cold storage in sterile water. Can J Microbiol 35: 1055–1060

5. Smith JE, McKay D, Molina R (1994) Survival of mycorrhizal fungal isolates stored in sterile water at two temperatures and retrieved on solid and liquid nutrient media. Can J Microbiol 40:736–742

6. Burdsall HH, Dorworth EB (1994) Preserving cultures of wood-decaying Basidiomycotina using sterile distille water in cryovials. Mycologia 86:275–280

7. Borman AM, Szekely A, Campbell CK, Johnson EM (2006) Evaluation of the viability of pathogenic filamentous fungi after prolonged storage in sterile water and review of recent published studies on storage methods. Mycopathologia 161:361–368

8. Richter DL (2008) Revival of saprotrophic and mycorrhizal basidiomycete cultures after 20 years in cold storage in sterile water. Can J Microbiol 54:595–599

9. Perrin PW (1979) Long-term storage of cultures of wood-inhabiting fungi under mineral oil. Mycologia 71:867–869

10. Johnson GC, Martin AK (1992) Survival of wood-inhabiting fungi stored for 10 years in water and under oil. Can J Microbiol 38:861–864

11. Sharma B, Smith D (1999) Recovery of fungi after storage for over a quarter of a century. World J Microbiol Biotechnol 15:517–519

12. Delcán J, Moyano C, Raposo R, Melgarejo P (2002) Storage of Botrytis cinerea using different methods. J Plant Pathol 84:3–9

13. Baskarathevan J, Jaspers MV, Jones EE, Ridgway HJ (2009) Evaluation of different storage methods for rapid and cost-effective preservation Botryosphaera species. N Z Plant Protect 62:234–237

14. Smith D (1998) The use of cryopreservation in the ex-situ conservation of fungi. CryoLetters 19:79–90

15. Raper KB, Alexander DF (1945) Preservation of molds by lyophil process. Mycologia 37:499–525

16. Antheunisse J (1973) Viability of lyophilized microorganisms after storage. Antonie Leeuwenhoek 39: 243–248

17. Hwang S-W, Kwolek WF, Haynes WC (1976) Investigation of ultra-low temperature for fungal cultures. III. Viability and growth rate of mycelial cultures following cryogenic storage. Mycologia 68:377–387

18. Bazzigher G (1962) Ein vereinfachtes Gefriertrocknungsverfahren zur Konservierung von Pilzkulturen. Phytopathol Z 45:53–56

19. Pertot E, Puc A, Kremser M (1977) Lyophilization of nonsporulating strains of the fungus Claviceps. Eur J Appl Microbiol 4:289–294

20. Tommerup IC (1988) Long-term preservation by L-drying and storage of vesicular arbuscular mycorrhizal fungi. Trans Brit Mycol Soc 90:585–591

21. Tan CS, Stalpers JA, van Ingen CW (1991) Freeze-drying of fungal hyphae. Mycologia 83:654–657

22. Croan SC, Burdsall HH Jr, Rentmeester RM (1999) Preservation of tropical wood-inhabiting basidiomycetes. Mycologia 91:908–916

23. Sundari SK, Adholeya A (1999) Freeze-drying vegetative mycelium of Laccaria fraterna and its subsequent regeneration. Biotechnol Tech 13:491–495

24. Tan CS, Vlug IJA, Stalpers JA, van Ingen CW (1994) Microscopical observations on the influence of the cooling rate during freeze-drying of conidia. Mycologia 86:281–289

25. Tan CS, van Ingen CW, Talsma H, van Miltenburg JC, Steffensen CL, Vlug IJA, Stalpers JA (1995) Freeze-drying of fungi, influence of composition and glass transition temperature of the protectant. Cryobiology 32:60–67

26. Ryan MJ, Smith D (2004) Fungal genetic resource centres and the genomic challenge. Mycol Res 108:1351–1362

27. Ryan MJ, Smith D (2007) Cryopreservation and freeze-drying of fungi employing centrifugal and shelf freeze-drying. Meth Mol Biol 368:127–140

28. Homolka L, Lisá L, Eichlerová I, Nerud F (2001) Cryopreservation of basidiomycete strains using perlite. J Microbiol Meth 47:307–313

29. Homolka L, Lisá L, Nerud F (2003) Viability of basidiomycete strains after cryopreservation, comparison of two different freezing protocols. Folia Microbiol 48:219–226

30. Homolka L, Lisá L, Nerud F (2006) Basidiomycete cryopreservation on perlite, evaluation of a new method. Cryobiology 52:446–453

31. Homolka L, Lisá L, Kubátová A, Váňová M, Janderová B, Nerud F (2007) Cryopreservation of filamentous micromycetes and yeasts using perlite. Folia Microbiol 52:153–157
32. Challen MP, Elliott T (1986) Polypropylene straw ampoules for the storage of microorganisms in liquid nitrogen. J Microbiol Meth 5:11–23
33. Hubálek Z (1996) Cryopreservation of microorganisms. Academia Publishing House, Prague
34. Hwang S-W (1960) Effects of ultra-low temperatures on the viability of selected fungus strains. Mycologia 52:527–529
35. Smith D (1983) Cryoprotectants and the cryopreservation of fungi. Trans Brit Mycol Soc 80:360–363
36. Ito T (1996) Preservation of fungal cultures at the Institute of Fermentation, Osaka (IFO). In: Samson RA, Stalpers JA, van der Mei D, Stouthamer AH (eds) Culture collections to improve the quality of life. The Netherlands and the World Federation for Culture Collections, Centraalbureau voor Schimmelcultures, Baarn, pp 210–211
37. Hwang S-W (1968) Investigation of ultra-low temperature for fungal cultures. I. An evaluation of liquid nitrogen storage for preservation of selected fungal cultures. Mycologia 60:613–621
38. Elliott TJ (1976) Alternative ampoule for storing fungal cultures in liquid nitrogen. Trans Brit Mycol Soc 67:545–546
39. Stalpers JA, de Hoog A, Vlug IJ (1987) Improvement of the straw technique for the preservation of fungi in liquid nitrogen. Mycologia 79:82–89
40. Hoffmann P (1991) Cryopreservation of fungi. World J Microbiol Biotechnol 7:92–94
41. Chandler D (1994) Cryopreservation of fungal spores using porous beads. Mycol Res 98:525–526
42. Belkacemi L, Barton RC, Evans EGV (1997) Cryopreservation of Aspergillus fumigatus stock cultures with a commercial bead system. Mycoses 40:103–104
43. Palágyi Z, Nagy Á, Vastag M, Ferency L, Vágvölgyi C (1997) Maintenance of fungal strains on cryopreservative-immersed porous ceramic beads. Biotechnol Tech 11:249–250
44. Feltham RKA, Power AK, Pell PA, Sneath PHA (1978) A simple method for storage of bacteria at -76° C. J Appl Bacteriol 44:313–316
45. Stummer BE, Zanker T, Scott ES (1999) Cryopreservation of airdried conidia of Uncinula necator. Austr Plant Pathol 28:82–84
46. Kitamoto Y, Suzuki A, Shimada S, Yamanaka K (2002) A new method for the preservation of fungus stock cultures by deepfreezing. Mycoscience 43:143–149
47. Singh SK, Upadhyay RC, Kamal S, Tiwari M (2004) Mushroom cryopreservation and its effect on survival, yield and genetic stability. CryoLetters 25:23–32
48. Leeson EA, Cann JP, Morris GJ (1984) Maintenance of algae and protozoa. In: Kirsop BE, Snell JJS (eds) Maintenance of microorganisms. Academic, London, pp 131–160
49. Leef J, Mazur P (1978) Physiological response of Neurospora conidia to freezing in the dehydrated, hydrated or germinated state. Appl Environ Microbiol 35:72–83
50. Morris GJ, Smith D, Coulson GE (1988) A comparative study of the changes in the morphology of hyphae during freezing and viability upon thawing for twenty species of fungi. J Gen Microbiol 134:2897–2906
51. Goos RD, Davis EE, Butterfield W (1967) Effect of warming rates on the viability of frozen fungus spores. Mycologia 59:58–66
52. Dahmen H, Staub T, Schwinn FT (1983) Technique for long-term preservation of phytopathogenic fungi in liquid nitrogen. Phytopathology 73:241–246
53. Smith D, Coulson GE, Morris GJ (1986) A comparatjve study of the morphology and viability of hyphae of Penicillium expansum and Phytophthora nicotianae during freezing and thawing. J Gen Microbiol 132:2013–2021
54. Jong SC, Davis EE (1978) Conservation of reference strains of Fusarium in pure culture. Mycopathologia 66:153–159
55. Hubálek Z (2003) Protectants used in the cryopreservation of microorganisms. Cryobiology 46:205–229
56. Homolka L (1976) On the problem of maintenance and cultivation of higher fungi. Folia Microbiol 21:189–190
57. Chvostová V, Nerud F, Homolka L (1995) Viability of wood-inhabiting basidiomycetes following cryogenic preservation. Folia Microbiol 40:193–197
58. Roquebert MF, Bury E (1993) Effect of freezing and thawing on cell membranes of Lentinus edodes, the Shiitake mushroom. World J Microbiol Biotechnol 9:641–647
59. Mata G, Salmones D, Pérez R, Guzmán G (1994) Behavior of some strains of the genus Pleurotus after different procedures for freezing in liquid nitrogen. Rev Microbiol Sao Paulo 25:197–200
60. Danell E, Flygh G (2002) Cryopreservation of the ectomycorrhizal mushroom Cantharellus cibarius. Mycol Res 106:1340–1342
61. Douds DD Jr, Schenck NC (1990) Cryopreservation of spores of vesicular-arbuscular mycorrhizal fungi. New Phytol 115:667–674
62. Stoychev I, Homolka L, Nerud F, Lisá L (1998) Activities of ligninolytic enzymes in some white-rot basidiomycete strains after recovering from cryopreservation in liquid nitrogen. Antonie Leeuwenhoek 73:211–214
63. Schipper MAA, Bekker-Holtman J (1976) Viability of lyophilized fungal cultures. Antonie Leewenhoek 42:325–328
64. Smith D, Ward SM (1987) Notes on the preservation of fungi. CAB International Mycological Institute, Slough, UK

65. Ryan MJ, Jeffries P, Bridge PD, Smith D (2001) Developing cryopreservation protocols to secure fungal gene function. CryoLetters 22:115–124

66. Voyron S, Roussel S, Munaut F, Varese GC, Ginepro M, Declerck S, Marchisio VF (2009) Vitality and genetic fidelity of white-rot fungi mycelia following different methods of preservation. Mycol Res 113:1027–1038

67. Homolka L, Lisá L, Eichlerová I, Valášková V, Baldrian P (2010) Effect of long-term preservation of different basidiomycetes on perlite in liquid nitrogen on their growth, morphological, enzymatic and genetic characteristics. Fungal Biol 114:929–935

68. Kerem Z, Hadar Y (1993) Effect of manganese on lignin degradation by *Pleurotus ostreatus* during solid-state fermentation. Appl Environ Microbiol 59:4115–4120

Long-Term Preservation of Fungal Cultures in All-Russian Collection of Microorganisms (VKM): Protocols and Results

3

Svetlana M. Ozerskaya, Natalya E. Ivanushkina,
Galina A. Kochkina, Svetlana S. Eremina,
Alexander N. Vasilenko, and Nadezhda I. Chigineva

Abstract

Results of successful preservation experience are given for the taxonomic groups of fungi preserved in All-Russian Collection of Microorganisms (VKM): the species names, conservation methods, storage time estimates.

Keywords

Fungi • Culture collection • Storage time • Survival • Lyophilization • Freeze-drying • Cryopreservation • Sterile soil

Introduction

Preservation and long-term storage of type, authentic, and other kinds of fungal cultures in a living state is of high importance both for the fundamental and practical mycology.

Long-time storage of strains is performed in microbiological culture collections (biological resource centers). Various methods of preservation of fungal cultures have been reported [1–3].

Freeze-drying (lyophilization) and cryopreservation methods are utilized for thousands of fungal strains in microbial collections all over the

S.M. Ozerskaya (✉) • N.E. Ivanushkina • G.A. Kochkina
S.S. Eremina • A.N. Vasilenko • N.I. Chigineva
G. K. Skryabin Institute of Biochemistry
and Physiology of Microorganisms, Russian Academy
of Science, prospekt Nauki, 5, Pushchino,
Moscow Region 142290, Russia
e-mail: smo@dol.ru

world [4–6]. Nevertheless, it is clear that the fungal strains of different species vary in ability to survive after the long-time storage preservation under laboratory conditions. Some of them are very difficult to maintain *ex situ*, whereas others could be easily and successfully preserved alive by using almost any conservation technique. Available information on the maximal time periods in which the reliable storage of different fungal species are ensured does not cover those for the diversity of fungi maintained in culture collections. This chapter presents the methods of cryopreservation, freeze-drying, and preservation in sterile soil that are utilized in VKM fungal collection, accompanied by data on maximal storage time registered. The methods take into consideration the special features of cultures preserved as well as the equipment used.

VKM fungal collection (All-Russian Collection of Microorganisms, Russia) was established in 1955 and has a long-term experience for preservation and storage of fungal

V.K. Gupta et al. (eds.), *Laboratory Protocols in Fungal Biology: Current Methods in Fungal Biology*,
Fungal Biology, DOI 10.1007/978-1-4614-2356-0_3, © Springer Science+Business Media, LLC 2013

cultures. Collection of filamentous fungi is currently composed of approximately 5,000 strains (545 genera, 1,450 species) belonging to species of the kingdoms Chromista (*Oomycetes*) and Fungi (*Zygomycetes, Ascomycetes,* and *Basidiomycetes*). The current use of different preservation methods for more than 3,800 strains maintained in VKM for more than 40 years was analyzed using a specially designed database. The database keeps the protocols of conservation methods, storage conditions, the calculated time of reliable storage, special requirements of growth, and other information related to the issue. Data presented in this chapter are derived from this database. The information on preservation methods is also available in VKM catalogue (http://www.vkm.ru/Catalogue.htm), in data sheet for each strain.

Cryopreservation of Filamentous Fungi

According to published data, the fast cooling rates followed by storage in liquid nitrogen at −196°C allow secure and long-term preservation of some fungal cultures [7]. However, the ability to resist damage by freezing and warming differs considerably among genera/species and depends on their particular features (presence and type of sporulation, chemical composition of cytoplasmic membrane and cell wall, physiological state, etc.). Selection of optimal cryoprotectants, rates of cooling and warming has enabled increasing the number and diversity of taxa preserved by this method [8, 9].

More than 70% filamentous fungal of VKM (2,714 strains belonging to 1,148 species and 405 genera) are stored using various cryopreservation protocols. Cultures with abundant sexual and nonsexual sporulation usually were preserved by using fast cooling rates followed by storage either in liquid nitrogen or in ultra-low temperature freezers at −80°C.

It was noticed that some cultures of *Zygomycetes* belonging to the genera *Mortierella, Basidiobolus, Coemansia,* and *Lobosporangium*

(syn. *Echinosporangium*) do not survive the ultrarapid freezing procedure even if they have abundant sporulation. Successful preservation of such strains was achieved by modification of the cryopreservation regime, for example using slow programmed freezing. The same method was used either for nonsporulating fungi or zoospore-forming ones (*Basidiomycetes, Oomycetes*).

According to our data, some part of strains of *Oomycetes* (20%), *Basidiomycetes* (4%), *Zygomycetes* (1%), and *Ascomycetes* (1%) did not survive cryopreservation at all freezing regimes and modification applied [10]. The strains most difficult to maintain belong to genera *Brevilegnia, Dictyuchus, Phytophthora,* and to some species of *Achlya* and *Saprolegnia*. Similar situations have also been seen with some species of *Basidiomycetes* (*Suillus, Amanita, Dictyophora, Mutinus,* etc.). They are usually maintained by subculturing and preservation under mineral oil.

It has been suggested that those microbial cultures that are able to survive the freezing and a short storage will permanently stay in the vital state after any length of storage [11]. According to our data this is not quite true: some strains of *Achlya colorata, Antrodia serialis, Armillaria cepistipes, Athelia rolfsii, Ceratellopsis equiseticola, Choanephora conjuncta, Clitocybe nuda, Coemansia aciculifera, Collybia butyracea, Conidiobolus thromboides, Exobasidium karstenii, E. splendidum, Kickxella alabastrina, Lactarius deliciosus, Marasmius oreades, Mortierella gamsii, M. humilis, Mycena pura, Phallus impudicus, Rhizoctonia solani, Sclerotium tuliparum, Suillus variegatus,* and *Ustilago scabiosae* have lost their ability to grow after 5–7 years of storage in liquid nitrogen, although they were in the vital state after 24 h of storage. The reason is not yet known. Nevertheless, the viability test showed that 350 strains of fungi remain alive after 19.5 years of storage (Table 3.1).

The cooling equipment being used in VKM is storage tanks "Bioproducts-0.5" with capacity of 500 liters of liquid nitrogen and ultra-low temperature freezers (−80°C, Sanyo, Japan).

Table 3.1 Storage time of VKM fungal cultures

Sr. No.	Name of species	Cryopreservation		Freeze-Drying		Soil	
		Number of strains	Storage time (years)	Number of strains	Storage time (years)	Number of strains	Storage time (years)
1	*Absidia blakesleeana* Lendner 1924	1	19.30	4	32.91	3	27.14
2	*A. coerulea* Bainier 1889	2	19.69	5	29.67	5	12.42
3	*A. cuneospora* G.F.Orr et Plunkett 1959	1	19.69	1	27.38	0	16.24
4	*A. cylindrospora* Hagem 1908			2	24.02	2	28.45
5	*A. glauca* Hagem 1908	4	19.37	10	36.13	5	15.58
6	*A. hyalospora* (Saito 1906) Lendner 1908	1	19.37	1	31.10	1	
7	*A. repens* van Tieghem 1878	1	19.24	1	22.36		
8	*A. spinosa* Lendner 1907	1	12.28	2	19.46	2	11.64
9	*A. bisexualis* Coker et Couch 1927	2	0.51				
10	*Achlya bonariensis* Beroqui 1969	1	0.16				
11	*A. colorata* Pringsheim 1882	2	6.33				
12	*A. intricata* Beneke 1948	1	0.15				
13	*A. sparrowii* Reischer 1949	1	1.11				
14	*Acladium curvatum* Bonorden 1851			1	32.40		
15	*Acremonium alternatum* Link 1809	1	0.58	2	27.04		
16	*A. arxii* W.Gams 1971	1	19.44	3	27.32		
17	*A. atrogriseum* (Panasenko 1964) W.Gams 1971	1	17.46	2	32.65		
18	*A. bacillisporum* (Onions et G.L. Barron 1967) W.Gams 1971			1	15.50		
19	*A. bactrocephalum* W.Gams 1971			3	24.26		
20	*A. berkeleyanum* (P.Karsten 1891) W.Gams 1982	1	19.49	3	30.12		
21	*A. biseptum* W.Gams 1971			1	25.30		
22	*A. breve* (Sukapure et Thirumalachar 1966) W.Gams 1971	3	19.29	4	27.78		
23	*A. cavaraeanum* (Jasevoli 1924) W.Gams 1971	1	19.92	1	6.05		
24	*A. charticola* (J.Lindau 1907) W.Gams 1971	2	19.56	3	25.98		
25	*A. crotocinigenum* (Schol-Schwarz 1965) W.Gams 1971			4	32.50		
26	*A. cymosum* W.Gams 1971	1	6.52	1	28.44		
27	*A. domschii* W.Gams 1971			2	28.36		
28	*A. egyptiacum* (J.F.H.Beyma 1933) W.Gams 1971	1	19.77	1	29.72		

(continued)

Table 3.1 (continued)

Sr. No.	Name of species	Cryopreservation		Freeze-Drying		Soil	
		Number of strains	Storage time (years)	Number of strains	Storage time (years)	Number of strains	Storage time (years)
29	A. gamsii Tichelaar 1971			1	7.06		
30	A. henneberii W.Gams 1971			1	31.29		
31	A. hyalinulum (Saccardo 1879) W.Gams 1971			1	3.23		
32	A. implicatum (J.C.Gilman et E.V.Abbott 1927) W.Gams 1975	2	19.75	3	27.42		
33	A. incrustatum W.Gams 1971	1	17.71	1	25.80		
34	A. killense Gruetz 1925	3	20.27	4	26.11		
35	A. lichenicola W.Gams 1971			1	24.26		
36	A. persicinum (Nicot 1958) W.Gams 1971	2	19.36	2	30.12		
37	A. polychromum (J.F.H.Beyma 1928) W.Gams 1971	2	19.92	3	25.34		
38	A. rutilum W.Gams 1971	1	8.01	3	19.10		
39	A. salmoneum W.Gams et Lodha 1975			3	2.70		
40	A. sclerotigenum (Moreau et R.Moreau 1941 ex Valenta 1948) W.Gams 1971	1	5.09	3	24.26		
41	A. strictum W.Gams 1971	14	19.79	20	39.59	1	9.71
42	A. tubakii W. Gams 1971			1	6.25		
43	Acrophialophora fusispora (S.B.Saksena 1953) Samson 1970	1	17.48	1	6.36		
44	Acrostalagmus albus Preuss 1851	1	19.58	1	18.18		
45	A. luteoalbus (Link 1809) Zare et al. 2004	2	19.34	8	27.50	3	13.56
46	Acrothecium robustum J.C.Gilman et E.V.Abbott 1927	1	19.28	1	22.40		
47	Actinomucor elegans (Eidam 1884) C.R.Benjamin et Hesseltine 1957			7	35.32	7	28.66
48	Agaricus arvensis Schaeffer 1774	2	20.03				
49	A. bisporus (J.Lange 1926) Imbach 1946	19	20.93				
50	Albonectria rigidiuscula (Berkeley et Broome 1875) Rossman et Samuels 1999	1	19.84	1	13.21	1	7.45
51	Alternaria alternata (Fries 1832) Keissler 1912	1	19.28	11	29.77		
52	A. brassicae (Berkeley 1836) Saccardo 1880	1	19.76				
53	A. brassicicola (Schweinitz 1832) Wiltshire 1947	1	19.27	2	24.42		
54	A. cheiranthi (Libert 1827) P.C.Bolle 1924	1	19.78	1	16.13		
55	A. chrysanthemi E.G.Simmons et Crosier 1965	1	19.76	1	24.95		

No.	Species					
56	A. cucumerina (Ellis et Everhart 1895) J.A.Elliott 1917 var. cucumerina	1	19.78			1
57	A. dauci (J.G.Kuehn 1855) J.W.Groves et Skolko 1944	1	19.78			
58	A. dianthicola Neergaard 1945	1	19.76			
59	A. geophila Daszewska 1912	1	12.56			
60	A. godetiae (Neergaard 1933) Neergaard 1945	1	12.56			
61	A. macrospora Zimmermann 1904	1	12.56	2	14.87	
62	A. multirostrata E.G.Simmons et C.R.Jackson 1968	1	17.79	1	11.28	
63	A. nobilis (Vize 1877) E.G. Simmons 2002	1	19.76			
64	A. radicina Meier et al. 1922	1	17.59	1	22.55	
65	A. raphani J.W.Groves et Skolko 1944	1	19.30	1	27.64	
66	A. solani Sorauer 1896	1	19.76	1	11.82	
67	A. tenuissima (Kunze 1818) Wiltshire 1933	1		2	8.64	
68	Amauroascus aureus (Eidam 1887) von Arx 1971	1	6.18	1	17.43	1
69	Amblyosporium botrytis Fresenius 1863	1	15.33	1	31.28	12.18
70	Amerosporium concinnum Petrak 1953	1	19.52	1	5.32	
71	Ampelomyces artemisiae (Voglino 1905) Rudakov 1979	1	12.57	1	11.42	
72	A. heraclei (Dejeva 1967) Rudakov 1979	1	12.18	1	13.21	
73	A. humuli (Fautrey 1890) Rudakov 1979	1		1	16.55	
74	A. polygoni (Potebnia 1907) Rudakov 1979	1	12.28	2	13.12	
75	A. quisqualis Cesati 1852	1	12.57	2	18.13	
76	A. ulicis (Adams 1907) Rudakov 1979	1	12.37	1	18.54	
77	A. uncinulae (Fautrey 1893) Rudakov 1979	1	12.57	1	10.41	
78	Anthurus archeri (Berkeley 1859) E. Fisch. 1886	1	1.64			
79	Aphanoascus fulvescens (Cooke 1879) Apinis 1968	1	20.37	1	20.62	
80	Aphanocladium album (Preuss 1848) W.Gams 1971	3	17.71	3	24.20	
81	Aphanomyces helicoides Minden 1915	1	9.93			
82	Apiospora montagnei Saccardo 1875	1		1	1.63	
83	Aplanes treleaseanus (Humphrey 1893) Coker 1927	1	15.93			
84	Aposphaeria caespitosa (Fuckel 1869) Jaczewski 1917	1	19.19	1	20.57	
85	Arachniotus aurantiacus (Kamyschko 1967) von Arx 1971	1	12.44	1	20.62	
86	Armillaria bulbosa (Barla 1887) Kill et Watling 1983	1	12.01			
87	A. gallica H.Marxmueller et Romagn. 1987	1	0.79			

(continued)

Table 3.1 (continued)

Sr. No.	Name of species	Cryopreservation		Freeze-Drying		Soil	
		Number of strains	Storage time (years)	Number of strains	Storage time (years)	Number of strains	Storage time (years)
88	A. mellea (Vahl 1792) Kummer 1871	3	19.93				
89	Arthrinium arundinis (Corda 1838) Dyko et Sutton 1981	1	12.57	3	10.13		
90	A. saccaricola F. Stevens 1917			1	16.91		
91	A. sphaerospermum Fuckel 1874	1	19.19	1	20.82	1	3.85
92	Arthrobotrys arthrobotryoides (Berlese 1888) J.Lindau 1907	1	19.30			1	3.91
93	A. cladodes Drechsler 1937	1	15.73	1	23.73	1	3.91
94	A. conoides Drechsler 1937	2	19.30	1	9.44	2	3.91
95	A. longa Mekhtieva 1973	1	10.64	1	9.42		
96	A. longispora Press 1853	1	15.63			1	3.91
97	A. oligospora Fresenius 1850	4	21.21	5	22.24	1	2.55
98	A. robusta Duddington 1951	1	19.40	1	27.06	1	3.91
99	A. superba Corda 1839	1	19.25	1	18.25	2	3.91
100	Ascochyta bohemica Kabat et Bubak 1905	1	20.32				
101	A. boltshauseri Saccardo 1891	1	20.32	1	27.54		
102	A. cucumeris Fautrey et Roumeguere 1891	1	20.32	1	27.91		
103	A. malvicola Saccardo 1878	1	19.21				
104	A. pinodes L.K.Jones 1927	1	19.21	1	22.41		
105	A. pisi Libert 1830	1	19.44	1	23.61	1	8.67
106	A. viciae Libert 1837	1	20.30	1	26.76		
107	Ascotricha chartarum Berkeley 1838	1	18.95	1	10.75		
108	Aspergillus alliaceus Thom et Church 1926			3	21.91	2	20.77
109	A. amylovorus Panasenko 1964 ex Samson 1979	1	12.45	1	14.96	1	20.73
110	A. aureolatus Muntanola-Cvetkovic et Bata 1964			1	9.61	1	9.58
111	A. awamori Nakazawa 1915			6	28.56	4	20.73
112	A. awamori Nakazawa 1915 var. fumeus Nakazawa et al. 1936			1	17.96	1	20.91
113	A. brasiliensis Varga et al. 2007			1	31.96	1	22.13
114	A. caespitosus Raper et Thom 1944			1	9.66	1	9.58
115	A. candidus Link 1809			5	38.93	5	20.87
116	A. carbonarius (Bainier 1880) Thom 1916			2	20.02	1	20.87
117	A. carneus Blochwitz 1933			1	16.12		

No.						
118	A. clavatus Desmazieres 1834		7	30.13	7	27.41
119	A. echinulatus (Delacroix 1893) Thom et Church 1926		1	9.60	1	9.58
120	A. ficuum (Reichardt 1867) Thom et Currie 1916		1	13.22	2	1.53
121	A. fischeri Wehmer 1907		4	27.20	3	27.41
122	A. flavipes (Bainier et R.Sartory 1911) Thom et Church 1926		4	31.64	4	20.90
123	A. flavus Link 1809		13	39.07	9	20.87
124	A. flavus Link 1809 var. columnaris Raper et Fennell 1965		1	26.98	1	8.69
125	A. foetidus Thom et Raper 1945		1	14.18	1	8.08
126	A. fumigatus Fresenius 1863		12	38.91		28.72
127	A. giganteus Wehmer 1901		2	16.87	2	17.90
128	A. heteromorphus Batista et H. Maia 1957		1	15.87		
129	A. insuetus (Bainier 1908) Thom et Church 1929		1	15.90	1	11.41
130	A. janus Raper et Thom 1944		1	36.04	1	20.92
131	A. japonicus Saito 1906		4	24.93	2	20.66
132	A. kanagawaensis Nehira 1951		2	39.62	2	16.06
133	A. melleus Yukawa 1911		3	39.08	1	1.91
134	A. niger van Tieghem 1867		26	27.30	18	27.41
135	A. niveus Blochwitz 1929		3	28.54	2	37.78
136	A. nutans McLennan et Ducker 1954		1	37.39	1	11.74
137	A. ochraceus G.Wilhelm 1877		8	24.89	7	20.89
138	A. oryzae (Ahlburg 1878) E.Cohn 1884		21	30.35	19	23.11
139	A. oryzae (Ahlburg 1878) E.Cohn 1884 var. effusus (Tiraboschi 1908) Y.Ohara 1951	12.45	1	12.34	1	9.58
140	A. pallidus Kamyschko 1963	1	1	18.59	1	21.90
141	A. parvulus G.Smith 1961		1	37.39	1	4.65
142	A. penicilliformis Kamyschko 1963		3	38.97	2	21.72
143	A. phoenicis (Corda 1840) Thom et Currie 1916		1	14.94	1	7.66
144	A. pseudodeflectus Samson et Mouchacca 1975		1	36.19		
145	A. puniceus Kwon-Chung et Fennell 1965		1	9.87	1	4.32
146	A. repens (Corda 1842) Saccardo 1882		9	38.67	8	27.41
147	A. restrictus G. Smith 1931		1	15.70	1	2.24
148	A. sclerotiorum G.A.Huber 1933		2	21.13	1	2.05
149	A. silvaticus Fennell et Raper 1955		1	27.29	1	8.08

(continued)

Table 3.1 (continued)

Sr. No.	Name of species	Cryopreservation		Freeze-Drying		Soil	
		Number of strains	Storage time (years)	Number of strains	Storage time (years)	Number of strains	Storage time (years)
150	A. subsessilis Raper et Fennell 1965	1	12.43	1	24.42	1	10.82
151	A. sulphureus (Fresenius 1863) Thom et Church 1926			1	17.11	1	22.65
152	A. sydowii (Bainier et R.Sartory 1913) Thom et Church 1926			6	25.86	4	21.01
153	A. tamarii Kita 1913			1	38.72	1	20.73
154	A. terreus Thom 1918	1	6.67	17	35.75	15	34.79
155	A. terreus Thom 1918 var. africanus Fennell et Raper 1955	1		1	14.73	1	7.74
156	A. terreus Thom 1918 var. aureus Thom et Raper 1945			1	14.73	1	7.22
157	A. terricola Marchal et É.J. Marchal 1893			3	17.41	3	20.92
158	A. terricola Marchal et É.J. Marchal 1893 var. americanus Marchal et É.J. Marchal 1921			1	21.47	1	7.74
159	A. tubingensis Mosseray 1934			1	7.19	1	
160	A. umbrosus Bainier et Sartory 1912			3	38.97	2	21.03
161	A. unguis (Weill et L.Gaudin 1919) Thom et Raper 1939	1	12.45	5	22.22	5	16.58
162	A. ustus (Bainier 1881) Thom et Church 1926			11	24.75	6	20.89
163	A. varians Wehmer 1897			1	18.57		
164	A. versicolor (Vuillemin 1903) Tiraboschi 1908			11	29.25	11	22.34
165	A. wentii Wehmer 1896			3	26.56	3	7.69
166	Athelia rolfsii (Curzi 1932) C.C.Tu et Kimbrough 1978	1	8.63				
167	Aureobasidium microstictum (Bubak 1907) W.B.Cooke 1962	1	19.84	1	26.42		
168	A. pullulans (de Bary 1866) G.Arnaud 1918 var. melanigenum Hermanides-Nijhof 1977	1	6.76	6	27.81	2	12.60
169	A. pullulans (de Bary 1866) G.Arnaud 1918 var. pullulans	1	20.20	10	31.03	1	3.76
170	Backusella circina J.J. Ellis et Hesseltine 1969			2	31.44		
171	B. lamprospora (Lendner 1908) Benny et R.K.Benjamin 1975	2	19.63	4	39.34	2	8.78
172	Bactridium equiseticola Milko et Dunaev			1	10.24		
173	Basidiobolus magnus Drechsler 1964	1	20.05				
174	B. meristosporus Drechsler 1955	1	20.20				
175	Beauveria bassiana (Balsamo-Crivelli 1835) Vuillemin 1912	3	20.39	6	30.68		
176	B. brongniartii (Saccardo 1892) Petch 1924	4	20.39	5	31.16	1	10.92

177	*Benjaminiella poitrasii* (R.K.Benjamin 1960) von Arx 1981	1	19.54	2	16.91	2	24.89
178	*Bionectria ochroleuca* (Schweinitz 1832) Schroers et Samuels 1997			1	26.61		
179	*Bipolaris australiensis* (M.B.Ellis 1971) Tsuda et Ueyama 1981	2	19.28	5	19.06	2	8.53
180	*B. bicolor* (Mitra 1931) Shoemaker 1959			1	6.61		
181	*B. cynodontis* (Marignoni 1909) Shoemaker 1959	1	19.11	3	25.93	1	8.43
182	*B. kusanoi* (Y.Nisikado 1928) Shoemaker 1959			1	7.04		
183	*B. nodulosa* (Berkeley et M.A.Curtis 1886) Shoemaker 1959			1	5.45		
184	*B. sorokiniana* (Saccardo 1890) Shoemaker 1959	1	19.34	3	19.77		
185	*B. spicifera* (Bainier 1908) Subramanian 1971			1	5.40		
186	*B. victoriae* (F. Meehan et H.C. Murphy 1946) Shoemaker 1959	1	17.46	1	16.28		
187	*Biscogniauxia nummularia* (Bulliard 1790) Kuntze 1891	1	19.31	1	0.66		
188	*Bispora antennata* (Persoon 1801) E.W. Mason 1953	2	19.15	2	27.07		
189	*B. betulina* (Corda 1838) S.Hughes 1958	1	17.44	2	8.96		
190	*B. effusa* Peck 1891	1	20.20	1	15.09		
191	*Bjerkandera adusta* (Willdenow 1787) P.Karsten 1879	2	8.31	2			
192	*Blakeslea trispora* Thaxter 1914	8	19.56	17	40.03	4	11.74
193	*Blumeriella jaapii* (Rehm 1907) Arx 1961	2	8.60	3	8.36		
194	*Botryosphaeria rhodina* (Berkeley et M.A.Curtis 1889) von Arx 1970	1	19.96	1	26.19		
195	*Botryosporium longibrachiatum* (Oudemans 1890) Maire 1903			1	2.01		
196	*Botryotinia narcissicola* (P.H.Gregory 1941) N.F.Buchwald 1949	1	18.95	1	5.39		
197	*B. polyblastis* (P.H.Gregory 1938) N.F.Buchwald 1949			1	19.35		
198	*Botryotrichum piluliferum* Saccardo et Marchal 1885	1	19.21	4	27.64		
199	*Botryoxylon geniculatum* (Corda 1839) Ciferri 1962			2	28.41		
200	*Botrytis aclada* Fresenius 1850	1	19.29	2	17.59		
201	*B. anthophila* Bondartsev 1913	1	19.19	1	22.52		
202	*B. bifurcata* J.H. Mill., Giddens et A.A. Foster 1958			1	6.88		
203	*B. cinerea* Persoon 1794	1	19.39	15	22.56		
204	*B. convallariae* (Klebahn 1930) Ondrej 1972 ex Boerema et Hamers 1988			2	3.56		
205	*B. convoluta* Whetzel et Drayton 1932	1	19.21	2	17.62		
206	*B. elliptica* (Berkeley 1881) Cooke 1901	1	19.39	1	19.39		

(continued)

Table 3.1 (continued)

Sr. No.	Name of species	Cryopreservation		Freeze-Drying		Soil	
		Number of strains	Storage time (years)	Number of strains	Storage time (years)	Number of strains	Storage time (years)
207	B. fabae Sardina 1929	1	19.11				
208	B. galanthina (Berkeley et Broome 1873) Saccardo 1886	1	19.21	1	9.62		
209	B. gladiolorum Timmermans 1941	1	19.29	2	23.47		
210	B. hyacinthi Westerdijk et J.F.H.Beyma 1928	1	19.19				
211	B. lutescens Saccardo et Roumeguere 1882	1	19.19				
212	B. squamosa J.C.Walker 1925	1	19.31				
213	B. tulipae (Libert 1830) Lind 1913	1	19.48	1	13.16		
214	Brachysporium nigrum (Link 1824) S. Hughes 1958			1	26.08		
215	Burgoa anomala (Hotson 1912) Goidanich 1937	1	19.86	1	8.98		
216	Byssochlamys nivea Westling 1909	2	19.42	2	30.50	1	22.83
217	Cadophora fastigiata Lagerberg et Melin 1928	1	19.39	2	21.47		
218	C. malorum (Kidd et Beaumont 1924) W. Gams 2000	4	19.74	4	19.27	1	
219	C. melinii Nannfeldt 1934	1	19.47	1	22.91		
220	Calcarisporium arbuscula Preuss 1851	1	1.93	3	28.45		
221	C. griseum Spegazzini 1902	1	7.62	2	27.88		
222	Calocera viscosa (Persoon 1794) Fries 1828	1	0.10				
223	Ceratellopsis aquiseticola (Boudier 1917) Corner 1950	1	12.06				
224	Ceratocystis paradoxa (Dade 1928) C.Moreau 1952	2	19.84	2	17.43	1	6.47
225	C. pilifera (Fries 1822) C.Moreau 1952	1	18.85	1	25.82		
226	Cercospora armoraciae Saccardo 1876	1	20.32	1	11.89		
227	C. beticola Saccardo 1876			2	15.29		
228	C. carotae (Passerini 1887) Kaznowski et Siemaszko 1929	1	20.32	1	12.70		
229	C. rosicola Passerini 1875	1	16.65				
230	C. violae Saccardo 1876			1	12.67		
231	Ceriporiopsis gilvescens (Bresadola 1908) Domanski 1963	1	12.57				
232	Cerrena unicolor (Bulliard 1788) Murrill 1903	1	12.58				
233	Chaetocladium brefeldii van Tieghem et G.Le Monnier 1873	1	17.58	2	33.55	1	16.98
234	C. jonesii (Berkeley et Broome 1854) Fresenius 1863	1	17.58	1	27.42		
235	Chaetocytostroma sp.			1	20.53		
236	Chaetomidium pilosum (C.Booth et Shipton 1966) von Arx 1975	1	14.44	1	18.28	1	3.11

No.	Name						
237	*Chaetomium amesii* Sergeeva 1965	1	19.27	1	31.01		
238	*C. angustispirale* Sergeeva 1956	1	19.27	1	17.37		
239	*C. aureum* Chivers 1912	2	20.50	2	15.45		
240	*C. crispatum* (Fuckel 1867) Fuckel 1870	1	19.27	1	20.34		
241	*C. elatum* Kunze 1817	3	20.50	3	27.68	1	11.03
242	*C. fieberi* Corda 1837			1	9.05	1	10.68
243	*C. funicola* Cooke 1873	1	19.27	1	26.15		
244	*C. globosum* Kunze 1817	7	19.42	11	31.21	3	10.68
245	*C. homopilatum* Omvik 1953	1	19.32	1	10.66		
246	*C. indicum* Corda 1840	1	19.27	2	17.66		
247	*C. megalocarpum* Bainier 1910	2	19.27	2	17.39		
248	*C. nozdrenkoae* Sergeeva 1961	1	18.85	1	35.25		
249	*C. perlucidum* Sergeeva 1956	1	19.27	1	35.29		
250	*C. semenis-citrulli* Sergeeva 1956	1	19.27	1	17.35		
251	*C. spirale* Zopf 1881	1	19.27	1	11.07		
252	*C. subaffine* Sergeeva 1961	1	19.27	1	35.29		
253	*C. subspirilliferum* Sergeeva 1960	1	20.37	1	25.61		
254	*C. trilaterale* Chivers 1912			1	11.80		
255	*Chaunopycnis alba* W. Gams 1979			1	2.01		
256	*Chloridium virescens* (Persoon 1797) W.Gams et Holubova-Jechova 1976 var. *caudigerum* (Hoehnel 1903) W.Gams et Holubova-Jechova 1976	1	19.66	1	21.35		
257	*Choanephora circinans* (H.Naganishi et N.Kawakami 1955) Hesseltine et C.R.Benjamin 1957	1	20.06	1	27.36		
258	*C. conjuncta* Couch 1925	2	5.81	1	8.64		
259	*C. cucurbitarum* (Berkeley et Ravenel 1875) Thaxter 1903	1	20.19	1	20.93		
260	*C. infundibulifera* (Currey 1873) Saccardo 1891	1	19.07	1	26.29		
261	*Chondrostereum purpureum* (Persoon 1794) Pouzar 1959	1	20.09				
262	*Chrysonilia sitophila* (Mont. 1843) Arx 1981			1	19.81	1	9.02
263	*Chrysosporium carmichaelii* Oorschot 1980			1	22.10		
264	*C. keratinophilum* D.Frey 1959 ex J.W.Carmichael 1962	1	9.56	2	19.07	1	5.95
265	*C. lobatum* Scharapov 1978			1	31.11	1	5.87
266	*C. lucknowense* Garg 1966	4	5.61	5	8.59		

(continued)

Table 3.1 (continued)

Sr. No.	Name of species	Cryopreservation		Freeze-Drying		Soil	
		Number of strains	Storage time (years)	Number of strains	Storage time (years)	Number of strains	Storage time (years)
267	C. merdarium (Link 1818 ex Greville 1823) J.W.Carmichael 1962			2	25.63	1	5.87
268	C. queenslandicum Apinis et R.G.Rees 1976	2	19.81	2	31.15	1	6.07
269	C. tropicum J.W.Carmichael 1962	1	9.08	1	24.13		
270	Circinella muscae (Sorokin 1870) Berlese et de Toni 1888	1	2.81	5	28.07	4	28.66
271	C. rigida G.Smith 1951	1	19.67	1	26.96	1	16.39
272	C. umbellata van Tieghem et G.Le Monnier 1873	1	19.67	1	6.91	1	19.35
273	Cladobotryum binatum Preuss 1851			1	29.01		
274	C. dendroides (Bulliard 1791) W.Gams et Hoozemans 1970			3	25.88		
275	C. multiseptatum de Hoog 1978			1	9.00		
276	C. varium Nees 1816	1	1.20	3	27.38		
277	C. verticillatum (Link 1809) S. Hughes 1958			2	28.41		
278	Cladophialophora chaetospira (Grove 1886) Crous et Arzanlou 2007			1	17.82		
279	Cladosporium aecidiicola Thuemen 1876	1	19.19	1	5.95		
280	C. brevicompactum Pidoplichko et Deniak 1941			2	27.73		
281	C. bruhnei Linder 1947			1	9.30		
282	C. cladosporioides (Fresenius 1850) G. A. de Vries 1952	2	19.51	15	37.72	3	6.72
283	C. colocasiae Sawada 1916	1	19.34	1	18.35		
284	C. cucumerinum Ellis et Arthur 1889			1	18.62		
285	C. elegantulum Pidoplichko et Deniak 1938			2	26.47	1	7.58
286	C. gossypiicola Pidoplichko et Deniak 1941			2	34.83		
287	C. halotolerans Zalar et al. 2007			1	13.43		
288	C. herbarum (Persoon 1794) Link 1816			31	36.58	8	5.76
289	C. macrocarpum Preuss 1848	1	18.99	3	17.54		
290	C. sphaerospermum Penzig 1882	1	19.29	7	26.65	2	14.36
291	C. straminicola Pidoplichko et Deniak 1938			1	26.05		
292	C. transchelii Pidoplichko et Deniak 1938			1	13.10		
293	Clavariadelphus pistillaris (Fries 1753) Donk 1933	2	2.16				
294	Claviceps paspali F.Stevens et J.G.Hall 1910	3	18.93				
295	C. purpurea (Fries 1823) Tulasne 1853	3	18.93	1	4.55		

296	Clitocybe odora (Bulliard 1784) P.Kummer 1871	1	0.01				
297	Clonostachys rosea (Link 1816) Schroers, Samuels, Seifert et W.Gams 1999	14	19.85	23	37.79	16	19.96
298	C. solani (Harting 1846) Schroers et W.Gams 2001	1	18.89	3	32.42		
299	Coemansia aciculifera Linder 1943	1	8.85				
300	Cokeromyces recurvatus Poitras 1950	2	20.14	3	27.13	1	22.74
301	Colletoconis aecidiophila (Spegazzini 1886) de Hoog et al. 1978	1	16.86	1	9.68		
302	Colletotrichum coccodes (Wallroth 1833) S.Hughes 1958			3	6.41		
303	C. gloeosporioides (Penzig 1880) Saccardo 1882	1	19.32	2	37.54		
304	C. lindemuthianum (Saccardo et Magnus 1878) Briosi et Cavara 1889			1	9.42		
305	C. musae (Berkeley et M.A.Curtis 1874) Arx 1957	1	19.30	1	24.90		
306	Collybia butyracea (Bulliard 1792) P.Kummer 1871	1	2.82				
307	Colpoma quercinum (Persoon 1796) Wallroth 1823			1	10.61		
308	Conidiobolus coronatus (Costantin 1897) Batko 1964	4	15.93				
309	C. thromboides Drechsler 1953	1	6.78				
310	Coniochaeta verticillata (van Emden 1973) Dania García, Stchigel et Guarro 2006	1	18.95	1	14.79		
311	Coniophora puteana (Schumacher 1803) P.Karsten 1868	4	12.01				
312	Coniothyrium concentricum (Desmazieres 1840) Saccardo 1878	1	18.99	1	12.63		
313	C. fuckelii Saccardo 1878			1	9.65		
314	C. hellebori Cooke et Massee 1886	1	17.68	1	16.15		
315	C. rosarum Cooke et Harkness 1882	1	19.39	2	16.28		
316	C. wernsdorffiae Laubert 1905	1	18.99	1	16.61		
317	Coprinus atramentarius (Bulliard 1783) Fries 1838	1	6.11				
318	C. comatus (O.F. Mueller 1780) Persoon 1797	2	12.05				
319	C. disseminatus (Persoon 1801) Gray 1821	1	3.82				
320	C. domesticus (Bolton 1788) Gray 1821	1	20.14				
321	C. ephemerus (Bulliard 1786) Fries 1838	1	18.86				
322	C. micaceus (Bulliard 1785) Fries 1838	2	18.82				
323	C. radians (Desmazieres 1828) Fries 1838	1	18.82				
324	C. sterquilinus (Fries 1821) Fries 1838	1	18.86				
325	Corynascus sepedonium (C.W.Emmons 1932) von Arx 1973	1	20.37	1	39.91		

(continued)

Table 3.1 (continued)

Sr. No.	Name of species	Cryopreservation		Freeze-Drying		Soil	
		Number of strains	Storage time (years)	Number of strains	Storage time (years)	Number of strains	Storage time (years)
326	Cryphonectria parasitica (Murrill 1906) M.E.Barr 1978	1	19.34	1	18.90		
327	Cunninghamella blakesleeana Lendner 1927	2	14.09			2	13.81
328	C. echinulata (Thaxter 1891) Thaxter 1905	7	19.42	15	35.33	12	29.23
329	C. homothallica Kominami et Tubaki 1952	1	20.07				
330	C. japonica (Saito 1905) Pidoplichko et Milko 1971	4	20.19	6	33.07	1	28.66
331	C. vesiculosa P.C.Misra 1966	1	14.09				
332	Curvularia clavata B.L.Jain 1962			1	0.65		
333	C. comoriensis Bouriquet et Jauffret 1955 ex M.B.Ellis 1966	1	17.79	1	13.83		
334	C. geniculata (Tracy et Earle 1896) Boedijn 1933	2	19.28	3	19.15		
335	C. inaequalis (Shear 1907) Boedijn 1933	3	19.74	3	20.46	1	4.33
336	C. lunata (Wakker 1898) Boedijn 1933	1	19.34	4	29.21		
337	C. protuberata Nelson et Hodges 1965			1	1.33		
338	Cyathus olla (Batsch 1783) Persoon 1801	1	0.46				
339	Cylindrium cordae Grove 1886			1	27.86		
340	Cylindrocarpon album (Saccardo 1877) Wollenweber 1917	1	17.66	1	29.98		
341	C. chlamydospora Schischkina et Tzanava 1973	1	19.96	1	11.89		
342	C. congoense J.A.Meyer 1958	1	19.54	1	18.45		
343	C. destructans (Zinssmeister 1918) Scholten 1964 var. destructans	1	15.66	1	14.82		
344	C. didymum (Hartig 1846) Wollenweber 1926	1	19.54	1	9.05		
345	C. gracile Bugnicourt 1939	1	15.58	2	11.58		
346	C. heteronema (Berkeley et Broome 1865) Wollenweber 1928	1	19.51	1	18.45		
347	C. ianthothele Wollenweber 1917 var. minus Reinking 1938	1	19.51				
348	C. magnusianum Wollenweber 1928	1	15.58	1	14.78	1	10.86
349	C. obtusisporum (Cooke et Harkness 1884) Wollenweber 1926	1	19.54	1	18.94	1	
350	C. peronosporae (Fautrey et Lambotte 1896) Rudakov 1981			1	32.16		
351	C. stilbophilum (Corda 1838) Rudakov 1981			2	29.51		
352	C. theobromicola C. Booth 1966			3	7.89		
353	Cylindrocephalum stellatum (Harz 1871) Saccardo 1886			1	27.87		
354	Cylindrophora alba Bonorden 1851			1	19.78		

No.	Species						
355	*C. hoffmannii* Daszewska 1912	1	17.71	1	28.06	1	5.65
356	*Dacrymyces stillatus* Nees 1817	1	6.62	1	23.58		
357	*Dactylaria dimorphospora* Veenbaas-Rijks 1973	1	16.08				
358	*Dactylellina asthenopaga* (Drechsler 1937) M. Scholler, Hagedorn et A. Rubner 1999	1	19.28	1	0.08		
359	*Daedalea quercina* (Linnaeus 1753) Persoon 1801	2	20.11	1	28.29		
360	*Dendrodochium toxicum* Pidoplichko et Bilai 1947	1	19.31	1	14.23	1	4.79
361	*Dendrostibella macrospora* W.Bally 1917	1	6.03				
362	*D. mycophila* (Persoon 1822) Seifert 1985			1	25.87		
363	*Dendryphion nanum* (Nees 1816) S.Hughes 1958			1	24.14		
364	*D. penicillatum* (Corda 1838) Fries 1849			2	15.59		
365	*Dichobotrys* sp.	1	19.26	1	18.88		
366	*Dichotomomyces cejpii* (Milko 1964) D.B. Scott 1970	1	18.95	1	19.72	1	1.51
367	*Dictyophora duplicata* (Bosc 1811) E.Fischer 1888	1	2.39				
368	*Dictyostelium discoideum* (Bosc 1811) E.Fischer 1888	1	5.74	1	10.34		
369	*Dictyuchus monosporus* Leitgeb 1869	1	12.56				
370	*Dicyma ampullifera* Boulanger 1897	1	19.19	1	12.36		
371	*D. olivacea* (Emoto et Tubaki 1970) Arx 1982	1	19.39	1	22.93		
372	*D. ovalispora* (S.Hughes 1951) Arx 1982	1	6.68	1	12.32		
373	*Didymopsis helvellae* (Corda 1854) Saccardo et Marchall 1885			1	28.44		
374	*Dimargaris bacillispora* R.K.Benjamin 1959	1	10.64	1	8.30		
375	*Dinemasporium strigosum* (Persoon 1801) Saccardo 1881	1	17.79	1	10.82		
376	*Diplocladium majus* Bonorden 1851			2	28.45		
377	*D. penicillioides* Saccardo 1886			2	27.21		
378	*Dipodascopsis tothii* (Zsolt 1963) L.R.Batra et Millner 1978			1	17.71		
379	*D. uninucleata* (Biggs 1937) L.R.Batra et Millner 1978 var. *uninucleata*	1	4.03	2	17.71		
380	*Dipodascus aggregatus* Francke-Grosmann 1952	1	18.86	1	17.71		
381	*Discula brunneotingens* E.I.Meyer 1953	1	20.30	1	14.78		
382	*D. pinicola* (Naumov 1926) Petrak 1927 var. *mammosa* Lagerberg et al. 1927	1	19.32	1	39.74		
383	*Dispira cornuta* van Tieghem 1875	1	12.31				
384	*Dissoacremoniella silvatica* Kirilenko 1970	1	19.43	1	22.90		

(continued)

Table 3.1 (continued)

Sr. No.	Name of species	Cryopreservation		Freeze-Drying		Soil	
		Number of strains	Storage time (years)	Number of strains	Storage time (years)	Number of strains	Storage time (years)
385	*Doratomyces purpureofuscus* (Schweinitz 1832) F.J.Morton et G.Smith 1963	1	19.38	1	20.31		
386	*D. stemonitis* (Persoon 1801) F.J.Morton et G.Smith 1963	1	19.39	3	24.58	1	5.68
387	*Drechmeria coniospora* (Drechsler 1941) W. Gams et H.-B. Jansson 1985			1	13.88		
388	*Drechslera avenacea* (M.A.Curtis ex Cooke 1889) Shoemaker 1959			2	22.21		
389	*D. biseptata* (Saccardo et Roumeguere 1881) M.J.Richardson et E.M.Fraser 1968	1	19.28	1	12.26	1	4.33
390	*D. campanulata* (Leveille 1841) B.Sutton 1976	1	19.37	2	5.32		
391	*D. poae* (Baudys 1916) Shoemaker 1962			1	18.20		
392	*Duddingtonia flagrans* (Duddington 1949) R.C.Cooke 1969	1	19.30				
393	*Echinobotryum rubrum* Sorokin ex Jaczewski 1917			1	25.93		
394	*Eladia saccula* (E.Dale 1926) G.Smith 1961	1	19.79	1	12.81		
395	*Emericella nidulans* (Eidam 1883) Vuillemin 1927			11	39.08	9	27.41
396	*E. quadrilineata* (Thom et Raper 1939) C.R. Benjamin 1955			3	26.90	2	11.42
397	*E. rugulosa* (Thom et Raper 1939) C.R. Benjamin 1955			4	37.09	4	16.74
398	*E. variecolor* Berkeley et Broome 1857			1	26.90	1	8.68
399	*Emericellopsis donezkii* Beliakova 1974	3	20.43	7	32.76		
400	*E. glabra* (J.F.H.Beyma 1940) Backus et Orpurt 1961	1	19.32	2	21.67		
401	*E. humicola* (Cain 1956) Gilman 1956	1	19.32	1	21.45		
402	*E. maritima* Beliakova 1970	1	14.87	1	14.81		
403	*E. minima* Stolk 1955	10	20.41	9	32.50		
404	*E. pallida* Beliakova 1974	1	20.43	1	22.93	1	9.64
405	*E. robusta* van Emden et W.Gams 1971			1	20.39		
406	*E. terricola* J.F.H.Beyma 1940	1	19.32	1	21.59		
407	*Engyodontium album* (Limber 1940) de Hoog 1978			3	11.56		
408	*Entomophthora dipterigena* (Thaxter 1888) Saccardo et Traverso 1891	1	13.91				
409	*E. pyriformis* Thoizon 1967	1	0.16				

No.	Species						
410	E. thaxteriana I.M.Hall et J.Bell 1963	2	12.05				
411	Entyloma gaillardianum Vanky 1982	1	12.34	1	1.64		
412	Epicoccum nigrum Link 1815	1	19.80	2	21.78		
413	Eremascus fertilis Stoppel 1907	1	18.85	1	5.73		
414	Eremothecium ashbyi Guilliermond 1935	3	17.31				
415	E. gossypii (S.F.Ashby et W.Nowell 1926) Kurtzman 1995	3	16.31				
416	Eupenicillium javanicum (J.F.H.Beyma 1929) Stolk et D.B.Scott 1967 var. javanicum	1	18.30	1	20.18	1	24.83
417	Eurotium amstelodami (Mangin 1909) Thom et Church 1926	1	3.99	8	38.97	8	32.38
418	E. chevalieri L. Mangin 1909			5	36.93	4	20.75
419	E. halophilicum C.M.Christensen et al. 1959			1	28.92		
420	E. herbariorum (F.H.Wiggers 1780) Link 1809	1	18.86	1	37.95	1	21.57
421	E. rubrum Jos. König et al. 1901			7	32.45	5	11.24
422	E. tonophilum Ohtsuki 1962	1	19.32	1	39.91	1	19.77
423	Evlachovaea kintrischica B.Borisov et Tarasov 1999	1	1.00	1	15.54		
424	Exobasidium bisporum Sawada 1950			1	1.64		
425	E. karstenii Saccardo et Trotter 1912	2	9.02				
426	E. myrtilli Siegmund 1879	1	8.17	1	1.64		
427	E. pachysporum Nannfeldt 1981	1	2.04				
428	E. vaccinii (Fuckel 1861) Woronin 1867	2	12.19	1	1.81		
429	E. warmingii Rostrup 1888	1	18.48	1	1.64		
430	Exophiala castellanii Iwatsu et al. 1999	1	20.38	2	16.34		
431	E. heteromorpha (Nannfeldt 1934) de Hoog et Haase 2003	1	19.53	1	21.63		
432	E. lecanii-corni (Benedek et Specht 1933) Haase et de Hoog 1999			1	7.93		
433	E. moniliae de Hoog 1977			1	7.93		
434	E. salmonis J.W.Carmichael 1966	1	17.77	1	5.95		
435	Exserohilum pedicellatum (A.W. Henry 1924) K.J. Leonard et Suggs 1974	1	19.46	1	18.88		
436	Farlowiella carmichaeliana (Berkeley 1836) Saccardo 1891			1	26.00		
437	Farrowia seminuda (L.M.Ames 1949) D. Hawksworth 1975			1	10.72		
438	Fennellomyces linderi (Hesseltine et Fennell 1955) Benny et R.K.Benjamin 1975	1	19.63	1	15.02	1	25.98

(continued)

Table 3.1 (continued)

Sr. No.	Name of species	Cryopreservation		Freeze-Drying		Soil	
		Number of strains	Storage time (years)	Number of strains	Storage time (years)	Number of strains	Storage time (years)
439	Fibroporia vaillantii (de Candolle 1815) Parmasto 1968	1	20.09				
440	Filobasidiella depauperata (Petch 1932) R.A.Samson et al. 1983	1	7.57				
441	Flammulina velutipes (Curtis 1777) Singer 1951	5	20.01				
442	Fomes fomentarius (Linnaeus 1753) Fries 1849	1	20.05				
443	Fomitopsis pinicola (Swartz 1810) P.Karsten 1889	4	20.03				
444	F. rosea (Albertini et Schweinitz 1805) P.Karsten 1881	1	20.09				
445	Fonsecaea pedrosoi (Brumpt 1922) Negroni 1936	1	19.71	1	16.12		
446	Funalia trogii (Berkeley 1850) Bondartsev et Singer 1941	2	20.05				
447	Fusarium agaricorum Sarrazin 1887	1	17.66	1	27.21		
448	F. aquaeductuum (Rabenhorst et Radlkofer 1863) Lagerheim 1891	1	0.19	2	33.19		
449	F. aquaeductuum (Rabenhorst et Radlkofer 1863) Lagerheim 1891 var. medium Wollenweber 1931			1	23.04	1	5.64
450	F. arthrosporioides Sherbakoff 1915			1	10.16		
451	F. avenaceum (Fries 1832) Saccardo 1886	2	19.98	3	33.20	2	14.34
452	F. avenaceum (Fries 1832) Saccardo 1886 var. herbarum (Corda 1839) Saccardo 1886	2	17.46	1	26.39		
453	F. cerealis (Cooke 1878) Saccardo 1886	1	2.19	1	22.49		
454	F. chlamydosporum Wollenweber et Reinking 1925			2	29.57		
455	F. concolor Reinking 1935	2	1.88	1	5.10		
456	F. culmorum (W.G.Smith 1884) Saccardo 1895	1	17.66	3	25.41	2	11.43
457	F. decemcellulare Brick 1908	2	17.46	2	33.20	2	9.66
458	F. episphaeria (Tode 1791) Snyder et Hansen 1945			1	15.12	1	6.65
459	F. epistroma (Hoehnel 1909) C.Booth 1971			2	32.43		
460	F. equiseti (Corda 1838) Saccardo 1886	2	19.26	6	29.63	2	23.56
461	F. expansum Schlechtendal 1824			1	29.47		
462	F. fujikuroi Nirenberg 1976	1	7.64	1	21.06	1	21.87
463	F. graminearum Schwabe 1839	2	15.66	4	38.91	2	7.09
464	F. heterosporum Nees et T. Nees 1818			2	29.86	1	7.09
465	F. heterosporum Nees et T. Nees 1818 var. pucciniophilum	1	17.71	1	23.19		

466	F. incarnatum (Roberge 1849) Saccardo 1886	3	16.56	4	28.49		
467	F. javanicum Koorders 1907	2	19.85	2	32.16	2	6.59
468	F. lateritium Nees 1816	3	19.78	6	32.78	1	21.85
469	F. merismoides Corda 1838	1	7.68	2	33.20		
470	F. merismoides Corda 1838 var. merismoides			1	0.57		
471	F. nivale (Fries 1849) Cesati 1860 ex Saccardo 1886	2	17.66	3	27.70	1	5.97
472	F. oxysporum Schlechtendal 1824	9	19.86	20	33.20	8	23.93
473	F. poae (Peck 1903) Wollenweber 1913	2	17.71	3	31.70	1	12.68
474	F. redolens Wollenweber 1913			3	14.98		
475	F. sambucinum Fuckel 1869	3	7.68	9	30.03	3	13.96
476	F. sambucinum Fuckel 1869 var. ossicola (Berkeley et M.A.Curtis 1875) Bilai 1955			1	1.16		
477	F. sarcochroum (Desmazieres 1850) Saccardo 1879	1	4.98	1	21.31		
478	F. solani (Martius 1842) Saccardo 1881	2	16.15	14	32.19	4	16.90
479	F. sporotrichioides Sherbakoff 1915	1	17.71	4	32.12	1	15.85
480	F. sporotrichioides Sherbakoff 1915 var. sporotrichioides	1	17.71	2	31.70	1	12.84
481	F. tricinctum (Corda 1838) Saccardo 1886	2	17.46	4	32.19	1	4.75
482	F. ventricosum Appel et Wollenweber 1913			3	29.13	2	12.68
483	F. verticillioides (Saccardo 1882) Nirenberg 1976	15	19.98	21	38.76	16	23.65
484	F. viride (Lechm.) Wollenweber 1917	1	17.66	1	23.96		
485	F. wolgense Rodigin 1942			1	33.02		
486	Fusicladium pomi (Fries 1825) Lind 1913	1	19.81				
487	Gabarnaudia betae (Delacroix 1897) Samson et W.Gams 1974	1	10.01	3	27.06		
488	Gaeumannomyces caricis J.Walker 1980	1	5.61				
489	Galactomyces geotrichum (E.E.Butler et L.J.Petersen 1972) Redhead et Malloch 1977	3	19.26	3	21.10	1	3.45
490	G. reessii (van der Walt 1959) Redhead et Malloch 1977	1	19.26	1	20.12	1	3.45
491	Ganoderma applanatum (Persoon 1799) Patouillard 1889	3	12.63				
492	Geastrum fimbriatum Fries 1829	1	0.47				
493	Geomyces pannorum (Link 1824) Sigler et J.W.Carmichael 1976	2	19.45	14	33.38		
494	Geosmithia lavendula (Raper et Fennell 1948) Pitt 1980			1	20.84	1	5.31
495	G. namyslowskii (K.M.Zalessky 1927) Pitt 1980			1	21.00		
496	Geotrichum amycelicum Redaelli et Ciferri 1935	1	18.85	1	11.15	1	15.07

(continued)

Table 3.1 (continued)

Sr. No.	Name of species	Cryopreservation		Freeze-Drying		Soil	
		Number of strains	Storage time (years)	Number of strains	Storage time (years)	Number of strains	Storage time (years)
497	G. bipunctatum Rolland et Fautrey 1894			1	28.38		
498	G. candidum Link 1809	19	20.39	23	31.16	1	1.59
499	G. fragrans (Berkhout 1923) Morenz 1960 ex Morenz 1964	3	19.85	4	31.67		
500	G. klebahnii (Stautz 1931) Morenz 1964	2	19.77	3	29.62		
501	Gibberella fujikuroi (Sawada 1917) Wollenweber 1931	2	19.31	2	5.25	1	27.17
502	G. zeae (Schweinitz 1821) Petch 1936			2	8.72	1	5.60
503	Gibellulopsis nigrescens (Pethybridge 1919) Zare. W. Gams et Summerbell 2007	2	19.31	5	32.31	2	4.80
504	Gilbertella persicaria (E.D.Eddy 1925) Hesseltine 1960	1	11.41	1	27.13	1	27.09
505	Gliocephalotrichum bulbilium J.J.Ellis et Hesseltine 1962			1	11.92		
506	Gliocladiopsis tenuis (Bugnicourt 1939) Crous et M.J. Wingfeld 1993			1	14.46		
507	Gliocladium album (Preuss 1851) Petch 1926			2	26.90		
508	G. ammoniphilum Pidoplichko et Bilai 1953	1	19.25	1	28.55	1	9.57
509	G. aurifilum (W. Gerard 1874) Seifert, Samuels et W. Gams 1985	1	0.54	1	14.41		
510	G. cholodnyi Pidoplichko 1931	2	16.14	2	26.27	2	9.13
511	G. comtus Rudakov 1981			1	30.99		
512	G. deliquescens Sopp 1912			1	17.27		
513	G. penicillioides Corda 1840			2	28.51		
514	G. viride Matruchot 1893	1	7.80	4	32.23		
515	Gliomastix cerealis (P.Karsten 1887) C.H.Dickinson 1968	2	7.80	3	25.16	1	10.97
516	G. inflata C.H.Dickinson 1968			2	23.17		
517	G. luzulae (Fuckel 1870) E.W. Mason 1953 ex S.Hughes 1958	1	6.22	3	29.44		
518	G. murorum (Corda 1839) S. Hughes 1958 var. felina (Marchal 1895) S.Hughes 1958	4	19.90	5	32.53	1	9.25
519	G. murorum (Corda 1838) S. Hughes 1958 var. murorum	6	19.88	10	33.64	1	9.49
520	Gloeophyllum odoratum (von Wulfen 1788) Imazeki 1943	1	12.39				
521	G. sepiarium (von Wulfen 1786) P.Karsten 1879	5	20.09				
522	Gongronella butleri (Lendner 1926) Peyronel et Dal Vesko 1955	2	19.69	6	32.12	1	5.42
523	G. lacrispora Hesseltine et J.J.Ellis 1961	1	15.37	1	18.86		

No.	Species					
524	Gonytrichum caesium Nees 1818			1	27.64	
525	G. macrocladum (Saccardo 1880) S.Hughes 1951	1	17.66	1	6.36	
526	Graphium putredinis (Corda 1839) S. Hughes 1958			1	23.88	
527	Grifola frondosa (Dickson 1785) Gray 1821	1	12.58			
528	Guepiniopsis buccina (Persoon 1801) L.L.Kennedy 1958	1	12.06			
529	Gymnoascus reessii Baranetzky 1872			1	36.56	
530	Hansfordia pulvinata (Berkeley et M.A.Curtis 1875) S.Hughes 1958	1	19.43	2	26.42	
531	H. triumfettae (Hahsford 1943) S. Hughes 1952			1	27.00	
532	Haplaria repens Bonorden 1851	1	12.55	1	14.08	
533	Haplographium delicatum Berkeley et Broome 1859	1	19.37	1	15.65	
534	Haplotrichum capitatum (Link 1809) Link 1824	2	19.56	2	28.92	
535	Hapsidospora milkoi Beliakova 1975	1	4.09	1	28.59	
536	Harposporium liliputanum Dixon 1952	1	19.28	1	17.19	
537	Harzia acremonioides (Harz 1871) Costantin 1888	1	19.47	3	27.73	
538	Harziella capitata Costantin et Matr. 1899			1	3.02	
539	Helicodendron tubulosum (Riess 1853) Linder 1929	1	19.29	1	11.44	
540	Helicostylum elegans Corda 1842	1	17.58	1	39.15	15.39
541	H. pulchrum (Preuss 1851) Pidoplichko et Milko 1971			2	27.11	16.63
542	Helminthosporium solani Durieu et Montagne 1849	1	19.59	1	9.56	
543	Hemicarpenteles ornatum (Subramanian 1972) Arx 1974			1	39.58	2.64
544	Hericium coralloides (Scopoli 1772) Persoon 1794	4	12.56			
545	H. erinaceus (Bulliard 1791) Persoon 1797	2	12.63			
546	Hesseltinella vesiculosa H.P.Upadhyay 1970	1	6.83	1	15.88	
547	Heterobasidion annosum (Fries 1821) Brefeld 1888	1	20.09			
548	Hirsutella thompsonii F.E. Fischer 1950			1	15.26	
549	Hormiactis alba Preuss 1851			1	23.26	
550	Hormoconis resinae (Lindau 1906) von Arx et G.A. de Vries 1973	2	19.47	12	34.15	17.28
551	Hormonema macrosporum L. Voronin 1986	1	19.80	1	26.36	
552	H. prunorum (Dennis et Buhagiar 1973) Hermanides-Nijhof 1977	1	16.92	1	27.42	
553	Humicola fuscoatra Traaen 1914	1	19.56	2	22.56	
554	H. grisea Traaen 1914	1	19.28	2	27.50	

(continued)

Table 3.1 (continued)

Sr. No.	Name of species	Cryopreservation		Freeze-Drying		Soil	
		Number of strains	Storage time (years)	Number of strains	Storage time (years)	Number of strains	Storage time (years)
555	*H. grisea* Traaen 1914 var. *thermoidea* Cooney et Emerson 1964			1	11.62		
556	*H. insolens* Cooney et R. Emerson 1964			1	11.62		
557	*Hymenochaete tabacina* (Sowerby 1796) Levielle 1846	1	19.55				
558	*Hyphozyma sanguinea* (C.Ramirez 1952) de Hoog et M.T.Smith 1981	1	19.28	1	20.45		
559	*H. variabilis* de Hoog et M.T.Smith 1981 var. *odora* de Hoog et M.T.Smith 1981	1	19.28	1	21.64		
560	*H. variabilis* de Hoog et M.T.Smith 1981 var. *variabilis*	1	19.28	1	20.45		
561	*Hypomyces ochraceus* (Persoon 1801) Tulasne et C. Tulasne 1865			2	18.58		
562	*Hypsizygus ulmarius* (Bulliard 1790) Redhead 1984	1	8.54				
563	*Inonotus dryophilus* (Berkeley 1847) Murrill 1904	1	19.41				
564	*I. obliquus* (Ach. ex Persoon 1801) Pilat 1942	1	8.58				
565	*I. rheades* (Persoon 1825) Bondartsev et Singer 1941	2	12.29				
566	*Irpex lacteus* (Fries 1818) Fries 1828	1	12.39				
567	*Isaria farinosa* (Holmskjold 1781) Fries 1832	3	19.56	5	22.64	3	18.38
568	*I. fumosorosea* Wize 1904	2	19.51	5	27.38	1	2.32
569	*Itersonilia perplexans* Derx 1948	2	6.72	1	9.95		
570	*Kickxella alabastrina* Coemans 1862	1	6.69	1	6.46		
571	*Kuehneromyces lignicola* (Peck 1872) Redhead 1984	1	12.07				
572	*K. mutabilis* (Schaeffer 1774) Singer et A.H.Smith 1946	2	12.39				
573	*Laccaria bicolor* (Maire 1937) P.D.Orton 1960	1	0.84				
574	*Lactarius helvus* (Fries 1821) Fries 1838	1	12.58				
575	*Laetiporus sulphureus* (Bulliard 1788) Murrill 1920	3	19.97				
576	*Lasiodiplodia theobromae* (Pat. 1892) Griffon et Maublanc 1909	1	7.36				
577	*Lecanicillium fungicola* (Preus 1851) Zare et W. Gams 2008	1	4.90	3	27.16		
578	*L. fusisporum* (W.Gams 1971) Zare et W.Gams 2001			1	25.03		
579	*L. lecanii* (Zimmermann 1898) Zare et W. Gams 2001	2	19.83	2	31.78		
580	*L. muscarium* (Petch 1931) Zare et W. Gams 2001	3	19.86	6	29.49		
581	*L. psalliotae* (Treschew 1941) Zare et W. Gams 2001	1	19.28	9	29.57		
582	*Leccinum scabrum* (Bulliard 1783) Gray 1821	1	12.07				

No.	Species	n	%	n	%	n	%
583	Lecythophora decumbens (J.F.H.Beyma 1942) E. Weber et al. 2002	1	20.38	1	21.28		
584	L. fasciculata (J.F.H.Beyma 1939) E. Weber et al. 2002	1	20.38	1	19.35		
585	L. hoffmannii (J.F.H.Beyma 1939) W. Gams et McGinnis 1983	1	20.51	2	26.59		
586	L. mutabilis (J.F.H. Beyma 1944) Gams et McGinnis 1983	1	20.38	1	25.14		
587	Lentinula edodes (Berkeley 1878) Pegler 1975	2	8.52				
588	Lentinus lepideus (Fries 1815) Fries 1825	2	20.09				
589	L. sulcatus Berkeley 1845	1	12.12				
590	L. tigrinus (Bulliard 1781) Fries 1825	3	20.96				
591	Lenzites betulina (Linnaeus 1753) Fries 1838	1	12.32				
592	Lepista luscina (Fries 1818) Singer 1951	1	0.21				
593	L. nuda (Bulliard 1790) Cooke 1871	2	8.25				
594	Leptographium lundbergii Lagerberg et Melin 1927			1	1.56		
595	Leptosphaeria coniothyrium (Fuckel 1870) Saccardo 1875	1	18.95	1	6.03		
596	Leucoagaricus leucothites (Vittadini 1835) M.M. Moser ex Bon 1977	1	3.04				
597	Linderina pennispora Raper et Fennell 1952	1	6.92	1	17.15		
598	Lobosporangium transversale (Malloch) M.Blackwell et Benny 2004	1	7.16				
599	Lycoperdon perlatum Persoon 1796	1	8.02				
600	L. pyriforme Schaeffer 1763	1	20.03				
601	Macrolepiota gracilenta (Krombholz 1836) Wasser 1978	1	12.20				
602	M. procera (Scopoli 1772) Singer 1948	1	12.31				
603	M. puellaris (Fries 1863) M.M.Moser 1967	1	12.06				
604	M. rhacodes (Vittadini 1833) Singer 1948	1	12.01				
605	Macrophoma mantegazziana (Penzig 1882) Berlese et Voglino 1886	1	0.97	1	11.56		
606	Magnusiomyces magnusii (F.Ludwig 1886) de Hoog et M.T. Smith 2004			1	24.07		
607	Malbranchea cinnamomea (Libert) Oorschot et de Hoog 1984			1	22.25		
608	Marasmius oreades (Bolton 1792) Fries 1836	1	8.29				
609	Mariannaea elegans (Corda 1838) Samson 1974	2	19.40	5	32.69	2	9.56
610	Melanconium apiocarpum Link 1825			1	6.27		
611	M. bicolor Nees 1817			1	5.84		

(continued)

Table 3.1 (continued)

Sr. No.	Name of species	Cryopreservation		Freeze-Drying		Soil	
		Number of strains	Storage time (years)	Number of strains	Storage time (years)	Number of strains	Storage time (years)
612	*Melanocarpus albomyces* (Cooney et R.Emerson 1964) von Arx 1975	2	18.60	1	34.85		
613	*Melanospora betae* Panasenko 1938	1	4.06	1	21.47		
614	*M. damnosa* (Saccardo 1895) Lindau 1897	1	19.92	1	9.99		
615	*M. kurssanoviana* (Beliakova 1954) Czerepanova 1962	1	16.04				
616	*M. phaseoli* Roll-Hansen 1948	1	0.12	1	21.13		
617	*Memnoniella echinata* (Rivolta 1884) Galloway 1933	2	20.53	2	12.53		
618	*Menispora ciliata* Corda 1837	1	19.23	1	16.73		
619	*Merimbla ingelheimense* (J.F.H. Beyma 1942) Pitt 1980	1	15.95	1	15.95	1	18.03
620	*Metarhizium anisopliae* (Metschnikoff 1879) Sorokin 1883	2	19.54	2	22.25	1	15.82
621	*Microascus cirrosus* Curzi 1930	1	19.31				
622	*M. trigonosporus* C.W.Emmons et B.O.Dodge var. *terreus* Kamyschko 1966	1	20.37	1	36.69		
623	*Microbotryum silenes-inflatae* (de Candolle 1815 ex Liro 1924) G.Deml et Oberwinkler 1982	3	12.22	1	1.84		
624	*M. violaceum* (Persoon 1797) G.Deml et Oberwinkler 1982	2	12.18				
625	*Microdiplodia pruni* Diedicke 1914	1	19.89				
626	*Microsphaeropsis olivacea* (Bonorden 1869) Höhnell 1917	1	19.39	1	12.05		
627	*Mirandina corticola* G.Arnaud 1952 ex Matsushima 1975	1	19.25	1	23.42		
628	*Monascus* sp.			1	5.27		
629	*Monilia brunnea* J.C.Gilman et E.V.Abbott 1927	1	15.21	1	27.24		
630	*M. diversispora* J.F.H.Beyma 1933	1	15.21	1	35.69		
631	*M. medoacensis* (Saccardo 1913) J.F.H. Beyma 1933	1	19.43	1	28.21		
632	*M. megalospora* (Berkeley et M.A.Curtis 1869) Saccardo 1886			1	26.72		
633	*M. shawi* P.Filho	1	19.56	1	29.68	1	9.18
634	*Moniliella suaveolens* (Lindner 1895 ex Lindner 1906) von Arx 1972 var. *nigra* (Burri et Staub 1909) de Hoog 1979	4	19.54	3	23.99		
635	*M. suaveolens* (Lindner 1895 ex Lindner 1906) von Arx 1972 var. *suaveolens*	1	18.95	1	10.76		
636	*Monilinia fructigena* (Aderhold et Ruhland 1905) Honey 1936	1	20.02				

No.	Species						
637	*Monocillium dimorphosporum* W.Gams 1971			1	17.12		
638	*M. indicum* S.B.Saksena 1955	1	16.31	1	15.10		
639	*M. nordinii* (Bourchier 1961) W.Gams 1971	1	19.83	1	22.59		
640	*M. tenue* W.Gams 1971			2	24.20		
641	*Monodictys levis* (Wiltshire 1938) S.Hughes 1958	1	17.77	2	21.23	1	25.87
642	*M. paradoxa* (Corda 1938) S.Hughes 1958	1	17.66	1	11.72		
643	*Monographella cucumerina* (Lindfors 1919) Arx 1984	1	3.81	2	17.75		
644	*Mortierella alliacea* Linnemann 1953	1	13.86				
645	*M. alpina* Peyronel 1913	2	13.93	2	21.97	2	21.91
646	*M. ambigua* B.S.Mehrotra 1963	1	11.73				
647	*M. angusta* Linnemann 1969	1	14.34				
648	*M. beljakovae* Milko 1973	1	12.12				
649	*M. bisporalis* (Thaxter 1914) Bjoerling 1936	2	11.19	1	2.74		
650	*M. capitata* Marchal 1891	1	13.86	1	21.28	1	13.30
651	*M. dichotoma* Linnemann 1936 ex W.Gams 1977	1	6.98	1	16.35	1	1.08
652	*M. elasson* Sideris et G.E.Paxton 1929	2	14.09				
653	*M. elongata* Linnemann 1941	2	15.97	1	4.40		
654	*M. exigua* Linnemann 1941	3	13.93	2	22.78		
655	*M. gamsii* Milko 1974	7	15.91				
656	*M. gemmifera* M.Ellis 1940	3	12.12	2	12.73	2	1.73
657	*M. globalpina* W.Gams et Veenbaas-Rijks 1976	1	13.86				
658	*M. globulifera* O.Rostrup 1916	2	13.14	1	2.76	1	1.61
659	*M. horticola* Linnemann 1941	2	13.99				
660	*M. humilis* Linnemann 1936 ex W.Gams 1977	5	14.09				
661	*M. hyalina* Harz 1871 var. *hyalina*	3	13.99	2	13.93	2	1.08
662	*M. jenkinii* (A.L.Smith 1898) Naumov 1935	3	13.28	3	32.53		
663	*M. lignicola* (G.W.Martin 1937) W.Gams et R.Moreau 1959	1	13.96	1	39.33	1	23.94
664	*M. longicollis* Dixon-Stewart 1932	2	19.42	4	35.06	4	28.66
665	*M. minutissima* van Tieghem 1878	4	13.99	1	14.31	1	1.00
666	*M. mutabilis* Linnemann 1941	1	6.84	1	14.53		
667	*M. nigrescens* van Tieghem 1878	1	3.17				
668	*M. oligospora* Bjoerling 1936	1	13.99	1	2.52	1	1.08
669	*M. parvispora* Linnemann 1941	5	19.07	5	22.80	5	0.79

(continued)

Table 3.1 (continued)

Sr. No.	Name of species	Cryopreservation		Freeze-Drying		Soil	
		Number of strains	Storage time (years)	Number of strains	Storage time (years)	Number of strains	Storage time (years)
670	*M. polycephala* Coemans 1863			1	21.31		
671	*M. pulchella* Linnemann 1941	1	13.99	1	6.83		
672	*M. pusilla* Oudemans 1902	1	16.82	1	7.16		
673	*M. reticulata* van Tieghem et G.Le Monnier 1873	1	6.98	1	39.26	1	24.34
674	*M. sarnyensis* Milko 1973	1	13.86				
675	*M. sclerotiella* Milko 1967	1	11.08				
676	*M. simplex* van Tieghem et G.Le Monnier 1873	1	11.95			1	20.46
677	*M. strangulata* van Tieghem 1875	1	3.79	1	21.01		
678	*M. stylospora* Dixon-Stewart 1932			1	15.95	1	25.98
679	*M. turficola* Y.Ling 1930	1	11.02				
680	*M. verticillata* Linnemann 1941	7	19.42	8	21.74	2	11.98
681	*M. zonata* Linnemann 1936 ex W.Gams 1977	1	13.95	1	5.09		
682	*M. zychae* Linnemann 1941	5	12.12	2	31.68	1	5.37
683	*Mucobasispora tarikii* Moustafa et Abdul-Wahid 1990			1	6.25		
684	*Mucor abundans* Povah 1917	1	17.79	1	29.37	1	5.99
685	*M. aligarensis* B.S.Mehrotra et B.R.Mehrotra 1969	1	20.18	1	23.28		
686	*M. amphibiorum* Shipper 1978	1	15.37	1	18.86		
687	*M. bacilliformis* Hesseltine 1954	1	5.84	1	28.79		
688	*M. bainieri* B.S.Mehrotra et Baijal 1963			1	32.41		
689	*M. circinelloides* van Tieghem 1875 var. *circinelloides*	4	19.54	17	33.63	13	28.04
690	*M. circinelloides* van Tieghem 1875 var. *griseocyanus* (Hagem 1908) Schipper 1976	1	11.98	3	22.46	3	24.30
691	*M. circinelloides* van Tieghem 1875 var. *janssenii* (Lendner 1907) Schipper 1976			7	32.18	6	27.75
692	*M. circinelloides* van Tieghem 1875 var. *lusitanicus* (Bruderlein 1916) Schipper 1976			8	35.15	3	28.44
693	*M. flavus* Bainier 1903	9	19.32	19	36.92	6	26.71
694	*M. fragilis* Bainier 1884	1	19.63	1	16.60	1	18.53
695	*M. fuscus* Bainier 1903	1	7.14	3	39.12	2	19.26
696	*M. genevensis* Lendner 1908	1	19.56	3	21.90		

697	M. guilliermondii Nadson et Philippow 1925	1	19.63	1	34.77	1	8.62
698	M. hiemalis Wehmer 1903 var. corticolus (Hagem 1910) Schipper 1973	1	0.14	3	18.48	1	9.59
699	M. hiemalis Wehmer 1903 var. hiemalis	4	19.67	17	40.19	5	28.13
700	M. hiemalis Wehmer 1903 var. luteus (Linnemann 1936) Schipper 1973			1	16.83	1	16.22
701	M. hiemalis Wehmer 1903 var. silvaticus (Hagem 1908) Schipper 1973			3	26.19	1	2.61
702	M. inaequisporus Dade 1937	1	19.63	1	14.72	1	0.09
703	M. indicus Lendner 1930	1	11.95	2	40.30	1	26.56
704	M. laxorrhizus Y.Ling 1930	3	19.58	5	31.36	2	20.52
705	M. microsporus Namyslowski 1910	1	19.56	1	19.44		
706	M. mousanensis Baijal et B.S.Mehrotra 1966	1	19.69	1	17.14	1	24.54
707	M. mucedo Linnaeus 1753	4	20.05	9	38.20	2	15.41
708	M. oblongiellipticus H.Naganishi, Hirahara et Seshita ex Pidoplichko et Milko 1971			1	34.63		
709	M. odoratus Treschew 1940	1	19.54	2	16.18		
710	M. piriformis A.Fischer 1892	4	18.69	5	27.32		
711	M. plasmaticus van Tieghem 1875	1	19.67	1	28.03	1	5.23
712	M. plumbeus Bonorden 1864	4	17.79	16	34.81	14	28.41
713	M. psychrophilus Milko 1971	1	20.14	1	14.88		
714	M. racemosus Fresenius 1850 var. chibinensis (Neophytova 1955) Schipper 1976	1	11.78	4	24.74	3	28.30
715	M. racemosus Fresenius 1850 var. racemosus	7	19.54	31	38.88	26	30.18
716	M. racemosus Fresenius 1850 var. sphaerosporus (Hagem 1908) Schipper 1970			3	32.34	2	28.30
717	M. ramosissimus Samoutsevitch 1927	1	19.63	1	17.67		
718	M. recurvus E.E.Butler 1952 var. indicus Baijal et B.S.Mehrotra 1965			1	19.87	1	18.75
719	M. recurvus E.E.Butler 1952 var. recurvus	1	19.63	1	15.92		
720	M. saturninus Hagem 1910			1	35.92	1	8.11
721	M. sinensis Milko et Beliakova 1971	1	6.83	2	25.58	2	23.61
722	M. strictus Hagem 1908			2	22.88	1	8.38
723	M. tuberculisporus Schipper 1978	1	19.63	1	20.64		

(continued)

Table 3.1 (continued)

Sr. No.	Name of species	Cryopreservation		Freeze-Drying		Soil	
		Number of strains	Storage time (years)	Number of strains	Storage time (years)	Number of strains	Storage time (years)
724	*M. ucrainicus* Milko 1971			1	17.38		
725	*M. variabilis* A.K.Sarbhoy 1965	1	19.58	1	14.04	1	19.25
726	*M. zonatus* Milko 1967	1	19.56	2	27.40	1	5.69
727	*M. zychae* Baijal et B.S.Mehrotra 1965 var. *zychae*	2	6.02	2	39.08		
728	*Mutinus caninus* (Hudson 1762) Fries 1849	1	12.30				
729	*M. ravenelii* (Berkeley et Curtis 1855) E.Fischer 1888	1	1.81				
730	*Myceliophthora fergusii* (Klopotek 1974) Oorschot 1977			1	1.98		
731	*M. lutea* Costantin 1892			1	29.47		
732	*M. thermophila* (Apinis 1962) van Oorschot 1977	2	20.49	3	20.37	1	19.25
733	*Mycena pura* (Persoon 1794) P.Kummer 1871	1	8.65				
734	*M. viscosa* Maire 1910	1	12.19				
735	*Mycocladus corymbifer* (Cohn 1884) Vanova 1991	6	19.54	18	29.84	17	28.66
736	*Mycogone cervina* Ditmar 1817	1	19.85				
737	*M. nigra* (Morgan 1895) C.N.Jensen 1912	1	19.53	4	27.93		
738	*M. rosea* Link 1809	1	19.37	4	23.30		
739	*Mycosticta cytosporicola* Frolov 1968	1	17.66	2	22.13		
740	*Mycotypha africana* R.O.Novak et Backus 1963			1	21.41		
741	*M. indica* P.M.Kirk et Benny 1985			1	13.81		
742	*Myrothecium cinctum* (Corda 1842) Saccardo 1886	2	19.28	2	16.86		
743	*M. roridum* Tode 1790	1	19.47	4	22.02		
744	*M. verrucaria* (Albertini et Schweinitz 1805) Ditmar 1813	3	19.82	4	31.60		
745	*Myxotrichum setosum* (Eidam 1882) G.F.Orr et Plunkett 1963			1	12.11		
746	*M. stipitatum* (Eidam 1882) G.F.Orr et Kuehn 1963	1	4.06	1	20.40		
747	*Nadsoniella nigra* Issatschenko 1914 var. *hesuelica* Lyakh et Ruban 1970	1	18.85	1	6.16		
748	*Nakataea sigmoidea* (Cavara 1889) Hara 1939	1	19.29				
749	*Nectria cosmariospora* Cesati et de Notaris 1863	2	18.93	2	23.42	2	3.45
750	*N. inventa* Pethybridge 1919			1	27.90		
751	*Nematogonum mycophilum* (Saccardo 1886) Rogerson et W. Gams 1981	1	0.54				

No.	Species	n	%	n	%	n	%
752	Neocosmospora vasinfecta E.F.Smith 1899 var. africana (von Arx 1955) Cannon et D.Hawksworth 1984	2	19.32	2	20.63		
753	Neonectria galligena (Bresadola 1901) Rossman et Samuels 1999	1	18.95				
754	Neoscytalidium dimidiatum (Penzig 1887) Crous et Slippers 2006	1	19.39				
755	Neottiosporra caricina (Desmazieres 1836) Hoehnel 1924			1	15.69		
756	Neovossia setariae (Ling 1945) Yu et Lou 1962	1	15.20				
757	Neurospora crassa Shear et B.O.Dodge 1927	5	18.94	8	33.87		
758	N. sitophila Shear et B.O.Dodge 1927	1	19.42	4	25.64	4	9.02
759	N. toroi F.L.Tai 1935	1	18.95	1	17.57	1	9.02
760	Newbya pascuicola M.C.Vick et M.W.Dick 2002	1	4.39				
761	Niesslia exilis (Albertini et Schweinitz 1805) G.Winter 1887	1	18.95	1	7.35	1	4.37
762	Nigrospora gossypii Jaczewski 1929	1	17.46	1	27.64		
763	Nigrospora oryzae (Berkeley et Broome 1873) Petch 1924	2	19.45	4	18.56	1	26.03
764	Nodulisporium verrucosum (J.F.H.Beyma 1929) G. Smith 1954	1	19.71				
765	Nomuraea rileyi (Farlow 1883) Samson 1974	1	9.84	1	12.70		
766	Ochrocladosporium elatum (Harz 1871) Crous et U. Braun 2007	1	12.57	1	27.80	1	7.91
767	Oedocephalum sp. (Berkeley et Broome 1873) Petch 1924	1	15.55	1	28.99	1	15.98
768	Oidiodendron cereale (Thuemen 1880) G.L.Barron 1962	1	0.18	4	27.05		
769	O. echinulatum G.L.Barron 1962	1	19.57	1	7.85		
770	O. tenuissimum (Peck 1894) S. Hughes 1958			1	1.95		
771	O. truncatum G.L.Barron 1962			3	8.70		
772	Olpitrichum sp.	1	20.45	1	19.95		
773	Oospora minor Delitsch 1943	1	15.64	1	8.14		
774	O. nicotianae Pezzolato 1899	1	19.75	2	33.00		
775	O. oryzae Ferraris 1902	1	19.94	1	1.05		
776	O. sajanica Ogarkov 1979	1	16.10	1	25.73		
777	O. sulphurea (Preuss 1852) Saccardo et Voglino 1886	2	19.54	2	20.49		
778	O. tenuis (P.Maze 1910) Berkhout 1923	1	19.94	1	18.62		
779	O. uvarum Karamboloff 1931	1	15.64				
780	O. variabilis (Lindner 1898) J.Lindau 1907	1	19.85	1	14.69		
781	Ophiostoma piceae (Münch 1907) Syd. et P. Syd. 1919			2	10.07		
782	Ostracoderma sp.	1	1.88				

(continued)

Table 3.1 (continued)

Sr. No.	Name of species	Cryopreservation		Freeze-Drying		Soil	
		Number of strains	Storage time (years)	Number of strains	Storage time (years)	Number of strains	Storage time (years)
783	Ovadendron sulphureo-ochraceum (J.F.H.Beyma 1933) Sigler et J.W.Carmichael 1976	1	19.77	1	29.32		
784	Paecilomyces borysthenicus B.A. Borisov et Tarasov 1997			1	6.32		
785	P. inflatus (Burnside 1927) J.W.Carmichael 1962			3	2.00		
786	P. lilacinus (Thom 1910) Samson 1974	4	19.45	12	31.69	5	19.37
787	P. marquandii (Massee 1898) S.Hughes 1951	3	19.54	4	31.29	2	18.38
788	P. puntonii (Vuill 1930) Nann. 1934	1	16.39	1	11.91		
789	P. variotii Bainier 1907	12	19.40	19	32.02	13	33.89
790	Papulaspora biformospora Kirilenko 1971	1	19.63	1	20.69		
791	Papulaspora sp.	1	19.30	1	23.48	1	3.20
792	Paraconiothyrium sporulosum (W. Gams et Domsch 1969) Verkley 2004	1	19.39	2	16.28		
793	Parasitella parasitica (Bainier 1884) Sydow 1903	1	7.18	2	22.32		
794	Passalora fulva (Cooke 1883) U. Braun et Crous 2003	1	18.99	1	22.51		
795	Paxillus panuoides (Fries 1818) Fries 1838	2	20.01				
796	Penicillium adametzii K.M.Zalessky 1927			2	35.16	3	33.89
797	P. albicans Bainier 1907			1	22.03	1	7.15
798	P. alicantinum C.Ramirez et A.T.Martinez 1980			1	10.67	1	12.20
799	P. anatolicum Stolk 1968			1	14.02	1	15.32
800	P. aragonense C.Ramirez et A.T.Martinez 1981			1	10.66	1	12.20
801	P. arenicola Chalabuda 1950			1	14.54	1	17.92
802	P. atramentosum Thom 1910					1	2.57
803	P. aurantioflammiferum C.Ramirez et al. 1980			1	10.67	1	12.20
804	P. aurantiogriseum Dierckx 1901	2	20.50	31	34.27	28	34.84
805	P. bilaiae Chalabuda 1950			1	15.01	1	18.52
806	P. brevicompactum Dierckx 1901	2	20.48	12	28.04	9	18.10
807	P. brunneum Udagawa 1959	1		1	10.59	1	12.07
808	P. camemberti Thom 1906			10	35.42	9	24.37
809	P. canescens Sopp 1912	2	20.48	12	36.88	10	39.72
810	P. capsulatum Raper et Fennell 1948	2	19.49	3	17.24	3	18.17

No.	Species					
811	P. castellonense C.Ramirez et A.T.Martinez 1981		1	17.69	1	12.20
812	P. chermesinum Biourge 1923		3	35.23	2	24.93
813	P. chrysogenum Thom 1910	19.34	29	35.75	22	43.04
814	P. cinerascens Biourge 1923		1	23.35	1	14.95
815	P. citreonigrum Dierckx 1901	18.94	10	25.87	10	33.89
816	P. citrinum Thom 1910	18.30	16	37.11	12	20.45
817	P. commune Thom 1910	2.51	6	20.26	6	18.03
818	P. cordubense C.Ramirez et A.T.Martinez 1981		1	10.65	1	12.20
819	P. corylophilum Dierckx 1901		1	28.97	1	15.77
820	P. cyaneum (Bainier et Sartory 1913) Biourge 1923 ex Thom 1930		1	21.73	1	9.56
821	P. daleae K.M.Zalessky 1927	2.26	1	5.73	1	18.10
822	P. decumbens Thom 1910	18.07	9	36.80	7	29.21
823	P. dierckxii Biourge 1923	18.30	6	20.27	6	20.45
824	P. digitatum (Persoon 1801) Saccardo 1881		3	17.59	2	21.28
825	P. diversum Raper et Fennell 1948		1	22.03	1	18.04
826	P. dodgei Pitt 1980		1	27.75	1	12.58
827	P. duclauxii Delacroix 1892		6	19.86	6	20.33
828	P. expansum Link 1809		5	30.96	5	38.34
829	P. fagi A.T.Martinez et C.Ramirez 1978		1	10.68	1	12.07
830	P. funiculosum Thom 1910	20.48	8	30.78	7	25.05
831	P. glabrum (Wehmer 1893) Westling 1911		9	34.77	9	24.90
832	P. gladioli Machacek 1928	20.44	2	35.58	2	12.07
833	P. glaucum Link 1805	20.50	1	15.50	1	25.04
834	P. grancanariae C.Ramirez et al. 1978		1	10.68	1	12.20
835	P. granulatum Bainier 1905		5	26.68	5	12.33
836	P. griseofulvum Dierckx 1901		6	19.61	5	24.88
837	P. herquei Bainier et Sartory 1912		3	37.41	3	22.37
838	P. hirsutum Dierckx 1901 var. hirsutum		1	2.77	1	5.41
839	P. hispanicum C.Ramirez et al. 1978		1	10.68	1	12.07
840	P. humuli J.F.H.Beyma 1937		1	19.85	1	24.92
841	P. ilerdanum C.Ramirez et al. 1980		1	26.66	1	12.20
842	P. indonesiae Pitt 1980		2	29.58	2	14.21

(continued)

Table 3.1 (continued)

Sr. No.	Name of species	Cryopreservation		Freeze-Drying		Soil	
		Number of strains	Storage time (years)	Number of strains	Storage time (years)	Number of strains	Storage time (years)
843	P. inflatum Stolk et Malla 1971			1	2.95		
844	P. insectivorum (Sopp 1912) Biourge 1923			1	15.05	1	25.04
845	P. islandicum Sopp 1912	1	2.23	3	20.87	3	12.07
846	P. italicum Wehmer 1894	1	14.09	3	20.32	3	20.35
847	P. janczewskii K.M.Zalessky 1927			8	27.39	8	27.37
848	P. jensenii K.M.Zalessky 1927			8	37.45	8	26.06
849	P. kirovogradum Beliakova et al.	1	18.21	1	5.52	1	6.07
850	P. lagena (Delitsch 1943) Stolk et Samson 1983	2	19.44	2	24.97		
851	P. lanosum Westling 1911			3	33.03	3	15.57
852	P. lapidosum Raper et Fennell 1948			3	23.28	4	24.86
853	P. lehmanii Pitt 1980			2	30.40	2	24.94
854	P. lineatum Pitt 1980			1	10.51	1	11.55
855	P. lividum Westling 1911			2	15.45	1	15.33
856	P. malacaense C.Ramirez et A.T.Martinez 1980			1	10.65	1	12.20
857	P. martensii Biourge 1923 var. moldavicum Solovei 1975		2.26	1	10.48	1	10.71
858	P. megasporum Orpurt et Fennell 1955			2	15.78	3	20.33
859	P. melinii Thom 1930	1	3.99	4	27.39	4	20.45
860	P. miczynskii K.M.Zalessky 1927			7	30.90	6	24.92
861	P. minioluteum Dierckx 1901	1	9.07	6	10.72	2	12.20
862	P. mirabile Beliakova et Milko 1972	1	4.01	1	16.95	1	9.56
863	P. mongoliae Beliakova et al.	1	17.98	1	23.82	1	6.07
864	P. multicolor Grigorieva-Manoilova et Poradielova 1915			1	19.35		
865	P. multicolor Novobranova 1972			1	12.94	1	15.09
866	P. murcianum C.Ramirez et A.T.Martinez 1981			1	10.65	1	12.20
867	P. novae-zeelandiae J.F.H.Beyma 1940	1	19.47	5	22.02	5	9.70
868	P. ochrochloron Biourge 1923			6	36.27	4	17.51
869	P. onobense C.Ramirez et A.T.Martinez 1981			1	10.69	1	12.20
870	P. ovetense C.Ramirez et A.T.Martinez 1981			1	10.67	1	12.20
871	P. oxalicum Currie et Thom 1915			4	31.32	5	24.94
872	P. palmense C.Ramirez et al. 1978			1	10.68	1	12.20

873	*P. paxilli* Bainier 1907			7	27.21	5	18.52
874	*P. phoeniceum* J.F.H.Beyma 1933			3	20.27	3	15.02
875	*P. piceum* Raper et Fennell 1948	1	20.53	2	34.66	3	24.87
876	*P. pinophilum* Thom 1910			2	37.80	2	26.35
877	*P. poltaviae* Beliakova et al.	1	18.21	1	5.51	1	6.07
878	*P. purpurogenum* Stoll 1904	1	9.25	9	36.78	7	24.87
879	*P. quercetorum* Baghdadi 1968	1	4.06	1	20.52	1	20.45
880	*P. raistrickii* G.Smith 1933			3	36.68	3	20.13
881	*P. resticulosum* Birkinshaw et al. 1942			1	26.72	1	17.75
882	*P. restrictum* J.C.Gilman et E.V.Abbott 1927	2	18.30	9	37.65	7	24.92
883	*P. roqueforti* Thom 1906	2	3.99	9	21.05	9	24.87
884	*P. roseopurpureum* Dierckx 1901			4	22.90	4	15.41
885	*P. rubrum* Stoll 1904	4	20.48	12	31.53	12	24.99
886	*P. rugulosum* Thom 1910	1	18.07	14	36.65	14	25.04
887	*P. sclerotiorum* J.F.H.Beyma 1937			6	25.87	5	18.09
888	*P. senticosum* D.B.Scott 1968			1	20.73		15.32
889	*P. severskii* Schechovtsov 1981			1	11.10	1	5.72
890	*P. simplicissimum* (Oudemans 1903) Thom 1930	2	18.30	18	27.37	17	24.92
891	*P. solitum* Westling 1911 var. *crustosum* (Thom 1930) Bridge et al. 1989	2	3.99	7	33.56	6	24.90
892	*P. solitum* Westling 1911 var. *solitum*	2	17.98	8	34.70	8	24.92
893	*P. solocongelatus* Beliakova et al.	1	17.98	1	7.99	1	6.07
894	*P. spinulosum* Thom 1910	2	4.01	16	24.81	14	24.93
895	*P. terraconense* C.Ramirez et A.T.Martinez 1980			1	10.65	1	12.79
896	*P. thomii* Maire 1917	1	2.26	8	21.38	8	24.90
897	*P. turbatum* Westling 1911			1	17.63	1	20.12
898	*P. turolense* C.Ramirez et A.T.Martinez 1981			1	10.70	1	12.20
899	*P. umbonatum* Sopp 1912			1	32.95		
900	*P. valentinum* C.Ramirez et A.T.Martinez 1980			1	10.66	1	12.20
901	*P. vanbeymae* Pitt 1980			1	27.09		
902	*P. variabile* Sopp 1912			5	34.08	3	15.15
903	*P. vasconiae* C.Ramirez et A.T.Martinez 1980			1	10.63	1	12.20
904	*P. velutinum* J.F.H.Beyma 1935			9	26.62	9	20.16

(continued)

Table 3.1 (continued)

Sr. No.	Name of species	Cryopreservation		Freeze-Drying		Soil	
		Number of strains	Storage time (years)	Number of strains	Storage time (years)	Number of strains	Storage time (years)
905	*P. verrucosum* Dierckx 1901			5	35.63	3	19.24
906	*P. verruculosum* Peyronel 1913	2	18.07	4	31.53	3	18.52
907	*P. vinaceum* J.C.Gilman et E.V.Abbott 1927			4	34.62	4	25.05
908	*P. viridicatum* Westling 1911			7	21.31	5	24.93
909	*P. vulpinum* (Cooke et Massee 1888) Seifert et Samson 1985			8	37.02	9	24.80
910	*P. waksmanii* K.M.Zalessky 1927	1	2.26	5	36.82	5	34.08
911	*P. westlingii* K.M.Zalessky 1927			1	15.51	1	9.56
912	*P. zacinthae* C. Ramírez et A.T. Martínez 1981			1	26.64	1	12.20
913	*Penidiella strumelloidea* (Milko et Dunaev 1986) Crous et U. Braun 2007			1	14.15		
914	*Perenniporia medulla-panis* (Jacquin 1778) Donk 1967	1	18.54				
915	*Periconia macrospinosa* Lefebvre et Aar.G. Johnson 1949	2	19.47	1	16.13		
916	*Pestalotia macrotricha* Klebahn 1914	1	19.27				
917	*P. pezizoides* de Notaris 1841	1	19.85	2	20.58		
918	*Petriella sordida* (Zukal 1890) G.L. Barron et J.C. Gilman 1961			1	10.23		
919	*Phaeococcomyces nigricans* (Rich et Stern 1958) de Hoog 1979			1	23.58		
920	*Phaeoisaria hippocrepiformis* Milko et Dunaev	1	19.38	1	5.26		
921	*Phallus hadriani* Ventenat 1798	1	15.06				
922	*P. impudicus* Linnaeus 1753	1	1.76				
923	*Phanerochaete sanguinea* (Fries 1828) Pouzar 1973	1	3.24				
924	*Phellinus igniarius* (Linnaeus 1753) Quelet 1886	1	20.05				
925	*P. lundellii* Niemelae 1972	3	20.01				
926	*P. populicola* Niemelae 1975	3	20.01				
927	*Philalophora atrovirens* (J.F.H.Beyma 1935) Schol-Schwarz 1970	1	20.38	1	25.35		
928	*P. bubakii* (Laxa 1930) Schol-Schwarz 1970	1	20.40	3	15.42		
929	*P. cyclaminis* J.F.H. Beyma 1942			1	6.36		
930	*P. lagerbergii* (Melin et Nannfeldt 1934) Conant 1937	1	19.37				
931	*P. melinii* (Nannf. 1934) Conant 1937			3	5.86		
932	*P. verrucosa* Medlar 1915	1	19.74	1	19.52		

No.	Species						
933	*Phlebia rufa* (Persoon 1801) M.P.Christiansen 1960	1	2.80				
934	*Phlebiopsis gigantea* (Fries 1815) Juelich 1978	2	20.03				
935	*Pholiota adiposa* (Batsch 1789) P.Kummer 1871	1	20.09				
936	*P. lenta* (Persoon 1801) Singer 1951	1	0.20				
937	*P. nameko* (T.Ito 1929) S.Ito et S.Imai apud S.Imai 1933	1	8.08				
938	*P. squarrosa* (Weigel 1771) P.Kummer 1871	1	12.19				
939	*Phoma betae* A.B.Frank 1892	1	17.46	2	24.02		
940	*P. destructiva* Plowright 1881			1	1.73		
941	*P. eupyrena* Saccardo 1879			1	5.21		
942	*P. glomerata* (Corda 1840) Wollenweber et Hochapfel 1936	1	19.30	6	25.47		
943	*P. hedericola* (Durieu et Mont. 1856) Boerema 1976			1	4.61		
944	*P. jolyana* Pirozynski et Morgan-Jones 1968 var. *circinata* (Kuznetzova 1971) Boerema et al. 1977	1	19.54	4	20.04		
945	*P. leveillei* Boerema et G.J.Bollen 1975			1	1.53		
946	*P. lingam* (Tode 1791) Desmazieres 1849			4	8.45		
947	*P. lycopersici* Cooke 1885			1	2.14		
948	*P. pinodella* (L.K. Jones 1927) Morgan-Jones et K.B. Burch 1987	1	20.52				
949	*P. pomorum* Thuemen 1879			1	7.21		
950	*P. sorghina* (Saccardo 1878) Boerema et al. 1973	1	19.45	2	20.14		
951	*P. tracheiphila* (Petri 1929) L.A. Kantschaweli et Gikaschvili 1948	1	12.55				
952	*Phomatospora* sp.	1	18.85	1	26.61		
953	*Phomopsis helianthi* Muntanola-Cvetcovic et al. 1981	1	8.01	1	5.68		
954	*Phycomyces blakesleeanus* Burgeff 1925	1	19.56	8	31.79	1	1.51
955	*P. nitens* (C.Agardh 1823) Kunze 1823	2	19.71	2	31.87		
956	*Phyllosticta pucciniospila* C. Massalongo 1900	1	11.44	1	23.98		
957	*Phytophthora cactorum* (Lebert et Cohn 1870) J.Schroeter 1886	1	17.04				
958	*P. capsici* Leonian 1922	2	15.93				
959	*P. cinnamomi* Rands 1922	4	13.13				
960	*P. cryptogea* Pethybridge et Lafferty 1919	1	0.18				
961	*P. drechsleri* Tucker 1931	3	4.16				
962	*P. megasperma* Drechsler 1931 var. *megasperma*	1	0.19				
963	*Pidoplitchkoviella terricola* Kirilenko 1975	1	19.44	1	15.30		

(continued)

Table 3.1 (continued)

Sr. No.	Name of species	Cryopreservation		Freeze-Drying		Soil	
		Number of strains	Storage time (years)	Number of strains	Storage time (years)	Number of strains	Storage time (years)
964	*Piedraia hortae* Fonseca et Leao 1928	1	19.34	1	15.70		
965	*P. hortae* Fonseca et Leao 1928 var. *paraguayensis* Fonseca et Leao 1928	1	19.34	1	7.90		
966	*P. sarmentoi* M.J.Pereira 1930	1	19.34	1	17.54		
967	*Pilaira anomala* (Cesati 1851) J.Schroeter 1886			1	26.00		
968	*P. caucasica* Milko 1970			1	16.20		
969	*P. moreaui* Y.Ling 1926	1	0.02	1	16.74		
970	*Pilobolus crystallinus* (F.H.Wiggers 1780) Tode 1784			1	15.64		
971	*P. longipes* van Tieghem 1878	1	3.68	1	15.64		
972	*P. umbonatus* Buller 1934			1	15.65		
973	*Piptoporus betulinus* (Bulliard 1786) P.Karsten 1881	3	12.39				
974	*Pirella circinans* Bainier 1882	1	19.32	1	27.13	1	14.40
975	*P. circinans* Bainier 1882 var. *volgogradensis* (Milko 1974) Benny et Schipper 1988			1	14.71		
976	*P. naumovii* (Milko 1970) Benny et Schipper 1992	1	19.30	1	15.35	1	25.59
977	*Pleurodesmospora coccorum* (Petch 1924) Samson, W. Gams et H.C. Evans 1980			1	28.25		
978	*Pleurophoma cava* (Schulzer 1871) Boerema 1996	1	19.80	3	23.72		
979	*Pleurotus cornucopiae* (Paulet 1793) Rolland 1910	1	20.01				
980	*P. eryngii* (De Candolle 1805) Quelet 1872	1	20.14				
981	*P. ostreatus* (Jacquin 1775) P.Kummer 1871	16	20.14				
982	*P. pulmonarius* (Fries 1821) Quelet 1872	2	8.15				
983	*Pochonia bulbillosa* (W. Gams et Malla 1971) Zare et W. Gams 2001	1	19.98	3	28.36		
984	*P. chlamydosporia* (Goddard 1913) Zare et W. Gams 2001	1	19.45	1	31.72		
985	*Polycephalomyces tomentosus* (Schrad. 1799) Seifert 1985	1	19.53	1	24.82		
986	*Polyscytalum pustulans* (M.N.Owen et Wakefield 1919) M.B. Ellis 1976	1	15.56	1	30.58		
987	*Preussia fleischhakii* (Auerswald 1866) Cain 1961			1	19.14		
988	*Protomyces macrosporus* Unger 1834	1	18.99				

S.M. Ozerskaya et al.

989	*Pseudallescheria boydii* (Shear 1922) McGinnis et al. 1982	2	19.42	2	21.30	1	3.42
990	*Pseudeurotium bakeri* C.Booth 1961	1	20.43	1	21.45		
991	*P. desertorum* Mouchacca 1971	1	18.86	1	14.48		
992	*P. ovale* Stolk 1955 var. *milkoi* Belyakova 1969	2	20.43	2	20.65		
993	*P. ovale* Stolk 1955 var. *ovale*	1	19.32	1	22.82		
994	*P. zonatum* J.F.H.Beyma 1937	9	20.37	9	36.92	1	3.81
995	*Pseudogymnoascus caucasicus* Cejp et Milko 1966	1	19.32	1	16.38		
996	*P. roseus* Raillo 1929	2	19.32	2	20.63		
997	*Puccinia albescens* (Greville 1824) Plowright 1888	1	18.50	1	5.32		
998	*P. bupleuri* F.Rudolphi 1829	1	18.50	1	1.84		
999	*P. suaveolens* (Persoon 1801) Rostrup 1869	1	12.22				
1000	*Pycnidiella resinae* (Ehrenberg 1818) Hoehnel 1915			1	6.69		
1001	*Pycnoporus cinnabarinus* (Jacquin 1776) Fries 1881	2	12.34				
1002	*Pyricularia grisea* Saccardo 1880	3	15.82	1	13.76		
1003	*Pyronema omphalodes* (Bulliard 1791) Fuckel 1870	1	18.95	1	17.28		
1004	*Pythium heterothallicum* W.A.Campbell et F.F.Hendrix 1968	1	5.90				
1005	*P. intermedium* de Bary 1881	1	6.50				
1006	*P. mamillatum* Meurs 1928	1	17.04				
1007	*P. oedichilum* Drechsler 1930	1	15.98				
1008	*P. paroecandrum* Drechsler 1930	1	8.99				
1009	*P. spinosum* Sawada 1926	1	0.15				
1010	*P. sylvaticum* W.A.Campbell et F.F.Hendrix 1967	1	0.03				
1011	*Radiomyces embreei* R.K.Benjamin 1960			2	17.40	2	24.86
1012	*R. spectabilis* Embree 1959	1	19.69	1	20.91	1	24.89
1013	*Ramichloridium biverticillatum* Arzanlou et Crous 2007			1	5.98		
1014	*Rhinocladiella atrovirens* Nannfeldt 1934	1	17.68	1	12.15		
1015	*Rhinotrichum aureum* Cooke et Massee 1889			1	28.38		
1016	*R. lanosum* Cooke 1871	1	19.45	1	21.52		
1017	*Rhizoctonia crocorum* (Persoon 1801) De Candolle 1815	1	19.30			1	5.60
1018	*R. solani* J.G.Kuehn 1858	5	19.90	1	5.53	3	10.96
1019	*R. tuliparum* (Klebahn 1905) Whetzel et J.M. Arthur 1924			1	14.94		
1020	*Rhizomucor miehei* (Cooney et R.Emerson 1964) Schipper 1978	1	19.67	1	21.41	1	17.16
1021	*R. pusillus* (Lindt 1886) Schipper 1978	3	19.69	4	28.96	4	22.48

(continued)

Table 3.1 (continued)

Sr. No.	Name of species	Cryopreservation		Freeze-Drying		Soil	
		Number of strains	Storage time (years)	Number of strains	Storage time (years)	Number of strains	Storage time (years)
1022	*R. tauricus* (Milko et Schkurenko 1970) Schipper 1978	1	19.63	2	20.82	2	24.69
1023	*Rhizopus microsporus* van Tieghem 1875 var. *chinensis* (Saito 1904) Schipper et Stalpers 1984	3	11.63	5	34.24	5	27.09
1024	*R. microsporus* van Tieghem 1875 var. *microsporus*	3	19.71	10	36.72	10	28.45
1025	*R. microsporus* van Tieghem 1875 var. *oligosporus* van Tieghem 1875			2	22.07	2	28.09
1026	*R. microsporus* van Tieghem 1875 var. *rhizopodiformis* (Cohn 1884) Schipper et Stalpers 1984			4	8.70		
1027	*R. oryzae* Went et Prinsen Geerligs 1895	2	19.71	26	38.13	24	30.35
1028	*R. stolonifer* (Ehrenberg 1818) Vuillemin 1902 var. *stolonifer*	3	19.71	23	33.61	14	28.66
1029	*Robillarda sessilis* (Saccardo 1878) Saccardo 1880			1	6.61		
1030	*Rosellinia mammiformis* (Persoon 1801) Cesati et de Notaris 1863	1	19.27	1	10.72		
1031	*Rozites caperata* (Persoon 1796) P.Karsten 1879	1	12.09				
1032	*Russula decolorans* (Fries 1821) Fries 1838	1	12.02				
1033	*R. grisea* Fries 1838	1	20.01				
1034	*R. velutipes* Velenovsky 1920	1	12.30				
1035	*R. vesca* Fries 1838	1	8.81				
1036	*Rutola graminis* (Desmazieres 1834) J.L. Crane et Schoknecht 1977	1	19.39	1	6.87		
1037	*Saksenaea vasiformis* S.B.Saksena 1953	1	19.05				
1038	*Saprochaete gigas* (Smit et L.Meyer 1928) de Hoog et M.T.Smith 2004	1	19.86	1	16.65		
1039	*Saprolegnia asterophora* de Bary 1860	1	15.19				
1040	*S. bielhamensis* (M.W.Dick 1969) Milko 1979	2	13.34				
1041	*S. ferax* (Gruithuisen 1821) Nees 1843	1	13.80				
1042	*S. litoralis* Coker 1923	1	13.09				
1043	*S. mixta* de Bary 1883	1	0.17				
1044	*S. terrestris* Cookson 1937 ex R.L.Seymour 1970	1	0.17				
1045	*S. unispora* (Coker et Couch 1923) R.L.Seymour 1970	1	0.17				

No.	Species	n	%	n	%	n	%
1046	Schizophyllum commune Fries 1815	3	20.09				
1047	Sclerotinia ricini G.H.Godfrey 1919	1	18.95	1	5.10		
1048	S. sclerotiorum (Libert 1837) de Bary 1884	2	19.31				
1049	Scopulariopsis acremonium (Saccardo 1882) Bainier 1907	1	19.31	1	23.60	1	4.86
1050	S. asperula (Saccardo 1882) Hughes 1958	1	19.25	1	26.22	1	9.56
1051	S. brevicaulis (Saccardo 1882) Bainier 1907	9	20.47	14	32.48	7	18.53
1052	S. brumptii Salvanet-Duval 1935	1	15.24	1	21.54		
1053	S. coprophila (Cooke et Massee 1887) W. Gams 1971			1	31.33		
1054	S. flava (Sopp 1912) F.J.Morton et G.Smith 1963	1	19.28	1	23.56	1	20.68
1055	S. halophilica Tubaki 1973	1	15.89				
1056	S. koningii (Oudemans 1902) Vuuillemin 1911	1	19.54				
1057	Sepedonium ampullosporum Damon 1952			1	6.85		
1058	S. macrosporum Saccardo et Cavara 1900	1	7.75	1	27.21	1	9.50
1059	Septoria lycopersici Spegazzini 1881	1	19.96				
1060	Serpula lacrymans (von Wulfen 1781) J.Schroeter 1888	2	18.87	3	24.26		
1061	Simplicillium lamellicola (F.E.W. Smith 1924) Zare et W. Gams 2001	1	19.56				
1062	Sordaria fimicola (Roberge ex Desmazières 1849) Cesati et de Notaris 1863	1	19.36	2	20.55	1	7.15
1063	Spadicesporium acrosporum V.N.Borisova et Dvoinos 1982	1	19.41	1	22.15		
1064	S.acrosporum-majus V.N.Borisova et Dvoinos 1982	1	19.41	1	11.45		
1065	S.bifurcatum V.N.Borisova et Dvoinos 1982	1	19.41	1	19.24		
1066	S.bifurcatum-majus V.N.Borisova et Dvoinos 1982	1	19.41	1	13.52		
1067	S.copiosum V.N.Borisova et Dvoinos 1982	1	19.41	1	13.52		
1068	S.persistens V.N.Borisova et Dvoinos 1982	1	19.41	1	22.24		
1069	S.ramosum V.N.Borisova et Dvoinos 1982	1	19.76	1	13.52		
1070	Sparassis crispa (von Wulfen 1781) Fries 1821	1	7.49				
1071	Sphaerellopsis filum (Bivona-Bernardi 1813–1816) Sutton 1977	1	12.66	1	5.76		
1072	Sphaeropsis malorum Peck 1883	1	19.98				
1073	S. sapinea (Fries 1823) Dyko et B. Sutton 1980	1	19.89				
1074	Sporodiniopsis dichotoma van Hoehnel 1903	1	15.89	1	15.95	1	10.80
1075	Sporormiella australis (Spegazzini 1887) S.I.Ahmed et Cain 1972			1	2.41	1	5.01

(continued)

Table 3.1 (continued)

Sr. No.	Name of species	Cryopreservation		Freeze-Drying		Soil	
		Number of strains	Storage time (years)	Number of strains	Storage time (years)	Number of strains	Storage time (years)
1076	S. intermedia (Auerswald 1868) S.I.Ahmed et Cain ex Kobayasi 1969			1	11.76		
1077	Sporothrix fungorum de Hoog et G.A. de Vries 1973			1	1.94		
1078	Sporotrichum aeruginosum Schweinitz 1886 var. microsporum Karsten 1905			1	32.35		
1079	S. bombycinum (Corda 1839) Rabenhorst 1844	1	19.29	3	29.59	1	27.59
1080	S. gorlenkoanum Kuritzina et Sizova 1967			1	23.64	1	20.27
1081	S. laxum Nees 1816	1	19.29	1	26.27		
1082	S. mycophilum Link 1818			1	27.68		
1083	S. pruinosum J.C.Gilman et E.V.Abbott 1927	6	19.81	8	31.16	1	5.84
1084	S. roseolum Oudemans et Beijerinck 1903	1	15.19	1	26.16		
1085	Stachybotrys chartarum (Ehrenberg 1818) S.Hughes 1958	1	19.30	11	26.67	2	9.56
1086	S. cylindrospora C.N.Jensen 1912			1	14.86		
1087	Stachylidium variabile Schaeffer et Saccardo			1	19.09		
1088	Stagonospora paludosa (Saccardo et Spegazzini 1879) Saccardo 1884	1	19.89				
1089	Stemphyliomma sp.			1	26.05	1	25.87
1090	Stemphylium botryosum Wallroth 1833	1	17.05	1	23.43		
1091	S. sarciniforme (Cavara 1890) Wiltshire 1938	1	17.79	3	23.43		
1092	Stenocarpella maydis (Berkeley 1847) B. Sutton 1980			1	11.57		
1093	Stephanoma sp.	1	16.23	1	16.03		
1094	Stereum hirsutum (Willdenow 1787) Persoon 1800	2	20.03				
1095	S. sanguinolentum (Albertini et Schweinitz 1805) Fries 1838	1	20.03				
1096	Stigmina carpophila (Leveille 1843) M.B. Ellis 1959	1	19.91				
1097	Stilbella bulbicola Hennings 1905	1	19.44	1	29.49		
1098	Stilbotulasnella conidiophora Bandoni et Oberwinkler 1982	1	12.54				
1099	Strobilomyces strobilaceus (Scopoli 1770) Berkeley 1851	1	15.10				
1100	Stropharia rugosoannulata Farlow ex Murrill 1922	1	12.20				
1101	Syncephalastrum racemosum Cohn ex J.Schroeter 1886	2	19.24	12	29.66	12	29.24
1102	Syncephalis cornu van Tieghem et G.Le Monnier 1873			1	20.64	1	24.09

No.	Species	n	%	n	%	n	%
1103	*S. nodosa* van Tieghem 1875			1	23.87	1	19.47
1104	*Taeniolella aquatilis* (Woronichin 1925) Milko 1985	1	19.31	1	9.74	1	11.55
1105	*Talaromyces emersonii* Stolk 1965			1	21.01		
1106	*T. flavus* (Kloecker 1902) Stolk et Samson 1972	1	18.07	4	15.44	4	24.86
1107	*T. luteus* (Zukal 1889) C.R. Benjamin 1955			8	37.03	7	18.17
1108	*T. stipitatus* (Thom 1935) C.R.Benjamin 1955			1	27.27	1	12.07
1109	*T. thermophilus* Stolk 1965			1	17.19	1	1.25
1110	*T. ucrainicus* Udagawa 1966	1	18.07	3	22.55	3	21.86
1111	*T. wormannii* (Kloecker 1903) C.R.Benjamin 1955			2	22.58	2	12.07
1112	*Taphrina bergeniae* Döbbeler 1979	1	18.99				
1113	*T. carnea* Johanson 1886	1	8.12				
1114	*T. deformans* (Berkeley 1857) Tulasne 1866	1	7.17				
1115	*T. pruni* (Fuckel 1861) Tulasne 1866	1	18.99				
1116	*Tetraploa aristata* Berkeley et Broome 1850	1	19.28				
1117	*Thamnidium elegans* Link 1809	1	17.71	3	27.08	3	12.14
1118	*Thamnostylum piriforme* (Bainier 1880) Arx et H.P. Upadhyay 1970	3	19.69	4	29.92	4	28.16
1119	*Thelebolus polysporus* (P.Karsten 1871) Otani et Kanzawa 1970	1	19.83	1	25.82		
1120	*Thermomyces ibadanensis* Apinis et Eggins 1966			2	13.79		
1121	*Thielavia appendiculata* Srivastava et al. 1966	1	14.44	1	12.14		
1122	*T. hyrcaniae* Nicot 1961	1	4.06	1	17.91		
1123	*T. inaequalis* Pidoplichko et al. 1973	3	18.93	3	20.63		
1124	*T. ovispora* Pidoplichko et al. 1973	3	19.32	3	18.92		
1125	*T. pallidospora* Pidoplichko et al. 1973			1	17.81		
1126	*T. terrestris* (Apinis 1963) Malloch et Cain 1972	1	20.50				
1127	*T. terricola* (J.C.Gilman et E.V.Abbott 1927) Emmons 1930	3	19.23	3	33.99		
1128	*T. terricola* (J.C.Gilman et E.V.Abbott 1927) Emmons 1930 var. minor (Rayss et Borut 1958) C.Booth 1961	1	18.86	2	19.98	1	3.14
1129	*Thielaviopsis basicola* (Berkeley et Broome 1850) Ferraris 1912	1	19.49	2	18.54		
1130	*Thysanophora canadensis* Stolk et Hennebert 1968	1	17.68	1	7.31		
1131	*T. penicillioides* (Roumeguere 1890) W.B.Kendrick 1961	2	19.19	5	26.17		
1132	*Tilachlidium pinnatum* Preuss 1851			1	29.01		
1133	*Tilletia caries* (de Candolle 1815) Tulasne et C.Tulasne 1847	1	12.54	1	1.84		

(continued)

Table 3.1 (continued)

Sr. No.	Name of species	Cryopreservation		Freeze-Drying		Soil	
		Number of strains	Storage time (years)	Number of strains	Storage time (years)	Number of strains	Storage time (years)
1134	*Tilletiopsis albescens* Gokhale 1972	1	16.39	1	9.92		
1135	*T. washingtonensis* Nyland 1950	3	18.85	3	29.65		
1136	*Tolypocladium cylindrosporum* W.Gams 1971	1	19.83	1	23.58		
1137	*T. geodes* W.Gams 1971			3	6.25		
1138	*T. inflatum* W.Gams 1971	2	19.26	4	22.72		
1139	*Trametes hirsuta* (von Wulfen 1788) Pilat 1939	3	12.58				
1140	*T. pubescens* (Schumacher 1803) Pilat 1939	2	20.05				
1141	*T. versicolor* (Linnaeus 1753) Lloyd 1921	1	11.24				
1142	*T. zonatella* Ryvarden 1978	2	8.25				
1143	*Tricellula aquatica* J.Webster 1959	1	16.21	2	22.18		
1144	*Trichaptum abietinum* (Persoon ex J.F. Gmelin 1792) Ryvarden 1972	2	20.09				
1145	*Trichocladium asperum* Harz 1871	1	19.54	1	15.64		
1146	*T. opacum* (Corda 1837) S.Hughes 1952	1	19.41	3	13.52		
1147	*Trichoderma album* Preuss 1851	1	15.66	3	32.91		
1148	*T. aureoviride* Rifai 1969	3	19.81	3	28.44	2	11.75
1149	*T. citrinoviride* Bissett 1984			1	1.92		
1150	*T. flavofuscum* (J.H.Miller et al. 1957) Bissett 1991			1	14.25		
1151	*T. hamatum* (Bonorden 1851) Bainier 1906	2	19.98	3	31.34	1	10.01
1152	*T. harzianum* Rifai 1969	7	19.54	12	32.01	4	10.27
1153	*T. koningii* Oudemans 1902	6	19.81	6	31.37	5	14.29
1154	*T. longibrachiatum* Rifai 1969	3	19.40	8	28.10	2	11.90
1155	*T. parceramosum* Bissett 1991			1	14.21		
1156	*T. polysporum* (Link 1816) Rifai 1969	3	19.29	6	29.00		
1157	*T. pseudokoningii* Rifai 1969	2	19.79	3	27.88	1	6.33
1158	*T. reesei* Simmons 1968	4	19.81	5	28.41	4	5.05
1159	*T. saturnisporum* Hammill 1970	1	0.54	2	14.25		
1160	*T. virens* (J.H. Miller, Giddens et A.A. Foster 1957) Arx 1987	2	18.89	2	32.33	2	18.38
1161	*T. virgatum* Rifai			1	14.21		
1162	*T. viride* Persoon 1801	15	20.45	22	37.88	12	20.68

No.	Name	n	%	n	%	n	%
1163	*T. viride* Persoon 1801 var. *kizhanicum* Krapivina 1975			1	28.08		
1164	*Trichosporiella cerebriformis* (G.A. de Vries et Kleine-Natrop 1957) W. Gams 1971	1	19.58	1	19.22		
1165	*Trichosporon dulcitum* (Berkhout 1923) Weijman 1979	1	19.75	1	27.05		
1166	*Trichosporum herbarum* Jaap 1916	1	0.54	1	27.82		
1167	*Trichothecium plasmoparae* Viala 1932	1	19.28	1	26.16		
1168	*T. roseum* (Persoon 1801) Link 1809	5	20.41	9	39.07	4	8.98
1169	*Trichurus spiralis* Hasselbring 1900	1	17.68	1	6.37		
1170	*Tritirachium oryzae* (Vincens 1923) de Hoog 1972	3	19.86	5	32.87	1	3.38
1171	*Truncatella angustata* (Persoon 1801) S.Hughes 1958	1	19.45	1	20.34		
1172	*Tympanosporium parasiticum* W.Gams 1974			1	30.21		
1173	*Ugola praticola* (Pidoplichko 1950) Stalpers 1984			1	15.83	1	8.47
1174	*Ulocladium alternariae* (Cooke 1871) E.G. Simmons 1967	1	20.52	2	8.62		
1175	*U. atrum* Preuss 1852	1	19.59	6	24.79	1	8.34
1176	*U. botrytis* Preuss 1851	1	19.54	9	29.83	1	25.87
1177	*U. chartarum* (Preuss 1851) E.G.Simmons 1967	3	19.44	6	24.42		
1178	*U. consortiale* (Thuemen 1876) E.G.Simmons 1967	1	19.47	3	24.07		
1179	*U. oudemansii* E.G.Simmons 1967	1	19.30	2	24.59		
1180	*Umbelopsis isabellina* (Oudemans 1902) W.Gams 2003	5	19.42	7	35.00	7	28.33
1181	*U. nana* (Linnemann 1941) Arx 1984	3	14.34	2	16.85	2	17.20
1182	*U. ramanniana* (Moeller 1903) W.Gams 2003	3	19.42	11	23.76	11	28.66
1183	*U. vinacea* (Dixon-Stewart 1932) Arx 1984	1	6.83	3	29.13	2	21.06
1184	*Ustilago cordae* Liro 1924	1	18.50	1	1.84		
1185	*U. cynodontis* (Hennings 1892) Hennings 1893	1	12.13	1	1.64		
1186	*U. filiformis* (Schrank 1793) Rostrup 1890	1	9.32	1	1.64		
1187	*U. hordei* (Persoon 1801) Lagerheim 1889	1	12.31	1	1.81		
1188	*U. maydis* (de Candolle 1815) Corda 1842	1	0.20	1	1.64		
1189	*U. perennans* Rostrup 1890	1	18.50	1	1.84		
1190	*U. vinosa* (Berkeley 1847) Tulasne et C.Tulasne 1847	1	12.25				
1191	*Venturia* sp.			1	19.39		
1192	*Verticillium albo-atrum* Reinke et Berthold 1879	2	19.31	2	15.60	1	4.79
1193	*V. aspergillus* Berkeley et Broome 1873			1	33.19		
1194	*V. cellulosae* Dasz. 1912			1	9.19		

(continued)

Table 3.1 (continued)

Sr. No.	Name of species	Cryopreservation		Freeze-Drying		Soil	
		Number of strains	Storage time (years)	Number of strains	Storage time (years)	Number of strains	Storage time (years)
1195	*V. cercosporae* Petrak et Ciferri 1932			1	18.56		
1196	*V. dahliae* Klebahn 1913	12	19.75	15	31.61	12	7.96
1197	*V. epiphytum* Hansford 1943			1	21.55		
1198	*V. fumosum* Seman 1968	1	19.25	1	23.56	1	4.74
1199	*V. insectorum* (Petch 1931) W.Gams 1971			1	14.98		
1200	*V. lecanii* (Zimmermann 1898) Viegas 1939	3	19.36	7	26.42		
1201	*V. leptobactrum* W.Gams 1971	2	8.84	1	15.14		
1202	*V. nigrescens* Pethybridge 1919			1	4.67		
1203	*V. nubilum* Pethybridge 1919	1	19.31	1	23.64	1	4.82
1204	*V. sulphurellum* Saccardo 1882	1	19.26	1	14.76		
1205	*V. tricorpus* I.Isaac 1953	1	19.31	1	28.32	1	4.97
1206	*V. villosum* Rudakov 1981	1		1	27.38		
1207	*Viennotidia humicola* (Samson et W.Gams 1974) P.F.Cannon et D.Hawksworth 1982			1	5.05	1	5.01
1208	*Volutella ciliata* (Albertini et Schweinitz 1805) Fries 1832	1	19.83	1	21.53		
1209	*Wallemia sebi* (Fries 1832) Arx 1970	2	19.59	2	13.22		
1210	*Wallrothiella subiculosa* Hoehnel 1912	3	20.10	1	21.53		
1211	*Wardomyces anomalus* Brooks et Hansford 1923			1	1.11		
1212	*Westerdykella dispersa* (Clum 1955) Cejp et Milko 1964	1	19.34	1	19.68		
1213	*W. multispora* (Saito et Minoura ex Cain 1961) Cejp et Milko 1964	1	3.99	1	19.94		
1214	*Xeromyces bisporus* L.R.Fraser 1953	1	18.85	1	15.48		
1215	*Xylobolus frustulatus* (Persoon 1801) Boidin 1958	1	19.99				
1216	*Zygorhynchus exponens* Burgeff 1924	3	20.10	4	19.46		
1217	*Z. heterogamus* (Vuillemin 1886) Vuillemin 1903			1	27.33	1	1.66
1218	*Z.macrocarpus* Y.Ling 1930			1	9.93		
1219	*Z. moelleri* Vuillemin 1903	3	19.69	5	25.09	2	14.60
1220	*Zygosporium echinosporum* Bunting et E.W.Mason 1941	1	19.54	1	20.24	1	1.00
1221	*Z. mycophilum* (Vuillemin 1910) Saccardo 1911	1	19.54	1	10.79		

Freeze-Drying of Filamentous Fungi

Currently, freeze-drying is used to preserve approximately 80% of filamentous fungi maintained in VKM (2,991 strains, belonging to 1,010 species and 303 genera). Fungi from different taxonomical groups (*Zygomycetes, Ascomycetes*— both teleo- and anamorph) able to produce dormant structures (spores, sclerotia, etc.) usually survive freeze-drying [12]. According to our data, from 87 to 92% strains of these fungal groups remain alive in this method. We noticed that 57% of freeze-dried cultures stored at 5°C for more than 20 years were in a vital state, and cultures of more than 190 species have been sustained for even 30–40 years of storage. Some species did not survive freeze-drying even when the sporulation is abundant, those are: *Conidiobolus coronatus*, *C. obscurus* (syn. *Entomophthora thaxteriana*), *C. thromboides* (syn. *Entomophthora virulenta*), *Erynia conica* (syn. *Entomophthora conica*), *Pandora dipterigena* (syn. *Entomophthora dipterigena*), *Cunninghamella homothallica*, *Cunninghamella vesiculosa*. Species of genus *Botrytis* (*B. cinerea*, *B. fabae*, and *B. gladiolorum*), forming only sclerotia as a dormant structure, remain in vital state in freeze-drying only for rather a short time—less than 10 years [10].

Nonsporulating microorganisms from *Oomycetes* and *Basidiomycetes* are not stored in VKM by freeze-drying, since sterile mycelium generally do not remain viable. However, some ectomycorrhizal fungi (e.g., *Laccaria laccata*) could be successfully lyophilized. For the positive result preliminary slow freezing (to −32°C) of fungal material is required [13].

The equipment used in VKM for freeze-drying is centrifugal freeze-dryer system Micromodulyo (Edwards, UK).

Drying in Sterile Soil of Filamentous Fungi

This simple and popular method for preservation of fungi was applied in the beginning of the twentieth century [14]. Species of *Aspergillus, Penicillium* can be maintained by this way more effectively than other micromycetes. According to T.P. Suprun [15] who investigated preservation of 78 *Penicillium* species (more than 1,000 strains) in sterile soil for 7–10 years, the best preserved strains were representatives of *Assymmetrica* section. Less effectively preserved species were *Biverticillata-Symmetrica* and the lowest effectiveness was observed with strains of the section *Monoverticillata*.

Species of *Zygomycetes* could be stored in soil for periods ranging from 6 months (*Cunninghamella elegans*) to 5 years (*Rhizopus stolonifer* var. *stolonifer* [syn. *R. nigricans*]) [16].

This method is also efficient for preservation of some human, animal, and plant pathogens with retaining their virulence [7]. For example, *Alternaria japonica* (syn. *A. raphani*), *Fusarium oxysporum*, and the species of *Septoria* (*S. avenae, S. nodorum, S. passerinii, S. tritici*) have retained their ability to infect a plant host after 2–5 years of storage [17–19]. Some degraded strains of micromycetes partly recuperated their lost qualities after preservation in soil [15].

Species of *Alternaria, Pseudocercosporella, Septoria* are genetically more stabile compared with *Fusarium*, therefore they did not show these kinds of changes and can be effectively preserved in soil [3].

Protocols

Protocol of Cryopreservation

Preparation of Cryovials (2.0 mL Externally Threaded Polypropylene, Nunc, Denmark)

- Labelled (6 for each culture) with an index, a collection number of a strain and a date of cryopreservation (month, year).
- Sterilized by autoclaving, at 121°C for 20 min.

Preparation of Cryoprotectant: 10% (v/v) Glycerol

- Pour 5 mL of glycerol into 12 mL glass tubes.
- Sterilized by autoclaving at 121°C for 20 min.
- Stored at +5°C for no longer than a month.

Preparation of Cultures

- Grow *sporulating* fungal cultures on slant agar under optimal growth condition and on suitable medium (www.vkm.ru).
- Wash off spores from agar surface with 5 mL of cryoprotectant.
- Titer of spores' suspension should be not less than 10^6 spores/mL.
- Grow *nonsporulating* fungal cultures on Petri dishes under optimal growth condition and on suitable medium with agar concentration 5% (w/v). Incubate culture to get a mature colony.
- Cut mycelial plugs (5 mm diameter) from vigorously growing colony part.

Filling of Vials

- For sporulating cultures add 0.2-mL aliquots of suspension to each cryovial using a Pasteur pipet. This procedure is carried out under sterile aerobic conditions.
- For nonsporulating cultures place 4 mycelial plugs into each cryovial using transfer needle, and then add 0.2 mL of cryoprotectant.

Fast Cooling Rates Regime of Freezing (~400 grad/min)

- Place the cryovials with cultures (spore suspension) in special containers; thoroughly fixate a position of every vial.
- Immediately place cryovials with cultures in the vapour phase of liquid nitrogen or in ultra-low temperature freezer.

Programmed Regime of Freezing

The First Protocol

- Place 18 cryovials with cultures (mycelial plugs) in a special container "NALGENE™" (Cryo 1°C Freezing Container, Cat. No. 5100-0001).
- Fill the container with 250 mL of isopropanol.
- Place the container in a mechanical freezer (−80°C). Temperature in this container decreased by 1°C/min.
- When temperature achieves −70°C, transfer cryovials in special container (thoroughly fixate a position of every vial), and place in the vapour phase of liquid nitrogen or in the ultra-low temperature freezer.

The Second Protocol

- Place 18 cryovials with cultures (mycelial plugs) in a special metal container, thermostatic inside by expanded polystyrene (container made in VKM).
- Place the container in the ultra-low temperature freezer (−70°C). It was empirically shown in VKM that temperature in this container decreased by 0.4°C/min.
- When temperature achieves −70°C, transfer cryovials in a special container (thoroughly fixate a position of every vial), and place in the vapour phase of liquid nitrogen or in the ultra-low temperature freezer.

Thawing

- Pull out the cryovial of container, in which it was stored—either in cryogenic tank or in an ultra-low temperature freezer.
- Warm the cryovial rapidly by immersion in a shaking (for increase in heat exchange) water bath (37°C) for 1–2 min.

Control of Viability

- To estimate the viability of fungal culture before cryopreservation, place either one volume of the suspension (0.2 mL) or 4 plugs under optimal growth conditions on a suitable medium. This procedure is carried out under sterile aerobic conditions.
- To estimate the viability of fungal culture after cryopreservation, sterilize the thawing cryovials surface by wiping with 70% (v/v) ethanol. Asepticaly transfer the contents (spore suspension or mycelial plugs) using a Pasteur pipet or a transfer needle onto a suitable growth medium. This procedure is carried out under sterile aerobic conditions.

Protocol of Freeze-Drying

Preparation of Ampoules

Glass tubes (gray glass, diameter 7 mm, length 110 mm):

- Wash successively with detergent, tap water, and distilled water.
- Dry.

- Plug loosely with cotton wool to a depth of 1 cm.
- Label with an index, a collection number of a strain, and a date of freeze-drying (month, year).
- Sterilize in dry oven at 160°C for 2 h.

Preparation of Lyoprotectant Agent

- Pour 5 mL 10% (v/v) skimmed milk into each 12-mL glass tube.
- Sterilized by autoclaving at 105°C for 30 min.
- Stored at +5°C for no longer than a month.

Preparation of Cultures

- Grow sporulating fungal cultures on slant agar under optimal growth condition and on suitable medium (www.vkm.ru).
- Wash off spores from agar surface with 5 mL of skimmed milk.
- Titer of spores' suspension should be more, than 10^6 spores/mL.

Filling of Ampoules

- Add 0.2-mL aliquots of suspension to each ampoule using a Pasteur pipet. This procedure is carried out under sterile aerobic condition.

The First Stage (Primary Drying)

- Transfer the ampoules to the spin freeze-drier.
- Freeze (temperature in a refrigerator −45°C) under the reduced pressure of the ambient gas during centrifugation (30 min).
- Dry via water sublimation (temperature of a freeze dryer is −45°C) in vacuum (from 4×10^{-2} to 6×10^{-2} mbar) till the moisture level achieves 5–10%.
- Duration of the first stage is 3 h.
- Switch off the vacuum pump after the first stage. The system is filled with gas; the ampoules are removed from the centrifuge.

Preparation of Ampoules for the Second Stage of Freeze-Drying

- Constrict ampoules to diameter of 2–3 mm using an air/gas torch with horizontal flame preventing overheating of cultures just below the cotton wool plug (approximately 50 mm from the ampoule bottom)

The Second Stage (Secondary Drying)

- Attach the constricted ampoules via rubber tubes to the manifold connecting with the vacuum pump.
- Drying (vacuum 100 mm) till the moisture level reach 2%.
- Duration of the second stage is 2.5 h.
- Seal the ampoules across the constriction using an air/gas torch.

Vacuum Control

- Immediately after sealing, test vacuum in ampoules using a high-voltage spark tester.

Control of Culture Viability

- To estimate the viability of fungal strains prior to freeze-drying one volume of the spore suspension (0.2 mL) place under optimal growth condition on suitable medium. This procedure is carried out under sterile aerobic conditions.
- To estimate the viability of fungal culture after freeze-drying, test ampoules after 24 h storage.
- This procedure is carried out under sterile aerobic conditions.
- Sterilize a control ampoule's surface with 70% ethanol and open ampoules using a cutter.
- Reconstitute the dried suspension with sterile tap water (0.2–0.3 mL) using a Pasteur pipet.
- After 30 min (when rehydration is complete) transfer suspension under optimal growth condition on suitable medium.

Storage

- Store the ampoules d at +5–8°C in the dark.

Protocol of Drying in Sterile Soil

Preparation of Sterile Soil

- Place 5 g of finely cultivated (garden) soil into 12-mL glass tube.
- Sterilized by autoclaving, at 121°C for 30 min for three consecutive days.

Preparation of Cultures

- Grow sporulating fungal cultures on slant agar under optimal growth condition and on suitable medium (www.vkm.ru).
- Wash off spores from agar surface with 5 mL of sterile tap water.
- Titer of spores' suspension should be not less than 10^6 spores/mL.

Soil Inoculation

- Add 1 mL spore suspension to glass tubes with sterile air dry soil (moisture is under 20%).
- Incubate at room temperature till soil dry up (near 1 month).
- Store in the refrigerator at 4–7°C.

Control of Viability

- Transfer a few grains of soil onto fresh agar medium, add a little water and incubate under optimal conditions.

Result

The real storage time estimates obtained in VKM are given in Table 3.1. They are not final data: the cultures are still being stored, and we expect to get longer storage times later on. Some cells of the table are empty; this is the case if the culture is not stored this method.

Conclusion

The conservation techniques used in VKM presents effective preservation of the stock of filamentous fungi from different taxonomic groups. The possibility and practical time estimates of secure long-term storage of fungal cultures belonging to 1,221 species and 424 genera was shown. The represented information could be used as a reference for researchers intending to maintain pure cultures of microorganisms for a long time. The data produced are also accessible online on the VKM Web site.

Acknowledgements We thank Ludmila Evtushenko for constructive comments on the manuscript.

This work was supported by grants of the program MCB of the Russian Academy of Science and the Ministry of Education and Science of the Russian Federation (N. 16.518.11.7035).

References

1. OECD best practice guidelines for biological resource centers. OECD; 2007. 115p.
2. Glyn NS, Day JGD (2007) Long-term *ex situ* conservation of biological resources and the role of biological resource centers. In: Day JG (ed) Methods in molecular biology, vol 368. Humana Press, Totowa, pp 1–14
3. Smith D (1998) Culture and preservation. In: Hawksworth DL, Kirsop BE (eds) Living resources for biotechnology. Filamentous fungi. Cambridge Univ. Press, UK, pp 75–99
4. Guidelines for the establishment and operation of collections of cultures of microorganisms. 3rd ed. WFCC; 2010. 19p. http://www.wfcc.info/pdf/Guidelines_e.pdf.
5. Smith D, Onions AHS (1994) The preservation and maintenance of living fungi, 2nd edn. AB International, Wallingford, UK, p 132
6. Nakasone KK, Peterson SW, Jong S-C (2004) Preservation and distribution of fungal cultures. In: Mueller GM et al (eds) Biodiversity of fungi. Inventory and monitoring methods. Elsevier/Academic Press, Amsterdam, pp 37–47
7. Ryan MJ, Smith D (2007) Cryopreservation and freeze-drying of fungi employing centrifugal and shelf freeze-drying. In: Day JG (ed) Methods in molecular biology, vol 368. Humana Press, Totowa, pp 127–140
8. Smith D, Thomas VE (1998) Cryogenic light microscopy and the development of cooling protocols for the cryopreservation of filamentous fungi. World J Microb Biotech 14:49–57
9. Sidyakina TM (1988) Methods of preservation of microorganisms. In: Veprintzev BN (ed) Konservatziya geneticheskikh resursov. ONTI NTzBI AN USSR, Pushchino, p 59 (In Russian)
10. Ivanushkina NE, Kochkina GA, Eremina SS, Ozerskaya SM (2010) Experience in using modern methods of long-term preservation of VKM fungi. Mikol Phytopathol 44:19–30 (In Russian)
11. Ryan MJ, Smith D, Jeffries P (2000) A decision-based key to determine the most appropriate protocol for the preservation of fungi. World J Microb Biotech 16:183–186
12. Milosevic MB, Medic-Pap SS, Ignatov MV, Petrovic DN (2007) Lyophilization as a method for pathogens long term preservation. Proc Nat Sci Matica Srska Novi Sad 113:203–210
13. Sundari SK, Adholeya A (1999) Freeze-drying vegetative mycelium of *Laccaria fraterna* and its subsequent regeneration. Biotechnol Tech 13:491–495

14. Raper KB, Fennell DI (1965) The genus *Aspergillus*. The Williams and Wilkins Company, Baltimore, p 686
15. Suprun TP (1965) Preservation of microscopic fungi in sterile soil. Mikrobiologiya 34:539–545 (In Russian)
16. Belyakova LA, Lavrova LN, Kudryavtzev VI (1967) Preservation of fungal cultures. Metody khranenija kollekcionnykh kul'tur mikroorganizmov. M.: Nauka, p 7–55 (In Russian)
17. Atkinson RG (1953) Survival and pathogenicity of *Alternaria raphani* after five years in dried soil cultures. Can J Bot 31:542–546
18. Hine RB (1962) Saprophytic growth of *Fusarium oxysporum* f. *niveum* in soil. Phytopathology 52:840–845
19. Shearer BL, Zeyen RJ, Ooka JJ (1974) Storage and behavior in soil of *Septoria* species isolated from cereals. Phytopathology 64:163–167

Fungal Specimen Collection and Processing

4

Anthonia O'Donovan, Vijai Kumar Gupta, and Maria G. Tuohy

Abstract

The study of fungi relies, in part, on the axenic culture of isolates. Because so many fungi are found in nature in close proximity to each other, and many other organisms, study of their structure and function relies on the ability to grow and maintain a pure culture of the fungi. This chapter describes the methodology used to isolate a fungus that is present in abundance in a soil sample and covers many techniques that are widely used in microbiology laboratories.

Keywords

Fungi • Isolation • Culture • Soil • Spread plate • Direct culture • Storage

A. O'Donovan (✉)
School of Natural Sciences, National University
of Ireland Galway, University Road, Galway, Ireland
e-mail: anthonia.odonovan@nuigalway.ie

V.K. Gupta
Molecular Glycobiotechnology Group,
Department of Biochemistry, School of Natural Sciences,
National University of Ireland Galway,
University Road, Galway, Ireland

Assistant Professor of Biotechnology, Department
of Science, Faculty of Arts, Science & Commerce,
MITS University, Rajasthan, India

M. G. Tuohy
Molecular Glycobiotechnology Group, Department
of Biochemistry, School of Natural Sciences,
National University of Ireland Galway,
University Road, Galway, Ireland

Introduction

Fungi, including yeasts and molds, are ubiquitous in nature and have been recovered from diverse, remote, and extreme environments. Approximately 100,000 fungal species have been described to date. Over the last decade, approximately 1,200 new species of fungi have been described in each year. It is now estimated that there may be from 1.5 to 5 million extant fungal species [1]. Fungi are eukaryotic microbes with a filamentous growth form. A wide variety of fungi can be isolated from soil and cultivated on media in the laboratory. Different media will encourage the growth of different types of microbes through the use of inhibitors and specialized growth substrates [2].

Fungi are often isolated using nonselective agar medium. This allows the isolation of the

V.K. Gupta et al. (eds.), *Laboratory Protocols in Fungal Biology: Current Methods in Fungal Biology*,
Fungal Biology, DOI 10.1007/978-1-4614-2356-0_4, © Springer Science+Business Media, LLC 2013

maximum number of fungal taxa from a sample (it should be noted that all media used in micro-biological laboratories are selective to some degree or another; truly nonselective media do not exist [3]). So-called nonselective media are only media and incubation conditions designed to isolate as large a part of the microbiota in soil as is possible. The least selective media today isolate maybe 5–15% of the fungal population of soils. The media used for fungi are different from the media recommended for bacteria [4].

Saprotrophic fungi can be subcultured on media containing nutrients appropriate to their growth and development. Several different types of media have been used successfully. The organic and mineral fractions of media are designed to supply nutrients similar to or commonly found in the environment of the fungus. Some commonly used media consist of fruit or vegetable extracts mixed with sugars and agar, and set in Petri dishes. Other commonly used materials include soil (soil agar), potato (PDA), tomato plus other vegetables (V8 JUICE Agar), Malt extract (MA), and dung (DUNG AGAR). These can be more highly defined by replacing the organic compo-nent with known organic materials including nutrient dextrose (NDY) and sabouraud dextrose (Sabouraud Agar; [3, 4]).

Specifically targeted fungi can be subcultured using selective media. Routine methods of pro-viding selective media include the addition of specific compounds required by an organism as a nutrient source or the omission of compounds required by most other organisms [5, 6]. Changing some physical properties of the media can also favor growth of specific organisms (e.g., lower-ing the pH allows fungal growth to prevail over bacterial growth). Incubation temperature is another common method to encourage growth of target organisms or to deter growth of contami-nating microbes (e.g., spore-forming fungi can be isolated by the use of their ability to withstand high temperature). Altering conditions such as water content and the presence or absence of light along with providing aerobic or anaerobic condi-tions may also help in promoting the growth of a particular organism [3, 4, 7].

Many media are available from commercial sources (e.g., from Difco or Oxoid), which have a more consistent quality than those prepared from the raw materials each time.

The inclusion of antibiotics in media is a com-monly used method to suppress the growth of bac-teria. As most antibiotics are denatured by heat, the antibacterial agents are usually added to sterilized molten agar, that is just above setting temperature (hand warm, but not hot to touch). Antibiotics such as penicillin, streptomycin, and tetracycline are commonly used, either alone or more commonly, in combination. Chloramphenicol can be auto-claved and is added during preparation [3, 7].

It is possible to design very selective media using combinations of the aforementioned tech-niques. Almost any physiological group of organ-isms can, in theory, be cultured selectively. A single stage isolation from a soil sample can be changed to a multistage isolation process by rep-lica plating. An example of this includes the transfer of individual colonies in their original orientation on the plate by pressing a strip of ster-ile tape onto the surface of the plate, removing a small sample of each colony and pressing it onto the surface of a fresh plate. This can be a different growth medium so that only a part of the original population is able to grow on the new medium. Using this technique, progressive selective media can be used to isolate microorganisms with com-binations of properties [3, 4, 8].

Some general observations:

- Fungi grow better on media with a high C:N ratio (bacteria grow better on low C:N ratio media).
- More fungi are isolated on low pH media (bacteria are isolated using neutral pH media).
- Spores from common fungi such as *Aspergillus*, *Mucor*, and *Penicillium* in soil grow rapidly on many fungal isolation media and prevent growth of the slower growing fungi.
- Many fungi produce antibiotic compounds on plates and prevent growth of other microorganisms.
- Some microbes are very difficult to culture at all on media in the laboratory. They may be very slow growing or very sensitive to compo-nents of normal media [9].

Our knowledge of soil fungi is derived primar-ily from dilution and plating techniques (which are described in the following sections). These methods have limitations in that they are biased in favor of rapidly growing and sporulating organisms [10, 11] and consequently most of the fungi identified by these techniques are Fungi Imperfecti (*Pencillium* and *Aspergillus*). It is extremely difficult to grow soil basidiomycetes on solid media in the lab [11], but their abundance in soil can be revealed by microscopy. Nevertheless, dilution plate techniques are still a popular and useful tool for studying the relative abundance of culturable populations.

Sample Collection

Fungi are ubiquitous in their occurrence but are inconspicuous because of the small size of their structures, and their cryptic lifestyles in soil, on dead matter, and as symbionts of plants, animals, or other fungi. Common sources for their isolation include lakes, river muds, and soils.

Soils are discontinuous heterogenous environments that contain large numbers of diverse organisms. The main sampling problems in soil microbiology are usually a result of the complexity of the medium being sampled. If a method requires a "generalized" sample of the soil, the problem is to determine what soil horizons to sample. How many samples are required (to estimate variability), exactly where to take the samples (to determine spatial variation), and how frequently samples should be taken (to determine temporal variation) also have to be determined and can greatly influence findings. These problems are often interrelated, for example, a larger sample subdivided into smaller ones after mixing is different to many small independent samples. The first will show experimental or procedural error, while the second will show differences due to natural field variation as well as procedural errors. Therefore, many factors are to be considered when sampling soil [9]. Microbial populations also vary considerably with soil type, soil depth, moisture levels, use of pesticides, climate, etc.

Once a sampling site is selected, collect the desired soil samples, ensuring no contamination of the samples. Store the samples at 4°C until

ready to be cultured. Samples should be cultured within 72 h of collection if possible to eliminate the storage effects on the microbial population.

Air dry the samples to remove moisture and to allow for sieving of the soil, which removes large particulate, litter and debris.

Materials

1. Soil.
2. Balance.
3. The media: Aureomycin-rose Bengal-glucose-peptone agar: glucose, peptone, potassium dihydrogen phosphate (KH_2PO_4), magnesium sulfate ($MgSO_4.7H_2O$), rose Bengal, agar, Aureomycin hydrochloride, and reagent grade water.
4. Hot water bath (45 °C) to keep molten agar.
5. Sterile saline blanks for dilution (9 and 90 mL) in dilution bottles and test tubes.
6. Sterile pipettes (P1000 and P200) and sterile tips.
7. Vortexer.
8. Spreaders for spread-plating (made from glass pipette).
9. 70% ethanol (in squirt bottles and dishes).
10. Bunsen burner.
11. Inoculating loops (for transferring colonies).

Methods

Sample Preparation

Sample Dilution
1. To prepare the soil sample for culturing, appropriate dilutions of a suspension must be prepared: Add 10 g of soil sample (air dried and sieved) to 90 mL of sterile saline. This will create a 1/10 dilution of the soil sample, and the container should be labeled appropriately.
2. Vortex the solution for at least 1 min. The intent of the stirring is to suspend the organisms in the solution.
3. Allow the soil particulate matter to settle for up to 15 min. Decant the liquid portion into test tubes labeled with the appropriate sample and the dilution (1/10).

4. Transfer 1 mL of the 1/10 dilution into a test tube containing 9 mL of sterile saline using a sterile pipette. This will create a 1/100 dilution of the sample. Vortex the test tube until the sample is thoroughly mixed. Ensure the sample is labeled with the appropriate sample and dilution.

5. Transfer 1 mL of the 1/100 dilution into a test tube containing 9 mL of sterile saline using a sterile pipette. This will create a 1/1,000 dilution of the sample. Vortex the test tube until the sample is thoroughly mixed. Ensure the sample is labeled with the appropriate sample and dilution.

6. Repeat the above steps (4–5) to create additional dilutions if required. Dilutions for fungal cultures normally need not exceed 1/10,000.

Preparing Plates

1. An example of one type of media is Aureomycin-rose Bengal-glucose-peptone agar. To make this agar, dissolve 10 g glucose, 5 g peptone, 1 g potassium dihydrogen phosphate (KH_2PO_4), 0.5 g magnesium sulfate ($MgSO_4.7H_2O$), 0.035 g rose Bengal, and 20 g agar in 800 mL reagent grade water. The pH should be 5.4.

2. Sterilize by autoclaving at 105 °C for 30 min. Then keep molten in a 45 °C waterbath.

3. Dissolve 70 mg Aureomycin hydrochloride in 200 mL reagent grade water, sterilize by filtration.

4. Add the sterilized Aureomycin hydrochloride solution to the cooled (42–45 °C) agar base.

5. Pour 25 mL molten agar into sterile petri dishes (100×15 mm) and allow agar to solidify. Poured plates can be stored at 4 °C for up to 4 weeks.

Note: Many media are available from commercial sources. Commercial sources have a more consistent quality than those prepared from the raw materials each time.

Culturing

Spread Plate

1. Using a sterile pipette transfer 100 μL of the sample or appropriate sample dilution onto the agar surface (if you do not know which dilution is suitable, then several need to be plated out). (Vortex sample before taking the aliquot.)

2. Use the spreader (bent glass rod), which has been flame-sterilized, to spread the dilution over the surface as evenly as possible to ensure proper distribution (Note: cool the spreader before spreading).

3. Flame-sterilize the rod between different dilution factors and media or if it comes into contact with any nonsterile surface. (If dilution factors are spread in order of decreasing dilution—that is, 1/10,000, 1/1,000, 1/100, etc.—spreaders need not be resterilized except with changes of media.) Allow the moisture to be absorbed into the agar before the incubation.

4. Label plates with date, organism, dilution factor, type of agar, and incubation temperature. Prepare triplicate plates for each dilution.

5. Invert plates and incubate at the selected temperatures. Observe plates after day 2 and until day 5/7. Slow growing fungi may not produce noticeable colonies until day 6 or 7.

6. After the microbial colonies are readily visible (2–7 days), count the number of colonies on each plate and calculate the average number of fungi in the soil sample. (Plates can be stored at 4 °C for not longer than 24 h before counting if required).

Results

Choose a plate with a dilution that gives the most reliable counts for your calculations. The optimal maximal number of colonies per plate is 100. If there are more than 150 colonies per plate the result is recorded as TNTC (too numerous to count) and a plate with a lower number of colonies (a lower sample dilution) is selected. If there are no colonies on any of the plates, the results are recorded as <1 for the highest dilution.

Note: Interpretation of cultured sample data is complicated by two factors:

1. Certain species only grow well on certain agar media, and, therefore, may be present but not culturable because an inappropriate agar was used. Other species are not culturable at all.

2. Fungi grow at different rates, so slow-growing fungi may be over-run by fast-growing fungi, and, therefore, not be apparent in the culture.

Therefore, to perform culturable sampling correctly, three or more types of agar may need to be used for each sample.

It is important to note that quantification of fungi is different to bacteria because a fungal colony can form from a spore (single cell), a cluster of spores, or from a mycelial fragment (containing more than one viable cell). It is assumed that each fungal colony developed in a laboratory culture originates from a single colony forming unit, which may or may not be a single cell.

Identification

Fungal colonies are identified to genus or species, if possible, using a combination of colony macroscopic characteristics (e.g., color, morphology, growth rate) and microscopic characteristics (e.g., spore shape, size, color, hypha morphology [12–14]). Referring to definitive texts and Web sites regarding fungal identification is recommended.

Isolate a Pure Culture

Many cultures may consist of multiple types of fungi. To isolate a single fungus, a method to specifically select the target fungus for further fungal growth can be chosen. An example of such a method is to select one colony if possible or sweep the end of a hyphal tip and aseptically plate the sample out onto a fresh agar plate. Selective plates may be used to culture specific fungi. Once the target fungus has been isolated in this manner, further culturing onto nonselective agar without antibiotic can be conducted to ensure a pure culture has been achieved.

Direct Culture Methods
Some fungi may not be detected using the soil dilution/spread plate method previously given. In this case it is possible to directly inoculate the fungal sample by directly positioning the soil

sample containing what appears to be one kind of fungus onto the agar: A small fragment of soil is placed on an agar plate and incubated. Then the emergent hyphal tips are subcultured. Transfer these samples to fresh agar plates and incubate to develop a pure culture.

Storage of Fungi

It is expensive to continually subculture isolated fungal samples. Furthermore, the fungi may mutate during continuous subculturing. Preservation of fungi is essentially reducing the metabolic rate to the slowest possible. This can be achieved in several different ways.

Storage in Water
Plugs of agar containing the fungal culture are placed in vials containing only sterile water. The vials are lidded and stored at 10 °C for up to 5 years. This is a remarkably useful and cheap technique. (This is not recommended for many Basidiomycota).

Cold Temperature
Place the isolate on a slope or plate and store at 4 °C. Some fungi stored at cold temperatures remain viable for up to 4 years.

Under Oil
The fungus is subcultured onto low sugar media (e.g., one-sixth strength NDY) in a lidded tube. The culture is incubated to allow the isolate to cover the surface of the agar. The fungus is then covered with sterile mineral oil. The tubes are kept at room temperature. Fungi may be stored for up to a year. The oil slows access of oxygen to the culture. However, it is crucial the culture is covered by at least 1 cm of oil at all times and that the lid is kept clean. Contaminating fungi can grow on the top of the tube if it has nutrients on it. Also, moderate changes in air pressure lead to the movement of air in and out of the tube. Contaminants present in the air can lodge on the surfaces of the lidded tube. Therefore, topping the oil to just below the lid is recommended. Flaming the surfaces before subculturing is essential [15].

Freeze Drying or Lyophilization

Cultures are placed in a lyophilization tube, cooled, and then freeze dried. Fungi preserved in this way can be stored for up to 20 years. The process is commonly used for spore-forming Deuteromycetes. It is less successful for non-spore-forming fungi.

Low-Temperature Storage

Some fungi can be stored for many years when placed in 10–25% sterile glycerol in water in sealed vials that are then placed in liquid nitrogen or stored at − 70 °C. The fungi may need to be taken in stages through the cooling process. It is important to experiment with the correct glycerol dilution as some fungi will dessicate in 10% glycerol [15].

References

 1. Hibbett DS, Ohman A, Glotzer D, Nuhn M, Kirk P, Nilsson RH (2011) Progress in molecular and morphological taxon discovery in Fungi and options for formal classification of environmental sequences. Fungal Biol Rev 25:38–47
 2. Hurst CJ, Crawford RL, Garland JL, Lipson DA, Mills AL, Stetzenbach LD (2007) Manual of environmental microbiology, 3rd edn. ASM Press, Washington, DC. 20036–2904, USA ISBN 978-1-55581-379-6
 3. Maier RM, Pepper IL, Gerb CP (2000) Environmental microbiology. Academic Press, San Diego, USA
 4. http://www.scribd.com/doc/16572739/PROTOZOA-AND-FUNGI-CULTURING-IN-THE-LAB
 5. Hunter-Cevera JC, Fonda ME, Belt A (1986) Isolation of cultures. In: Demain AL, Solomon NA (eds) Manual of industrial microbiology and biotechnology. American Society for Microbiology, Washington, DC, pp 3–23
 6. Seifert KA (1990) Isolation of filamentous fungi. In: Labeda DP (ed) Isolation of biotechnological organisms from nature. McGraw-Hill Publishing Co., New York, pp 21–51
 7. Collins CH, Grange JM, Lyne PM, Falkinham JO (2004) Microbiological methods. Hodder Arnold, London, UK. ISBN 9780340808962
 8. Nautiyal CS, Dasgupta S, Varma A, Oelmüller R (2007) Screening of plant growth-promoting rhizobacteria advanced techniques in soil microbiology. Springer, Berlin, Heidelberg
 9. http://wvlc.uwaterloo.ca/biology447/modules/module8/soil/chapter3Soil446.htm
10. Bonito G, Isikhuemhen OS, Vilgalys R (2010) Identification of fungi associated with municipal compost using DNA-based techniques. Bioresour Technol 101:1021–1027
11. Thorn RG, Reddy CA, Harris D, Paul EA (1996) Isolation of saprophytic basidiomycetes from soil. Appl Environ Microbiol 62:4288–4292
12. de Hoog GS, Guarro J, Gene J, Figueras MJ (2004) Atlas of clinical fungi. Centraalbureau voor Schimmelcultures, Utrecht, The Netherlands
13. Pitt JI, Hocking AD (1997) Fungi and food spoilage. Academic press, Sydney
14. Samson RA, Hoekstra ES, Frisvad JC (2004) Introduction to food and airborne fungi. Centraalbureau voor Schimmelcultures, Utrecht, The Netherlands
15. http://bugs.bio.usyd.edu.au/learning/resources/Mycology/Growth_Dev/axenicCulture.shtml

Chemical and Molecular Methods for Detection of Toxigenic Fungi and Their Mycotoxins from Major Food Crops

S. Chandra Nayaka, M. Venkata Ramana,
A.C. Udayashankar, S.R. Niranjana,
C.N. Mortensen, and H.S. Prakash

Abstract

Mycotoxins are the secondary metabolites produced by certain molds on a wide range of agricultural commodities and are closely related to human and animal food chains. Mycotoxins are capable of causing disease in humans and other animals, and their detection is largely dependent on the sample matrix and the type of fungus causing their contamination. The strict regulations on trade of contaminated grains and seeds and other produce in industrial countries lead to economic burdens on farmers. In developing countries, the situation is aggravated where regulations may be nonexistent or not enforced and where consumption of home-grown cereals leads to a wide exposure to toxins. Important mycotoxins that occur quite often in food are deoxynivalenol/nivalenol, trichothecenes, zearalenone, ochratoxin A fumonisins, and aflatoxins. High concentrations of mycotoxins such as aflatoxins are consumed by humans in areas of the world with higher-than-average levels of liver cancer, childhood malnutrition, and disease. This chapter introduces rapid, robust, and user-friendly protocols currently applied in the identification of toxigenic fungi and important mycotoxins.

S.C. Nayaka • A.C. Udayashankar
S.R. Niranjana • H.S. Prakash (✉)
Department of Studies in Biotechnology, Asian Seed
Health Centre, University of Mysore Manasagangotri,
Mysore, Karnataka 570 006, India
e-mail: hsp@appbot.uni-mysore.ac.in

M.V. Ramana
Microbiology Division, Defence Food Research
Laboratory, Siddarthanagar, Mysore,
Karnataka 570 011, India

C.N. Mortensen
Department of Agriculture and Ecology, University
of Copenhagen, Thorvaldsensvej 40, DK-1871
Frederiksberg C Copenhagan, 2630 Taastrup, Denmark

V.K. Gupta et al. (eds.), *Laboratory Protocols in Fungal Biology: Current Methods in Fungal Biology*,
Fungal Biology, DOI 10.1007/978-1-4614-2356-0_5, © Springer Science+Business Media, LLC 2013

Keywords

Mycotoxins • Fungi • Polymerase chain reaction (PCR) • Thin-layer chromatography (TLC) • High-performance liquid chromatography (HPLC)

Introduction

The term "mycotoxin" combines the Greek word for fungus "mykes" and the Latin word "toxicum" meaning poison. Mycotoxins have received considerable attention, especially over the last few decades. The problem related to mold damage and the hazard of consuming damaged grains have been recognized since historical times. The term "mycotoxin" is usually reserved for the toxic chemical products formed by a few fungal species that readily colonize crops in the field or after harvest and thus pose a potential threat to human and animal health through the ingestion of food products prepared from these commodities [1].

The possibility of human diseases occurring as a result of the consumption of mold-damaged rice and wheat was raised in Japan and other Asian countries during the first half of this century. Awareness of risks from eating over-wintered millet was reported in the USSR [2]. However, the serious worldwide concern about mycotoxins began in the early 1960s after it was discovered in the United Kingdom that Turkey "X" disease is caused by aflatoxins. More than 300 mycotoxins have been identified, although only around thirty with toxic properties that are genuinely of concern for human beings or animals were reported by the French Agency of Food, Environmental and Occupational Health and Safety [3]. Some mycotoxins are rather rare in occurrence; others—such as aflatoxin, ochratoxin, fumonisins, and trichothecenes—are quite common in some years. The molds primarily responsible for producing mycotoxins are *Aspergillus, Fusarium,* and *Penicillium* spp. Occurrence of mycotoxins in food and animal feed often exhibits a geographical pattern; for example, *Aspergillus* species meet optimal conditions in tropical and subtropical regions, whereas *Fusarium* and *Penicillium* species are adapted to the moderate climate of North America and Europe [4]. The toxins can be produced in major food crops like, maize, wheat, sorghum, rice, soybeans, peanuts, and other food and feed crops in the field, during transportation, or improper storage. Moreover, in animals consuming contaminated feed, mycotoxins can deposit in different organs and also subsequently affect food of animal origin (e.g., meat, eggs, milk, and milk products). Worldwide trade with food and feed commodities has resulted in a wide distribution of contaminated material [5]. One of the characteristics of mycotoxins is that they can exude toxic properties in minute quantities; thus, sensitive and reliable methods are required for their detection and quantification, which generally involves sophisticated sampling, sample processing, extraction, and assay techniques.

Different methods have been applied in the detection and quantification of mycotoxins from food and feed samples, including ELISA (enzyme-linked immunosorbent assay), immunoaffinity cartridge, solid-phase ELISA, and selective adsorbent mini-column procedures [6]. TLC (thin-layer chromatography) and HPLC (high-performance liquid chromatography) [7] are more accurate for quantification of mycotoxins in food and feedstuffs' produce. Under practical storage conditions monitoring for the occurrence of fungi is often conducted; however, in practice, it is difficult to distinguish several toxigenic fungal species from their close relatives, and accurate identification based on traditional methods is very difficult owing to their genetic variation and high morphological similarity.

The conventional scheme of isolation and identification of toxigenic fungi from food samples is cumbersome and requires skilled personnel to achieve proper identification. Even with taxonomical expertise, identification is commonly difficult regarding some fungi genus that contains a large number of closely related species [8]. Robust DNA-based tools often offer accurate, rapid, and sensitive identification and characterization of species (e.g., *Fusarium*) that belong to a complex genus [9]. The application of molecular biology techniques is an alternative to cumbersome and time-consuming conventional culture methods for precise identification of toxigenic fungal species before they can enter the food chain. The polymerase chain reaction (PCR) assay has allowed rapid, specific, and sensitive detection of toxigenic species without the need for prior growth of the organisms. The traditional molecular markers are mainly based on ribosomal DNA, β-tubulin, and calmodulin genes or have been based on anonymous DNA sequences. These DNA sequences are obtained from an unbiased sampling of genomic DNA, and these may or may not contain functional genes involved in toxin production [10].

Developing markers from anonymous sequences requires comparative analyses among related species of DNA profiles generated from randomly amplified fragments by using RAPD (random amplified polymorphic DNA) or AFLP (amplified fragment length polymorphism). In the last decade, numerous PCR assays have been developed for rapid detection and differentiation of toxigenic and nontoxigenic fungi from major commodities by using specific genes associated with mycotoxin biosynthesis [11].

Some Important Mycotoxins

Aflatoxins (*Aspergillus* spp.)

Aflatoxins are chemical derivatives of difuran-coumarin, mainly produced by *Aspergillus flavus*/*A. parasiticus*. Aflatoxins have been implicated in subacute and chronic effects in humans. These effects include primary liver cancer, chronic hepatitis, jaundice, hepatomegaly, and cirrhosis through repeated ingestion of low levels of aflatoxin. It is also considered that aflatoxins may play a role in a number of diseases, including

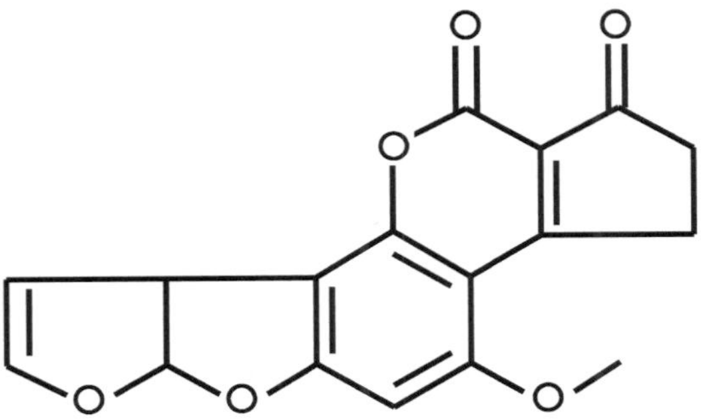

Fig. 5.1 Structure of aflatoxin B$_1$

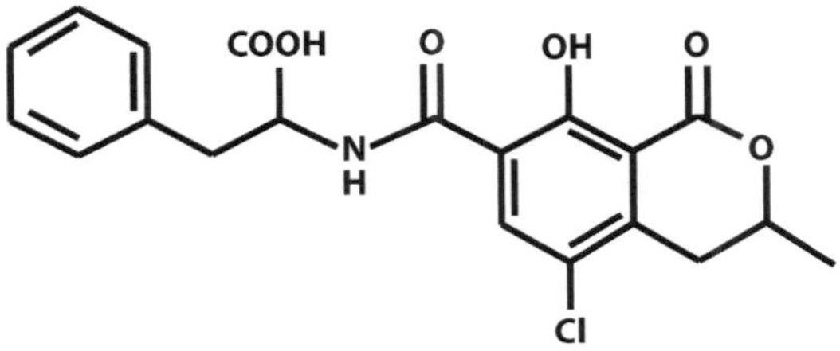

Fig. 5.2 Structure of ochratoxin A

Reye's syndrome, kwashiorkor, and hepatitis [12]. Aflatoxins can also affect the immune system. *A. flavus* infects many of our food crops, such as nuts, grains, and culinary herbs. Primary economic concerns are infestations that occur in corn and peanuts. The major aflatoxins consist of aflatoxins B1 (Fig. 5.1), B2, G1, and G2.

Ochratoxin

Ochratoxin A is the most important and most commonly occurring structurally related group of compounds; it is often abbreviated to OTA or OA (Fig. 5.2). Ochratoxin A is the major mycotoxin of this group, and it is an innately fluorescent compound produced primarily by *Aspergillus ochraceus* and *Penicillium verrucosum* [13]. Ochratoxin A is a potent toxin that affects mainly the kidneys, in which it can cause both acute and chronic lesions. Ochratoxin A is a potent teratogen in mice, rats, hamsters, and chickens, and a nephrotoxic effect has been demonstrated in all mammalian species.

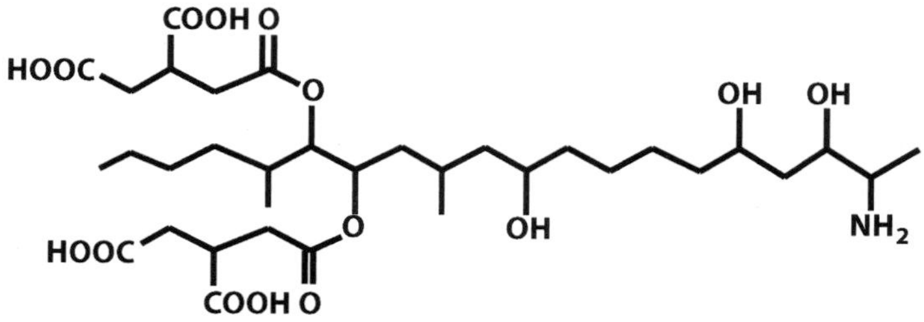

Fig. 5.3 Structure of
fumonisin B₁

Fumonisins

The fumonisins are a group of nonfluorescent
mycotoxins—FB₁ (Fig. 5.3), FB₂, and FB₃ being
the major entities—produced primarily by
Fusarium verticillioides and *F. proliferatum* [13],
Fusarium nygamai, as well as *Alternaria alternata* f. sp. *lycopersici.* Fumonisins are thought to
be synthesized by condensation of the amino
acid alanine into an acetate-derived precursor.
Numerous species-specific diseases have been
attributed to fumonisin-contaminated feed, including leukoencephalomalacia in horses and pulmonary edema and hydrothorax in swine [14]. These
compounds have been shown to have carcinogenic potential in animal models and are the only
known inhibitors of ceramide kinase, a key
enzyme involved in inflammatory cascades.

Deoxynivalenol

Deoxynivalenol, also known as DON or vomi-
toxin, is one of about 150 related compounds
known as the trichothecenes that are mainly produced by *Fusarium graminearum* and, in some
geographical areas, by *F. culmorum* (Fig. 5.4)
[15]. These two species are important plant
pathogens and cause Fusarium head blight in
wheat and Gibberella ear rot in maize. Toxicity of
deoxynivalenol is characterized by vomiting,
particularly in pigs, feed refusal, weight loss, and
diarrhea. A study reporting human food poison-
ing by infected wheat containing deoxynivalenol
in India showed a range of symptoms, including
abdominal pains, dizziness, headache, throat
irritation, nausea, vomiting, diarrhea, and blood
in the stool [16].

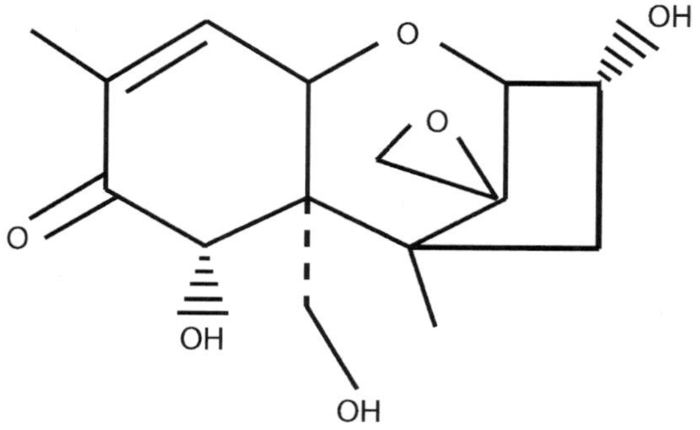

Fig. 5.4 Structure of deoxynivalenol

The potential presence of toxins in the food
supply means that expensive testing and remedial
actions are necessary to assure that they do not
reach dangerous levels in our food. This testing
and losses in crop quality and yield associated
with these fungal diseases are estimated to cost
agriculture billions of dollars annually, and the
presence of fungal toxins in our crops places the
competitiveness of our agricultural exports at
risk. The presence of mycotoxins is unavoidable;
therefore, testing of raw materials and products is
required to keep our food and feed safe. The presence of mycotoxins in food crops is a serious and
common quality problem that has become more
obvious as a result of the research of recent years.
Several chemical and biological detection systems exist for the determination of mycotoxins.
Biological assays were used when analytical and
methods were not available for routine analysis
because biological assays are qualitative and
often are nonspecific and time-consuming. Various
analytical methods for mycotoxin analysis have
been developed, such as TLC, HPLC, and HPTLC
[7]. Many research laboratories have also adopted
molecular detection methods for rapid and accurate

detection of toxigenic molds from major food crops.

This chapter presents information on the general protocols adopted for detection of important toxigenic *Aspergillus*, *Penicillium*, and *Fusarium* species by PCR and their mycotoxins mainly by TLC and HPLC.

Materials

(See Note 1)

DNA Extraction

1. Sterile distilled water.
2. 1× lysis buffer (made freshly approximately every 2 weeks): 100 mM Tris–HCl pH 7.4 (37°C), 100 mM EDTA, 6% SDS, 2% β-mercaptoethanol.
3. Vortex mixer.
4. Water bath.
5. Phenol: chloroform (1:1).
6. Microcentrifuge.
7. Disposable polypropylene microcentrifuge tubes: 1.5 mL conical; 2 mL screw-capped.
8. 3 M ammonium acetate.
9. Isopropanol.
10. Tris EDTA: 10 mM Tris–HCl pH 7.5 (25°C), 0.1 mM EDTA.
11. Phenol solution equilibrated with 10 mM Tris–HCl pH 8, 1 mM EDTA (Sigma-Aldrich, Gillingham, UK).
12. Ethanol (70%).
13. Agarose (e.g., molecular biology grade agarose from Sigma-Aldrich, USA).
14. TBE buffer: 50 mM Tris, 50 mM boric acid, 1 mM EDTA. Dilute when needed from a 10× stock.
15. Ethidium bromide: 0.5 mg/mL stock.
16. Gel loading mixture—40% (w/v) sucrose, 0.1 M EDTA, 0.15 mg/mL bromophenol blue.
17. Horizontal electrophoresis equipment (e.g., Biorad Wide Mini Sub Cell).

18. UV transilluminator and camera suitable for photographing agarose gels (e.g., Syngene Gene Genius Bioimaging System).

Basic Equipment Required for Polymerase Chain Reaction

1. Thermal cycler.
2. Micropipettes.
3. Agarose gel electrophoresis unit.
4. Centrifuge.
5. UV-gel documentation chamber.
6. Eppendorf tubes.

PCR Reagents

1. Agarose.
2. dNTP mix.
3. PCR buffer.
4. $MgCl_2$.
5. Template DNA.
6. Taq DNA polymerase.
7. Oligonucleotide primers.
8. Ethidium bromide.
9. TBE buffer (Tris–EDTA–boric acid buffer).

PCR Assay Set-up

Prepare a master mix (for 50 µL reaction) containing the following:

1. 5 µL 10× PCR buffer.
2. µL MgCl2 (25 mM)×(number of reactions+1).
3. µL dNTP mix (1.25 mM).
4. 30 µL sterile distilled H_2O.
5. Dispense the master mix at 43 µL per PCR tube.
6. Dispense primers in pairs at 2.5 µL per tube.
7. Dispense template at 1 µL per tube.
8. Prepare the Taq solution by diluting the appropriate amount of stock Taq DNA polymerase to 1 unit/µL with sterile distilled H_2O.
9. Dispense diluted Taq polymerase at 1 µL per tube.

General PCR Conditions

After setting up the reaction, the following conditions can be used for proper amplification of target genes:

1. Initial denaturation: 94°C/5 min.
2. Denaturation: 94°C/1 min.
3. Annealing temperature: 52–60°C/1 to 1.30 min.
4. Extension temperature: 72°C/1 to 2 min.
5. Repeated for 30 to 35 cycles.
6. Final extension temperature: 72°C/5 to 10 min.
   After successful amplification load 5 μL of amplified PCR products to agarose gel (1%) in containing 1% ethidium bromide and visualize under UV.

Materials Required for Chromatography-Based Method

1. Silica gel-coated plates (Merck chemicals).
2. Solvents, analytical grade and HPLC grade (Sigma-Aldrich, USA).
3. Purified mycotoxin standards (Sigma-Aldrich, USA).
4. Micropipettes (Eppendorf 1–1,000 μL).
5. Chromatography chamber (CMAG).
6. Spray reagents.
7. UV scanner (CMAG).
8. Mechanical cup shaker (CMAG).
9. Conical flasks (250 mL).
10. Filter papers.
11. Separating funnels.
12. HPLC system with UV and fluorescent detectors (Hitachi F-4500).
13. C18/C8 cartridges.
14. Immunoaffinity/Solid phase extraction columns for clean up.

Methods

DNA Fingerprinting Methods

The methods detailed as follows describe the general protocol for obtaining pure DNA from fungal culture, food, and grain samples [17]. The volumes

and number of tubes used per sample may need to be varied, depending on the type of sample and the quantity of mycelia being processed.

Extraction of DNA from Pure Cultures of Fungi [18]

1. Take small pinch of mycelium into micro-centrifuge tubes.
2. Add 500 μL of lysis buffer, and macerate the mycelium with the help of a sterile glass rod.
3. Add 50 μL of 10% SDS, vortex, and incubate at 65°C for 10 min.
4. Add 500 μL phenol: chloroform (1:1) to each tube and vortex briefly.
5. Centrifuge 5 min at 10,000 rpm in the microfuge, then carefully transfer as much as possible of the top aqueous layer to a clean tube. Do not disturb the debris at the interface.
6. Add 40 μL 6 M ammonium acetate, 700 μL isopropanol, invert gently to mix, and spin for 2 min, and incubate the mixture at −20°C for 1–2 h.
7. Centrifuge at 10,000 rpm for 10 min and discard the supernatant.
8. Add 300 μL cold 70% ethanol, centrifuge for 2 min, and discard the supernatant.
9. Centrifuge for 10 s and remove the remaining liquid with a micropipette.
10. Allow the pellet to dry for 20 min in a fume hood and then resuspend it in 50 μL TE or sterile distilled water.

Extraction of DNA from Contaminated Food Samples (e.g., Maize)

This method has been used to prepare DNA from contaminated food grains at several laboratories. The steps are as follows:

1. Ground the contaminated food grains.
2. Add 400 μL of lysis buffer, vortex, and incubate at 65°C for 10 min.
3. Add 500 μL phenol:chloroform (1:1) to each tube and vortex briefly.

4. Centrifuge 15 min at 12,000 rpm in the microfuge, then carefully transfer as much as possible of the top aqueous layer to a clean tube. Do not disturb the debris at the interface.

5. 40 μL 6 M ammonium acetate, 600 μL isopropanol, invert gently to mix and spin for 2 min. Incubation of the mixture at −20°C for 10–60 min before centrifugation may improve recovery of DNA but can result in reduced purity of the sample. Remove the supernatant.

6. Add 50 μL RNase in TE to the pellet and incubate at 37°C for 15 min.

7. Pool into 200 μL samples or add 150 μL TE, and then add 200 μL phenol, vortex, and centrifuge at 10,000 rpm for 6 min.

8. Transfer carefully the top aqueous layer to a fresh tube.

9. Add 10 μL 6 M ammonium acetate, 600 μL isopropanol, and incubate at −20°C for 10 min.

10. Centrifuge at 12,000 rpm for 10 min and discard the supernatant.

11. Add 800 μL cold 70% ethanol, centrifuge at 10,000 rpm for 2 min, and discard the supernatant.

12. Centrifuge at 10,000 rpm for 15 s and decant the remaining liquid with a micropipette.

13. Allow the pellet to dry for 20 min in a fume hood and then resuspend it in 50–200 μL TE buffer.

Polymerase Chain Reaction

During the last few decades great advances have been made in molecular diagnostic technology, especially in the development of rapid and sensitive methods for the detection of plant pathogenic fungi [19]. A number of DNA-based techniques that have been developed include restriction fragment length polymorphism, pulse field gel electrophoresis, and PCR. PCR has been gaining popularity mainly because of its ease of application compared to other DNA-based techniques.

There are already many examples of PCR-based assays developed for the detection of fungi in plant pathology, but the reports on their use in specific detection of toxigenic fungi are limited.

Many mycotoxin biosynthetic pathway genes are present within gene clusters, and some of these appear to have undergone horizontal transfer from one species to another and are now present in several species [7]. Regions of homology within mycotoxin biosynthetic gene from the different species can be used to develop primers to detect the presence of the relevant mycotoxigenic species. This strategy was successfully applied for aflatoxin producers [20], trichothecene-producing fungi [18], fumonisin-producing *Fusarium* species [18], and also for producers of ochratoxin [21]. PCR-based detection has been applied as an alternative assay, replacing cumbersome and time-consuming microbiological and chemical methods for detection and identification of the most serious pathogenic and mycotoxin producers in the fungal genera *Fusarium*, *Aspergillus*, and *Penicillium* spp. (Table 5.1).

Polymerase Chain Reaction-Based Detection of Aflatoxigenic Fungal Species

Target gene and primers: *Nor1*— 5′CGCTACGCCGGCACTCTCGGCA3′ (forward) and 5′TGGCCGCCAGCTTCGACACTC3′ (reverse). Amplicon size 400 bp.

Reaction conditions:

1. Initial denaturation: 94°C/5 min.
2. Denaturation: 94°C/1 min.
3. Annealing temperature: 58°C/1 min.
4. Extension temperature: 72°C/1 min. Repeated for 30 cycles.
5. Final extension temperature: 72°C/5 min.
6. After successful amplification, load 5 μL of amplified PCR products to agarose gel (1%) containing 1% ethidium bromide and visualize under UV (Fig. 5.5).

Polymerase Chain Reaction-Based Detection of Ochratoxigenic Fungi

Target gene and primers: *pks1*—5′AGT CTTCGCTGGGTGCTTCC3′ (forward) and 5′ AGCACTTTTCCCTCCATCTATCC3′ (reverse). Amplicon size 630 bp.

Table 5.1 Primer sequences developed for metabolic pathway genes for the detection of toxigenic fungi

S. No.	Toxin	Target gene	Primer sequence 5'–3'	Tm (°C)	Amplicon size (bp)	Reference
1	Trichothecenes	Tri5	GAGAACTTTCCCACCGAATAT GATAAGGTTCAATGAGCAGAG	58	450	[18]
		Tri6	GATCTAAACGACTATGAATCACC GCCTATAGTGATCTCGCATGT AGA GCC CTG CGA AAG(C/T) ACT GGT GC	58	541	
2	Fumonisins	Fum5	GTC GAG TTG TTG ACC ACT GCG CGT ATC GTC AGC ATG ATG TAG C	62	845	[22]
		Fum13	AGTCGGGGTCAAGAGCTTGT TGCTGAGCCGACATCATAATC	58	998	
3	Aflatoxins	Aflr1	CGC GCT CCC AGT CCC CTT CAT T CTT GTT CCC CGA GAT GAC CA	65	1,032	[20]
		Ver1	GCC GCA GGC CGC GGA GAA AGT GGT GGG GAT ATA CTC CCG CGA CAC AGCC	65	537	
		Nor1	ACCGCTACGCCGGCACTCTCGGCAC GTTGGCCGCCAGCTTCGACACTCCG	65	400	
		Omt1	GTG GAC GAA CCT AGT CCG ACA TCAC GTC GGC GCC ACG CAC TGG GTT GGGG	65	797	
4	Ochratoxin	Pks1	AGTCTTCGCTGGGTGCTTCC AGCACTTTTCCCTCCATCTATCC	56	550	[21]

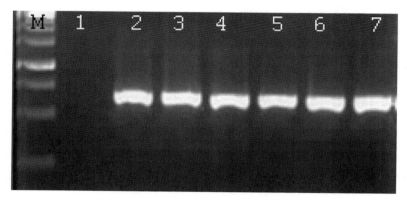

Fig. 5.5 Detection of aflatoxin-producing *Aspergillus* species targeting *aflR* gene (400 bp). Lane M, 1-kb DNA ladder; lane 2, negative control; lanes 3–8 aflatoxigenic *Aspergillus* spp

Reaction conditions:
1. Initial denaturation: 94°C/5 min.
2. Denaturation: 94°C/1 min.
3. Annealing temperature: 56°C/1 min.
4. Extension temperature: 72°C/1 min. Repeated for 30 cycles.
5. Final extension temperature: 72°C/5 min.
6. After successful amplification, load 5 μL of amplified PCR products to agarose gel (1%) containing 1% ethidium bromide and visualize under UV (Fig. 5.6).

Polymerase Chain Reaction-Based Detection of Tricothecene-Producing *Fusarium* Species

Target gene and primers: *tri6*—5'GATCTAAACGACTATGAATCACC3' (forward) and 5'GCCTATAGTGATCTCGCATGT3' (reverse). Amplicon size 446 bp.
Reaction conditions:
1. Initial denaturation: 94°C/5 min.
2. Denaturation: 94°C/1 min.

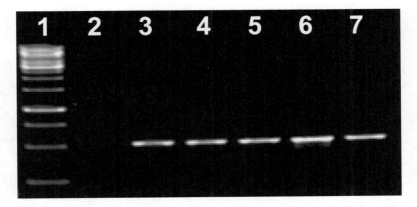

Fig. 5.6 Detection of ochratoxin-producing fungi targeting PKS gene (630 bp). Lane 1, 1-kb DNA ladder; lane 2, negative control; lanes 3–5, OTA-producing *Aspergillus*; lanes 6 and 7, OTA-positive strains of *Penicillin* spp

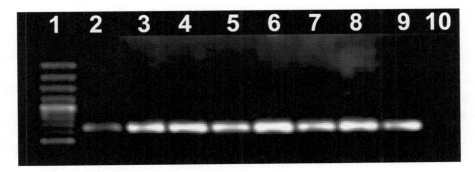

Fig. 5.7 Detection of trichothecene-producing *Fusarium* spp. by targeting *tri6* gene (440 bp). Lane1, 1-kb DNA ladder; lanes 2–9, positive strains of *Fusarium* spp.; lane 10, negative control

3. Annealing temperature: 56°C/1 min.
4. Extension temperature: 72°C/1 min. Repeated for 30–35 cycles.
5. Final extension temperature: 72°C/8 min.
6. After successful amplification, load the PCR amplicons into ethidium bromide- containing agarose gel and visualize bands under UV (Fig. 5.7).

Polymerase Chain Reaction-Based Detection of Fumonisins Producing *Fusarium* Species

Target gene and primers: *Fum13*— 5'AGTCGGGGTCAAGAGCTTGT3' (forward) and 5'TGCTGAGCCGACATCATAATC3' (reverse). Amplicon size 998 bp.

Reaction conditions:
1. Initial denaturation: 94°C/5 min.
2. Denaturation: 94°C/1 min.
3. Annealing temperature: 58°C/1 min.
4. Extension temperature: 72°C/1.30 min. Repeated for 30 to 35 cycles.
5. Final extension temperature: 72°C/8 min.
6. After successful amplification, load 5 μL of amplified PCR products to agarose gel (1%) containing 1% ethidium bromide and visualize under UV (Fig. 5.8).

Chromatography Methods

Sample Extraction and Clean-up for Mycotoxins Analysis

During the chromatographic methods the determination step is usually preceded by a number of

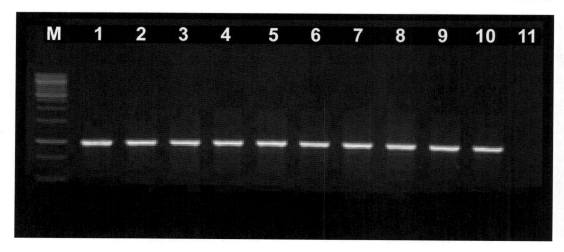

Fig. 5.8 PCR amplification *fum 13* gene (998 bp) of toxigenic *F. verticillioides* and *F. proliferatum.* Lane M, 1-kb DNA marker; lanes 1–5, *F. verticillioides* standard strains; lanes 5–10, *F. proliferatum* isolates; lane 11, nontoxigenic *F. verticillioides* isolate

operations such as sampling, sample preparation, extraction, and clean-up. The reliability of the results obtained by these procedures is highly dependent on the efficiency of these steps. A large number of components that are originally present in the sample must be reduced, and interfering compounds that show the same behavior in the chromatographic column must be removed as much as possible [7].

Conventional techniques such as column chromatography and liquid–liquid extraction usually require high amounts of solvent, are time-consuming, tedious to apply, and expertise is needed. Therefore, new approaches have been investigated to simplify the extraction and clean-up procedures. A number of clean-up columns have been developed that are used after the conventional extraction step. These procedures make use of different principles (immunoaffinity columns, solid phase extraction, ion exchangers, and others), but all have in common that they are commercially available and are easy to use. They have the additional advantages that less solvent is required and sample preparation can be speeded up considerably. The immunoaffinity columns (IACs) reveal high selectivity, as only the analyte is retained on the column and can then be eluted easily after a rinsing step in order to remove interfering components. Clean-up procedures are used for the removal of interfering compounds such as lipids, carbohydrates, and proteins [23].

IACs for clean-up purposes have become increasingly popular in recent years because they offer high selectivity. IACs are easy to use, and their application for purification of samples that are contaminated with several mycotoxins has already been well investigated. Because mycotoxins are low weight molecules, they are only immunogenic if they are bound to a protein carrier. If this problem is overcome, specific antibodies can be produced and bound to an agarose, sepharose, or dextran carrier. The mycotoxin molecules bind selectively to the antibodies after a preconditioning step, and subsequent to a washing step the toxin can be eluted with a solvent, causing antibody denaturation. Interfering substances do not interact and the column is therefore washed to remove the matrix [24].

Thin-Layer Chromatography

TLC was a very popular technique to separate and detect mycotoxins. TLC is the most commonly utilized test because more than one mycotoxin can be detected from each test sample. TLC is based on the separation of compounds by their migration on a specific matrix with a specific solvent. The distance that a compound will travel is a unique identifier for specific compounds and a retention factor (R_f) has been determined for most mycotoxins. As with any detection system, a positive control containing purified mycotoxins must be run in parallel to ensure accuracy. For mycotoxin assays, silica gel TLC, with both precoated and self-coated plates, can be used. Detection and identification procedures have been specifically developed for each single mycotoxin, making use of molecular properties or reactions with spray reagents [25].

High-Performance Liquid Chromatography

HPLC or high-pressure liquid chromatography became available for the analysis of foodstuffs in the early seventies and gained importance in the determination of mycotoxins, particularly when several types of column packings and detectors became available. HPLC is the method of choice because it offers the advantages of good resolution, high degree of precision, reproducibility, and sensitivity. HPLC methods are mainly used for the final separation of matrix components and detection of the analyte of interest. Nowadays, HPLC methods are widespread, because of their superior performance and reliability compared with TLC. HPLC methods have been developed for almost all major mycotoxins in cereals and other agricultural commodities. Reversed phase (RP) chromatography is most commonly used for the determination of mycotoxins in agricultural samples—for example, a C_8 or C_{18} hydrocarbon phase with mixtures of polar solvents (e.g., water: methanol or water:acetonitrile combinations). Detection is mainly performed using diode array detection; alternatively, fluorescence detection (FLD), which utilizes the emission of light from molecules that have been excited to higher energy levels by absorption of electromagnetic radiation, is employed. FLD features superior sensitivity, although frequently derivatization of the analyte has to be performed in order

to make the detection possible at all or enhance the sensitivity even further [26]. A short summary for the determination of the common toxins of *Aspergillus* and *Fusarium* spp. by chromatographic method is presented in the following section.

Detection of Aflatoxin by Thin-Layer Chromatography

A number of methods have been developed for the determination of aflatoxins by TLC. Silica plates are most commonly in use, with a number of solvent systems based on chloroform and small amounts of methanol or acetone. However, a shift can be observed to less toxic and more environmentally friendly mixtures (e.g., toluene/ethylacetate or acetone/isopropanol).

Extraction and Clean-up

Place 50 g of finely ground sample in a wide-mouth polypropylene screw-cap bottle with 100 mL of chloroform and water mixture (1:1) and place on a wrist-action shaker for 30 min. Allow the contents to settle, and a 10- to 25-mL aliquot of the solvent extract is filtered through four layers of Whatman filter paper. Dry the organic layer by rotary evaporator and reconstitute the compound in 3–4 mL of methanol. Load onto a preconditioned C18 clean-up column. Wash the column with 5 mL of phosphate buffered saline and then elute with chloroform–methanol (97:3) mixture. Eluants are allowed to dry for a few minutes and re-dissolved in suitable solvents for TLC analysis, as recommended by the supplier.

Thin-Layer Chromatography

Spot the sample (0.5–2 mL) by using a capillary tube on TLC plate. The spot should be as small and compact as possible, with a distance of 1–2 cm from the edges of the plate and between the spots.

1. Place TLC plate in chromatography chamber and run until solvent front is 2–3 cm from top of the plate (approximately 30–45 min).
2. Developing solvent: 80% benzene or toluene, 15% methanol, 5% acetic acid.

Table 5.2 R_f values and visible color of aflatoxin

Mycotoxin	R_f value Solvent system(B:M:A)	Color Visible light	UV light Long wave	Short wave	Color after spray treatment Long-wave ultraviolet light
Aflatoxin B1	0.14	Yellow	Green	Faint green	Blue
Aflatoxin B2	0.20		Blue	Faint blue	Pink
Aflatoxin G1	0.23		Blue	Faint blue	Pink

Table 5.3 R_f value and visible color of Ochratoxin A

Mycotoxin	R_f value Solvent system (TEF)	Color Visible light	UV light Long wave	Short wave	Color after spray treatment Long-wave ultraviolet light
Ochratoxin A	0.55	Yellow	Green	Green	Faint blue

3. Observe the TLC plate under UV scanner at 256 nm compared with standard toxin. Toxins were visualized in visible or ultraviolet light, before and after the plate was sprayed with freshly prepared mixture of 0.5 mL of *p*-anisaldehyde in 85 mL of methanol containing 10 mL of glacial acetic acid and 5 mL of concentrated sulfuric acid and then heated at 130°C for 10 min. The 5- to 10-min heating time was better for fluorescence development (Table 5.2).

Detection of Ochratoxin by Thin-Layer Chromatography

OTA detection by TLC can be performed by spotting samples and spikes onto a SG-60 plate and development with a mixture of toluene/methanol/acetic acid or toluene/ethyl acetate/formic acid. Under long-wavelength UV light OTA will appear blue–green at a retention value of 0.55.
1. Extraction and clean-up: As mentioned above.
2. Spot the sample (0.5–2 mL) by using a capillary tube on TLC plate. The spot should be as small and compact as possible with a distance of 1–2 cm from the edges of the plate and between the spots.
3. Developing solvent: 90% toluene, ethyl acetate, formic acid (5:4:1, v/v/v)
4. Observe the TLC plate under UV scanner at 256 nm compared with standard toxin.

5. Observation: Toxins were visualized in visible or ultraviolet light, before and after the plate was sprayed with a freshly prepared mixture of 0.5 mL of *p*-anisaldehyde/silver chloride in 85 mL of methanol containing 10 mL of glacial acetic acid and 5 mL of concentrated sulfuric acid and then heated at 130°C for 10 min. The 5- to 10-min heating time was better for fluorescence development (Table 5.3).

Detection of DON by Thin-Layer Chromatography

TLC is still common, and with the introduction of high-performance TLC (HPTLC) and scanning instruments, separation efficiency and precision have increased. Reagents (e.g., sulfuric acid or para-anisaldehyde) are necessary to visualize the only short wavelength absorbing DON. Other spray reagents include 4-para-nitrobenzyl-pyridine or nicotinamide in combination with 2-acetyl-pyridine) or $AlCl_3$, which is the most useful visualization reagent for DON. Typical detection limits by TLC are in the range of 20–300 ng/g.
1. Extraction and clean-up: 50 g of finely ground sample were placed in a wide-mouth polypropylene screw-cap bottle with 100 mL of a methanol–water mixture (1:1) and placed on a wrist-action shaker for 30 min.
2. The contents were allowed to settle, and a 10- to 25-mL aliquot of the solvent extract was

Table 5.4 R_f value and visible color of DON

Mycotoxin	R_f value	Color			Color after spray treatment	
			UV light			Long-wave
	Solvent system (CMW)	Visible light	Long wave	Short wave	Visible light	ultraviolet light
Deoxynivalenol	0.45	Yellow	Brown	Brown	Brown	Brick red

filtered through Whatman 4 filter paper and extracted with 25 mL of ethyl acetate.

3. Ethyl acetate was completely evaporated by rotary evaporation and the pellet resuspended in 3 to 4 mL of acetone–water (1:1).

4. Extracts were passed through preconditioned C18 column, and elutes were dried and reconstituted with 1 mL of a methanol–water mixture (1:1).

5. These extracts were used for TLC analysis.

6. Thin-layer chromatography: Spot the sample (0.5–2 mL) by using a capillary tube on TLC plate. The spot should be as small and compact as possible, with a distance of 1–2 cm from the edges of the plate and between the spots.

7. Place TLC plate in chromatography chamber and run until solvent front is 2–3 cm from top of plate (approximately 30–45 min).

8. Developing solvent: chloroform: methanol: water (9:1:0.2).

9. Observe the TLC plate under UV scanner at 256 nm compare with standard toxin. Toxins were visualized in ultraviolet, before and after the plate was sprayed with a freshly prepared mixture of 0.5 mL of p-anisaldehyde/silver chloride in 85 mL of methanol and then heated at 130°C for 20 min (Table 5.4).

Detection of Fumonisins by Thin-Layer Chromatography [27]

TLC is the simplest and most frequent screening method used for detection of fumonisins, but like all other methods, extraction and clean-up make a major contribution to accuracy and precision of obtained data. Derivatization is necessary before fluorescent detection can be performed, because fumonisins do not contain a chromophore to exhibit radiation. Reversed phase TLC (on C_{18} modified silica plates) has also been employed with acidic vanillin or fluorescamine/sodium borate buffer as a spray reagent.

1. Extraction and clean-up: 50 g of finely ground sample were placed in a wide-mouth polypropylene screw-cap bottle with 100 mL acetonitrile: water (1:1) and placed on a wrist-action shaker for 30 min.

2. The contents were allowed to settle, and a 10- to 25-mL aliquot of the solvent extract was decanted and filtered using Whatman 4 paper.

3. A C18 clean-up column was preconditioned with 5 mL of methanol followed by 5 mL of 1% aqueous potassium chloride (KCl). Two milliliters of the filtrate was combined with 5 mL 1% aqueous KCl and applied to the column.

4. The column was washed with 5 mL 1% aqueous KCl followed by 2 mL acetonitrile: 1% aqueous KCl (1:9), and the eluants were discarded.

5. The fumonisins were eluted with 4 mL acetonitrile: water (7:3), and the column eluant was evaporated to dryness under a stream of air on a heating module for TLC analysis.

6. Thin-layer chromatography: The sample residue was dissolved in 100 μL acetonitrile: water (1:1).

7. 10 μL was spotted on a C_{18} TLC plate along with 10-μL fumonisins standards (5, 10, and 100 ppm) dissolved in acetonitrile: water (1:1).

8. Observation: The TLC plate was developed in methanol: 1% aqueous KCl (3:2), air dried, and sprayed with 0.1 M sodium borate buffer (pH 8–9) followed by fluorescamine (0.4 mg/mL in acetonitrile). After 1 min, the plate was sprayed with 0.01 M boric acid: acetonitrile (40:60). The plate was then air dried at room temperature and examined under long-wave UV light. Fumonisin levels were estimated by visual comparison with standards (Table 5.5).

Table 5.5 Rf value and visible color of fumonisins

Mycotoxin	R_f value Solvent system (M:Kcl)	Color after spray treatment Long-wave ultraviolet light
Fumonisins	0.5 (FB1) 0.1 (FB2)	Bright yellowish-green fluorescent bands

Detection of Aflatoxins by High-Performance Liquid Chromatography

1. Instrument: Liquid chromatography methods for the determination of aflatoxins in foods by reversed-phase HPLC (Hitachi F-4500). The emitted light is detected at 435 nm after excitation at 365 nm. Stationary phases for HPLC usually include C_{18} material, with mobile phases being mixtures of water, methanol, or acetonitrile. A fluorescence detector and a suitable data system are required to provide sensitive and specific detection and quantification of aflatoxins.
2. Solvents: All solvents shall be of HPLC grade, and all reagents should be analytical grade.
3. Extraction: A ground sample (20 g) is extracted with a methanol–water (7:3) mixture (80 mL). Corn and wheat samples will be kept in a vibrating shaker for 15–30 min. Extracts are filtered immediately after extraction through filter paper. After filtration the sample is evaporated to dryness at 40°C in a rotary evaporator.
4. Clean-up by IAC: The use of IACs is now well established in aflatoxin determination. MycoSep® (Romer Labs, Union, MO) columns, which remove matrix components efficiently and can produce a purified extract within a very short time, are also available. Conventional clean-up with silica columns has also been reported [16].
5. Standard preparation: Aflatoxin B1, B2, G1, and G2 standard can be purchased from private companies (Sigma-Aldrich, USA). Each of aflatoxins was diluted in methanol to 1 mg/mL solution of G2, 1 mg/mL of B2 10 mg/mL of G1, and 10 mg/mL of B1. 100 μL aliquot of each aflatoxin solution was then combined in

a 2-mL glass vial and mixed well. This mixture was further diluted in series to 100,000 folds in water: methanol (7:3 v/v) and used as the standard solution.

6. Chromatography conditions: Column: Hypersil GOLD®, 3 μm, 100×2.1 mm; Flow Rate: 800 μL/min λex: 365 nm λem: 455 nm; Mobile Phase: Water: Methanol (7:3 v/v) (isocratic elution); Column Temperature: 40°C; Injection Volume: 10 μL of the prepared standard solution; Analytes: aflatoxin B1 and aflatoxin B2. The instruments will be controlled and the data analyzed using the suitable data system. No step changes of the excitation and emission wavelengths will be used during the run.
7. Observation: Aflatoxins fluoresce strongly on illumination with 365-nm ultraviolet light. Figure shows the fluorescence chromatogram of the two common aflatoxins with an excitation wavelength of 365 nm and an emission wavelength of 435 nm.

Detection of Ochratoxins by High-Performance Liquid Chromatography

1. Instruments: The liquid chromatograph equipped with quaternary pump, autoinjector with a stainless steel reverse phase 150×4.6 mm, 3-mm particle size C18 Supelco HPLC column (Supelco, USA). A fluorescence detector and a suitable data system are required to provide sensitive and specific detection and quantification of ochratoxins derivatized with OPA/mercaptoethanol.
2. Solvents: All solvents shall be of HPLC grade, and all reagents should be analytical grade.
3. Extraction: Sample extraction is generally performed with a mixture of water and organic solvents depending on the type of matrix. An IUPAC/AOAC method for the determination of OTA in barley uses a mixture of $CHCl_3$ and H_3PO_4 [28]; for green coffee, $CHCl_3$ is only employed [29]. For determination in wheat, a number of extraction solvents are used, including mixtures of toluene/HCl/$MgCl_2$, CHCl3/ethanol/acetic acid, and dichloromethane/H_3PO_4.

4. Clean-up by IAC: The use of IACs is now well established in ochratoxin determination. The extract is forced through the column, and ochratoxins are bound to the antibody. Five milliliters of the final extract, corresponding to 5% (v/v) of the original material, was placed into the IAC. The sample was allowed to pass though the column at a flow rate of 2–3 mL/min. Slowly elute the bound ochratoxin from the column using 1.5 mL of desorption solution; allow this to pass through the column by gravity and collect in a sample vial.

5. Standard preparation: Ochratoxin will be purchased from private companies (Sigma-Aldrich, USA). Ochratoxin was diluted 50 µg/mL in benzene: acetic acid (99:1). 50 µL aliquot of each solution was then combined in a 2-mL glass vial and mixed well. This mixture was further diluted in series to 100,000 folds in acetonitrile: water (7:3 v/v) used as the standard solution.

6. Chromatography conditions: Reversed phase HPLC approach with a C_{18} column [21]. Flow Rate: 800 µL/min λex: 365 nm λem: 455 nm; and an acidic buffer (acetic acid) in an acetonitrile/water mixture as a mobile phase. Column Temperature: 40°C; Injection Volume: 10 µL of the prepared standard solution; Analytes: Ochratoxin A.

7. The instruments will be controlled and the data analyzed using the suitable data system. No step changes of the excitation and emission wavelengths will be used during the run.

8. Observation: Quantify the ochratoxin A concentration by comparing the sample peak area to that of a standard.

Detection of Deoxynivalenol by High-Performance Liquid Chromatography

1. Instruments: The liquid chromatograph equipped with quaternary pump, autoinjector, and UV detector was used with a stainless steel reverse phase 150×4.6 mm, 3 mm particle size C18 Supelco HPLC column.

2. Solvents: All solvents shall be of HPLC grade, and all reagents should be analytical grade.

3. Extraction: Place 10 g of the ground sample into the ultraturax and then add 40 mL of distilled water and 2 g of polyethylene glycol. The mixture is stirred for 1 min. The extract is filtered through a fluted filter and then through a microfiber filter.

4. Clean-up by immunoaffinity chromatography: Place 1 mL of the final extract into the IAC. Use 10 mL of redistilled water for column washing. The elution of DON is conducted with 1 mL of methanol. The elution solvent is removed by a gentle stream of nitrogen and re-dissolved in 300 µL mobile phase.

5. Standard preparation: DON purchased from private companies (Sigma-Aldrich) is diluted to 200 µg/mL in ethyl acetate: methanol (95:5). 50 µL. Aliquot of the solution is then combined in a 2 mL glass vial and mixed well. Serially dilute this mixture to 1,000 folds in methanol: water (7:3 v/v) used as the standard for HPLC.

6. Chromatography condition: Samples of 50 µL are injected into the HPLC column and heated to 30°C. The used mobile phase consisted of a methanol: water solution (8:2 v/v). The flow rate is of 0.6 mL/min. Deoxynivalenol is determined at a wavelength of 218 nm by using UV detector.

7. Observation: Quantify the deoxynivalenol concentration by comparing the sample peak area to that of a standard.

Detection of Fumonisins by High-Performance Liquid Chromatography

1. Instrument: HPLC system consisting of an isocratic pump capable of a flow rate of 1 mL/min and a suitable injector capable of 10 µL injections. Columns containing C_{18}- or C_{8}-modified silica packing material of 3- to 5-mm particle size. A fluorescence detector and a suitable data system are required to provide sensitive and specific detection and quantification of fumonisins derivatized with OPA/mercaptoethanol.

2. Solvents: All solvents will be used of HPLC grade and all regents should be analytical grade.

3. Extraction: Place finely ground sample (25 g) into a container suitable for centrifuging (250-mL polypropylene centrifuge bottle). Add 100 mL extraction solvent (methanol–water, 3:1) and homogenize the contents for 3 min. Centrifuge the container at 10,000 rpm for 10 min at 4°C. Filter the supernatant through a Whatman 4 filter paper.

4. Solid phase extraction (SPE) cartridges: Sample extracts are generally cleaned up on SPE columns containing strong anion exchange material. For optimal simultaneous handling of cartridges, the use of a commercial SPE manifold is recommended.

5. Standard preparation: Fumonisin standards are prepared in acetonitrile: water (1:1) and stored at 4°C. Stock solutions of individual fumonisins standards of concentration 250 μg/mL are used, from which a working standard is prepared containing 50 μg/mL of each analog. Derivatize standards by mixing 25 μL working standard with 225 μL OPA reagent at the base of a small test tube. Inject 10 μL into the HPLC using a standardized time of 1–2 min between the addition of OPA reagent and injection.

6. Chromatography conditions: The HPLC mobile phase is a mixture of methanol and 0.1 M sodium dihydrogen phosphate in water. For most reversed-phase columns, a solvent composition of 75% to 80% methanol will be required. The pH of the mixture is adjusted to 3.35 with o-phosphoric acid and filtered through a 0.45-mm membrane filter.

7. OPA reagent: OPA reagent for derivatizing the fumonisins is prepared by dissolving 40 mg OPA in 1 mL of methanol and diluting with 5 mL of 0.1 M disodium tetraborate.

8. Observation: Quantify the fumonisins' concentration by comparing the sample peak area to that of a standard.

Summary

Many agricultural commodities are vulnerable to attack by fungi that produce mycotoxins. Detection of mycotoxins and toxin-producing fungi from food and feeds are very essential. In the present chapter we discussed available techniques for detection and quantification of major mycotoxigenic fungi and their toxins from agricultural produce. The standard methods varied from lab to lab and toxin to toxin and also from commodity to commodity. International agencies such as International Union of Pure and Applied Chemistry (IUPAC), Association of Official Analytical Chemists (AOAC), and The European Mycotoxin Awareness Network have developed their own methodologies for detection of mycotoxins from different food matrices. In conclusion, a broad range of techniques for practical analysis and detection of a wide spectrum of mycotoxins are available. This chapter presented some recent developments in scientific and technological basis analytical methods that offer flexible and broad-based methods for analysis of toxins and toxigenic fungi.

Notes

DNA Extraction and Polymerase Chain Reaction Conditions

1. Make use of suitable microbiological aseptic technique when working with DNA. Wear gloves to prevent nuclease contamination from the surface of the skin. Use sterile, disposable plasticware and automatic, aerosol-resistant pipettes reserved for DNA work.

2. Wipe pipettes with Dnase-removal solutions when transitioning between handling crude extracts to handling more purified material.

3. Equilibrated phenol can typically be purchased from commercial sources. Alternatively, you can equilibrate it yourself. There are also commercial sources of phenol and chloroform mixed together and equilibrated. The pH is important because chromosomal DNA will end up in the phenol phase if the pH is acid (around pH 5).

4. Phenol and chloroform should be used in a hood. Phenol is a dangerous substance that will burn you if it gets on your skin. Always wear gloves and be careful. A solution of

PEG 400 is recommended for first aid. Phenol is both a systemic and local toxic agent.

5. DNA should be kept frozen in a non-frost-free freezer. DNA should not be allowed to defrost between uses, as this will break long molecules.

6. Make a PCR master mix for 50 μL reaction containing DNA and PCR ingredients. After setting up the reaction, specific reaction conditions can be used for proper amplification of target genes. After successful amplification, the PCR amplicons can be stained with ethidium bromide-containing agarose gel and bands can be visualized under UV.

Mycotoxin Analysis

7. A laboratory or part of a laboratory should be reserved for mycotoxin analysis only and the work confined to that area. The bench top should be of a nonabsorbent material, such as formica, for example (Whatman Benchkote can also be used, but it must be removed and destroyed after use), and should be screened from direct sunlight.

8. Analyses should be performed in a well-ventilated laboratory, preferably under an efficient extraction hood, and fume cupboard facilities should be available.

9. Many of the solvents used are highly flammable and have low flash points. Bunsen burners, electric fires, and sparking apparatus such as centrifuges should not be used in the same laboratory. The amount of flammable solvents in the laboratory should be kept to a minimum and stored in a fire-resistant cupboard or bin.

10. Swab accidental spills of toxin with 1% NaOCl bleach, leave 10 min, and then add 5% aqueous acetone. Rinse all glassware exposed to aflatoxins with methanol, add 1% NaOCl solution, and after 2 h add acetone to 5% of the total volume. Allow a 30-min reaction and wash thoroughly.

11. Weighing and transferring mycotoxins in dry form should be avoided; they should be dissolved in a solvent. The electrostatic nature of a number of the mycotoxins in dry form results in a tendency for them to be easily dispersed in the working area and to be attracted to exposed skin and clothes.

12. Containers of mycotoxin standard solutions should be tightly capped, and their weights may be recorded for future reference before wrapping them in foil and storing them in a freezer.

13. During the grinding and weighing of samples, there is a risk of absorbing toxin either through the skin or by the inhalation of dust. There is also the risk of developing allergic reactions due to spores and organic material. These risks should be minimized by working under an extraction hood, by good hygiene, and by wearing protective clothing and masks.

14. Glassware and TLC plates should be decontaminated by soaking for 2 h in a 1% sodium hypochlorite solution. After this time an amount of acetone equal to 5% of the total volume of the bleach bath should be added, and the glassware soaked for an additional 30 min. Spraying of TLC plates must be carried out in an efficient fume cupboard or spray cabinet. Always ensure that this equipment is working before commencing use. When viewing chromatograms under UV light the eyes should be protected by UV filter or by wearing protective spectacles.

References

1. Speijers GJA, Speijers MHM (2004) Combined toxic effects of mycotoxins. Toxicol Lett 153:91–98
2. Bhat RV (1988) Mold deterioration of agricultural commodities during transit: problems faced by developing countries. Int J Food Microbiol 7(3):219–225
3. Michelangelon. Pascale (2009) Detection methods for mycotoxins in cereal grains and cereal products. ISPA, CNR 117:15–25
4. Jajic I, Verica J, Glamocic D, Abramovic B (2008) Occurrence of deoxynivalenol in maize and wheat in Serbia. Int J Mol Sci 9:2114–2126
5. Bhat, R, Vasanthi S (1999) Mycotoxin contamination of foods and feeds: overview, occurrence and economic impact on food availability, trade, exposure of farm animals and related economic losses. Third Joint FAO/WHO/UNEP International Conference on Mycotoxins, Tunis, Tunisia, 3–6 Mar 1999

6. mycotoxins\Mycotoxins in Grain AgriPinoy_net.mht, http://helica.com/food-safety/mycotoxins/?gclid= CIL4rJS-qLMCFVF96wodC0cA5g

7. Richard JL, Lyon RL, Fichtner RE, Ross PF (1989) Use of thin layer chromatography for detection and high performance liquid chromatography for quantitating gliotoxin from rice cultures of *Aspergillus fumigatus* fresenius. Mycopathologia 107:145–151

8. Chandra Nayaka S, Udaya Shankar AC, Niranjana SR, Ednar GW, Mortensen CN, Prakash HS (2010) Detection and quantification of fumonisins from *Fusarium verticillioides* in maize grown in southern India. World J Microbiol Biotechnol 26:71–78

9. Chandra NS, Wulff EG, Udayashankar AC, Niranjana SR, Mortensen NM, Prakash HS (2011) Prospects of Molecular markers for *Fusarium* diversity. Appl Microbiol Biotechnol 90:1625–1639

10. Sweeney MJ, Dobson AD (1999) Molecular biology of mycotoxins biosynthesis. FEMS Microbiol Lett 175(2):149–163

11. Niessen L (2007) Current trends in molecular diagnosis of ochratoxin A producing fungi. In: Rai M (ed) Mycotechnology—present status and future prospects. I.K. International Publishing House, New Delhi, India, p 320

12. http://www.mycotoxins.org/. Accessed 20 Sep 2011

13. CAST (2003) Mycotoxins—risks in plant, animal and human systems. Task Force Report No. 139. Ames, IA: Council for Agricultural Science and Technology, pp. 1–191.

14. Bennett JW, Klich M (2003) Mycotoxins. Clin Microbiol Rev 16:497–516

15. Richard JL (2000) Mycotoxins—an overview. In: Richard JL (ed) Romer Labs' guide to mycotoxins, vol 1. Romer Labs, Union, MO, pp 1–48

16. http://www.mycotoxins.org/ Accessed 18 Sep 2011

17. Fredricks DN, Caitlin S, Amalia M (2005) Comparison of six DNA extraction methods for recovery of fungal DNA as assessed by quantitative PCR. J Clin Microbiol 43(10):5122–5128

18. Ramana MV, Balakrishna K, Murali HS, Batra HV (2011) Multiplex PCR-based strategy to detect contamination with mycotoxigenic *Fusarium* species in rice and fingermillet collected from southern India. J Sci Food Agric 91:1666–1673

19. McCartney HA, McCartney JF, Bart AF, Elaine W (2003) Molecular diagnostics for fungal plant pathogens. Pest Manag Sci 59(2):129–142

20. Shapira R, Paster N, Eyal O, Menasherov M, Mett A, Salomon R (1996) Detection of aflatoxigenic molds in grains by PCR. Appl Environ Microbiol 62:3270–3273

21. Dao HP, Mathieu F, Lebrihi A (2005) Tow primer pairs to detect OTA producers by PCR method. Int J Food Microbiol 104:61–67

22. Bluhm BH, Flaherty JE, Cousin MA, Woloshuk CP (2002) Multiplex polymerase chain reaction assay for the differential detection of trichothecene- and fumonisin-producing species of *Fusarium* in cornmeal. J Food Prot 65:1955–1961

23. Cichna MM (2001) New strategies in sample clean-up for mycotoxin analysis. World Mycotoxin J 4(3):203–215

24. Visconti A, Pascale M (1998) Determination of zearalenone in corn by means of immunoaffinity clean-up and high-performance liquid chromatography with fluorescence detection. J Chromatogr A 818:133–140

25. Pittet A (2001) Natural occurrence of mycotoxins in foods and feeds: a decade in review. In: de Koe WJ (ed) Mycotoxins and phycotoxins in perspective at the turn of the millennium. Wageningen, The Netherlands, pp 153–172

26. Irena KC, Helena P (2009) An overview of conventional and emerging analytical methods for the determination of mycotoxins. Int J Mol Sci 10:62–115

27. Rottinghaus GE, Coatney CE, Minor HC (1992) A rapid, sensitive thin layer chromatography procedure for the detection of fumonisin B_1 and B_2. J Vet Diagn Invest 4(3):326–329

28. Ghali R, Hmaissia-khlifa K, Ghorbel H, Maaroufi K, Hedili A (2009) HPLC determination of ochratoxin A in high consumption Tunisian foods. Food Control 20:716–720

29. Anonymous (2000) Official methods of analysis. 17th ed. Chapter. 49. In: Williams H (ed) Official Methods of Analysis of the Association of Official Analytical Chemists. AOAC INTERNATIONAL Methods No. 990.3, Gaithersburg, MD, USA, pp 20–22

Identification Key for the Major Growth Forms of Lichenized Fungi

Jeyabalan Sangeetha and Devarajan Thangadurai

Abstract

Lichens are classified as cup fungi, mainly under the phylum of Ascomycota and rarely as Basidiomycota. The body of the lichen is called a thallus, in which the mycobionts and photobionts are stratified in separate layers. Based on their characteristics and thalli, lichens are classified into four main groups. This chapter consists of a pair of parallel and opposing statements that can be compared and will help to identify several genera of foliose, fruticose, crustose, and squamulose lichens.

Keywords

Lichenized fungi • Foliose • Fruticose • Crustose • Squamulose • Identification keys

Introduction

Kingdom Fungi comprises seven major phyla, such as Ascomycota, Basidiomycota, Blastocladiomycota, Chytridiomycota, Glomeromycota, Microsporidia, and Neocallimastigomycota [1–3]. The total number of fungi is estimated at 700,000 to 1.5 million species. Members of Ascomycota are commonly known as sac fungi.

J. Sangeetha (✉)
Department of Zoology, Karnataka University,
580003, Dharwad, Karnataka, India
e-mail: drsangeethajayabalan@gmail.com

D. Thangadurai
Department of Botany, Karnataka University,
Dharwad, 580003 Karnataka, India

They are the largest phylum of fungi, containing about 32,000 named species. The fungal symbionts in the majority of lichens belong to the Ascomycota phylum, and a few belong to Basidiomycota. Among the identified fungal species, about 17,500 lichens are recorded [4–7]. The number of undescribed species of lichenized fungi has been estimated at roughly 10,000. These fungi form meiotic spores called ascospores, which are enclosed in a special sac-like structure called an ascus. However, some members of Ascomycota do not reproduce sexually and do not form asci or ascospores. These members are assigned to Ascomycota based upon morphological or physiological similarities to ascus-bearing taxa and, in particular, by phylogenetic comparisons of DNA sequences [8–11]. In addition to lichens, this phylum includes morels (e.g., *Morchella deliciosa*, *M. elata*), a few mushrooms and truffles, single-celled yeasts, and many filamentous fungi living as saprotrophs, parasites, and mutualistic symbionts [12–15].

Structurally, lichens are the strangest of all forms of fungi and are dual organisms formed by a symbiotic relationship of a fungus (mycobiont) with an algae or cyanobacterium (photobiont) [16,17]. Some factors, like substratum chemistry, stability, and longevity, and light and moisture availability, are affecting the presence and abundance of lichens. Some substrata that support lichen growth include rock surface, woody plant bark, wood, soil, microhabitats, and broad evergreen leaves in the humid tropics [3,18]. Lichens appear in a variety of colors, like white, grey, brown, orange, brilliant red, or yellow [19–21]. Lichen morphology is quite different from that of other fungi [22,23]. Each lichen thallus has a complete microscopic feature with unique characteristics [17]. From an ecological perspective, lichen may be even more complex than free-living bacteria and nonsymbiotic fungi. Longevity and considerable environmental stress tolerance seem to be the major features shared by all ascomycetes [24]. The basic lichen body is called a thallus, and it consists of fungal hyphae that are organized into various tissues and thread-like fungal cells. Lichens consist of four layers: the top surface, which is a layer of tightly packed thallus called upper cortex; an algal layer, where the photobionts lives; the medulla, which is bound with loose hyphae; and another protective covering, the lower cortex [25,26]. In lichen, the algae and the fungi can reproduce individually but to form recognizable lichen, viable fungal spores have to unite with the appropriate algal spores to form new lichen. The surface of the lichen has three important structures: fruiting body, vegetative powdery, and hard-bodied peg-like stalk [27,28].

The detection and identification of lichens are based on the conventional method of observation of physical characteristics of spores, hyphae, arrangement thallus, life cycle, and relationship with the host cell. An identification key can be prepared for each genus by using these observations. Normally, an identification key is a printed or computer-aided document that provides the details of the lichen and helps in the identification.

An identification key is a single-access key, and it will work with fixed sequence of identification steps, each with multiple alternatives, the choice of which determines the next step. Some lichens can be identified with a few notes, and others may require more notes for identification. This key will assist the investigator in identifying an unknown specimen to its genus. The amount of information presented in the key limits the effectiveness of the key. Hence, the key should be as comprehensive as possible; inclusion of too much information in a single point makes the key difficult to follow. The keys are organized in groups of options in the middle column, and these are numbered in ascending order in the left-hand column (1a, 1b, 2a, 2b, etc.). Having first decided which key to use, start at the top and choose the most appropriate option to describe the specimen in question (e.g., 1a or 1b). Description of the option will lead to either a possible solution in the form of a name or the number of a further option, which will be in the right-hand column. Moving down with the key to the number indicated earlier once again offers either a possible solution or a further pair of options to choose and so on until the search ends with no further options offered [26,29]. The identification obtained from a key should be viewed as only a suggestion of the specimen's real identity. Full identification requires comparison of the specimen with some authoritative source and further well-developed molecular techniques [30].

This chapter has two primary objectives. The first is to provide protocol for the collection and isolation of lichenized fungi. The second is to provide comprehensive identification keys to the main growth forms of lichens.

Materials

1. Autoclave
2. Incubator
3. Laminar air flow chamber
4. Centrifuge
5. Dissecting microscope
6. Compound microscope
7. Inverted microscope

8. Microscopic slide
9. Distilled water
10. Petroleum jelly
11. Ethanol
12. 4 % distilled water agar medium: agar, 4 g; distilled water, 100 mL
13. Malt/yeast extract medium (MY medium): malt extract, 20 g; yeast extract, 2 g; agar, 20 g; distilled water, 1,000 mL
14. Lilly and Barnett's medium (LB medium): glucose, 10 g; asparagine, 2.0 g; KH_2PO_4, 1.0 g; $MgSO_4 \cdot 7H_2O$, 0.5 g; $Fe(NO_3)_3 \cdot 9H_2O$, 0.2 mg; $ZnSO_4 \cdot 7H_2O$, 0.2 mg; $MnSO_4 \cdot 4H_2O$, 0.1 mg; thiamine, 0.1 mg; biotin, 5 µg; distilled water, 1,000 mL
15. Lactophenol cotton blue

Methods

Lichens can be isolated from ascospores, conidia, isidia, soredia, and thallus fragments. Lichens show maximum growth at 15–20 °C under the pH of 5–6. Lichens can be cryopreserved so that they remain viable for extended periods of time [16,17,31]. Spot tests (see Note 1) can be done with sodium hypochlorite (C), potassium hydroxide (K), potassium iodide (I), or paraphenylenediamine (P) to identify the color reaction of lichen substances (see Note 2). For laboratory cultures the most useful methods are isolating lichens from discharged spores and thallus fragments. When mycobionts cannot be isolated from the spores, other parts like conidia, isidia, and soredia may be applicable for isolation [10,26,32].

Spore Discharge Method

1. Collect thalli from the field and clean it (see Note 3).
2. Remove the apothecia or perithecia from the thallus.
3. Place it into the dishes containing distilled water and wash it.
4. Blot dry to remove excess water.
5. Fix these structures to the bottom of a Petri dish with petroleum jelly.

6. Pour 4 % water agar medium on the lid of the Petri dish and allow it to solidify (see Note 4).
7. Replace the top cover of the Petri dish with new covers containing freshwater agar medium.
8. Observe spores discharged onto the water agar medium under inverted microscope.
9. Keep discharged spores on a sterile glass slide containing medium.
10. Keep the slides in a Petri dish in humid conditions.
11. Observe continuously for the germination of spores and mycelia growth, using lactophenol cotton blue stain under microscope (see Note 5).
12. After germination, excise small blocks of the agar containing spores and transfer to the culture tubes or Petri dishes containing MY medium and LB medium.
13. Note all measurements and match with the lichen identification key for the identification of the isolated lichen.

Isolation of Mycobionts from Thallus

1. Using sterilized cutter or blade, cut pieces from a fresh thallus.
2. Store in small test tubes containing water or on wet filter paper at 15 °C. After 2 weeks a new medullary hyphae will be elongated.
3. Excise a portion of the newly elongated hyphae and transfer to a test tube containing fresh culture medium (MY medium or LB medium).
4. Prepare sufficient number of replicates for confirmation.
5. Identify the isolated lichen with the list of lichen identification keys.

Key for the Identification of Lichenized Fungi

Based on the growth forms, lichens are grouped into four main categories: foliose, fruticose, crustose, and squamulose. Basic growth forms used for identification of lichen specimens are thallus adenation or attachment to the substrate, lobe width, color, soredia, isidia, cilia, ahizines, tomentum, veins, and pores. Pairs of keys are constructed based on the contrast characters. Each key is followed until one reaches a match for the description of the specimen. Keys to the major growth forms are as follows [20,25,26,28,33–36]:

1. *Foliose Lichens.* Thallus foliose, leaf-like with branching lobes, adnate to suberect or umbilicate with a central umbilicus below; thalli are flattened, having upper and lower cortex; the lobes can be narrow or broad; rhizines present.
2. *Fruticose Lichens.* Thallus fruticose or shrubby, richly divided, rounded or flattened in cross section, decumbent or tufted to pendulous, and attached at the base or free growing; thallus filamentous forming small tufts or mats; no lower cortex and rhizines.
3. *Crustose Lichens.* Thallus crustose, closely attached to the substrate and lacking a lower cortex and rhines but sometimes with a lobed margin; or thallus consisting of small crowded squamulose.
4. *Squamulose Lichens.* Thallus horizontally spreading, blister-like, squamulose; lacking lower cortex and rhizines; the lower layer is white; some lichens produce fruiting structure called podetium, an erect, hollow stalk.

Identification Key to the Foliose Lichens

Certain lichens are leaf-like and composed of lobes usually strap-shaped and live on leaves, sometimes as parasites. These special leaf-living lichens are known as foliose lichens. Foliose are loosely attached to the substrate and have an upper and lower cortex. Foliose lichens grow slowly on culture medium. The identification keys for the foliose lichens are listed in Table 6.1 [16,22,35,37–45].

Identification Key to the Fruticose Lichens

Fruticose lichens are mostly three-dimensional. Most are branched, the thallus is attached to the substratum at one point, and the remaining major portion is either growing erect or hanging. Thalli may be pendulous strands or hollow, upright stalks. Presence of soredia and isidia is also important for identification. The key for the identification of fruticose lichens is listed in Table 6.2 [16,25,35,38–40,43,45,46].

Identification Key to the Crustose Lichens

The thallus in crustose lichen is closely attached to the substrate and may not be removed from it without destruction of the substrate. The thallus usually lacks lower cortex and rhizines. About 75 % of the lichens are crustose and found in ice-free areas of the highest mountains. Compared to other lichens, crustose lichens can grow faster on culture medium. Crustose lichens can tolerate extreme conditions, such as exposure to a rock surface, and also demonstrate a clearly defined growth type. Crustose lichens have two types of fruiting body: open disc-shaped apothecia and flask-shaped perithecia. These characters play important roles in identification of crustose. The identification key for the crustose lichens is given in Table 6.3 [22,25,38,39,43,44,46–53].

Identification Key to the Squamulose Lichens

Sometimes, crustose lichens develop blister-like "squamules" where the areolae are enlarged in their upper part and become partially free from the substrate. Such lichens are known as squamulose lichens. Here the lichen thallus is in the form of minute lobes. Their form is similar to that of crustose lichens in that they possess an upper cortex but no lower cortex. The squamulose are often developed in rock surfaces in hot and arid regions of the world. A key for the identification of squamulose lichens is listed in Table 6.4 [23,28,35,38,39,43,46,54–59].

Table 6.1 Identification keys for foliose lichens

1a	Thallus squamulose; photobiont cyanobacterium 2	
b	Thallus gelatinous, homomerous, without arachnoid; photobiont *Nostoc*..3	
2a	Photobiont cells arranged in chains; cortex more than one layer thick; hypothallus present.........11	
b	Photobiont cells arranged in small groups; lacking a distinct hypothallus; brown and smooth upper cortex..5	
3a	Thallus with pseudoparenchymatic cortex of one cell layer..4	
b	Thallus with perithecia..5	
4a	Thallus >0.5 mm thick; spores simple .... *Physma*	
b	No thallospores and anthraquinones 7	
5a	Thallus attached by umbilicus .... *Thyrea*	
b	Thallus isidiate or papillate...6	
6a	Thallus attached by umbilicus 7	
b	Thallus blackish, attached on rock...9	
7a	Thallus surface grey to brown or dark brown...8	
b	Thallus green; apothecia lecanorine...*Rhizoplaca*	
8a	Perithecia embedded in thallus, immersed in the thallus, producing simple, hyaline ascospores .... *Dermatocarpon*	
b	Apothecia lecideoid; ascocarps absent; thallus dark color ....*Umbilicaria*	
9a	Thallus surface yellow or orange..10	
b	Thallus surface dark brown or greenish grey or yellowish grey...15	
10a	Thallus surface yellow; K- negative; ascospores simple; apothecia absent 11	
b	Thallus surface orange; K+dark violet; ascospores polarilocular...13	
11a	Thallus loosely appressed, lobes rounded with concentric ridges, not gelatinous, photobiont bluegreen algae; apothecia often with thalline margin 59	
b	Thallus lobes elongate, without concentric ridges; photobiont green algae...............................12	
12a	Thallus soridiate, greyish yellow or bright yellow to greenish; K-; pigment mainly in cortex; apothecia sessile; narrow lobes hardly >1 mm wide ....*Candelaria*	
b	Thallus lobes >1 mm wide..50	
13a	Ascospores bilocular...14	
b	Ascospores tetralocular..*Teloschistes*	
14a	Upper and lower cortex prosoplectenchymatic; with elongate lumina not wider than walls..........*Josefpoeltia*	
b	Upper cortex paraplectenchymatic; lower cortex prosoplectenchymatic ....*Xanthomendoza*	
15a	Lower side smooth; with or without rhizines 16	
b	Lower side tomentose; with or without cortex, with or without rhizines.....................................53	
16a	Upper surface of lobes with white; rounded or elongate pseudocyphellae 17	
b	Upper surface of lobes without pseudocyphellae 20	
17a	Thallus yellowish to greenish grey; with usnic acid in the cortex.................*Flavopunctelia*	
b	Thallus whitish grey; with atranorin in the cortex...18	
18a	Pseudocyphellae with reticulate pattern...43	
b	Pseudocyphellae rounded; without reticulate pattern 19	
19a	Lobes ascending; > 1 cm wide; lower side black ....*Cetrelia*	
b	Lobes appressed to substrate; < 1 cm wide; lower side white.............................*Punctelia*	
20a	Cilia present on the lobe margins...21	
b	Cilia absent from the margins...34	
21a	Cilia with inflated base..22	
b	Cilia without inflated base 23	
22a	Upper side of thallus yellowish green; usnic acid present..*Relicina*	
b	Upper side of thallus whitish grey; atranorin present...*Bulbothrix*	
23a	Marginal rhizines or cilia white or grey, with perpendicular side branchlets; ascospores one-septate 45	
b	Marginal rhizines or cilia black, unbranched; ascospores simple and colorless................................24	

(continued)

Table 6.1 (continued)

24a	Rhizines one or more times dichotomously branched	43
b	Rhizines unbranched	25
25a	Thallus lobes linear, with parallel margins, dichotomously branched	26
b	Thallus lobes elongate or short and wide; irregularly branched	28
26a	Lower side brown, without rhizines; terricolous lichen with upright lobes	40
b	Lower side black; with rhizines; predominantly epiphytic with spreading lobes	27
27a	Lobes regularly dichotomously branched; apothecia concave when young	*Everniastrum*
b	Lobes irregularly branched; apothecia flat to convex	*Cetrariastrum*
28a	Thallus <2 mm wide; flattened	29
b	Thallus >5 mm wide; ascending	31
29a	Cilia robust, tapering; upper surface white	*Canomaculina*
b	Cilia slender, not tapering; upper surface grey	30
30a	Medulla yellow to orange	*Myelochroa*
b	Medulla white	*Parmelinopsis*
31a	Underside near lobe tips without rhizines	41
b	Underside rhizinate to the margin	32
32a	Lobe tips appressed to substratum; rhizines with uniform length; shorter towards the lobe tips	*Parmotremopsis*
b	Lobe tips ascending; rhizines with variable length; dimorphic	33
33a	Underside near margin brown; with scattered rhizines of variable length	*Rimelia*
b	Underside near margin pale; with dense; short rhizines mixed with scattered longer ones	*Rimeliella*
34a	Rhizines absent	35
b	Rhizines present	40
35a	Thallus lobes hollow; with cavity between medulla and black lower cortex	36
b	Thallus lobes compact	37
36a	Lobes with large pore; thallus with perforations on upper side; soralia laminal; atranorin, sticitic and constictic acids	*Menegazzia*
b	Lobes without large pores; thallus without perforations on upper side; soralia terminal or subterminal	*Hypogymnia*
37a	Thallus lobes erect; attached near to the base; lower side near the tip often white; epiphytic	
b	Thallus attached to the substrate; lower side black; epilithic	38
38a	Upper side yellowish or greenish grey, with usnic acid; ascospores hyaline, simple	*Psiloparmelia*
b	Upper side pale to dark grey, without usnic acid; ascospores grey, 1-3-septate; widespread	39
39a	Thallus K+yellow; cortex containing atranorin; apothecia with dark hypothecium; thallus lobes elongate and laterally confluent	46
b	Thallus K-; cortex lacking atranorin; apothecia with pale hypothecium; thallus lobes not laterally confluent	*Hyperphyscia*
40a	Thallus pale yellowish on both sides; lower side with pseudocyphellae; rhizines scarce	*Cetraria*
b	Thallus not pale yellowish on both sides; lower side without pseudocyphellae	41
41a	Rhizines absent in marginal zone, scarce and restricted to small patches; thallus lobes wide, > 10 mm wide	*Parmotrema*
b	Rhizines present up to lobe margins, regularly spread all over the surface; thallus lobes usually >2 mm wide, rather narrow and deeply dissected	42
42a	Rhizines frequently dichotomously branched	*Hypotrachyna*
b	Rhizines mostly unbranched	43
43a	Apothecia completely black, without grey thalline margin	*Pyxine*
b	Apothecia with grey thalline margin	44
44a	Ascospores one-septate, grey to brown; thallus closely appressed to the substrate; lower side whitish or black	45
b	Ascospores simple, hyaline; thallus mostly loosely appressed to the substrate; lower side brown to black	49

(continued)

Table 6.1   (continued)

45a	Thallus ascendant; cortex prosoplectenchymatic, longitudinally arranged hyphae; upper surface faintly longitudinally striate 84
b	Thallus flattened; cortex paraplectenchymatic or prosoplectenchymatic; hyphae not longitudinally arranged; upper surface not striate 46
46a	Apothecia with dark hypothecium; thallus closely appressed to substrate with adjoined lobes, with divaricatic acid .... *Dirinaria*
b	Apothecia with pale hypothecium; thallus less closely appressed to substrate and clearly foliose, without divaricatic acid 47
47a	Upper surface pale grey, K+yellow; medulla white .... *Physcia*
b	Upper surface dark grey, K-negative; medulla orange to red .... *Phaeophyscia*
48a	Thallus upper surface whitish grey, with atranorin; without usnic acid 49
b	Thallus upper surface yellowish grey to greenish grey; without atranorin; with usnic acid 52
49a	Lower surface pale brown to brown; ascospores < 10 μm long 50
b	Lower surface dark brown to black; ascospores > 10 μm long 51
50a	Thallus closely appressed to rock .... *Paraparmelia*
b	Thallus loosely appressed to bark .... *Pseudoparmelia*
51a	Pycnoconidia bifusiform, < 10 μm long .... *Canoparmelia*
b	Pycnoconidia curved, > 10 μm long .... *Parmeliopsis*
52a	Thallus lobes with rounded tips, > 2 mm wide, flat to concave; epiphytic; lower surface black .... *Flavoparmelia*
b	Thallus lobes with slightly incised tips, 0.5-10 mm wide, flat to convex; lower surface pale brown .... *Xanthoparmelia*
53a	Lobes with tomentose or felty lower side; veins, rhizines or cilia present 54
b	Lobes with naked, much longer than wide, smooth lower side; lacking long cilia; underside pigmented; veins and rhizines absent, apothecia present; cilia present .... *Heterodermia*
54a	Thallus linear, lower side with thick tomentose layer .... *Anzia*
b	Thallus irregularly lobed, lower side with thin or interrupted tomentose layer or with felty lower side 55
55a	Thalli with usually large, 10-30 mm wide lobes, ascending or loosely appressed; photobiont green algae 56
b	Thallus lobes <5 mm wide, closely appressed; upper surface tomentose; on lower side tomentum often restricted to marginal patches; Photobiont blue-green algae; apothecia present .... *Erioderma*
56a	Lower side pale brown tomentose, with large mottled spots; cyphellae or pseudocyphellae present 57
b	Lower side continuously tomentose or with larger delimited patches without tomentum; cyphellae or pseudocyphellae absent 58
57a	Cyphellae present; white spots on lower side with prominent, raised margin .... *Sticta*
b	Pseudocyphellae present, yellow spots on lower side without raised margin .... *Pseudocyphellaria*
58a	Thallus larger not sharply delimited; lobes broad and rounded; veins absent; pseudocyphellae absent in upper side; epiphytic .... *Lobaria*
b	Thallus wider or narrower, vein-like raised bands on lower side, thallus bright green; black cephalodia present on the surface .... *Peltigera*
59a	Apothecia on upper surface, often absent .... *Leioderma*
b	Apothecia present on lower side at the tip of lobules .... *Nephroma*

Table 6.2 Key for identification of fruticose lichens

1a	Dimorphic, composed of fruticose and crustose or squamulose part; thallus containing algae or not; ascocarp or basidiocarp present or absent...	2
b	Uniformly fruticose; thallus containing algae and ascocarps	11
2a	Thallus without algae; producing basidiocarps, subulate or mushroom-shaped	3
b	Thallus with algae; producing ascocarps, branched	6
3a	Fruticose thallus subulate...	4
b	Fruticose thallus mushroom-shaped .... *Omphalina*
4a	Photobiont containing thallus crustose	5
b	Photobiont containing thallus squamulose.. *Lepidostroma*
5a	Fruticose thallus <2 cm long.. *Clavulinopsis*
b	Fruticose thallus <2 mm long; photobiont *Nostoc*...*Massalongia*
6a	Thallus crustose; podetia not branched, with single apothecium	7
b	Thallus squamulose; podetia branched, without apothecium	10
7a	Apothecia terminal, disc-like to globose	8
b	Apothecia convex, covering thallus...*Cetraria*
8a	Cephalodia present on crustose thallus ....*Pilophorus*
b	Cephalodia absent	9
9a	Apothecia terminal pink; cephalodia and phyllocladia absent .... *Baeomyces*
b	Apothecia whitish to pink; with convex disc and thin margin...*Dibaeis*
10a	Fruticose thallus branches hollow, branched; without ascocarps; cilia present...	14
b	Fruticose thallus branches solid, unbranched; with terminal ascocarp; cilia absent..................... *Phyllobaeis*
11a	Thallus branches hollow; without arachnoid	12
b	Thallus branches solid, with arachnoid..	20
12a	Thallus surface glossy, pale brown to brown; frequent perforations or pseudocyphellae...........................	13
b	Thallus surface dull, whitish, greenish or greyish; perforations absent or restricted to axils of ramifications	14
13a	Branches <2 mm thick; true perforations present.. *Cladia*
b	Branches <2 mm thick; pseudocyphellae present...	16
14a	Thallus branches with squamules	19
b	Thallus branches without squamules	15
15a	Epiphytic; pseudocyphellae present...	16
b	Pseudocyphellae absent	17
16a	Thallus branches thin, < 1 mm wide; whitish.. *Oropogon*
b	Thallus branches thick, > 1 mm wide; greenish grey	24
17a	Thallus richly branched and ascending..*Dendriscocaulon*
b	Thallus unbranched; without anastomoses	18
18a	Thallus, whitish grey, subulate with smooth surface ....*Thamnolia*
b	Thallus greenish grey to brown, with smooth or tomentose surface	19
19a	Thallus surface slightly arachnoid, without cortex ....*Cladina*
b	Thallus hallow; surface smooth, with thin cortex..*Cladonia*
20a	Thallus branches with tough axial strand; ascospores simple ....*Usnea*
b	Thallus branches without tough central strand	21
21a	Epiphytic...	22
b	Epilithic...	34
22a	Thallus branches flattened; thallus greenish grey or white..	23
b	Thallus branches cylindrical; flattened sections; thallus bluish grey or greenish to whitish yellow	26
23a	Thallus <0.2 mm long, stiff hairs composed of bundled hyphae; apothecium disc orange, K+dark violet .... *Seirophora*
b	Thallus >0.2 mm long; not hairy; apothecium disc not orange	24
24a	Thallus greenish grey, > 2 cm long, pendant	32

(continued)

Table 6.2   (continued)

b	Thallus whitish, < 2 cm long, erect; surface smooth	37
25a	Thallus greenish grey to brown; with green algae; branches flattened; with green algae	26
b	Thallus bluish grey to black, basal branches whitish; branches cylindrical; with blue-green algae	30
26a	Thallus branches with granular to tomentose surface, grey to blue-grey	27
b	Thallus branches with smooth surface, greenish grey to brown	28
27a	Thallus branching dendroid with flattened terminal	35
b	Thallus branching more irregular	*Tornabea*
28a	Thallus branches flattened, composed of apothecia	*Polystroma*
b	Thallus branches cylindrical	29
29a	Thallus brown; branches <0.5 mm wide, slender and branched	38
b	Thallus greenish grey; branches >0.5 mm	40
30a	With scattered tomentum; no apothecia; thallus with multilayered cortex, not gelatinous	*Dendriscocaulon*
b	Without tomentum; apothecia present; thallus with single layered cortex or without cortex, gelatinous	31
31a	Richly branched; main branches <0.1 mm wide; with tomentum; cortex composed of a single cellular layer; algae *Scytonema*, gelatinous	*Polychidium*
b	Sparingly branched; main branches >0.1 mm wide, without tomentum; no cellular cortex; algae *Nostoc* in chains, not gelatinous	*Lempholemma*
32a	On green algae	33
b	On blue-green algae	41
33a	Fruticose thallus with granular, coralloid, cylindrical or squamulose, phyllocladia; with green algae	34
b	Fruticose thallus without phyllocladia; with blue-green algae	35
34a	Surface of cephalodia and phyllocladia present and smooth; apothecia terminal, lateral or absent	*Stereocaulon*
b	Surface of phyllocladia tomentose	*Leprocaulon*
35a	On soil	36
b	On rock or mossy tree trunks	38
36a	Thallus whitish yellow, with loose, long, and slender branches	*Alectoria*
b	Thallus whitish, with dense, short, and blunt branches	37
37a	Thallus branches flattened, never with ascocarps	*Siphula*
b	Thallus branches coralloid; mazaedium present	*Acroscyphus*
38a	Thallus brown; branches rounded; < 0.5 mm thick	*Bryoria*
b	Thallus greenish grey to grey; branches flattened, > 0.5 mm thick	39
39a	Ascocarps black with mazaedium	*Bunodophoron*
b	Ascocarps yellow to grey without mazaedium	40
40a	Thallus and ascocarps grey; cortex of palissadic structure; without cartilaginous strands	*Roccella*
b	Thallus greenish grey, apothecial disc yellowish; cortex of strongly conglutinated, periclinal hyphae; cartilaginous strands present	*Ramalina*
41a	Thallus not gelatinous, white or grey	*Peltula*
b	Thallus gelatinous, black	42
42a	Lobes flat, unbranched, > 1 cm long	*Jenmania*
b	Lobes rounded and branched, < 1 cm long	*Ephebe*

Table 6.3 Identification key for crustose lichens

1a	Ascocarp present	2
b	Ascocarp absent; Apothecia present	5
2a	Perithecia present	130
b	Lirellae present	169
3a	Thallus crustose undifferentiated; perithecia spherical within thalloid warts	13
b	Thallus crustose poorly developed; perithecia embedded in verrucose around ostiole	13
4a	Conidangia present	188
b	Conidangia absent	193
5a	Apothecia covered by powdery masses of ascospores, ripening above the asci	6
b	Apothecia covered by ascospores; thallus with soredia; spores globose	*Srangospora*
6a	Apothecia pin-shaped, with up to 2 mm long, thin stalk	7
b	Apothecia sessile or with a short stalk	9
7a	Ascospores simple, pale brown or colorless	8
b	Ascospores 1-3 septate; brown color	*Arthonia*
8a	Apothecia dark brownish, often covered by colored pruina; ascospores pale brown	*Chaenotheca*
b	Apothecia reddish brown; hyaline	*Sclerophora*
9a	Apothecia without thalline margin	10
b	Apothecia with thalline margin	11
10a	Apothecia with distinct proper margin, clearly delimited; ascospores ripening and coloring in the asci	*Pyrgidium*
b	Apothecia without proper margin, indistinctly delimited; ascospores ripening and coloring inside the asci	12
11a	Ascospores one-septate	38
b	Ascospores two-septate	12
12a	Ascospores 7-9 x 5-6 μm; excipulum wide, with corona-like extension	*Nadvornikia*
b	Ascospores 9-13 x 4.5-6.5 μm; excipulum without corona-like extension	188
13a	Thallus white to pale grey to blackish; paraphyses unbranched; hymenium hyaline	210
b	Thallus yellowish to brownish; paraphyses slender; acicular hyaline	210
14a	Hymenium I+blue, throughout or around the asci; ascospores I-, simple or variously septate, with equal cells; paraphyses straight	15
b	Hymenium I-; asci I-; paraphyses unbranched	109
15a	Ascospores grey to brown at old, usually one-septate	16
b	Ascospores persistently hyaline, variously septate	17
16a	Epiphytic on moss or plant debris	17
b	Epiphytic on rock	19
17a	Ascospore wall with septa thin	19
b	Ascospore wall with septa thickened	18
18a	Apothecia lecideine; ascospore wall thickening along lateral outer walls	19
b	Apothecia lecanorine or biatorine; ascospore wall thickening near apex and septa	20
19a	Apothecia lecanorine with thalline margin	20
b	Apothecia lecideine without thalline margin	21
20a	Thallus margin lobate	*Dimelaena*
b	Thallus margin not lobed	22
21a	Spores with halo, muriform	86
b	Spores without halo, transversely one-, rarely three-septate	22
22a	Ascospore septa thin, lumina edged	*Buellia*
b	Ascospore septa thickened, lumina rounded	23
23a	Apothecia lecideine; Ascospores thin-walled, pointed poles	*Hafellia*
b	Apothecia biatorine; ascospores thickened polar walls, not pointed	*Rinodina*

(continued)

Table 6.3 (continued)

24a	Ascospores simple	25
b	Ascospores muriform	81
25a	Apothecia lecanorine, margin as of thallus color	26
b	Apothecia lecideine or biatorine, without distinct margin	39
26a	Ascospores <20 μm long	27
b	Ascospores >20 μm long	56
27a	Ascospores >8 in each ascus	28
b	Ascospores <8 in each ascus	31
28a	Thallus coarse-areolated, brown; apothecia immersed	*Acarospora*
b	Thallus continuous or granular-areolated, grey; apothecia sessile	29
29a	Thallus yellow; epiphytic on rock	31
b	Thallus greenish; epiphytic on other substrates	30
30a	Asci of Lecanora-type, with distinct I+pale axial mass in tholus	*Maronina*
b	Asci of Fuscidea-type, with continuously I+blue axial mass in tholus	*Maronea*
31a	Ascocarps immersed in thallus; ascus apex I	*Aspicilia*
b	Ascocarps sessile in thallus; ascus apex I+blue	32
32a	Hypothecium pale brown to brown	33
b	Hypothecium colorless	34
33a	Hymenium colorless	*Vainionora*
b	Hymenium pale purplish	*Tephromela*
34a	Thallus bright yellow	*Candelariella*
b	Thallus white or grey	35
35a	Conidia pleurogenous; cupular, thick-walled in excipulum	*Protoparmelia*
b	Conidia acrogenous; cupular thick-walled in excipulum absent	*Lecanora*
36a	Ascus apex I+pale blue; spores >8/ascus	37
b	Ascus apex I+strongly blue; spores 4-8/ascus; ascocarps immersed in thallus warts	37
37a	Ascocarps immersed in thallus warts; ascospores <4/ascus; epiphytic or on moss; cephalodia absent	*Megaspora*
b	Ascocarps lecanorine; ascospores >8/ascus; flat, rounded cephalodia present	*Placopsis*
38a	Ascocarps immersed in thallus warts; ascospores thick-walled (5 μm thick), > 60 μm long	39
b	Ascocarps lecanorine; ascospores thin-walled (2 mm thick), < 50 μm long	*Ochrolechia*
39a	Apothecial disc orange to bright red, K+purplish to dark violet	40
b	Apothecial disc usually brown or black	41
40a	Apothecial disc bright red, K+purplish	*Pyrrhospora*
b	Apothecial disc orange, K+dark violet; margin black	*Bahianora*
41a	Ascospores >16 per ascus, small and globose	*Piccolia*
b	Ascospores <16 per ascus, not globose	42
42a	Paraphyses anastomosing; exciple absent; apothecia globular	43
b	Paraphyses not anastomosing; exciple present; apothecia often flat	45
43a	Hymenial gelatine absent	*Vezdaea*
b	Hymenial gelatine present	44
44a	Thallus yellow to green, completely sorediate; apothecia colorless	*Psilolechia*
b	Thallus grey to brown, partially sorediate; apothecia variously colored	64
45a	Asci without tholus, I	*Schaereria*
b	Asci with distinct, I+blue tholus	46
46a	Asci with weakly I+blue tholus	47
b	Asci with strongly I+blue tholus	50
47a	Epiphytic or on organic detritus	48
b	Saxicolous; ascus apex variable	51

(continued)

Table 6.3 (continued)

48a	Epiphytic; central amyloid tube in ascus apex present	*Malcolmiella*
b	On organic detritus; no distinct amyloid structure in ascus apex	49
49a	Thallus grey; apothecia pink	57
b	Thallus dark brown; apothecia black	*Placynthiella*
50a	Apothecia pink, usually < 1 mm wide, with thin proper margin	*Trapelia*
b	Apothecia dark brown, 1-3.5 mm wide, with thick, prominent proper margin	*Ainoa*
51a	Central tube in ascus apex present	*Porpidia*
b	Small apical cap in ascus apex present	*Lecidea*
52a	Asci with I+ pale axial mass in tholus	53
b	Asci with I+ blue-staining tholus	54
53a	Axial mass with Lecanora-type asci; apothecia dark brown to black; thallus whitish, often K+ yellow or C+ orange	*Lecidella*
b	Axial mass conical; apothecia variously colored; thallus reactions different	55
54a	Asci with I+ weakly blue-staining tholus; apothecia brownish to blackish discolored; thallus grey, granular, K-, C+ red	*Trapeliopsis*
b	Asci with I+ strongly blue-staining apical layer; apothecia dark brown to black; thallus brown	*Fuscidea*
55a	Apothecia with distinct margin same as thallus color	56
b	Apothecia with margin same as disc color or without distinct margin	62
56a	Apothecia immersed, lacerate, erect margin; disc grey to pale brown, often white-pruinose	57
b	Apothecia sessile with constricted base, with entire or crenulate margin	58
57a	Paraphyses thick and conspicuously septate	*Phlyctidia*
b	Paraphyses slender, simple	*Phlyctella*
58a	Ascospores with thin septa	59
b	Ascospores with swollen septa	60
59a	Spores one-septate; disc yellow to brown, K	*Lecania*
b	Spores many-septate; disc red, K+ purplish	*Haematomma*
60a	Spores one-septate, > 15 μm wide; septa thinner than lumina; disc grey, K	*Megaloblastenia*
b	Spores many-septate; septa thicker than lumina; disc yellow to brownish, mostly K+ purplish	61
61a	Ascospores one- to three-septate	63
b	Ascospores > three-septate	81
62a	Ascospores one-septate	63
b	Ascospores many-septate	69
63a	Ascospores 2-4/ascus, > 30 μm long; hymenium guttulate	34
b	Ascospores 8/ascus, < 30 μm long; hymenium unclear	65
64a	Ascospores 2/ascus, > 40 μm long; hymenium clear	*Lopezaria*
b	Ascospores 2-8/ascus, > 30 μm long; hymenium inspersed with oil-like droplets	65
65a	Ascospores with thick septa; disc K+ dark violet	*Caloplaca*
b	Ascospores with thin septa	66
66a	Apothecia without distinct margin	67
b	Apothecia with distinct margin	67
67a	Asci with ocular chamber surrounded by I+ weakly staining, rounded axial mass; ascospores not halonate; exciple with well-defined cortical and medullary parts	69
b	Asci without rounded, I+ weakly staining axial mass around ocular chamber; ascospores not halonate; exciple compact	68
68a	Ascus tholus containing a conical, I+ weakly staining axial mass around ocular chamber; ascospores halonate	*Catinaria*
b	Ascus tholus containing a tubular, I+ strongly staining; apothecia yellowish; ascospores not halonate	*Catillaria*

(continued)

6 Identification Key for the Major Growth Forms of Lichenized Fungi   103

Table 6.3 (continued)

69a	Ascospores large, broadly ellipsoid, 75-135 x 30-40 µm, single	81
b	Ascospores < 10 µm wide, long fusiform to filiform, 8/ascus	70
70a	Ascospores with thick septa, up to 3-septate; disc mostly K+dark violet	*Caloplaca*
b	Ascospores with thin septa, up to 3-septate; disc K-	71
71a	Apothecia with weak inapparent margin; apothecia convex to globose; paraphyses branched	*Micarea*
b	Apothecia with prominent margin; disc flat; paraphyses unbranched; sessile	72
72a	Ascus tholus conical, I+weakly staining axial mass around ocular chamber	73
b	Ascus tholus containing a tubular, I+strongly staining	77
73a	Ascospores fusiform, < 5 times as long as wide	74
b	Ascospores acicular, > 5 times as long as wide	76
74a	Apothecia with brown lid-like appendage on the margin	*Auriculora*
b	Apothecia without lid-like appendage	75
75a	Ascus tholus with rounded, I+weakly staining axial mass around ocular chamber	*Megalaria*
b	Ascus tholus with pointed, I+weakly staining axial mass around ocular chamber	*Biatora*
76a	Excipulum paraplectenchymatic; asci with rounded axial mass	*Bacidina*
b	Excipulum prosoplectenchymatic; asci with conical axial mass	*Bacidia*
77a	Excipulum with byssoid outer layer	*Byssoloma*
b	Excipulum smooth outside	78
78a	Apothecia <0.5 mm side, soon convex; exciple paraplectenchymatic	*Fellhanera*
b	Apothecia usually >0.5 mm wide, flat; exciple prosoplectenchymatic	79
79a	Campylidia absent; tubular structure in ascus apex distinctly stained over its whole length	*Mycobilimbia*
b	Campylidia present; tubular structure in ascus apex widened and less distinct towards the tip	80
80a	Ascospores fusiform; dark brown ellipsoidal; one septate thallus distinct	*Cyphelium*
b	Ascospores needle-shaped to cylindrical; campylidia absent; pale brown with greenish	*Microcalicium*
81a	Apothecia yellow to red, K+dark purple	82
b	Apothecia not yellow to red, K	83
82a	Thallus white or grey	*Brigantiaea*
b	Thallus greenish yellow	*Letrouitia*
83a	Hymenium inspersed with oil-like droplets; spores single	89
b	Hymenium clear; spore number per ascus various	84
84a	Ascus tholus containing a tubular I+strongly staining; no campylidia present; paraphyses unbranched	*Bapalmui*
b	Ascus tholus without tubular, I+structure, often with wide ocular chamber; campylidia present; paraphyses strongly branched	85
85a	Ascomata with tomentose margin	*Lasioloma*
b	Ascomata with smooth margin	86
86a	Campylidia absent	87
b	Campylidia present	88
87a	Ascospores with gelatinous sheet	*Rhizocarpon*
b	Ascospores without gelatinous sheet	*Schadonia*
88a	Campylidia consisting of a thalloid tube and a short, brownish; ocular chamber of ascus wide; excipulum paraplectenchymatic	190
b	Campylidia consisting of a large, greyish; ocular chamber of ascus wide; excipulum paraplectenchymatic	189
89a	Ascocarps compound with several punctiform discs in usually raised areas which are differently colored; ascospores transversely septate	90
b	Ascocarps simple, with single disc, at age sometimes deformed; ascospores transversely muriform	96
90a	Thallus felt-like, greenish	95
b	Thallus with compact upper layer	91

(continued)

Table 6.3 (continued)

(continued)

Table 6.3   (continued)

112a	Apothecial margin rounded to flat; discs tiny, rarely >0.5 mm wide	200
b	Apothecial margin lacerate, forming slips which cover the disc in part; discs often several mm wide	194
113a	Apothecia with raised thalline margin, discs invisible through thallus splits, pale, often white-pruinose	200
b	Apothecia exerted above thallus, without thalline margin; discs widely exposed, brownish	114
114a	Apothecia compound with elongated discs level with the margin; ascospores hyaline, transversely septate	119
b	Apothecia simple	115
115a	Ascospores hyaline, muriform	*Gyrostomum*
b	Ascospores grey, bacillary	116
116a	Paraphyses branched and anastomosing throughout	117
b	Paraphyses unbranched, with indistinct transverse connections	119
117a	Thallus without bristles	*Gyalideopsis*
b	Thallus with white or black bristles or hyphophores	118
118a	Apothecia with constricted base and prominent margin; bristles usually black; hyphophores absent	*Tricharia*
b	Apothecia widely adnate and with indistinct margin; bristles usually white, when present; hyphophores often present	188
119a	Ascospores acicular, 4 µm wide	120
b	Ascospores ovoid to fusiform, generally over 4 µm wide	121
120a	Epiphytic; thallus grey; apothecia very elongated, tubular, forming 2 mm long stalks, brown	*Gomphillus*
b	Terrestrial; thallus yellow; apothecia discoid, black	*Arthrorhaphis*
121a	Apothecia immersed; ascospores grey, ovoid and muriform	*Diploschistes*
b	Epiphytic; apothecia sessile; ascospores various	122
122a	Apothecia persistently immersed, with flat disc, often >1 mm large, elongate; margin lacerate	123
b	Apothecia initially immersed, finally sessile with constricted base, with concave disc, mostly <1 mm large, always rounded; margin entire	125
123a	Margin inapparent, granular; paraphyses with thin transverse connections; asci I+ pale blue, with thin apex	*Phlyctidia*
b	Margin lacerate; paraphyses simple; asci I-, with thickened apex	124
124a	Margin rounded to flat; discs rounded to elongate, often several mm long	170
b	Margin lacerate; discs rounded to slightly elongate	*Chroodiscus*
125a	Apothecia forming coralloid-branched structures, greenish; ascospores 5-7-septate; hymenium usually absent	*Polystroma*
b	Apothecia not proliferating	126
126a	Apothecia with carbonized excipulum covered by pale pruina; ascospores muriform	*Ramonia*
b	Apothecia not with carbonized margin	127
127a	Hymenium I+(pale) blue	128
b	Hymenium I	130
128a	Photobiont blue-green alga	*Bryophagus*
b	Photobiont Trentepohlia-like	129
129a	Spores 8/ascus, muriform	*Gyalecta*
b	Spores 8-16/ascus, transversely septate	*Cryptolechia*
130a	Apothecia pale yellow to orange; ascospores one-septate	*Dimerella*
b	Apothecia brown; ascospores bacillar-pluriseptate	*Gyalidea*
131a	Ascospores simple	132
b	Ascospores septate	133
132a	Ascospores 16-20/ascus	*Kalbiana*
b	Ascospores 8/ascus	133

(continued)

Table 6.3 (continued)

133a	Ascospores thick-walled, with spines; paraphyses persistent	*Monoblastia*
b	Ascospores thinwalled, smooth; paraphyses disappearing in an early stage	*Verrucaria*
134a	Ascospores brown or grey-brown	135
b	Ascospores persistently colorless	144
135a	Ascospore with thin septa and edged lumina	136
b	Ascospore with thick septa and rounded lumina	139
136a	Asci ovoid with narrow ocular chamber	137
b	Asci subcylindrical with wide ocular chamber	138
137a	Ascospores with 1-3 transverse septa; ascocarps simple	*Mycomicrothelia*
b	Ascospores transversely septate to muriform; ascocarps mostly multi-chambered	144
138a	Ascospores one-septate	*Clypeopyrenis*
b	Ascospores muriform	*Anthracothecium*
139a	Ascospores one-septate	140
b	Ascospores transversely septate only, with 3 septa	141
140a	Ascospores without pigment granules in endospore	*Distopyrenis*
b	Ascospores with pigment granules in endospore	*Granulopyrenis*
141a	Ascospores three-septate, < 50 μm long	142
b	Ascospores three-septate, > 70 μm long	*Architrypethelium*
142a	Paraphyses absent; algae present in hymenium	*Staurothele*
b	Paraphyses present; no algae in hymenium	143
143a	Spores without longitudinal grooves, with prominent median transverse septum	*Pyrenula*
b	Spores with longitudinal grooves, with prominent median longitudinal septum	*Sulcopyrenula*
144a	Ascospores with thickened septa and rounded lumina; ascocarps often compound	145
b	Ascospores with thin septa and edged lumina; ascocarps usually simple	153
145a	Ascospores transversely septate	146
b	Ascospores muriform	150
146a	Thallus poorly developed; ascocarps naked at maturity, never aggregated in pseudostromata	*Pseudopyrenula*
b	Thallus well developed; ascocarps immersed in pseudostromata	147
147a	Ostioles free, apical	148
b	Ostioles fused to form a compound ascocarp	*Astrothelium*
148a	Ascospores 1-septate, > 45 μm long, with needle-shaped crystals in the wall	*Megalotremis*
b	Ascospores more than 2-septate, < 45 μm long, without crystals	149
149a	Paraphyses branched; ascus apex with narrow ring surrounding a small ocular chamber; wall thickening of ascospores pronounced at the edges	*Trypethelium*
b	Paraphyses unbranched; ascus apex with wide apical ring and wide ocular chamber; wall thickening of the ascospore more equal	*Lithothelium*
150a	Ostiole apical; jigsaw puzzle-like hyphae	151
b	Ostiole lateral	152
151a	Ascocarps in brown shiny pseudostromata, K + red; pseudostroma wall composed of brown	*Bathelium*
b	Ascocarps not in brown pseudostromata, K-; wall not composed of brown	*Laurera*
152a	Ostioles free	*Campylothelium*
b	Ostioles fused	*Cryptothelium*
153a	Ascus tip thin, truncate; paraphyses unbranched	154
b	Ascus tip thickened with an ocular chamber, rounded; paraphyses often branched	157
154a	Ascocarps with subapical whorl of black bristles	*Trichothelium*
b	Ascocarps without bristles	155
155a	Ascospores transversely septate; asci with chitinoid apical ring	*Porina*
b	Ascospores muriform; asci without chitinoid apical ring	156
156a	Medulla white	*Clathroporina*
b	Medulla yellow	*Myeloconis*

(continued)

Table 6.3 (continued)

157a	Thallus gelatinous; spores one-septate	*Pyrenocollema*
b	Thallus subcuticular; grey or green, spores variously septate	158
158a	Paraphyses absent; hymenial gelatine I+reddish	159
b	Paraphyses present, persistent; hymenial gelatine I-	160
159a	Spores transversely septate	*Thelidium*
b	Spores muriform	*Polyblastia*
160a	Paraphyses unbranched; macroconidia usually present, cylindrical, septate; Photobiont *Cephaleuros*	*Strigula*
b	Paraphyses branched; macroconidia more or less lacking	161
161a	Ascocarps simple	162
b	Ascocarps multilocular	168
162a	Ascospores transversely septate	163
b	Ascospores muriform	167
163a	Ascospores ovoid-fusiform, 1-5-septate, > 4 μm wide	164
b	Ascospores filiform, 6-10-septate, 2 μm wide	*Celothelium*
164a	Ascospores ovoid-fusiform, 1-septate, rarely finally 3-septate	165
b	Ascospores fusiform, 3-11-septate, > 4 μm wide	*Polymeridium*
165a	Lower ascospore cell shorter; lichenized; microconidia globose to ellipsoid; macroconidia present; ostiole often lateral	*Anisomeridium*
b	Lower ascospore cell longer; nonlichenized; microconidia bacillar; macroconidia lacking; ostiole apical	166
166a	Paraphyses slender, without refractive bodies near the septa; asci clavate	*Arthopyrenia*
b	Paraphyses short-celled, with refractive bodies near the septa; asci obpyriform	*Naetrocymbe*
167a	Asci with indistinct apical thickening; ascospores 8 in each ascus; algiferous thallus present	*Helenella*
b	Asci with pronounced apical thickening; ascospores 2 in each ascus; nonlichenized	*Julella*
168a	Paraphyses indistinct, with many oil droplets	*Mycoporum*
b	Paraphyses distinct, without oil droplets	169
169a	Ascospores with 1-3 transverse septa; asci rather cylindrical; paraphyse cell ends with refractive bodies	*Tomasellia*
b	Ascospores submuriform, with enlarged end cells; paraphyse cells without refractive bodies	*Exiliseptum*
170a	Paraphyses branched; hymenium I+red or blue; ascospore lumina not rounded, with slightly rounded edges, septa I-; asci with rather thin tholus with small ocular chamber surrounded by a small I+blue ring	171
b	Paraphyses unbranched; hymenium I-, rarely I+pale blue; ascospore lumina rounded, lentiform, septa often I+blue or violet; asci with thick tholus, completely I-	175
171a	Ascocarp walls conspicuous and carbonized	172
b	Ascocarp walls indistinct, not carbonized; Asci ovoid to globose, with strongly thickened tip	174
172a	Ascocarps immersed in thallus with thick thalline carbonized margin; hymenium always gelatinous	173
b	Ascocarps exerted with prominent carbonized margin; hymenium not gelatinous	174
173a	Ascospores muriform, hyaline; ascocarps short and exerted with thick thallus margin	*Helminthocarpon*
b	Ascocarps variously septate, often brown at maturity; ascocarps level with thallus, with thin thalline margin	*Sclerophyton*
174a	Lirellae with closed gaping labiae; asci with I+blue cap in tip, which extends laterally; spores transversely septate	*Opegrapha*
b	Lirellae with wide open disc; asci with tiny I+blue cap in tip; spores transversely septate	*Lecanographa*
175a	Hymenium gelatinous; ascospores muriform	*Arthothelium*
b	Hymenium gelatinous; ascospores transversely septate	*Arthonia*
176a	Ascocarps stellate	177
b	Ascocarps single	180

(continued)

Table 6.3 (continued)

(continued)

Table 6.3 (continued)

201a	Medulla pink throughout; with glossy, short, clavate isidia	203
b	Medulla red throughout, exposed along the margins	*Cryptothecia*
202a	Thallus greenish, glossy, with scattered soralia-like, yellow spots	*Myeloconis*
b	Thallus grey, dull, with raised, dense, yellow soralia	*Megalospora*
203a	With soradi fine or corsea	204
b	With schizidia	206
204a	Thallus C+red	*Pertusaria*
b	Thallus C	205
205a	Soredia fine; with stictic acid (P+orange, K+orange)	*Thallotrema*
b	Soredia coarse; with hypoprotocetraric acid (P-, K-)	206
206a	Schizidia accumulated in groups, shortly stalked	207
b	Schizidia arising single, leaving scattered, round scars on thallus	207
207a	Thallus bluegrey, pruinose	208
b	Thallus greenish, not pruinose	208
208a	Schizidia small, <0.2 mm wide; with protocetraric acid (P+red)	209
b	Schizidia >0.5 mm wide; with psoromic acid (P+yellow, K-)	*Ocellularia*
209a	Isidia cylindrical; with psoromic acid (P+yellow)	*Myriotrema*
b	Isidia gradually tapering; no lichen substances (P	*Thelotrema*
210a	Medulla I+blue; globuse 8 elongate, multiseptate	*Belonia*
b	Medulla I-; Spores 8 globuse elipsoid	*Biatorella*

Table 6.4 Identification key for squamulose lichens

1a	Perithecia present	2
b	Apothecia present	5
2a	Ascospores simple or transversely septate; no algae present in hymenium	3
b	Ascospores muriform	4
3a	Ascospores simple; Squamules large; no algae in the hymenium	*Catapyrenium*
b	Ascospores transversely septate	5
4a	Algae present in hymenium; cortical cells of thallus smooth-walled; spores muriform, brown	28
b	Algae absent in hymenium; cortical cells of thallus finely papilose; spores simple, colorless	6
5a	Squamules rounded, whitish with raised margin; perithecia absent	*Normandina*
b	Squamules usually elongated and greenish, without raised margin	7
6a	Photobiont green algae	24
b	Photobiont bluegreen algae	*Psoroglaena*
7a	Photobiont bluegreen algae	8
b	Photobiont green algae	13
8a	Thallus not gelatinous, closely appressed to substrate; heteromerous, lower cortex lacking or weakly developed; with distinct cortex and medulla	9
b	Thallus umbilicate, reddish brown, not sorediate or isidiate; apothecia immersed in warts	*Phylliscum*
9a	Ascospores >100 per ascus; photobiont cells in tetrads or single	*Phyllopeltula*
b	Thallus blue-grey to brown, epiphytic	10
10a	Thallus with distinct, often tomentose, prothallus; apothecia with thalline margin	*Parmeliella*
b	Thallus without distinct prothallus; not umbilicate; apothecia without thalline margin	11
11a	Apothecial with thalline margin; photobiont bluegreen algae, cells in densely winding chains; ascospores simple, hyaline; isidiae or papillae bluish	12
b	Apothecia without thalline margin; ascospores simple, hyaline; corex composed of longitudinally arranged hyphae; thallus with concentric ridges	*Coccocarpia*
12a	Asci with I+blue apical plug; thallus brownish; with white spots on lobe margins	*Fuscopannaria*
b	Asci without I+blue apical plug; thallus usually blue-grey	*Pannaria*

(continued)

Table 6.4 (continued)

13a	Upper surface of squamules byssoid, lacking cortex	14
b	Upper surface of squamules smooth, with cortex, pruinose	15
14a	Epiphytic; thallus squamules connected into rosette-like thalli, algiferous	*Crocynia*
b	Thallus squamules widely scattered, without algae	*Cyphellostereum*
15a	Tomentose prothallus present	16
b	No tomentose prothallus	19
16a	Squamules with an upper and lower cortex comprised of a thin layer of cubic cells	30
b	Squamule cortex otherwise, lower cortex usually absent	17
17a	Apothecium margin concolorous with thallus; with algae	*Physcidia*
b	Apothecium margin not concolorous with thallus, without algae	18
18a	Ascospores generally >25 μm, transversely many-septate	*Squamacidia*
b	Ascospores generally <25 μm, simple or one-septate	*Phyllopsora*
19a	Thallus yellow to orange	20
b	Thallus pale grey to greenish grey or brownish	22
20a	Thallus lobes ascending, 2 mm long, convex; medulla yellow; apothecia black, globose; thallus K + dark violet	*Xanthopsorella*
b	Thallus K- or K + weakly reddish	21
21a	Apothecia lecanorine, yellow	*Candelina*
b	Apothecia lecideine, black	27
22a	On soil	23
b	On rock	29
23a	Squamules whitish grey on both sides, elongate and erect; cortical cells of thallus finely papillose; squamules delicate, lacerate	28
b	Squamules more pale on lower side, appressed to substrate; ascocarps present	25
24a	Cortical cells of thallus finely papillose; squamules delicate, lacerate	*Agonimia*
b	Squamules pale grey-green, orbicular, with concentric wrinkles, usually with a raised sorediate margin; spores 5 septate	*Normandina*
25a	Ascospores simple	26
b	Ascospores transversely septate; apothecia lecideine	*Toninia*
26a	Apothecia lecanorine	33
b	Apothecia lecideine	27
27a	Squamules whitish grey on both sides, erect	28
b	Squamules greenish or brownish above, pale below	29
28a	Cortical cells of thallus finely papillose; squamules delicate	*Agonimia*
b	Thallus of appressed squamules;, lacerate, pale perithecial wall	35
29a	Squamules elongated	*Pseudohepatica*
b	Squamules rounded to moderately elongated and incised, mostly <5 mm long	30
30a	Thallus thin, corticate with a thin layer of cubic cells on both sides	*Eschatogonia*
b	Thallus without cortex	31
31a	Ascospores simple	32
b	Ascospores transversely septate	34
32a	Thallus C + red, with labriform soralia	*Hypocenomyce*
b	Thallus C-; apothecia lecanorine	33
33a	Ascospores acicular; with capitate soralia	*Bacidiopsora*
b	Ascospores fusiform to bacillary; without capitate soralia	*Psorella*
34a	Ascospores with many in each ascus, < 10 μm long	*Acarospora*
b	Ascospores 8 in each ascus, > 20 μm long	*Placopsis*
35a	Thallus appressed squamules, 1-3 μm across, brown, pale perithelial wall	*Staurothele*
b	Thallus erect squamules; perithelial wall brown to black	*Endocarpon*

Notes

1. Note that K+ or C+ denotes a positive color reaction and K- or C- indicates that there is no color change. A very small amount is enough, and it can be put in eye-dropper bottles and kept in the fridge when not in use.
2. Sodium hypochlorite (C) is common bleach; it is best to use bleach without any additives. Avoid getting it on your clothes.
3. Potassium hydroxide KOH (K) or alternatively sodium hydroxide is used as a 10 % solution. Care should be taken while preparing this chemical as it is highly caustic in the concentrated form.
4. Clean and freeze the material for future use. Before use allow a few hours for equilibration.
5. Placing the media in the upper lid limits contamination. Discharged spores will be attached to the agar surface either singly or groups. For single-spore isolation reduce the discharge time or increase the distance between the ascocarp and the water agar medium. Approximate spore discharge time is 24 h.
6. In some lichens, spores germinate within 1 day after dispersal.

References

1. Hibbett DS, Binder M, Bischoff JF, Blackwell M, Cannon PF, Eriksson O et al (2007) A higher-level phylogenetic classification of the Fungi. Mycol Res 111:509–547
2. Schessler A, Schwarzott D, Walker C (2001) A new fungal phylum, the Glomeromycota: phylogeny and evolution. Mycol Res 105:1413–1421
3. Wearing J (2010) Fungi. Crabtree Publishing Company, New York
4. Feuerer T, Hawksworth DL (2007) Biodiversity of lichens, including a worldwide analysis of checklist data based on Takhtajan's Floristic regions. Biodivers Conserv 16:85–98
5. Mueller GM, Schmit JP (2007) Fungal biodiversity: what do we know? What can we predict? Biodivers Conserv 16:1–5
6. Kirk PM, Cannon PF, Minter DW, Stalpers JA (2008) Dictionary of the fungi. CAB International, Wallingford, UK
7. Lumbsch HT, Ahti T, Termann S, AmoDePaz G, Aptroot A, Arup U et al (2011) One hundred new species of lichenized fungi: a signature of undiscovered global diversity. Phytotaxa 18:1–127
8. Porter TM, Schadt CW, Rizvi L, Martin AP, Schmidt SK, Scott-Denton L et al (2008) Widespread occurrence and phylogenetic placement of a soil clone group adds a prominent new branch to the fungal tree of life. Mol Phylogenet Evol 46:635–644
9. Blackwell M, Spatafora JW (2004) Fungi and their allies. In: Bills GF, Mueller GM, Foster MS (eds) Biodiversity of fungi: inventory and monitoring methods. Elsevier Academic Press, Amsterdam, pp 18–20
10. Lutzoni F, Kauff F, Cox CJ, McLaughlin D, Gail C, Dentinger B et al (2004) Assembling the fungal tree of life: progress, classification, and evolution of subcellular traits. Am J Bot 91:1446–1480
11. James TY, Kauff F, Schoch CL, Matheny PB, Hofstetter V, Cox CJ et al (2006) Reconstructing the early evolution of fungi using a six-gene phylogeny. Nature 443:818–822
12. Bland J (1971) Forests of Lilliput: the realm of mosses and lichens. Prentice-Hall, Upper Saddle River, NJ
13. Alexopoulos CJ, Mims CW, Blackwell M (1996) Introductory mycology. John Wiley & Sons, New York
14. Schadt CW, Martin AP, Lipson DA, Schmidt SK (2003) Seasonal dynamics of previously unknown fungal lineages in tundra soils. Science 301:1359–1361
15. Progovitz RF (2003) Black mold your health and your home. The Forager Press, New York, pp 48–52
16. Ahmadjian V (1993) The lichen symbiosis. John Wiley & Sons, New York, pp 8–29
17. Yoshimura I, Yamamoto Y, Nakano T, Finnie J (2002) Isolation and culture of lichen photobionts and mycobionts. In: Kranner I, Beckett RP, Varma A (eds) Protocols in lichenology: culturing, biochemistry, ecophysiology and use in biomonitoring. Springer-Verlag, Berlin, Heidelberg, New York, pp 3–33
18. Mueller GM, Bills GF, Foster MS (2004) Biodiversity of fungi: inventory and monitoring methods. Academic, San Diego
19. Orangea A (1989) *Macentina stigonemoides* (Verrucariaceae), a new lichenized species from Great Britain and Ireland. Lichenologist 21:229–236
20. Adler MT (1990) An artificial key to the genera of the Parmeliaceae (Lichenes, Ascomycotina). Mycotaxon 38:331–347
21. Brodo IM, Sharnoff SD, Sharnoff S (2001) Lichens of North America. Yale University Press, New Haven
22. Nash TH (1996) Lichen biology. Cambridge University Press, Cambridge, UK, pp 1–36
23. Purvis W (2000) Lichens. Smithsonian Institution Press, Washington, DC
24. Dobson F (1979) Lichens: an illustrated guide. Richmond Publishing Co., Mexico City
25. Hale ME, Cole M (1988) Lichens of California. University of California Press, Berkeley, CA
26. Goward T, McCune B, Meidinger D (1994) The Lichens of British Columbia illustrated keys. Part I. Foliose and squamulose species. Victoria, British Columbia. Ministry of Forests Research Program, Canada

27. Alstrup V, Olech M (1993) Lichenicolous fungi from Spitsbergen. Pol Polar Res 14:33–42

28. Nash TH (2002) Lichen Flora of the Greater Sonoran Desert Region: the pyrenolichens and most of the squamulose and macrolichens. Arizona State University, Mesa, AZ

29. Jordan M (2004) The encyclopedia of fungi of Britain and Europe. Francis Lincoln Ltd., London, UK, pp 18–23

30. Vitt DH, Marsh E, Bovey RB (1998) Mosses, lichens and ferns of Northwest North America. Lone Pine Publishing, Edmonton, Alberta, Canada

31. Yamamoto Y, Mizuguchi R, Yamada Y (1985) Tissue cultures of *Usnea rubescens* and *Ramilina yasudae* and production of usnic acid in their cultures. Agric Biol Chem 49:3347–3348

32. Nguyen TT, Joshi Y, Lucking R, Wang X, Dzung NA, Koh Y, Hur J (2010) Notes on some new records of Foliicolous lichens from Vietnam. Taiwania 55:402–406

33. Purvis OW (1992) The lichen flora of Great Britain and Ireland. British Lichen Society, London, UK

34. Bungartz F, Rosentreter R, Nash TH (2002) Field guide to common epiphytic macrolichens in Arizona. Arizona State University, Tempe

35. McCarthy PM, Malcolm WM (2004) Key to the genera of Australian macrolichens. Australian Biological Resources Study. http://www.anbg.gov.au/abrs/lichenlist/introduction.htmL. Accessed 01 Sep 2011

36. de Vries B, de Varies I (2008) Common lichens of Cypress Hills in Interprovincial Park Saskatchewan, Canada. Fish and Wildlife Branch, Ministry of Environment, Canada

37. Smith AL (1975) Lichens. Richmond Publishing Co., Mexico City

38. Thomson JW (1997) American Arctic lichens. University of Wisconsin Press, Madison

39. Case J (2002) Key to lichens of the Alberta Prairies and Parklands. Permanantly stored at the Lichen Key Archive. http://www.toyen.uio.no/botanisk/lav/LichenKey/index.htm. Accessed 08 Aug 2011

40. Ovstedal DO, Smith RIL (2001) Lichens of Antarctica and South Georgia: a guide to their identification and ecology. Cambridge University Press, Cambridge, UK

41. Søchting U, Frödén P (2002) Chemosyndromes in the lichen genus *Teloschistes* (Teloschistaceae, Lecanorales). Mycol Prog 3:257–266

42. Sinha GP, Elix JA (2003) A new species of *Hypogymnia* and a new record in the lichen family Parmeliaceae (Ascomycotina) from Sikkim, India. Mycotaxon 87:81–84

43. Sipman H (2005) Identification key and literature guide to the genera of Lichenized Fungi (Lichens) in the Neotropics. http://www.bgbm.org/BGBM/STAFF/Wiss/Sipman/keys/eokeyC.htm. Accessed 09 Sep 2011

44. Wetmore CM (2004) The isidiate corticolous *Caloplaca* species in North and Central America. Bryologist 107:284–292

45. Aptroot A (2004) Key to the macrolichens and checklist of the lichens and lichenicolous fungi of New Guinea. http://www.myco-lich.com/Lichenology-in-Iran/identifications/useful-ke. Accessed 01 Sep 2011

46. Wetmore C (2005) Keys to the Lichens of Minnesota. http://www.myco-lich.com/Lichenology-in-Iran/identifications/useful-ke. Accessed 01 Sep 2011

47. Weber WA (1962) Environmental modification and the taxonomy of the crustose lichens. Sv Bot Tidskr 56:293–333

48. Hertel H (1998) Problems in monographing Antarctic crustose lichens. Polarforschung 58:65–67

49. Tibell L (1996) Phaeocalicium (Mycocaliciaceae, Ascomycetes) in Northern Europe. Ann Bot Fennici 33:205–221

50. Awasthi DD (1991) A key to microlichens of India, Nepal and Srilanka. Bibl Lichenol 40:1–337

51. Upreti DK, Chatterjee S (2004) Lichen genus Tephromela in India. Geophytology 32:47–52

52. Westberg M (2010) The identity of Candelariella canadensis. Lichenologist 42:19–22

53. McCarthy PM (2001) Key to the genera of crustose pyrenocarpous lichens in Australia. 2001. Australian Biological Resources Study. http://www.anbg.gov.au/abrs/lichenlist/introduction.htmL. Accessed 05 Sep 2011

54. Jørgensen PM (2001) Four new Asian species in the lichen genus *Pannaria*. Lichenologist 33:297–302

55. McCune B, Meidinger DV (1994) The Lichens of British Columbia: foliose and squamulose species. Ministry of Forests, Research Branch, British Columbia, Canada

56. Aptroot A (1991) A conspectus of *Normandina* (Verrucariaceae, lichenized Ascomycetes). Willdenowia 21:263–267

57. Awasthi DD (2000) Lichenology in Indian Subcontinent. Shiva Offset Press, Dehra Dun, India

58. McCarthy PM (2010) Checklist of the lichens of Australia and its Island Territories. Australian Biological Research Society, Canberra. http://www.anbg.gov.au/abrs/lichenlist/introduction.htmL. Accessed 05 Sep 2011

59. Yamamoto Y, Kinoshita Y, Yoshimura I (2002) Culture of thallus fragments and redifferentiation of lichens. In: Kranner I, Beckett RP, Varma A (eds) Protocols in lichenology: culturing, biochemistry, ecophysiology and use in biomonitoring. Springer-Verlag, Berlin, Heidelberg, New York, pp 34–46

Microscopic Methods for Analytical Studies of Fungi

De-Wei Li

Abstract

Optical microscopy is essential in mycological research and analytical studies of fungi. This chapter describes techniques and procedures commonly used for microscopic studies of fungi and analytical studies of airborne fungal spores with an optical microscope.

Keywords

Fungi • Optical microscopy • Fungal spores • Fungal morphology • Freehand sectioning • Freezing microtome • Mounts • Stains • Bright field • Phase contrast • Differential interference contrast

Introduction

In the last decade, the development of molecular technology has advanced fungal systematics in an unprecedented and revolutionary way. It has allowed mycologists to examine the phylogenetic relationships of fungal taxonomic groups at a molecular level from a new perspective. More importantly, mycologists can examine the natural essence of these organisms, rather than view them from artificial and superficial perspectives. Such a development has led to significant changes at all taxonomic ranks of fungal classification. It has also led to the discovery of "Cryptomycota" (or Rozellida), a new phylum arguably in Mycota or even a new Kingdom in biology [1]. The final placement of "Cryptomycota" is subject to debate at present. One thing for sure is that this group of organisms is a new clade on the evolutionary tree of life and it has not been discovered before. However, morphology-based fungal taxonomy is an undisputable fundamental basis of modern molecular fungal systematics. Traditional morphology-based taxonomy will still be imperative to complete a global inventory of Fungi [2]. Fungal molecular systematics will not replace morphology-based fungal systematics in the foreseeable future; rather, the two schools are supplementary to each other. The International Code of Nomenclature for algae, fungi, and plants (Melbourne Code) requires morphological Latin or English diagnosis for valid new fungal taxon descriptions [3]. Microscopic observation and analysis of fungi is still a simple, economic, and efficient way to morphologically study, characterize, and identify fungi. Many mycologists still

De-Wei Li (✉)
Valley Laboratory, The Connecticut Agricultural Experiment Station, 153 Cook Hill Road, Windsor, CT 06095, USA
e-mail: dewei.li@ct.gov

V.K. Gupta et al. (eds.), *Laboratory Protocols in Fungal Biology: Current Methods in Fungal Biology*, Fungal Biology, DOI 10.1007/978-1-4614-2356-0_7, © Springer Science+Business Media, LLC 2013

enjoy observing the beauty of fungi and are very excited to see an undescribed species under a microscope. In the last 10 years, an average of 1,196 new species of fungi were described in each year [2]. Among these newly described species, a large number of fungi (74.4%, 8,895 species) were described based on only morphological characters without DNA sequence data from 1999 to 2009 [2]. At this time, fungal morphology and microscopy still have practical and scientific value in studying fungi. Microscopic methods are widely employed in fungal research and commercial laboratories, which serve the indoor mold as well as the indoor air quality industries.

Fungi are very diverse in their morphology. Thus, microscopic observation of their reproductive structures, such as spores and their arrangement, conidiophores, conidiogenesis, acervuli, picnidia, ascomata, basidiomata, and sporangiophores, is very important in the classification of fungi and in the diagnosis of various infections of plants, human beings, and animals.

In this chapter we mainly discuss optical microscopy, which is commonly used in observation and analysis of fungi. Other microscopes used in fungal research, such as scanning electronic microscope (SEM), transmission electronic microscope (TEM), and confocal microscope, and so on, are not covered in this chapter. A number of reference books can be consulted for techniques for using these microscopes [4, 5].

Procedures and Techniques for Microscopic Studies of Fungi

Specimen Preparation

Pretreatment

When dried herbarium specimens are used for study, it is necessary to rehydrate these specimens by placing a drop of water directly on the fruiting bodies of Ceolomycetes or Ascomycetes for a few minutes [6]. Alternatively, one may place a dried specimen on sterile wet filter paper in a Petri dish or a moisture chamber and rehydrate it overnight at room temperature. Another simple pretreatment method is to add a drop of 5–10%

potassium hydroxide (KOH) solution to rehydrate, soften, and clear fungal material prior to staining. The pretreatment will make superficial fruiting bodies of fungi much easier to be removed from substrates or specimens. It will also make fruiting bodies embedded in plant materials much easier to section.

When sporulation is not present or fruiting bodies are not mature in/on fresh plant materials, the fresh specimen can be pretreated by placing it on sterile wet filter paper in a Petri dish and incubating it for two to several days at room temperature to assist sporulation or maturation of fruiting bodies.

For fungi already growing on artificial media and fresh, living materials with sporulating fungal structures, such pretreatment is not necessary.

Sectioning

Freehand Sectioning

Freehand sectioning of fungi growing in plant tissues often provides an adequate method for rapid and inexpensive microscopic observation of their structures, if done properly. This very simple technique often leads to high-quality images for publication.

Equipment and Materials

- A good light source, desk lamp or a reading light.
- A comfortable position for yourself, e.g., sitting in front of a bench or desk.
- A number of *double-edged razor blades*, not single-edged ones.
- A small Petri dish filled with water.
- A beaker filled with water.
- A small paintbrush (#1).
- The material to be sectioned.

Procedure

Hold the fungal material to be sectioned between the thumb and index finger of one hand; a razor blade held by the other hand is drawn across the material with the edge towards the operator. The razor slides on the index finger of the hand holding the fungal material. Wet the upper surface of the blade edge with water before sectioning. As the sections are cut, they should slide into water and float in it. The sections in the water are removed with a small brush onto a glass slide for observation under a microscope. Freehand sectioning needs to be practiced a number of times before satisfactory sections (smooth, even, and thin) can be cut.

If the fungal material is small or soft, it should be placed in a small Styrofoam block or chip made out of packing material with a slit or pith cut on the top to hold the fungal material or use two pieces tied together with an elastic band or a piece of tape to hold the fungal material. The holding material should be able to hold the fungal material firmly, but not too rigid and hard to crush the fungal material.

A double-edged razor blade is crucial for successful freehand sectioning. The low angle of the edges of a double-edged razor blade allows one to cut subject materials into thin sections. The blade will become dull by slicing the materials several times. Thus, it is necessary to move to an unused portion of the edge of the blade after cutting a few sections. When full lengths of both edges of the blade are used, the blade is no longer useable.

To practice freehand sectioning, it is better to start with herbaceous plant materials. The stem should be <5 mm in diameter and relatively soft. Trim the materials to the size and length that are easy to handle in your hand.

Sectioning with a Freezing Microtome

A number of brands/designs of microtomes are available for fungal research, such as sled microtome, rotary microtome, cryomicrotome (freezing microtome), ultramicrotome, vibrating microtome, saw microtome, and laser microtome.

The freezing microtome is commonly used. Nag Raj [7] described a freestanding freezing microtome (a modified Reichert freezing microtome) in detail in his monumental book *Coelomycetous Anamorphs with Appendages-Bearing Conidia*. Sections are cut to 10–20 μm thick and transferred to water from the cutting edge of the blade with a fine needle or a fine camel or goat hair brush [8, 9].

For information about operating a microtome and preparation of samples, consult the manufacturer's manual.

Sample Mounting

Tapelift Mounts

To tapelift is to use a small piece of 2×2 to 5×5 mm clear/transparent tape to gently touch a colony so that fungal structures are removed from the colony by the sticky side of the tape with a pair of tweezers. The tools needed for tapelift are a roll of clear tape mounted on a tape dispenser, a pair of tweezers, and a pair of scissors.

Procedure

1. Examine the Petri dish under a stereo microscope to locate the area or colony of interest.
2. Flame tweezers and scissors.
3. Cut off a piece of tape with desired size with the scissors and tweezers.
4. Place the sticky side of the tape on the colony with the tweezers and press it against the colony gently.
5. Move the tape out of the Petri dish and flop its sticky side up.
6. Place it on a slide.
7. Add a drop of 70% alcohol over it and let it evaporate for 2–3 min.
8. Add a drop of water, lactic acid, or mounting agent of choice with staining agent.
9. Place a coverslip on it.
10. Place it under a compound microscope.
11. Take photomicrographs, if necessary.

It is a rather simple, convenient, and efficient method to prepare wet mounts for microscopic observation with a reduced disturbance to fungal structures. Many commercial laboratories are routinely using this method. However, the disadvantage of this method is that the tape itself reduces the resolution and clarity of the field and results in reduced quality of images. By choosing the proper tape and careful preparation, this method still can produce quality photomicrographs, which are high enough for peer-reviewed publications.

The quality of tape is the foremost important factor for the quality of the wet mount. A number of brands of clear, transparent, ultra-transparent, or super clear tapes (12.7–19 mm wide) are available on the market. Titan Ultra Clear Tape (59212 95882) is considered the best one (Seifert, personal communication; 2010); unfortunately, this brand is not available in the United States. The author was able to purchase Titan Ultra Clear Packing Tape (5 cm wide), which yielded satisfactory, but not super, results. Among the different types of tape the author has tested with water, 85% lactic acid, or 0.1% lacto-fuchsin, the best one is Moore Crystal Clear Tape (No. CCD-134-C, Moore Push Pin Co., Wyndmoor, PA). The fresh mounts provide rather good clarity and resolution. The observation should be finished in one half to one hour. This tape will shift the color to the slight bluish side. White balance can easily correct its color when color photomicrographs are taken. This tape is not available in stores, but it can be ordered online. It is sold for torn book and blueprint repair or restoration. Mainstays Crystal Clear Tape (imported by Walmart Canada, Mississauga, Ontario) is rather satisfactory in resolution and clarity. Scotch Transparent Tapes 600 (12.7 and 19 mm wide) (3 M Co., St. Paul, MN) did not produce photomicrographs suitable for publication owing to the mosaic background induced by its uneven glue coating. This tape is not recommended. Try several brands yourself with the mounting media of your choice and choose one which gives the most satisfactory results.

Scotch Crystal Clear Tape (Ultra Clear-Permanent) has not been tested. Packing clear tapes (2.5–3.75 cm wide) sold in post offices are much thicker than other tapes. Their thickness deteriorates resolution significantly, and they are not suitable for microscopic fungal studies.

Wet Mounts Using a Needle or a Cutting Tool

The other method to prepare a wet mount for microscopic observation is to use a needle (an inoculating needle, an insect pin no. 4, a sewing needle, a miniature scalpel, or a custom-made cutting tool with a small chisel point, etc.) to cut a small piece of colony including fungal reproductive structures (conidiophores or ascomata) [10].

The points of needles or cutting edges of scalpel should be so sharp so as to cut the colony without too much disturbance to fungal structures. The area to be cut on a colony should have mature conidiophores or ascomata, but not excessive conidia. It is helpful to use a stereo microscope to examine the colony and determine which area to cut first for microscopic observation. It is rather difficult to wash excessive conidia from a cut colony sample on a slide. When a piece of colony sample with excessive conidia is picked, it would be better to prepare another one. Float the cut colony sample from the needle tip onto a slide with a drop of 70% alcohol. When the alcohol evaporates, apply a drop of lactic acid or a mounting medium of choice, with or without staining agent, depending on the fungus and the optics to be used for observation. For example, it is not necessary to use a staining agent for dematiaceous hyphomycetes; water or lactic acid would be adequate.

Squash or Tease Mounts

This technique is useful for observing conidiophores and conidiogenesis in conidiomata (avervuli or picnidia) and arrangements of asci and ascospores in ascomata as well as the tissue structures of these fruiting bodies. Often a wet mount is prepared without sectioning for a quick identification.

Under a stereo microscope, one or two superficial conidiomata or ascomata can be removed from substrates with a sharp pin without difficulty. Dip the tip of the pin in water so that a droplet of water at the tip of the pin can be used to retrieve and transfer conidiomata or ascomata to a drop of water on a slide. Place a coverslip over the conidiomata or ascomata on the slide. Press the coverslip gently with a needle or a rubber eraser and move it sideways to obtain a reasonably good squash mount of well-separated conidiogenous cells or asci.

Excessive plant material attaching to fungal fruiting bodies prevents making clean and clear slides for observation. Rehydrating dried specimens often assists in the separation of fungal fruiting bodies from plant materials.

Teasing the tissues of the conidiomata or ascomata with two fine insect pins will produce similar, but frequently inferior results. When teasing is used to remove excessive plant tissues from the fungal fruiting bodies, it may improve the quality of the mounts.

Staining and Mounting

Whether to use a staining agent and what kind of staining agent to be used is a decision each mycologist must make based on the fungus under study, the key diagnostic characters of the fungus to be observed, and microscopy to be used.

It may require application of a staining agent to observe colorless, transparent, or lightly colored fungi or fungal structures and improve the contrast and differentiation of the fungal structures in the samples to be examined. For a pigmented fungus, a mounting medium providing a neutral background without altering the color or the chemical composition of the specimen is sufficient and satisfactory for microscopic observation, such as water and lactic acid.

The function of a proper mounting medium is to increase contrast by providing a refractive index higher than that of a glass slide and by providing a refractive index close to that of the specimen/sample to achieve optimal transparency. A mounting medium chemically compatible with the specimen is preferable. A stain or a mounting medium should retain its optimal condition, not drying out during microscopic study or storage. Many stains and mounting media are available for fungal microscopic studies, water, KOH, lactic acid, lacto-fuchsin, lacto-cotton blue, modified lacto-cotton blue (MLCB), and lacto-phenol blue are most commonly or routinely used by mycologists and lab technicians (see Note 1). A number of publications provide detailed information on the stains and mounting media [11–13]. For information on special agents, such as nuclear staining agents, and the stains to differentiate fungi in animal tissues, see references [11, 13, 14].

It should be pointed out that some mycologists employ a special staining agent, mounting media, or the technique they developed for their microscopic studies on certain taxonomic groups. It is necessary to check their publications or monographs for the detailed information on how they prepare their specimens for their studies. It often saves time by following these methods.

Malloch [15] indicated that some fungi are hydrophobic and difficult to "wet," even with a wetting agent. He recommended adding a drop or two of 95% ethanol to the fungal material for a few seconds. Before the alcohol is completely evaporated, adding a drop of a mounting medium of your choice will improve the quality of wet mounts.

Some staining media widely used in the past contain lacto-phenol, such as lacto-phenol blue [11]. Lacto-phenol cotton blue solution is used in some research [16]. These staining agents are no longer recommended because of health concerns for phenol, which is readily absorbed through the skin and known to result in some detrimental health effects from long-term exposure [13, 17]. Thus, these staining agents are not covered in this chapter.

Microscopy: Principles and Application

Optical microscope is an indispensable instrument to a mycologist. Microscopes really extend the vision of mycologists and biologists into a microcosmic world to study fungi microscopically. Microscopy is using microscopes to observe samples, specimens, or structures that are invisible to the unaided eye.

Both stereo (dissecting) and compound microscopes are necessary instruments for fungal observation and analyses.

The objective of the microscopy is to produce a quality image showing more details of the specimen within its designed capacity.

Stereo Microscopes
The stereo microscope, also known as a dissecting microscope, is an optical microscope variant designed for low-magnification observation of the surface of a sample/specimen using incident light illumination rather than transillumination with magnification range of ×5 to ×60. It employs

two separate optical paths, with two objectives and two eyepieces providing slightly different viewing angles to the left and right eyes so as to produce a three-dimensional view of the sample/specimen under the microscope.

There are two major types of magnification systems in stereo microscopes: fixed magnification and zoom magnification systems. In the former one primary magnification is achieved by a pair of objective lenses with different discrete magnifications. The zoom magnification is able to continuously change magnification within a designed range of ×5 to ×60. Recently, newer versions of stereo microscopes have been improved with motor controlled zooms integrated with digital imaging systems.

The observation of fungal materials under stereo microscope is essential before a compound microscope is used for microscopic examination [18]. The significance of using a stereo microscope is often not fully appreciated. A number of morphological characters of fruiting bodies, conidiophores, and conidial orientations on the fungal colonies on natural substrates or culture media are diagnostic under a stereo microscope. For instance, the orientation of conidial chains and the conidial mass shapes of *Aspergillus* and *Penicillium* are diagnostic features for identifying some fungi of these genera to species [10, 19]. The morphological features of fruiting bodies are critical for identifying most fungi to genera or to species, such as *Alternaria*, *Aspergillus*, *Chaetomium*, *Penicillium*, and powdery mildews. These characteristics are undisturbed for observation under a stereo microscope, but can be lost or at least altered due to disturbance in a wet mount preparation. The patterns of sporulation and conidial chains are very important characters to differentiate species of *Alternaria* [20, 21]. The stereo microscope is used to observe and locate fungal materials on the plant tissues or on the culture media. Removal of conidiomata and ascomata from plant materials or colonies with a needle or a piece of tape are conducted under a stereo microscope with an increased change of success of picking right fungal materials to prepare wet mounts for observation under a compound microscope. It is often used to prepare the

samples or subsamples for further observation under a compound microscope.

It is relatively easy to use a stereo microscope. Observation starts from low power and proceeds to high power. Stereo microscopes are normally equipped with an internal light source projecting light from upper back and bottom. For better quality of photomicrography, an external fiber optic light source is sometimes necessary to enhance the characters with better lighting from an optimal angle. Occasionally, simultaneous lighting from above and below is necessary to observe certain fungal structures, such as conidial orientation from a culture.

Compound Microscopes

A compound microscope is a microscope that uses multiple lenses (4–6 objectives) to collect light from the sample and then a pair of eyepiece lenses to focus the light into the eye or camera with a magnification range of ×40 to ×2000.

Slayter [22] indicated that the quality of the image produced by a compound microscope depends on the magnification, the resolution of the microscope, and the contrast produced in the image.

Microscope objectives are characterized by two parameters: magnification and numerical aperture [23]. Magnification is achieved by objectives (object lenses) and eyepieces (ocular lenses) chosen for your microscope with a magnifying range of ×4 to ×100. For most compound microscopes, a set of objectives ranges from four to six lenses. The objective lenses are available with magnifications of ×4, ×10, ×20, ×40, ×60, ×100 (oil immersion lens). Eyepieces of ×10, ×15, and ×20 are available. A pair of ×10 eyepieces is most commonly used. The total magnification = objective × eyepiece.

Numerical aperture ranges from 0.10 to 1.25, corresponding to focal lengths of approximately 40–2 mm, respectively. Numerical apertures can be achieved with oil immersion as high as 1.6, at which the highest resolution can be reached.

The resolution of an optical microscope is the shortest distance between two points on a specimen that can be distinguished as separate entities [24]. The resolving power of a microscope is the

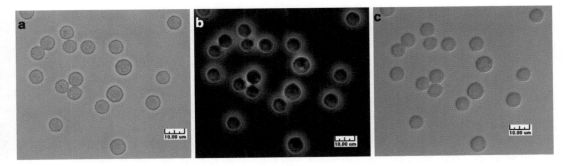

Fig. 7.1 Spores of *Arcyria* sp. captured in the same viewing field with bright field (**a**), phase contrast (**b**), and differential interference contrast (**c**)

most important technical specification or feature of the optical system and determines the ability to differentiate between fine details of a specimen. The maximal resolution for an optical compound is 0.2 μm (200 nm) [25]. The wavelengths of visible spectrum of light are from 390 to 750 nm. The resolution of an image is limited by the wavelength of light used to illuminate the sample. When objects in the specimen are much smaller than the wavelength of the light, they do not interfere with the waves so that these objects are indistinguishable.

Using a microscope with a more powerful magnification does not increase its resolution beyond 0.2 μm; it just enlarges the image. However, objects closer than 0.2 μm will become blurry and only be observed as an indistinguishable entity.

Contrast is the difference in visual properties that makes an object distinguishable from other objects and the background. Contrast is determined by the difference in the color and brightness of the object and other objects within the same field of view under a microscope.

The contrast of a fungal sample is determined by several factors: type of specimen, mounting medium, degree of optical aberration, proper setting of aperture diaphragms, contrast mechanism employed, and characteristics of the detector. Proper usage of contrast can improve the quality of photomicrographs.

Illumination Techniques

Three illumination techniques—bright field, phase contrast, and differential interference contrast (DIC)—commonly used in microscopic studies of fungi (Fig. 7.1).

Figure 7.1 illustrates the differences of spores of *Arcyria* sp. in three images captured in the same viewing field under differing contrast illumination techniques: bright field, phase contrast, and DIC.

Bright Field Microscopy

Bright field microscopy is the simplest and the most basic illumination technique (see Fig. 7.1a). This illumination technique uses transmitted visible light. The advantages of the technique are the simplicity and the minimal sample preparation required. Its limitations are the low contrast of most biological samples and the low resolution due to the blur of out of focus material. It often has difficulty to observe colorless or transparent fungal spores and other structures clearly due to low contrast, especially septation without aid of staining.

Despite its limitations, the majority of the laboratories providing services to Indoor Air Quality and Indoor Mold Industry use bright field illumination according to the survey conducted by the task force of AIHA (unpublished data), but the colorless specimens/spores are routinely stained to increase the contrast in these labs.

For a research laboratory, the choice of illumination techniques relies on the quality to show the characteristic of fungal structure to be observed.

Phase contrast and differential interference techniques make observation of unstained biological materials possible.

Phase Contrast

Phase contrast is a widely used technique showing differences in refractive index as difference in contrast [26]. It allows observing colorless or transparent biological materials or certain fungal structures, such as septation in colorless spores without the aid of staining agents. It was developed by Frits Zernike, a Dutch physicist, in the 1930s during his study of diffraction gratings. He was awarded the Nobel Prize in 1953 for his invention of the phase contrast microscope [27].

It necessitates a special set of objectives with phase rings in the back focal plane of the objective. A matching set of phase rings are placed in the condenser and these rings are marked for their setting position. The phase objective is required to match the specified phase setting on the turret condenser to enable best observation possible. The major limitations of phase contrast technique are halo artifacts around objects and the requirement of thin specimen preparations (see Fig. 7.1b) [28].

Differential Interference Contrast

DIC (also known as Nomarski interference contrast) was created and introduced to microscopy by George Nomarski, a Polish physicist and optics theoretician in 1950s [29]. DIC is an excellent technique, but much more expensive for rendering contrast in colorless or transparent specimens. Contrast is very good and the condenser aperture can be used fully open, thereby reducing the depth of field and maximizing resolution. DIC uses a special prism (Nomarski prism, Wollaston prism) in the front focal plane of the condenser and one in the rear focal plane of the objective [30]. It also needs a polarized light source which is generated by two polarizing filters installed in the light path, one below the condenser (the polarizer), and the other above the objective (the analyzer). This technique is widely used to study live biological specimens and unstained tissues. Under DIC technique, biological specimens including fungi show a pseudo-color that deviates from their natural color.

The technique produces a three-dimensional effect of the specimen's surface which is actually a monochromatic shadow-cast image showing the gradient of optical paths for both high and low spatial frequencies in the specimen (see Fig. 7.1c) [31]. A DIC turret condenser usually has a bright field illumination setting as well as the DIC setting to match each objective. A DIC turret condenser in some models has a bright field and a set of phase illumination settings. DIC produces much higher resolution images than phase contrast and bright field.

Compound Microscope Calibration, Adjustment, Use, and Care

All compound microscopes should be equipped with an ocular micrometer or micro-grid for measuring fungal structures during microscopic observation, if necessary. The ocular micrometer must first be calibrated with each objective before it can be used. Each microscope must be individually calibrated routinely at least once a year. For the same magnification on the microscopes of different brand, or even different models of the same brand, the measurements by micrometer with the lenses may vary slightly. The ocular micrometer in the eyepiece is measured against a stage micrometer or a graduated slide with each objective. The length of each division of the ocular micrometer is calculated, recorded, and documented for each objective. The calibration result for each objective should be recorded and taped on the microscope for quick reference.

A microscope should be used properly. It involves proper alignment and setting-up. The first thing is to understand your microscope and to know the features and functions as well as the limits of the microscope. The aperture should be set based on the objective lens to be used. For instance, when using a ×60 objective lens, the aperture should be closed down to the 0.3 position on a bright field microscope to achieve the optimal contrast. A wide-open aperture will result in low contrast and wash out the details in the image against an over bright background. The field of view should be centered. Use your microscope with care. Start with your lens on the lowest magnification to locate the area that you want to observe and move to higher magnification lens in a sequence. All objectives are parfocal and parcentered by design to minimize refocusing, when switching lenses. Understand the depth of focus of each objective on your microscope. It determines your workable space associated your specimen and objective. Never crush your lens into a slide when you focus or switch objectives. Such an action will scratch the lens. A dirty lens should be cleaned with lens paper and lens cleaning solution. Immersion oil should be cleaned from the ×100 objective, when observation with the oil lens is complete.

Eyepieces on a microscope should be adjusted individually so that both eyes of an observer will focus on the same plane (parfocal). If two eyes are focused on different focus planes, it is easy to lead to eye fatigue. To some individuals it may result in dizziness or headache. Make yourself comfortable ergonomically by adjusting the position and height of your microscope and chair to avoid ergonomic fatigue or injury. A chair that is adjustable in height and with back support is preferred.

Observation and Measurement of Fungi under Light Microscope

Microscopic observation can be used in many areas of mycological studies, such as fungal taxonomy, identification, fungal development/biology, pathogen and host relationship, etc. Two factors, proper preparation of samples (slide) and proper usage of microscope determine whether a microscopic study can obtain the data necessary for your objectives. Take detailed notes when you make your observations, especially a key or diagnostic character or a detail that is not able to be captured by photomicrography with sufficient quality. All photos should be annotated in your notebook including the digital filing information

where the photo(s) is saved. All photomicrographs should be annotated with scale bars. If your system does not have the feature to insert a scale bar to a photo, the magnification or all magnifying factors should be recorded in details. Always save your photos to a folder with a clear and meaningful title so that you can retrieve them later. How often we experience frustration by the difficulties of retrieving a photo due to hazy memories!

When it is necessary to measure fungal structures during microscopic observation, it is recommended to measure enough numbers of a structure under study. Scan the whole sample first and make sure to measure individuals that represent the full range of a population. Randomly choose the individual subject to measure. Avoid any bias in the process of measurement. Make 30 measurements for each structure, if possible, so that a statistical data analysis can be conducted later. Keep in mind that the best resolution of a light microscope is 0.2 μm. Accuracy of a measurement <0.2 μm is meaningless. This is the reason why for new taxon description, all measurements are rounded up to closest 0.5 μm or whole number in some publications.

Photomicrographic System and Usage

Drent [32] indicated that digital photomicrographic systems are dominant in image capturing for microscopy which demands high resolution, retaining true color and proper usage of limited light conditions. Digital cameras can capture images with quality comparable to traditional film photography. It is the norm to use digital photomicrographs in academic papers for publishing in peer-reviewed journals including mycological journals. Photographic software programs make editing and post image-capture process a much easier operation. Making a photoplate is no longer a time-consuming and laborious process. A simple plate with four to six images can be done within an hour. Digital images can be captured, stored, and retrieved much more easily than traditional photomicrography with 35-mm film. The photomicrograph system is either mounted on the trinocular or the ocular tube. The ocular setup is more economical.

Color shift under a microscope can be corrected by white balance or by a neutral density filter slide [33].

We have to be careful to edit the images that are for scientific research. Adjusting contrast and brightness, correcting shifted color, and cropping out excessive edges are acceptable, but any Photoshop alterations to the contents of images for aesthetic and other reasons are questionable such as removing and replacing background. Such an action may compromise the integrity of scientific research.

Capturing a good image with a high resolution, a proper contrast, and true color relies on two factors or skills: (1) using the photographic system properly, and (2) preparing a good wet mount with your research fungal material. Read the manual of your photographic system and become familiar with the functions and features of your gear prior to operation. You will avoid unnecessary frustration and improve your photomicrographic skill. You may enjoy it during the process.

TIFF, GIF, and JPG are the common file formats used for taking photos, especially JPG. These formats are able to provide the quality necessary for publication without any problem. The key requirement for publication is ≥ 300 dpi. For a photo to be published at a print size of 4 by 4 in., the photo digital size is 1.44 mb at 300 dpi resolution. Medium quality setting for a camera with photo size 0.5–1.5 mb is adequate. A quality photo is determined by being in focus and having good resolution, proper contrast, nice clarity, and neutral color. Size of a photomicrograph does not need to be over 2 mb in most cases. RAW is not necessary, unless the photomicrograph needs to be enlarged to poster size, or you need the quality of professional photography. Working with large size photos will significantly slow down your computer and your work progress or you would need to acquire a very powerful personal computer with high quality and speed. Storage and digital filing for larger photos could be another issue, as they will fill up your hard drive space rather quickly.

Microscopy for Airborne Fungal Spore Analysis

Analysis of airborne fungal spores provides important information to the indoor and airborne fungi industry and IAQ professionals to help their indoor mold investigation. Both qualification and enumeration of fungal spores are included in the analysis. Analysis of airborne fungal spores is often referred to as "spore count." It is covered here as a separate section to discuss its procedures.

Lab materials for spore counts are similar to those described in previous sections. No other special material is needed.

For spore count analysis a compound microscope is commonly equipped with $\times 20, \times 40, \times 60$, and $\times 100$ objectives [34, 35]. Objectives of $\times 4$ and $\times 10$ are rarely chosen for a microscope for spore count. A $\times 20$ lens can serve for locating sample trace, scanning the whole trace, and identifying large spores and hydrophilic fungal spores, such as *Stachybotrys*, *Memnoniella*, *Chaetomium*, *Alternaria*, and *Epicoccums*. Objectives of $\times 40$ or $\times 60$ are used for overall spore enumeration and identification [36–39]. It is the preference of a lab or an analyst to decide which one to use. More and more labs tend to use a $\times 60$ lens for better observation of details of fungal structures and not significantly reduced viewing area and depth of field. The $\times 100$ objective is occasionally used to observe fine, diagnostic characters or structures on small spores (<6 μm). A number of research labs used a $\times 100$ objective for their research [39–42]. Such a practice is not economical for a commercial lab because of the time restraints.

The Environmental Microbiology Proficiency Analytical Testing task force of American Industrial Hygienist Association conducted a survey several years ago. The survey found that the majority of accredited environmental laboratories conduct fungal spore counts under magnifications of $\times 200$ to $\times 600$ and a few laboratories use $\times 1000$.

For spore count the ocular micrometer is used to measure the size of fungal spores and the dimensions of sample traces. It can also be used as a guidepost during the analysis of the sample.

Fungal Spore Identification

It is not difficult to identify common fungal spores to genus level, such as *Alternaria*, *Cladosporium*, *Pithomyces*, *Epicoccum*, *Ganoderma*, *Chaetomium*, *Fusarium*, *Curvularia*, *Torula*, *Nigrospora*, *Spegazzinia*, etc. It is a challenge to identify some spores to genus level. This is the reason why the *Aspergillus*/*Penicilium*-like group is used to accommodate colorless, ellipsoid, ovoid, or subglobose spores that often belong to more than a half dozen of different genera, *Aspergillus*, *Pacilomyces*, *Penicilium*, *Acremonium*, *Geosmithia*, and *Merimbla*. For the spores of Basidiomycota and Ascomycota are often identified as basidiospores and ascospores. Among these spores, a few may be able to be identified to genus, such as *Ganoderma* and *Leptosphaeria*. There is no doubt that it is necessary to conduct more research on fungal spore morphology and their taxonomic values for fungal spore identification. Possession of a good collection of mycological literature is vital to fungal spore identification.

Procedures to Identify and Quantify Spore Traps

Ideally, 100% of a sample trace should be analyzed. Analysis of spore count samples cover the traces ranging from 25% to 100%. Only a small number of labs analyze their samples by covering 100% of the trace. A majority analyzes a portion of the sample by subsampling in the trace due to time or economic restraints. For a normal spore load, it is minimal to analyze 25% of the trace. Occasionally the analyzed area might be reduced to <25%, if a spore load is extraordinarily high (>800 spores/sample) and has a relatively even spore distribution. Be extremely cautious when <25% of the trace is analyzed. This practice is allowed in some commercial labs with quality-control measures in place. For research this practice should be avoided.

A number of methods are available for examining spore trap samples, such as transverse pass, and longitudinal pass methods, zigzag field, and random fields (Fig. 7.2). These methods

illustrated in Fig. 7.2 are straightforward. The methods have two different approaches: using the entire field of view or using a micrometer to guide passes and the counting process. The micrometer method has its obvious advantage. Because the view area is much narrower than the width of the field of view, the analyst does not need to turn his/her eyes to read the entire field of view. It is much easier to focus on the area within the micrometer. This method normally imposes less occupational stress on eyes [43].

For pass methods when spores/conidia touch the edge of the micrometer that is guiding the counting pass, always count those that touch the left edge of the micrometer and exclude any that touch the right edge (Fig. 7.3) [43].

For random and zigzag field methods, randomly scan across the trace (3–5 fields per pass) in a zigzag pattern. The number of fields counted should be equivalent to 25% of the trace area or more. Count all spores within the entire field of view. Try to scatter fields randomly over the entire trace. For spores present on the perimeter of the field, count only the ones attaching or on the perimeter on the left-hand side (Fig. 7.4), not the ones on the right-hand side. Move the field to the left so as to reveal the entire spore for a more precise identification. This should be done when all spores in the field of view are analyzed except for the ones on the left edge. Thus, the spores with the partial view can be identified.

Collecting Reference Data and Calculating Trace and Coverage Area

- With a calibrated microscope, the full length of the ocular micrometer under ×40, ×60, and ×100 lenses can be calculated by multiplying by 100. The length of the objective you choose to use will be the width of passes covering through the samples, if pass method is used. If field-of-view method is used, the diameters of field of view should be measured.

- Obtain dimensions of the adhesive band for the sample deposit for each sample type from the manuals or from the manufacturers. For example, the trace size of Air-O-Cell is 1.055 mm ×14.4 mm (15.19 mm^2); Burkard samplers, 14 mm ×2 mm (28 mm^2); and Allergenco samplers, 14.5 mm ×1.1 mm (16 mm^2).

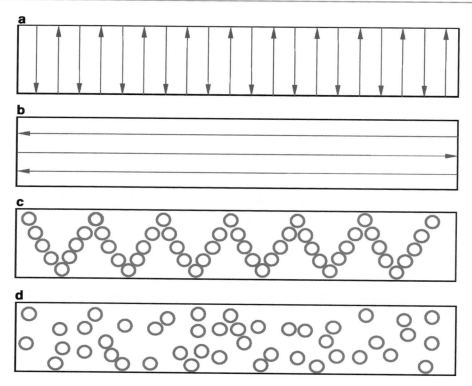

Fig. 7.2 Fungal sample trace reading methods. (**a**) Transverse passes. (**b**) Longitudinal passes. (**c**) Zigzag fields. (**d**) Random fields

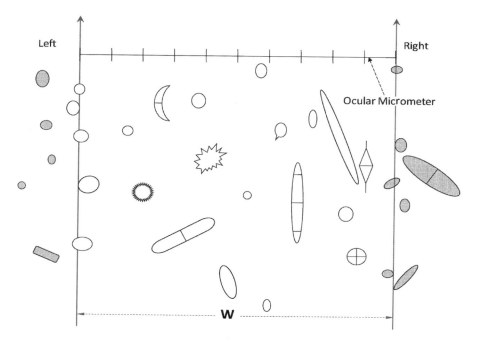

Fig. 7.3 Spore-counting criteria for the pass methods. Only count the spores touching the left boundary line of the micrometer pass. Do not count the ones touching the right boundary line. Two vertical solid arrows indicate the direction of a pass guided by the ocular micrometer. W is the width of the micrometer and counting pass. The unshaded spores are counted, and the shaded spores are not counted

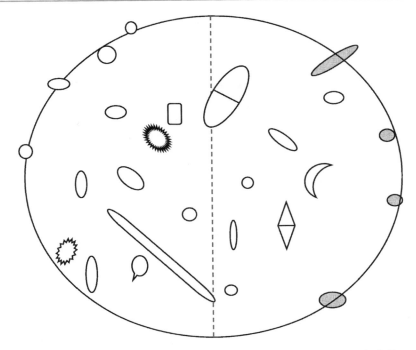

Fig. 7.4 Spore-counting criteria of random and zigzag field methods. The spores (*unshaded*) on the left half of the perimeter of the field of view and other unshaded spores are counted, and those (*shaded*) on or touching the right half are not counted

- Use the data collected in steps 1 and 2 to calculate the number of passes necessary to cover the entire trace for using pass method. The diameter of the field of view is used to calculate the number of fields needed to cover the entire trace, if using the field-of-view method. Using these results, the numbers of passes or field-of-views necessary to cover a portion of trace area, such as 30%, can be calculated.

Procedure of Analyzing Samples Collected with Air-O-Cell Cassettes or Similar Devices

Sample Preparation

- Check the sample label against the Chain of Custody sheet.
- Fill up the data sheet with client information, sample information (number, time, location, air volume, samplers).
- Remove or cut the seal.
- Pry the cassette open with a weighing spatula or a penny.

- Mark the trace with a permanent marker.
- Apply a little bit of clear nail polish to each of four corners of the sample.
- Remove the coverslip with a pair of tweezers and turn it over.
- Place the sample on a slide with trace facing up and allow the nail polish to anchor the sample on the slide.
- Add a drop of staining agent on the sample.
- Place a 22 × 22 mm coverslip on the sample.
- Write down the lab ID number and the sample ID number on the right-hand side of the slide.

Sample Analysis

- Place the prepared slide on the stage of the microscope.
- Scan the whole trace with the ×20 objective. If *Stachybotrys*, *Memnoniella*, and *Chaetomium* spores are found during scanning, start the first pass/field-of-view from that location. When the first pass/field-of-view is completed, move to the end of the slide and start the 2nd pass/field-of-view.

- Move the slide to the marking place as the starting point.
- Switch to ×40 or ×60 objective.
- Start to count and identify spores with a multiple channel counter following the reading methods you have chosen. For less common spores that are not listed on the counter, mark each of the spores on the data sheet directly. Use a micrometer to guide the passes or use field-of-view to count spores.
- Write down the location with two stage rulers when a spore has to be measured. Move the spore to the center of the field to measure. Once the measurement is complete, use the coordinates to return the location where

counting stopped to resume the counting process.
- Take notes on the information of spores in clumps, presence of conidiophores, hypha/mycelium, dust load, animal hair, arthropod parts, pollen grains, etc.
- When the other end of the trace is reached, write down the numbers of each fungal taxon counted on the data sheet.
- Enter the data to a computer and calculate the concentrations of airborne fungi.
- Generate a lab report, when analyses of all samples from the same project are complete.

The equation for calculating spore concentration is as follows:

$$\text{Spores} / m^3 = \text{total spores} \times (\text{number of fields in the whole trace/number of fields counted}) \times (1{,}000 \, L / (\text{air flow rate} \times \text{sampling time})).$$

Procedure for Analyzing Samples Collected with Allergenco Sampler or Similar Devices

This sampler can collect twelve discrete samples on the same slide.

Sample Preparation
- Remove sample slide from the slide box.
- Check the sample label against the Chain of Custody sheet and make sure the information matches. Check the number of samples on the slide and the sequence of the sample collection. Mark the sample which is the first one collected.
- Fill up the data sheet with client information, sample information (number, time, location, air volume, samplers).
- Mark one end of all traces on the slide with a permanent marker
- Add two drops of staining agent on the sample slide, if 12 samples were collected on the same slide.
- Place a 40×22 mm coverslip on the sample. For ≤6 traces on a slide, a 22×22 coverslip will be used.

- Write down the lab ID number and the sample ID number on the right-hand side of the slide.

Sample Analysis
- Same as the samples for Air-O-Cell. Choose the longitudinal pass method to analyze multiple samples on the same slide.

Procedure of Analyzing Samples Collected with Burkard Sampler or Similar Devices

The Burkard 7-day Recording Volumetric Spore Sampler (Burkard Manufacturing Co., Rickmansworth, UK or Burkard Scientific Uxbridge, Middlesex, UK) collects airborne samples on Melinex tape coated with adhesives. The tape is attached to a slow rotating drum. The drum rotates 2 mm/h. The flow rate of this sampler is 10 L/min. This is a great sampler for outdoor aeromycological studies. It is too bulky for use indoors.

Sample Preparation
- Mark the beginning end of the Melinex tape.
- Remove the entire tape from the drum carefully with a pair of fine forceps.

- Place the tape on the cutting block (or a glass plate).
- Cut the tape with a sharp razor blade or a surgery scalpel into seven pieces. Each piece covers 24 h of sampling (48 mm in length).
- Prepare the slides by writing the date, starting time of the samples (sampler # if using more than one trap) onto the slides with a permanent marker.
- Lay a bead of 10% Gelvatol in distilled water, about 45 mm long, down the center of the slide. This works well as an adhesive to hold the tape on the slide. The exposed segment of tape can then be rolled onto the Gelvatol, using fine pointed forceps, taking care not to get bubbles under the tape. Once in place, the tape can be readjusted to make sure it is adhered perfectly straight. The start of the tape should consistently be placed at the same end of the slide.
- Wait 24 h to allow the Gelvatol to dry. At this stage, the slides can be stored for later mounting and analysis.
- Mounting the slides. Place 5–7 drops of melted glycerin jelly (in a 50°C hot water bath) on the exposed tape and carefully cover the tape with a 22×50 mm coverslip. Warming the slides on a slide warmer at 48°C. Press the coverslip gently to spread the glycerin jelly evenly.

Sample Analysis

- Put a sample slide under a microscope.
- Identify and count fungal spores using one single longitudinal traverse, or 12 bi-hourly transverse traverses depending on the objectives of the investigation or research.
- Remaining steps are the same steps as for Air-O-Cell. Choose longitudinal pass method to analyze multiple samples on the same slide.

Procedure of Analyzing Samples Collected with 25-mm Mixed Cellulose Ester (MCE) Filter Cassette

- Remove the MCE filter from the cassette with the filter forceps and mount onto a slide.
- Apply a clearing agent (such as triacetin) containing a dye to the filter. This process

should be conducted in a clean area (i.e., in a hood or a clean Petri dish) to avoid contamination.
- Wait approximately 10–30 min for the clearing process.
- Add a mounting agent.
- Place a 25×25 mm coverslip gently over the sample.
- Count 25–50% or a fixed area of the filter. Or draw a fine cross on the back of the slide and randomly choose one quarter of the sample to analyze.

For 37-mm filter cassettes, they are too big to mount in a slide directly. Cut the filter into four equal pieces with an alcohol-wiped scalpel blade by cutting a cross over it. Randomly choose one piece and mount it on a slide for clearing and staining. The rest of the steps are the same as for previous samples.

Notes

Commonly Used Mounting and Staining Agents/Reagents

Water

Water is routinely used as a mounting medium in microscopic observation/examination in mycological research. It serves two purposes: (1) rehydration of the fungal samples/specimens [18], if necessary and (2) retention of the natural color of the fungus and its structures [7, 44]. When observing living and fresh materials, application of water as a mounting agent will retain fungi's natural shape, size, and color of the structures so that these characters can be accurately recorded. It is essential to mount fungal materials in water to observe mucoid, noncellular appendages or mucoid layer on ascospores. These glutinous structures and noncellular appendages are invisible in lactic acid. The disadvantage of water as a mounting agent is that water evaporates rather quickly. It requires adding water regularly to the wet mount slide to avoid the material drying out. If the observation is not complete, the wet mount may be sealed with nail polish. Some fungi are hydrophobic. Adding a wetting agent to the water (e.g., Triton X 100, Tween 80, or Span 60) will improve the quality of the wet mounting and avoid formation of air bubbles among the fungal structures under the coverslip on the slides [15]. However, water is not regularly used as a mounting medium for microscopic analysis in commercial laboratories.

Lactic Acid (85%)

This chemical is commonly used as a mounting medium in microscopic observation of fungi. It retains the natural color of fungal structures with minimal alteration. It desiccates rather slowly. A wet mount without application of any sealant can keep its quality for several weeks before noticeable deterioration occurs. If a wet mount slide is sealed with nail polish or other sealant, the slide can be kept for several months. When distorted fungal materials in their morphological characters due to dehydration are mounted in lactic acid and left on a lab bench for overnight to 24 h, the lactic acid is able to rehydrate the fungal materials and results in recovery of morphology of some fungal structures.

Gelatinous fungal structures and noncellular appendages are invisible in lactic acid. If the fungus under observation may develop mucoid structures, it should be mounted in water to make sure that the mucoid structures will not be missed.

When lactic acid mounting does not produce the quality you need, a drop of a staining agent can be added to a side of the coverslip to allow the staining agent to diffuse inside the coverslip without disturbing the mounting. A piece of paper towel can be placed at the opposite side of the coverslip to draw out excessive mounting medium and to expedite the diffusion process.

Lacto-Fuchsin

Carmichael [45] published this agent (dissolve 0.1 g acid fuchsin in 100 mL 85% anhydrous lactic acid). It is a widely used stain and mounting medium. It turns the cytoplasmic elements a deep pink. Dhingra and Sinclair [11] considered it to be superior to cotton blue. To obtain the best observation, freshly prepared mounts should be used, especially for photomicrography. Because

it increases the contrast of some fungal structures and provides excellent clarity, it can be used for temporary and semipermanent mounts for most taxonomic groups of fungi [13]. It is very useful to observe septation in spores and conidiogenesis of hyaline fungi. For dematiaceous hyphomycetes or fungi with unique color/pigments, it is not necessary to use this agent or any other staining agents, since the fungi will lose their natural color. This agent can be diluted by adding lactic acid to the solution to obtain a lighter pink color to fungal materials.

Cotton Blue in Lactic Acid

This is a widely used staining-mounting medium by staining the cytoplasm of fungal cells. The composition of this staining medium includes only two ingredients: 0.01 g cotton blue and 100 mL 85% lactic acid. The procedure to make this medium is as follows: add lactic acid to a beaker with a magnetic stirrer; place the beaker on a hot plate; and add cotton blue powder to the lactic acid, which is being heated and stirred until the cotton blue has dissolved. After the solution is cool, filter out the undissolved particles [13]. The solution can be diluted with 85% lactic acid, if a lighter blue color is required. Permanent slides can be made with this staining-mounting medium by applying two coats of nail polish to seal them.

Lactic Acid in Glycerol

To make this mounting medium, mix lactic acid, glycerol, and distilled water in the ratio of 1:2:1 (v/v/v) [13].

Modified Lacto-Cotton Blue

The ingredients of this staining agent are: 3 mL cotton blue stock (1.0 g cotton blue and 99 mL 85% lactic acid), 250 mL glycerol, 100 mL 85% lactic acid, and 50 mL deionized water [46]. The stain retains the natural color of pigmented fungal materials, such as spores/conidia and stains hyaline and thin-walled spores and other structures subtly. It is developed for observing airborne fungal spores collected on a slide, which includes spores of multiple fungal taxa without overstaining the sample. Occasionally, the resultant color may be found to be too light for certain spores. The contrast and clarity produced by this agent are not as good as lacto-fuchsin. Sime et al. [46] indicated that this stain is able to rehydrate desiccated and deformed spores so as to regain their original shapes. The rehydration process takes several hours or overnight. No doubt, the rehydration will assist in identification of these; otherwise, deformed fungal spores under a microscope may hinder identification.

Polyvinyl Alcohol in Lactic Acid (PVLG)

It is a permanent mounting medium. It has a minimal distorting effect on fungal material and sets rapidly. It requires a 2- to 3-day time period for clearing of spore content in mounted slides when used as a clearing agent. It is composed of 1.66 g polyvinyl alcohol, 1 mL glycerol, 10 mL lactic acid, and 10 mL distilled water [11]. To make this agent, mix all liquid ingredients in a dark bottle first; then add the polyvinyl alcohol to the well-mixed liquid ingredients. The polyvinyl alcohol dissolves slowly. Thus, a water bath at 70–80°C is necessary to hold the bottle until the solution becomes clear in 4–6 h. This solution can be stored in a dark bottle for a year without deterioration of the quality.

Melzer's Reagent

This reagent is composed of 1.5 g potassium iodide, 0.5 g iodine, 20 mL distilled water, and 22 g chloral hydrate [13]. Melzer's is routinely used in a microscope slide preparation for observing amyloid reactions in fruiting bodies of Ascomycota and Basidiomycota [44, 47]. It is not a staining agent. It is a reagent to trigger amyloid reactions in certain fungi resulting in color change.

- Melzer's-positive reaction (amyloid)—test material turns blue, bluish gray to black.
- Dextrinoid reaction (pseudoamyloid)—test material turns brown to reddish-brown.
- Melzer's-negative (inamyloid)—test material does not show color change, or turn to faintly yellowish-brown.

The amyloid reaction includes two subtypes of reactions:

- Euamyloid reaction—test material turns blue without potassium hydroxide (KOH)-pretreatment.
- Hemiamyloid reaction—test material turns red in Lugol's solution, but shows no reaction in Melzer's reagent; when pretreated with KOH, it turns blue in both reagents [48].

Melzer's reagent is the most widely used reagent in studies of *Agaricales* and *Boletales*.

- KOH (5–10%), potassium hydroxide [13].
- Deionized water 1,000 mL.
- Potassium hydroxide 50–100 g.

KOH is not a staining agent. It is used to soften and clear fungal tissues as a simple and quick method in microscopic preparations. It is rather advantageous to examine thick, gelatinous/mucoid material or specimens containing keratinous material, such as skin scales, nails, or hair. The cellular material and background keratin in these samples are digested by this reagent to reveal the fungal material by improving visibility of fungal structures under a compound microscope. KOH is also an agent used for determining chemical reactions of fleshy macrofungi (mushrooms).

KOH is a very strong corrosive reagent. Extreme care should be taken when using it or preparing this reagent. Proper personal protection gear will help avoid personal injury.

Acknowledgments I am grateful to Dr. James LaMondia for his pre-submission review and to Diane Riddle for her editorial assistance.

References

1. Jones MDM, Forn I, Gadelha C, Egan MJ, Bass D, Massana R et al (2011) Discovery of novel intermediate forms redefines the fungal tree of life. Nature 474(7350):200–3. doi:10.1038/nature09984
2. Hibbett DS, Ohman A, Glotzer D, Nuhn M, Kirk P, Nilsson HR (2011) Progress in molecular and morphological taxon discovery in Fungi and options for formal classification of environmental sequences. Fungal Biol Rev 25:38–47
3. Norvell LL (2011) Fungal nomenclature. 1. Melbourne approves a new code. Mycotaxon 116 (1):481–90.

4. Ross LE, Dykstra M (2003) Biological electron microscopy: theory, techniques and troubleshooting. Springer, New York
5. Pawley JB (2006) Handbook of biological confocal microscopy, 3rd edn. Springer, New York
6. Rossman AY, Samuels GJ, Rogerson CT, Lowen R (1999) Genera of Bionectriaceae, Hypocreaceae and Nectriaceae (Hypocreales, Ascomycetes). Stud Mycol 42:1–248
7. Nag Raj TR (1993) Coelomycetous anamorphs with appendages-bearing conidia. Mycologue Publications, Waterloo, ON
8. Michaelides J, Hunter L, Kendrick B, Nag Raj TR (1979) Icone Generum Coelomycetum. Supplement—Synoptic key to 200 Genera of Coelomycetes. Biol. Ser. University of Waterloo, Waterloo
9. Nag Raj TR (1981) Coelomycete systematic. In: Cole GT, Kendrick B (eds) The biology of conidial fungi. Academic Press, New York, pp 43–84
10. Pitt JJ. A laboratory guide to common Penicillium species. Food Science Australia; 2000
11. Dhingra OD, Sinclair JB (1985) Basic plant pathology methods. CRC Press, Boca Raton
12. Li DW, Yang C, Harrington F (2007) Microscopic analytical methods for fungi. In: Yang C, Heinsohn P (eds) Sampling and analysis of indoor microorganisms. John Wiley & Sons, Hoboken, NJ, pp 75–103
13. Kirk PM, Cannon PF, Minter DW, Stalpers JA (2008) Dictionary of the fungi, 10th edn. CABI, Wallingford, UK
14. Smith MB, McGinnis MR (2008) Diagnostic histopathology. In: Hospenthal DR, Rinaldi MG (eds) Infectious disease: diagnosis and treatment of human mycoses. Humana Press, Totowa, NJ, pp 37–51
15. Malloch D (1997) Moulds, under the microscope, mounting media. University of Toronto. www.botany.utoronto.ca/ResearchLabs/MallochLab/Malloch/Moulds/Examination.html. Accessed 18 Dec 2005
16. Li W, Zhuang W (2009) New species and new Chinese records of Dothideomycetes. Mycotaxon 106:413–418
17. Sigma-Aldrich, (2000) Material safety data sheet P4161 Phenol ACS reagent. Sigma Chemical Co, St. Louis, MO, p 6
18. Barr ME (1987) Prodromus to class loculoascomycetes. Hamilton I. Newell, Inc., Amherst, MA
19. Klich MA (2002) Identification of common Aspergillus species. CBS, Ultrecht, Netherlands
20. Simmons EG (1999) Alternaria themes and variations (226–235) classification of citrus pathogens. Mycotaxon 70:263–323
21. Zhang TA (2003) Flora fungorum sinicorum, vol 16. Science Press, Beijing
22. Slayter EM (1973) Optical microscope. In: Gray P (ed) The encyclopedia of microscopy and microtechniques. Van Nostrand Reinhold Co, New York, pp 382–389
23. Piston DW (1998) Choosing objective lenses: the importance of numerical aperture and magnification in digital optical microscopy. Biol Bull 195:1–4
24. Davidson MW (2011) Resolution. microscope U: the source for microscope education. http://www.microscopyu.com/articles/formulas/formulasresolution.html. Accessed 13 Jun 2011
25. Morris KJ (1995) Modern microscopic methods of bioaerosol analysis. In: Cox CS, Wathes CM (eds) Bioaerosols handbook. Lewis Publishers, Boca Raton, pp 285–316
26. Spring KR, Davidson MW (2011) Specialized microscope objectives. http://www.microscopyu.com/articles/optics/objectivespecial.html. Accessed 13 Jun 2011
27. Van Berkel K, Van Helden A, Palm L (1999) Frits Zernike 1888–1966. A history of science in the Netherlands. Survey, themes and reference. Brill, Leiden, pp 609–611
28. Murphy DB, Hoffman R, Spring KR, Davidson MW (2011) Specimen contrast in optical microscopy. www.microscopyu.com/articles/formulas/specimen-contrast.html Accessed 13 Jun 2011
29. Allen RD, David GB, Nomarski G (1969) The Zeiss-Nomarski differential interference equipment for transmitted-light microscopy. Z Wiss Mikrosk Technik 69:193–221
30. Pluta M (1846) Nomarski's DIC microscopy: a review. Proc SPIE 1994:10–25
31. Centonze-Frohlich V (2008) Phase contrast and differential interference contrast (DIC) microscopy. J Vis Exp 17:844
32. Drent P (2011) Digital imaging-new opportunities for microscopy. Nikon. http://www.microscopyu.com/articles/digitalimaging/drentdigital.html. Accessed 13 Jun 2011
33. Kim JY, Kim JW, Seo SH, Kye YC, Ahn HH (2011) A novel consistent photomicrography technique using a reference slide made of neutral density filter. Microsc Res Tech 74:397–400
34. Aylor DE (1995) Vertical variation of aerial concentration of Venturia inaequalis ascospores in an apple orchard. Phytopathology 85:175–181
35. Muilenberg ML (1989) Aeroallergen assessment by microscopy and culture. Immunol Allergy Clin North Am 9:245–268
36. Calderon C, Lacey J, McCartney HA, Rosas I (1995) Seasonal and diurnal variation of airborne basidiomycete spore concentrations in Mexico City. Grana 34:260–268
37. Li DW, Kendrick B (1996) Functional and causal relationships between indoor and outdoor airborne fungi. Can J Bot 74:194–209
38. Oh JW, Lee HB, Lee HR, Pyun BY, Ahn YM, Kim KE et al (1998) Aerobiological study of pollen and mold in Seoul, Korea. Allergol Int 47:263–270
39. Sterling M, Rogers C, Levetin E (1999) An evaluation of two methods used for microscopic analysis of airborne fungal spore concentrations from the Burkard spore trap. Aerobiologia 15:9–18
40. Gilliam MS (1975) Periodicity of spore release in Marasmius rotula. Mich Botanist 14:83–90
41. Hestbjerg H, Dissing H (1995) Studies on the concentration of Ramularia beticola conidia in the air above

sugar beet fields in Denmark. J Phytopathol 143: 269–273

42. Vismer HF, Cadman A, Terblanche APS, Dames JF (1995) Airspora concentrations in the Vaal Triangle: monitoring and potential health effects. 2, Fungal spores. S Afr J Sci 9:408–411

43. Lacey J, Venette J (1995) Outdoor air sampling techniques. In: Cox CS, Wathes CM (eds) Bioaerosols handbook. Lewis Publishers, Boca Raton, pp 407–471

44. Dennis RWG (1981) British ascomycetes. J. Cramer, Vaduz

45. Carmichael JW (1955) Lacto-fuchsin: a new medium for mounting fungi. Mycologia 47:611

46. Sime AD, Abbott LL, Abbott SP (2002) A mounting medium for use in indoor air quality spore-trap analyses. Mycologia 94:1087–1088

47. Largent D, Johnson D, Watling R (1977) How to identify mushrooms to genus III: microscopic features. Mad River Press, Arcata, CA

48. Baral HO (1987) Lugol's solution/IKI versus Melzer's reagent: hemiamyloidity, a universal feature of the ascus wall. Mycotaxon 29:399–450

Scanning Electron Microscopy for Fungal Sample Examination

8

Eduardo Alves, Gilvaine Ciavareli Lucas,
Edson Ampélio Pozza,
and Marcelo de Carvalho Alves

Abstract

This chapter presents general and specific methods for preparing fungi, plant, and seed tissues infected with fungi, and nematodes parasited by fungi using scanning electron microscopy (SEM). Conventional methods with chemical fixation as well as cryofixation are included, with details for subsequent steps of sample preparation to get optimal fungal specimens for visualization and examination through SEM.

Keywords

Scanning electron microscopy • SEM of seed-borne fungi • SEM of plant pathogenic fungi • SEM of nematophagous fungi • Ultra-structure • Methods for SEM

E. Alves
Department of Phytopathology, Ultra-structural Analysis and Electron Microscopy Laboratory—Ufla, Lavras, Minas Gerais 37200-000, Brazil

G.C. Lucas
Department of Phytopathology, Ultra-structural Analysis and Electron Microscopy Laboratory—Ufla, Lavras, Minas Gerais 37200-000, Brazil

Departamento de Fitopatologia, Universidade Federal de Lavras, Caixa postal 3037, Lavras, Minas Gerais 37200-000, Brazil

E.A. Pozza
Departamento de Fitopatologia, Universidade Federal de Lavras, Caixa postal 3037, Lavras, Minas Gerais 37200-000, Brazil

M. de Carvalho Alves (✉)
Department of Soil and Rural Engineering, Campus of the Federal University of Mato Grosso, Av. Fernando Correa da Costa, 2367, Boa Esperança, Cuiaba, Mato Grosso 78060900, Brazil
e-mail: marcelocarvalhoalves@gmail.com

Introduction

Fungi are small, generally microscopic, eukaryotic, usually filamentous, branched, spore-bearing organisms that lack chlorophyll [1]. They can be obligate parasites, nonobligate parasites, or biotrophs. Owing to these characteristics these microorganisms can develop several interactions with plants, animals, or the environment and can be used to produce food and enzymes for industrial processes [2].

Scanning electron microscopy (SEM) has caused a revolution in the study of the microscopic world. Its advantages include bi-dimensional aspect of the images with high depth of field; large increase in magnitude from 10 to 1,000,000 times; rapid image processing, digitalization, and acquisition; ease of preparation of samples and operation; and accessible costs [3]. The images generated by SEM are also used to enlarge the possibilities of teaching–learning interaction based on the application of virtual reality techniques for visualization of these microorganisms [4] and to study details of fungi taxonomy [5].

When using SEM, the goal is to investigate the external features of a specimen. However, the SEM can be used to probe internal cellular detail removing the overlying material, by fracturing, cutting, or tearing the specimen [6]. The objective of the observation will also affect how the specimens must be prepared.

The SEM is a helpful tool to study fungi, their interaction with other organisms, and their use in industrial processes. It permits study of several aspects of the morphology, such as surface details, fungi parasitism, and saprophytism. This tool has been used since the beginning of the 1970s to study these microorganisms [7], and to date has produced a lot of knowledge about fungi and their interactions.

SEM analysis of biologic material, like fungi, requires optimal preparation of the samples. The prevention of degeneration processes and changes in the material is necessary during microscopic observation. Both problems would lead to the formation of undesirable artifacts.

Some methods to prepare study fungi and their interactions with others organisms are presented in this chapter. Most of these protocols are used frequently in the Ultra-structural Analysis and Electron Microscopy Laboratory, a multiuser facility at the Federal University of Lavras, Brazil. The goal is to present the protocol applied to fungi studies, with greater coverage about specimen preparation and SEM application. For further information about the application and operation of the SEM, see references [3, 6, 8].

Materials

These materials are used to prepare fungi for SEM observation. The necessary materials can vary depending on the protocol used.

1. Razor blades and scalpel.
2. Specimen collection supplies (microcentrifuge tubes, pipettes, dishes (3 and 9 cm),

containers, centrifuge tube (15 and 50 mL), Erlenmeyer flasks, and tissue culture plates).
3. Fixative solution 1: modified Karnovsky's solution (glutaraldehyde 2.5%, paraformaldehyde 2.0% in cacodylate sodium buffer 0.05 M, pH 7.2, $CaCl_2$ 0.001 M).
4. Buffer cacodylate 0.05 M.
5. Fixative solution 2: aqueous 1% osmium tetroxide solution 0.05 M pH 7.2.
6. Distilled water or sterile distilled water.
7. Acetone dehydration series (25, 50, 75, 90, and 100%).
8. Ethanol dehydration series (25, 50, 75, 90, 95, and 100%).
9. 30% Glycerol.
10. Poly-L-lysine-coated coverslip with diameter of 13 mm.
11. Double-stick tape (3 M).
12. Double-stick carbon tape.
13. Aluminum stubs.
14. Aluminum foil.
15. Critical point dryer.
16. Sputter coater.
17. Desiccator container.
18. Silica gel.
19. Liquid nitrogen.
20. Styrofoam box (about 10×15 cm).
21. Plastic box a little bit small than Styrofoam box.
22. Parafilm.
23. Coverslip 13 mm of diameter and 22 mm square.
24. Ariel (a commercially available washing powder—Proctor and Gamble).
25. Ultramicrotome or microtome.
26. Cryo-transfer system.
27. Stereomicroscope.
28. SEM.

Methods

Routine Protocol

The method presented below is for preparing fungi or yeasts from solid media or from any substrate for routine observation using conventional approaches. This method enables the study of fungi on the surface of the substrate,

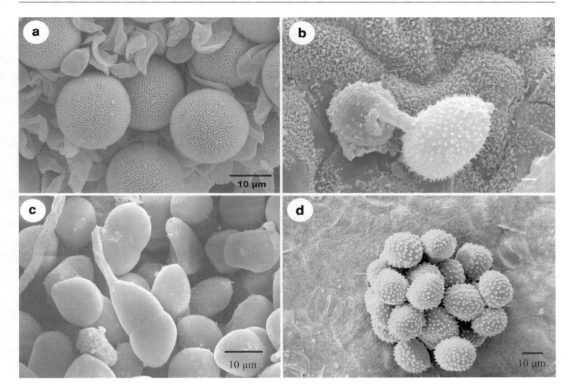

Fig. 8.1 Scanning electron micrographs of rust fungi. (**a**) *Puccinia nakanishiki* urediniospores (lemon grass rust). (**b**) *Phakopsora pachyrhizi* germinated urediniospore with appressorium (soybean rust). (**c**) *Prospodium* *bicolor* teliospore (*Tabebuia* sp. rust). (**d**) *Hemileia vastatrix* urediniospore (coffee rust). Specimens were prepared as described in section "Routine Protocol"

the morphology of hyphae, spores production, pre-invasion events of the pathogenesis in plants (Figs. 8.1 and 8.2), and interaction between fungi and others organisms such as nematodes and insects.

1. Squares or discs with about 0.5 cm of a solid medium with mycelia of a fungus or yeast species (after the incubation period necessary to produce the structure that you would like to see) or a piece of leaves, roots, stems or others parts from infected plant or organisms must be immersed in a microcentifuge tube with fixative solution (modified Karnovsky's fixative 2.5% glutaraldehyde–2.5% paraformaldehyde, 0.05 M cacodylate buffer, $CaCl_2$ 0.001 M) at pH 7.2 and kept for 24 h or more in a refrigerator (see Note 1).

2. Then the specimens are washed in cacodylate buffer (three times, for 10 min each wash).

3. Post-fix in 1% osmium tetroxide aqueous solution in water for 1–4 h at room temperature.

4. Rinse three times in distilled water.

5. Follow dehydration in crescent series of acetone solutions (25, 50, 75, 90, and 100%, once for concentrations up to 90% and twice for the 100% concentration) for 10 min each.

6. Afterwards, the samples are transferred to a critical point dryer to complete the drying process with carbon dioxide as a transition fluid.

7. The specimens obtained are mounted on aluminum stubs (same time could be necessary to use stereomicroscope), with a double-stick carbon tape pasted on a film of aluminum foil (see Note 2). Take care to keep the area to be observed upward.

8. Coat with gold in a sputter. Keep the material in desiccator with silica gel until observing.

9. Observe in a SEM.

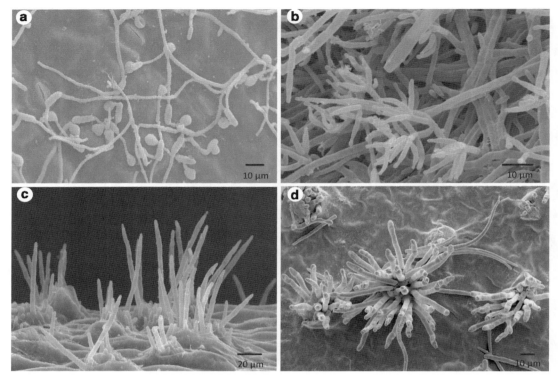

Fig. 8.2 Scanning electron micrographs of plant patho-genic fungi. (**a**) *Colletotrichum gloeosporioides* conidia germinated with appressoria on coffee leaf. (**b**) *Verticillium* sp. mycelia on culture medium. (**c**) *C. gloeosporioides* acervuli on coffee leaf. (**d**) *Cercospora coffeicolla* conid-iophores and conidia on coffee leaf. Specimens were pre-pared as described in section "Routine Protocol"

Sample Preparation to Observe Fungi Inside Plant Tissue or Inside a Fungal Fruiting Body

It is possible to use SEM to see internal cellular detail inside a fungal fruiting body or inside plant tissue infected by fungi, if overlying material is removed. This material can be removed by frac-turing, cutting, or tearing into the specimen and removing part of the tissue that is covering the fungi inside the material. This can be accom-plished in the following three ways.

Cutting the Tissue in Liquid Nitrogen (Cryofracture)

This technique permits one to get smooth and clean sections that can become possible to see inside some materials. The process is simple and could be done in two ways.

Glycerol Method

This is a simple method that needs just a few materials to be developed and it is possible to get good results (Figs. 8.3 and 8.4).

1. Fix the sample as described at step 1 in section "Routine Protocols."
2. Move the material to 30% glycerol (see Note 3) at least 30 min.
3. Transfer the material for a container with liquid nitrogen. Wait for the specimen deep to the bottom of the container.
4. Take each specimen one-to-one with a forceps and put over a metal surface, inside a plastic container with liquid nitrogen placed inside a styrofoam box. Immediately, fracture the speci-men in many parts as possible with a scalpel.
5. Take the parts fractured and put in a microcen-trifuge tube with distilled water.
6. Follow steps 3–9 in section "Routine Protocol."

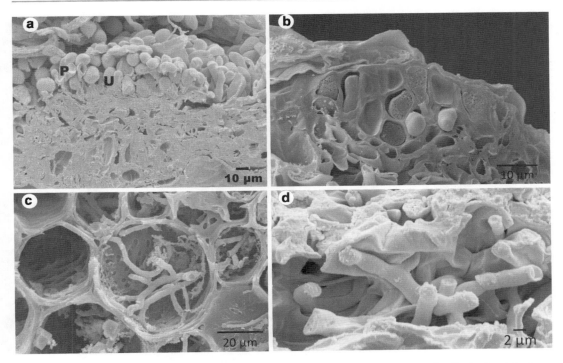

Fig. 8.3 Scanning electron micrographs of plant pathogenic fungi inside plant tissue. (**a**) Cross section of grape leaf with uredium of the rust fungus *Phakopsora euvitis*. *U*-urediniospore; *P*-paraphysis. (**b**) Cross section of soybean leaf with telium of the rust fungus *Phakopsora pachyrhizi*. (**c**) Cross section of coffee leaf petiole showing *C. gloeosporioides* hyphae inside the cell. (**d**) Hyphae of *P. euvitis* inside grape leaf sponge parenchyma. Specimens were prepared as described in section "Glycerol Method"

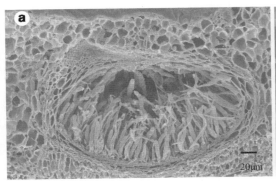

Fig. 8.4 Scanning electron micrographs of the fruiting body of *Diaporthe phaseolorum* f. sp. *Meridionali* in soybean stem. (**a**) General view of the ascocarp. (**b**) Detail of asci with ascospores. Specimens were prepared as described in section "Glycerol Method" (Courtesy of Regiane Medici)

Ethanol Method

This method is a simple, rapid, and inexpensive way to observe fine detail inside the tissue. It is adapted from reference [9], as follows:

1. Fix tissues in 2% glutaraldehyde/0.1 M cacodylate buffer, pH 7.2.
2. Wash in 0.1 M cacodylate buffer + 5% sucrose for 30 min.
3. Cut tissue into strips 1 mm × 4 mm (for leaves) or small parts for fungi fruiting bodies.
4. Post fix in 2% OsO_4/0.1 M cacodylate buffer, for 2 h.
5. Wash in 0.1 M cacodylate + 5% sucrose for 30 min.
6. Dehydrate in ethyl alcohol, 70, 95, 100% (2×), 15 min each.

 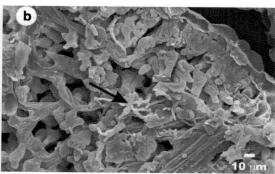

Fig. 8.5 Scanning electron micrographs of rust soybean fungus (*Phakopsora pachyrhizi*) inside soybean leaves tissue. (**a**) Intern face of epidermis. *Arrow* hyphae; *C*-sponge parenchyma cell. (**b**) Sponge parenchyma with hyphae of the fungus. *Arrow* hyphae. Specimens were prepared as described in section "Removing the Leaf Epidermis" (Courtesy of Elisandra B. Zambenedetti Magnani)

7. While submerged in the final change of absolute alcohol, insert material into Parafilm sleeves (make by rolling 2 cm strips of Parafilm around an applicator stick). Crimp sleeves shut on both ends (see Note 4).

8. Take this material with forceps and held under liquid nitrogen until frozen.

9. The tissue which is visible through the Parafilm and frozen ethanol is fractured with a single-edge razor blade or scalpel held in a hemostat and pre-cooled in liquid nitrogen.

10. Pick the fractured fragments up with forceps and return to a fresh volume of absolute ethanol.

11. Remove the Parafilm sleeve.

12. Then the specimen is critical point dried using liquid CO_2 as a transition fluid.

13. The fractured face (which is distinctly smoother and shinier when viewed in a stereomicroscope than the other surface) is oriented upward and secured to a specimen stub.

14. The specimens are coated with layer of vaporized gold and can be observed in a SEM.

Preparing Thick Sections

One alternative process to observe inside the tissue is to make thick sections in ultramicrotome (1–2 μm). Those sections could be assembled in coverslip or on stubs and submitted to a treatment to remove the resin and then coated with gold [10].

1. In one ultramicrotome, cut thick section (1 μm) from a tissue in a resin block.

2. Take the sections with a gold wire loop to transfer to a glass coverslip measuring 13 mm of diameter. Allow the sections to dry flat against the glass.

3. Put a drop of resin solvent and gradually replace the solvent by drawing off solvent from one side with filter paper while you add fresh solvent at the other side with a pipette.

4. After 20 min, examine the sections in a stereomicroscope. When the traces of resin are removed, wash the sections with methyl alcohol by filter paper/pipette method. Allow the coverslip to dry in a dust-free place.

5. Put the coverslip on stub covered with a film of aluminum foil using a double-stick tape.

6. Coat it in the sputter for examination in the SEM.

Removing the Leaf Epidermis

This method could be used to see the fungal colonization inside the leaf blade in the internal part of the epidermis and in the parenchyma (Fig. 8.5) [11]. It is modified from the leaf fracture method [12].

1. Leaves from the species in study are inoculated, in random marked places, with a 50 μL

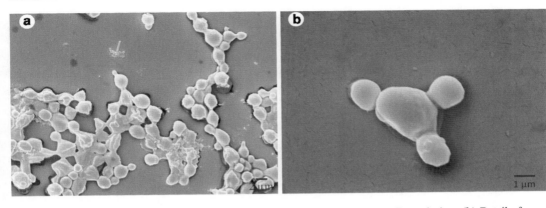

Fig. 8.6 Scanning electron micrographs of yeasts on polystyrene membrane. (**a**) General view. (**b**) Detail of yeast cells. Specimens were prepared as described in section "Use of Polystyrene Surface to Study Yeasts in SEM"

droplet of the spore suspension from the fungus tested. They are kept in dew chamber for 24 h.

2. The inoculated places from leaves are harvested after a time necessary to start the colonization (this time can vary from 12 to 96 h, depended on the fungus), and cut into 3 × 3 mm squares.

3. The collected material is prepared as described up to step 7 (critical point dry) in section "Routine Protocol."

4. After mounting the specimen on stubs, take a piece of double-stick tape and put over the specimens. Push softly and then make a fast upward movement to remove the epidermis.

5. Place the stripped part on a new stub and coat with gold in a sputter with both specimens (it will form two parts: the stripped epidermis and the parenchyma part that was kept on the surface of the first stub).

6. Observe in a SEM.

Use of Proteases to Clean and Reveal Details of Interfaces Between Plant Cells and Fungal Structures

This technique could be used to remove all cytoplasmic remnants from sectioned host cells to examine morphology of fungal haustoria, inside leaf cells as well as inner surface of the host cells, adapted from references [13, 14]. This technique

was associated successfully with the process described in section "Glycerol Method."

1. Firstly, cut the plant tissue as described for glycerol liquid nitrogen technique (steps 1–3).

2. Put the material in a 5% (w/v) aqueous solution of Ariel (a commercially available washing powder—Proctor and Gamble) or other washing powder contented protease and incubate overnight (approximately 14 h) at 30°C.

3. Following incubation, the material is removed from Ariel solution and washed for 30 min in multiple changes of distilled water.

4. Follow steps 3–9 of the section "Routine Protocol."

Use of Polystyrene Surface to Study Yeasts in SEM

This technique was developed to study bacterial adhesion [15]; however, it is used to study adhesion, polysaccharides production, and morphology of yeasts (Fig. 8.6) grown in liquid culture or fermentation medium in SEM (see Note 5). This protocol is split into three major steps:

1. Prepare covered glass slides (PCGS) following the method described as follows [16]. Take one-half of a polystyrene Petri dish and dissolve in 50 mL of amyl acetate. Clean glass slides are dipped into the polystyrene solution and dried overnight in a laminar flow hood to

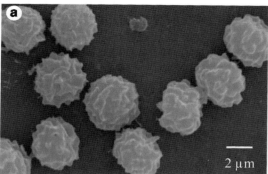

Fig. 8.7 Scanning electron micrographs of *Aspergillus* spp. spores on coverslip with poly-L-lysine. (**a**) *A. tubingensis*. (**b**) *A. japonicus*. Specimens were prepared as described in section "Study of Yeasts or Fungi Spores on Poly-L-Lysine Coverslips" (Courtesy of Daiani Maria da Silva)

remove traces of the solvent and to prevent surface contamination.

2. Sterilization of the PCGS inside a 50-mL centrifuge tube (Falcon tubes) for 20 min at 120°C.
3. In aseptic conditions, transfer 25 mL of liquid medium and inoculate the yeast.
4. The centrifuge tube is transferred into an Erlenmeyer, which is incubated inside optimal conditions and time necessary for yeast growing in rotator agitation.
5. After that, remove the polystyrene surface from the slide by passing a blade along the edges of the slide, cut the area with the yeast, and put it into a dish (3-cm Petri dish) with fixative solution described in step 1 of section "Routine Protocol." Keep the polystyrene membrane immersed in fixative solution with the surface of the yeast facing upward.
6. Follow steps 2–4 in section "Routine Protocol."
7. Follow dehydration in increasingly more concentrated ethanol solutions (25, 50, 75, 90, 95, and 100%, once for concentrations up to 95% and twice for the 100% concentration) for 10 min each. The dehydration needs to be in ethanol series because the acetone is solvent for polystyrene.
8. Follow steps 6–9 in section "Routine Protocol."

Study of Yeasts or Fungi Spores on Poly-L-Lysine Coverslips

This technique can be used to study yeast in liquid culture or fungi spore morphology and is adapted from reference [6] and processed as follows (Fig. 8.7).

1. Prepare circular coverslips (13 mm of diameter) coated with poly-L-lysine. Use a swab to pass poly-L-lysine on coverslips and wait 10 min to dry.
2. For adhering yeast cells or fungi spores, transfer 50 µL of suspension (yeast cells or spores) to a coated coverslip and wait for 10 min. Protect the slide from the evaporation.
3. With a fine-tip pipette take out the liquid and gently put in 100 µL of the fixative solution described in section "Routine Protocol" and follow steps 2–9. Take care when pouring off the previous solution to put on a new one, and do not let the specimen dry out. Keep the cell or spore layer uppermost at all times.

Fixation of Fungal Structures with Osmium Tetroxide Vapor

This methodology can be used for imaging soft fungi that can lose their spores (such as *Aspergillus* and *Penicillium* species) (Fig. 8.8),

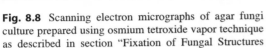

Fig. 8.8 Scanning electron micrographs of agar fungi culture prepared using osmium tetroxide vapor technique as described in section "Fixation of Fungal Structures with Osmium Tetroxide Vapor." (**a**) *Aspergillus* sp. (**b**) *Penicillium* sp.

for entomopathogenic fungi on insects [17], exudation of conidia from fruiting bodies or present any types of alterations due to conventional methods. It is modified from reference [18].

1. Prepare the material to be processed (it can be an agar plug from a fungal sporulation area). Put with care on a glass surface and keep the fungal layer uppermost.
2. Put the specimens inside a 9-cm Petri dish. Use a piece of moisturized filter paper to make a dew chamber.
3. In a fume hood, place 1 mL of 2% glutaraldehyde solution in an opened container and put it inside the Petri dish with the specimens. Keep it closed at room temperature (22–24°C) for 2 h.
4. Move the specimens for other Petri dish, place 1 mL of 2% osmium tetroxide solution in an opened container, and close the dish.
5. Keep the Petri dish covered with aluminum foil inside the fume hood for 12–48 h (see Note 6).
6. After fixation the material must be kept inside a desiccator with silica gel for 1–2 days. The OsO_4 vapor fixation combination with silica gel desiccation can reduce disruption satisfactorily, but specimens were not as well preserved as with conventional method.
7. The specimens obtained are mounted on aluminum stubs, with a double-stick carbon tape pasted on a film of aluminum foil.
8. Coat with gold in a sputter.
9. Observe in a SEM in no more than 3 days.

Desiccator Method

In our lab we sometime have a problem keeping fungi spores (such as: *Aspergillus* (Fig. 8.9) and *Penicillium* species) attached to conidiophores to be observed in SEM. We have observed that the largest loss of spore has occurred at critical point dry process. Then we developed the protocol below with good results.

1. Take some plug from Petri dish content, a sporulated fungal culture. Remove as much underlying agar as possible.
2. Place some excised plug in a small Petri dish (3 cm in diameter) with the fixative solution described in section "Routine Protocol" and follow the steps 2–5.
3. Afterward, the samples are transferred to a desiccator containing silica gel to complete the drying process.
4. Follow steps 7–9 in section "Routine Protocol."

Ultrasonication to Remove Fungi Appressoria

Some pathogenic fungi, before invasion of a plant, produce an appressorium. Under this structure is formed a narrow hypha *peg* used for penetration. To study the hole in the plant tissue that the fungus has used to penetrate through to the epidermis it is necessary to remove the

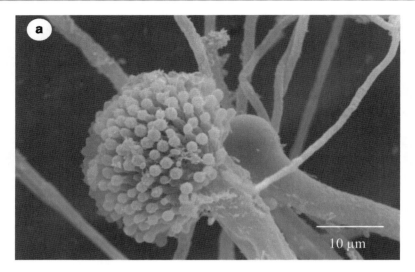

Fig. 8.9 Scanning electron micrograph of agar *Aspergillus* sp. culture prepared using the technique described in section "Desiccator Method"

apressorium. The technique presented below is used with this objective (Fig. 8.10) [11].

1. Leaves from the species in study are inoculated, in random marked places, with a droplet of the spore suspension from the tested fungus. They are kept in dew chamber for 24 h.
2. The inoculated places from leaves are harvested after a time necessary to start the appressoria formation (this time can vary depended on the fungus), and cut into 5×5 mm squares.
3. Put the material in microcentrifuge tube with fixative solution (modified Karnovsky's fixative 2.5% glutaraldehyde–2.5% paraformaldehyde, 0.05 M cacodylate buffer, $CaCl_2$ 0.001 M) at pH 7.2 and kept for 24 h or more.
4. Then specimens are washed in cacodylate buffer (three times, for 10 min each wash).
5. Move the material to a beaker with cacodylate buffer and put the beaker into an ultrasound. Sonicating the material for 10 s.
6. Follow steps 3–9 described in the section "Routine Protocol."

Fixation with Tannic Acid for Fungi

In this protocol the tannic acid is used as a mordent for osmium tetroxide, according to the methodology modified from reference [19]. It can improve the fixation and increase the secondary electron emission by specimen surface. It could be necessary for samples with low conductivity.

1. Place the fungal tissue in 2.0% glutaraldehyde aqueous solution with 0.2% of tannic acid and keep for 3 h.
2. Move the sample for a new container with 2.0% glutaraldehyde aqueous solution with 2% of tannic acid and keep for more 3 h.
3. Rinse in distilled water three times for 10 min each.
4. Post-fix in 1% OsO_4 aqueous solution for 1 h.
5. Rinse in distilled water for three times and follow steps 5–9 described in section "Routine Protocol."

Cryofixation and Freeze-Substitution Technique for Preparing Fungal Specimens for Scanning Electron Microscopy

This technique could be used to improve the quality of the fixation process [20]. This process provides a much-improved structural and biochemical preservation of cell that is close to its living state relative to aqueous fixation protocols. It can be developed using many expensive

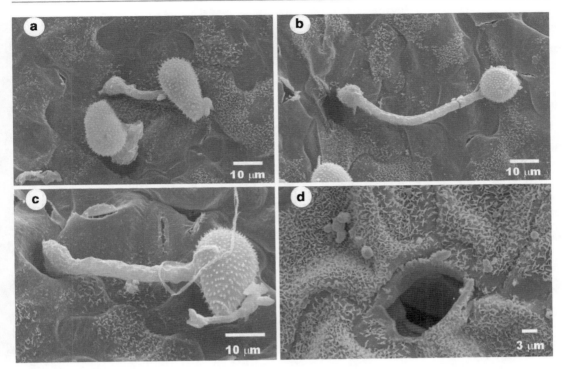

Fig. 8.10 Scanning electron micrographs of rust soybean fungus (*Phakopsora pachyrhizi*) on soybean leaves. (**a**) General view of urediniospore in germination forming appessoria. (**b**) General view of urediniospore in germination with appessorium removed. (**c**) General view of ured-iniospore in germination with appessorium not removed totally yet. (**d**) Detail of a hole seen after removal of the appressorium. Specimens were prepared as described in section "Ultrasonication to Remove Fungi Appressoria" (Courtesy of Elisandra B. Zambenedetti Magnani)

machines, but the following procedure is a simple and economic method:

1. Prepare a solution 1 with 1% tannic acid and 1% glutaraldehyde in anhydrous EM-grade acetone in a polypropylene cryovials and one solution 2 with 1% osmium tetroxide in anhydrous EM-grade acetone and put both in a −85°C freezer for 1 h prior to the cryofixation.

2. Take some agar plugs or pieces of plant tissue with fungal structures and put in 30% glycerol and keep for 30 min.

3. Take the plugs and deep one-to-one in styrofoam box with liquid nitrogen and wait for cooling.

4. In a hood fume collect the plugs and transfer to a vial with the solution 1 kept in a box with ice and transfer to a −85°C freezer and leave there for 48–72 h.

5. Wash the specimen three times, at low-temperature (−85°C freezer), with 100% acetone (P.A.) for 15 min.

6. Transfer the sample to solution 2 in the same condition of step 4.

7. Wash again for three times in acetone as developed in step 5.

8. Transfer the vial to −20°C in a freezer for 2 h.

9. Transfer the sample to a 4°C refrigerator for 2 h.

10. Afterward, the samples are transferred to a critical point dryer to complete the drying process with carbon dioxide as a transition fluid.

11. Follow steps 7–9 in section "Routine Protocol."

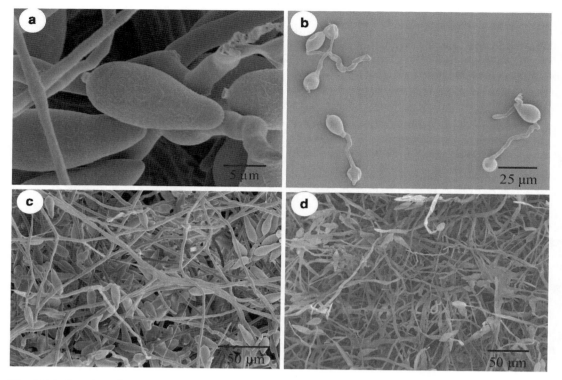

Fig. 8.11 Scanning electron micrographs of fungus *Magnaporthe grisea* prepared using cryo-chamber system as described in section "Use of Cryochamber in SEM to Study Fungi." (**a**) Detail of conidia and conidiophores. (**b**) Conidia germinating on a plastic coverslip. (**c, d**) Image of two cultures of the fungus to show the difference to produce spores between the strains using this technique (Images taken in Bioimaging Lab at University of Delaware—USA advised by Dr. Kirk J. Czymmek)

Use of Cryochamber in SEM to Study Fungi

The accessory for SEM, called cryo-transfer, permits to observe samples at low temperature. It is possible to get excellent preservation and fast preparation of delicate samples to be observed in SEM, which could not be possible using other techniques avoiding artifact formation. For example, this method could be used to measure the ability to produce spore by fungi in agar plates [20, 30] or to observe activities of nematophagous fungi in soil [21] and the fungi spore germination and penetration on plants or artificial hydrophobic surfaces (Fig. 8.11).

1. Agar plugs from a medium of fungi culture, pieces of leaves, artificial hydrophobic membranes or nematodes with fungi in a small plastic container (see Note 7) are taken mounted using a double-stick carbon tape to a metal stub and plunged into liquid nitrogen for cryofixation for 20 s. It is possible to use Tissue Tak with carbon graphite to paste the specimens on stubs.

2. Frozen specimens are placed in a cryo-chamber of a cryo-transfer system maintained at −170°C and transferred to a cold specimen stage of a SEM.

3. Sublimation of superficial frozen water is performed by heating and maintaining the cold specimen stage at −60°C for 10 min.

4. The specimens are sputter-coated with gold (approximately 30 nm in thickness) in the cryo-chamber.

5. Observe with the electron microscope at 20 kV.

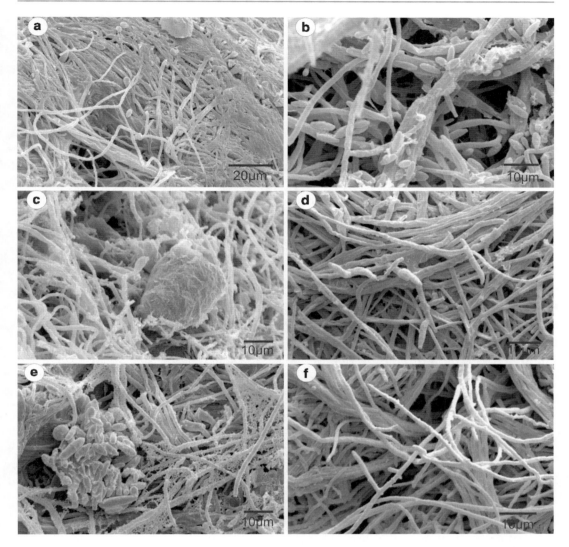

Fig. 8.12 Scanning electron micrographs of fungi in seeds prepared using technique described in section "Scanning Electron Microscopy as a Complementary Methodology to Identify Seed-Borne Fungi." (**a**, **b**) *Fusarium oxysporum*, presenting microconidia, (**c**) false heads of microconidia formed on short conidiophores along the hyphae, (**d**) terminal and intercalary phialides, (**e**) clusters of conidia and (**f**) hyphae

Scanning Electron Microscopy as a Complementary Methodology to Identify Seed-Borne Fungi

Techniques of observation of fungi in seeds with light microscope and stereomicroscopy can be supplemented by alternative methods with greater precision, such as SEM (Fig. 8.12). This topic describes the material and protocols to prepare seeds for studying seed-borne fungi in SEM.

Material

1. Materials for blotter test incubation method for seed health analysis.
2. Eight repetitions of 50 seeds for each species.

3. Absorbent paper.
4. Petri dishes or equivalent containers.
5. Transparent lid for light.
6. NUV light.
7. Freezer.
8. Sodium hypochlorite solution at 1% concentration.
9. 2,4-D salt (sodium 2,4-dichloro-phenoxyacetate) at 5–10 ppm concentration.
10. Incubator with temperature and light regulation.
11. Agar 0.2%.
12. Laminar flow hood.
13. Autoclave.
14. Dishwasher.
15. Microscopes, compound and stereomicroscope.
16. Mannitol or other osmotic compound at osmotic potentials of −0.6 to −1.0 MPa.

Protocol for Blotter Test Incubation Method for Preparing Seeds for Health Analysis in SEM [5]

1. Seeds are submitted to the standard blotter test. The blotter test preparation is conducted in accordance with the International Seed Testing Association criteria [22]. For instance, it will be presented a protocol for maize, common bean and cotton seeds, but other seeds could be prepared following this protocol with slight alterations.

2. Seeds of maize are incubated initially 24 h in room condition, then kept 24 h in a freezer (−20°C), followed by 5 days under NUV light, with 12 h photoperiod, at 20°C. To prevent seed germination of common bean and cotton during the incubation period, the substrate paper must be moistened with 2,4-D salt solution, 5–10 ppm, and 0.2% agar to prevent seeds from rolling during handling.

3. After that, seeded dishes are placed under NUV light with 12-h photoperiod at 20°C for 7 days.

4. Seeds are individually examined with a stereomicroscope from 30 to 80 magnification and slide mounts are prepared for observation using light microscope, in order to confirm fungi identity.

5. If the doubt about the fungi is kept then the seed is prepared for SEM as described in section "Routine Protocol."

Protocol for Blotter Test Incubation Method with Water Restriction for Preparing Seeds for Health Analysis in SEM [5]

This technique to get seeds to be prepared for SEM is similar to that from the preceding protocol, with the addition of the water restriction technique [23].

1. Lots of healthy seeds are chosen in order to ensure the occurrence of the inoculated fungal species.

2. The pure colony of each pathogen is scraped with a Drigalsky loop in order to obtain a suspension of conidia and mycelium with 1 mL solution.

3. The solution is transferred to Petri dishes with 15 cm in diameter, with potato dextrose agar (PDA)+mannitol medium, at −1.0 MPa. Mannitol concentrations are obtained through Van't Hoff equation [24].

4. The Petri dishes are incubated in chamber with 20°C and photoperiod of 12 h, for 5 days.

5. The seeds are disinfected in a solution of sodium hypochlorite 1% per 3 min, then washed with sterile water and dried in the shade per 24 h. Then, 35 g of seeds of cotton (*Gossypium hirsutum*), soybean (*Glycine max*) and common bean (*Phaseolus vulgaris*) are distributed in single layer on the colony of *Colletotrichum gossypii* var. *cephalosporioides*, *Colletotrichum truncatum* and *Colletotrichum lindemuthianum*, respectively.

6. A suspension of conidia with concentration of 1×10^6 conidia/mL is sprayed on the single layer of seeds to ensure the inoculation.

7. After that, the plates returned to the chamber of 20°C with photoperiod of 12 h, per 144 h.

8. Follow steps 4 and 5 in the section "Protocol for Blotter Test Incubation Method for Preparing Seeds for Health Analysis in SEM."

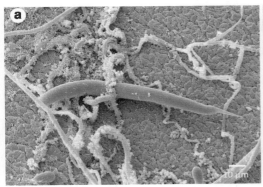
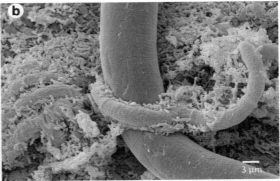
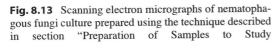

Fig. 8.13 Scanning electron micrographs of nematophagous fungi culture prepared using the technique described in section "Preparation of Samples to Study Nematophagous Fungi." (**a**) General view of the fungus trapped the nematode. (**b**) Detail of a trap

Preparation of Samples to Study Nematophagous Fungi

This protocol describes the process to prepare nemathophagous fungi for SEM from in vitro trails. It is modified from reference [25] (Fig. 8.13).

1. In a fume hood transfer an agar disk of nematophagous fungus from single spore culture for a Petri dish with medium and put the dish into a growth chamber at 25°C, in dark for 5 days.
2. Pour on 1 mL of a nematode suspension (about 100 specimens) on the dish and incubate again under same conditions. Observe if 50 or more nematodes are captured, then pour on fixative solution described in the section "Routine Protocol" to cover the mycelia.
3. Keep in refrigerator for 72 h to complete the fixation.
4. Then the dish is washed in cacodylate buffer (three times, for 10 min each wash).
5. Agar disks are taken from the medium carefully, moved for a vial and post-fix in 1% osmium tetroxide aqueous solution in water for 1–4 h at room temperature.
6. Follow steps 4–9 in section "Routine Protocol."

Preparation of Samples to Study Nuclei and Chromosomes of Fungi in SEM

This technique is important to determine whether fungal nuclei and chromosomes have structures distinct from those of plants and animals, and give information about the manner of chromatin compaction into condensed chromosomes during nuclear division in fungi. The protocol presented her is for *Cochliobolus heterostrophus* and *Neurospora crassa* based on reference [26] but this technique can be used for other fungi with some changes as reviewed in reference [27].

1. Take 200 mL droplet of conidial suspension (1.5×10^5/mL) containing 30 mM hydroxyurea (HU) and 3% (w/v) glucose and incubated to germinate on a glass coverslip under optimal conditions for the fungus.
2. After washing with distilled water to remove HU, 200 mL of 3% glucose is added to the germinated conidia adhering to the slide and incubation is resumed for 30 min.
3. After washing away the medium with water, germinated conidia on the slide are incubated in newly added 200 mL Vogel's medium containing 50 mg/mL thiabendazole for more than 2 h. Thiabendazole treatment is used to arrest nuclear division at metaphase.

4. After completing incubation, the germlings of fungus is rinsed with water and treated with 0.05% glutaraldehyde (buffered with 50 mM cacodylate, 2 mM $MgCl_2$, pH 7.2) for 30 s.

5. The slide is then immersed in the fixative (99.5% methanol:glacial acetic acid = 7:3) for 2 h at room temperature. Bursting of the germling cells occurred upon immersion in the fixative to discharge the nuclei and chromosomes from cells and spread them on the glass slide surface.

6. Follow steps 1–9 in section "Routine Protocol," with a slight difference at step 7, where the coverslip is put over the double-stick tape on stub.

Other Methods to Grow and to Take Fungi Material for Preparing Samples for SEM

Using Microculture of Fungi in Coverslip

1. In a Petri dish, place two sheets of filter paper and pour on distilled water to prepare a moisture chamber.

2. Take a glass slide for microscopy and put two agar disks, with the fungi mycelia took from the edge of a good fungal culture. Put these disk about 1.5 cm away one from the other with the mycelia up.

3. Place a coverslip on the two disks. Cover the dish, and incubate under optimal conditions for the fungus. Wait the fungus colonization on the coverslip.

4. Remove the coverslip with the attached mycelia and spores and put in a small Petri dish (3 cm), with fixation solution from section "Routine Protocol." Keep the mycelia layer uppermost.

5. Follow steps 2–9 in section "Routine Protocol," with a slight difference at step 7, where the coverslip is put over the double-stick tape on stub.

Using Filter for Mycelia Preparation

This method could be used to take mycelia and conidia from a fungus growing in an agar media for preparation for SEM. We have used this technique to evaluate the effect of fungicides, plant extracts, essential oils, or other substances on mycelial and conidia of fungi [28].

1. The substance that will be evaluated is transferred through an agar disk covered with mycelia of fungus. The dish is transferred into a growth chamber in optimal conditions for the fungus to grow.

2. After fungus have grown, pour distilled water on the dish and prepare a hyphal and conidial suspension, using a spatula to remove the mycelia and filter in Millipore membrane PTFE hydrophilic of pore size 0.45 μm and diameter of 13 mm.

3. Take the Millipore membrane and prepare for SEM as described in section "Routine Protocol."

Using Cellophane Membrane to Prepare Fungal Sample

This technique could be used to study the hyphal interactions of one biological control fungi with one plant pathogenic fungi or for collect hyphae from one fungus to study morphologic characteristic modified from reference [29].

1. A cellophane membrane is placed on an agar media in a Petri dish.

2. An agar disk covered with mycelia of antagonistic fungus is put on one end of the dish and the plant pathogenic fungus on the other. The fungi are placed to grow in a growth chamber under optimal conditions for both fungi.

3. After growth, the cellophane from the interaction area is cut, fixed, and prepared for SEM following steps 5–8, as described in section "Use of Polystyrene Surface to Study Yeasts in SEM." If the goal is to observe just one fungus then a good area (with characteristic structures of the fungus) could be selected in any part of the cellophane.

Notes

1. The fungi must be in Petri dishes on an appropriate media to obtain typical and good growth and sporulation. When cutting the agar disk, remove as much underlying agar as possible.

Other fixative solution could be used, but in our laboratory we have used this solution as standard. The sample could be kept in this solution, in a refrigerator, for a long time (6 months or more depending on the type of material) for SEM preparation.

2. In our laboratory, in order to make it easier to clean the stubs, we are using a piece of aluminum foil to cover this support. An aluminum foil square (about 2.5×2.5 cm) is cut and placed on the stub. It is a good idea to make certain that the foil is affixed well.

3. Glycerol is a cryoprotector, and we have used it because it is cheaper and easier to use. 20% sucrose and 15% dextran could be used.

4. The Parafilm sleeves are used so that the section does not move.

5. Stainless steel, plastic, or glass surface chips of 10×20 mm could be used to study yeast in place of the polystyrene membrane. The chips can be placed inside an Erlenmeyer with liquid culture and kept under agitation for a time necessary to grow the yeast. Collect the chips and prepare as described in section "Use of Polystyrene Surface to Study Yeasts in SEM."

6. The time depends on the size and type of the material. The fixation process occurs slowly owing to the osmium vapor. If the material becomes black, it is a good signal. If it does not turn black, it could be necessary to change the osmium tetroxide solution as often as is necessary.

7. Fill a small plastic container (ca. $1.5 \times 0.5 \times 0.3$ cm) with 3 mm soil layers. The fungus is added to the soil together with nematodes. The specimens are incubated in a moisture chamber with optimal temperature, for 3–5 days. After incubation, the plastic is frozen and prepared as described in step 1. With cryo-SEM, it is possible to study soil colonization and nematode parasitism by the introduced fungus, as well as the formation and action of traps.

Acknowledgements The authors gratefully acknowledge CNPq (Conselho Nacional de Desenvolvimento Científico e Tecnológico), CAPES (Coordenadoria de Aperfeiçoamento de Pessoal de Nível Superior) and FAPEMIG (Fundação de Amparo à Pesquisa do Estado de Minas Gerais—Brazil) for financial support to Ultra-structural analysis and electron microscopy laboratory of the Ufla, Brazil.

References

1. Agrios GN (2005) Plant pathology. Academic, New York, 922p
2. Alexopoulos CJ, Mims CW, Blackwell M (1996) Introductory mycology. Wiley, New York
3. Bozzola JJ, Russell LD (1999) Electron microscopy. Jones and Bartlett Publishers, Boston
4. Sforza PM, Eisenback JD (2001) The virtual nematode. In: ISPP-IT symposium covering instructional technology in plant pathology, New Zealand, 2001
5. Alves MC, Pozza EA (2009) Scanning electron microscopy applied to seed-borne fungi examination. Microsc Res Tech 72:482–488
6. Bozzola JJ (2007) Conventional specimen preparation techniques for scanning electron microscopy of biological specimens. In: John K, Clifton NJ (eds) Electron microscopy: methods and protocols. Humana Press, New York, pp 449–466
7. Ito Y, Nozawa Y, Setoguti T (1970) Examination of several selected fungi by scanning electron microscope. Mycopathologia 41:299–305
8. Mims CW (1991) Using electron microscopy to study plant pathogenic fungi. Mycologia 83:1–19
9. Humphreys WJ, Spurlock BO, Johnson JS (1975) Transmission electron microscopy of tissue prepared for scanning electron microscopy by ethanol cryofracturing. Stain Technol 50:119–125
10. Winborn WB (1976) Removal of resin from specimens for SEM. In: Hayat MA (ed) Principles and techniques of scanning electron microscopy, vol 5. Van Nostrand Reinhold, New York, pp 21–35
11. Zambenedetti Magnani EB, Alves E, Araujo DV (2007) Eventos dos Processos de Pré-Penetração, Penetração e Colonização de Phakopsora pachyrhizi em Folíolos de Soja. Fitopatol Bras 31:156–160
12. Hughes FL, Rijkenberg FHJ (1985) Scanning electron microscopy of early infection in the uredial stage of *Puccinia sorghi* in *Zea mays*. Plant Pathol 34:61–68
13. Richardson BA, Mims CW (2001) A simple technique for the removal of plant cell protoplasm to facilitate scanning electron microscopy of fungal haustoria and plant cell wall features. Micros Today 9:14–15
14. Honegger R (1985) Scanning electron microscopy of the fungus-plant interface: a simple preparative technique. Trans Br Mycol Soc 84:530–553
15. Alves E, Pascholati SF, Leite B (2001) Formation of bacterial biofilms on the surface of polystyrene: a methodology for the scanning electron microscope. In: XVIII Congresso da Sociedade Brasileira de Microscopia e Microanalise, 2001, Águas de Lindóia-SP. Sociedade Brasileira de Microscopia e

Microanálise, Acta Microscopica, São Carlos-SP, 2001, pp 155–156

16. Leite B, Nilcholson RL (1992) Mycosporine-alanine: a self-inhibitor of germination from the conidial mucilage of *Colletotrichum graminicola*. Exp Mycol 16:76–86

17. Moino A Jr, Alves SB, Lopes RB, Oliveira PM, Neves J, Pereira RM et al (2002) External development of the entomopathogenic fungi *Beauveria bassiana* and *Metarhizium anisopliae* in the subterranean termite *Heterotermes tenuis*. Sci Agric 59:267–273

18. Kim KW (2008) Vapor fixation of intractable fungal cells for simple and versatile scanning electron microscopy. J Phytopathol 156:125–128

19. Kunoh H, Ishizaki H, Nakaya K (1977) Cytological studies of early stages of powdery mildew in barley and wheat leaves: (II) significance of the primary germ tube of *Erysiphe graminis* on barley leaves. Physiol Plant Pathol 10:191–199

20. Howard RJ, Bourett TM, Duncan KE (2000) Impact of cryo-techniques on cytological studies of plant pathogenic fungi and their hosts. In: Proceedings microscopy and microanalysis, Philadelphia, 2000, pp 674–675

21. Jansson HB, Persson C, Odeslius R (2000) Growth and capture activities of nematophagous fungi in soil visualized by low temperature scanning electron microscopy. Mycologia 92:10–15

22. Machado JC, Langerak CJ (2002) General incubation methods for routine seed health analysis. In: International Seed Testing Association (ed) Seed-borne fungi: a contribution to routine seed health analysis. International Seed Testing Association, Zurich, pp 48–59

23. Machado JC, Guimarães RM, Vieira MGGC, Souza RM (2004) Use of water restriction technique in seed pathology. In: Seed testing international, vol 128. Switzerland, pp 12–16

24. Salisbury FB, Ross CW (1991) Plant physiology. Wadsworth, Belmont

25. Martinelli PRP, Santos JM (2010) Microscopia eletrônica de varredura de fungos nematófagos associados a *Tylenchulus semipenetrans* e *Pratylenchus Jaehni*. Biosci J 26:809–816

26. Tsuchiya D, Koga H, Taga M (2004) Scanning electron microscopy of mitotic nuclei and chromosomes in filamentous fungi. Mycologia 96:208–210

27. Wieloch W (2006) Chromosome visualization in filamentous fungi. J Microbiol Methods 67:1–8

28. Melo IS, Faull JL (2004) Scanning electron microscopy of conidia of *Thichoderma stromaticum*, a biocontrol agent of witches broom disease of cocoa. Braz J Microbiol 35:330–332

29. Elad Y, Chet I, Boyle P, Henis Y (1983) Parasitism of *Trichoderma* spp. on *Rhizoctonia solani* and *Sclerotium rolfsii*—scanning electron microscopy and fluorescence microscopy. Phytopathology 73:85–88

30. Jeon J, Goh J, Yoo S, Chi M, Choi J, Rho H et al (2008) A putative MAP kinase kinase kinase, MCK1, is required for cell wall integrity and pathogenicity of the rice blast fungus, Magnaporthe oryzae. Mol Plant Microbe Interact 21:525–534

High-Resolution Imaging and Force Spectroscopy of Fungal Hyphal Cells by Atomic Force Microscopy

9

Biplab C. Paul, Hui Ma, Laelie A. Snook,
and Tanya E.S. Dahms

Abstract

Various forms of microscopy applied to image fungal hyphae have been limited by either diffraction or the ability to image viable specimens. The advent of the atomic force microscope offered the opportunity to image live fungal hyphae under ambient conditions at very high resolution. The force spectroscopy capabilities of the microscope facilitate physical and chemical characterization of the fungal cell surface. In this chapter, we describe the detailed protocols that have allowed high-resolution imaging and force spectroscopy of fungal spores, germinants, and hyphae for both fixed and viable specimens.

Keywords

Adhesion • Atomic force microscopy • Force spectroscopy • Cell wall ultrastructure • Hyphae • Spore • Viscoelasticity

Introduction

Conventional imaging techniques for probing the ultrastructure of the fungal cell surface—scanning electron microscopy (SEM), for example—require rigorous sample preparation, which can alter the native state of the cell. The development of cryo-SEM allowed cell imaging without the need for critical point drying and gold coating, but since electron microscopy requires vacuum, the imaging of cells under ambient conditions remained illusive. The advent of atomic force microscopy (AFM) presented the opportunity to overcome these limitations and provide the highest possible topographic and lateral resolution of cell surfaces with little or no sample preparation. In addition to its ability to provide ultrastructural information comparable to that obtained by cryo-SEM, AFM can also be used to probe chemical and physical surface characteristics. Integration of AFM with optical microscopes compensates for the microscope's relatively small field of view (100×100 μm). AFM has become a very popular

B.C. Paul • T.E.S. Dahms (✉)
Department of Chemistry and Biochemistry,
University of Regina, Regina, SK, Canada S4S 0A2
e-mail: tanya.dahms@uregina.ca

H. Ma
Department of Chemistry, National University of
Singapore, Singapore

L.A. Snook
Department of Human Health and Nutritional Sciences,
50 Stone Road East, Guelph, Ontario, N1G 2W1 Canada

V.K. Gupta et al. (eds.), *Laboratory Protocols in Fungal Biology: Current Methods in Fungal Biology*,
Fungal Biology, DOI 10.1007/978-1-4614-2356-0_9, © Springer Science+Business Media, LLC 2013

tool in microbiology for observing cell morphology, cell surface ultrastructure, chemical and physical properties of microbes and eukaryotic cells.

The heart of the AFM instrument is the microfabricated tip on the end of a metal-coated cantilever. Three piezoelectric micropositioners (X, Y, and Z) maintain a constant distance between the tip and sample (Z), as the tip is raster scanned (X, Y) over the sample surface (tip scanning) or vice versa (sample scanning). The optical lever, consisting of a laser reflected from the top surface of the cantilever into a four quadrant photodetector, is the most common AFM detection method [1]. Changes in topography and tip–sample interactions give rise to cantilever deflection or lateral movement registered by the photodetector as a vertical or horizontal displacement, respectively. This signal is sent to the computer for data collection and to the electronic control unit (ECU), which adjusts the height (Z) of the AFM tip through a fast feedback loop. Vertical and horizontal displacement of the laser in the photodiode is registered as topography and tip–sample interactions, respectively. For rough samples, lateral force images are convoluted with topographic signals as the tip encounters large changes in the slope, giving rise to the "edge effect," [2] which improves contrast and helps delineate surface feature boundaries for live samples [3].

Immobilization of the sample onto a solid surface is mandatory for imaging cells, fixed or viable, by AFM. Otherwise, raster scanning of the AFM tip can dislodge the sample. Imaging in noncontact mode can alleviate this problem, but it comes at the cost of resolution [3]. Therefore, methods described in this chapter relate to contact mode imaging of both fixed and viable cells.

In contact mode AFM imaging, specimen immobilization will prevent movement during scanning. There are various immobilization techniques, including physical trapping [4], electrostatic interaction by coating glass coverslips [5], and chemical tethering of the cell to the substrate [6]. Early studies by Kasas and Ikai [7] used physical trapping to immobilize cells on filter paper, while more recent studies have introduced a micro-patterned surface for this purpose [8]. The latter methods are appropriate for spherical cells such as certain bacteria or yeast. Most rod-shaped bacteria imaged by AFM [9] have been immobilized through electrostatic interactions or by culturing cells on chemically functionalized solid surfaces. The different surface characteristics of conidia (hydrophobic) and hyphae (hydrophilic) make it difficult to use a chemical method to adhere the entire fungal body to a solid support. Thus, prior to developing a method for imaging live filamentous fungal hyphae [4], the majority of AFM studies had been restricted to spores or yeast [10]. Zhao et al. [11] determined the elasticity of both hydrated and dehydrated *Aspergillus nidulans* by electrostatically immobilizing fixed cells on poly-L-lysine-coated coverslips; however, imaging growing filamentous fungi requires the continuous supply of media, which is not possible with this method. Ma et al. [3] devised a method to image live fungal hyphae by AFM. Dague et al. [12] recently imaged the dynamic germination of *Aspergillus fumigatus* spores immobilized on filter paper by AFM and monitored the effect of antimicrobial agents on germination, but their work has not been extended to live hyphae. Here we present sample preparation techniques for fungal spores, fixed hyphae, and live hyphae, and the methods used for their AFM imaging and analysis by force spectroscopy (FS).

Materials

1. Complete medium [13, 14].
2. Sterilized distilled water.
3. Ultrapure water (18 MΩ, Thermo Scientific).
4. Ethanol (95%, Fisher Scientific).
5. Acetone (Fisher Scientific).
6. Hydrogen peroxide (Fisher Scientific).
7. Sulfuric acid (Fisher Scientific).
8. Octadecyltrichlorosilane (($CH_3(CH_2)_{17}SiCl_3$) $NaNO_3$, Sigma-Aldrich).
9. Compressed air duster (Grand and Toy).
10. Incubator (Queue).
11. Kim wipes (Kimberly-Clark).
12. Filter paper (Whatman, USA).
13. Glass coverslips (22 mm × 22 mm, Fisher Scientific).
14. Petri dish (100 mm × 15 mm, polystyrine, disposable, sterile Petri dish, VWR International).

15. Dialysis tubing (Spectra/Pore® MWCO-8000).
16. Triton-X-100 (Sigma-Aldrich).
17. Cantilevers (Ni$_3$N$_4$ AFM tip; spring constant: 0.05–0.5 nN/nm; tip radius: 10 nm; Bruker).
18. Standard sample (AFM Calibration Grating, Bruker).

Experimental Methods

Here we describe the sample preparation and microscopy methods developed for AFM imaging and force spectroscopy of fungal spores, and germinating and mature fungal hyphae.

Preparation of Fungal Spores for Imaging

Previous studies reported that the spore surface is composed of hydrophobic rodlet (RodA) layers composed of hydrophobin proteins [15], making the cell surface hydrophobic [16]. Since the glass substrate surface is hydrophilic, the hydrophobic spore would not favorably adhere to the glass coverslip. For this reason, the glass surface was chemically modified with silanes to facilitate spore adherence to coverslips through hydrophobic interactions. Physical confinement in a porous membrane by filtering the spore suspension can also be used to image spores by AFM [12], but with the drawback of having the spores exposed to water, which can lead to surface modification through water imbibition and swelling [4]. Thus, imaging spores under dry conditions is the only way to examine its native ultrastructure and physio-chemical surface properties. Steps involved in silanization, all of which are conducted in a fumehood, are described in the following section.

Cleaning Coverslips

Fungal spores are approximately 1–3 μm in diameter [17]; therefore, contaminants on the glass coverslip surface will lead to imaging artifacts and also can adversely affect the silanization

process required to adsorb spores onto the glass surface. Prior to coating, coverslips must be cleaned.

1. Dip coverslips in 1 M HCl for 2 min (see Note 1).
2. Wash coverslips in deionized water and let air dry.
3. Soak coverslips in Piranha solution (5 mL of 30% H$_2$O$_2$ + 15 mL of 18 M H$_2$SO$_4$) for 1 h (see Note 2).
4. Remove the coverslip and wash with copious amounts of deionized water.
5. Dip coverslip in methanol for 2 min and let air dry.
6. Dip coverslip in acetone for 2 min and let air dry.
7. Store the clean coverslips in a dust-free container.
8. Clean coverslips can be sterilized with 70% ethanol or by autoclaving.
9. Clean coverslips can be coated with either poly-L-lysine or a silanization reagent.
10. Coated coverslips can be sterilized with 70% ethanol (see Note 3).

Silanization

1. Immerse glass coverslips in a solution of 2% octadecyltrichlorosilane (see Note 4).
2. Rinse coverslip in 100% hexanes (3×) with gentle shaking, 1 min each.
3. Air dry over night in the fume hood.

Spore Sample Preparation

1. Touch the silanized coverslip to the conidiating culture (spores are produced after 3–4 days culture), so that a thin layer of conidia adhere to the coverslip.
2. Incubate the sample for 1 h at room temperature, allowing spores to settle and form a strong hydrophobic interaction with the coverslip.
3. Remove any nonadhering spores or dust using compressed air. Nonadhering spores on the coverslip will make AFM imaging extremely difficult.

The sample is now ready for AFM imaging in contact mode.

Preparing Fixed and Dehydrated Fungal Hyphae for Imaging

A. nidulans is a filamentous fungus that forms mycelium when it grows for 2–3 days. The height of the mycelium is beyond the limit of many AFM Z-piezo ranges (10–100 μm). While it is possible to view hyphae at low resolution by CCD or light microscopy and visually line up the AFM tip on the surface of structures that exceed the AFM Z piezo limit [18], it is easier to probe a sample by AFM with less height variability. Therefore, two-dimensional hyphal samples were grown for imaging by AFM, whereby hyphae were grown for a short period of time between two laterally aligned coverslips.

Formaldehyde Fixation

1. Prepare spore suspensions by pipetting 10 μL of sterile water onto the conidiating culture.
2. Collect the water from the culture that now contains spores and add it to 90 μL of sterile water in an Eppendorf tube.
3. Mix the spore suspension by vortex.
4. Serially dilute the spore suspension to reduce the spore count (~1/100 final dilution).
5. Add 20 μL of the spore suspension to 980 μL of complete medium (CM) and mix thoroughly by vortex.
6. Pipette 200 μL of the spore suspension onto a sterile, clean glass coverslip laying in a sterile Petri dish.
7. Place a second sterile, clean glass coverslip on top of the first one very carefully so that the medium remains between the two coverslips.
8. Add 2 mL of liquid CM to the Petri dish area surrounding the coverslips to avoid dehydration. Ensure that the additional 2 mL of liquid media added does not come into contact with the coverslips.
9. Incubate the sample at 37 °C for 16 h.
10. Carefully remove the top coverslip and wick off the remaining media from the bottom coverslip using a Kimwipe, being careful not to touch the mycelia.
11. Wash the bottom coverslip with 50 mM warm (37 °C) phosphate buffer (dilute

30.75 mL of 1 M K_2HPO_4 and 19.25 mL of 1 M KH_2PO_4 to 1 L with 18 MΩ water).
12. Treat with fixative solution (see Note 5) (200 μL; 3.7% formaldehyde, 2% of Triton-X-100, in 50 mM phosphate buffer, pH 7.0) and incubate for 10 min.
13. Wash carefully (3×) with deionized water by depositing 100 μL at a time on one corner of the coverslip, allowing the solution to disperse, and wicking the solution from the other side of the coverslip with a Kimwipe.
14. Dry the sample at room temperature in a covered Petri dish.

Fixation by OsO4

To compare AFM and SEM images, samples can be fixed with OsO_4 and then critical point dried and gold-coated [3]. A detailed protocol is given below.

1. Grow hyphae on coverslips following steps 1–10, described in section "Formaldehyde Fixation."
2. Attach the coverslip to the lid of a Petri dish using double-sided tape and invert each plate over approximate 200 μL of OsO_4 (4% aqueous) in the Petri dish. Incubate for at least 2 h at RT until the highly volatile OsO_4 completely penetrates the hyphae (agar is dark from Os accumulation).
3. Freeze the fixed cells by plunging into cold anhydrous acetone (60 mL, −80 °C) and dehydrate for at least 4 h.
4. Then warm the sample slowly to room temperature (2 h at −20 °C, 2 h at 4 °C, 2 h at RT) to avoid condensation.
5. Dehydrate the sample (Emitech K850 Critical Point Dryer; Quorum Technologies, UK). In this method acetone is exchanged for liquid CO_2 (in a pressure chamber), which is heated through its phase transition temperature, becoming gas at the critical point.
6. The sample can then be imaged, or gold sputter coated (Emitech SC7620 Sputter Coater; Quorum Technologies, UK) for direct comparison with traditional SEM data. Gold-coated samples are more robust, and so can be imaged for several days after preparation.

7. Mount the coverslip on the AFM sample holder for imaging. Gold-coated samples, with a harder surface, offer superior contact mode AFM images.

Preparing Viable Fungal Hyphae

Many microbes have been imaged live by AFM; however, viable filamentous fungi have not been as well explored with this method. Many previous studies focus on unicellular, spherical yeast, which are amenable to physical trapping for AFM imaging and force spectroscopy [19–21]. Fixed and rehydrated *A. nidulans* fungal hyphae and spores were probed by FS and imaged [11, 22], and the surface dynamics of *A. fumigatus* spores have been probed in liquid medium during germination [16]. However, to date only our laboratory has studied the cell wall ultrastructure and physical properties of viable hyphae, from spore to germling [4, 12, 23, 24]. Below we describe the steps used to prepare samples of live germlings and hyphae for AFM imaging.

1. Place a piece of sterile (boiled) dialysis tubing membrane on a Petri dish containing agar media.
2. Inoculate the membrane with the fungal spore suspension (1 µL of 1/100 dilution). (Follow step 1 of section "Spore Sample Preparation" for spore suspension.)
3. Incubate at 37 °C for 16 h.
4. Following adequate hyphal growth, remove the membrane from the Petri dish for transfer to the AFM.

Imaging Spores and Fixed Hyphae by AFM

1. After the 30-min warm-up, mount the sample onto the AFM stage using double-sided tape (see Note 6).
2. Locate the specimen visually with the CCD camera or optical microscope (see Note 7).
3. Bring the AFM cantilever very close (~50 µm) to the sample surface, but far enough away to

avoid crashing the tip into the surface and breaking the cantilever.

4. For instruments with an optical lever design, the laser must be aligned on the portion of the cantilever directly above the AFM tip. Ideally, this region would correspond to an optimal signal (see Note 8).
5. Adjust the mirror between the laser and detector to give an optimum signal in the four-quadrant photodetector, which monitors the position of the AFM cantilever and thus tip response.
6. Many AFM instruments allow the sensor response, which is a plot of distance vs. cantilever deflection, to be tested. For a properly aligned laser and a new AFM cantilever, the sensor response should be linear. If not, disengage the tip, readjust the laser and mirror, and then reengage the tip. If it is not possible to obtain a linear sensor response, it is likely that the cantilever is old, and it is worth using a new cantilever.
7. Use the line scan function in topography mode to evaluate instrument feedback. Each time the same line of the sample is scanned, the topographic patterns should overlap. If not, adjust the input gains, scan speed and other parameters until this is achieved.
8. Once instrument feedback has been established, which is crucial for AFM imaging, a large field of view (100×100 µm) can be imaged at low resolution (200×200 pixels) to identify spores or hyphae of interest (see Note 9).
9. The zoom function allows a smaller field of view (<500 nm $\times 500$ nm) for imaging the surface of single spores or hyphae at higher resolution (up to $1,000 \times 1,000$ pixels).

These methods have been used to resolve 10 nm hydrophobin rodlets on the native spore surface (Fig. 9.1a) and 25 nm subunits on the hyphal surface of viable specimens (Fig. 9.1b) [4].

AFM Imaging of Live Hyphae

After growing hyphae on dialysis tubing membrane following the steps described in the section

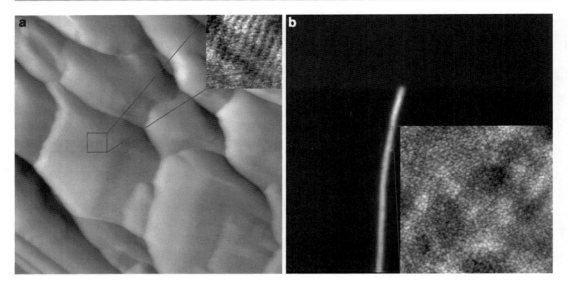

Fig. 9.1 Atomic force microscopy (AFM) images of *A. nidulans* spore and fixed hyphal cell surface. (**a**) High-resolution AFM image (1.5 μm × 1.5 μm, 1,000 × 1,000 pixel) reveals the surface structure of a dry spore; and *inset* is the image of the 10 nm-wide rodlets. (**b**) AFM image (80 μm × 80 μm) of hyphae fixed with OsO_4 and critical point dried at low resolution (200 × 200 pixel). The *inset* (1.5 μm × 1.5 μm, 1,000 × 1,000 pixel) shows the ultrastructure of the same hyphal cell wall surface resolving 25 nm surface subunits. Images were adapted and reprinted with permission from reference [4]

"Preparing Viable Fungal Hyphae," the sample is mounted on the AFM and imaged following the steps described below:

1. Attach the dialysis tubing membrane to a glass coverslip with double-sided tape.
2. Insert a small piece of filter paper into the dialysis membrane tube (see Note 10).
3. Mount the coverslip onto the sample holder of the AFM with double-sided tape.
4. Add media dropwise to the filter paper to deliver media nutrients through the membrane to the hyphae during imaging (see Note 11).
5. Make sure that the dialysis tubing remains moist; otherwise it will dry out into a highly convoluted structure that is too rough for AFM imaging (Fig. 9.2) (see Note 12).
6. Follow the steps described in steps 2–8 of the section "Imaging Spores and Fixed Hyphae by AFM" for imaging live hyphae.

Force Spectroscopy of Live and Fixed Hyphae

AFM is not only a tool for nanometer-scale imaging, but can also be used for force spectroscopy, in which the tip approaches and retracts from the sample surface. A force curve is generated by plotting cantilever deflection vs. tip distance from the surface in the Z direction, and this is converted to force vs. distance using the spring constant of the cantilever. Newer instruments have combined the raster scanning and force spectroscopy capabilities to produce force maps. The same principles apply, but rather than probing only a single point beneath the tip, force curves can be collected over a large surface area of the cell. When the AFM tip exerts a small amount of force, the degree of cell indentation will depend on the viscoelasticity of both the cell and cantilever spring constant [25]. The slope of

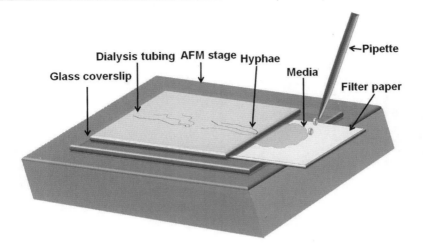

Fig. 9.2 Schematic of the assembly for AFM imaging of hyphae on dialysis tubing. Dialysis tubing is fixed on a coverslip with double-sided tape then placed on the AFM stage. Media is delivered from underneath the dialysis tubing through a piece of filter paper. More sophisticated environmental chambers on new AFMs can be used for media delivery

the approach curve into the cell surface in conjunction with appropriate models can be used to measure the spring constant of the cell, cell wall, or envelope, while the last portion of the retraction cycle reflects tip–sample adhesion. For acquisition of force spectroscopic data for both live and fixed hyphae in specific regions, it is necessary to image the sample first using the method described above, followed by the steps given below:

1. Determine the spring constant of the cantilever based on its resonant frequency [26], or calibrate the AFM cantilever using the method described by Gibson et al. [27] to ensure accurate force data (see Note 13).
2. After mounting the sample on the AFM stage, steps described in the section "Imaging Spores and Fixed Hyphae by AFM" should serve as a guide for imaging.
3. Switch to force spectroscopy or force mapping mode, allowing measurement at any given point or across the entire surface, respectively (see Note 14).
4. For *A. nidulans* hyphal samples, the tip is initialized 1 μm from the sample surface.
5. Adjust the approach speed of the AFM tip during force curve acquisition, which will vary with each AFM system. For force mapping, the approach speed will determine the force map acquisition time [28].
6. Image and acquire FS data for the sample surface, substrate surface, and a hard surface as a reference.

Determination of Hyphal Viscoelasticity

Viscoelasticity can report on whole cell turgor and cell wall integrity. The spring constant of the hyphal cell, K_w can be determined from the following equation using the slope, S, of the approach force–distance curve (Fig. 9.3, section b-c). A force curve from a hard surface is used to determine the slope of the approach portion, designated S_h.

$$K_w = K_c S / (S_h - S) \qquad (9.1)$$

where K_c is the spring constant of the cantilever used to acquire the force curve of the hyphal sample determined according to step 1 in section "Force Spectroscopy of Live and Fixed Hyphae."

Once K_w has been calculated, the viscoelasticity of the hyphal cell wall can be calculated according to the following equation [11]:

$$E = 0.80(K_w / h)(R / h)^{1.5} \qquad (9.2)$$

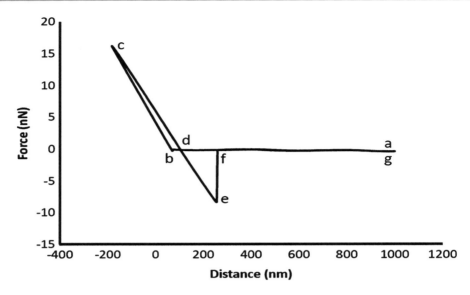

Fig. 9.3 Force curve from the live hyphal cell surface of *A. nidulans*. Representative force vs. distance curve of live *A. nidulans* hyphae. The slope of the b–c region can be used to determine the mechanical properties of cell using appropriate models. The e–f region describes adhesion between AFM tip and sample surface in pN, which will depend on the chemical properties of AFM tip and sample

where h is the cell-wall thickness measured by TEM and R is the radius of hyphal cell measured by AFM or TEM.

Such models hold for the conditions for which they have been developed, in this case hyphae with intact cell walls and relatively small surface subunits. However, the model may no longer fit data from hyphae with compromised cell walls [13].

Determining Surface Adhesion

The adhesion force between tip and sample depends on the chemical nature of the tip and sample surface. Adhesion force measurements are useful to determine the relative hydrophobicity, hydrophilicity, or electrostatic properties of the sample [29]. If chemically functionalized tips are used, the force of binding between the two molecules (FS) or the surface distribution of a particular molecule (FM) can be determined. The last portion of the retract curve is used to determine the adhesion force in pN between the sample and AFM tip as the distance between points e and f of the force–distance curve (see Fig. 9.3). Adhesion values can provide valuable insight on carbohydrate remodeling, for instance as a function of polarized hyphal growth [4] or cell wall biosynthetic enzyme mutations [13].

Notes

1. If preparing a large number of coverslips, a home-built rack can be used for dipping, since it is important to keep coverslips separate to ensure both proper cleaning and subsequent coating of the surface.
2. *Caution*: Piranha solution is extremely caustic and volatile. Work in a fume hood double-gloved.
3. Coated coverslips should not be autoclaved, as the coating will degrade at high temperatures.
4. Immersion of the coverslip in octadecyltrichlorosilane allows formation of covalent bonds between silane and silicate groups of the glass surface, resulting in the formation of a self-assembled hydrocarbon monolayer on the coverslip surface. Hence, the coverslip surface is transformed from being hydrophilic to hydrophobic.

5. Mix 10 mL of 37% formaldehyde and 0.2 mL of 1.7 M Triton-X-100 and dilute to 100 mL with 50 mM potassium phosphate buffer.

6. For instruments with optical lever detection, the laser will increase cantilever temperature over time, so it is best to align the laser onto the end of the cantilever to allow it to stabilize (~30 min). The piezoelectric scanners perform best when warm (many labs who use AFM for QC find images improve over time, Asylum, personal communication), so it is worthwhile to allow the AFM to raster scan for at least 30 min with the tip out of feedback prior to imaging.

7. It is very useful to first examine AFM samples by light microscopy to visualize the position of spores, germlings, hyphae, or mycelia. Areas of interest can be marked underneath the glass coverslip using a thin-tipped marker to make them visible under the CCD camera. It is very difficult to view spores with the CCD camera (×200), so a large area of the coverslip (100 × 100 μm) must first be imaged at low resolution (200 × 200 pixels) to locate spores. If the AFM is mounted on a light microscope with ×40 or ×60 objectives, at this magnification (×400–×600) it will be possible to view the spores, eliminating the need to image large areas at low resolution or premark areas of interest on the coverslip using a separate light microscope.

8. It can be difficult to position the laser on the cantilever by viewing through a CCD camera. One way to tell that the red laser is actually hitting the surface of the cantilever is if a diffraction pattern appears below the cantilever.

9. Look for an area on the coverslip with limited three-dimensional growth and adequate space between hyphae to allow access to a single hypha. Make sure the mycelium is not positioned beneath the free end of cantilever. The AFM head can be tilted (front down, back up) slightly to avoid the latter situation which will ultimately prevent the tip from engaging in feedback.

10. Cut the filter paper to a width slightly smaller than that of the dialysis tubing, and to a length slightly longer, such that part of the filter paper remains outside the membrane for media delivery.

11. During live imaging, 2 drops (~10 μL) of media must be delivered to the sample every 30 min to keep the hyphae viable. Care should be taken when delivering media, since excess liquid on the membrane surface will create extreme capillary forces that interfere with AFM feedback. As a result, it can be difficult to establish and maintain feedback while imaging live samples, requiring AFM imaging parameters to be adjusted more frequently.

12. During live hyphal imaging, if the AFM tip is pulled to the surface of the dialysis tubing by capillary forces, feedback will be destabilized, so extra care should be taken during media delivery. In some cases, reducing the set point can also solve this problem.

13. Varying geometry and characteristics of AFM probes may cause image distortion and overestimation of the lateral size (X and Y) of small features. To minimize experimental error, AFM probes can be calibrated using an AFM calibration grating with known dimensions, or nanometer-scale gold spheres [4, 30]. Accurate determination of physical parameters such as hardness, friction or adhesion requires an accurate cantilever spring constant, which can be derived from the spectrum of its thermal vibration in air.

14. Note that most modern instruments use a mode called force mapping, which combines force spectroscopy with raster scanning, creating a map of the sample surface describing tip–sample interactions and sample compression. Force spectroscopy still remains a useful method for fast surface analysis.

References

1. Allison DP, Mortensen NP, Sullivan CJ, Doktycz MJ (2010) Atomic force microscopy of biological samples. Wiley Interdiscip Rev Nanomed Nanobiotechnol 2:618–634

2. Nie H, Walzak MJ, McIntyre NS, El-Sherik AM (1999) Application of lateral force imaging to enhance

topographic features of polypropylene film and photo-cured polymers. Appl Surf Sci 144–145:633–637

3. Ma H, Snook LA, Kaminskyj SG, Dahms TE (2005) Surface ultrastructure and elasticity in growing tips and mature regions of *Aspergillus* hyphae describe wall maturation. Microbiology 151:3679–3688

4. Meyer LR, Zhou X, Tang L, Arpanaei A, Kingshott P, Besenbacher F (2010) Immobilisation of living bacteria for AFM imaging under physiological conditions. Ultramicroscopy 110:1349–1357

5. Schar-Zammaretti P, Ubbink J (2003) The cell wall of lactic acid bacteria: surface constituents and macromolecular conformations. Biophys J 85:4076–4092

6. Bearinger J, Dugan L, Wu L, Hill H, Christian A, Hubbell J (2009) Chemical tethering of motile bacteria to silicon surfaces. Biotechniques 46:209–216

7. Kasas S, Ikai A (1995) A method for anchoring round shaped cells for atomic force microscope imaging. Biophys J 68:1678–1680

8. Miyahara Y, Mitamura K, Saito N, Takai O (2009) Fabrication of microtemplates for the control of bacterial immobilization. J Vac Sci Technol 27:1183–1185

9. Doktycz MJ, Sullivan CJ, Hoyt PR, Pelletier DA, Wu S, Allison DP (2003) AFM imaging of bacteria in liquid media immobilized on gelatin coated mica surfaces. Ultramicroscopy 97:209–216

10. Dufrêne YF (2010) Atomic force microscopy of fungal cell walls: an update. Yeast 27:465–471

11. Zhao L, Schaefer D, Xu H, Modi SJ, LaCourse WR, Marten RM (2005) Elastic properties of the cell wall of *Aspergillus nidulans* studied with atomic force microscopy. Biotechnol Prog 21:292–299

12. Dague E, Alsteens D, Latgé JP, Dufrêne YF (2008) High-resolution cell surface dynamics of germinating *Aspergillus fumigatus* conidia. Biophys J 94:656–660

13. Paul BC, El-Ganiny AM, Abbas M, Kaminskyj SG, Dahms TE (2011) Quantifying the importance of galactofuranose in *Aspergillus nidulans* hyphal wall surface organization by atomic force microscopy. Eukaryot Cell 10:646–653

14. Kaminskyj S (2001) Fundamentals of growth, storage, genetics and microscopy in *Aspergillus nidulans*. Fungal Genet Newsl 48:25–26

15. Hess WM, Stocks DL (1969) Surface characteristics of *Aspergillus* conidia. Mycologia 61:560–571

16. Girardin H, Paris S, Rault J, Bellon-Fontaine MN, Latgé JP (1999) The role of the rodlet structure on the physicochemical properties of *Aspergillus* conidia. Lett Appl Microbiol 29:364–369

17. Latgé JP (1999) *Aspergillus fumigatus* and aspergillosis. Clin Microbiol Rev 12:310–340

18. Wyatt HDM, Ashton NW, Dahms TE (2008) Cell wall architecture of *Physcomitrella patens* is revealed by atomic force microscopy. Botany 86:385–397

19. Dague E, Bitar R, Ranchon H, Durand F, Yken HM, Francois JM (2010) An atomic force microscopy analysis of yeast mutants defective in cell wall architecture. Yeast 27:673–684

20. Heinisch JJ, Dufrêne YF (2010) Is there anyone out there? Single-molecule atomic force microscopy meets yeast genetics to study sensor functions. Integr Biol (Camb) 2:408–415

21. Dupres V, Dufrêne YF, Heinisch JJ (2010) Measuring cell wall thickness in living yeast cells using single molecular rulers. ACS Nano 4:5498–5504

22. Zhao L, Schaefer D, Marten RM (2005) Assessment of elasticity and topography of *Aspergillus nidulans* spores via atomic force microscopy. Appl Environ Microbiol 71:955–960

23. Ma H, Snook LA, Tian C, Kaminskyj SG, Dahms TE (2006) Fungal surface remodelling visualized by atomic force microscopy. Mycol Res 110:879–886

24. Kaminskyj SG, Dahms TE (2008) High spatial resolution surface imaging and analysis of fungal cells using SEM and AFM. Micron 39:349–361

25. Arnoldi M, Kacher CM, Bauerlein E, Radmacher M, Fritz M (1998) Elastic properties of the cell wall of *Magnetospirillum gryphiswaldense* investigated by atomic force microscopy. Appl Phys A66:613–614

26. Cleveland J, Manne S (1993) A nondestructive method for determining the spring constant of cantilevers for scanning force microscopy. Rev Sci Instrum 64:403–405

27. Gibson TC, Weeks LB, Lee RIJ, Abell C, Rayment T (2001) A nondestructive technique for determining the spring constant of atomic force microscope cantilevers. Rev Sci Instrum 72:2340–2343

28. Kim KS, Lin Z, Shrotriya P, Sundararajan S, Zou Q (2008) Iterative control approach to high-speed force-distance curve measurement using AFM: time-dependent response of PDMS example. Ultramicroscopy 108:911–920

29. Dufrêne YF (2000) Direct characterization of the physicochemical properties of fungal spores using functionalized AFM probes. Biophys J 78:3286–3291

30. Arnsdorf FM, Xu S (1994) Atomic (scanning) force microscopy in cardiovascular research. J Cardiovasc Electrophysiol 7:639–652

Vitaly Erukhimovitch and Mahmoud Huleihel

Abstract

Reliable and rapid identification of the fungal pathogens that cause plant diseases is playing an important role in their control strategies. The available methods for identification of fungi are time-consuming and not always very specific. Fourier-transform infrared microscopy is proving to be a reliable and sensitive method for detection of molecular changes in cells. Fungi pathogens display typical infrared spectra that differ from spectra of substrate material such as potato, which make it is possible to detect and identify such pathogens directly from the infected tissue.

In addition, although different strains of the same fungi species display very similar infrared spectra, there are specific spectral differences between them that might be successfully used, with the assistance of advanced statistical methods, for the identification of these fungal strains.

Keywords

Fungal pathogens • Fungal detection • Fungi • Fourier-transform infrared microscopy • Spectral biomarkers • Potato

Introduction

Fungal pathogens are considered one of the most common causes of severe diseases in various plants. Infection with fungal pathogens can lead, in many cases, to a great deal of economic damage [1]; consider, for instance, *Colletotrichum coccodes*, a major pathogen of potato and tomato [2]. Infected seed tubers are a major source of contamination of field soils and storage areas. In many cases, it is difficult to detect low levels of contamination in the early stages of infection [3]. Early identification enables one to precisely target

V. Erukhimovitch
Ben-Gurion University of the Negev,
Analytical Equipment Unit, POB 653,
Beer-Sheva 84105, Israel
e-mail: evitaly@bgu.ac.il

M. Huleihel (✉)
Department of Virology and Developmental Genetics,
Ben-Gurion University of the Negev, POB 653,
Beer-Sheva 84105, Israel
e-mail: mahmoudh@bgu.ac.il

V.K. Gupta et al. (eds.), *Laboratory Protocols in Fungal Biology: Current Methods in Fungal Biology*,
Fungal Biology, DOI 10.1007/978-1-4614-2356-0_10, © Springer Science+Business Media, LLC 2013

a pathogen with the most effective treatment, thereby preventing large economic damage. Most commercially available identification systems are based on the physiological and nutritional characteristics of fungi. Such identification systems are usually time-consuming (2–4 weeks) and not always very specific. Polymerase chain reaction (PCR)-based methods developed for the detection and identification of plant pathogenic fungi are rapid and sensitive [4, 5]. Primers, designed to conserve regions of the internal transcriber spacer regions within ribosomal gene clusters, have been used to detect and identify plant pathogenic fungi [6]. Although this method is promising, it is not yet in large-scale use and is expensive.

Fourier-transform infrared (FTIR) spectroscopy is considered valuable because of its sensitivity, rapidity, low expense, and simplicity. These factors, together with the large information already known about spectral peaks obtained from FTIR spectra of living cells [7], make it an attractive technique for detection of pathogens. This technique was used for the detection and characterization of cancer cells [8, 9], cells infected with viruses [10], and microorganisms, including some fungi [11–16].

Materials

1. Zinc sellenide crystals.
2. Sterile distilled water.
3. Vortex.
4. Water bath.
5. Incubator shaker.
6. Microcentrifuge.
7. Pipators.
8. Sterile tips (in different sizes) for the appropriate pipators.
9. Disposable polypropylene microcentrifuge tubes (2 mL screw-capped).
10. Potato Dextrose Agar (Difco).
11. Potato Dextrose Broth medium (Difco).
12. Fungi: Different fungi strains were obtained from ATCC for examination by FTIR microscopy.
13. Infected tissues (such as potatoes) with known strains of fungi.

Method

Procedures Used for Preparation of Fungi Samples

FTIR microscopy method is applied in our studies for the detection and identification of different fungal species and strains that are isolated and purified from culture media or directly examined on the infected tissues.

1. Fungi growing on solid media: The examined fungi are cultivated for several days (3–6 days) on Potato Dextrose Agar (PDA, Difco) at 27 °C. Small aliquots of fungi are picked up from the growing fungi on the agar with a bacteriological loop, suspended in 500 μL of sterile distilled H_2O, pelleted by centrifugation at 2,000 rpm for 5 min. Each pellet is washed twice with sterile distilled H_2O (by suspending it with 500 μL of sterile distilled H_2O and pelleting by centrifugation at 2,000 rpm for 5 min) and suspended with 50 μL of sterile distilled H_2O.

2. Fungi growing in liquid media: The examined fungi are cultivated and identified using classical microbiological techniques [2]. Briefly, samples of the fungi are cultivated and maintained in Potato Dextrose Broth media (Difco). These cultures are grown for 3–10 days at continuous shaking conditions and at a temperature of 27 °C (the growth time depends on the fungi species).

3. Samples of these fungi are purified from these media by spinning about 1 mL of medium containing fungi at 2,000 rpm for 5 min, washing twice with sterile distilled H_2O and the pellet is suspended in appropriate volume (about 50 μL) of sterile distilled H_2O.

4. Direct examination of fungi from infected tissue: Small epidermal aliquots of samples are scratched from the infected (or uninfected) areas on the surface of tissue (such as potato tubers), suspended in 500 μL of sterile distilled H_2O, pelleted by centrifugation at 1,000 rpm for 5 min. Each pellet is washed twice with H_2O and re-suspended with 50 μL of sterile distilled H_2O.

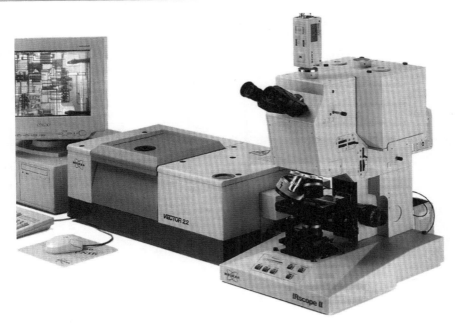

Fig. 10.1 A representative photo of the FTIR microscopy instrument used in this study

Preparation of Slides for FTIR Microscopy Examination

Since ordinary glass slides exhibit strong absorption in the wavelength range of interest to us, zinc sellenide crystals, which are highly transparent to IR radiation, are used. A drop of 5 μL of the obtained suspension (containing the fungal sample as described above in steps 1–3) is placed on the zinc sellenide crystal, air dried for 20–30 min at room temperature (or for 5–10 min by air drying in a laminar flow) until all the water had evaporated, and then examined by FTIR microscopy.

FTIR Spectra Measurement

FTIR measurements are performed in transmission mode with a liquid nitrogen-cooled MCT detector of FTIR microscope (Bruker IRScope II) coupled to an FTIR spectrometer (Bruker Equinox model 55/S, OPUS software) (Fig. 10.1). Figure 10.2 demonstrates a principal scheme of FTIR microscope. The spectra are obtained in the wave number range of 600–4,000 cm^{-1} in the

mid-IR region. A spectrum is taken as an average of 128 scans to increase the signal/noise ratio, and the spectral resolution was at 4 cm^{-1} with Backman Harris 4-Term adopization function. Since the obtained samples are heterogeneous in many cases, appropriate regions are chosen by FTIR microscope out of different impurity (salts, medium residuals, etc.). The optimal aperture used in this study was 100 μm, since this aperture gave the best signal/noise ratio. At lower apertures, the quality of the spectra is bad owing to high levels of noise. In addition, at apertures lower than 20 μm, there is diffraction of the IR beam. Baseline correction and normalization are obtained for all the spectra by OPUS software. Baseline correction is performed by the rubber band as follows: each spectrum is divided up into ranges of equal size. In each range, the minimum y-value is determined. The baseline is then created by connecting the minima with straight lines. Starting from "below" a rubber band stretched over this curve constituted the baseline. The baseline points that do not lie on the rubber band are discarded. Normalization is performed by a vector method, as follows. The average y-value of

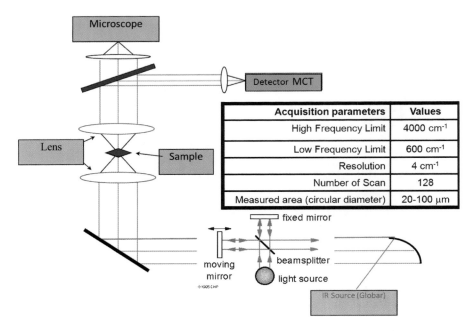

Fig. 10.2 A representative diagram of the sample measurement by FTIR microscopy with different parameters used in these measurements

the spectrum is first calculated. This average value is then subtracted from the spectrum so that the middle of the spectrum was pulled down to $y = 0$. The sum of the squares of all the y-values is then calculated, and the spectrum is divided by the square root of this sum. The vector norm of the resulting spectrum is 1. Peak positions are determined using second derivation.

FTIR Spectra Analysis

The obtained spectra are analyzed for specific regions which show distinct differences between normal uninfected and infected tissues with fungi. Also, similar analysis is done for the differentiation between different fungi strains.

For instance, Fig. 10.3 shows the FTIR spectra of both uninfected and infected samples obtained from potato tubers. Although there is a very high similarity between the spectra of the infected and the uninfected control tissues, some of the characteristic fungal bands appeared in the spectra of the infected tissue samples (while

they are missing in the spectra of the control uninfected tissue samples) as follows:

1. A significant band is found in all examined infected samples at $1,545 \, cm^{-1}$ (see Fig. 10.3b), but it is missing in the control uninfected samples.
2. A peak at $1,405 \, cm^{-1}$ is found in infected potato samples, but it is missing in control uninfected samples (see Fig. 10.3c).

These markers seem to be very characteristics of the fungal spectra compared to the spectra of uninfected potato epidermal samples as shown above (see Fig. 10.3).

Statistical Analysis

Cluster Analysis

The obtained spectral results of infected and control uninfected potato tissues are classified using cluster analysis. Cluster analysis (CA) is an unsupervised technique that examines the interpoint distances between all the samples and presents that information in the form of a two-dimensional

Fig. 10.3 FTIR spectra of purified fungi (*Colletotrichum*) grown in standard medium, control uninfected potato samples, and potato samples infected with fungi obtained by scratching technique. The obtained spectra were examined in various regions: (**a**) of 600–2,000 cm⁻¹, (**b**) 1,480–1,600 cm⁻¹, and (**c**) 1,340–1,460 cm⁻¹

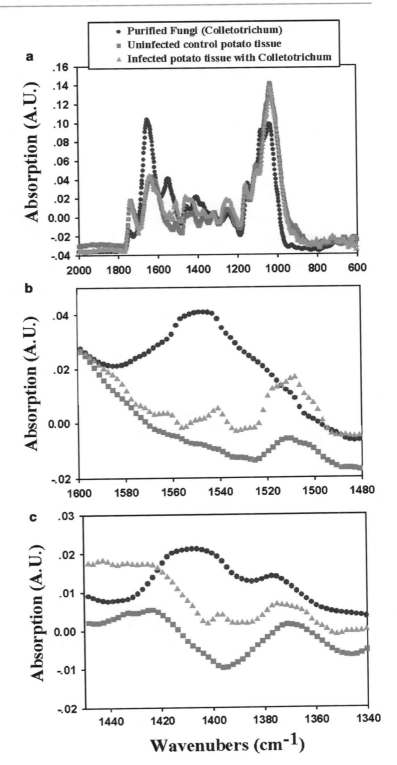

plot known as a dendrogram. These dendrograms present the data from high-dimensional row spaces in a form that facilitates the use of human pattern recognition abilities. To generate the dendrogram, CA methods form clusters of samples based on their nearness in row space. A common approach is to initially treat every sample as a cluster and to join the closest clusters together. This process is repeated until only one cluster remains. Cluster analysis was performed according to Ward's algorithm by OPUS software.

Principal Component Analysis

Principal component analysis (PCA) is a standard tool in modern data analysis [17, 18]. It is a common approach for the reduction of dimensionality. In the transformed space the data are uncorrelated, but not statistically independent. It is widely used in identification problems, with the assumption that the most separable directions are those with the highest variance. This is frequently the case, but not always. It is easy to show a scenario with most separable direction with much lower variance than the maximal variance one.

Basically, PCA is a mathematical algorithm that reduces the dimension of the problem dealt with. In other words, instead of using many variables, the variability in the data is described using only a few principal components [19].

The first linear combination is called the first principal component (PC1), and contains in region III approximately 62.5% of the variance. The second principal component (PC2) accounts for most of the residual variance and is perpendicular to the first one. The subsequent principal components obey the same rules. This method allows the reduction of our spectra to 36 variables in the lower and higher wave number regions that account for almost 100% of the variance [20].

Notes

1. It is important to dry completely the examined samples because water spectral band may overlap important spectral bands of the tested sample. Therefore, air drying in a laminar flow or with a fan might be helpful and recommended.
2. The 5-μL drop of sample (fungi) should be placed on the zinc sellenide crystal as a concentrated drop and then slightly spread on the crystal.
3. When choosing by microscope the aperture of the sample to be scanned, it is important to choose an aperture with confluent fungal cells.
4. Be careful not to choose possible contaminants such as salts, rather than fungi, for scanning.
5. Fungi are usually constructed of vast numbers of threads called hyphae, which are tangled together, rather than arranged in an organized structure. This morphology makes it difficult to dissolve fungi in water or spread them on surfaces. FTIR-microscopy measurements require the smearing of the measured specimen on the crystal surface, thus presenting a serious challenge to the fungi measurement. The fungi were torn into small pieces and mixed as evenly as possible within the distilled water before taking and placing the sample on the zinc sellenide crystals.
6. After placing the sample containing the fungi on the zinc sellenide crystals, try to spread the drop as carefully as possible in order to obtain homogeneous, thin (about 20 μm) layers.
7. When the samples placed on zinc sellenide crystal for FTIR microscopy examination are obtained from fungi growing on solid or in liquid media, the optimal signal/noise ratio is achieved using an aperture of 100 μm. However, when the samples are scratched from areas on the surface of tissue, part of these pieces may be with a radius smaller than 100 μm. Therefore, in these cases only large pieces (than 100 μm) should be selected for examination with 100-μm aperture. If smaller pieces are examined, it is necessary to decrease the aperture size in order to get spectra only from the examined sample without unrelated around regions.

References

1. Agrios GN (1997) Plant pathology. Academic, New York
2. Tsror (Lahkim) L, Erlich O, Hazanovsky M (1999) Effect of *Colletotrichum coccodes* on potato yield, tuber quality, and stem colonization during spring and autumn. Plant Dis 83:561–565
3. Bang U (1986) Effects of planting potato tubers attacked by *Phoma exigua* var. *foveata* on yield and contamination of progeny tubers. Potato Res 29:321–331
4. Chu PWG, Waterhouse PM, Martin RR, Gerlach WL (1989) New approaches for the detection of microbial plant pathogens. Biotechnol Genet Eng Rev 7:45–111
5. Errampalli D, Saunder J, Cullen D (2001) A PCR-based method for detection of potato pathogen, Helminthosporium solani, in silver scurf infected tuber tissue and soils. J Microbiol Methods 44:59–68
6. White TJ, Bruns T, Lee S, Taylor J (1990) Amplification and direct sequencing of fungal ribosomal RNA genes for phylogenetics. In: Innis MA, Gelfand DH, Sninsky JJ, White TJ (eds) PCR protocols, a guide to methods and applications. Academic, San Diego, pp 315–322
7. Diem M, Boydstom-White S, Chiriboga L (1999) Infrared spectroscopy of cells and tissues: shining light onto a novel subject. Appl Spectrosc 53: A148–A161
8. Rigas B, LaGuardia K, Qiao L, Bhandare PS, Caputo T, Cohenford MA (2000) Infrared spectroscopic study of cervical smears in patients with HIV: implications for cervical carcinogenesis. J Lab Clin Med 35:26–31
9. Huleihel M, Erukhimovitch V, Talyshinsky M, Karpasas M (2002) Spectroscopic characterization of normal primary and malignant cells transformed by retroviruses. Appl Spectrosc 56:640–645
10. Salman A, Erukhimovitch V, Talyshinsky M, Huleihil M, Huleihel M (2002) FTIR-spectroscopic method for detection of cells infected with herpes viruses. Biopolymers 67:406–412
11. Naumann D, Helm D, Labischinski H (1991) Microbiological characterizations by FT-IR spectroscopy. Nature 351:81–82
12. Gordon SH, Jones RW, McClelland JF, Wicklow DT, Greene RV (1999) Transient infrared spectroscopy for detection of toxigenic fungi in corn: potential for on-line evaluation. J Agric Food Chem 47:5267–5272
13. Mariey L, Signolle JP, Amiel C, Travert J (2001) Discrimination, classification, identification of microorganisms using FTIR spectroscopy and chemometrics. Vib Spectrosc 26:151–159
14. Maquelin K, Kirschner C, Choo-Smith LP, Ngo-Thi NA, Vreewijk V, Stammler M et al (2003) Prospective study of the performance of vibrational spectroscopies for rapid identification of bacterial and fungal pathogens recovered from blood cultures. J Clin Microbiol 41:324–329
15. Erukhimovitch V, Tsor L, Hazanovsky M, Talyshinsky M, Mukmanov I, Souprun Y et al (2005) Identification of fungal phyto-pathogens by Fourier-transform infrared (FTIR) microscopy. J Agric Technol 1:145–152
16. Naumann A, Navarro-Gonzalez M, Peddireddi S, Kues U, Polle A (2005) Fourier transform infrared microscopy and imaging: detection of fungi in wood. Fungal Genet Biol 42:829–835
17. Camastra F, Vinciarelli A (2008) Machine learning for audio, image and video analysis. Springer, London
18. Duda RO, Hart PE, Stork DG (2001) Pattern classification, 2nd edn. Wiley, Hoboken
19. Zwielly A, Gopas J, Brkic G, Mordechai S (2009) Discrimination between drug-resistant and non-resistant human melanoma cell lines by FTIR spectroscopy. Analyst 134:294–300
20. Diem M, Griffith P, Chalmers J (2008) Vibrational spectroscopy for medical diagnosis. Wiley, Chichester

Diagnosis of Parasitic Fungi in the Plankton: Technique for Identifying and Counting Infective Chytrids Using Epifluorescence Microscopy

Télesphore Sime-Ngando, Serena Rasconi, and Mélanie Gerphagnon

Abstract

Fungal epidemics, especially in the form of parasitic chytrids, are omnipresent in aquatic environments, infecting diverse organisms. Major target hosts are algae, primarily diatoms, chlorophytes, and colonial or filamentous cyanobacteria. Chytrids are also called "zoosporic" organisms because their life cycle includes dispersal forms, that is, uniflagellate zoospores, and host-associated infective sporangia. They are considered relevant not only for the evolution of their hosts but also for the population dynamics and successions of phytoplankton communities, thus representing an important ecologically driving force in the food web dynamics. However, ecological knowledge of microscopic fungal parasites in aquatic environments is weak, compared to terrestrial ecosystems. We propose a routine protocol based on size fractionation of pelagic samples and the use of the fluorochrome calcofluor white (which binds to β-1,3 and β-1,4 polysaccharides) for diagnosing, identifying, and counting chitinous fungal parasites (i.e., the sporangia of chytrids). The protocol offers a valid method for the quantitative ecology of chytrid epidemics in aquatic ecosystems and food web dynamics.

Keywords

Direct counting method • Fungi • Sporangia • Parasitism • Environmental samples

T. Sime-Ngando (✉) • M. Gerphagnon
UMR CNRS 6023, Université Blaise Pascal,
Clermont II, 24 Avenue des Landais, BP 80026,
63171 Aubière Cedex, France
e-mail: telesphore.sime-ngando@univ-bpclermont.fr

S. Rasconi
Department of Biology, University of Oslo,
Blindernvn. 31, Oslo, 0371, Norway

Introduction

Fungal infections are recurrent in aquatic ecosystems [1, 2]. The most described aquatic fungi in freshwater ecosystems belong to Chytridyomycota (or chytrids). Chytrids infect a wide variety of hosts, including fishes, eggs, zooplankton, and other aquatic fungi but especially phytoplankton.

V.K. Gupta et al. (eds.), *Laboratory Protocols in Fungal Biology: Current Methods in Fungal Biology*,
Fungal Biology, DOI 10.1007/978-1-4614-2356-0_11, © Springer Science+Business Media, LLC 2013

Typical phytoplankton hosts include prokaryotes and eukaryotes, primarily large size diatoms and filamentous species [3]. Associated chytrids are external eucarpic parasites that produce a specialized rhizoidal system within host cells, that is, the diet conveying system that leads to the formation of the chitinous fruit bodies: the sporangium. This parasitic stage produces numerous uniflagellate spores, the zoospores, which constitute the dissemination phase of the life cycle [4].

Various approaches have been used to study fungal parasites but routine techniques for reliably identifying and counting these organisms are missing in the context of aquatic microbial ecology [5, 6]. So far, observations of parasitic fungi were obtained by using phase contrast light microscopy with live or Lugol's iodine preserved samples [7–9]. Such conventional microscopy allows observation of fungal sporangia or similar forms (especially in laboratory cultures), but is a poor approach for characterizing chytrid parasites in natural samples, at the complex community level. For example, a simple light microscopy observation of fungal rhizoidal systems, that is, a pertinent criterion for identifying chytrids [4, 10, 11], is very difficult. This situation may help explain the confusion of chytrids with protistan flagellates such as choanoflagellates or other bacterivorous flagellates in the group of Bicosoeca, which are attached to phytoplankton but do not harm their host [5–7].

Earlier studies on chytrids were restricted to morphological descriptions and focused on few species [12–15]. Electron microscopy was used to describe different life stages and the ultrastructural cytology of fungal zoospores and spore differentiation [16–18], providing the basis for chytrid taxonomy [19, 20]. Studies on pelagic chytrids started in the British lakes [21], and different authors have provided descriptions of morphological characters [22–26]. However, few attempts have been made to include the related parasitism pathway in the aquatic food web dynamics, and to understand environmental factors that trigger epidemics as well [27]. Some authors have also investigated the effects of parasitism on the growth of algal host species and on the genetic structure of infected populations [28].

Parasites are thus considered relevant not only for the evolution of their hosts but also for the population dynamics such as successions of phytoplankton communities, and for structuring microbial communities in general [9, 29]. Moreover, chytrids can represent interesting key intermediates in the food chain [30, 31]. The nutrients from infected large-size algae that could not be fed directly by zooplankton can be transferred from sporangium to grazers via fungal zoospore production. Fungal zoospores have suitable dimensions and represent a valuable food source for zooplankton [32]. The activity of zoosporic fungi and the related biogeochemical processes can thus be crucial in matter and energy transfer in aquatic systems [29]. Methodological limitations for the study of the ecological dynamics of chytrid populations can be overcome with epifluorescence microscopy coupled to a specific fluorochrome targeting molecular tracers (i.e., some types of polysaccharides) of the fungal chitinous structures, including sporangium and the rhizoidal system.

The protein stain fluorescein isothiocyanate (FITC) and, in particular, the chitin stain calcofluor white (CFW), were suggested as good markers that offer useful tools for the investigation of fungal dynamics in aquatic samples [33]. CFW binds to β-1,3 and β-1,4 polysaccharides such as those found in cellulose or in chitin, which commonly occur in the fungal cell wall [1, 2]. It fluoresces when exposed to UV light and is currently used in clinical mycology for direct microscopic examination of skin scrapings, hairs, nails, and other clinical specimen for fungal elements [34, 35]. In contrast to FITC, CFW penetrates into infected host cells and is more efficient for the observation of the complete rhizoidal system of parasites, that is, a pertinent criterion for chytrid identification [4, 10, 11].

The main objective of this chapter is to provide, in a simplified step-by-step format, a routine protocol based on size-fractionation of pelagic samples and the use of the fluorochrome calcofluor white for diagnosing, identifying, and counting fungal parasites (i.e., sporangia of chytrids) within phytoplanktonic communities [3], together with practical advices on how to apply the method.

Materials

1. 25-μm nylon filter.
2. 0.2-μm filters.
3. High-performance concentration/diafiltration system. As an example, we use the system Amicon model DC 10LA (Epernon, France) equipped with a reusable hollow fiber cartridge (0.2 μm cutoff, surface area of 0.45 m²).
4. 36% Formaldehyde.
5. Calcofluor white ($C_{40}H_{44}N_{12}O_{10}S_2$ Fluorescent Brightener 28; Sigma catalog no. F3543).
6. 10 N NaOH.
7. Balance.
8. Distilled water.
9. 15 and 0.2-mL tubes.
10. Glass slides and coverslips.
11. Epifluorescence microscope equipped with appropriate UV filter sets and Neofluar objective lens (optional).

Methods

Concentrations of Cells (See Note 1)

1. Pass the sample (ca. 20 L) through the 25-μm pore size nylon filter (see Note 2).
2. Collect large phytoplankton cells in the >25 μm size fraction by washing the filter with 40 mL of 0.2 μm filtered lake water.
3. Fix the concentrate sample with formaldehyde (2% final conc.), before staining and analysis.
4. Concentrate nanoplanktonic cells in the <25 μm size fraction (i.e., the 20 L filtrate) ca 20× by ultrafiltration to a volume of approximately 1 L, entry pressure 0.9 bar.
5. Fix about 180 mL of the ultrafiltrate retentate with formaldehyde (2% final conc.), before staining and analysis.

Preparation of Calcofluor Stock Solution

1. Weigh 35 mg of Calcofluor White into a 15-mL tube.

2. Add 7 mL of sterile distilled water and 2–3 drops of 10 N NaOH (to increase pH to 10–11). Calcofluor does not dissolve well in neutral solutions.
3. Dissolve the calcofluor.
4. Adjust the volume to 10 mL by adding sterile distilled water.
5. Distribute the stock solution in 0.2-mL tubes and store in a light-proof tube at −20 °C.

Staining and Visualization

1. In the dark, stain aliquots (about 200 μL) of concentrated and fixed materials by adding 1–2.5% (vol/vol) of CFW stock solution directly in solution for 10 min.
2. Mount drop (5–10 μL) of the stained samples between glass slides and coverslips for observations and counting.
3. In a dark room, examine the slides under an epifluorescence microscope equipped with an appropriate set of filters and objective lens. Shift between white and UV light to visualize and determine parasites and phytoplankton cells, and check the viability of the host cell, e.g., presence of chloroplast.
4. Applied a standard procedure for microscopic counting (see Notes 3 and 4).

Notes

1. Different approaches were tested to concentrate samples: the total community approach and the size-fractionated community approach. For the former approach, 180 mL of experimental samples were fixed with formaldehyde (2% final conc.) and aliquots were concentrated in three different ways: (1) by simple gravity, following Utermöhl's [36] method before staining the chyrids; (2) by vacuum pressure on two different filters before staining directly onto filters; and (3) by vacuum pressure on the same two types of filters but after staining in solution.

   For the Utermöhl method, 100 mL of fixed samples were settled for at least 24 h. For each of the two filter-vacuum pressure methods, 10 mL×2 of fixed samples were filtered onto polycarbonate white filters (pore size 0.6 µm, catalog no. DTTP02500, Millipore) and nuclepore polycarbonate black filters (pore size 0.8 µm, catalog no. 110659, Whatman), by using gentle vacuum (<0.2 bar or 20 kPa). For the total community approach using the classical Utermöhl [36] method, visualization of fungal parasites was very difficult and most of the time practically impossible for all the stain concentrations tested. The main reason was that staining directly in the Utermöhl chamber resulted in very poor-quality specimens of parasites observed in any given sample. Other disadvantages of the procedure include the relatively long sedimentation time and the difficulty of increasing the volume analyzed.

The alternative total community approaches based on vacuum pressure concentrations on polycarbonate filters, that is, white (0.6-µm-pore-size) and black (0.8-µm-pore-size) filters, yielded similar quality images of fungal parasites, either when CFW staining was done before (i.e., in solution) or after (i.e., on filters) concentrating phytoplankton host cells onto filters. However, substantial differences were noted depending both on the type of the filter and on the concentration of the stain. In general, for the two types of filters, high levels of background noises were obtained when using CFW at final concentrations of 3, 10, or 20%, precluding any accurate assessment of numerical and phenotypic characteristics of both host cells and their fungal parasites. Staining with 1% CFW final concentration substantially improved the viewing of chytrids on filters, with an increasing contrast from the white DTTP Millipore to the black Whatman filters. However, none of the membrane-retaining approaches yielded satisfactory images of morphological and cellular features of the host cells (e.g., presence of chloroplast, viability of the host cell). Accordingly, the proposed protocol is based on the size-fractionation approach using 1–2.5% vol/vol CFW final concentration (from the stock solution), which substantially enhanced the observational results.

2. The approach is efficient since it is based on the concentration of large initial volumes and size-partitioning of samples, a step that we judged necessary in order to yield good analytic images of infectious sporangia for accurate diagnosis and identification of parasites. In addition, this approach yielded satisfactory images of morphological and cellular features of the host cells, for phytoplankton identification based on phenotypic features and viability of the host cell, through the integrity of cell wall and the presence of chloroplasts, which are fundamental parameters to assess the intensity of the disease. We consider this protocol optimal for the diagnosis and quantitative assessment of phytoplanktonic chytrid infections in natural samples. Finally, the approach was designed to freeze-conserve particulate DNA samples for quantifying the propagule stages (i.e., zoospores) of chytrids via FISH targeting of specific rRNA oligonucleotide probes (see Chap. 5).

3. To estimate the infectivity parameters of ecological interest in phytoplankton population, several algorithms are used according to formula proposed by Bush et al. [37] These parameters include the prevalence of infection (Pr), that is, the proportion of individuals in a given phytoplankton population having one or more sporangia or rhizoids, expressed as $Pr (\%) = [(N_i/N) \times 100]$, where N_i is the number of infected host cells and N the total number of host cells. The second parameter is the mean intensity of infection (I), calculated as $I = N_p/N_i$, where N_p is the number of parasites and N_i is the number of the infected individuals within a host population.

4. We propose a third parameter concerning the prevalence of infection of cells in colonial (or filamentous) species (Pr_{CF}). $Pr_{CF} (\%) = [(N_i/N) \times 100]$, where N_i is the number of infected host cells in parasitized colonies (or filaments) and N the total number of parasitized host colonies (or filaments).

Acknowledgements SR and MG were supported by PhD Fellowships from the French Ministère de la Recherche et de la Technologie (MRT). This study receives grant-aided support from the French ANR Programme Blanc # ANR 07 BLAN 0370 titled DREP: Diversity and Roles of Eumycetes in the Pelagos.

References

1. Sigee DC (2005) Freshwater microbiology. Wiley, Chichester, UK
2. Tsui CKM, Hyde KD (2003) Freshwater mycology. Fungal Diversity Press, Hong Kong
3. Rasconi S, Jobard M, Jouve L, Sime-Ngando T (2009) Use of calcofluor white for detection, identification, and quantification of phytoplanktonic fungal parasites. Appl Environ Microbiol 75:2545–2553
4. Sparrow FK (1960) Aquatic phycomycetes. University of Michigan Press, Ann Arbor
5. Lefevre E, Bardot C, Noel C, Carrias J-F, Viscogliosi E, Amblard C, Sime-Ngando T (2007) Unveiling fungal zooflagellates as members of freshwater picoeukaryotes: evidence from a molecular diversity study in a deep meromictic lake. Environ Microbiol 9:61–71
6. Lefèvre E, Roussel B, Amblard C, Sime-Ngando T (2008) The molecular diversity of freshwater picoeukaryotes reveals high occurrence of putative parasitoids in the plankton. PLoS One 3:e2324
7. Kudoh S, Takahashi M (1990) Fungal control of population-changes of the planktonic diatom Asterionella-formosa in a shallow eutrophic lake. J Phycol 26:239–244
8. Sen B (1987) Fungal parasitism of planktonic algae in Shearwater I. Occurrence of Zygorhizidium affluens Canter on Asterionella formosa Hass. in relation to the seasonal periodicity of the alga. Arch Hydrobiol 76:129–144
9. Van Donk E, Ringelberg J (1983) The effect of fungal parasitism on the succession of diatoms in Lake Maarsseveen I (The Netherlands). Freshwater Biol 13:241–251
10. Canter HM (1950) Fungal parasites of the phytoplankton I. Studies on British chytrids X. Ann Bot 14:263–289
11. Canter HM, Lund JWG (1951) Studies on plankton parasites. III. Examples of the interaction between parasitism and other factors determining the growth of diatoms. Ann Bot 15:359–371
12. Ingold CT (1940) Endocoenobium Eudorinae gen. et sp. nov., a chytridiaceous fungus parasitizing Eudorina elegans ehrenb. New Phytol 39:97–103
13. Huber-Pestalozzi G (1944) Chytridium Oocystidis (spec. nova?) ein Parasit auf Oocystis lacustris Chodat. Aquat Sci 10:117–120
14. Canter HM (1953) Annotated list of British aquatic chytrids. Trans Br Mycol Soc 36:278–303
15. Canter HM, Lund JWG (1969) The parasitism of planktonic desmids by Fungi. Osterr Bot Z 116: 351–377
16. Beakes GW, Canter HM, Jaworski GHM (1992) Comparative ultrastructural ontogeny of zoosporangia of Zygorhizidium affluens and Z. planktonicum, chytrid parasites of the diatom Asterionella formosa. Mycol Res 96:1047–1059
17. Beakes GW, Canter HM, Jaworski GHM (1992) Ultrastructural study of operculation (discharge apparatus) and zoospore discharge in zoosporangia of Zygorhizidium affluens and Z. planktonicum, chytrid parasites of the diatom Asterionella formosa. Mycol Res 96:1060–1067
18. Beakes GW, Canter HM, Jaworski GHM (1993) Sporangium differentiation and zoospore fine-structure of the chytrid Rhizophydium planktonicum, a fungal parasite of Asterionella formosa. Mycol Res 97:1059–1074
19. Powell MJ (1978) Phylogenetic implications of the microbody-lipid globule complex in zoosporic fungi. Biosystems 10:167–180
20. Barr DJS (1992) Evolution and kingdoms of organisms from the perspective of a mycologist. Mycologia 84:1–11
21. Cook WRI (1932) An account of some uncommon British species of the Chytridiales found in algae. New Phytol 31:133–144
22. Reynolds N (1940) Seasonal variations in Staurastrum paradoxum eyen. New Phytol 39:86–89
23. Canter HM, Lund JWG (1948) Studies on plankton parasites. I. Fluctuations in the numbers of Asterionella formosa Hass. in relation to fungal epidemics. New Phytol 47:238–261
24. Canter HM, Lund JWG (1953) Studies on plankton parasites. II. The parasitism of diatoms with special reference to lakes in the English Lake District. Trans Br Mycol Soc 36:13–37
25. Canter HM (1972) A guide to the fungi occurring on planktonic blue-green algae. In: Desikachary TV (ed) Taxonomy and biology of blue-green algae. University of Madras, Madras, India, pp 145–159
26. Pongratz E (1966) De quelques champignons parasites d'organismes planctoniques du Léman. Aquat Sci 28:104–132
27. Gleason FH, Kagami M, Lefevre E, Sime-Ngando T (2008) The ecology of chytrids in aquatic ecosystems: roles in food web dynamics. Fungal Biol Rev 22:17–25
28. De Bruin A, Ibelings BW, Kagami M, Mooij WM, van Donk E (2008) Adaptation of the fungal parasite Zygorhizidium planktonicum during 200 generations of growth on homogeneous and heterogeneous populations of its host, the diatom Asterionella formosa. J Eukaryot Microbiol 55:69–74
29. Rasconi S, Jobard M, Sime-Ngando T (2011) Parasitic fungi of phytoplankton: ecological roles and implications for microbial food webs. Aquat Microb Ecol 62:123–137
30. Masclaux M, Bec A, Kagami M, Perga ME, Sime-Ngando T, Desvilettes C, Bourdier G (2011) Food quality of anemophilous plant pollen for freshwater zooplankton. Limnol Oceanogr 56:939–946

31. Gleason FH, Kagami M, Marano AV, Sime-Ngando T (2009) Fungal zoospores are valuable food resources in aquatic ecosystems. Inoculum 60:1–3

32. Kagami M, von Elert E, Ibelings BW, de Bruin A, van Donk E (2007) The parasitic chytrid, *Zygorhizidium* facilitates the growth of the cladoceran zooplankter, *Daphnia* in cultures of the inedible alga, *Asterionella*. Proc Biol Sci 274:1561–1566

33. Müller U, Sengbush P (1983) Visualization of aquatic fungi (Chytridiales) parasitizing on algae by means of induced fluorescence. Arch Hydrobiol 97:471–485

34. Hageage GJ, Harrington BJ (1984) Use of calcofluor white in clinical mycology. Lab Med 15:109–112

35. Harrington BJ, Hageage GJ (2003) Calcofluor white: a review of its uses and application in clinacal micology and parasitology. Lab Med 34:361–367

36. Utermöhl H (1958) Zur Vervollkommung der quantitative Phytoplankton Methodik. Mitt Int Verein Limnol 9:1–38

37. Bush AO, Lafferty KD, Lotz JM, Shostak AW (1997) Parasitology meets ecology on its own terms: Margolis et al. revisited. J Parasitol 83:575–583

Fungal Cell Wall Analysis

Pilar Pérez and Juan C. Ribas

Abstract

Fungal cell wall is a rigid structure mainly composed of polysaccharides (up to 90 %) and glycoproteins. It is essential for survival of the fungal cells, because it protects them against bursting caused by internal turgor pressure and against mechanical injury. Because of its absence in mammalian cells, it is an attractive target for antifungal agents. Thus, for various reasons, it might be important to know how the cell wall is synthesized, and how to analyze its composition. We provide here information about in vitro analysis of the biosynthetic activities of the main fungal wall and describe some methods for rapid analysis of cell wall composition by using specific enzymatic degradations. We also describe some additional methods that can be occasionally used to analyze fungal wall properties or composition. These methods provide powerful tools to evaluate changes in fungal cell walls and will be useful for screening new compounds for antifungal activity that might cause inhibition of cell wall biosynthesis and/or alter the structure of the fungal cell wall.

Keywords

Cell wall • Polysaccharides • Glucan • Chitin • Mannan • Antifungal drugs

Introduction

The fungal wall is responsible for the cell shape, provides mechanical protection, and supports the internal osmotic pressure of fungal cells. In addition, it acts as a filter for large molecules, and its rigid structure is useful for penetration into and colonization of insoluble substrates. The cell wall is also the surface of interaction between pathogenic fungi and their host. Indeed

P. Pérez (✉) • J.C. Ribas
Instituto de Biología Funcional y Genómica (IBFG)
CSIC, Universidad de SalamancaC, Zacarías González
s/n 37007, Salamanca, Spain
e-mail: piper@usal.es

V.K. Gupta et al. (eds.), *Laboratory Protocols in Fungal Biology: Current Methods in Fungal Biology*,
Fungal Biology, DOI 10.1007/978-1-4614-2356-0_12, © Springer Science+Business Media, LLC 2013

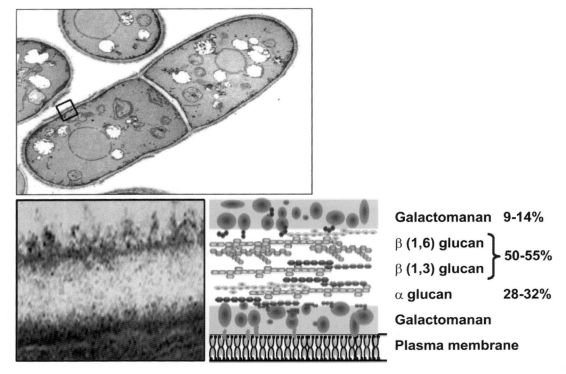

Fig. 12.1 Structure and composition of fungal cell wall. Transmission electron micrograph (TEM) of a fission yeast cell. A TEM detail of the cell wall is presented in the *lower panel* with a scheme of the organization and com- position of the two main cell wall layers—electron dense and electron transparent layers—and the inner plasma membrane-bound glycoproteins

the host defense response is usually directed against the cell wall. This structure is not simply a rigid exoskeleton but has the elasticity necessary to permit morphological changes during fungal growth and life cycle.

To build the walls, fungal cells need to synthesize wall components, export them across the plasma membrane, and assemble them outside the cell. The wall is composed basically of polysaccharides (70–90%) and glycoproteins (10–30%). Although composition varies among fungal species, and may even vary within a single fungal isolate, depending upon the growth conditions, most walls have a common structure [1]. When observed by transmission electron microscopy (TEM) the cell walls show a dark external layer formed by glycoproteins and an internal layer more transparent to the electrons, which mainly contains fibrillar polysaccharides (Fig. 12.1). The major fungal wall fibrillar components are: glu-

cose homopolymers, β(1,3)-D-glucan with some β(1,6) branches, that constitutes 48–54% of total cell wall polysaccharides; chitin, a β(1,4)-N-acetylglucosamine polymer; and α(1,3) (1,4)-D-glucan. Chitin accounts for only 1–2% of yeasts wall [2, 3], whereas filamentous fungi, such as *Neurospora* or *Aspergillus*, contain 10–20% chitin in their walls [1]. In both yeasts and filamentous fungi, chitin forms microfibrils from interchain hydrogen bonding that have enormous tensile strength and significantly contribute to the overall integrity of the cell wall [4].

The wall polysaccharides are formed at the plasma membrane by synthase enzymes and extruded into the periplasmic space where they bind to each other [5–7]. The linkages among the different components, which results in a tightly linked network, are generated by transglycosylation [8, 9] and are responsible for the mechanical strength of the cell wall [5–7, 10].

The formation and remodeling of the cell wall involves several biosynthetic pathways and the concerted actions of numerous gene products within the fungal cell. Many of the genes involved in cell wall synthesis or regulation have been cloned by complementation of mutants altered in wall structure or defective in the biosynthesis of cell wall components. Those mutants were isolated in many different ways, reflecting the complexity of functions involved in cell wall integrity and cell viability. Moreover, many of the genes and enzymes critical for assembly and biogenesis of fungal walls remain unidentified or poorly characterized. The main studies on fungal wall composition and biosynthesis have been performed in *Saccharomyces cerevisiae* [3, 11] but can be extended to other fungi.

Cell Wall Components

Glucan

Glucan is the main structural polysaccharide of the wall, and it represents 50–60% of this structure's dry weight. The majority of glucan polymers are composed of glucose units with β(1,3) bonds (65–90%), although there are also some β(1,6), β(1,3)(1,4) and β(1,4) glucans. Usually the main backbone is β(1,3)-D-glucan with β(1,6) branches (Fig. 12.2). The β(1,3)-D-glucan is synthesized by a complex of enzymes known as glucan synthases located in the plasma membrane. These enzymes catalyze the formation of linear glucan chains composed of, approximately, 1,500 β(1,3)-bound glucose residues. In these linear chains, new glucose units bind, forming β(1,6) branches in variable proportion depending on the organism—from almost linear to highly branched—which can bind to other glucans, to chitin or to glycoproteins, providing a great mechanical resistance to the wall, which is essential to maintain the fungal cell integrity (see Fig. 12.2).

The genes coding for the putative β(1,3)-D-glucan synthase catalytic subunit were initially identified in *S. cerevisiae* and named *FKS1* and *FKS2* [12, 13]. The Fks protein family of β(1,

3)-D-glucan synthase is very well conserved in fungi and plants. Orthologs of these genes have been described in the main fungal genera such as *Schizosaccharomyces*, *Candida*, *Aspergillus*, *Cryptococcus* or *Pneumocystis* [1, 3, 14]. Besides the catalytic subunit, fungal glucan synthases (GS) require GTP-bound Rho1 GTPase for their activity [15, 16]. This family of enzymes use uridine-diphospho-glucose (UDP-Glc) as substrate and catalyze the reaction 2 UDP-Glc → [Glc-β-1,3-Glc].

A second β-linked glucan contained in most fungal walls is the β(1,6)-glucan. This polymer is shorter than β(1,3)-glucan, it does not form a fibrillar structure, and acts as a flexible glue by forming covalent cross-links to β(1,3)-glucan, chitin, and glycoproteins [6].

Some fungi contain α(1,3)(1,4)-glucan in their cell wall. However, the corresponding in vitro α(1,3)-glucan synthase activity has not been described yet. A putative catalytic subunit was first described in *Schizosaccharomyces pombe* [17, 18]. Ags1/Mok1 is a multidomain integral membrane protein with a predicted domain highly similar to starch synthase in the inner side and another domain similar to α-amylase and other proteins implicated in glycogen metabolism in the outer side. *S. pombe* contains five genes coding Ags/Mok proteins, and genomes of other fungi, including several human fungal pathogens in which cell wall α-glucan accounts for around 35% of the total wall polysaccharides, contain sequences of predicted proteins homologous to these genes [19, 20].

Chitin

Chitin is a β(1,4)-linked homopolymer of N-acetylglucosamine present in the cell walls of all fungi studied to date with the exception of *S. pombe*. Chitin represents 1–2% of the dry weight of the yeast cell wall whereas in the filamentous fungi it can reach up to 10–20% [1]. Chitin is synthesized from N-acetylglucosamine units by the enzyme chitin synthase (CS) that deposits microfibrils of chitin outside of the plasma membrane. This family of enzymes use uridine-d

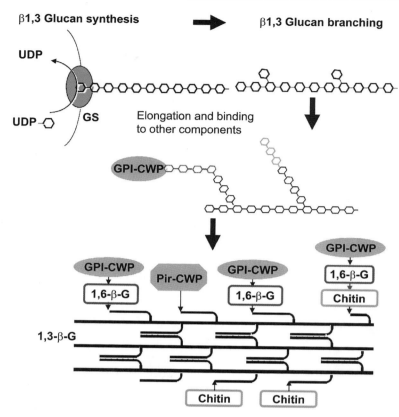

Fig. 12.2 Schematic representation of the synthesis and organization of the β-glucans forming the central core of the fungal cell wall. Linear β(1,3)-glucan chains are initially synthesized in the plasma membrane by the glucan synthase. Then cell wall transglycosidases form β(1,6)-branched β(1,3)-glucan that is also linked to β(1,6)-glucan through the side chains and to chitin via β(1,4) linkages. Proteins are covalently attached to β(1,6)-glucans (GPI-CWP) through the GPI remnant or to β(1,3)-glucans (PIR-CWP) through a glutamine residue following a transglutaminase reaction

iphosphate-N-acetylglucosamine (UDP-GlcNAc) as substrate and catalyze the reaction 2 UDP-GlcNAc → [GlcNAc-β-1,4-GlcNAc]. Chitin biosynthesis has been mainly studied in *S. cerevisiae*, which has three chitin synthases (CS1–3) responsible for the synthesis of chitin [21] at different times and places during cell growth. The number of chitin synthase genes varies from 1 to 20 according to the fungal species. The large family of chitin synthase (CS) enzymes fall into seven classes according to the evolution of their amino acid sequences [22]. The multiplicity of enzymes suggests that they have redundant roles in chitin synthesis and makes it difficult to find functional significance to the different classes [23].

Glycoproteins

Glycoproteins represent 30–50% of the dry weight of the *S. cerevisiae* or *Candida* walls, and around 20% of the dry weight of *S. pombe* and the filamentous fungi walls. These wall proteins have diverse functions, participating in the maintenance of the cellular form, taking part in adhesion processes, transmitting signals to cytoplasm, and remodeling the components of the wall. The glycoproteins present in the cell wall are extensively modified with both N- and O-linked carbohydrates, predominantly or exclusively formed by mannose residues known as mannan. In some cases, the mannan backbone presents single residues or side chains of different sugars, galactomannan, rham-nomannan, glucogalactomannan, rhamnogalacto-mannan, etc. [1, 24, 25]. Most cell wall proteins are attached through a glycosylphophatidyl inositol (GPI) remnant to β(1,3)-glucan or chitin, via a branched β(1,6)-glucan linker [11].

Analysis of the Cell Wall Synthases

We will present here the methods described for in vitro measurements of the enzymatic activities responsible for the biosynthesis of the two main structural wall polysaccharides: β(1,3)-glucan synthase (GS) and chitin synthase (CS). Both are integral membrane proteins, localized in the plasma membrane with their catalytic sites facing the inner side of the membrane.

β(1,3)-Glucan Synthase (GS)

The original method to detect the in vitro β(1,3)-glucan synthase activity was described nearly three decades ago [4, 26, 27]. This method has been modified, simplified and improved but essentially the basis of the protocol remains unaltered [28, 29].

Membrane Extracts Preparation

The source of enzyme activity is a crude membrane extract partially purified from the total cell extract.

1. Cell cultures (100 mL) are collected at early log-phase (A_{600} 0.7–1.0) and centrifuged 5 min at 4 °C, 3,000×g (5,000 rpm in a GSA-type rotor).
2. Cells are suspended in 30–40 mL cold buffer A (50 mM Tris–HCl pH 7.5, 1 mM EDTA, and 1 mM β-mercaptoethanol, centrifuged 5 min at 4 °C, 3,000×g (6,000 rpm in a SS34-type rotor), transferred to a 1.5-mL tube (with screw cap), washed with 1 mL cold buffer A (1 min at 16,000×g, 13,200 rpm) and resuspended in 100 μL cold buffer A containing 50 μM GTPγS (GTPγS is more stable than GTP, not hydrolysable and therefore, better GS activator). Glucan synthase is very labile

and GTPγS is very useful to preserve the enzyme activity.

3. Cells are broken with glass beads (0.5 mm diameter, filling in all the liquid with glass beads and discarding by gentle drop out the excess of beads not entrapped by liquid capillarity) in a FastPrep apparatus (Q-Biogene, MP Biomedicals, Thermo Scientific) during 15 s at a speed of 6.0 and at 4 °C if possible. Alternatively, the cells can be broken in glass tubes with glass beads by 6 or 7 cycles of vortexing for 30 s and cooling down in ice for another 30 s. Due to enzyme instability, the rest of process must be done at 4 °C and the sample must be kept on ice.
4. Broken material and glass beads are diluted with 30 mL buffer A. Beads and cell debris are removed by low speed centrifugation, 5 min at 4 °C and 3,000×g.
5. The supernatant is then centrifuged at 36,000–38,000×g for 30 min at 4 °C and the membrane pellet is resuspended carefully by using a glass stick and a vortex to extend the membrane material throughout the entire bottom surface of the tube. Then 25 μL of buffer A containing 33% glycerol and 50 μM GTPγS is added and the membranes are homogenized by vortexing with the glass stick throughout the tube surface. The process is repeated with another 25, 50, and 50 μL of the same buffer until the membranes are homogenized in 100–150 μL and stored at −80 °C.
6. A homogeneous emulsion of membrane extract is critical for a reproducible GS assay. The amount of protein is quantified by using the Bradford dye-binding assay (Bio-Rad) with bovine serum albumin as standard. The protein concentration of the enzyme extract is usually kept at 3–5 mg/mL.

GS Assay
The GS mixture contains 5 mM UDP-[^{14}C]-glucose (200 cpm/nmol) (PerkinElmer), 150 μM GTP or GTPγS, 0.75% bovine serum albumin, 2 mM EDTA, 75 mM Tris–HCl pH 8.0, and 5 μL of enzyme extract (15–25 μg of protein) in a total volume of 40 μL. The correct amount of protein in the assay is critical: higher protein concentration

does not result in proportional increase of GS activity; therefore, the relative GS activity per milligram of protein decreases.

The reaction mixture is incubated for 30–60 min at 30 °C and stopped by addition of 1 mL of 10% trichloroacetic acid (TCA). The samples are kept at least 30 min at 4 °C, filtered in Whatman GF/C glass fiber filters, and washed three times with 1 mL of 10% TCA and twice with 1 mL of ethanol. The filters are placed into vials, 2 mL of liquid scintillation is added and the radioactivity of the filters is measured in a Beckman scintillation counter. One unit of GS activity is the amount of enzyme that catalyzes the incorporation of 1 μmol of glucose into glucan per min at 30 °C. The specific activity is expressed as milliunits per mg of protein and the reactions are always performed in duplicate. The GS enzyme is very labile and therefore, the data for each GS assay must be reproducible and calculated from at least three to four independent experiments.

When the membrane extract is obtained in the absence of GTPγS, we can measure basal and maximal GS activity by omitting or adding GTP to the reaction mixture. Alternatively, the GS assay may be done at pH 7.0–7.5 and may contain 25–30 mM potassium or sodium fluoride [4, 12, 30–37] and 0.5% Brij-35 [38, 39].

The detergents 2% Tergitol NP-40 (with 2 M NaCl) or 1.0% CHAPS are used for GS fractionation and solubilization of the regulatory subunit [12, 15, 35, 40–43], and 0.5% CHAPS, 0.1% cholesteryl hemisuccinate, or 0.1–0.2% CHAPS, 0.5–1.12% octyl glucoside are used for partial solubilization of GS microsomal fractions [30, 32, 34, 35, 44].

The GS reaction product can be confirmed to be β(1,3)-glucan, not a contaminant product such as glycogen, by degradation with Zymolyase-100T (AMS Biotechnology) or Kitalase (Wako Pure Chemical Industries). Zymolyase-100T is a preparation partially purified by affinity chromatography from *Arthrobacter luteus* that contains β(1,3)-glucanase, protease, and mannanase activities, but it does not contain β(1,6)-glucanase and α-glucanase activities. Similarly, Kitalase is a preparation from

Rhizoctonia solani with β(1,3)-glucanase, protease, hemicellulase, pectinase, and amylase activities, but it does not contain α(1,3)-glucanase activity. Kitalase is also named as Lysing enzymes from *Rhizoctonia solani* (Sigma-Aldrich) or as Yeast Lytic Enzyme from *Rhizoctonia solani* (MP Biomedicals).

The degradation mixture contains the GS reaction product (40 μL), either 20 μg Zymolyase-100T, 50 mM citrate-phosphate pH 5.6 or 25 μg Kitalase, 50 mM potassium-acetate pH 5.0, and either 0.2% Tween 20 or Triton X-100 (1% detergent produces lower degradation) in a volume of 300 μL. The mixture is incubated 15–24 h at 30 °C with shaking, stopped with 10% TCA and processed as a standard GS assay. Zymolyase or Kitalase degradation in the absence of detergent is not complete, with a 15–20% residual product, likely due to the protection conferred by membrane vesicles. Other enzyme complexes result in only partial degradation of the reaction product.

Microtiter-Based Fluorescence Assay

This method has been described as an alternative to the use of UDP-[^{14}C]-glucose for the GS assay [39]. This method takes advantage of substituting radioactive substrate for the fluorochrome aniline blue that is specific for linear β(1,3)glucan. Aniline binding is proportional to the amount of linear glucan, and it can be measured in a microplate fluorescence reader (excitation at 400 nm, emission at 460 nm). The GS mixture (50 μL total volume, 100 μg of enzyme protein) is similar to that described above except that UDP-[^{14}C]-glucose is omitted and it may contain 0.5% Brij-35. The reactions are performed in microtiter plate wells at 30 °C for 30–60 min and stopped with 10 μL of 6 N NaOH. The glucan product is solubilized by heating at 80 °C for 30 min followed by the addition of 210 μL of aniline blue mix (1 mL contains 400 μL of 0.1% aniline blue, 210 μL of 1 N HCl, and 590 μL of 1 M glycine/NaOH pH 9.5). The plate is incubated at 50 °C for 30 min and at room temperature for another 30 min to allow reaction with the fluorochrome. Then, the fluorescence is quantified in a fluorescence reader. Linear β(1,3)-glucans (such as pachyman, curdlan, or yeast glucan dissolved in 1 N NaOH at 80 °C for 30 min) are used as standards, in a reaction mixture containing the same components except the membrane extract.

In Situ GS Assay Using Permeabilized Whole Cells

This method takes advantage of being a more direct enzymatic assay, omitting the steps of membrane extract preparation. In addition, this method can be applied for GS, chitin synthase, or other membrane-bound or cytosolic enzymes. The enzyme activity in permeabilized cells can yield similar or higher activities than those in cell extracts [45, 46]. The procedure is as follows:

1. Early log-phase cells are collected by centrifugation (1,500 × g, 5 min) and suspended in 40 mM EDTA, 100 mM β-mercaptoethanol (3.5 mL per g of cells wet weight).

2. The cells are incubated at 30 °C for 30 min with shaking, collected by centrifugation at 3,000 × g for 5 min, washed with 5 mL of 1.2 M sorbitol, resuspended (7 mL per g of cells) in 50 mM citrate phosphate pH 6.3, 1 mM EDTA, 1.2 M sorbitol, and incubated at 30 °C for another 30 min with shaking. Similar result can be obtained keeping the cells on ice for 30 min without shaking. Then, the cells are centrifuged at 3,000 × g for 5 min, suspended in 30 mL of cold 50 mM Tris–HCl pH 7.5 for osmotic shock, kept on ice for 5 min and centrifuged at 13,000 × g for 5 min.

3. The cells are suspended (1–1.5 mL per g of cells) in the buffer of the reaction assay of choice (50 mM Tris–HCl pH 7.5, 33% glycerol for GS and CS assays) and stored at −80 °C for weeks. Under those conditions, more that 90% of the cells are permeabilized (measured by staining with methylene blue).

Alternatively, the cells can be collected, washed with ice-cold water, suspended in cold 50 mM Tris–HCl pH 7.5, 1 mM EGTA, 1 mM β-mercaptoethanol, 0.5 mM phenylmethylsulfonyl fluoride (PMSF), 33% glycerol, and kept for 10 min in ice. 3.5 M glycerol can be replaced by 1.2 M glycerol, sorbitol, mannitol, or 1 M KCl. The cells are washed and suspended (1 g of cells wet weight per mL) in cold buffer without glycerol or the corresponding osmolyte [38]. In another protocol

the cells are collected, washed with 50 mM Tris–HCl pH 7.5, 1 mM EDTA, 1 mM DTT, 33% glycerol, resuspended in the same buffer and permeabilized with 2% toluene/methanol (1.1) at 22 °C for 5 min. The cells are washed twice at 4 °C with cold buffer and resuspended in cold buffer (1 g of cells wet weight per mL) [47].

The GS assay contains the same mixture than that of the membrane assay except that the 5 μL of enzyme extract is replaced by 5 μL of permeabilized cells (5 μg of cells wet weight), in a total volume of 40 μL.

Chitin Synthase (CS)

This protocol was developed for *S. cerevisiae* chitin synthase enzymes CSI, CSII, and CSIII [48]. The in vitro activities of other fungi may be different and the protocols may need to be adapted.

Membrane Extracts Preparation

As for the GS, the source of enzyme activity is a crude membrane extract. The protocol for membrane extract preparation is similar to that described for GS except that the buffer is 50 mM Tris–HCl pH 7.5. Similarly, the membranes are resuspended in the same buffer containing 33% glycerol.

CS Assay

CS is a zymogenic enzyme that must be degraded partially to show its maximal activity. As mentioned previously, three CS activities have been described in *S. cerevisiae*, corresponding to three different proteins. A method to determine the three activities in the same membrane preparation by the use of several modifications in the reaction conditions has been described [48]. Most of the in vitro CS activity corresponds to CSI. The reaction mixture contains 37 mM Tris–HCl pH 7.5, 4.8 mM magnesium acetate, 5–10 μL of membrane suspension (up to 20 μL) and 2 μL of trypsin at the optimal concentration for activation (0.1–2.0 mg/mL) in a total volume of 40 μL. The mixture is incubated 15 min at 30 °C and the proteolysis is stopped by adding 2 μL of soybean trypsin inhibitor at a concentration 1.5 times that of the used trypsin solution. The tubes are placed on ice and made 1 mM in UDP-[^{14}C]-GlcNAc (400 cpm/nmol) and 32 mM in GlcNAc in a total volume of 46 μL. Samples are incubated for 30–60 min at 30 °C, stopped with 1 mL of 10% TCA, and processed as for the GS assay (see previous). The specific activity is expressed as ηmoles of GlcNAc incorporated per hour and mg of protein.

For CSII and CSIII, the reaction mixture before proteolysis contains 32 mM Tris–HCl pH 8.0, 5 mM cobalt acetate, 20 μL of membrane suspension, 2 μL of trypsin, and 1.1 mM in UDP-[^{14}C]-GlcNAc (400 cpm/nmol) in a total volume of 46 μL. For CSIII, the reaction mixture also contains 5 mM nickel acetate. After proteolysis is stopped, the mixture is made 32 mM in GlcNAc in a volume of 50 μL, incubated for 90 min at 30 °C, and stopped with 1 mL of 10% TCA.

Although the CSI assay detects the three activities, CSII and CSIII are minor contributors and therefore they do not alter significantly the value of CSI activity. However, they can be calculated based in the inhibitor effect of Ni^{2+} and Co^{2+}. Ni^{2+} is a powerful inhibitor of CSI and II but has little effect on CSIII and Co^{2+} stimulates CSII and III but inhibits CSI. In summary:

CSI activity (total CS−CSII+III): CS assay minus CS assay (+Co^{2+}).
CSIII activity: CS assay (+Co^{2+}, +Ni^{2+}).
CSII activity (CSII+III−CSIII): CS assay (+Co^{2+}) minus CS assay (+Co^{2+}, +Ni^{2+}).

Microtiter-Based Fluorescence Assay

As for GS, an alternative to the use of UDP-[^{14}C]-GlcNAc for the CS assay has been described [49]. The CS reaction is similar but the radioactive UDP-[^{14}C]-GlcNAc is omitted. The wells of a microtiter plate are coated with wheat germ agglutinin (WGA) which binds with high affinity and specificity to chitin. The procedure involves the binding of the synthesized chitin to the WGA-coated surface. Then, horseradish peroxidase–WGA conjugate is added to the mixture. The WGA of the conjugate will bind to the chitin previously fixed in the well. The horseradish peroxidase activity is measured at 600 nm, and

the amount of chitin is calculated using acid-solubilized chitin as standard. This method is suitable for the three CS activities.

In Situ CS Assay Using Permeabilized Whole Cells

Similar to the procedure described previously for GS [38, 45–47].

Analysis of Cell Wall Polysaccharides

Different methods can be used to analyze the cell wall polysaccharide composition. All the methods, either for a precise analysis or for a rapid estimation of cell wall polymers, require a separation of the wall from the rest of the cell components. In general, the current methods have been adapted for a simple, accurate and rapid analysis of wall polysaccharides and are all based on labeling and fractionation of cell wall polysaccharides using chemical and enzymatic procedures. Basically, the methods available have been established using *S. cerevisiae* and *S. pombe* models, although the techniques used can easily be adapted for any organism.

Radioactive Labeling and Fractionation of the Cell Walls

A basic procedure to quantify the cell wall polymers consists in ^{14}C-glucose labeling and fractionation of the cell wall polysaccharides as follows (Fig. 12.3):

Cell Wall Labeling

1. Exponentially growing cultures are adjusted to 5×10^6 cells/mL, 5–7 mL are supplemented with [U-^{14}C]-glucose (3 μCi/mL) (Hartmann Analytic) and incubated for at least one doubling time (3 h) to allow ^{14}C incorporation into the cell. One doubling time means 50% labeling of cell material. If a stronger labeling is required, the cultures can be incubated for longer times, the ^{14}C-glucose can be increased (up to 18 μCi/mL), and the glucose concentration in the culture medium can be reduced to 1 or 0.5%.

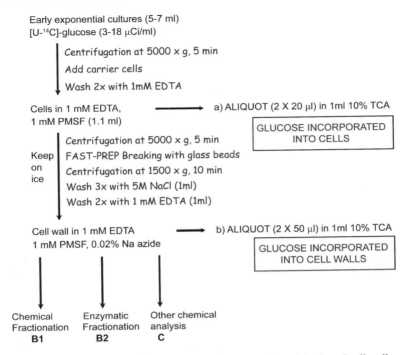

Early exponential cultures (5-7 ml)
[U-¹⁴C]-glucose (3-18 μCi/ml)

Centrifugation at 5000 × g, 5 min
Add carrier cells
Wash 2x with 1mM EDTA

Cells in 1 mM EDTA, ⟶ a) ALIQUOT (2 X 20 μl) in 1ml 10% TCA
1 mM PMSF (1.1 ml)

GLUCOSE INCORPORATED
INTO CELLS

Keep on ice
Centrifugation at 5000 × g, 5 min
FAST-PREP Breaking with glass beads
Centrifugation at 1500 × g, 10 min
Wash 3x with 5M NaCl (1ml)
Wash 2x with 1 mM EDTA (1ml)

Cell wall in 1 mM EDTA ⟶ b) ALIQUOT (2 X 50 μl) in 1ml 10% TCA
1 mM PMSF, 0.02% Na azide

GLUCOSE INCORPORATED
INTO CELL WALLS

Chemical   Enzymatic   Other chemical
Fractionation   Fractionation   analysis
B1   B2   C

Fig. 12.3 Overview of [U-¹⁴C]-glucose radioactive labeling and purification of the fungal cell wall

2. Cells (1–2×10^7 cells/mL) are harvested by centrifugation ($5,000 \times g$ for 5 min).

3. Centrifuged cells from exponentially growing cultures without radioactive glucose (300 μL of a concentrate of 10^9–10^{10} cells/mL) are added to minimize lost material.

4. Cells are washed twice with 1 mM EDTA, transferred to 1.5–2.0 mL tubes and resuspended in 1.1 mL of 1 mM EDTA, 1 mM PMSF.

5. Total glucose incorporation is monitored in two 50 μL aliquots added to 1 mL of cold 10% TCA and kept at 4 °C for at least 30 min. Then aliquots are filtered through a fiberglass filter Whatman GF/C, washed three times with 1 mL of 10% TCA, twice with 1 mL of ethanol, and counted in a liquid scintillation counter. Eventually, the samples can be stored in 10% TCA at 4 °C and analyzed with the samples obtained in further steps.

6. The remaining cells (1.0 mL) are centrifuged ($5,000 \times g$ 5 min), resuspended in 100 μL of 1 mM EDTA, 1 mM PMSF, filled with cold glass beads (0.5 mm diameter) to completely cover the cell suspension and broken in a FastPrep homogenizer (Q-Biogene, MP Biomedicals, Thermo Scientific) during 3× 20 s pulse at a speed of 6.0 and at 4 °C. Complete cell lysis is confirmed by microscopic observation.

7. The broken cells are collected and the glass beads are washed twice with 1 mM EDTA, 1 mM PMSF to collect all residual material.

8. The broken material is centrifuged at $1,500 \times g$ for 5 min, washed three times with 5 M NaCl, then again twice with 1 mM EDTA.

9. The cell wall pellet is resuspended in 1.1 mL of 1 mM EDTA, 1 mM PMSF, 0.02% Na azide, and heated at 100 °C for 5 min, to inactivate the intrinsic hydrolytic cell wall enzymes that would interfere the wall analysis.

Total radioactivity incorporated into the cell wall is monitored in two 50-μL aliquots that are added to 1 mL of cold 10% TCA and processed as described previously.

Cell Wall Alkali-Fractionation and Analysis

The most common method for cell wall fractionation is that used for *S. cerevisiae* [50, 51], which can easily be adapted for other organisms. This method allows the separation of cell wall β-glucan into alkali-soluble and alkali-insoluble fractions. The alkali-soluble fraction contains β(1,3)-glucan, mannan, and some β(1,6)-glucan; and the alkali-insoluble fraction contains chitin and β(1,3)-glucan β(1,6)-glucan linked to the chitin.

1. The cell wall suspension (1 mL) is extracted twice with 6% NaOH for 90 min at 80 °C and centrifuged at $1,500 \times g$ for 5 min.
2. The alkali-extracted supernatant is divided into four aliquots of 250 μL.
   2.1. Two alkali-extracted aliquots are used to precipitate the mannan with Fehling´s reagent [52] as follows: unlabeled purified mannan from *S. cerevisiae* (Sigma) is added to the supernatant as carrier (0.1 mL from a stock of 50 mg/mL in water). Fehling´s reagent (2 mL) is then added to the samples, mixed, and left overnight at 4 °C to precipitate the mannan. Fehling's reagent is freshly prepared for each experiment by adding one volume of reagent B (3.5% $CuSO_4$) to one volume of reagent A (17.3% potassium sodium tartrate dissolved in 12.5% KOH). After centrifugation at $1,500 \times g$ for 10 min, the pellet is washed with Fehling´s reagent and solubilized in 20–40 μL of 6 N HCl (drop by drop and mixing until completely solubilized). Then, 100 μL of 50 mM Tris–HCl pH 7.5 is added and the solution is transferred to a vial with 2 mL of liquid scintillation. The tube is washed twice with 100 μL of buffer to collect the residual mannan, which is added to the vial containing liquid scintillation and analyzed (total mannan fraction).
   2.2. The other two alkali-extracted aliquots are precipitated with 2 volumes of ethanol, allowed to dry, dissolved in 100 μL of water, collected (washing the tube twice) and analyzed with liquid scintillation as the previous samples (mannan + alkali soluble glucan). The difference between both fractions is the alkali-soluble glucan (β1,3 + β1,6).
3. The alkali-insoluble residue is washed with water several times by centrifugation until it reaches a neutral pH, then suspended in 1.3 mL of water and divided into six aliquots of 200 μL.
   3.1. The radioactivity of two aliquots is counted directly (alkali insoluble glucan + chitin).
   3.2. Two aliquots are incubated for 24–36 h at 30 °C with 25 μg Zymolyase-100T in 50 mM citrate-phosphate pH 5.6, 0.02% Na azide in a volume of 300 μL, and two aliquots are processed similarly but without Zymolyase as a control. After incubation, the four samples are centrifuged and the pellets are washed twice, resuspended in 100 μL of buffer, added to 1 mL of 10% TCA, and processed by filtration in Whatman GF/C glass fiber filters and scintillation counting as described above. The residue remaining after Zymolyase digestion is the chitin fraction. Treatment of the alkali-insoluble fraction with recombinant chitinase from *Pyrococcus furiosus* (Wako Pure Chemical Industries) is not needed because the remaining residue is the alkali-insoluble glucan (β1,3 + β1,6), which can be obtained as a difference between fractions.

Alkaline extraction is a widespread procedure for cell wall analysis of many organisms [1, 25, 53–59]. However, the data are only reproducible when maintaining the extraction conditions; the proportions of alkali-soluble and insoluble fractions can change depending on the alkali concentration, temperature, and incubation time. In fact, a process such as storing the cell walls at −20 °C results in total alkali solubility of the cell wall under conditions similar to those described above, showing only a small residue coincident with the chitin fraction.

Chemical fractionation of *S. pombe* cell wall polysaccharides is similar to that of *S. cerevisiae* [60], although the cell wall composition is different and therefore the results differ. The alkaline extraction procedure is rather harsh in *S. pombe*, due to the absence of the chitin responsible for the alkali-insoluble maintenance of some β(1,3)-glucan, and to a considerably smaller amount of β(1,6)-glucan than in *S. cerevisiae* [61, 62]. Therefore, alkaline extraction causes solubilization of nearly all *S. pombe* cell wall polymers and as a result, more gentle methods involving enzymes capable of specifically digesting one polymer without altering the others, described below, yield more accurate results.

Cell Wall Enzymatic Fractionation and Analysis

This protocol was adapted for *S. pombe* cell wall. The most commonly used procedure permits quantification of the three major cell wall polymers, β-glucans, α-glucans and galactomannoproteins. Once cell wall labeling and purification has been performed as described above, the procedure is as follows:

1. Half of the cell wall material (500 µL) is divided in:

    1.1. Two 100 µL aliquots that are incubated for 24–36 h at 30 °C with shaking with 25 µg Zymolyase-100T in 50 mM citrate-phosphate pH 5.6, 0.02% Na azide in a volume of 300 µL, and two aliquots are processed without enzyme as control. A similar option is the incubation with 100–200 µg of Kitalase in 50 mM potassium-acetate pH 5.0, 0.02% Na azide.

    1.2. Two 100 µL aliquots are incubated for 24–36 h at 30 °C with 100 units of Quantazyme (MP Biomedicals) or 100 units of recombinant β-1,3-Glucanase Yeast Lytic Type (Wako Pure Chemical Industries) in 50 mM potassium phosphate monobasic pH 7.5, 60 mM β-mercaptoethanol, 0.02% Na azide in a volume of 300 µL.

    1.3. Two 50 µL aliquots are processed without enzyme as control.

2. After incubation, the samples are centrifuged (16,000 × *g* for 3 min). The supernatant is removed and 2 × 50 µL aliquots are counted directly with 2 mL of liquid scintillation. 1 mL of 10% TCA is added to the pellet and the radioactivity incorporated is determined by filtration and liquid scintillation counting as described previously. As mentioned, Zymolyase-100T and Kitalase contain β(1,3)-glucanase, protease, mannanase and other activities but not α-glucanase activity. Therefore, the residue obtained after Zymolyase-100T or Kitalase digestion is considered α-glucan and the supernatant, β-glucan plus galactomannan. Quantazyme and the β-1,3-Glucanase Yeast Lytic Type are two recombinant endo-β(1,3)-glucanases capable of digesting β(1,3)-glucan without degrading the β(1,6)-glucan or α-glucan [5, 6]. Therefore, the residue obtained after Quantazyme or similar recombinant β(1,3)-glucanase treatment is considered α-glucan plus β(1,6)-glucan and galactomannan, and the supernatant is β(1,3)-glucan.

3. Half of the cell wall material (500 µL) is divided in two aliquots for galactomannan quantification. The wall is solubilized in alkali (6% NaOH) by adding 250 µL of 12% NaOH and heating at 80 °C for 1 h. Then, the galactomannan is precipitated from the alkali-solubilized aliquots with the Fehling's reagent and quantified as described previously for *S. cerevisiae* cell walls.

Besides Zimolyase 100T, Kitalase and Quantazyme, there is an ample variety of enzymes and enzyme complexes commercially available that can be used for the cell wall determination of a specific fungal organism. These enzymes can provide information either individually or in combination. In the later case, depending on the cell wall polysaccharides composition and linkages, the order of enzymes can be important for maximal degradation of each enzyme. In addition, the enzymes with exohydrolytic activity present poor or null activity against the cell wall polysaccharides and therefore, all the enzymes to be tested should contain endohydrolytic activity. Some of these enzymes are:

• *β(1-3;1-4)*-D-*glucan hydrolases* such as that from *Bacillus subtilis* (Biosupplies). This endoglucanase specifically hydrolyzes β-D-glucans containing both β(1,3) and β(1,4)-D-glucosidic linkages in linear sequence. It does not hydro-lyze β(1,3)-glucans or β(1,4)-glucans.

- *α(1,4)-glucanases or α-amylases*. These enzymes are endoglucanases that hydrolyze α(1,4)-D-glucosidic linkages in polysaccha-rides containing three or more α(1,4)-linked D-glucose units. Examples of α-amylases that can be used are α-amylase from *Bacillus subti-lis* (heat stable, Sigma), from *Bacillus licheni-formis* (Termamyl-120, heat stable up to 90 °C, Sigma), from *Aspergillus oryzae* (Taka-Diastase, Taka-Amylase A, Sigma), from *Rhizopus* sp. (Merck Calbiochem), from por-cine pancreas (Sigma), from human pancreas (Merck Calbiochem) or from human saliva (Sigma, Merck Calbiochem). No additional activity for these enzymes has been reported.

- *Chitinases*. These enzymes hydrolyze internal linkages in the chitin chain, a linear polymer of β(1,4)-N-Acetyl-D-glucosamine units. A recombinant Chitinase (Wako Pure Chemical Industries) from *Pyrococcus furiosus* is a ther-mostable and powerful enzyme. Chitinases from *Trichoderma viride* (Sigma) and from *Streptomyces griseus* (Sigma) are combina-tion of exo and endochitinases, but with less efficient hydrolytic activity.

- *Chitosanases*. These enzymes catalyze the hydrolysis of β(1,4) linkages between D-glu-cosamine (GlcN-GlcN) residues in chitosan. Chitosanases from *Streptomyces* sp. (Sigma; Merck Calbiochem) are available enzymes.

- *Mannanases*. These are α-mannosidases that cleave terminal α(1,2), α(1,3) and/or α(1,6)-linked mannose residues in mannan. The sup-plied enzymes do not contain contaminant protease or glycosidase activities. α(1,2;1,3)-mannosidase, recombinant from *Xanthomonas manihotis* (Merck Calbiochem) cleaves termi-nal α(1,2) and α(1,3)-linked mannose residues. Its activity is efficient and reproducible. α(1,6)-mannosidase, recombinant from *X. manihotis* (ProZyme, AMS Biotechnology), cleaves ter-minal α(1,6)-linked mannose residues. It is recommended for use after digestion with α(1,2;1,3)-mannosidase for increasing the degradation efficiency. Other available enzymes are α(1,2;1,6)-mannosidase and

α(1,2;1.3;1,6)-mannosidase from *Canavalia ensiformis*, Jack bean (Sigma, ProZyme, AMS Biotechnology), and α(1,2)-mannosidase from *Aspergillus saitoi* (ProZyme).

- *Proteases* such as Proteinase K, recombinant, from *Tritirachium album* (Roche Applied Science); Protease S, recombinant thermo-stable from *Pyrococcus furiosus* (Sigma); Turbo3C protease, recombinant, from human rhinovirus 3C protease; and aminopeptidase T, recombinant, thermostable from *Thermus aqualicus* (Wako Pure Chemical Industries).

- *Enzymatic complexes*. They can be used indi-vidually or in combination with other enzymes. The enzymatic composition of some complexes is well characterized although they must be used cautiously because they may contain other not-tested activities. The enzymatic activities of other complexes are not characterized and therefore can only be used to test whether the present and absent activities are helpful for a specific cell wall analysis. We already mentioned Zymolyase-20T and 100T (AMS Biotechnology) and Kitalase (Wako Pure Chemical Industries, City Chemical) also sold as Lysing Enzymes from *Rhizoctonia solani* (Sigma) or as Yeast Lytic Enzyme from *Rhizoctonia solani* (MP Biomedicals). Uskizyme (Wako Pure Chemical Industries) is a preparation from *Trichoderma* sp. with β(1,3)-glucanase, cel-lulase, protease, and chitinase activities. Westase (Cosmo Bio) is a preparation from the liquid culture supernatant of *Streptomyces rochei*. This complex contains mainly β(1,3)-glucanase and β(1,6)-glucanase activities according to the specifications sheet of the manufacturer, but minor or absent activities are not reported. Driselase (Sigma, Kyowa Hakko Kogyo) is a crude powder from *Basidiomycetes* sp. containing laminarinase, xylanase and cellulase activities. Glucanex or lysing enzymes from *Trichoderma harzianum* (Sigma), previously known as Novozyme-234 (Novozymes Corp., discontinued) contains β-glucanase, cellulase, protease, and chitinase activities. It also contains α(1,3)-glucanase activity, although it is not reported in the specifications sheet. Viscozyme (Sigma, Novozymes Corp.) is a multi-enzyme complex from *Aspergillus* sp. containing a wide range of carbohydrases, including arabanase, cellulase, β-glucanase, hemicellulase, and xylanase.

Other Cell Wall Chemical Fractionation and Analysis

Other methods that require more time and effort consist of combinations of enzymatic cell wall degradations (and dialysis), chemical degradations, and analytical techniques that permit the determination of the degradation products. These methods usually give more precise information about the type of bonds between the units forming the polymers. Common chemical degradations include alkali solubilization, acid hydrolysis, periodate oxidation, Smith degradation, borohydride reduction, β-elimination, carboxymethylation, and permethylation. Common analytical techniques include determination of reducing sugars, of total sugars, of glucose, of glucosamine, methylation analysis, gas–liquid chromatography, mass spectrometry, paper chromatography, gel filtration (size-exclusion) chromatography, ion-exchange chromatography, thin-layer chromatography (TLC), affinity chromatography, high-performance anionic-exchange chromatography (HPAEC), nuclear magnetic resonance (NMR) spectroscopy, and X-ray diffraction [6, 7, 53, 57, 60–70].

Colorimetric Determination of the Cell Wall Chitin

This is a colorimetric method useful to evaluate the amount of chitin in the cell wall [71]. This method is less precise than [^{14}C] labeling and measurement but it is faster and less toxic.

1. Cell cultures (200 mL) are harvested at early log-phase (A$_{600}$ of 1.0, approximately 200 mg wet weight), washed twice with water, and the wet cell pellet is weighed.
2. Cells are resuspended in 1.0 mL water and a volume corresponding to 100 mg of cells is transferred to glass tubes.

3. The walls of 100 mg cells are extracted in 1 mL of 6% KOH at 80 °C for 90 min and the suspension is cooled down, neutralized with 100 µL glacial acetic acid, transferred to a 1.5 mL tube (washing the glass tube with water) and centrifuged 1 min at 16,000 × *g*.
4. The insoluble cell wall material is washed three times with water, resuspended in 600 µL of 50 mM potassium phosphate pH 7.5, and incubated with 1 unit (5 µL) of recombinant chitinase from *Pyrococcus furiosus* (Wako Pure Chemical Industries) at 85 °C for 2 h. Then, the degraded cell wall material is incubated with 25 µL of Glusulase (PerkinElmer) at 37 °C for 1 h, stopped at 100 °C for 1 min and centrifuged 1 min. Glusulase degrades the chitobiose and small chitin fragments formed by the endo-chitinase treatment into GlcNAc monomers.
4. The amount of GlcNAc is quantified colorimetrically by the Reissig method [72] using different concentrations of GlcNAc (0, 0.02, 0.04, 0.06, 0.08, and 0.1 µmol) as standards. Each reaction contains 250 µL of water, 250 µL of 270 mM potassium tetraborate pH 9.5 and 0, 2, 4, 6, 8 and 10 µL of 10 mM GlcNAc. The samples contain 250 µL of 270 mM potassium tetraborate pH 9.5, 200 or 150 µL of water and 50 or 100 µL respectively of the degraded cell wall material. The samples are boiled (100 °C) during 8 min, cooled down in ice/water and 3 mL of Reissig reagent is added and mixed. The Reissig reagent must be prepared freshly and contains 1 g of 3,5-diamino-benzaldehyde, 1.25 mL of 37% HCl and glacial acetic acid up to 100 mL. The reactions are done in duplicate. The samples are incubated 40 min at 37 °C and the generated color is measured at 585 nm (quartz cuvettes).

Cell Wall β(1,6) Glucan Determination

The most common method for β(1,6)glucan determination consists of alkaline extraction of the cell wall and analysis of the alkali-insoluble β(1,6)-glucan [73, 74], although part of the β(1,6)-glucan is in the alkali-soluble fraction [63, 75].

1. Isolated cell walls from 50 mL cultures are obtained as described in page 9 and Fig 12.3, and extracted three times with 1 mL of 3% NaOH at 75 °C for 1 h (removes mannoproteins and alkali-soluble glucan).

2. The extracted walls are washed once with 1 mL of 100 mM Tris–HCl pH 7.5, once with 1 mL of 10 mM Tris–HCl pH 7.5, resuspended in 1 mL of 10 mM Tris–HCl pH 7.5 and incubated with 1 mg of Zymolyase-100T at 37 °C for 16 h. Approximately 90% of the glucose-containing carbohydrate is released into the supernatant. Zymolyase releases the β(1,6)-glucan to the supernatant.

3. The insoluble material is removed by centrifugation ($13,000 \times g$, 15 min) and the supernatant is dialyzed against water (6–8 kDa pore size) for 16 h. The carbohydrate retained after dialysis is the amount of β(1-6)-glucan. Total carbohydrate of each alkali-insoluble fraction (dialysis-retained, Zymolyase-soluble and Zymolyase-insoluble) is measured as hexose by the borosulfuric acid or phenol-sulfuric methods [76, 77]. The alkali-insoluble β(1,6)-glucan is determined as percentage of total carbohydrate (the sum of both the Zymolyase-soluble and insoluble fractions). The alkali-insoluble β(1,3)-glucan is determined as the amount of Zymolyase-soluble material before dialysis subtracted from the amount of alkali-insoluble β(1,6)-glucan.

The alkali-soluble β(1,6)glucan can be detected by immunodetection [75, 78]. The alkali-soluble extract is spotted onto nitrocellulose, dried, blocked with 5% nonfat milk in TBST (10 mM Tris pH 8.0, 150 mM NaCl, 0.05% Tween 20) and probed with anti-β(1,6)-glucan antibody [70]. After antibody binding, the membrane is washed three times with TBST, probed with anti-rabbit IgG horseradish peroxidase secondary antibody, washed again three times with TBST, and visualized with an ECL detection kit (Amersham). The alkali-soluble β(1,3)glucan can also be detected with the same procedure by using commercial monoclonal anti-β(1,3)-glucan antibodies (Biosupplies) and anti-mouse secondary antibody. The amount of β(1,6)glucan or β(1,3)glucan is quantified in dots by using different concentrations of (β(1,6)glucan (pustulan) or β(1,3)-glucan (pachyman, curdlan, or laminarin) as standards, respectively.

β(1,6)-glucan can also be determined by high-performance liquid chromatography (HPLC), gas chromatography, mass spectrometry and NMR analysis of oligosaccharides after alkaline extraction and/or enzymatic degradation of the cell wall polymers.

Determination of Cell Wall Proteins

The cell wall proteins are glycoproteins highly modified with O- and N-linked oligosaccharides, predominantly or exclusively formed by mannose residues known as mannan. In some cases, the mannan backbone presents single residues or side chains of different sugars, galactomannan, rhamnomannan, glucogalactomannan, rhamnogalactomannan, etc. [1, 24, 25].

Most proteins found in the cell wall are water- or detergent-soluble and are usually secreted to the medium. Some few cell wall proteins are covalently linked to the polysaccharides and can be divided into two groups, proteins covalently attached to β(1,3)glucan (Pir proteins) through a glutamine residue, an alkali-labile linkage that can be extracted by a mild alkali treatment (30 mM NaOH at 4 °C for 16 h); and proteins covalently attached by a GPI anchor to the β(1,6)-glucan of a β(1,6)/β(1,3)glucan core that can be removed by β(1,3)glucanases or β(1,6)glucanases [1, 59, 79–82].

The SDS-soluble proteins are extracted by hot SDS-mercaptoethanol treatments, twice with 50 mM Tris–HCl pH 7.8–8.0, 2% SDS, 100 mM EDTA, and 40 mM β-mercaptoethanol for 5–15 min at 100 °C [70, 83, 84]. The proteins are concentrated and the SDS removed by precipitation with 9 volumes of cold acetone at −20 °C for 2 h. The proteins are dried, resuspended and analyzed by SDS-PAGE or Western blot.

The covalently attached cell wall proteins can be purified by three methods [82, 84]:

1. The proteins covalently attached to β(1,3)glucan are released by treatment with 30 mM

NaOH at 4 °C for 16 h. The reaction is stopped by adding acetic acid to neutrality followed by dialysis.

2. The GPI-attached proteins are released by degradation of the GPI anchor by treatment with undiluted pyridine hydrofluoride at 24 °C for 16 h. HF-pyridine is removed by dialysis. Treatment with recombinant endo-β(1,6)-glucanase (noncommercial) also releases the GPI proteins.

3. Both groups of covalently attached proteins can be released by treatment with recombinant endo-β(1,3)-glucanase (MP Biomedicals; Wako Pure Chemical Industries), followed by dialysis.

The mannan oligosaccharides can be analyzed directly or after protease treatment by enzymatic treatment with different mannosidases. As the glycoproteins are highly glycosylated and this may interfere with the electrophoresis protein analysis, the N-linked oligosaccharides can be removed by treatment with endoglycosidase H (Boehringer, Roche, New England Biolabs) [70, 85]. The proteins can be analyzed by SDS-PAGE or Western blot with specific antibodies or with the lectin concanavalin A (Sigma) that specifically binds to the mannan region [81, 86]. Additionally, the cell wall proteins can be biotin-labeled directly in the cell prior to cell wall purification, with the biotinylation reagent Sulfo-NHS-LC-Biotin (Pierce) in 50 mM potassium phosphate pH 8.0 by incubation for 90 min on ice. The proteins are analyzed by Western blot and visualized with streptavidin-horseradish peroxidase conjugate (Pierce) [87, 88].

The mannan can be analyzed by incorporation of radioactive orthophosphate into N-mannosylated glycoproteins. The cell wall is purified and extracted with SDS to obtain part of the mannan. Then the insoluble material is treated with β(1,3)glucanases to release the rest of the mannan. The phosphate is bound as mannose-6-phosphate to both soluble and covalently linked cell wall mannoproteins [83, 89]. The phosphorylated cell wall oligosaccharides can be characterized after hydrolysis in trifluoracetic acid (TFA) by Quaternary aminoethyl (QAE)-Sephadex A50 chromatography, Bio-Gel P2 chromatography, HPAEC, electrospray ionization tandem mass spectrometry (ESI-MS-MS), and methylation analysis with gas chromatography–mass spectrometry [83].

Nonquantitative Methods for Analysis of Cell Wall Polysaccharides

Other methods can also be used in order to detect cell wall differences in structure or composition without the need of precise polymer quantification. Among them, those more used are briefly described as follows.

Sensitivity of Cells to Enzymatic Degradation

This method is used for a rough analysis of the cell wall state and is used to corroborate other results that suggest an altered cell wall [29, 90]. It is also used as a screening for mutations affecting the cell wall. Common enzymes used in this procedure are Zymolyase-100T in 50 mM citrate/phosphate buffer pH 5.6, Kitalase in 50 mM potassium-acetate pH 5.0, or Glucanex (Sigma) in any of these buffers. Zymolyase-100T and Kitalase mainly degrade β(1,3)-glucan and mannoproteins, whereas Glucanex degrades the entire cell wall. Increased sensitivity of the cells to these enzymes can be due to different causes such as a decrease in the amount of cell wall β-glucan or an increase in the permeability of the cell wall to the enzymes, which can be caused by a decrease in the cell wall surface glycoproteins or an altered cell wall structure. A specific cell wall-related mutant strain may show different sensitivities to Zymolyase, Kitalase and Glucanex, depending on the importance that the α-glucan may have in maintaining cell integrity in this mutant strain. The sensitivity to degradation using α(1,3)-glucanase has never been assayed due to the lack of a commercially available enzyme, although the purification of recombinant α(1,3)-glucanases (mutanases) from *Penicillium purpurogenum* and *Trichoderma harzianum* has been described [91].

Hypersensitivity or Resistance to Cell Wall Biosynthesis Inhibitors

This method can be used to detect changes in the cell wall composition or structure, and to detect mutations in the enzymes involved in cell wall synthesis; it is also used in the analysis of genetic interactions between genes that might be specifically related to cell wall biosynthesis and cell integrity. The main inhibitors are:

- Calcofluor white used to stain chitin. Calcofluor also binds to linear $\beta(1,3)$glucan with high affinity when chitin is not present, as in the case of *S. pombe* [92]. The binding of this dye perturbs the wall structure at low concentration and halts cell growth at high concentration.
- Echinocandins (caspofungin, micafungin and anidulafungin, available for clinical use) are a family of antifungal drugs that specifically inhibit both, the in vitro $\beta(1,3)$-glucan synthase activity and the in vivo $\beta(1,3)$-glucan synthesis [28, 93–97]. Other families of $\beta(1,3)$-glucan synthase inhibitors like papulacandins and the acidic terpenoid enfumafungin are also used [29].
- Polyoxins and Nikkomycins are chitin synthase inhibitors [23, 94, 96] or recently discovered $\beta(1,6)$glucan synthase inhibitors can also be used for specific assays [98, 99].
- 2-deoxi-D-glucose can be used as competitor of glucose for assays of cell wall biosynthesis defects.

Fluorescence Microscopy

Direct observation of the cell wall using microscopy techniques is an important method for cell wall studies. Fluorochromes, lectins, and antibodies that stain a specific cell wall polysaccharide can be used for cell wall fluorescence microscopy analysis. The most commonly used are:

- Calcofluor white (Sigma) (25 µg/mL final concentration) stains the cell wall chitin and in its absence specifically stains linear $\beta(1,3)$glucan.

Similarly, aniline blue (0.5 mg/mL final concentration) (Biosupplies), specifically stains linear $\beta(1,3)$glucan, although its affinity for the cell wall $\beta(1,3)$glucan of growing poles is lower than that of Calcofluor.

- Lectins against the mannan of cell wall glycoproteins such as Concanavalin A, which binds mannose residues, the lectin from *Bandeiraea simplicifolia* that recognizes specifically terminal galactose residues or wheat germ agglutinin (WGA) that recognize specifically chitin. They can be used for immunofluorescence analysis bound to fluorescein isothiocyanate (FITC) (Sigma). Lectins are added (200 µg/mL) to the cells, in culture medium or in phosphate buffer solution (PBS) if the cells are fixed. After 15 min in dark, cells are washed and resuspended for microscopy observation. For the observation of chitin, since it is internal to the cell wall, WGA-FITC binding and visualization requires a previous mild alkali extraction of the cell wall. The same lectins, labeled with colloidal gold, can be used for immunoelectron microscopy.
- Antibodies raised against specific proteins or polymers can also be used (see below). These antibodies are combined with secondary fluorochrome-labeled antibodies for immunofluorescence studies.

Electron Microscopy (EM)

The techniques for EM are numerous and have advanced greatly during the last years:

Transmission Electron Microscopy

Cells fixed with glutaraldehyde have been used to study the cell wall and septum structures of fungal cells. This technique permits observation of the cell wall as a three-layered structure of polysaccharides with different electron densities (see Fig. 12.1) [92, 100–109]. Negative staining for TEM with uranyl acetate is an alternative method for visualization of fibrils network and other cell wall structures [110–114].

Scanning Electron Microscopic (SEM)

Cells fixed with glutaraldehyde are used to observe the cell surface clearly and with high fidelity. An improved technique is ultra-high-resolution low-voltage SEM (UHR-LVSEM) [103, 110–113, 115–118]. SEM microscopy can be coupled with lectins or antibodies labeled with colloidal gold particles. This technique served to detect a cell wall surface completely filled with particles specific for the mannan carbohydrate of glycoproteins [116], and to detect the β(1,3)-glucan in the bud scars of *S. Cerevisiae* [1].

Atomic Force Microscopy (AMF)

This technique is used to measure the mechanical properties of the fungal cell wall macromolecules [1, 113, 119, 120].

Cryoscanning and Cryosectioning Electron Microscopy

This technique uses cryofixation to physically immobilize the specimen [113]. An improved method combines high-pressure freezing with ultra-low temperature and low-voltage SEM (ULT-LVSEM). It is useful to analyze fractured and coated cell samples, allowing the observation of a fine plane and the ultrastructure of both external and internal cell components [121].

Transmission Immunoelectron Microscopy (IEM)

Cells are fixed with paraformaldehyde plus glutaraldehyde. This technique has improved with the method of cryofixation by high-pressure freezing followed by freeze-substitution to retain the antigenicity [122]. Specific mono- and polyclonal antibodies in combination with secondary antibodies conjugated with 10-nm gold particles have been used to locate the different types of β-glucans [122]. The currently described antibodies against different wall polysaccharides are:

- Anti-β(1,3)-glucan [123] (Biosupplies), a murine monoclonal antibody without cross-reactivity with β(1,4)-glucans or β(1,3;1,4)-glucans.

- Anti-β(1,3;1,4)-glucan (Biosupplies). It is a murine monoclonal antibody without cross-reactivity with β(1,3)-glucans [124].
- Anti-β(1,6)-glucan. This rabbit antiserum specifically recognizes this polymer in *S. cerevisiae* [70, 125] and *S. pombe* [122] but is not commercially available.
- Anti-β(1,6)-branched-β(1,3)-glucan [122, 126] is a rabbit antiserum obtained against grifolan (a type of β(1,6)-branched-β(1,3)-glucan) and does not reacts with linear β(1,3)-glucan. It is suggested that the hapten site of the antibody is the monoglucosyl β(1,6)-glucan-branched moiety of β(1,3)-glucan. It is not commercially available.
- Anti-α(1,3)-glucan. A polyclonal anti-α(1,3)-glucan has been used to analyze the cell wall α(1,3)-glucan but is not commercially available [64, 127]. Additionally, the monoclonal IgM antibody MOPC-104E (Sigma, Abcam) has been used to detect the α(1,3)-glucan in *Histoplasma capsulatum* yeast cell walls. This antibody specifically recognizes α(1,3)-glucan because only α(1,3)-linked and not β-linked glycosyl polysaccharides can block the antibody and because the cell walls lacking α(1,3)-glucan of an avirulent strain of *Histoplasma capsulatum* are not recognized by the antibody [128–130].
- Anti-GFP murine monoclonal antibody (JL-8 anti-GFP; BD Biosciences, Sigma) has been used to detect GFP-fused cell wall or plasma membrane proteins with secondary gold-labeled anti-mouse secondary antibodies [131].

The lectins mentioned previously, Concanavalin A, which binds mannose residues, and wheat germ agglutinin (WGA), which binds GlcNAc residues, are also used for IEM studies when labeled with colloidal gold particles [109, 132–134].

Acknowledgements We thank D. Posner for language revision. This work was supported by grants BFU2010-15641 and BIO2009-10597 from the Dirección General de Investigación, MICINN, Spain, and grant CSI038A11-2 from the Junta de Castilla y León, Spain.

References

1. Latge JP (2007) The cell wall: a carbohydrate armour for the fungal cell. Mol Microbiol 66:279–290
2. Cabib E, Bowers B, Sburlati A, Silverman SJ (1988) Fungal cell wall synthesis: the construction of a biological structure. Microbiol Sci 5:370–375
3. Lesage G, Bussey H (2006) Cell wall assembly in Saccharomyces cerevisiae. Microbiol Mol Biol Rev 70:317–343
4. Cabib E, Kang MS (1987) Fungal 1,3-β-glucan synthase. Methods Enzymol 138:637–642
5. Kapteyn JC, Ram AF, Groos EM, Kollar R, Montijn RC, Van Den Ende H et al (1997) Altered extent of cross-linking of β1,6-glucosylated mannoproteins to chitin in Saccharomyces cerevisiae mutants with reduced cell wall β1,3-glucan content. J Bacteriol 179:6279–6284
6. Kollar R, Reinhold BB, Petráková E, Yeh HJ, Ashwell G, Drgonová J et al (1997) Architecture of the yeast cell wall. β(1-6)-glucan interconnects mannoprotein, β(1-3)-glucan, and chitin. J Biol Chem 272:17762–17775
7. Kollar R, Petrakova E, Ashwell G, Robbins PW, Cabib E (1995) Architecture of the yeast cell wall. The linkage between chitin and β(1-3)-glucan. J Biol Chem 270:1170–1178
8. Cabib E, Blanco N, Grau C, Rodriguez-Pena JM, Arroyo J (2007) Crh1p and Crh2p are required for the cross-linking of chitin to β(1-6)glucan in the Saccharomyces cerevisiae cell wall. Mol Microbiol 63:921–935
9. Cabib E, Farkas V, Kosik O, Blanco N, Arroyo J, McPhie P (2008) Assembly of the yeast cell wall. Crh1p and Crh2p act as transglycosylases in vivo and in vitro. J Biol Chem 283:29859–29872
10. Kapteyn JC, Montijn RC, Vink E, de la Cruz J, Llobell A, Douwes JE, Shimoi H, Lipke PN, Klis FM (1996) Retention of Saccharomyces cerevisiae cell wall proteins through a phosphodiester-linked β-1,3-/β-1,6-glucan heteropolymer. Glycobiology 6:337–345
11. Klis FM, Boorsma A, De Groot PW (2006) Cell wall construction in Saccharomyces cerevisiae. Yeast 23:185–202
12. Douglas CM, Marrinan JA, Li W, Kurtz MB (1994) A Saccharomyces cerevisiae mutant with echinocandin-resistant 1,3-β-D-glucan synthase. J Bacteriol 176:5686–5696
13. Mazur P, Morin N, Baginsky W, El-Sherbeini M, Clemas JA, Nielsen JB et al (1995) Differential expression and function of two homologous subunits of yeast 1,3-β-D-glucan synthase. Mol Cell Biol 15:5671–5681
14. Pérez P, Ribas JC (2004) Cell wall analysis. Methods 33:245–251
15. Arellano M, Duran A, Perez P (1996) Rho1 GTPase activates the (1-3)β-D-glucan synthase and is involved in Schizosaccharomyces pombe morphogenesis. EMBO J 15:4584–4591
16. Qadota H, Python CP, Inoue SB, Arisawa M, Anraku Y, Zheng Y et al (1996) Identification of yeast Rho1p GTPase as a regulatory subunit of 1,3-β-glucan synthase. Science 272:279–281
17. Katayama S, Hirata D, Arellano M, Pérez P, Toda T (1999) Fission yeast α-glucan synthase Mok1 requires the actin cytoskeleton to localize the sites of growth and plays an essential role in cell morphogenesis downstream of protein kinase C function. J Cell Biol 144:1173–1186
18. Hoschstenbach F, Klis FM, Van den Ende H, Van Donselaar E, Peters PJ et al (1998) Identification of a putative alpha-glucan synthase essential for cell wall construction and morphogenesis in fission yeast. Proc Natl Acad Sci USA 95:9161–9166
19. Grun CH, Hochstenbach F, Humbel BM, Verkleij AJ, Sietsma JH, Klis FM et al (2005) The structure of cell wall alpha-glucan from fission yeast. Glycobiology 15:245–257
20. Vos A, Dekker N, Distel B, Leunissen JA, Hochstenbach F (2007) Role of the synthase domain of Ags1p in cell wall alpha-glucan biosynthesis in fission yeast. J Biol Chem 282:18969–18979
21. Cabib E, Roh DH, Schmidt M, Crotti LB, Varma A (2001) The yeast cell wall and septum as paradigms of cell growth and morphogenesis. J Biol Chem 276:19679–19682
22. Roncero C (2002) The genetic complexity of chitin synthesis in fungi. Curr Genet 41:367–378
23. Lenardon MD, Munro CA, Gow NA (2010) Chitin synthesis and fungal pathogenesis. Curr Opin Microbiol 13:416–423
24. Bowman SM, Free SJ (2006) The structure and synthesis of the fungal cell wall. Bioessays 28:799–808
25. Leal JA, Prieto A, Bernabe M, Hawksworth DL (2010) An assessment of fungal wall heteromannans as a phylogenetically informative character in ascomycetes. FEMS Microbiol Rev 34:986–1014
26. Shematek EM, Braatz JA, Cabib E (1980) Biosynthesis of yeast cell wall. I. Preparation and properties of β(1-3)glucan synthetase. J Biol Chem 255:888–894
27. Shematek EM, Cabib E (1980) Biosynthesis of yeast cell wall. II. Regulation of β(1-3)glucan synthetase by ATP and GTP. J Biol Chem 255:895–902
28. Ishiguro J, Saitou A, Durán A, Ribas JC (1997) cps1+, a Schizosaccharomyces pombe gene homolog of Saccharomyces cerevisiae FKS genes whose mutation confers hypersensitivity to cyclosporin A and papulacandin B. J Bacteriol 179:7653–7662
29. Martins IM, Cortés JCG, Muñoz J, Moreno MB, Ramos M, Clemente-Ramos JA et al (2011) Differential activities of three families of specific β(1,3)glucan synthase inhibitors in wild-type and resistant strains of fission yeast. J Biol Chem 286:3484–3496
30. Abe M, Nishida I, Minemura M, Qadota H, Seyama Y, Watanabe T et al (2001) Yeast 1,3-β-glucan synthase activity is inhibited by phytosphingosine localized to the endoplasmic reticulum. J Biol Chem 276:26923–26930

31. El-Sherbeini M, Clemas JA (1995) Nikkomycin Z supersensitivity of an echinocandin-resistant mutant of *Saccharomyces cerevisiae*. Antimicrob Agents Chemother 39:200–207

32. Inoue SB, Takewaki N, Takasuka T, Mio T, Adachi M, Fujii Y et al (1995) Characterization and gene cloning of 1,3-β-D-glucan synthase from *Saccharomyces cerevisiae*. Eur J Biochem 231:845–854

33. Kelly R, Register E, Hsu MJ, Kurtz M, Nielsen J (1996) Isolation of a gene involved in 1,3-β-glucan synthesis in *Aspergillus nidulans* and purification of the corresponding protein. J Bacteriol 178:4381–4391

34. Mazur P, Baginsky W (1996) *In vitro* activity of 1,3-β-D-glucan synthase requires the GTP-binding protein Rho1. J Biol Chem 271:14604–14609

35. Mol PC, Park HM, Mullins JT, Cabib E (1994) A GTP-binding protein regulates the activity of (1,3)-β-glucan synthase, an enzyme directly involved in yeast cell wall morphogenesis. J Biol Chem 269:31267–31274

36. Thompson JR, Douglas CM, Li W, Jue CK, Pramanik B, Yuan X et al (1999) A glucan synthase *FKS1* homolog in *Cryptococcus neoformans* is single copy and encodes an essential function. J Bacteriol 181:444–453

37. Wood RL, Miller TK, Wright A, McCarthy P, Taft CS, Pomponi S et al (1998) Characterization and optimization of *in vitro* assay conditions for (1,3)β-glucan synthase activity from *Aspergillus fumigatus* and *Candida albicans* for enzyme inhibition screening. J Antibiot (Tokyo) 51:665–675

38. Sestak S, Farkas V (2001) *In situ* assays of fungal enzymes in cells permeabilized by osmotic shock. Anal Biochem 292:34–39

39. Shedletzky E, Unger C, Delmer DP (1997) A microtiter-based fluorescence assay for (1,3)-β-glucan synthases. Anal Biochem 249:88–93

40. Ribas JC, Díaz M, Durán A, Pérez P (1991) Isolation and characterization of *Schizosaccharomyces pombe* mutants defective in cell wall (1-3)β-D-glucan. J Bacteriol 173:3456–3462

41. El-Sherbeini M, Clemas JA (1995) Cloning and characterization of *GNS1*: a *Saccharomyces cerevisiae* gene involved in synthesis of 1,3-β-glucan *in vitro*. J Bacteriol 177:3227–3234

42. Kang MS, Cabib E (1986) Regulation of fungal cell wall growth: a guanine nucleotide-binding proteinaceous component required for activity of (1,3)-β-D-glucan synthase. Proc Natl Acad Sci U S A 83:5808–5812

43. Kondoh O, Tachibana Y, Ohya Y, Arisawa M, Watanabe T (1997) Cloning of the *RHO1* gene from *Candida albicans* and its regulation of β-1,3-glucan synthesis. J Bacteriol 179:7734–7741

44. Schimoler-O'Rourke R, Renault S, Mo W, Selitrennikoff CP (2003) *Neurospora crassa FKS* protein binds to the (1,3)β-glucan synthase substrate, UDP-glucose. Curr Microbiol 46:408–412

45. Crotti LB, Drgon T, Cabib E (2001) Yeast cell permeabilization by osmotic shock allows determination of enzymatic activities *in situ*. Anal Biochem 292:8–16

46. Aimanianda V, Clavaud C, Simenel C, Fontaine T, Delepierre M, Latge JP (2009) Cell wall β-(1,6)-glucan of *Saccharomyces cerevisiae*: structural characterization and *in situ* synthesis. J Biol Chem 284:13401–13412

47. Frost DJ, Brandt K, Capobianco J, Goldman R (1994) Characterization of (1,3)-β-glucan synthase in *Candida albicans*: microsomal assay from the yeast or mycelial morphological forms and a permeabilized whole-cell assay. Microbiology 140:2239–2246

48. Choi WJ, Cabib E (1994) The use of divalent cations and pH for the determination of specific yeast chitin synthetases. Anal Biochem 219:368–372

49. Lucero HA, Kuranda MJ, Bulik DA (2002) A nonradioactive, high throughput assay for chitin synthase activity. Anal Biochem 305:97–105

50. Castro C, Ribas JC, Valdivieso MH, Varona R, del Rey F, Durán A (1995) Papulacandin B resistance in budding and fission yeasts: isolation and characterization of a gene involved in (1,3)β-D-glucan synthesis in *Saccharomyces cerevisiae*. J Bacteriol 177:5732–5739

51. Roncero C, Valdivieso MH, Ribas JC, Duran A (1988) Isolation and characterization of *Saccharomyces cerevisiae* mutants resistant to Calcofluor white. J Bacteriol 170:1950–1954

52. Algranati ID, Behrens N, Carminatti H, Cabib E (1966) Mannan synthetase from yeast. Methods Enzymol 8:411–416

53. Gopal PK, Shepherd MG, Sullivan PA (1984) Analysis of wall glucans from yeast, hyphal and germ-tube forming cells of *Candida albicans*. J Gen Microbiol 130:3295–3301

54. Gorka-Niec W, Perlinska-Lenart U, Zembek P, Palamarczyk G, Kruszewska JS (2010) Influence of sorbitol on protein production and glycosylation and cell wall formation in *Trichoderma reesei*. Fungal Biol 114:855–862

55. Bartnicki-Garcia S (1999) Glucans, walls, and morphogenesis: On the contributions of J. G. H. Wessels to the golden decades of fungal physiology and beyond. Fungal Genet Biol 27:119–127

56. Perez P, Garcia-Acha I, Duran A (1983) Effect of papulacandin B on the cell wall and growth of *Geotrichum lactis*. J Gen Microbiol 129:245–250

57. Sietsma JH, Wessels JG (1977) Chemical analysis of the hyphal wall of *Schizophyllum commune*. Biochim Biophys Acta 496:225–239

58. Tomazett PK, Felix CR, Lenzi HL, de Paula Faria F, de Almeida Soares CM, Pereira M (2010) 1,3-β-D-Glucan synthase of *Paracoccidioides brasiliensis*: recombinant protein, expression and cytolocalization in the yeast and mycelium phases. Fungal Biol 114:809–816

59. Latge JP (2010) Tasting the fungal cell wall. Cell Microbiol 12:863–872

60. Manners DJ, Meyer MT (1977) The molecular structures of some glucans from the cell walls of *Schizosaccharomyces pombe*. Carbohydr Res 57:189–203

61. Magnelli PE, Cipollo JF, Robbins PW (2005) A glu-canase-driven fractionation allows redefinition of *Schizosaccharomyces pombe* cell wall composition and structure: assignment of diglucan. Anal Biochem 336:202–212

62. Sugawara T, Takahashi S, Osumi M, Ohno N (2004) Refinement of the structures of cell-wall glucans of *Schizosaccharomyces pombe* by chemical modification and NMR spectroscopy. Carbohydr Res 339: 2255–2265

63. Magnelli P, Cipollo JF, Abeijon C (2002) A refined method for the determination of *Saccharomyces cerevisiae* cell wall composition and β-1,6-glucan fine structure. Anal Biochem 301:136–150

64. Sugawara T, Sato M, Takagi T, Kamasaki T, Ohno N, Osumi M (2003) In situ localization of cell wall α-1,3-glucan in the fission yeast *Schizosaccharomyces pombe*. J Electron Microsc (Tokyo) 52:237–242

65. Bush DA, Horisberger M, Horman I, Wursch P (1974) The wall structure of *Schizosaccharomyces pombe*. J Gen Microbiol 81:199–206

66. Cabib E, Durán A (2005) Synthase III-dependent chitin is bound to different acceptors depending on location on the cell wall of budding yeast. J Biol Chem 280:9170–9179

67. Manners DJ, Masson AJ, Patterson JC (1973) The structure of a β-(1-3)-D-glucan from yeast cell walls. Biochem J 135:19–30

68. Manners DJ, Masson AJ, Patterson JC, Bjorndal H, Lindberg B (1973) The structure of a β-(1-6)-D-glucan from yeast cell walls. Biochem J 135:31–36

69. Fontaine T, Simenel C, Dubreucq G, Adam O, Delepierre M, Lemoine J et al (2000) Molecular organization of the alkali-insoluble fraction of *Aspergillus fumigatus* cell wall. J Biol Chem 275:27594–27607

70. Montijn RC, van Rinsum J, van Schagen FA, Klis FM (1994) Glucomannoproteins in the cell wall of *Saccharomyces cerevisiae* contain a novel type of carbohydrate side chain. J Biol Chem 269:19338–19342

71. Bulawa CE, Slater M, Cabib E, Au-Young J, Sburlati A, Adair WL Jr, Robbins PW (1986) The *S. cerevisiae* structural gene for chitin synthase is not required for chitin synthesis *in vivo*. Cell 46:213–225

72. Reissig JL, Storminger JL, Leloir LF (1955) A modified colorimetric method for the estimation of N-acetylamino sugars. J Biol Chem 217:959–966

73. Boone C, Sommer SS, Hensel A, Bussey H (1990) Yeast *KRE* genes provide evidence for a pathway of cell wall β-glucan assembly. J Cell Biol 110:1833–1843

74. Dijkgraaf GJ, Brown JL, Bussey H (1996) The *KNH1* gene of *Saccharomyces cerevisiae* is a functional homolog of *KRE9*. Yeast 12:683–692

75. Dijkgraaf GJ, Abe M, Ohya Y, Bussey H (2002) Mutations in Fks1p affect the cell wall content of β-1,3- and β-1,6-glucan in *Saccharomyces cerevisiae*. Yeast 19:671–690

76. Badin J, Jackson C, Schubert M (1953) Improved method for determination of plasma polysaccharides with tryptophan. Proc Soc Exp Biol Med 84:289–291

77. McKelvy JF, Lee YC (1969) Microheterogeneity of the carbohydrate group of *Aspergillus oryzae* α-amylase. Arch Biochem Biophys 132:99–110

78. Lussier M, Sdicu AM, Shahinian S, Bussey H (1998) The *Candida albicans KRE9* gene is required for cell wall β-1,6-glucan synthesis and is essential for growth on glucose. Proc Natl Acad Sci U S A 95:9825–9830

79. Ecker M, Deutzmann R, Lehle L, Mrsa V, Tanner W (2006) Pir proteins of *Saccharomyces cerevisiae* are attached to β-1,3-glucan by a new protein-carbohydrate linkage. J Biol Chem 281:11523–11529

80. de Groot PW, Ram AF, Klis FM (2005) Features and functions of covalently linked proteins in fungal cell walls. Fungal Genet Biol 42:657–675

81. de Groot PW, Yin QY, Weig M, Sosinska GJ, Klis FM, de Koster CG (2007) Mass spectrometric identification of covalently bound cell wall proteins from the fission yeast *Schizosaccharomyces pombe*. Yeast 24:267–278

82. Klis FM, Brul S, De Groot PW (2010) Covalently linked wall proteins in ascomycetous fungi. Yeast 27:489–493

83. Mrsa V, Ecker M, Strahl-Bolsinger S, Nimtz M, Lehle L, Tanner W (1999) Deletion of new covalently linked cell wall glycoproteins alters the electrophoretic mobility of phosphorylated wall components of *Saccharomyces cerevisiae*. J Bacteriol 181:3076–3086

84. de Groot PW, de Boer AD, Cunningham J, Dekker HL, de Jong L, Hellingwerf KJ et al (2004) Proteomic analysis of *Candida albicans* cell walls reveals covalently bound carbohydrate-active enzymes and adhesins. Eukaryot Cell 3:955–965

85. Carotti C, Ragni E, Palomares O, Fontaine T, Tedeschi G, Rodriguez R et al (2004) Characterization of recombinant forms of the yeast Gas1 protein and identification of residues essential for glucanosyltransferase activity and folding. Eur J Biochem 271:3635–3645

86. Kapteyn JC, ter Riet B, Vink E, Blad S, De Nobel H, Van Den Ende H et al (2001) Low external pH induces *HOG1*-dependent changes in the organization of the *Saccharomyces cerevisiae* cell wall. Mol Microbiol 39:469–479

87. Mrsa V, Seidl T, Gentzsch M, Tanner W (1997) Specific labelling of cell wall proteins by biotinylation. Identification of four covalently linked O-mannosylated proteins of *Saccharomyces cerevisiae*. Yeast 13:1145–1154

88. Mrsa V, Tanner W (1999) Role of NaOH-extractable cell wall proteins Ccw5p, Ccw6p, Ccw7p and Ccw8p (members of the Pir protein family) in stability of the *Saccharomyces cerevisiae* cell wall. Yeast 15:813–820

89. Ballou L, Hernandez LM, Alvarado E, Ballou CE (1990) Revision of the oligosaccharide structures of yeast carboxypeptidase Y. Proc Natl Acad Sci U S A 87:3368–3372

90. Calonge TM, Nakano K, Arellano M, Arai R, Katayama S, Toda T et al (2000) *Schizosaccharomyces pombe* rho2p GTPase regulates cell wall α-glucan biosynthesis through the protein kinase pck2p. Mol Biol Cell 11:4393–4401

91. Fuglsang CC, Berka RM, Wahleithner JA, Kauppinen S, Shuster JR, Rasmussen G et al (2000) Biochemical analysis of recombinant fungal mutanases. A new family of α1,3-glucanases with novel carbohydrate-binding domains. J Biol Chem 275:2009–2018

92. Cortés JC, Konomi M, Martins IM, Munoz J, Moreno MB, Osumi M et al (2007) The (1,3)β-D-glucan synthase subunit Bgs1p is responsible for the fission yeast primary septum formation. Mol Microbiol 65:201–217

93. Bal AM (2010) The echinocandins: three useful choices or three too many? Int J Antimicrob Agents 35:13–18

94. Chapman SW, Sullivan DC, Cleary JD (2008) In search of the holy grail of antifungal therapy. Trans Am Clin Climatol Assoc 119:197–215; discussion 215–216

95. Shao PL, Huang LM, Hsueh PR (2007) Recent advances and challenges in the treatment of invasive fungal infections. Int J Antimicrob Agents 30:487–495

96. Vicente MF, Basilio A, Cabello A, Pelaez F (2003) Microbial natural products as a source of antifungals. Clin Microbiol Infect 9:15–32

97. Varona R, Pérez P, Durán A (1983) Effect of papulacandin B on β-glucan synthesis in Schizosaccharomyces pombe. FEMS Microbiol Lett 20:243–247

98. Kitamura A, Higuchi S, Hata M, Kawakami K, Yoshida K, Namba K et al (2009) Effect of β-1,6-glucan inhibitors on the invasion process of Candida albicans: potential mechanism of their in vivo efficacy. Antimicrob Agents Chemother 53:3963–3971

99. Kitamura A, Someya K, Hata M, Nakajima R, Takemura M (2009) Discovery of a small-molecule inhibitor of β-1,6-glucan synthesis. Antimicrob Agents Chemother 53:670–677

100. Biely P, Kovarik J, Bauer S (1973) Cell wall formation in yeast. An electron microscopic autoradiographic study. Arch Microbiol 94:356–371

101. Feldmesser M, Kress Y, Mednick A, Casadevall A (2000) The effect of the echinocandin analogue caspofungin on cell wall glucan synthesis by Cryptococcus neoformans. J Infect Dis 182:1791–1795

102. Johnson BF, Yoo BY, Calleja GB (1973) Cell division in yeasts: movement of organelles associated with cell plate growth of Schizosaccharomyces pombe. J Bacteriol 115:358–366

103. Osumi M, Sato M, Ishijima SA, Konomi M, Takagi T, Yaguchi H (1998) Dynamics of cell wall formation in fission yeast, Schizosaccharomyces pombe. Fungal Genet Biol 24:178–206

104. Roh DH, Bowers B, Schmidt M, Cabib E (2002) The septation apparatus, an autonomous system in budding yeast. Mol Biol Cell 13:2747–2759

105. Schmidt M, Bowers B, Varma A, Roh DH, Cabib E (2002) In budding yeast, contraction of the actomyosin ring and formation of the primary septum at cytokinesis depend on each other. J Cell Sci 115:293–302

106. Coluccio A, Bogengruber E, Conrad MN, Dresser ME, Briza P, Neiman AM (2004) Morphogenetic pathway of spore wall assembly in Saccharomyces cerevisiae. Eukaryot Cell 3:1464–1475

107. Bowers B, Levin G, Cabib E (1974) Effect of polyoxin D on chitin synthesis and septum formation in Saccharomyces cerevisiae. J Bacteriol 119:564–575

108. Cabib E, Sburlati A, Bowers B, Silverman SJ (1989) Chitin synthase 1, an auxiliary enzyme for chitin synthesis in Saccharomyces cerevisiae. J Cell Biol 108:1665–1672

109. Shaw JA, Mol PC, Bowers B, Silverman SJ, Valdivieso MH, Durán A et al (1991) The function of chitin synthases 2 and 3 in the Saccharomyces cerevisiae cell cycle. J Cell Biol 114:111–123

110. Konomi M, Fujimoto K, Toda T, Osumi M (2003) Characterization and behaviour of α-glucan synthase in Schizosaccharomyces pombe as revealed by electron microscopy. Yeast 20:427–438

111. Osumi M, Yamada N, Kobori H, Taki A, Naito N, Baba M et al (1989) Cell wall formation in regenerating protoplasts of Schizosaccharomyces pombe: study by high resolution, low voltage scanning electron microscopy. J Electron Microsc (Tokyo) 38:457–468

112. Konomi M, Ishiguro J, Osumi M (2000) Abnormal formation of the glucan network from regenerating protoplasts in Schizosaccharomyces pombe cps8 actin point mutant. J Electron Microsc (Tokyo) 49:569–578

113. Osumi M (1998) The ultrastructure of yeast: cell wall structure and formation. Micron 29:207–233

114. Kopecka M, Fleet GH, Phaff HJ (1995) Ultrastructure of the cell wall of Schizosaccharomyces pombe following treatment with various glucanases. J Struct Biol 114:140–152

115. Sipiczki M, Yamaguchi M, Grallert A, Takeo K, Zilahi E, Bozsik A et al (2000) Role of cell shape in determination of the division plane in Schizosaccharomyces pombe: random orientation of septa in spherical cells. J Bacteriol 182:1693–1701

116. Osumi M, Yamada N, Yaguchi H, Kobori H, Nagatani T, Sato M (1995) Ultrahigh-resolution low-voltage SEM reveals ultrastructure of the glucan network formation from fission yeast protoplast. J Electron Microsc (Tokyo) 44:198–206

117. Kobori H, Yamada N, Taki A, Osumi M (1989) Actin is associated with the formation of the cell wall in reverting protoplasts of the fission yeast Schizosaccharomyces pombe. J Cell Sci 94:635–646

118. Bernard M, Latge JP (2001) Aspergillus fumigatus cell wall: composition and biosynthesis. Med Mycol 39:9–17

119. Dufrene YF (2010) Atomic force microscopy of fungal cell walls: an update. Yeast 27:465–471

120. Zhao L, Schaefer D, Xu H, Modi SJ, LaCourse WR, Marten MR (2005) Elastic properties of the cell wall of Aspergillus nidulans studied with atomic force microscopy. Biotechnol Prog 21:292–299

121. Osumi M, Konomi M, Sugawara T, Takagi T, Baba M (2006) High-pressure freezing is a powerful tool for visualization of *Schizosaccharomyces pombe* cells: ultra-low temperature and low-voltage scanning electron microscopy and immunoelectron microscopy. J Electron Microsc (Tokyo) 55:75–88

122. Humbel BM, Konomi M, Takagi T, Kamasawa N, Ishijima SA, Osumi M (2001) *In situ* localization of β-glucans in the cell wall of *Schizosaccharomyces pombe*. Yeast 18:433–444

123. Meikle PJ, Bonig I, Hoogenraad NJ, Clarke AE, Stone BA (1991) The location of (1-3)-β-glucans in the walls of pollen tubes of *Nicotiana alata* using a (1-3)-β-glucan-specific monoclonal antibody. Planta 185:1–8

124. Meikle PJ, Hoogenraad NJ, Bonig I, Clarke AE, Stone BA (1994) A (1->3,1->4)-beta-glucan-specific monoclonal antibody and its use in the quantitation and immunocytochemical location of (1->3,1->4)-beta-glucans. Plant J 5:1–9

125. Montijn RC, Vink E, Muller WH, Verkleij AJ, Van Den Ende H, Henrissat B et al (1999) Localization of synthesis of β1,6-glucan in *Saccharomyces cerevisiae*. J Bacteriol 181:7414–7420

126. Adachi Y, Ohno N, Yadomae T (1994) Preparation and antigen specificity of an anti-(1-3)-β-D-glucan antibody. Biol Pharm Bull 17:1508–1512

127. Reese AJ, Yoneda A, Breger JA, Beauvais A, Liu H, Griffith CL et al (2007) Loss of cell wall α(1-3) glucan affects *Cryptococcus neoformans* from ultrastructure to virulence. Mol Microbiol 63:1385–1398

128. Rappleye CA, Eissenberg LG, Goldman WE (2007) *Histoplasma capsulatum* α-(1,3)-glucan blocks innate immune recognition by the b-glucan receptor. Proc Natl Acad Sci U S A 104:1366–1370

129. Eissenberg LG, Moser SA, Goldman WE (1997) Alterations to the cell wall of *Histoplasma capsulatum* yeasts during infection of macrophages or epithelial cells. J Infect Dis 175:1538–1544

130. Kugler S, Schurtz Sebghati T, Groppe Eissenberg L, Goldman WE (2000) Phenotypic variation and intracellular parasitism by *Histoplasma capsulatum*. Proc Natl Acad Sci U S A 97:8794–8798

131. Gonzalez M, Goddard N, Hicks C, Ovalle R, Rauceo JM, Jue CK et al (2010) A screen for deficiencies in GPI-anchorage of wall glycoproteins in yeast. Yeast 27:583–596

132. Cabib E, Bowers B, Roberts RL (1983) Vectorial synthesis of a polysaccharide by isolated plasma membranes. Proc Natl Acad Sci U S A 80:3318–3321

133. Molano J, Bowers B, Cabib E (1980) Distribution of chitin in the yeast cell wall. An ultrastructural and chemical study. J Cell Biol 85:199–212

134. Roberts RL, Bowers B, Slater ML, Cabib E (1983) Chitin synthesis and localization in cell division cycle mutants of *Saccharomyces cerevisiae*. Mol Cell Biol 3:922–930

Histopathological Technique for Detection of Fungal Infections in Plants

Vijai Kumar Gupta and Brejesh Kumar Pandey

Abstract

Microscopic examination of the interaction between pathogenic fungi and their host plants has been instrumental in deciphering the biology of this relationship and can serve as a useful diagnostic tool. In this chapter, we describe the technique of fixing fungal infections of plant samplings for histopathological experiments. Toluidine blue O' staining methods coupled with stereoscopic microscopy are used to scan the infection structures of the fungus *Fusarium* spp. and host response in *Psidium guajava* L. root tissues.

Keywords

Fungal infections • Histopathological experiments • Microscopy • Staining techniques • TOLUIDINE blue O' • *Fusarium* spp. • *Psidium guajava* L.

Introduction

The ability to observe the growth of fungal structures in host tissues under the microscope is an important tool in the study of plant pathogenesis. Over the years many staining techniques that highlight fungal structures in plant tissues have been reported. In particular, technologies such as stereoscopic microscopy have enhanced our ability to visualize hyphae in plant tissue [1–4].

The use of certain staining techniques can facilitate considerably microscopic observations and experimental research on plant pathology

V.K. Gupta (✉)
Molecular Glycobiotechnology Group, Department of Biochemistry, School of Natural Sciences, National University of Ireland Galway, University Road, Galway, Ireland

Assistant Professor of Biotechnology, Department of Science, Faculty of Arts, Science & Commerce, MITS University, Rajasthan, India
e-mail: vijai.gupta@nuigalway.ie; vijaifzd@gmail.com

B.K. Pandey
Molecular Plant Pathology Laboratory, Central Institute for Subtropical Horticulture, Indian Council of Agricultural Research, P.O.kakori, Rahmankhera, Lucknow, UP 227 017, India

V.K. Gupta et al. (eds.), *Laboratory Protocols in Fungal Biology: Current Methods in Fungal Biology,*
Fungal Biology, DOI 10.1007/978-1-4614-2356-0_13, © Springer Science+Business Media, LLC 2013

by allowing plant and fungal tissues to be differentiated. More specifically, staining can aid examination of fungal colonization and infection processes, such as differentiating hyphae in life cycles that involve a transition from a biotrophic to a necrotrophic phase. Staining of specific tissues also can simplify identification of fungal inoculum or hyphal presence in asymptomatic plant tissue. The effectiveness of a particular staining technique can vary greatly depending on the particular fungus and plant species. Toluidine blue O' has been used to stain and identify callose deposition produced by host plants in response to intracellular infection of plant cells by fungi in some plant–fungus interactions [5]. Toluidine blue O' staining techniques was applied to examine the infection structures of the fungus *Fusarium* in root tissues wilt-infected guava plants. The usefulness of this staining method was based on the visual contrast between host plant tissue and fungal hyphae provided by polychromatic dye and resolution, and the relative ease of preparation and use [6]. This study describes an improved method for fixation of sampling of fungal-infected plant parts, and staining and observation of fungal infections in plant tissue for histopathological visualization.

Materials

1. Sterilized water
2. 0.1% $HgCl_2$
3. Glass slides
4. Whatman filter paper no. 41
5. Formaldehyde
6. Glacial acetic acid
7. Alcohol
8. Xylene
9. Paraffin wax
10. Toluidine blue O'
11. DPX-mount

12. Microprocessor-based automatic tissue processor (Electra, YSl 104, Yorko)
13. Microtome (MICROM—HM 350)
14. Stereoscopic microscope (Leica—LEITZ—DM RBE)

Methods

The methods presented in the following sections describe general procedures for fixation, staining, and microscopy of fungal infections of plant samplings. Modifications that may be needed to fix the sampling properly from different types and sources of material are also described.

Killing and Fixation

Roots samples were collected from wilt-affected and healthy plants. Root pieces 2–4 cm long were cut and surface sterilized using 0.1% $HgCl_2$, washed two to three times in sterilized water, and the excess water absorbed on Whatman filter paper 41. Then samples were kept in formaldehyde: acetic acid: alcohol (5 mL: 5 mL: 90 mL) for a minimum of 48 h (see Note 1).

Dehydration

The samples were processed with the alcohol:xylene series (as per the flow chart depicted in Fig. 13.1) using an automatic tissue processor (Yorko) (see Note 2).

Infiltration and Embedding

The samples were embedded in melted paraffin wax (54–56 °C) for at least 4–8 h in order to completely replace the xylene with paraffin wax in a square-shaped block (see Note 3).

Alcohol 30% for 30 min

↓

Alcohol 50% for 30 min

↓

Alcohol 70% for 30 min

↓

Alcohol 80% for 30 min

↓

Alcohol 95% for 30 min

↓

Absolute alcohol for 30 min

↓

Alcohol: xylene (25:75) for 30 min

↓

Alcohol: xylene (50:50) for 30 min

↓

Alcohol: xylene (75:25) for 30 min

↓

Pure xylene for 30 min

Fig. 13.1 Samples processed with the alcohol:xylene series. This process removes the water from the plant tissues and facilitates sectioning

Sectioning

Section (10 μm thick) cutting was done using a microtome. Blocks were prepared in paraffin wax and thin sections 10 μm thick were cut with the help of a microtome (MICROM—HM 350S). At least 20 slides were prepared for each sample (see Note 4).

Staining and Mounting

The sections were stained in 0.1% aqueous toluidine blue O' and were mounted in DPX after

bringing them to xylene through the alcohol:xylene series. The detailed procedure is given in the flow diagram depicted in Fig. 13.2 (see Note 5).

Microscopy and Imaging

Samples were mounted in 50% (v/v) DPX mount and viewed under a stereoscopic microscope (Leica—LEITZ DM RBE) using a Hoya CM500S filter (IR cut-off 650 nm). Images were captured using a CCD camera with a Bayer Array RGB filter for brilliant pictures (Interline transfer frame readout CCD—ICX252AQ) and Leica DFC Twain and Leica Image Manager analysis software (soft microscopy with imaging control software system).

Notes

1. The FAA solution is prepared based on the type of material, that is, soft tissue, moderate tissue, or hard tissue (use 25% ethanol for very delicate material, 50% for normal use, and 70% ethanol for very tough material). The samples were left in the FAA solution at least 48 h or until they were processed further. This depends on the hardness of the tissue.
2. This process removes the water from the plant tissues and facilitates sectioning.
3. Infiltration and embedding of the material was done in paraffin wax to remove the xylene from the tissues. The blocks of wax were prepared in L molds in which the material was embedded.
4. Sectioning of the material was done with the automatic microtome (MICROM—HM 350S).
5. Sections were stained in 0.1% aqueous toluidine blue O' and were mounted in DPX after bringing them to xylene through the alcohol:xylene series as described by Jensen [7]. The samples were examined for anatomical details as per the technique described by Pandey [3].

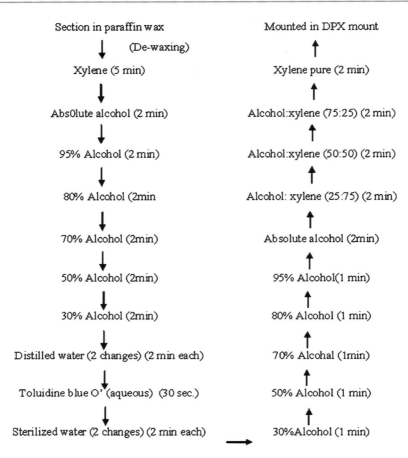

Fig. 13.2 Sections were stained in 0.1% aqueous toluidine blue O' and were mounted in DPX after bringing them to xylene through the alcohol:xylene series

Acknowledgements The authors are very grateful to Director, CISH; Head, Department of Crop-Protection, Central Institute of Subtropical Horticulture (CISH), Lucknow; and Prof. Shakti Baijal, Ex-Dean, FASC, MITS University, Rajasthan for providing the necessary research grants.

References

1. Johansen DA (1940) Plant microtechnique. McGraw-Hill, New York
2. Meyberg M (1988) Selective staining of fungal hyphae in parasitic and symbiotic plant-fungus associations. Histochemistry 88:197–199
3. Pandey BK (1984) Studies of chickpea blight caused by *Ascophyta rafiei* (Pass) Labr. with special reference to survival in crop debris [PhD thesis]. Department of Plant Pathology. G. B. Pant University of Agriculture & Technology, Pantnagar, UP, India
4. Saha DC, Jackson MA, Johnson-Cicalese JM (1988) A rapid staining method for detection of endophytic fungi in turf and forage grasses. Phytopathology 78:237–239
5. Gupta VK, Misra AK, Pandey BK (2012) Histopathological changes during wilting in guava root. Arch Phytopathol Plant Protect 45(5):570–573 Available from http://www.tandfonline.com/doi/abs/10.1080/03235408.2011.588047
6. Knight NL, Sutherland MW (2001) A rapid differential staining technique for *Fusarium pseudograminearum* in cereal tissues during crown rot infections. Plant Pathol Online. First-doi:10.1111/j.1365-3059.2011.02462.x
7. Jensen WA (1962) Botanical histochemistry: principals and practices. WH Freeman, San Francisco, London

Development of Media for Growth and Enumeration of Fungi from Water

Segula Masaphy

Abstract

Fungi are found in water resources as natural primary inhabitants or as secondary inhabitants that enter the water source as contaminants. Many of the fungi in water resources can be directly harmful to human, animal, and plant health, or cause problems in food processing and preparation, or by producing biofilms in water-distribution systems. Hence, water fungi are of concern for consumers. The ability to detect fungi in water sources is therefore important with respect to minimizing the risk of contamination and for safety-management protocols. However, there is no one uniform method for determining fungal load in water. Various new molecular-based methods are being developed to analyze water resources, but the traditional colony-based ones are still the methods of choice for enumeration and characterization of fungal populations in water. Recent developments in those methods for water mycological quality examination, particularly with regard to media composition, are presented.

Keywords

CFU • Detection • Enumeration • Fungi • Membrane filtration • Biofilm • Treated/untreated water • Routine analyses

Introduction

An awareness of the importance of fungi in water destined for human consumption has emerged in recent years. Although still limited, the number of publications on this topic is rising, demonstrating the presence of a range of fungal species, some of which are known to be directly pathogenic to humans, cause allergic reactions, or have harmful effects due to their production and release of toxins into the water [1–3]. Fungi are also suspected of contributing to negative organoleptic qualities in drinking water [4], and to biofilm production in distribution systems [5]. Plant-pathogenic and food-spoilage fungi have also been found in treated and untreated water [2].

S. Masaphy (✉)
Applied Microbiology and Mycology Lab,
MIGAL, Galilee Technology Center, POB 831, 11016,
South Industrial Area, Kiryat Shmona, Israel
e-mail: segula@migal.org.il

V.K. Gupta et al. (eds.), *Laboratory Protocols in Fungal Biology: Current Methods in Fungal Biology*,
Fungal Biology, DOI 10.1007/978-1-4614-2356-0_14, © Springer Science+Business Media, LLC 2013

In terms of human health, there is relatively little information on the role played by fungal contamination of water in illnesses and infections in the general public. However, the greater concern is related to immunosuppressed patients, who may be infected by drinking water, bath water, or recreational water bodies [6]. Hence, studies have been aimed at examining the mycological quality of hospital water systems, to assess the risk of fungal transmission to patients and possible infection. Potentially health-disruptive fungi have been detected in tap water [7–12], water-distribution systems [10, 13], bottled water [14, 15], bathing water [16], swimming pools [17, 18], surface water sources (rivers, streams, canals, lakes, and ponds) [2, 7], and groundwater [19].

Species of potentially human-pathogenic yeasts and molds recovered from drinking water include *Aspergillus fumigatus* [20], *Fusarium* [21], *Penicillium* [14], *Aureobasidium pullulans* (found in saunas) [22], *Absidia, Mucor, Candida* [23], *Trichoderma viride* [9], and *Chaetomium globosum* [4]. Various potentially pathogenic fungi, including dermatophytes, have also been isolated from swimming pools, such as *Cladosporium* spp., *Penicillium* spp., *Aspergillus* spp., *Rhizopus* spp., *Fusarium* and *Trichophyton rubrum, Mucor* spp. and *Candida albicans* [17, 18, 24], among others.

With the increasing concern about fungal contamination in consumed water, fungal examination of different water sources is on the rise, and it is recommended that water resources be monitored for fungal contamination as part of water-safety management programs. However, the field of mycological water quality, including methodologies and regulations, is far less established than the fields of bacteriological water quality and mycological food quality. In 1975, the methods for fungal detection in water and wastewater were still only tentative, as laid out in the fourteenth edition of the Standard Methods for the Examination of Water and Wastewater [25]. Even today, on a global scale, there are different regulations and methodologies in place for mycological water safety, and no uniform or standardized method for determining the mycological quality of potable water resources has been recognized.

Water is very heterogeneous with regard to the fungal diversity found within it. Fungi can be found in water resources as natural primary inhabitants or as secondary inhabitants that have entered the water source as contaminants. As summarized by Hageskal et al. [26] primary inhabitants are those that are adapted to aquatic environments, belonging to the phylum Chytridiomycota. Secondary inhabitants are all other fungi, which enter natural water bodies from the air, soil, and wastewater. The survival and proliferation of these latter fungi depend on the characteristics of the water, that is, nutrition load, temperature, pH, other microorganisms, and, in some cases, the presence of disinfectants (such as chlorine in swimming pools) [27].

The origin of the water source and its designated use also vary (e.g., tap water for drinking, bottled water, swimming pool water, recreational water, or wastewater), and these need to be considered in the establishment and application of a detection method. Each of these water bodies may support different fungal loads and diversities. It is clear that no one defined fungal indicator can give information on all of the different types of damage that can be expected from each water source. It is important to detect the individual fungus that is causing a problem and identify it, but quantifying the total fungal load may give a faster and clearer estimation of the degree of fungal contamination, so that appropriate action can be taken.

Traditionally, quantification of fungi in water sources has been based on culturing the colonies on general nutritional media, or culturing specific fungi on more selective media, and presenting the fungal load as CFU (colony forming units) in a certain volume of water. Although it is agreed that not all fungal species in water will grow on a specific medium under particular growth conditions, culturing and enumeration of as many members of the fungal community in the water source as possible can give a good estimation of the level of fungal contamination. Although efforts are being made today to develop rapid molecular procedures for the detection of fungal contamination (e.g., PCR-based protocols, ergosterol-content determination, the use of gene probes, protein probes, and mass spectrometry), culturing fungal colonies on agar media is still the method of choice to enumerate and isolate fungi of interest. This chapter focuses on examinations of mycological water quality using the CFU approach, and on the recent developments in media used for this purpose.

Fungal Culturing Approach

Historically, the examination of fungi in water bodies has consisted of adopting analytical methods and media used to examine fungal loads in foods. Hence, the "pour plate" and "spread plate" techniques, used with solid products, were recommended for water examinations as well. Later, the "membrane filtration" (MF) technique was also recommended for examinations of fungi in water.

Techniques

Spread Plate Technique

This is a well-established technique for mycological examinations of food and environmental samples. This and the pour plate technique, both established and used for different types of food and soil samples, were also adopted for water examination. In this method, a water sample is spread by glass or plastic spreader on the surface of agar medium in a Petri dish. With a suitable medium and incubation conditions, this technique enables the growth of a range of fungal species, as most fungi are aerobic. The advantage of this method is that the colonies can be differentiated by appearance, and the cultured fungi of interest can be further isolated and identified separately. The disadvantage of this method is sample loss on the spreader, and the low species diversity growing on the agar surface in comparison to the pour plate technique.

Pour Plate Technique

In this technique, the water sample is placed in an empty Petri dish, to which molten medium is then added. After mixing the two, the medium is allowed to solidify, and the culture is incubated for fungal growth. This method is well suited to the enumeration of fungi since there is no loss of sample as in the spread plate technique. In addition, this technique allows the germination and growth of colonies inside the agar medium, where each colony is surrounded by a homogeneous microclimate in terms of nutrient, oxygen, and moisture levels. This allows for higher fungal diversity, as fungi can grow on the surface or at different depths in the medium. The disadvantage is that it becomes difficult to isolate the colony of interest if that colony is growing inside the medium.

Both the spread plate technique and the pour plate technique are simpler and less costly than the MF technique, but require higher quantities of medium, and more space for plates incubation.

MF Technique

Today, this is the most commonly used technique in routine microbiological quality examinations of low-turbidity water (drinking, bottled, swimming pool, and bathing water). The MF technique was initially developed as an alternative to other methods for bacteriological analyses of water samples in the late 1950s [28, 29]. In 1975, it was adopted as a standard procedure for bacterial water examination in the eleventh edition of the SMEWW [30]. It was only in 1976 that Qureshi and Dutka [31] examined the MF technique for the recovery of fungi from water, comparing different brands of MFs for this purpose. In 1978, it was used for the enumeration of *Candida* in natural water [32]. The procedure is based on filtering a volume of water through a MF, and the fungal units (conidia, chlamydospores, and other hyphal units) are trapped on the membrane surface. The filter is then transferred onto nutritional agar medium and the fungal colonies are left to develop on the filter surface. In their comprehensive work on the optimization of methodologies for fungal recovery from water, Kinsey et al. [2] showed that this technique recovers a lower number of different taxa than the direct spread plate technique. In a work conducted in our lab (S. Masaphy, unpublished data), the recovery of fungi (*Penicillium* sp., *Aspergillus* sp., *Candida albicans,* and

Saccaromyces sp.) artificially inoculated into bottled water, tap water, and swimming pool water was similar using MF and spread plate techniques in terms of fungal units. The MF technique has several advantages, especially in examining water with low microbiological loads. Moreover, this technique enables examining a large number of samples in a short period of time.

For all three of the aforementioned methods, successful fungal recovery is related mainly to selecting the right medium.

Media

Early media used for the detection of fungi in water sources were similar to those used for the detection of fungi in other products, since many of the fungi in water actually reach the water body from the surrounding environment. Today, more specific media for water examination are being developed, taking into consideration the low-nutritional conditions of the water matrix and the use of the MF technique. In general, media supporting a broad range of fungal taxa while restricting linear expansion of the fungal colonies and simultaneously inhibiting bacterial growth are preferred for the detection and enumeration of fungi in environmental samples. Different media are used, according to the type of water source and the aim of the fungal detection.

Nutritional characteristics are the most influential factor in the suitability of a particular medium for the recovery of water fungi. To detect a wide range of fungal propagules present in the water source, a nutrition-rich medium, with the addition of an antibacterial agent, is used, such as Sabourand dextrose agar (SDA) [33]. However, low-nutrition media have also been recommended for fungal recovery from water [2]. Comparing poor and rich media, half-strength corn meal agar (CMA/2) was recommended by Kinsey et al. [2] for routine fungal examination since it provided good results, with recovery of higher fungal diversity, and is inexpensive. The rich SDA medium supported higher fungal counts, but mostly from common hypomycete

(Fungi Imperfecti) genera such as *Penicillium* and *Aspergillus,* whereas CMA/2 supported other genera [2].

Medium pH is also important. Generally, fungi tend to grow in more acidic media than bacteria; hence, many of the mycological media are adjusted to be more acidic, thereby supporting fungal growth while inhibiting bacterial growth. This is even more relevant when fungi are being detected in food products that may themselves be acidic. Thus, due to adoption of media from the food discipline, some officially recommended media are already acidic, such as modified aureomycin-rose bengal-glucose-peptone agar (MARGPA), which has a pH of 5.4 [34]. Other recommended media, such as SDA, corn meal medium, and dichloran-18% glycerol (DG18), have pH values between 5.6 and 6 [35]. In 1962, Mossel et al. [36] showed better recovery of molds and yeasts from foods using media with more neutral pH containing an antibacterial agent than with media based on acid pH alone. In a recent work in our lab [37], we also showed that for recovery of fungi from a range of fungus-inoculated water sources, rose bengal-chloramphenicol (RBC) medium with pH 7.2 was superior to MARGPA.

Another important consideration in fungal recovery is their rapid growth. Unlike most bacterial colonies, filamentous fungi tend to form expansive colonies, which may cover small colonies of slower germinating or growing fungi and yeast. Therefore, chemicals that inhibit hyphal growth need to be added to limit overgrowth of the fast-growing colonies. Dichloran and rose bengal were shown to perform this function and were incorporated into a nutrition-rich medium to restrict the linear expansion of hyphal growth [38–41]. In 1973, Jarvis [42] developed and used rose bengal-chlortetracycline medium, and in 1979, King et al. [43] showed that introducing dichloran and rose bengal together (with reduced rose bengal concentration) allowed greater recovery of molds. More recently, DG18 has been recommended in water examinations and widely used by Hageskal et al. [9, 10]. This medium was developed for xerophilic fungi from foods [44], and was recommended by Samson et al. [35] as a general medium for the isolation and enumeration of fungi in food with water activity (a_w) >90. Askun et al. [45] compared DG18 and RBC medium for fungal examination in raisins and obtained higher fungal species diversity with RBC, although both media gave similar results for total fungal counts. We compared RBC medium with MARGPA for the recovery of fungi from different water sources and found RBC to be superior.

As to the antibacterial agent, Korburger and Rodgers [46] showed the positive effect of adding antibiotic to the medium on enumeration of yeasts and molds, and today, as mentioned above, media for detection of fungi in water samples include a wide spectrum of different antibiotics, such as chlortetracycline (auromycin) and streptomycin [34, 47], in addition to the hyphal-restricting agent. We found that the antibiotic chloramphenicol is simplest to use as it is autoclavable.

Assessing and Counting

The observed fungal colonies are counted and referred to as CFU. There are two important issues to consider. First, filamentous fungi tend to spread over the medium, overlapping other slow-growing fungi. As mentioned, to overcome this problem, rose bengal or dichloran are incorporated into the medium. However, the concentration of the added compounds is important, as it can limit fungal growth too severely. The second issue involves observation of the fungal colonies on the medium surface. Some of the fungi are colored due to colored spores, whereas others appear pale and are difficult to observe. Upon using rose bengal to limit the overgrowth of fungi, we found that it also strongly improves the colony count: the filamentous fungi and yeast colonies tend to absorb the rose bengal, giving them a sharper color and reducing the need for optical magnification [37]. This is especially true when MF technique is used, as it is difficult to observe light-colored colonies on the white filter.

Procedures

Media

Some of the more common media used for water fungal detection and enumeration are presented here.

1. Rose bengal-chloramphenicol (RBC) agar (commercially available). Add 5 g peptone, 10 g glucose, 1 g K_2HPO_4, 0.5 g $MgSO_4\cdot7H_2O$, 0.05 g rose bengal, and 15.5 g agar to 1 L distilled water. Heat to dissolve with stirring, and autoclave. Adjust final pH to 7.2 ± 0.2.

2. Dichloran-RBC (DRBC) agar. Add 5 g peptone, 10 g glucose, 1 g K_2HPO_4, 0.5 g $MgSO_4\cdot7H_2O$, 0.002 g dichloran, 0.025 g rose bengal, 0.1 g chloramphenicol, and 15 g agar to 1 L distilled water. Adjust pH to 5.4–5.8.

3. Aureomycin-rose bengal-glucose-peptone agar (ARGPA). Add 5 g peptone, 10 g glucose, 1 g KH_2PO_4, 0.5 g $MgSO_4\cdot7H_2O$, 0.035 g rose bengal, and 20 g agar to 800 mL distilled water. Prepare separately: dissolve 70 mg chlorotetracycline (aureomycin hydrochloride) in 200 mL distilled water, filter-sterilize. Add to the cooled (42–45 °C) melted agar medium. Adjust final pH to 5.4.

4. Modified ARGPA (MARGPA). Add 5 g peptone, 10 g glucose, 1 g KH_2PO_4, 0.5 g $MgSO_4\cdot7H_2O$, 0.035 g rose bengal, and 20 g agar to 800 mL distilled water. Prepare separately: dissolve 200 mg chlorotetracycline in 200 mL distilled water and filter-sterilize. Add to cooled (42–45 °C) melted agar medium. Adjust final pH to 5.4.

5. Dichloran-18% glycerol agar (DG18). Add 5 g peptone, 10 g glucose, 1 g K_2HPO_4, 0.5 g $MgSO_4\cdot7H_2O$, 1 mL dichloran (0.2% in ethanol), 220 g glycerol, 0.1 g chloramphenicol, and 15 g agar to 1 L distilled water. Adjust to pH 5.4–5.8.

6. Neopeptone-glucose-rose bengal-auromycin. Add 5 g neopeptone, 10 g glucose, 3.5 mL rose bengal solution (1 g/100 mL), and 20 g agar to 1 L distilled water. Separately filter-sterilize chlorotetracycline or tetracycline

(1 g/150 mL water). Add 5 mL of this solution to 1 L agar solution immediately before use. Adjust to pH 6.5.

7. Czapek Dox (CZ) agar. This medium is recommended for *Aspergillus, Penicillium,* and similar fungi, among others, but not for total fungal recovery. Add 30 g saccharose, 3 g NaNO$_3$, 1 g K$_2$HPO$_4$, 0.5 g MgSO$_4$·7H$_2$O, 0.5 g KCl, 0.01 g FeSO$_4$, 0.5 g KCl, and 15 g agar to 1 L distilled water. Adjust to pH 7.3.

8. SDA. Add 10 g mycological peptone, 40 g glucose, and 15 g agar to 1 L distilled water. Adjust to pH 5.4–5.8.

If dehydrated commercial medium is being used, preparation should be as per the manufacturer's recommendations. Otherwise, weigh each of the medium components into 1 L or 800 mL sterile distilled water, heat while stirring on a hotplate to near boiling until the agar is dissolved and the medium is homogeneous. Autoclave (121 °C for 15 min). Cool medium to 45 °C, and then pour into the plates. If autoclave-sensitive antibiotic is to be used, add 200 mL of the filter-sterilized antibiotic at relevant concentration to 800 mL cooled medium, mix by stirring, and then pour into the plates. Keep at 45 °C, and adjust pH as required with HCl or NaOH.

5. Screw-cap bottles for dilutions, 100–500 mL volume
6. Borosilicate glass flasks, 250–1,000 mL volume
7. Sterile pipettes, glass or plastic, of appropriate volumes
8. Graduated cylinder, 100–1,000 mL
9. Sterile L-shaped glass rod or plastic disposable spreader rod
10. Petri dishes (either 50 or 90 mm), sterile, plastic
11. Forceps, with smooth tips, to handle filters without damaging them
12. Membrane filtration unit: filter funnel manifolds and filter manifolds (47 mm)
13. Membrane filter: 0.45-μm pore size white hydrophobic mixed cellulose acetate membrane filter, grid marked, 47 mm, preferably presterilized
14. Water bath maintained at 50 °C for tempering agar medium
15. Incubator maintained at 15, 20, or 25 °C, 90% relative humidity
16. Vortex
17. Heating stirrer
18. Colony counter with magnifying glass

Materials

1. Sterile distilled water
2. Ethanol or methanol in wide-mouth container for flame-sterilization of the forceps
3. 0.1 M NaOH and 0.1 M HCl
4. Dilution buffer composed of:
   (a) Potassium dihydrogen phosphate (KH$_2$PO$_4$) solution: Weigh 17 g KH$_2$PO$_4$ into 250 mL sterile water. Mix to complete dissolution. Adjust to pH 7.2 with 1 N NaOH, bring to 500 mL with sterile water.
   (b) MgCl$_2$ solution: Weigh 40.55 g of MgCl$_2$·6H$_2$O into 500 mL sterile water. Mix to complete dissolution.
   (c) Working dilution buffer: Transfer 1.25 mL from solution (a) and 5 mL from solution (b) into 1 L sterile distilled water. Sterilize solution before use.

Techniques

Fungi tend to spread unevenly in water bodies. It is therefore important to mix the water in the sample bottle vigorously prior to examination. When using a low volume of sample, it is recommended that several repeats (optimally five) be examined [34].

Spread Plate Technique

Streak 0.1–0.05 mL water sample onto the center of pre-solidified agar medium (10–20 mL medium) in a 90-mm Petri dish using a sterile pipette. Spread the sample with a spreading rod (available as disposable plastic rods, but glass rods, which can be ethanol-sterilized, are preferred). Streak back and forth across the plate, working up and down several times to distribute the fungal units as evenly as possible. Use a single rod per sample. Cover the plate and wait several minutes before inverting it and incubating.

Pour Plate Technique

Water sample (usually 1 mL, but as little as 0.1 mL can be used) is added to an empty 90-mm Petri dish. Then, 10–15 mL of molten propionate medium is poured at 45 °C. The medium should not be poured directly on the water sample. Mix water sample and medium and let the mixture solidify. In some cases, antibiotics may be added to the Petri dish as well before pouring the molten medium. Incubate plates noninverted.

MF Technique

Shake the sample bottle to distribute the fungal units uniformly. Filter the sample (10–1,000 mL, ideally 100 mL) through the MF. Rinse the sides of the funnel with 20–30 mL sterile dilution water. Turn off the filtration system vacuum and aseptically remove the MF from the filter base using sterile forceps. Overlay the MF on the agar medium surface in a 50-mm Petri dish. Close the dish and incubate, either inverted or noninverted. If the fungal counts exceed 80CFU/filtered volume, lower volumes of water should be used, or the water source can be diluted 1:10 with dilution water and filtered.

Incubation Conditions

Petri dishes are either incubated inverted or non-inverted. Incubating the plate in the inverted position prevents dripping of the condensed water onto the agar medium surface. Incubation is performed in the dark to avoid overproduction of conidia that might spread and recontaminate the agar medium surface. In some cases, however, incubation in the light, but not direct sunlight, is preferred to increase (colored) conidiation. The incubation temperature is usually relatively low (15–20 °C) to avoid overgrowth by fast-growing fungi, allowing the slower fungi to germinate and grow as well. In this case, a longer incubation may be needed to recover higher counts. It is recommended that the plate be observed as soon as possible. However, for the purpose of standardization,

it is recommended that the results be read after defined incubation periods. The SMEWW [34] recommends incubation of spread plates at 15 °C for 7 days, or 20 °C for 5–7 days. Slow-growing fungi may not produce noticeable colonies until 6 or 7 days of incubation. For the pour plate method, it is recommended that plates be incubated at 20–24 °C for 3–7 days. For the MF technique, the recommendation is 15 °C for 5 days or 20 °C for 3–7 days. In all cases, plates should be incubated in a humid (90–95% RH) atmosphere.

Counting

The count can provide an estimate of cultivable fungi extracted from the water sample. All filamentous fungi and yeast colonies may be counted together or separately. The number of fungi in the water sample is calculated as CFU/mL of water. The calculation should take into account the dilution factor. If counting cannot be performed immediately, the culture plate can be kept at 4 °C for 24 h. It is advisable to count colonies in plates that have the optimal number of 20–150 colonies per 90-mm plate for the spread plate technique, and up to 300 colonies for the pour plate technique. For the MF technique, it is suggested that a magnifying binocular microscope be used to count all of the colonies, which may be hard to see on the white filter background. Ideal plates for counting should have 20–80 colonies per filter.

References

1. Paterson RRM, Kelley J, Kinsey G (1997) Secondary metabolites and toxins from fungi in water. In: Morris R, Gammie A (eds) Proceedings of the 2nd UK Symposium on Health-Related Water Microbiology. University of Warwick, Coventry, UK, pp 78–94
2. Kinsey GC, Paterson RR, Kelley J (1999) Methods for the determination of filamentous fungi in treated and untreated waters. J Appl Microbiol (Symp Suppl) 85:214S–224S
3. Kelley J, Kinsey G, Paterson R, Brayford D (2003) Identification and control of fungi in distribution systems. AWWA Research Foundation and American Water Works Association, Denver, CO

4. Kikuchi T, Kadota S, Suehara H, Nishi A, Tsubaki K (1981) Odorous metabolites of a fungus, *Chaetomium globosum* KINZE ex FR. Identification of geosmin, a musty-smelling compound. Chem Pharm Bull 29:1782–1784

5. Doggett MS (2000) Characterization of fungal biofilms within a municipal water distribution system. Appl Environ Microbiol 66:1249–1251

6. Anaissie EJ, Penzak SR, Dignani MC (2002) The hospital water supply as a source of nosocomial infections. A plea for action. Arch Intern Med 162:1483–1492

7. Niemi RM, Knuth S, Lundström K (1982) Actinomycetes and fungi in surface waters and in potable water. Appl Environ Microbiol 43:378–388

8. Gonçalves AB, Paterson RRM, Lima N (2006) Survey and significance of filamentous fungi from tap water. Int J Hyg Environ Health 09:257–264

9. Hageskal G, Knutsen AK, Gaustad P, de Hoog GS, Skaar I (2006) Diversity and significance of mold species in Norwegian drinking water. Appl Environ Microbiol 72:7586–7593

10. Hageskal G, Gaustad P, Heier BT, Skaar I (2007) Occurrence of moulds in drinking water. J Appl Microbiol 102:774–780

11. Russell R, Paterson M, Hageskal G, Skaar I, Lima N (2009) Occurrence, problems, analysis and removal of filamentous fungi in drinking water. In: De Costa P, Bezerra P (eds) Fungicides: chemistry, environmental impact and health effects. Nova Science Publishers, New York

12. Paterson RRM, Lima N (2005) Fungal contamination of drinking water. In: Lehr J, Keeley J, Lehr J, Kingery TB III (eds) Water encyclopedia. Wiley, New York, pp 1–7

13. Kelley J, Paterson R, Kinsey G, Pitchers R, Rossmoore H (1997) Identification, significance and control of fungi in water distribution systems. In: Water Technology Conference Proceedings, US Public American Water Works Association; 1997 Nov 9–12. Denver, CO

14. Cabral D, Fernández Pinto VE (2002) Fungal spoilage of bottled mineral water. Int J Food Microbiol 72:73–76

15. Yamaguchi MU, Pontello Rampazzo RC, Yamada-Ogatta SF, Nakamura CV, Ueda-Nakamura T, Dias Filho BP (2007) Yeasts and filamentous fungi in bottled mineral water and tap water from municipal supplies. Brazil Arch Biol Technol 50:1–9

16. Cheung WHS, Chang KCK, Hung RPS (1991) Variations in microbial indicator densities in beach waters and health-related assessment of bathing water quality. Epidemiol Infect 106:329–344

17. Brandi G, Sisti M, Paparini A, Gianfranceschi G, Schiavano GF, De Santi M et al (2007) Swimming pools and fungi: an environmental epidemiology survey in Italian indoor swimming facilities. Int J Environ Health Res 17:197–206

18. Buot G, Toutous-Trellu L, Hennequin C (2010) Swimming pool deck as environmental reservoir of Fusarium. Med Mycol 48:780–784

19. Göttlich E, van der Lubbe W, Lange B, Fiedler S, Melchert I, Reifenrath M et al (2002) Fungal flora in groundwater-derived public drinking water. Int J Hyg Environ Health 205:269–279

20. Anaissie EJ, Costa SF (2001) Nosocomial aspergillosis is waterborne. Clin Infect Dis 33:1546–1548

21. Anaissie EJ, Kuchar RT, Rex JH, Francesconi A, Kasai M, Muller FM et al (2001) Fusariosis associated with pathogenic *Fusarium* species colonization of a hospital water system: a new paradigm for the epidemiology of opportunistic mold infections. Clin Infect Dis 33:1871–1878

22. Metzger WJ, Patterson R, Fink J, Semerdjian R, Roberts M (1976) Sauna-takers' disease. Hypersensitivity due to contaminated water in a home sauna. J Am Med Assoc 236:2209–2211

23. Muittari A, Kuusisto P, Virtanen P, Sovijärvi A, Grönroos P, Harmoinen A et al (1980) An epidemic of extrinsic allergic alveolitis caused by tap water. Clin Allergy 10:77–90

24. Papadopoulou C, Economou V, Sakkas H, Gousia P, Giannakopoulos X, Dontorou C et al (2008) Microbiological quality of indoor and outdoor swimming pools in Greece: investigation of the antibiotic resistance of the bacterial isolates. Int J Hyg Environ Health 211:385–397

25. American Public Health Association (APHA), American Water Works Association (AWWA), Water Environment Federation (WEF) (1975) Standard methods for the examination of water and wastewater, 14th edn. APHA Inc., Washington, DC

26. Hageskal G, Lima N, Skaar I (2009) The study of fungi in drinking water. Mycol Res 113:165–172

27. Bobichon H, Dufour-Morfaux F, Pitort V (2009) In vitro susceptibility of public indoor swimming pool fungi to three disinfectants. Mycoses 36:305–311

28. Ehrlich R (1955) Technique for microscopic count of microorganisms directly on membrane filters. J Bacteriol 70:265–268

29. Jannasch HW (1958) Studies on planktonic bacteria by means of a direct membrane filter method. J Gen Microbiol 18:609–620

30. American Public Health Association (APHA), American Water Works Association (AWWA), Water Environment Federation (WEF) (1960) Standard methods for the examination of water and wastewater, 11th edn. APHA Inc., Washington, DC

31. Qureshi AA, Dutka BJ (1976) Comparison of various brands of membrane filters for their ability to recover fungi from water. Appl Environ Microbiol 32:445–447

32. Buck JD, Bubucis PM (1978) Membrane filter procedure for enumeration of *Candida albicans* in natural waters. Appl Environ Microbiol 35:237–242

33. Arvanitidou M, Kanellou K, Constantinides TC, Katsouyannopoulos V (1999) The occurrence of fungi in hospital and community potable waters. Lett Appl Microbiol 29:81–84

34. American Public Health Association (APHA), American Water Works Association (AWWA), Water

Environment Federation (WEF) (2005) Standard methods for the examination of water and wastewater, 21st edn. APHA Inc., Washington, DC, pp 153–161. Chapter 9, Detection of fungi (9610)

35. Samson RA, Hoekstra ES, Frisvad JC (2004) Introduction to food- and airborne fungi, 7th edn. Centraalbureau voor Schimmelcultures, Utrecht

36. Mossel DAA, Visser M, Mengerink WJH (1962) A comparison of media for the enumeration of moulds and yeasts in foods and beverages. Lab Pract 11:109–112

37. Albaum S, Masaphy S (2009) Comparison of rose Bengal-chloramphenicol and modified aureomycin-rose Bengal-glucose-peptone agar as media for the enumeration of molds and yeasts in water by membrane filtration techniques. J Microbiol Methods 76:310–312

38. Kramer CL, Pady SM (1961) Inhibition of growth of fungi on rose Bengal media by light. Trans Kansas Acad Sci 64(2):110–116

39. Ottow JCG (1972) Rose bengal as a selective aid in the isolation of fungi and actinomycetes from natural sources. Mycologia 64:304–315

40. Henson OE (1981) Dichloran as an inhibitor of mold spreading in fungal plating media: effects on colony diameter and enumeration. Appl Environ Microbiol 42:656–660

41. Bragulat MR, Abarca ML, Castella G, Cabanes FJ (1995) Dyes as fungal inhibitors effect on colony enumeration. J Appl Bacteriol 79:578–582

42. Jarvis B (1973) Comparison of an improved rose Bengal-chlortetracycline agar with other media for the selective isolation and enumeration of moulds and yeasts in food. J Appl Bacteriol 36:723–727

43. King D, Hocking AD, Pitt JI (1979) Dichloran-rose Bengal medium for enumeration and isolation of molds from foods. Appl Environ Microbiol 37:959–964

44. Hocking AD, Pitt JI (1980) Dichloran-glycerol medium for enumeration of xerophilic fungi from low-moisture foods. Appl Environ Microbiol 39:488–492

45. Askun T, Eltem R, Taskin E (2007) Comparison of rose-Bengal chloramphenicol agar and dichloran glycerol agar (DG18) for enumeration and isolation of moulds from raisins. J Appl Biol Sci 1:71–74

46. Korburger JA, Rodgers MF (1978) Single or multiple antibiotic-amended media to enumerate yeasts and moulds. J Food Protect 41:367–369

47. Douglas KA, Hocking AD, Pitt JI (1979) Dichloran-rose Bengal medium for enumeration and isolation of molds from foods. Appl Environ Microbiol 37:959–964

Sabouraud Agar for Fungal Growth

15

Janelle M. Hare

Abstract

This article describes the history, theory, and use of Sabouraud agar for isolating and growing fungi. This includes an explanation of the role of each ingredient, the instructions for making this medium, variations upon the basic recipe, and the various recipes that are commercially available.

Keywords

Sabouraud agar • Fungi • Growth • Dermatophytes • Sabouraud agar (Modified) • Sabouraud agar Emmons • Glucose

Introduction

Sabouraud agar is one of the oldest and most commonly used media for isolating and growing fungi. It is effective in selectively isolating fungi from environmental samples such as air and soil, maintaining pure fungal cultures, and growing fungi to distinguish and identify different species, especially dermatophytes, by color and appearance. This article describes the theory behind the use of Sabouraud agar, the role of the ingredients in the medium, the preparation and use of the medium, observations about the variation in names and ingredients that can prove a source of confusion, as well as visual results in the growth.

History

Sabouraud agar medium was developed by the French dermatologist Raymond J. A. Sabouraud (pronounced sah-bū-rō′) in the late 1800s to support the growth of fungi, particularly dermatophytes [1, 2]. Sabouraud's medical investigations focused on bacteria and fungi that cause skin lesions, and he developed many agars and techniques to culture pathogens such as dermatophytes and *Malassezia*. The long incubation period (multiple weeks) of dermatophytes and the need to avoid bacterial contamination while culturing them was the driving force behind the development of this medium. Sabouraud also sought to provide a medium that would yield reliable results for fungal identification *across* laboratories. He recommended that all mycologists detail their exact media formulations and sources of ingredients as well as the temperatures and times of specimen incubation, in order to standardize observations

J.M. Hare (✉)
Department of Biology & Chemistry,
Morehead State University, 150 University Blvd,
Morehead, KY 40351, USA
e-mail: jm.hare@moreheadstate.edu

and reduce media-derived sources of differences in appearance [3].

Ironically, given Sabouraud's original desire to standardize the construction of fungal media, there are currently many sources of confusion and variation in both the names and ingredients associated with Sabouraud agar, also called Sabouraud's agar (abbreviated either SDA or SAB). Because of the old-fashioned use of the term "dextrose" to refer to D-glucose, the medium has been referred to as Sabouraud dextrose agar as well as Sabouraud glucose agar, the term most appropriate and consistent with standard chemical nomenclature [3]. Finally, a more recent modification of Sabouraud agar by Emmons is called either Sabouraud agar (Modified), or Sabouraud agar, Emmons [4]. Many of the historical details behind these names and ingredient variations are described in Odds' excellent review article [3].

Theory

Sabouraud agar is a selective medium that is formulated to allow growth of fungi and inhibit the growth of bacteria. The available means of inhibiting bacterial growth in Sabouraud's pre-antibiotic era was an acidic medium (pH 5.6). Currently, the addition of antibiotics or antimicrobials to the acidic medium is used to inhibit bacterial growth (and sometimes saprophytic fungi, depending on the particular antimicrobial used).

Sabouraud agar medium is complex and undefined, but contains few ingredients. Peptones, as soluble protein digests, are sources of nitrogenous growth factors that can vary significantly according to the particular protein source. The most variation is present in the source and method of these protein digests. Both Difco and BBL brand Sabouraud agars use pancreatic digests of casein as their peptone source, but they and other vendors also use a combination of pancreatic digest of casein and peptic digest of animal tissues. Sabouraud's original formulation contained a peptone termed "Granulée de Chassaing," which is no longer available. Mold morphology can vary slightly based on the peptones used, but

pigmentation and sporulation can be consistent if one uses a consistent method of medium prepared with the ingredients from the same source each time. Researchers should also explicitly describe the commercial or laboratory-prepared ingredients used in their medium.

Sabouraud originally used the sugar maltose as an energy source, and although this medium is still commercially available, glucose (formerly referred to as dextrose) is currently used most frequently. Glucose is present at the high level of 4% in Sabouraud's formulation to assist in vigorous fermentation and acid production by any bacteria present, inhibiting later bacterial growth [5].

In 1977, Emmons formulated an alternative version of Sabouraud's agar, which contains half the amount of glucose (2%) and a neutral pH of 6.8–7.0. The neutral pH seems to enhance the growth of some pathogenic fungi, such as dermatophytes. Agar concentrations ranging from 1.5% to 2.0% are found in commercial preparations of Sabouraud agar in both the original formula and Emmons modification, and serve to solidify the medium in tube and plate medium.

Materials

Sabouraud agar can be either made from individual ingredients (Tables 15.1 and 15.2), purchased either as dehydrated powder that must be dissolved in water, autoclaved, and dispensed, or as prepared medium that can be purchased in tube, plate, or broth form from a variety of commercial sources such as Becton Dickinson/Difco, Remel, or BBL. Various antimicrobials can be added to either the original recipe (Sabouraud agar), or Sabouraud agar, Modified/Emmons (see Table 15.2).

1. Deionized, distilled water.
2. Autoclave.
3. Graduated cylinder, 1,000 mL.
4. Erlenmeyer flask (2 L if making 1 L of medium).
5. Analytical balance (if using antimicrobial agents).
6. Balance for weighing media ingredients.
7. Stir bar.
8. Stirring hotplate.

Table 15.1 Ingredients for Sabouraud agar and Sabouraud agar (Emmons)

Ingredient	Sabouraud agar (g/L)	Sabouraud agar (Emmons) (g/L)
Pancreatic digest of casein	10	10
Glucose	40	20
Agar	15–20	15–20

Table 15.2 Antimicrobial and other additives to Sabouraud agar

Ingredient	Amount (per liter)	Notes
Chloramphenicol[a]	50 mg	Dissolve in 10 mL 95% ethanol
Cycloheximide[a]	0.5 g	Dissolve in 2 mL acetone
Gentamicin	50 mg	Dissolve in 5 mL water
Lecithin	0.7 g	Add directly with other powdered medium ingredients before autoclaving
Tween 80	5 g	
Olive oil		Spread 0.1 mL sterile olive oil on surface of plate

[a]Add these to molten, autoclaved media once it has been tempered in water bath to 45–50 °C

9. Slant tube rack for holding media tubes after autoclaving to solidify with a slanted surface.
10. Pancreatic digest of casein.
11. Glucose.
12. Chloramphenicol.
13. Gentamicin sulfate.
14. Cycloheximide.
15. Tween 80 (polysorbate 80).
16. Lecithin.
17. Olive oil, sterilized by autoclaving.
18. Sterile glass test tubes with caps.
19. Sterile Petri dishes, 100 mm diameter.

Method of Sabouraud Agar Preparation

1. Combine all ingredients, except any antimicrobials to be used, in ~900 mL of deionized water in a graduated cylinder while stirring with a magnetic stir bar.
2. Adjust to pH 5.6 with hydrochloric acid and adjust final volume to 1 L.
3. Transfer contents to a 2-L flask and boil on a heating/stirring plate while stirring, for 1 min.
4. Cover opening of flask loosely with aluminum foil and autoclave 15 min at 121 °C under pressure of 15 lb/in².
5. Cool to ~45–50 °C (roughly until one can support the flask underneath with an ungloved hand). If the antimicrobials chloramphenicol or cycloheximide are to be added, aseptically add them at this point and swirl medium gently (see Variations on Standard Sabouraud Agar).
6. Pour into Petri dishes or tubes and leave at room temperature overnight to solidify and dry. When pouring plates, fill each Petri dish with at least 25 mL of medium to allow for medium dehydration during the longer incubation period required for fungi. If preparing tubes, slant the rack of covered tubes immediately after pouring in a slant tube rack, either at a 5° or 20° slant.
7. Store all media at 4 °C, regardless of whether they contain antimicrobials (see Table 15.1).

Method of Sabouraud Agar, Emmons Modification Preparation

1. Combine all ingredients, except any antimicrobials to be used, in ~900 mL of deionized water in a graduated cylinder while stirring with a magnetic stir bar.
2. Adjust to pH 6.8–7.0 with hydrochloric acid and adjust final volume to 1 L.
3. Transfer contents to a 2-L flask and boil on a heating/stirring plate while stirring, for 1 min.
4. Cover opening of flask loosely with aluminum foil and autoclave 15 min at 121 °C under pressure of 15 lb/in².

5. Cool to ~45–50 °C (roughly until one can support the flask underneath with an ungloved hand). If the antimicrobials chloramphenicol or cycloheximide are to be added, aseptically add them at this point and swirl medium gently (see Variations on Standard Sabouraud Agar).
6. Pour into Petri dishes or tubes and leave at room temperature overnight to solidify and dry. Fill each Petri dish with at least 25 mL of medium to allow for medium dehydration during the longer incubation period required for fungi. If preparing tubes, slant the rack of covered tubes immediately after pouring in a slant tube rack, either at a 5° or 20° slant.
7. Store all media at 4 °C, regardless of whether they contain antimicrobials.

Variations on Standard Sabouraud Agar

Either Sabouraud agar or its Emmons version can be made more selective by adding antimicrobials (see Table 15.2). Antimicrobials commonly used are the aminoglycoside gentamicin, which inhibits gram-negative bacteria; chloramphenicol, which inhibits a wide range of gram-positives and gram-negatives; and/or cycloheximide, which inhibits primarily saprophytic fungi, but not dermatophytes or yeasts [6, 7]. Chloramphenicol and gentamicin are used at 50 mg/L (50 mg of chloramphenicol dissolved in 10 mL of 95% ethanol before adding to molten medium) and cycloheximide at 0.5 g/L (0.5 g dissolved in 2 mL of acetone before adding to molten medium) [8]. Chloramphenicol and cycloheximide should only be added after media has been autoclaved and then cooled to ~45–50 °C (see step 5 in Method of Sabouraud Agar Preparation). Gentamicin may be added to the medium ingredients before autoclaving.

Lecithin and Tween 80 are added to Sabouraud agar (see Table 15.2) that is used in monitoring environmental surfaces that may have been treated with antiseptics and quaternary ammonium compounds, as these additives neutralize the cleaning compounds [9]. Sterile olive oil can be spread on the surface of Sabouraud agar plates to grow lipophilic *Malassezia* species [10] (see Table 15.2).

Methods of Inoculation and Incubation

Sabouraud agar plates can be inoculated by streaking for isolation, as with standard bacteriological media, by exposing the medium to ambient air, or by tamping clinical sample material (hair, skin scrapings, etc.) onto the surface of the agar medium. When growing cultures in tubes, the caps should be screwed on loosely to admit air, as dermatophytes and most molds are obligate aerobes. Isolation of fungi is performed on plates, while slants are primarily used for maintaining pure, or stock, cultures once isolated. If using selective Sabouraud media, a control plate/tube without antimicrobials should also be inoculated for comparison. Typically, molds are incubated at room temperature or slightly warmer (25–30 °C), yeasts are incubated at 28–30 °C or both 30 and 37 °C if suspected to be dimorphic fungi.

Incubation times will vary, from approximately two days for the growth of yeast colonies such as *Malassezia*, to 2–4 weeks for growth of dermatophytes or dimorphic fungi such as *Histoplasma capsulatum*. Indeed, the incubation time required to acquire fungal growth is one diagnostic indicator used to identify or confirm fungal species. Dermatophytes in particular show characteristic incubation times ranging from 5 to 7 days (some *Epidermophyton* or *Microsporum* species) to 3–4 weeks for some *Trichophyton* species [11]. Cultures should be examined twice weekly and be held for 4–6 weeks before being reported as negative if infection by systemic agents such as *Histoplasma*, *Blastomyces*, or *Coccidioides* is suspected.

Results

Depending on the antimicrobials used, different types of microorganisms and groups of fungi may grow on Sabouraud agar (Table 15.3). Typically, saprophytic fungi are inhibited by cycloheximide and/or chloramphenicol, but yeasts and dermatophytes grow well in their presence. Conversely, even Sabouraud agar is unable to support the growth of a few dermatophytes in the absence of additives. For example, some *Trichophyton*

Table 15.3 Expected growth of various microbes on Sabouraud agar containing antimicrobials

Microbe	Growth on SAB+CAM[a]	Growth on SAB+CHX[b]
Candida albicans	Yes	Yes
Cryptococcus neoformans	Yes	No
Aspergillus niger	Yes	No
Trichophyton mentagrophytes	Yes	Yes
Microsporum audouinii	Yes	Yes
Blastomyces dermatitidis	Yes (mold phase at 25 °C) No (yeast phase at 37 °C)	Yes (mold phase at 25 °C) No (yeast phase at 37 °C)
Histoplasma capsulatum	Yes (mold phase at 25 °C) No (yeast phase at 37 °C)	Yes (mold phase at 25 °C) No (yeast phase at 37 °C)
Rhizopus spp.	Yes	No
Sporothrix schenckii	Yes	Yes
Penicillium roqueforti	Yes	No
Escherichia coli	No	No

[a]SAB+CAM=Sabouraud agar plus chloramphenicol
[b]SAB+CHX=Sabouraud agar plus cycloheximide

species require additional growth factors, such as thiamine and inositol (*T. verrucosum*) or nicotinic acid (*T. equinum*), and may not grow well, if at all, on Sabouraud agar [12]. *T. mentagrophytes* and *T. rubrum*, however, grow well on Sabouraud agar. Similarly, the growth of *Malassezia* species is significantly impaired without the addition of olive oil overlaid on the surface of a Sabouraud agar plate [10] (see Table 15.3).

Mold morphology should be observed on both the top (obverse) and bottom (reverse) surfaces, as differences can be seen on each surface.

Variation from lot to lot as well as between commercial vendors of Sabouraud agar can significantly impact the qualitative and quantitative growth of fungi. One study comparing five different commercial preparations of Sabouraud glucose agar observed significant differences in the quantitation of yeasts as well as the color of *Aspergillus* colonies; however, the dermatophytes yielded reliably similar appearances on the five media sources tested [13].

Notes

Fungi often produce spores that are easily dispersed into the laboratory upon opening of plates. Plates should be incubated with the lid on the top (as opposed to the typical practice of inverting microbiological plates for incubation) to avoid spreading spores when the plates are opened. After growth, plates should be wrapped in Parafilm to maintain them securely closed for storage and/or transport. Plate or tube cultures should be opened only within a class II biological safety cabinet to avoid contamination of laboratory spaces with fungal spores, possible infection of individuals by pathogenic fungi, or induction of allergic responses.

Because the growth of large numbers of fungi can pose a potential infection hazard, measures must also be taken to prevent infection of laboratory researchers. Note that some fungi are biosafety level one (BSL-1), whereas most are BSL-2 [14]. The American Society for Microbiology strongly recommends that environmental enrichment experiments should only be performed in BSL-2 laboratories. The following precautions apply to the use of Sabouraud agar:

1. Direct environmental samples (e.g., soil, water) that are known to contain infectious organisms should be handled according to the biosafety level of that infectious agent.
2. Cultures of enriched microorganisms derived from environmental samples should be handled using BSL-2 precautions.
3. Mixed, enriched, or pure cultures of microorganisms from environmental samples with a significant probability of containing infectious agents should be manipulated in a class II biosafety cabinet, if available.

4. Researchers should be aware if they work in regions with endemic fungi capable of causing systemic infections, and should avoid environmental isolations. Some safe fungi for student experimentation and handling include the molds *Penicillium camemberti* and *P. roqueforti* (used in making cheeses), *Rhizopus stolonifer* (used in making tempeh), *Aspergillus* species (except *A. fumigatus* and *A. flavus*), the yeasts *Saccharomyces cerevisiae*, *Rhodotorula rubrum*, and *Neurospora crassa*.

Acknowledgements I thank my colleague Ted Pass for helpful comments on this chapter.

References

1. Sabouraud R (1896) La question des teignes. Ann Dermatol Venereol (series 3) 7:87–135
2. Sabouraud R (1896) Recherche des milieux de culture propres a la différenciation des espèces trichophytiques a grosse spore. In: Les trichophyties humaines. Masson et Cie, Paris, pp 49–55
3. Odds FC (1991) Sabouraud('s) agar. J Med Vet Mycol 29:355–359
4. Emmons CW, Binford CH, Utz JP, Kwon-Chung KJ (1977) Culture media. In: Medical mycology, 3rd edn. Lea & Febiger, Philadelphia, p 535
5. Jarrett L, Sonnenwirth AC (1980) Gradwohl's and parasitic infections, 7th edn. American Public Health Association, Washington, DC
6. McDonough ES, Ajello Georg LK, Brinkman S (1960) In vitro effects of antibiotics on yeast phase of Blastomyces dermatitidis and other fungi. J Lab Clin Med 55:116
7. Lorian V (2005) Antibiotics in laboratory medicine. Lippincott, Williams & Wilkins, Baltimore
8. Hungerford LL, Campbell CL, Smith AR (1998) Veterinary mycology laboratory manual. Iowa State University Press, Ames
9. Curry AS, Graf JG, McEwen GN Jr (1993) CTFA microbiology guidelines. The Cosmetic, Toiletry and Fragrance Association, Washington, DC
10. Kwon-Chung KJ, Bennett JE (1992) Infections caused by *Malassezia* species. In: Medical mycology. Lea & Febiger, Philadelphia, pp 70–182
11. Robert R, Pihet M (2008) Conventional methods for the diagnosis of dermatophytosis. Mycopathologia 166:295–306
12. Georg LK, Camp LB (1957) Routine nutritional tests for the identification of dermatophytes. J Bacteriol 74:113–121
13. Brun S, Bouchara JP, Bocquel A, Basile AM, Contet-Audonneau N, Chabasse D (2001) Evaluation of five commercial Sabouraud gentamicin-chloramphenicol agar media. Eur J Clin Microbiol Infect Dis 20:718–723
14. Centers for Disease Control and Prevention. Biosafety in microbiological and biomedical laboratories (BMBL). 5th edn. Section VIII-B: Fungal agents, pp 170–181. Available from: http://www.cdc.gov/biosafety/publications/bmbl5/BMBL.pdf

A Method for the Formation of *Candida* Biofilms in 96 Well Microtiter Plates and Its Application to Antifungal Susceptibility Testing

Christopher G. Pierce, Priya Uppuluri, and Jose L. Lopez-Ribot

Abstract

Fungal infections are an increasing threat to an expanding population of immunocompromised patients. Of these, candidiasis remains the most common, now representing the third to fourth most prevalent infection in US hospitals. *Candida albicans* remains the major causative agent of candidiasis. Most manifestations of candidiasis are associated with biofilm formation on either host tissues or implanted biomaterials (i.e., catheters), which carries important negative consequences, as cells within biofilms show dramatically increased levels of antifungal drug resistance and protection from host defenses. Here we describe a rapid and robust model for the formation of *C. albicans* biofilms *in vitro* using 96 well microtiter plates, which can also be easily adapted for antifungal susceptibility testing. The read-out is colorimetric, based on the reduction of a tetrazolium salt (XTT) by metabolically active cells. This method simplifies biofilm formation, democratizes biofilm research, and provides a framework for the standardization of antifungal susceptibility testing of fungal biofilms.

Keywords

Candida • Candidiasis • Biofilms • Antifungals • Fungi • Susceptibility testing

Introduction

Since microorganisms can be diluted to a single cell and studied in pure culture, most investigations in the field of microbiology have typically involved the use of free-living (planktonic) cells in liquid cultures, leading to the "dogmatic" and almost universal consideration of microorganisms as unicellular life forms. However, during the last few decades there has been an increasing recognition of

C.G. Pierce • P. Uppuluri • J.L. Lopez-Ribot (✉)
Department of Biology, South Texas Center for
Emerging Infectious Diseases, The University of Texas
at San Antonio One UTSA Circle, San Antonio,
TX 78249, USA
e-mail: jose.lopezribot@utsa.edu

the role that microbial biofilms play both in nature and during disease [1, 2]. Biofilms are defined as structured microbial communities that are attached to a surface and encapsulated within a self-produced matrix. Biofilm formation carries important negative clinical consequences, because cells in biofilms, in stark contrast with their free-floating planktonic counterparts, are notoriously resistant to most antibiotics [3].

Candidiasis remains the most common fungal infection in hospitalized patients, now representing the third to fourth most frequent nosocomial infection worldwide [4–8]. *C. albicans* remains the main etiologic agent of candidiasis, although other species are on the rise [9]. The increase in the frequency of candidiasis in the last few decades is associated with the increase in use of a variety of medical implant devices in which *Candida* can form biofilms. These include, among others, different types of catheters (i.e., urinary, intravascular), endotracheal tubes, intracardiac devices, neurosurgical shunts, and prosthetic joints [10, 11]. Other manifestations of candidiasis, such as denture stomatitis and oropharyngeal candidiasis, also have a biofilm etiology [12].

From the clinical point of view, the most salient feature of *Candida* biofilms is their high levels of resistance against conventional antifungal agents, particularly azoles and polyenes [13–15]. However, newer antifungal agents, such as the echinocandins and liposomal formulations of amphotericin B, display increased efficacy against fungal biofilms [16–18]. Also, biofilms provide a safe haven where microorganisms are protected from host immune defenses and from which they can disseminate to colonize and infect distal sites [11]. In other instances, biofilm formation is directly responsible for failure of a contaminated device, very often necessitating removal. It is important to note that the commonly used Clinical and Laboratory Standards Institute broth microdilution techniques for antifungal susceptibility testing are based on the use of planktonic populations and will not enable prediction of the drugs' efficacy against fungal biofilms, which underscores the importance of developing standardized techniques for biofilm formation and to determine

the effectiveness of different antifungal agents and regimens against fungal biofilms.

Traditionally, most models for the formation of fungal biofilms are cumbersome, requiring the use of specialized equipment, expert handling, and long processing times. Moreover, since relatively few equivalent biofilms can be produced at the same time, the majority of these methods do not allow for high throughput screening [19]. Here, we describe a fast and highly reproducible method for the formation of multiple equivalent biofilms on the bottom of wells of microtiter plates, coupled with a colorimetric method that measures the metabolic activity of cells within the biofilm based on the reduction of 2,3-bis(2-methoxy-4-nitro-5-sulfo-phenyl)-2H-tetrazolium-5-carboxanilide (XTT). Although the method was initially developed for *C. albicans* and other *Candida* spp., it can also be easily adapted for other biofilm-forming fungal species of clinical interest, such as *Cryptococcus neoformans* and *Aspergillus fumigatus* [20]. The method can be used to examine multiple parameters and factors influencing biofilm formation, to estimate the biofilm-forming ability of multiple fungal isolates and mutant strains, and to perform antifungal susceptibility testing of fungal biofilms.

Materials

1. Clinical or laboratory strains of *C. albicans* (or other fungal species).
2. Yeast peptone dextrose (YPD) (1% w/v yeast extract, 2% w/v peptone, 2% w/v dextrose, 1.5% agar) or Sabouraud-dextrose agar plates or slants for subculturing *Candida* isolates.
3. YPD liquid medium (1% w/v yeast extract, 2% w/v peptone, 2% w/v dextrose).
4. RPMI-1640 without sodium bicarbonate supplemented with L-glutamine and buffered with 165 mM morpholinepropanesulfonic acid. From now on this medium will be referred to simply as RPMI 1640.
5. Sterile phosphate buffered saline, PBS (10 mM phosphate buffer, 2.7 mM potassium chloride, 137 mM sodium chloride, pH 7.4) (Sigma-Aldrich, St Louis, MO, USA).

6. Haemocytometer (Hausser Scientific, Horsham, PA, USA).

7. Polystyrene, flat-bottomed, 96 well microtiter plates.

8. 2,3-bis(2-methoxy-4-nitro-5-sulfo-phenyl)-2H-tetrazolium-5-carboxanilide (XTT). The XTT saturated solution is prepared at 0.5 g/L in sterile Ringer's lactate or PBS. This solution is light sensitive, so it should be covered with aluminum foil during preparation. The solution is filter-sterilized using a 0.22-μm-pore size filter (the filtration step will leave yellow residues on the filter, but this does not constitute a problem). Aliquot into 10 mL working volumes, and store at −70 °C, wrapped in aluminum foil.

9. Menadione, prepared as a 10-mM stock solution in 100% acetone. Aliquot into smaller volumes (about 50 μL) and store at −70 °C.

10. Vortex mixer.

11. Microtiter plate reader with 490-nm optical filter.

12. Bright field inverted microscope.

13. Multichannel pipette.

14. Antifungal drugs. Most drugs are initially solubilized in DMSO, but some antifungals are also soluble in water. If needed, concentrated stock solutions (i.e., 1 mg/mL) of the antifungals can be aliquoted into smaller volumes and stored at −70 °C until required.

Methods

The methods described below summarize the formation of *C. albicans* biofilms and procedures for antifungal susceptibility testing of cells within biofilms.

Formation of Candida Biofilms on 96 Well Microtiter Plates (Fig. 16.1)

1. *Candida* isolates are typically stored as glycerol stocks. From these stocks (or from a fresh culture if a recent clinical isolate) streak a loopful of cells onto a plate containing YPD agar or Sabouraud-dextrose agar and incubate overnight at 37 °C.

2. Flasks containing YPD liquid medium (typically 20 mL of medium in a 150-mL flask) are inoculated with a loopful of cells from the stock cultures and incubated overnight in an orbital shaker (150–180 rpm) at 30 °C. *C. albicans* should grow in the budding–yeast form under these conditions (check under the microscope).

3. Centrifuge the liquid cultures (approximately 3,000 g for 5 min at 4 °C), remove supernatant, and wash twice in sterile PBS (by resuspending the pellet in approximately 20 mL of ice-cold buffer, vortexing vigorously, followed by centrifugation as described previously).

4. Resuspend the final pellet of cells in approximately 20 mL of RPMI 1640 medium that has been prewarmed to 37 °C.

5. Prepare 1:100 and/or 1:1,000-fold dilutions in the same medium and count using a haemocytometer on a bright field microscope with a 40× objective lens. Calculate the volumes needed to prepare a suspension of cells at a final density of 1.0×10^6 cells/mL in RPMI 1640. The total volume needed will depend on the total number of wells (or plates) that need to be seeded for biofilm formation.

6. Use as many 96 well microtiter plates as needed according to the experimental design. We recommend performing a minimum of 2–4 replicates (entire rows or columns of the microtiter plate[s]) for each condition (i.e., isolate, strain, antifungal concentration) to be tested. Pipette 100 μL of the standardized inoculum into selected wells of the microtiter plate(s). Ideally, leave wells in column 12 on each plate empty, and use these wells as negative background controls during subsequent analyses. If multiple rows in the same plate, or the entire plate, or if multiple plates are to be seeded with the same fungal isolate, the use of a multichannel pipette is strongly recommended for this and successive steps.

7. After initial seeding, cover the microtiter plate with its original lid, seal with parafilm, and place inside an incubator. Incubate statically for 24–48 h at 37 °C.

8. After biofilm formation, use a multichannel pipette to aspirate the medium carefully so as not to touch and disrupt the biofilm. Wash the

Fig. 16.1 Schematic diagram of the protocol for the formation of *C. albicans* biofilms on 96 well microtiter plates

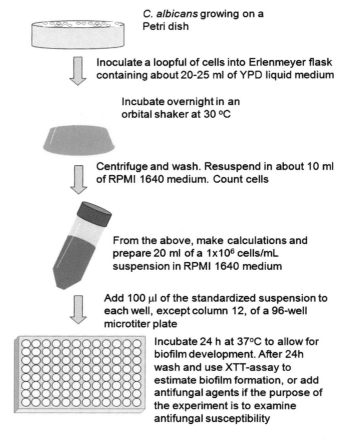

C. albicans growing on a Petri dish

Inoculate a loopful of cells into Erlenmeyer flask containing about 20-25 ml of YPD liquid medium

Incubate overnight in an orbital shaker at 30 °C

Centrifuge and wash. Resuspend in about 10 ml of RPMI 1640 medium. Count cells

From the above, make calculations and prepare 20 ml of a 1×10^6 cells/mL suspension in RPMI 1640 medium

Add 100 µl of the standardized suspension to each well, except column 12, of a 96-well microtiter plate

Incubate 24 h at 37°C to allow for biofilm development. After 24h wash and use XTT-assay to estimate biofilm formation, or add antifungal agents if the purpose of the experiment is to examine antifungal susceptibility

plates (using a multichannel pipette or an automatic plate washer) three times in sterile PBS (200–300 mL per well) to remove nonadherent cells that remain in the wells. After each wash the microtiter plates should be drained in an inverted position by blotting with paper towels to remove any residual PBS. Biofilms are now ready to be processed for XTT (to determine the extent of biofilm formation) or alternatively to be treated with antifungals for susceptibility testing.

At this point after biofilm formation, if the main purpose of a particular set of experiments is to estimate the biofilm-forming ability, the extent of biofilm formation, or to determine biofilm-forming kinetics, the XTT/menadione reagent can be added and the resulting color read using a microtiter plate reader.

Methods for Antifungal Susceptibility Testing Against Candida Biofilms on 96 Well Microtiter Plates

The following steps describe the preparation of antifungal agents for testing and the challenging of pre-formed biofilms with antifungal agents.

1. From stock solutions or powder prepare a final "high" working solution in RPMI 1640 medium of each antifungal to be tested. Typical high concentrations are 1,024 µg/mL for fluconazole, and 16 µg/mL for both amphotericin B and caspofungin. Other concentrations can be used for different agents.

2. Using a multichannel pipette, add 200 µL of the high working concentration of antifungal to the corresponding wells on column 1 of each microtiter plate containing the *C. albicans*

biofilms. Be careful not to touch or otherwise disrupt the biofilms.

3. Add 100 µL of RPMI 1640 to each well in columns 2–11.

4. Remove 100 µL of antifungal agent from the wells of column 1 and add to the adjacent wells in column 2 (already containing 100 µL of medium). Mix the contents well by pipetting up and down and remove the tips.

5. Repeat moving right until the wells of column 10, after which the final 100 µL volume from the wells of column 10 after mixing is discarded. In this way, a series of doubling dilutions of your agent(s) of interest have been created; from most concentrated in wells of column 1 to least concentrated in wells of column 10. Unchallenged biofilms in column 11 will serve as positive controls, and empty wells in column 12 will serve as negative controls.

6. Cover the plates with their lids, seal with parafilm, and incubate for 24–48 h at 37 °C. Other incubation times/conditions may be used depending on the experimental design.

Use of the XTT-reduction Assay to Estimate Fungal Cell Viability After Treatment with Antifungals

The XTT assay relies on the measurement of metabolic activity of cells within the biofilm based on the reduction of XTT, which yields a water-soluble formazan-colored product that can be measured using a microtiter plate reader. We (and others) have previously shown that the XTT-reduction assay shows excellent correlation between cellular density and metabolic activity, thus providing a semiquantitative measurement of biofilm formation [19, 21]. The XTT assay is nondestructive, and requires minimal post-processing of samples in stark contrast with other methods (i.e., cell counts).

1. After the incubation period wash the plates 3 times with PBS.

2. Thaw as many tubes containing 10 µL of the XTT solution as required for the experimental design (one per plate). To each tube, add 1 µL

of the stock solution of menadione to achieve a final menadione concentration of 1 µM (for uniformity, if multiple plates are processed at the same time, we recommend pooling all the resulting XTT/menadione tubes into a single solution in a clean sterile container).

3. Using a multichannel pipette add 100 µL of XTT/menadione solution to each well containing a pre-washed biofilm as well as to negative control wells for the measurement of background XTT-colorimetric levels.

4. Cover the plates in aluminum foil and incubate in the dark for 1–2 h at 37 °C.

5. Uncover the plates. Using a multichannel pipette remove 80 µL of the resulting colored supernatant from each well and transfer into the corresponding wells of a new microtiter plate, and read the plate(s) in a microtiter plate reader at 490 nm.

6. From the resulting colorimetric readings (and after subtracting the corresponding values for negative controls from wells in column 12 containing XTT only), calculate the sessile minimum inhibitory concentrations SMIC50 and SMIC80, which are the antifungal concentrations at which a 50 or 80% decrease in XTT readings are detected in comparison to the control biofilms formed in the absence of antifungal drug (in this case, values for column 11).

Notes

1. Always use proper microbiological handling techniques and universal precautions to handle microorganisms. Follow Institutional Guidelines Regarding Biosafety.

2. Please note that the seeding densities provided here have been optimized for the formation of biofilms on wells of microtiter plates, as quorum-sensing mechanisms play an important role in biofilm formation. Cell densities that are too high or too low will likely result in poor biofilms.

3. After initial seeding of the plates the time of incubation for biofilm formation may be varied depending on the specific objectives of the

study: for example, if the main objective of the study is examination of initial adherence, the incubation time can be reduced to 2–4 h, other investigators may want to study "fully mature" biofilms after 72–96 h incubation, or perhaps study the kinetics of biofilm formation for which multiple plates can be seeded at the same time and then individual plates can be processed after different incubation times (i.e., 0, 2, 4, 6, 8, 12, 24, and 48 h).

4. There are multiple washing steps during the entire protocol. It is critical to preserve biofilm integrity during these washing steps. Normally the biofilms are strongly attached to the bottom of the wells, and these washing procedures should not disrupt the preformed biofilms. If at the end of the washing procedures wells are observed with clearly disrupted biofilm layer at the bottom (normally this is visible by the naked eye), then these wells should be excluded from the analyses. This is one of the main reasons why we recommend performing sufficient replicates for each condition tested.

5. If all the methods have been followed properly, biofilms formed on the bottom of the wells should be visible by the naked eye, simply by looking at the underside of the microtiter plate. In addition, an inverted microscope can be used to examine morphological details of the formed biofilms. Images can be captured if such microscope is equipped with a camera(s) that would allow for image acquisition.

6. If testing experimental and/or new agents with unknown activity against biofilms, we recommend starting with high concentrations of the drug, normally up to 100–1,000 times higher than one would use against planktonic populations.

7. SMIC results are typically presented in a tabular fashion (i.e., multiple antifungals against multiple strains or isolates) or, alternatively, results for each individual strain against each antifungal can be presented as a graph by plotting percent inhibition (or colorimetric readings) versus antifungal concentration.

Acknowledgments Biofilm-related work in the laboratory has been funded by Public Health Service grants numbered R21DE017294 and R21AI080930 from the National Institute of Dental & Craniofacial Research and the National Institute for Allergy and Infectious Diseases (to Lopez-Ribot). Pierce is supported by a predoctoral fellowship, 51PRE30004, from the American Heart Association. Uppuluri is supported by a postdoctoral fellowship, 10POST4280033, from the American Heart Association. The content is solely the responsibility of the authors and does not necessarily represent the official views of the NIDCR, the NIAID, the NIH, or the AHA.

References

1. Costerton JW, Cheng KJ, Geesey GG, Ladd TI, Nickel JC, Dasgupta M et al (1987) Bacterial biofilms in nature and disease. Annu Rev Microbiol 41:435–464
2. Donlan RM (2002) Biofilms: microbial life on surfaces. Emerg Infect Dis 8:881–890
3. Donlan RM, Costerton JW (2001) Biofilms: survival mechanisms of clinically relevant microorganisms. Clin Microbiol Rev 15:167–193
4. Banerjee SN, Emori TG, Culver DH, Gaynes RP, Jarvis WR, Horan T et al (1991) Secular trends in nosocomial primary bloodstream infections in the United States, 1980–1989. National Nosocomial Infections Surveillance System. Am J Med 91: 86S–89S
5. Beck-Sague C, Jarvis WR (1993) Secular trends in the epidemiology of nosocomial fungal infections in the United States, 1980–1990. National Nosocomial Infections Surveillance System. J Infect Dis 167: 1247–1251
6. Edmond MB, Wallace SE, McClish DK, Pfaller MA, Jones RN, Wenzel RP (1999) Nosocomial bloodstream infections in United States hospitals: a three-year analysis. Clin Infect Dis 29:239–244
7. Wisplinghoff H, Bischoff T, Tallent SM, Seifert H, Wenzel RP, Edmond MB (2004) Nosocomial bloodstream infections in US hospitals: analysis of 24,179 cases from a prospective nationwide surveillance study. Clin Infect Dis 39:309–317
8. Wright WL, Wenzel RP (1997) Nosocomial *Candida*. Epidemiology, transmission, and prevention. Infect Dis Clin North Am 11:411–425
9. Pfaller M, Diekema DJ (2007) Epidemiology of invasive candidiasis: a persistent public health problem. Clin Microbiol Rev 20:133–163
10. Kojic EM, Darouiche RO (2004) *Candida* infections of medical devices. Clin Microbiol Rev 17:255–267
11. Ramage G, Martinez JP, Lopez-Ribot JL (2006) *Candida* biofilms on implanted biomaterials: a clinically significant problem. FEMS Yeast Res 6:979–986
12. Ramage G, Tomsett K, Wickes BL, Lopez-Ribot JL, Redding SW (2004) Denture stomatitis: a role for

Candida biofilms. Oral Surg Oral Med Oral Pathol Oral Radiol Endod 98:53–59

13. Jabra-Rizk MA, Falkler WA, Meiller TF (2004) Fungal biofilms and drug resistance. Emerg Infect Dis 10:14–19

14. Ramage G, Mowat E, Jones B, Williams C, Lopez-Ribot J (2009) Our current understanding of fungal biofilms. Crit Rev Microbiol 35:340–355

15. Ramage G, Saville SP, Thomas DP, Lopez-Ribot JL (2005) *Candida* biofilms: an update. Eukaryot Cell 4:633–638

16. Bachmann SP, VandeWalle K, Ramage G, Patterson TF, Wickes BL, Graybill JR et al (2002) In vitro activity of caspofungin against *Candida albicans* biofilms. Antimicrob Agents Chemother 46:3591–3596

17. Kuhn DM, George T, Chandra J, Mukherjee PK, Ghannoum MA (2002) Antifungal susceptibility of *Candida* biofilms: unique efficacy of amphotericin B lipid formulations and echinocandins. Antimicrob Agents Chemother 46:1773–1780

18. Ramage G, VandeWalle K, Bachmann SP, Wickes BL, Lopez-Ribot JL (2002) In vitro pharmacodynamic properties of three antifungal agents against pre-formed *Candida albicans* biofilms determined by time-kill studies. Antimicrob Agents Chemother 46:3634–3636

19. Ramage G, Van de Walle K, Wickes BL, Lopez-Ribot JL (2001) Standardized method for in vitro antifungal susceptibility testing of *Candida albicans* biofilms. Antimicrob Agents Chemother 45:2475–2479

20. Pierce CG, Uppuluri P, Tristan AR, Wormley FL Jr, Mowat E, Ramage G et al (2008) A simple and reproducible 96-well plate-based method for the formation of fungal biofilms and its application to antifungal susceptibility testing. Nat Protoc 3:1494–1500

21. Nett JE, Cain MT, Crawford K, Andes DR (2011) Optimizing a *Candida* biofilm microtiter plate model for measurement of antifungal susceptibility by tetrazolium salt assay. J Clin Microbiol 49:1426–1433

Screening for Compounds Exerting Antifungal Activities

17

Jean-Paul Ouedraogo, Ellen L. Lagendijk,
Cees A.M.J.J. van den Hondel, Arthur F.J. Ram,
and Vera Meyer

Abstract

There is a strong demand for the discovery of new antifungal drugs. More
and more human and plant pathogenic fungi develop resistance against
currently used drugs and therefore do not respond to antifungal treatments.
As humans and fungi are both eukaryotic cells in which many molecular
processes are conserved, compounds that have antifungal activity are also
often toxic for humans. To circumvent this, it is important to develop
methods and screens for the identification of compounds that specifically
kill fungi but do not affect men and the environment. In this chapter, we
describe methods to screen compounds for their ability to prevent growth
of the filamentous fungus *Aspergillus niger*, and to monitor whether these
compounds are fungicidal and whether they switch on the *agsA* reporter
system, which is representative for cell wall or cell membrane stress.

Keywords

Antifungal • Fungicide • *Aspergillus* • Cell wall • Cell membrane •
Susceptibility assay • Azoles • Polyenes • Echinocandins

Introduction

Both the plasma membrane and the cell wall of
fungi contain components that are unique to the
fungal kingdom. Hence, drugs that interfere with
the biosynthesis of these components are likely
to be fungal-specific. Azoles, polyenes, and
echinocandins are three groups of drugs that are
used nowadays to treat fungal infections [1].
Azoles inhibit ergosterol synthesis in fungi,
which is the cholesterol equivalent of animal
membranes. The four currently used azoles
include fluconazole, itraconazole, voriconazole,

J.-P. Ouedraogo • V. Meyer
Department Applied and Molecular Microbiology, Berlin
University of Technology, Institute of Biotechnology,
Gustav-Meyer-Allee 25, Berlin 13355, Germany

E.L. Lagendijk • C.A.M.J.J. van den Hondel
• A.F.J. Ram (✉)
Department of Molecular Microbiology
and Biotechnology, Leiden University, Sylviusweg 72,
Leiden 2333 BE, The Netherlands
e-mail: a.f.j.ram@biology.leidenuniv.nl

V.K. Gupta et al. (eds.), *Laboratory Protocols in Fungal Biology: Current Methods in Fungal Biology,*
Fungal Biology, DOI 10.1007/978-1-4614-2356-0_17, © Springer Science+Business Media, LLC 2013

and posaconazole; these drugs block ergosterol biosynthesis by inhibiting the activity of the cytochrome P450 lanosterol demethylase [2]. Fenpropimorph is a morpholine fungicide that also inhibits ergosterol biosynthesis and protects plants against pathogenic fungi [3]. Polyenes are amphiphatic drugs that strongly bind to ergosterol and create channels, thereby disrupting the integrity of fungal membranes. The most often used polyene, amphotericin B, is effective against several pathogenic fungi; however, its use is restricted because of detrimental side effects on mammalian cells [1].

In addition to the cell membrane, fungal cells are surrounded by a cell wall, which is essential for the fungus to withstand the internal turgor pressure. The fungal cell wall is composed of chitin, beta-glucans, and mannosylated proteins. Depending of the fungal species, polysaccharides (e.g., alpha-glucans, galactomannans) can be present as well [4, 5]. Some of these components are covalently linked and connected to each other to ensure the rigidity and strength of the fungal cell wall. Echinocandins are currently the only class of antifungals that target the biosynthesis of the fungal cell wall. The tree echinocandins used in medicine (caspofungin, micafungin, and anidulafungin) inhibit the function of the (1,3)-beta-D-glucan synthase, which is an essential enzyme for fungal cell wall biosynthesis [6].

Whether a fungal cell is directly killed by a drug is dependent on a variety of factors. Both the concentration of the drug as well as the intrinsic resistance of the specific fungus are important factors that determine drug sensitivity. At nonlethal concentrations, the drugs can trigger stress responses that can make the fungal cell more resistant towards the drugs. It has been well established that the addition of drugs that interfere with cell membrane or cell wall biosynthesis trigger the cell wall stress response pathway [7–9]. The pathway is partially conserved in fungi such as the yeasts Saccharomyces cerevisiae and Candida albicans and the filamentous fungi such as Aspergillus fumigatus and Aspergillus niger [10–12]. We have previously shown that the induction of the agsA gene from A. niger, which encodes a putative (1,3)-alpha-D-glucan synthase, is a very suitable and specific reporter to monitor fungal cell wall stress [7–11].

In the following sections, we describe microtiter- and microscopic-based methods to identify compounds that are fungicidal and, moreover, induce the agsA reporter. These methods allow the set up of high-throughput approaches to identify potential drugs that very specifically disrupt fungal-specific mechanisms essential for survival.

Materials

1. Glucose (50 %): For 1 L: Boil 500 mL Milli-Q (MQ) in a 1,000-mL beaker on a heated magnetic stirrer. Slowly add 500 g of D(+)-Glucose anhydrous. After glucose has been dissolved, let the solution cool down to RT, add MQ up to 1 L and autoclave.
2. ASPA+N (50×): For 1 L: Add 297.5 g (3.5 M) $NaNO_3$, 26.1 g (0.35 M) KCl, and 74.8 g (0.55 M) KH_2PO_4 to 600 mL MQ in a 1-L cylinder. When all salts are dissolved, set pH to 5.5 with KOH (use 5 M KOH). Add MQ up to 1 L and autoclave.
3. ASPA-N (50×): For 1 L: Add 26.1 g (0.35 M) KCl and 74.8 g (0.55 M) KH_2PO_4 to 600 mL MQ in a 1-L cylinder. When dissolved, set pH to 5.5 with KOH. Add MQ up to 1 L and autoclave.
4. $MgSO_4$ (1 M): For 1 L: Add 246.5 g $MgSO_4 \cdot 7H_2O$ to 600 mL MQ in a 1-L cylinder. When all salts are dissolved, add MQ up to 1 L and autoclave.
5. Trace element solution (1,000×): For 1 L: Add 10 g (26.9 mM) EDTA, 4.4 g (15.3 mM) $ZnSO_4 \cdot 7H_2O$, 1.01 g (5.1 mM) $MnCl_2 \cdot 4H_2O$, 0.32 g (1.3 mM) $CoCl_2 \cdot 6H_2O$, 0.315 g (1.3 mM) $CuSO_4 \cdot 5H_2O$, 0.22 g (0.18 mM) $(NH_4)_6 Mo_7O_{24} \cdot 4H_2O$, 1.11 g (10 mM) $CaCl_2$ and 1.0 g (3.6 mM) $FeSO_4 \cdot 7H_2O$ to 600 mL MQ. When dissolved, set pH to 4.0 with NaOH (use 1 M NaOH; 40 g/l) and HCl (use 1 M HCl; 75 mL 37 % hydrochloric acid/l), fill MQ up to 1 L and autoclave. (see Note 1)
6. Vitamin solution (1,000×): For 100 mL: Add 100 mg thiamin-HCl, 100 mg riboflavin, 100 mg nicotinamide, 50 mg pyridoxine, 10 mg pantotenic acid, 2 mg biotin to 50 mL of warm MQ (about 50–60 ° C) in a 100-mL

cylinder. When all vitamins are dissolved, add MQ up to 100 mL, sterilize by filtration, and store at 4 °C under dark conditions.

7. Minimal medium (MM): For 500 mL: Add under sterile conditions to 480 mL of sterile MQ: 10 mL of 50 % glucose, 10 mL of 50×ASPA+N, 1 mL of 1 M MgSO$_4$, and 500 μL of 1,000× trace element solution. For MM+agar, autoclave 480 mL of MQ with 7.5 g of agar (Scharlau) and add all components after autoclaving under sterile conditions.

8. 2× Minimal medium (2× MM): 2 % glucose, 2× ASPA+N, 4 mM MgSO$_4$, 2× trace element solution, 0.06 % yeast extract.

9. Complete medium (CM): For 500 mL: Add 0.5 g casamino acids, 2.5 g yeast extract and if required, 7.5 g agar to 480 mL of MQ and autoclave. Afterwards, add under sterile conditions: 10 mL of 50 % glucose, 10 mL of 50× ASPA+N, 1 mL of 1 M MgSO$_4$, 500 μl of 1,000× trace element solution.

10. Saline solution: For 1 L: Add 9 g (0.9 % w/v) NaCl to 900 mL MQ in a 1-L cylinder. When NaCl is dissolved, add MQ up to 1 L and autoclave.

11. YPD medium: 0.3 % yeast extract, 1 % bactopeptone, 2 % glucose.

12. Myracloth (Calbiochem, La Jolla, CA, USA).

13. Cotton sticks (Hecht).

14. Flat bottom 96 well plate (transparent, Sarstedt, Newton, NC, USA).

15. V-bottom 96 well plate (Sarstedt, Newton, NC, USA).

16. Polystyrol 96 well plate, black (Greiner, Monroe, NC, USA).

17. Multichannel pipette (20–300 μL) (Rainin, Woburn, MA, USA).

18. Microtiter plate reader (e.g., Victor 3, PerkinElmer, Waltham, MA, USA).

19. Incubator (Heraeus, Thermo Scientific, Waltham, MA, USA).

20. Inverted microscope (Leica ICC50).

21. Microscope counting chamber.

22. SYTOX-Green (Invitrogen, Paisley, UK).

23. Fluorescence microscope allowing both light and fluorescence imaging (GFP).

Methods

Inhibitory Testing Using Growth Assays

This method can be used to generally test growth inhibitory effects of compounds towards fungi like *A. niger*. To study growth inhibition, the optical density of the cultures is followed and visualized by microscopic means. Figure 17.1 depicts growth inhibition of *A. niger* in a 96 well plate incubated with different concentrations of the antifungals caspofungin and fenpropimorph.

1. Prepare spore solution of *A. niger* wild-type strain as follows: Streak spores from a single colony on a CM agar plate and incubate until the plate is abundantly covered with sporulated mycelium (3–6 days, 25–37°).

2. In order to harvest spores from CM agar plate, add 10 mL of saline solution to the plate and carefully release spores by scraping over the surface plate with a sterile cotton stick.

3. Pipette spore solution from the plate into a sterile 15-mL tube. If required, remove mycelial debris (vegetative mycelium, conidiophores) by filtration through a sterile myracloth filter.

4. Count spores using a microscope counting chamber.

5. Prepare a spore solution with the final titer of 7.5×10^5 spores/mL (see Note 2).

6. Start the growth inhibition assay with letting the spores germinate first: Fill each well of a flat-bottom 96 well plate with 30 μL sterile MQ, 50 μL 2× MM and 20 μL *A. niger* spore solution. Close the plate with a lid and incubate for 7 h at 30 °C.

7. Prepare a compound stock plate for efficient and fast addition of the compounds by using a V-bottom 96 well plate: Prepare serial dilutions of the compounds and add each 40 μL per well. Every compound should be tested in triplicate. Pipette also 40 μL water or any other solvent used as negative control to at least three wells. Store the compound plate at 4 °C until germination is finished.

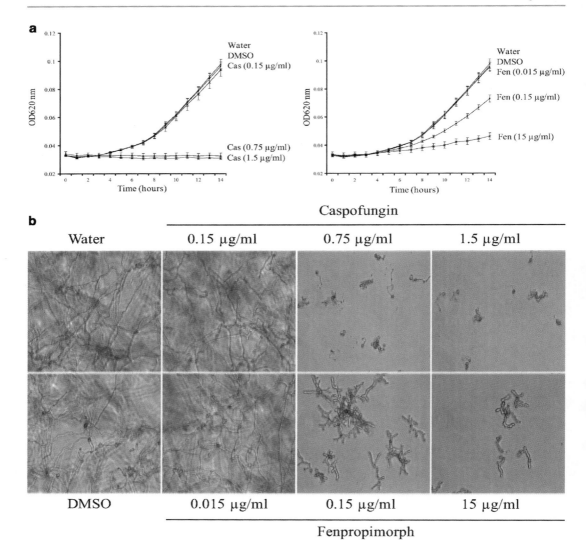

Fig. 17.1 Growth inhibition of *A. niger* incubated with different concentrations of the antifungals caspofungin (cas) and fenpropimorph (fen). Note that DMSO is the solvent for both compounds. (**a**) Optical density (OD_{620}) measured by a plate reader. (**b**) Microscopic pictures taken with an inverted microscope after 24 h of growth

8. Add 45 µL sterile MQ and 75 µL 2× MM to each well of the germination plate using a multichannel pipette.
9. Transfer 30 µL of the compounds (and controls) from the compound plate to the germination plate using a multichannel pipette.
10. Incubate the growth inhibition plate at 30 °C and record the kinetics of growth every hour by measuring the optical density at 620 nm (see Fig. 17.1a). Incubate for a maximum of 24 h (see Note 3).

11. Visualize growth of *A. niger* in the 96 well plate of each condition via microscopy (see Fig. 17.1b).

Cell Membrane Susceptibility Testing Using a Sytox Green Assay

The integrity of cell membrane can easily be carried out using the SYTOX-Green assay [13]. SYTOX Green is a high-affinity nucleic acid stain that can penetrate cells having compromised cell membranes but does not cross intact membranes. The method given below describes the procedure for testing the susceptibility of cell membranes of filamentous fungi and yeast towards potential antifungals [8].

1. Obtain fresh spores or cells of the fungal strains under investigation and count them using a microscope counting chamber.
2. Inoculate 10^2 spores or 10^5 yeast cells in black polystyrol 96 well plates containing 150 µL YPD medium and incubate at 28 °C for 20–40 h for filamentous fungi and for 12–16 h for yeast cells (see Note 4).
3. Add 0.2 µM final concentration of SYTOX-Green for filamentous fungi and 1 µM for yeast cells and place the plate into the dark (see Note 5).
4. Add 25 µL of the antifungal under investigation using serial dilutions.
5. Continue cultivation in the dark at 28 °C. Measure the kinetics of fluorescence formation in minute intervals up to 2 h using a microtiter plate reader at an excitation wavelength of 480 nm and an emission wavelength of 530 nm (see Note 6).
6. Calculate the relative fluorescence values by subtracting the fluorescence values of a culture incubated only with SYTOX-Green without an antifungal compound.

Cell Wall and Cell Membrane Susceptibility Testing Using an *agsA*::GFP Assay

The *agsA* gene coding for (1,3)-alpha-D-glucan synthase is specifically induced in response to compounds that interfere with cell wall or cell membrane integrity of *A. niger* [7, 11]. The *agsA* gene is therefore an excellent and fungal-specific marker for detecting cell surface integrity. Note that bacteria, yeasts (except *Schizosaccharomyces pombe*), plants, and mammals do not have a (1,3)-alpha-D-glucan synthase. To study the effect of compounds on (1,3)-alpha-D-glucan synthesis, two *A. niger* reporter strains, containing either a

cytoplasmatically (strain JvD1.1) or nuclear (strain RD6.47) targeted *gfp* gene under the control of the *agsA* promoter, can be used.

1. Obtain fresh spore solutions from the reporter strains JvD1.1 (expressing PagsA-GFP) and RD6.47 (expressing PagsA-H2B-GFP) as described previously.
2. Inoculate 2×10^4 conidia from the reporter strains in flat-bottom 96 well plate (Sarstedt) containing 100 µL 2× CM.
3. Incubate for 6 h at 37 °C.
4. After spore germination, add 100 µL of a two-fold dilution series for each antifungal compound to individual wells. The effect of each compound shall be tested for at least three to four different concentrations. Include respective negative (water or other solvent) and positive (caspofungin) controls.
5. After adding the compound solution, place the microtiter plates for 3 more hours at 30 °C (see Note 7).
6. Discard the medium by inverting the microtiter plate and analyze germlings that are adherent to the bottom of each well by fluorescence microscopy (see Note 8). Compounds which induce *agsA* expression will induce a strong GFP fluorescence even if germ tube elongation is inhibited. A wild-type *A. niger* strain shall always be used as a negative control because *agsA* expression will be naturally induced after prolonged cultivation.

Notes

1. The color of the 1,000× trace element solution is green when freshly made. After autoclaving, the color changes from green to purple within 2 weeks.
2. Spore solutions of *A. niger* can be stored at 4 °C. However, all assays described work best if the spore solution used is not older than 2 weeks.
3. The maximum time for incubation is 24 h, because the wells dry out owing to medium evaporation.

4. YPD is a complete medium for both yeast and filamentous fungi. Cultivate cells until the culture reaches the mid-logarithmic growth phase. The time required is strain-dependent.

5. It is very important that all experiments involving SYTOX Green are performed in the dark. 0.2 μM and 1 μM SYTOX Green are the optimal concentration for filamentous fungi and yeast, respectively.

6. Permeabilized mycelia or cells respond with increasing fluorescence already after a few minutes of incubation with the antifungal [14].

7. It is important to cultivate at 30 °C as higher temperatures negatively interfere with GFP folding.

8. Use a 40× objective. For GFP images, use a fixed exposure of, for example, 2 s. Process images using Adobe Photoshop.

References

1. Ostrosky-Zeichner L, Casadevall A, Galgiani JN, Odds FC, Rex JH (2010) An insight into the antifungal pipeline: selected new molecules and beyond. Nat Rev Drug Discov 9:719–727

2. Odds FC, Brown AJ, Gow NA (2003) Antifungal agents: mechanisms of action. Trends Microbiol 11:272–279

3. Marcireau C, Guilloton M, Karst F (1990) In vivo effects of fenpropimorph on the yeast *Saccharomyces cerevisiae* and determination of the molecular basis of the antifungal property. Antimicrob Agents Chemother 34:989–993

4. Latge JP (2007) The cell wall: a carbohydrate armour for the fungal cell. Mol Microbiol 66:279–290

5. Klis FM, Ram AF, De Groot PWJ (2007) A Molecular and genomic view of the fungal cell wall. In: Howard J, Gow NAR (eds) The mycota: biology of the fungal cell VIII. Springer-Verlag, Berlin, Heidelberg, pp 97–120

6. Kartsonis NA, Nielsen J, Douglas CM (2003) Caspofungin: the first in a new class of antifungal agents. Drug Resist Updat 6:197–218

7. Meyer V, Damveld RA, Arentshorst M, Stahl U, van den Hondel CA, Ram AF (2007) Survival in the presence of antifungals: genome-wide expression profiling of *Aspergillus niger* in response to sublethal concentrations of caspofungin and fenpropimorph. J Biol Chem 282:32935–32948

8. Ouedraogo JP, Hagen S, Spielvogel A, Engelhardt S, Meyer V (2011) Survival strategies of yeast and filamentous fungi against the antifungal protein AFP. J Biol Chem 286:13859–13868

9. Agarwal AK, Rogers PD, Baerson SR, Jacob MR, Barker KS, Cleary JD et al (2003) Genome-wide expression profiling of the response to polyene, pyrimidine, azole, and echinocandin antifungal agents in *Saccharomyces cerevisiae*. J Biol Chem 278:34998–35015

10. Levin DE (2005) Cell wall integrity signaling in *Saccharomyces cerevisiae*. Microbiol Mol Biol Rev 69:262–291

11. Damveld RA, vanKuyk PA, Arentshorst M, Klis FM, van den Hondel CA, Ram AF (2005) Expression of *agsA*, one of five 1,3-alpha-D-glucan synthase-encoding genes in *Aspergillus niger*, is induced in response to cell wall stress. Fungal Genet Biol 42:165–177

12. Valiante V, Jain R, Heinekamp T, Brakhage AA (2009) The MpkA MAP kinase module regulates cell wall integrity signaling and pyomelanin formation in *Aspergillus fumigatus*. Fungal Genet Biol 46:909–918

13. Thevissen K, Terras FR, Broekaert WF (1999) Permeabilization of fungal membranes by plant defensins inhibits fungal growth. Appl Environ Microbiol 65:5451–5458

14. Theis T, Marx F, Salvenmoser W, Stahl U, Meyer V (2005) New insights into the target site and mode of action of the antifungal protein of *Aspergillus giganteus*. Res Microbiol 156:47–56

Fluorescence *In Situ* Hybridization of Uncultured Zoosporic Fungi

Télesphore Sime-Ngando, Marlène Jobard, and Serena Rasconi

Abstract

Recently, molecular environmental surveys of the eukaryotic microbial community in lakes have revealed a high diversity of sequences belonging to uncultured zoosporic fungi. Although they are known as saprobes and algal parasites in freshwater systems, zoosporic fungi have been neglected in microbial food web studies. Recently, it has been suggested that zoosporic fungi, via the consumption of their zoospores by zooplankters, could transfer energy from large inedible algae and particulate organic material to higher trophic levels. However, because of their small size and their lack of distinctive morphological features, traditional microscopy does not allow the detection of zoosporic organisms such as chytrids in the field. We have designed an oligonucleotidic probe specific to Chytridiales (i.e., the largest group of the true-fungal division of Chytridiomycota) and provide simplified step-by-step protocols for its application to natural samples using both the classical monolabeled-FISH and the CARD-FISH approaches, for the assessment of uncultured zoosporic fungi and other zoosporic microbial eukaryotes in natural samples.

Keywords

Fungi • Zoosporic fungi • Sporangia • Spores • Fluorescence *in situ* hybridization (FISH) • Classical monolabeled-FISH • Catalyzed reporter deposition-FISH (CARD-FISH) • Environmental samples

T. Sime-Ngando (✉)
UMR CNRS 6023, Université Blaise Pascal,
Clermont II, 24 Avenue des Landais, BP 80026,
Aubière 63171 Cedex, France
e-mail: telesphore.sime-ngando@univ-bpclermont.fr

M. Jobard
LMGE UMR CNRS, U.F.R. SCIENCES ET
TECHNOLOGIES, 24 Avenue des Landais, BP 80026,
Aubière, 63171 Cedex, France

S. Rasconi
Department of Biology, University of Oslo,
Blindernvn. 31, Oslo 0371, Norway

Introduction

Recent molecular surveys of microbial eukaryotes have revealed overlooked, uncultured environmental fungi with novel putative functions [1–3], among which zoosporic forms (i.e., chytrids) are the most important in terms of diversity, abundance, and functional roles, primarily as infective parasites of phytoplankton

V.K. Gupta et al. (eds.), *Laboratory Protocols in Fungal Biology: Current Methods in Fungal Biology*,
Fungal Biology, DOI 10.1007/978-1-4614-2356-0_18, © Springer Science+Business Media, LLC 2013

[4, 5] and as valuable food sources for zoo-plankton via massive zoospore production, particularly in freshwater lakes [6–8]. However, owing to their small size (2–5 μm), their lack of distinctive morphological features, and their phylogenetic position, traditional microscopic methods are not sensitive enough to detect fungal zoospores among a mixed assemblage of microorganisms. Chytrids occupy the most basal branch of the kingdom Fungi, a finding consistent with choanoflagellate-like ancestors [9]. These reasons may help explain why both the infective (i.e., sporangia) and disseminating (i.e., zoospores) life stages of chytrids have been misidentified in previous studies as, respectively, phagotrophic sessile flagellates (e.g., choanoflagellates, bicosoecids) and as "small undetermined" cells. These cells often dominate the abundance of free-living heterotrophic nanoflagellates (HNFs) and are considered the main bacterivores in aquatic microbial food webs [2, 10]. Their contribution ranges from 10 to 90% of the total abundance of HNFs in pelagic systems (see review in ref. [11]). Preliminary data have shown that up to 60% of these unidentified HNFs can correspond to fungal zoospores [12], establishing the HNF compartment as a black box in the context of microbial food web dynamics [4]. A recent simulation analysis based on a Lake Biwa (Japan) inverse model indicated that the presence of zoosporic fungi leads to (1) an enhancement of the trophic efficiency index, (2) a decrease of the ratio detritivory/herbivory, (3) a decrease of the percentage of carbon flowing in cyclic pathways, and (4) an increase in the relative ascendency (indicates trophic pathways more specialized and less redundant) of the system [13]. Unfortunately, because specific methodology for their detection is not available, quantitative data on zoosporic fungi are missing.

Sporangia and the associated rhizoidal system are characterized by a chitinaceous wall (a common fungal structure element for many species) that can be targeted by specific fluorochromes such as calcofluor white [14]. In contrast, because the chitinaceous wall springs out after zoospore encystment, chytrid zoospores completely lack cell wall and chitin, precluding any simple use of fluorochromes for their quantitative assessment in natural environments [12]. Molecular approaches, primarily fluorescence *in situ* hybridization (FISH), offer an alternative for quantitatively probing both chytrid sporangia and zoospores in nature. FISH method is based on the detection of targeted nucleic acid sequences by the use of oligonucleotide probes labeled by a fluorochrome, usually Cy3 [15, 16]. One of the major limitations of FISH-based methods for natural samples is the autofluorescence interference from autotrophic organisms. During the past few years, numerous efforts have been made to improve the sensitivity of monolabeled probes for FISH assay, including the use of brightener fluorochromes [17] or of signal amplification with reporter enzymes [18]. Of particular interest is the hybridization method using horseradish peroxidase (HRP)-labeled probes activated by fluorescent tyramide (also known as catalyzed reporter deposition, CARD-FISH), which is very efficient in overcoming the interference from natural fluorescence [19]. The method is based on the fact that each HRP-labeled probe catalyzes the deposition of many labeled tyramides, so that numerous fluorescent molecules are introduced at the hybridization site, resulting in net fluorescence signal amplification, compared to the classical Cy3-monolabeled FISH probes. [20]

The main objective of this chapter is to provide, in a simplified step-by-step format, classical FISH and CARD-FISH protocols for the identification and quantitative assessment of uncultured zoosporic fungi and other zoosporic microbial eukaryotes in natural aqueous environments (cf. 12), together with practical advices on how to apply the methods.

Materials

1. Gloves (should be worn when manipulating most of the following materials).
2. 0.6-μm pore size polycarbonate white filters (e.g., catalog no. DTTP02500, Millipore, Billerica, MA, USA).

3. Appropriate Cy3-labeled oligonucleotidic probe and its reverse complement stored at −20 °C (see Note 1).

4. Sodium dodecyl sulfate (SDS).

5. Formaldehyde 37%.

6. FISH hybridization buffer (0.9 M NaCl, 20 mM Tris–HCl (pH 7.2), 0.01% SDS) containing 30% formamide and 2.5 ng μL^{-1} of Cy3-labeled probe (see Note 2).

7. Washing buffer — 20 mM Tris–HCl (pH 7.2), 5 mM EDTA, 0.01% SDS, 112 mM NaCl [21].

8. Appropriate filtration columns equipped with a peristaltic pump.

9. DAPI — 4.6-diamidino-2-phenylindole.

10. Glass slides and coverslips.

11. Nonfluorescent immersion oil.

12. Epifluorescence microscope equipped with appropriate filter sets (blue and UV) and neofluar objective lens (optional).

13. CARD-FISH hybridization buffer: 30% deionized formamide, 0.9 M NaCl, 20 mM Tris–HCl (pH 7.5), 0.01% SDS, and 10% blocking reagent (e.g., Roche Diagnostics/Boehringer).

14. Appropriate oligonucleotide probe labeled with HRP (in our case, commercially synthesized by Biomers, Germany).

15. TNT buffer — 0.1 M Tris–HCl (pH 7.5), 0.15 M NaCl, and 0.05% Tween 20.

16. TSA mixture — (1:1) of 40% dextran sulphate (Sigma-Aldrich, St. Louis, MO, USA) and 1× amplification diluent (PerkinElmer LAS, Waltham, MA, USA).

17. Fluorescein isothiocyanate coupled with tyramide (1×, Perkin-Elmer LAS).

Methods

Classical FISH Probing (see Note 3)

Fix experimental samples with 2% formaldehyde, vol:vol final concentration. The fixation step is facultative and can be avoided when observations are made without delay.

1. Filter-collect appropriate volumes (×3 replicates) of cultures, enriched cultures, or natural samples containing zoosporic organisms onto 0.6-μm pore size polycarbonate white filters (see Note 4) by using gentle vacuum (<20 kPa).

2. In the dark, pour the filters with targeting fungal zoospores and sporangia and perform hybridization in the standard FISH hybridization buffer (containing 30% formamide and 2.5 ng μL^{-1} of the Cy3-labeled oligonucleotide probe) for 3 h at 46 °C (see Note 5).

3. Use the reverse complement probe in a negative control to check for the autofluorescence interference from fungi and other natural plankton present in natural samples.

4. After hybridization, thoroughly rinse the filters in the washing buffer for 30 min at 48 °C.

5. Counterstain the filters in the dark at room temperature for 5 min with DAPI 0.5 μg mL^{-1}, and repeat the washing step.

6. Mount the filters between glass slides and coverslips using appropriate nonfluorescent immersion oil (see Note 6). At this stage, mounted filters can be conserved at −20 °C until microscopic observation.

7. In a dark room, examine the filters under an epifluorescence microscope equipped with appropriate set of filters and objective lens. Shift between blue and UV light to distinguish between Cy3 stain and DAPI, use different convenient magnifications for sporangia and zoospores, and apply a standard procedure for microscopic counting.

CARD-FISH Probing (see Note 7)

Perform steps 1 and 2 in Classical FISH Probing.

1. In the dark, pour the filters with targeting fungal zoospores and sporangia and perform hybridization in the CARD-FISH hybridization buffer (containing 30% formamide and 2.5 ng μL^{-1} of HRP labeled oligonucleotide probe) for 3 h at 35 °C (see Note 5).

2. After hybridization, thoroughly rinse the filters in the washing buffer for 2 × 20 min at 37 °C.

3. Equilibrate samples to increase enzyme activity in TNT buffer at room temperature for 15 min.

4. Perform signal amplification by 30-min incubation in TSA mixture, to which fluorescein isothiocyanate coupled with tyramide was added (1:50 vol/vol).

5. Transfer filters in two successive 5-ml TNT buffer baths at 55 °C for 20 min, in order to stop the enzymatic reaction and remove the dextran sulphate.

6. Follow steps 6 and 7 in Classical FISH Probing.

Notes

1. We propose to used a probe named Chyt1061 (sequence 5′>3′: CATAAGGTGCCGAACAA GTC), because of the sequence position (1,061 base pairs) on *Saccharomyces cerevisiae* small-subunit rDNA molecule (GenBank accession no. J01353). According to Behrens and collaborators [22], this position provides a good accessibility for FISH probing. There were two mismatches in the middle of the probe with sequences of chytridiales species (cf. 12), which did not result in loss of positive signal. Chyt1061 was designed *in silico* for targeting fungal species in the order Chytridiales, which is the largest order of the division Chytridiomycota (chytrids), mainly represented by phytoplanktonic parasites in aquatic environments [12]. The design was based on the alignment of rDNA sequences of Chytridiales obtained from GenBank (http://www.ncbi.nlm. nih.gov/), together with 106 sequences derived from 18S rDNA PCR surveys of freshwater picoeukaryotes conducted in French Lakes Pavin, Godivelle, and Aydat [12]. Distinct rDNA sequence unique to target organisms was localized and imported in Primer3 software (http://fokker.wi.mit.edu/primer3/input.htm) in order to design a probe with size between 18 and 27 bases, probe melting temperature (Tm) between 57 and 63 °C, and GC percentage at about 50%. The probe was analyzed for potential complementarities and no dimers or hairpins were found using Netprimer software (http://www.premierbiosoft.com/netprimer/ netprlaunch/netprlaunch.html). The probe was commercially synthesized by MWG-Biotech Company (Germany) and labeled with the fluorochrome Cy3 for classical FISH or application to environmental samples using the CARD-FISH approach [12].

2. In case you design your own probe because of the increasing availability of sequences in the database, hybridization stringency should be tested and validated using appropriate positive and negative cultures, before application to natural samples. In the absence of laboratory cultures, our probe Chyt1061 was evaluated from an adaptation of an alternative approach called clone-FISH, known from prokaryotes [23]. This approach is based on the genetic modification of a clone of *Escherichia coli* by inserting plasmid vector containing the target rDNA sequence. In our adaptation of the approach, cells of *E. coli* clone BL21 star were genetically transformed by inserting plasmid vector containing rDNA sequence from several different target fungal cells. Specific plasmid inserts came from freshwater lake surveys of picoeukaryote 18S rDNA and fungal 18S-ITS rDNA as well (cf. 12).

3. The specificity of the designed probe should be checked both *in silico* by using a basic local alignment search tool (e.g., BLAST, [24]), and in vivo by screening of clone libraries with classical FISH (the clone-FISH approach can be used here). In our case, clones containing rRNA gene inserts from different eukaryotes closely related to microorganisms of interest, and negative controls as well, were FISH-targeted following the protocol described in this chapter (i.e., with 30% formamide in the hybridization buffer). In addition, the in vivo transcription of the 18S rRNA gene insert was induced with IPTG (1 mM) for 1 h. The designed probe or its reverse complement probe was used, depending on the orientation of insertion into the vector, i.e. 3′→5′ or 5′→3′ way downstream the T7 promoter (cf. 12).

4. In vivo tests using the clone-FISH approach (see Note 1) yield the best fluorescence signal when hybridization was performed at 30% formamide concentration in the hybridization buffer. Assuming an increase of the effective

hybridization temperature of 0.5 °C per 1% of added formamide, the melting temperature (Tm) of the probe Chyt1061 was experimentally calculated at 61 °C [12].

5. This protocol is suitable for cultures and enriched cultures (i.e., concentrates of targeted zoosporic organisms during host blooms) (cf. 12). However, the FISH resolution for fungal images and species identification based on sporangium features is poor, compared to the calcofluor approach, which is more appropriate for the identification of zoosporic organisms [14]. In addition, The Cy3-monolabeled FISH probing of natural samples clearly showed that the fluorescence of targeted chytrid zoospores may be quite similar to the autofluorescence from natural picoautotrophs. That is why CARD-FISH is more appropriate for environmental samples.

6. For natural waters, the appropriate volume depends on the trophic status of the natural waters, whether the sampling period corresponds to a bloom period, and on the nature of the phytoplankton species. Zoosporic parasites are much more abundant when large-size phytoplankton hosts such as diatoms or filamentous cyanobacteria develop [5, 14]. In oligotrophic waters, concentration of natural samples could be required before harvesting targeted organisms onto polycarbonate filters [15].

7. One filter corresponding to one sampling time point can be cut in pieces before hybridization when several probes are used.

8. The mounting medium should not be fluorescent and will minimize the fading of fluorochromes. An example of mountant is a solution composed of 50% glycerol, 50% phosphate buffered saline (0.05 M Na2HPO4, 0.85% NaCl, pH 7.5), and 0.1% *p*-phenylenediamine (made fresh daily from a frozen 10% aqueous stock solution; Sigma-Aldrich, St. Louis, MO, USA).

9. The CARD-FISH protocol is suitable for natural samples, primarily in oligotrophic waters (see Note 3), where its application improves the detection and the recognition of chytrid, because of the enhanced signal conferred by HRP-labeled probes, compared to monolabeled

oligonucleotides. In addition, the choice of fluorescein as stain (emission in the green spectrum at 520 nm) significantly reduces the interference from natural fluorescence of autotrophic organisms, thereby preventing the use of the deductive approach based on a double counting of the same sample (i.e., with and without hybridization) [25]. However, similar to the simple FISH approach, the CARD-FISH resolution for fungal images and species identification based on sporangium features is poor, compared to the calcofluor approach, which is more appropriate for the identification of zoosporic organisms [14].

Acknowledgements M. Jobard and S. Rasconi were supported by Ph.D. Fellowships from the Grand Duché du Luxembourg (Ministry of Culture, High School, and Research) and from the French Ministère de la Recherche et de la Technologie (MRT), respectively. This study receives grant-aided support from the French ANR Programme Blanc # ANR 07 BLAN 0370 titled DREP: *Diversity and Roles of Eumycetes in the Pelagos.*

References

1. Jobard M, Rasconi S, Sime-Ngando T (2010) Diversity and functions of microscopic fungi: a missing component in pelagic food webs. Aquat Sci 72:255–268
2. Lefèvre E, Bardot C, Noël C, Carrias JF, Viscogliosi E, Amblard C et al (2007) Unveiling fungal zooflagellates as members of freshwater picoeukaryotes: evidence from a molecular diversity study in a deep meromictic lake. Environ Microbiol 9:61–71
3. Monchy S, Jobard M, Sanciu G, Rasconi S, Gerphagnon M, Chabe M et al (2011) Exploring and quantifying fungal diversity in freshwater lake ecosystems using rDNA cloning/sequencing and SSU tag pyrosequencing. Environ Microbiol 13(6):1433–1453. doi:10.1111/j.1462-2920.2011.02444.x. Epub 2011 Mar 9
4. Gachon C, Sime-Ngando T, Strittmatter M, Chambouvet A, Hoon Kim G (2010) Algal diseases: spotlight on a black box. Trends Plant Sci 15:633–640
5. Rasconi S, Jobard M, Sime-Ngando T (2011) Parasitic fungi of phytoplankton: ecological roles and implications for microbial food webs. Aquat Microb Ecol 62:123–137
6. Gleason FH, Kagami M, Marano AV, Sime-Ngando T (2009) Fungal zoospores are valuable food resources in aquatic ecosystems. Inoculum (Suppl Mycologia) 60:1–3
7. Kagami M, Von Elert R, Ibelings BW, de Bruin A, Van Donk E (2007) The parasitic chytrid, *Zygorhizidium*, facilitates the growth of the clado-

ceran zoosplankter, *Daphnia*, in cultures of the inedible alga, *Asterionella*. Proc Soc Biol 274:1561–1566

8. Kagami M, Helmsing NR, Van Donk E (2011) Parasitic chytrids could promote copepod survival by mediating material transfer from inedible diatoms. In: Sime-Ngando T, Niquil N (eds) Disregarded microbial diversity and ecological potentials in aquatic systems. Springer, Heidelberg, pp 49–54

9. James TY, Letcher PM, Longcore JE, Mozley-Standridge SE, Porter D, Powell MJ et al (2006) A molecular phylogeny of the flagellated fungi (Chytridiomycota) and description of a new phylum (Blastocladiomycota). Mycologia 98:860–871

10. Lefèvre E, Roussel B, Amblard C, Sime-Ngando T (2008) The molecular diversity of freshwater picoeukaryotes reveals high occurrence of putative parasitoids in the plankton. PlosOne 3(6):e2324

11. Sime-Ngando T, Lefèvre E, Gleason FH (2011) Hidden diversity among aquatic heterotrophic flagellates: ecological potentials of zoosporic fungi. In: Sime-Ngando T, Niquil N (eds) Disregarded microbial diversity and ecological potentials in aquatic systems. Springer, Heidelberg, pp 5–22

12. Jobard M, Rasconi S, Sime-Ngando T (2010) Fluorescence *in situ* hybridization of uncultured zoosporic fungi: testing with clone-FISH and application to freshwater samples using CARD-FISH. J Microbiol Methods 83:236–243

13. Niquil N, Kagami M, Urabe J, Christaki U, Viscogliosi E, Sime-Ngando T (2011) Potential role of fungi in plankton food web functioning and stability: a simulation analysis based on Lake Biwa inverse model. In: Sime-Ngando T, Niquil N (eds) Disregarded microbial diversity and ecological potentials in aquatic systems. Springer, Heidelberg, pp 65–79

14. Rasconi S, Jobard M, Jouve L, Sime-Ngando T (2009) Use of calcofluor white for detection, identification, and quantification of phytoplanktonic fungal parasites. Appl Environ Microbiol 75:2545–2553

15. Amann RI, Ludwig W, Schleifer KH (1995) Phylogenetic identification and *in situ* detection of individual microbial cells without cultivation. Microbiol Rev 59:143–169

16. Baschien C, Manz W, Neu TR, Marvanová L, Szewzyk U (2008) *In situ* detection of freshwater fungi in an alpine stream by new taxon-specific fluorescence *in situ* hybridization probes. Appl Environ Microbiol 74:6427–6436

17. Glöckner FO, Amann A, Alfreider R, Pernthaler J, Psenner R, Trebesius K et al (1996) An *in situ* hybridization protocol for detection and identification of planktonic bacteria. Syst Appl Microbiol 19:403–406

18. Schönhuber W, Zarda B, Eix S, Rippka R, Herdman M, Ludwig W et al (1999) *In situ* identification of cyanobacteria with horseradish peroxidase-labeled, rRNA targeted oligonucleotide probes. Appl Environ Microbiol 65:1259–1267

19. Schmidt B, Chao J, Zhu Z, DeBiasio RL, Fisher G (1997) Signal amplification in the detection of single-copy DNA and RNA by enzyme-catalyzed deposition (CARD) of the novel fluorescent reporter substrate Cy3.29-tyramide. J Histochem Cytochem 45:365–373

20. Not F, Simon N, Biegala IC, Vaulot D (2002) Application of fluorescent *in situ* hybridization coupled with tyramide signal amplification (FISH-TSA) to assess eukaryotic picoplankton composition. Aquat Microb Ecol 28:157–166

21. Pernthaler J, Glöckner FO, Schönhuber W, Amann R (2001) Fluorescence *in situ* hybridization (FISH) with rRNA-targeted oligonucleotide probe. Methods Microbiol 30:207–226

22. Behrens S, Rühland C, Inácio J, Huber H, Fonseca Á, Spencer-Martins I et al (2003) *In situ* accessibility of small-subunit rRNA of members of the domains *Bacteria*, *Archaea*, and *Eucarya* to Cy3-Labeled Oligonucleotide Probes. Appl Environ Microbiol 69:1748–1758

23. Schramm A, Fuchs BM, Nielsen JL, Tonolla M, Stahl DA (2002) Fluorescence *in situ* hybridization of 16S rRNA gene clones (Clone-FISH) for probe validation and screening of clone libraries. Environ Microbiol 4:713–720

24. Altschul SF, Madden TL, Schäffer AA, Zhang J, Zhang A, Miller W et al (1997) Gapped BLAST and PSI-BLAST: a new generation of protein database search programs. Nucleic Acids Res 25:3389–3402

25. Lefèvre E, Carrias J-F, Bardot C, Amblard C, Sime-Ngando T (2005) A preliminary study of heterotrophic picoflagellates using oligonucleotidic probes in Lake Pavin. Hydrobiologia 55:61–67

Jeyabalan Sangeetha and Devarajan Thangadurai

Abstract

In the past, conventional identification of fungi relied on the combination of morphological and physiological properties. In recent years, morphological studies, supplemented with staining techniques and biochemical methods, still play an important role in the overall identification of fungi in the molecular era. In most instances, these tools are widely used to determine the correct identity of yeasts and molds at the genus and species levels.

Keywords

Identification • Molds • Yeasts • Staining techniques • Biochemical methods

Introduction

In general, fungal identification requires greater visual acuity than bacteria. Unlike other important microorganisms such as bacteria and viruses, the identification of fungi heavily relies on morphological criteria. The characteristics of fungal structure are identified by observing colonial growth both macroscopically and microscopically. These morphological features and other classical methods that are routinely used in classification are also useful in fungal identification. The correct identity of fungal taxa is of great practical relevance in clinical mycology, plant pathology, biodeterioration, and biotechnology. A recent review illustrates the inability to identify fungi at the species, or even at the genus, level in many cases [1]. Many species of fungi, ascomycetes, basidiomycetes, and zygomycetes in particular, have different microscopic and macroscopic characteristics in each stage of their life cycle. Moreover, they are synonymous to each other with many names used to describe the same organism [2].

The ever-increasing number of yeasts and molds that are frequently impossible to identify using morphological criteria due to lack of sporulation has driven the need for the design and development of rapid and robust biochemical and molecular identification tools [3–7]. In recent years, the ability to accurately and reproducibly

J. Sangeetha (✉)
Department of Zoology, Karnataka University, 580003, Dharwad, Karnataka, India
e-mail: drsangeethajayabalan@gmail.com

D. Thangadurai
Department of Botany, Karnataka University, 580003, Dharwad, Karnataka, India

identify fungi has been greatly enhanced through automated biochemical methods and comparative DNA analysis [8–14]. The procedures for the identification of yeasts and molds are different. Identification of molds to the species level has been difficult, because of the amount of experience required to accurately identify these filamentous fungi. In many cases, molds are identified based upon colony and microscopic characteristics. The microscopic structures and morphological features like type, size, shape, and arrangement of spores, and size, color, and septation of hyphae usually provide definitive identification for molds. However, species-level mold identification has not been held to the same standards as bacterial and yeast identification. In yeast, morphological criteria and biochemical tests are generally used to determine genus and species level, respectively.

Direct microscopic examination of the fungal specimen provides a clear view and valuable information about the fungal structure. Yeasts and molds can be identified based on a combination of macroscopic, microscopic, and biochemical analysis. Macroscopic evaluation of fungi provides information about the probable region of the presence of fungi in a specimen, whereas microscopic examination reveals the important features like type and color of hyphae, conidia, septae, spores, and also the concentration of fungi. A definitive fungal stain is an important tool that is needed to begin the identification of fungal specimens. Detection of fungi using direct microscopy with various stains is quick, simple, and can be optimized through the use of ready-to-use staining solutions. The selection of staining method is primarily based on the sample used. A number of stains are routinely available to visualize fungi; some of these are special fungal stains and others are more general in use [15]. Achieving a successful identification comes from the use of appropriate stains and further microscopic examination. The major growth forms of the fungi that help in identification after staining are the yeast cells, hyphae, pseudohyphae, arthroconidia, chlamydoconidia, and endosporulating spherules [16, 17]. On the other hand, visualization of fungi in smear often helps in identification; in some circumstances, it is essential to establish further study on fungi. Even more, the evaluation of staining is useful when multiple organisms are cultured [18].

Additionally, fungal identification requires biochemical tests that will distinguish genera among families and species among genera. Further, strains within a single species are usually distinguished by genetic or immunological criteria. The identity of certain fungi depends on their ability or inability to grow in the absence of nutritional substances such as carbon and nitrogen sources. In addition, several routine biochemical tests, including urease production and proteolysis, are available for the identification of many molds. Moreover, the presence of various enzymes as determined by the biochemical tests is also useful in identifying fungi. When grown in selective liquid or solid media, fungi ferment carbohydrates and produce acids, alcohols, gases, and metabolic and enzymatic products in patterns characteristic of their genus and/or species. These fermentation products are commonly used in the differential identification of fungi [19].

Most frequently, these instant and incubated biochemical tests and other "expert systems" monitor the aptitude of the isolate to assimilate and ferment various sugar and nitrogen sources. In addition, the identification of yeasts has now been regularized by a variety of commercially available strips and kits that can be used to examine rapidly the absorption of carbon as well as nitrogen. The most reliable commercially available yeast identification kits are API 20C AUX, ATB 32C, MicroScan, and Vitek systems. In general, biochemical tests are not important in identifying molds as they are specific for yeasts and dermatophytes [20]. Many yeast-like fungi such as the genera *Geotricum* and *Trichosporon,* which form arthrospores, require a series of biochemical tests for their definitive identification. Most recently, fungal DNA analysis has been considered a powerful identification tool that requires specialized equipment and is impractical in much of routine laboratory work. Although there is an increasing move towards molecular diagnostic approaches, often use of the more easily available and still fundamental staining techniques and other biochemical tests are the choice of mycologists for much of the day-to-day fungal identification. The goal of this chapter is to acquaint mycologists with various staining techniques and differential biochemical tests available for detecting fungal specimens.

Materials

See Note 1.

1. A pure culture of 24- to 48-h old yeast cells (growing on Sabouraud or other nonselective agar)
2. Microscopic slides
3. Sterile saline or water
4. Flame source
5. Fume hood
6. A straight nichrome wire with a long handle for stabbing inoculation
7. A bent nichrome wire with a long handle for handling the mycelia types of fungal cultures
8. Pair of short, stiff teasing needles helpful in pulling apart dense masses of mycelium on the slide for better microscopic examination
9. Scalpel with blades
10. Pair of forceps
11. Light microscope, phase contrast microscope, and fluorescence microscope with filters
12. 20% KOH-Glycerol solution—20 g KOH, 20 mL glycerol, 80 mL distilled water (see Note 2)
13. Lactophenol Cotton Blue stain—20 mL lactic acid, 20 mL phenol, 40 mL glycerol, 0.05 g cotton blue or aniline blue. Dissolve phenol in lactic acid, glycerol, and distilled water, finally add cotton blue and mix well
14. India ink
15. Crystal violet
16. Gram's iodine
17. Ethanol
18. Absolute alcohol
19. Safranin
20. Giemsa stain
21. Wright stain—9.0 g (0.3% w/v) powdered Wright's stain, 1.0 g (0.033% w/v) powdered

Giemsa stain, 90 mL glycerin, and 2,910 mL absolute acetone-free methanol (see Note 3)
22. Weigert's solution A—0.6 g ferric chloride, 100 mL distilled water, 0.75 mL hydrochloric acid
23. Weigert's solution B—1.0 g hematoxylin, 100 mL 95% ethanol; combine equal parts of solution A and B
24. Weigert's iron hematoxylin solution A—1.0 g of hematoxylin, 100 mL 95% ethyl alcohol
25. Weigert's iron hematoxylin solution B—4.0 mL 29% aqueous ferric chloride, 95 mL distilled water, 1.0 mL concentrated hydrochloric acid
26. Weigert's iron working solution—mix equal parts of Weigert's iron solution A and B
27. Acridine orange
28. Schiff's reagent—0.025 g pararosaniline, 100 mL distilled water, sulfur dioxide gas (see Note 4)
29. Aldehyde fuchsin—2 g basic fuchsin, 1 mL paraldehyde, 1 mL concentrated hydrochloric acid, 200 mL 70% ethanol; ripen at room temperature for 48–72 h
30. Chromic acid—5 g chromic trioxide, 500 mL distilled water (see Note 5)
31. Metanil yellow—0.25 g metanil yellow, 100 mL distilled water, 0.25 mL glacial acetic acid
32. 15% Potassium hydroxide solution—15 g potassium hydroxide, 20 mL glycerol, 80 mL distilled water; store at 25 °C and discard if precipitation occur
33. 0.1% Calcofluor White (CFW) solution
34. Mucicarmine stain—1.0 g carmine, 0.5 g aluminum chloride (anhydrous), 2 mL distilled water (see Note 6)
35. Formal–ethanol mixture—10 mL 40% formaldehyde, 90 mL absolute alcohol
36. 5% Periodic acid (see Note 7)
37. Basic fuchsin solution
38. Sodium metabisulphate solution—1.0 g sodium metasulphite, 10 mL hydrochloric acid, 190 mL distilled water (see Note 8)
39. 0.2% Light green solution—0.2 g light green, 100 mL distilled water, 0.2 mL glacial acetic acid

40. Xylene (see Note 5)
41. 1% Sodium metabisulphite
42. Hexamine, preheated in a water bath to 56 °C for 1 h
43. Ferric chloride
44. 5% Sodium thiosulphate solution
45. 5% Silver nitrate, store in dark-colored bottle at 4 °C (see Note 9)
46. 3% Aqueous methenamine (hexamethylene tetramine) (see Note 9)
47. Aniline–acetic acid (see Note 5)
48. Thiosemicarbazide
49. Schmorl's solution—30 mL ferric chloride (1% aqueous), 4 mL potassium ferric cyanide, 6 mL distilled water; make immediately before use and do not reuse
50. Mellor bleach solution A—1% potassium permanganate
51. Mellor bleach solution B—1% sulphuric acid
52. Mellor bleach working solution—mix Mellor bleach solution A and B
53. Mellor bleach solution C—1% oxalic acid
54. Ammonium silver stock solution—25 mL 10% silver nitrate, add ammonium hydroxide drop by drop, until solution precipitates and clears again
55. Ammonium silver working solution—12.5 mL ammonium silver stock solution, 37.5 mL distilled water (see Note 10)
56. 0.1% Gold chloride
57. Neutral red stain
58. Sulfation reagent—45 mL glacial acetic acid, 15 mL concentrated sulphuric acid (see Note 11)
59. Toluidine blue O—3 g toluidine blue O, 60 mL distilled water, 2 mL concentrated hydrochloric acid, 140 mL absolute ethyl alcohol
60. Basal salt medium
61. 0.1% Carbohydrate
62. Bromocresol purple
63. Sugar disc
64. Whatman filter paper
65. Sugars
66. Hot-air oven
67. Yeast Nitrogen Base (YNB) medium
68. Molten agar
69. Yeast Carbon Base (YCB) medium
70. Peptone disc
71. Potassium nitrate
72. Durham's tube
73. Skim milk agar
74. Casein agar
75. Microgranular cellulose
76. Inoculation chamber
77. Autoclave
78. Petriplates
79. Incubator
80. Mycosel agar with cyclohexamide plate
81. 10% Tween solution
82. Cellulolysis Basal Medium (CBM)—5 g $C_4H_{12}N_2O_6$, 1 g KH_2PO_4, 0.5 g $MgSO_4 \cdot 7H_2O$, 0.1 g yeast extract, 0.001 g $CaCl_2 \cdot 2H_2O$, 1000 mL distilled water
83. 0.5% Esculin
84. 0.5% Arbutin
85. 2% Ferric sulfate
86. Sabouraud agar plates
87. Christensen's urea agar
88. API 20C Yeast Identification Kit
89. Biomerieux Vitek Yeast Biochemical Card
90. Abbott Yeast Identification Kit
91. Colorimeter (Vitek colorimeter, 52–1210)
92. Filling stand (e.g., Vitek 52–0700)
93. Sterile wooden applicator sticks
94. Sterile Pasteur or plastic pipettes (5 mL)
95. Squeeze bottle
96. Water bath (48–50 °C)
97. 0.45–0.50% Saline
98. Sterile tubes
99. Fine-tip markers
100. 0.05% Noble agar
101. Sterile inoculating loop
102. Vortex mixer
103. Polyester film

Methods

Wet Mount Techniques

Fungal specimens can be visualized using wet mount techniques through suspension of culture in either water or saline, mixed with alkali to dissolve background material [21] or mixed with a combination of alkali and contrasting dye (e.g., lactophenol cotton blue or India ink) [22, 23].

The dyes nonspecifically stain the fungal material, which increases contrast with the background and permits examination of the detailed structures. A variation is the India ink method, in which the ink darkens the background rather than the fungi.

Potassium Hydroxide Wet Mount

Potassium hydroxide (KOH) is used to dissolve proteinaceous material and facilitate detection of fungal elements that are not affected by strong alkali solution. It is a strong alkali used as a clearing agent to observe fungi in a wet mount preparation. The concentration of KOH is usually based on the specimen that is being used. Normally, 10–20% KOH is used; occasionally, 40% is used when the specimen is not cleared by 10–20% of KOH. In this method, the fungal structures, such as hyphae, large yeasts (*Blastomyces*), spherules, and sporangia, are well distinguished. In unstained preparations (KOH without stain), the fungal structures may be enhanced by using a phase contrast microscope [18, 24]. The clearing effect throughout the specimen can be accelerated by gently heating the KOH preparation.

Visualization of fungi can be further enhanced by the addition of dyes to the preparation. This method is quick, simple, and inexpensive [25, 26].

1. Place a large drop of KOH solution with a Pasteur pipette.
2. Transfer small quantity of the culture with a loop or the tip of a scalpel into the KOH drop.
3. Put a clean coverslip over the drop gently so that no air bubble is trapped.
4. Clearing can be hastened by gentle heating of the slide, but it is best avoided.
5. Observe under 20× and 40× objective of light or phase contrast microscope.
6. Look for budding yeast cells; branching hyphae; type of branching; and the color, separation, and thickness of hyphae (see Note 12).

Lactophenol Cotton Blue Wet Mount

Lactophenol cotton blue (LCB) is a mounting medium commonly used in microbiology laboratories for preparing mounts of fungal cultures. LCB is used as both mounting fluid and stain. In this method, phenol will kill the organisms, and the lactic acid preserves fungal structures; chitin in the fungal cell wall is stained by the cotton blue. It can be used alone or in conjunction with KOH. Library slides may be made by allowing the mount to dry for 3 weeks and then sealing with collodion [24, 27, 28].

1. Put a large drop of LCB with a Pasteur pipette.
2. Transfer a small quantity of the culture to the drop.
3. Tease the culture well with teasing needles, so as to get a uniform spread.
4. Put on a coverslip gently to avoid entrapment of air bubbles.
5. Examine under the 20× and 40× objectives of light microscope.
6. Observe the morphological features carefully.
7. Fungal elements will stain deep blue against a clear pale-blue background.

India Ink Wet Mount

India ink can be added to specimens to provide dark background that will highlight hyaline yeast cells and capsular material. This method is used to detect microorganisms that are surrounded by capsules. The dye is excluded by the capsule, creating a clear halo around the yeast cell. It is a rapid method for the preliminary detection and identification of specimens containing species of *Cryptococcus* [29].

1. Add a small drop of India ink on a smear.
2. Place a coverslip over the smear and press it gently to obtain a thin mount.
3. If India ink is too thick (dark), dilute it by 50% with saline.
4. Allow the preparation to stand for few minutes to settle.
5. Scan under low power in reduced light; switch to high power, if necessary.
6. Organisms possessing a capsule appear highly refractile, surrounded by a clear zone against a dark background.

Staining Techniques

Direct microscopic examination without stain lacks sensitivity, especially when hyphae are sparse in the specimen. A variety of differential stains are commonly used like Gram, Giemsa, Wright stain, toluidine blue O, and Weigert's iron hematoxylin to stain fungi [30]. The sensitivity of microscopic examination is improved when fungus-enhancing stains like Mayer's mucicarmine, periodic acid Schiff, Gomori's methenamine silver, acridine orange fluorescent, calcoflour white, thiosemicarbazide, Fontana-Masson, and Gridley's stains are used. Since the stain is immediately taken up by the fungal cell wall in the scraping, the staining usually becomes much brighter after 5–10 min [31]. Some specimens need alkali pretreatment. In that case, it is important to make sure that they do not react for a long period of time; otherwise, a gelatinous consistency will form, and the specimen should be neutralized with 10% lactic acid before staining and adjusted to pH 3.0–5.0 [32]. The method of preparation of smear for staining is as follows:

1. Take a clean grease-free glass slide.
2. Place a large drop of saline solution.
3. Transfer a small quantity of the culture with a loop or the tip of a scalpel into the saline drop.
4. Make a smear over the surface of the slide.
5. Fix by heat, if necessary.

Gram Staining

Gram stain is a key starting point to identify microbial species. The stain differentiates membrane structures between gram-positive and gram-negative microorganisms. Gram-positive microbes have a thick cell wall made up of peptidoglycan (50–90%), which are stained purple by crystal violet, whereas gram-negative microbes have a thinner layer (10% of cell wall), which are stained pink by the counter-stain safranin [24, 30, 33–36].

1. Apply two drops of crystal violet on smear for 30 s.
2. Wash with tap water.
3. Add two drops of Gram's iodine for 30 s.
4. Repeat step 2.
5. Add 95% ethanol.
6. Repeat step 2.
7. Add two drops of safranin.
8. Repeat step 2.

9. Observe in microscope under oil immersion.
10. Yeasts are gram-positive, but poorly stained; *Cryptococcus neoformans* is a notable exception (gram-negative).

Giemsa Staining

A variety of "Romanowsky-type" stains with mixtures of methylene blue and azure eosin compounds have been used successfully for many years on diverse fungi with various procedures and modifications. Giemsa stain is a member of the Romanowsky group of stains, which are defined as being the black precipitate formed from the addition of methanol [37]. In this stain, eosin ions are negatively charged and stain basic components of cells orange to pink. It was also originally designed to incorporate cytoplasmic (pink) staining with nuclear (blue) staining and fixation as a single step for smears and thin films. This stain has widely been used to examine *Pneumocystis jiroveci*, *Rhinosporidium seeberi*, and *Histoplasma capsulatum* [38–40].

1. Flood the smear with methyl alcohol and leave for 3–5 min for fixation.
2. Add prepared Giemsa stain and leave for 45 min.
3. Wash slide thoroughly with running tap water.
4. Blot dry with absorbent paper.
5. Observe under oil immersion.
6. Look for intracellular budding yeasts; fungi stain with purplish-blue.

Wright Staining

The Wright stain is an alcoholic solution of methylene blue, azure A, thionin, and eosin Y. Methyl groups are activated and react with charged components of the cell to produce coloration. It is used to detect blood parasites, viral and chlamydial inclusion bodies, yeast cells, and species of *Pneumocystis*. Eosin ions are negatively charged and stain basic components of cells orange to pink, whereas other dyes stain acidic cell structures to various shades of blue to purple [41, 42].

1. Cover the smear with freshly filtered Wright stain and leave for 1–3 min.
2. Without removing the stain, pour on buffer solution (pH 6.4).

3. Gently mix buffer and stain; upon proper mixing, metallic green sheen (green scum) rises to the surface of the fluid.
4. Leave for 3 min or longer.
5. Wash the slide gently with flowing tap water and wipe the bottom of the slide with a clean filter paper.
6. Air-dry the slide and observe under the microscope.
7. Intracellular yeast cells are typically stain blue and species of *Pneumocystis* stain purple.

Weigert's Iron Hematoxylin Staining

This stain can be used with fixatives that include polyvinyl alcohol, sodium acetate, and formalin. The staining method involves application of haemalum, which is a complex formed from aluminum ions and oxidized hematoxylin. This stains nuclei of cells blue. Counterstain eosin Y may also be used to color other structures in various shades of red, pink, and orange [28, 43–47].

1. Add staining solution on a smear and leave for 1–2 h.
2. Rinse with tap water.
3. Add 1% HCl.
4. Add 70% ethanol.
5. Repeat step 2.
6. Counterstain with eosin Y, if necessary.
7. Dehydrate with ethanol.
8. Clear with xylene and observe under microscope.
9. Yeast cells stain blue–gray to black.

Acridine Orange Staining

Acridine orange is a fluorochromatic dye that binds to nucleic acids of fungi. Under UV light, acridine orange stains RNA and single-stranded DNA orange, while double-stranded DNA appears green. At neutral pH, fungi and cellular materials stain reddish orange. At acid pH, fungi remain reddish orange but background material stains greenish yellow [31, 48–53].

1. Add Weigert's iron hematoxylin on smear for 5 min.
2. Wash well with tap water.
3. Place few drops of acridine orange solution for 2 min.
4. Repeat step 2.

5. Observe smear on the fluorescence microscope.
6. Fungi stain bright orange and the background appears greenish yellow.

Gridley Staining

Gridley staining method is used to identify fungi, based on Bauer chromic acid leucofuchsin stain with the addition of Gomori's aldehyde fuchsin stain and metanil yellow as counterstains. Against a yellow background, hyphae, conidia, yeast capsules, elastin, and mucin appear in different shades of blue to purple. It can be used to identify *Rhiosporidium seeberi* and *Histoplasma capsulatum* [28, 47, 51, 54, 55].

1. Place chromic acid on smear for 1 h.
2. Wash well with tap water.
3. Treat with sodium metabisulphite bleach for 1 min.
4. Repeat step 2.
5. Rinse with distilled water.
6. Place in Schiff's reagent for 20 min.
7. Repeat step 2.
8. Rinse with 70% ethanol.
9. Place in aldehyde fuchsin for 30 min.
10. Rinse off excess with 95% ethanol.
11. Repeat step 2.
12. Counterstain with metanil yellow for 1 min.
13. Rinse well with distilled water.
14. Dehydrate and observe under fluorescence microscope.
15. Fungi show purple color with yellow background (see Note 13).

Calcoflour White Staining

Calcoflour White (CFW) stain is used to detect fungal elements, particularly *Pneumocystis* species. The fluorophore shows a high affinity for chitin-forming hydrogen bonds with free hydroxyl groups and stains fungal cell walls blue. The use of CFW staining requires the addition of KOH, which helps to dissolve keratinized particles and emulsify solid, viscous material and enhance the visualization of fungal elements in microscopic examination. Positive results are indicated by a bright green to blue fluorescence using a fluorescence microscope (see Note 14) [56]. A bright yellow-green fluorescence is observed when collagen or elastin is present. KOH-CFW preparations may be preserved for several days at 4 °C. This stain can be used to identify *Fusarium solani*, *Aspergillus fumigatus,* and *Candida albicans* [24, 56–59].

1. On smear add a drop of 15% KOH and a drop of the CFW solution or mix in equal volumes before processing.
2. Mix and place a coverslip over the material.
3. If necessary, allow the KOH preparation to remain at room temperature (25 °C) for a few minutes until the material has been cleared; the slide may be warmed to speed up the clearing process.
4. Observe the slide by UV microscopy.
5. Fungal cell walls fluorescence apple green to blue.

Mayer's Mucicarmine Staining

Mucicarmine is a red stain that contains aluminum chloride and carmine. Aluminum is believed to form a chelation complex with the carmine and change the molecule to a positive charge, allowing it to bind with the acid substrates of low density, such as mucins. It is used to detect mucin-secreting fungi and capsules of *Cryptococcus neoformans* and *Rhinosporidium seeberi*. It will also stain the walls of the spores and the inner surface of the sporangia. However, the cell walls of yeasts and *Blastomyces dermatitis* may stain weakly with mucicarmine [55, 60–64].

1. Stain the smear with a working solution of Weigert's hematoxylin for 7 min.
2. Wash well in tap water.
3. Add metanil yellow for 1 min.
4. Repeat step 2.
5. Place in mucicarmine stain for 45 min.
6. Rinse quickly in distilled water.
7. Dehydrate in 95% ethanol and absolute alcohol (two changes of each).
8. Clear with two changes of xylene.
9. Mount in DPX and view under the microscope.
10. Mucopolysaccharide capsule stain deep rose to red, nuclei are black, and the other debris stain yellow.

Periodic Acid–Schiff Staining

Periodic acid–Schiff (PAS) reactions are effective stains for demonstrating fungal elements of essentially all fungi. Periodic acid attacks some carbohydrates containing 1,2-glycol or OH group with the conversion of this group to 1,2-aldehydes, which then react with the fuchsin-sulfurous acid to form the magenta color [25, 32]. Identification of fungal elements can be enhanced if a counterstain such as light green is used. Species of *Coccidioides*, *Cryptococcus*, *Histoplasma*, *Candida, Malassezia*, and *Aspergillus* can be stained with this stain [65].

1. Immerse the smear in ethanol for 1 min.
2. Place 5% periodic acid for 5 min.
3. Wash gently in running tap water.
4. Place basic fuchsin for 2 min.
5. Repeat step 3.
6. Add sodium metabisulphite (0.5%) for 3–5 min.
7. Repeat step 3.
8. Counterstain with dilute aqueous light green (0.2%) for 2 min.
9. Dehydrate with 70%, 80%, 95%, 100% ethanol and xylene, each for 2 min.
10. Observe under microscope.
11. Fungi stain bright pink-magenta or purple against green background when light green is used as a counterstain.

Grocott-Gomori Methenamine Silver Staining

Grocott-Gomori methenamine silver (GMS) staining is preferred for screening degenerated and nonviable fungi because it provides better contrast. The fungal cell wall contains mucopolysaccharides that are oxidized by GMS to release aldehyde groups, which later react with silver nitrate. Silver nitrate is converted to metallic silver, which becomes visible in the silver stains; this is useful in detecting fungal elements. Fungi stain in black against a pale-green background. *Pneumocystis jiroveci*, *Cryptococcus neoformans*, *Coccidiodes immitis*, *Histoplasma capsulatum*, *Aspergillus fumigatus,* and *Candida albicans* can be detected by this staining technique [18, 24, 55, 64, 66, 67].

1. Add two drops of absolute ethanol for 5 min.
2. Wash in distilled water.
3. Flood the smear with 4% chromic acid for 45 min.
4. Repeat step 2.
5. Add 1% sodium metabisulphite for 1–2 min.
6. Repeat step 2.
7. Add working solution of hexamine (smear becomes dark brown).
8. Wash with distilled water or if smear turns black, wash with 0.1% ferric chloride.
9. Add 5% sodium thiosulphate for 2 min.
10. Repeat step 2.
11. Wash with 1% light green solution for 1 min.
12. Dry and view under oil immersion.
13. The slide with fungal elements stains black; inner part of micelle or hyphae stains pink with background in pale green.

Double-Oxidation Thiosemicarbazide, Schmorl

The hydrazine group (H_2NNH-) combines with any aldehyde generated by periodic acid oxidation. The thiocarbamyl group ($-CSNH_2$) is a more powerful reducing agent than aldehydes and rapidly reduces ferricyanide to ferrocyanide, which immediately forms a prussian blue deposit at the site. The mallory bleach lightens background staining and improves contrast. It may also produce some aldehyde, which is removed in step 6. This method is widely used to identify fungal colonies in tissues [39, 68].

1. Do mallory bleach (place mellory working solution for 10 min and rinse with water; then place mallory solution C for 2 min until the tissue is bleached).
2. Wash with a running tap water.
3. Oxidize in periodic acid for 10–20 min.
4. Repeat step 2.
5. Place aniline-acetic for 30 min.
6. Repeat step 2.
7. Repeat step 3.
8. Repeat step 2.
9. Place into thiosemicarbazide for 10 min.
10. Repeat step 2 to remove all traces of thiosemicarbazide.
11. Place into freshly made Schmorl's solution for 10 min.

12. Repeat step 2.
13. Optionally, counterstain with eosin.
14. Repeat step 2.
15. Dehydrate with ethanol, clear with xylene, and mount with a resinous medium.
16. Fungi stain with blue, nuclei with red, and background is pink.

Fontana-Masson Staining

The Fontana-Masson (FM) stain can be used to detect the presence of melanin in cell walls of dematiaceous fungi such as species of *Bipolaris*, *Curvularia*, *Exophiala,* and *Phialophora*. FM stain is often believed to be a diagnostic tool to differentiate dematiaceous fungi from *Aspergillus* sp. and some Zygomycetes. Also, it is particularly useful for distinguishing capsule-deficient *Cryptococcus neoforamans* from *Histoplasma capsulatum* and *Blastomyces dermatitis*. Melanin has the ability to reduce solutions of ammonical silver nitrate to metallic silver without the use of an external reducing agent. The intensity and amount of staining may reflect differences in melanin deposition owing to growth rate, age, availability of precursors, or loss of pigment staining associated with hyphal death and destruction. Extent of stain intensity and its distribution in fungal elements in tissue were evaluated by means of intensity; for example, dark brown (strong intensity), medium brown (moderate intensity), and pale brown (weak intensity) [69].

1. Treat smear with ammonical silver nitrate solution for 20 min at 60 °C.
2. Check microscopically after 15 min and repeat step 1 if necessary.
3. Wash well in distilled water.
4. Tone with 0.1% gold chloride for 2 min.
5. Repeat step 3.
6. Fix in 2% aqueous sodium thiosulphate for 2 min.
7. Repeat step 3.
8. Counterstain with neutral red stain for 1 min.
9. Repeat step 3.
10. Rapidly dehydrate well in absolute alcohol, clear, and mount.
11. Observe under microscope.

12. All dematiaceous fungi show strong intensity (black); species such as *Bipolaris, Exophiala, Fonsecaea,* and *Phialophora* are darkly pigmented because of melanin.

Toluidine Blue O Staining

This stain is primarily used for the detection of *Candida albicans, Rhinosporidium seeberi,* and *Pneumocystis carinii.* Background staining is removed by sulfation reagent. Yeast cells get stained differentially and are difficult to distinguish from *Pneumocystis* cells. The stain can be replaced with specific fluorescent stains. Toluidine blue O gives polychromatic staining for all the fungal structures (such as conidia, germ tubes, haustoria, and hyphae) as well as cells [55, 64, 70–73].

1. Add sulfation reagent for 10 min.
2. Wash with tap water.
3. Add toluidine blue O for 3 min.
4. Add 95% ethyl alcohol, absolute ethyl alcohol, and xylene, each for 10 s for decolorizing.
5. Place a coverslip on the slide.
6. Observe with 20× and 40× objectives.
7. Fungi stain reddish blue to dark purple on light-blue background.

Manual Biochemical Methods

Biochemical tests have been used to classify and identify various groups of fungi. Because fungi grow rapidly in pure culture, it is possible to use biochemical methods to identify and classify them [74]. The classical broth methods were originally developed by Wickerham for utilization and fermentation testing of yeasts [75–77]. Biochemical methods like utilization of carbon and nitrogen, fermentation of carbon, and enzyme activity like caseinase, cellulase, gelatinase, glucosidase, fatty acid esterase, lipase, urease, and so forth are currently in use to assist in the differentiation of fungi.

Utilization of Carbon Sources

Assimilation tests are extremely important in the taxonomy of yeasts. They should be performed before tests involving chemical analysis of the fungus for the very simple reason that they are easier to perform and generally require no specialized apparatus. An assimilation test is performed based on the fact that nutritional factors are capable of differentiating fungi. Storage of carbohydrates fulfills multiple functions in fungi; they not only constitute a source of carbon and energy but also protect fungi against a variety of environmental stresses, such as desiccation and frost. Fungal metabolism dominates the assimilation of exogenous carbohydrates into tissues. Several sugars can be used as a carbon source. In physiologic characterization, tests for the ability of a fungi to utilize various carbon substrates (see Note 15) as the sole source of carbon by employing a basal medium such as yeast nitrogen base that contains ammonium sulfate (universal nitrogen source), vitamins, amino acids, and trace elements are required for growth of yeasts. Assimilation tests are read by growth and turbidity. For each test or organism, negative controls without a carbon source should be maintained [78–82].

Liquid Auxanographic Method

1. Prepare basal medium with bromocresol purple (0.5 g/L).
2. Adjust pH to 5.4 by adding NaOH or HCl.
3. Sterilize at 121 °C for 20 min.
4. Add filter-sterilized 1% (w/v) selected carbohydrate.
5. Pour the medium into test tubes.
6. Inoculate with fungal mycelium.
7. Incubate at 20 °C for 14 days.
8. Change in color of the medium to orange or yellow is taken as positive, whereas a change to pink or purple is negative.

Pour Plate Auxanographic Method

1. Sugar discs can be obtained commercially or prepared manually (steps 2–5).
2. Punch 6-mm diameter discs from Whatman no. 1 filter paper.
3. Sterilize the discs by placing them in a hot-air oven for 1 h.
4. Allow to cool, and then add one drop of 10% filter-sterilized sugar solution to each disc.
5. Dry the disc at 30 °C in incubator and store at 0 °C.

6. Prepare yeast nitrogen base (YNB) medium.
7. Prepare a yeast suspension from 24- to 48-h old culture in 2 mL of YNB by adding heavy inoculum.
8. Add this suspension to the 18 mL sterilized Molten agar mix well.
9. Pour the entire medium into Petri dish.
10. Allow the media to solidify at room temperature.
11. Now place the various carbohydrate-impregnated discs onto the surface of the agar plate.
12. Incubate at 30 °C for 4–7 days.
13. Positive reactions can be noted by growth and color change around the disc.

Utilization of Nitrogen Sources

Simple basal media to which single nutrients (vitamins, amino acids) could be added will be used as the bases of these tests. For testing various nitrogen substrates (see Note 16) as the sole source of nitrogen, one can use yeast carbon base that contains glucose (universal carbon source) and the vitamins, amino acids, and trace elements required for growth. In general, if yeast can utilize nitrate, it can also use nitrate as a nitrogen source. For each test or organism, negative controls without nitrogen source should be maintained [83–89].

Liquid Auxanographic Method

1. Prepare basal medium along with bromocresol purple (0.5 g/L).
2. Adjust pH to 4.5 with HCl or NaOH.
3. Sterilize the medium at 121 °C for 20 min.
4. Add filter-sterilized nitrogen compound (2 g/L).
5. Inoculate with fungal mycelium.
6. Incubate at 20 °C for 4–7 days.
7. Hyphal mat in liquid media is the positive result.

Pour Plate Auxanographic Method

1. Peptone discs can be obtained commercially or prepared manually (steps 2–5).
2. Punch 6-mm diameter discs from Whatman no. 1 filter paper.

3. Sterilize the discs by placing them in a hot-air oven for 1 h.
4. Allow to cool, and then add one drop of 3% filter-sterilized potassium nitrate or peptone solution to each disc.
5. Dry the disc in 30 °C in incubator and store at 0 °C.
6. Prepare yeast carbon base (YCB) medium.
7. Prepare a yeast suspension from 24- to 48-h old culture in 2 mL of YCB by adding heavy inoculum.
8. Add this suspension to the 18 mL sterilized molten agar and mix well.
9. Pour the entire medium into Petri dish.
10. Allow the media to solidify at room temperature.
11. Now place the various nitrate-impregnated discs onto the surface of the agar plate.
12. Incubate at 30 °C for 4–7 days.
13. Positive reactions can be noted by growth and color change around the disc.

Carbohydrate Fermentation

This method is a powerful tool for definitive characterization and taxonomy of yeasts. Carbohydrate fermentation tests whether a certain microbe has the capability to ferment different carbohydrates. Fungi able to ferment a particular sugar are also able to assimilate the same sugar; however, the reverse is not always true. To test fermentative abilities, a different basal medium is employed. Normally, 2% sugar solution is added to the basal medium in a test tube that also contains an inverted Durham tube in order to observe production of CO_2 and ethanol as the by-products of sugar fermentation. Because most yeasts are also able to assimilate ethanol as the sole source of carbon, it is necessary to incubate assimilation tests separately from fermentation tests, as ethanol vapor produced by fermentation can dissolve in assimilation tests and cause false-positive results [90–98].

1. Prepare basal medium and sterilize at 121 °C for 20 min.
2. Add filter-sterilized sugar at the concentration of 2% (w/v) to the medium aseptically.
3. Pour the medium to the test tubes.

4. Insert inverted single sterile Durham's tube in each and close the lid.
5. Incubate at 20 °C for 7 days.
6. Gas accumulation in Durham's tube is indicative of a positive result.

Casein Hydrolysis

Caseinase is an exoenzyme that is secreted outside of the cells into the surrounding media. It has the ability to break down milk protein, called casein, into small peptides and individual amino acids for their energy use or as building material. The hydrolytic reaction creates a clear zone around the cell as the casein protein is converted to soluble and transparent end products, like small chains of amino acids, dipeptides, and polypeptides. This test can be used to identify some species of yeast and fungi like *Citeromyces matritensis, Aspergillus dimorphicus, A. ochraceus, Fusarium illudens, F. moniliforme, F. solani, Penicillium citrinum, P. brevicompactum, P. chrysogenum, P. fellutanum,* and *P. waksmanii* [99, 100].

1. Prepare Petri plates with autoclaved skim milk agar or casein agar in sterile conditions.
2. Inoculate fungal mycelia onto the center of the plate and incubate at 20 °C for 14 days.
3. Examine for the presence of a clear zone.
4. The appearance of a clear zone around the fungal colony is the positive result.

Cellulose Hydrolysis

Cellulase is produced chiefly by fungi, bacteria, and protozoans that catalyze cellulolytic activity. In the most familiar case of cellulose activity, the enzyme complex breaks down cellulose to β-glucose. This type of cellulase is produced by symbiotic bacteria and fungi in the ruminating chambers of herbivores. Three hydrolytic enzymes, such as five endo-1,4-β-glucanases, one exo-1,4-β-glucanase, and one or several 1,4-β-glucosidases, are involved in cellulolysis. This can be used to identify *Sporotrichum poulverulentum, Trichoderma viride, Aspergillus niger, Peziza* sp.*, Fusarium* sp., and *Penicillium* sp [101–103].

1. Prepare basal salt medium with the addition of 1% (w/v) microgranular cellulose (sigma) and 1.2% (w/v) agar.

2. Autoclave at 121 °C for 30 min, disperse into Petri dishes.
3. Inoculate fungi onto the center of the agar plates.
4. Incubate at 20 °C for 14 days.
5. Appearance of a clear zone around the fungal colony is a positive result.

Cyclohexamide Resistance

This technique is to confirm the presence of a possible dimorphic fungus or dermatophytes. Determination of the resistance of isolates to cyclohexamide is useful when screening cultures for *Blastomyces dermatitidis, Coccidioides immitis, Epidermophyton floccosum, Histoplasma capsulatum, Microsporum* sp., *Paracoccidioides brasiliensis, Sporothrix schenckii,* and *Trichophyton* sp. All these fungi will grow in the presence of cycloheximide at 30 °C or less, while fungal species such as *Absidia, Aspergillus, Mucor, Rhizopus, Scedosporium* and many more are inhibited by cyclohexamide [104–106].

1. Inoculate a small portion of mold colony onto mycosel agar with cyclohexamide and Sabouraud agar plates.
2. Incubate at 30 °C for 7–10 days.
3. Observe for the growth of the colonies on plate.
4. Growth on Sabouraud agar and mycosel agar indicates resistant, and growth on Sabouraud agar but no growth on mycosel agar indicates sensitive. Repeat the test if there is no growth on Sabouraud agar or mycosel agar.

Fatty Acid Esterase Activity

Fatty acid esters are cleaved by enzymes including esterase, cutinase, and lipases, which can release free fatty acids from several sources, including lipids, phospholipids, sterol esters, waxes, cutin, and suberin. In this method, *Rhizopus circinans, R. microspores, Fusarium oxysporum, R. boreas, R. thermosus, R. usamii, R. stolonifer, R. fusiformis,* and *Pseudomonas cepacia* are screened by its enzyme activity [107–112].

1. Prepare basal medium with bromocresol purple as indicator at pH 5.4.
2. Prepare 10% tween solution.

3. Autoclave the medium and tween solution separately at 121 °C for 30 min.
4. Add the tween solution in a ratio of 1:9 by volume to the cooled medium.
5. Pour the medium into petri dishes.
6. Inoculate the fungal mycelia onto the center of the medium.
7. Incubate at 20 °C for 14 days.
8. Change in the color of the medium to purple is a positive result.

Gelatin Hydrolysis

Gelatin, a protein derived from collagen, is too large to enter the cell as a whole and, hence, the exoenzyme gelatinase cleaves gelatin to polypeptides and then further degrades polypeptides to amino acids, which are taken up and utilized by the fungi. This test can be performed to differentiate between fungi that produce gelatinase and those that do not produce the enzyme [113–116].

1. Inoculate culture into a nutrient gelatin tubes with a straight needle.
2. Incubate for 48 h at 22 °C.
3. Observe for liquification.
4. Liquification of the gelatin is a positive test for gelatin hydrolysis.

β-Glucosidase Activity

The hydrolysis of cellobiose to glucose is achieved by β-glucosidase. β-glucosidase is predominantly a cell associated or intracellular enzyme in many fungi. This enzyme is ubiquitous among cellulolytic fungi producing hydrolytic endonucleases or cellobiohydrolases. Activity of β-glucosidase can be detected by growth of the test fungi on agar containing esculin (6,7-dihydroxycoumarin 6-glucosidase) or arbutin (hydroquinone-β-D-glucopyranoside) as the sole carbon source. Splitting of the substrate by the enzyme yields glucose and a coumarin product that react with iron sulfate to produce a black color in the growth medium [117–120].

1. Sterilize cellulolysis basal medium (CBM) supplemented with 0.5% (w/v) esculin or arbutin and 1.6% (w/v) agar.
2. Add 1 mL of sterile 2% (w/v) aqueous ferric sulfate solution per 100 mL of CBM agar.

3. Pour the medium to the Petri dishes.
4. Inoculate the fungal culture into the medium.
5. Incubate at 25 °C in darkness for 5 days.
6. Development of black color in the medium indicates the production of β-glucosidase.

Lipase Activity

Lipase catalyzes the hydrolysis of the ester bonds of triacylglycerols, thereby releasing free fatty acids. Lipolytic activity has been associated with survival and pathogenicity of several fungal species. It can be used to detect species of *Malassezia* and *Candida* [121–124].

1. Prepare Sabouraud agar medium and sterilize at 121 °C for 30 min.
2. Add 0.1% (v/v) n-tributylin to the medium and pour into the Petri dishes.
3. Inoculate fungal mycelia onto the surface of the medium and incubate at 20 °C for 7 days.
4. Occurrence of clearance in the medium is a positive result.

Urease Activity

Urea may be hydrolyzed by some fungi with the help of urease so that the ammonia that is liberated can be used as a nitrogen source. Phenol red indicator is added to the broth or solid medium at pH 8.4, and the media will turn from red color to pink color owing to ammonia production. Urea hydrolysis is primarily used to distinguish *Trichophyton mentagrophytes, T. sulphureum, T. tonsurans,* and *T. verrucosum* from *T. gallianae* and *T. rubrum* and to identify *Cryptococcus neoformans* [125–128].

1. Prepare Christensen's urea agar medium and sterilize at 121 °C for 30 min.
2. Pour the medium into Petri dishes and inoculate fungal mycelium onto the center of the medium.
3. Incubate the inoculated tubes at 25–30 °C for 8 days.
4. Examine the slants every 2–3 days for color change.
5. Change in the color of the medium to pink or purple can be taken as a positive result, while a change to orange or yellow is considered as a negative result.

Automated Biochemical Methods

Conventional methods of identifying fungi require as long as 14–28 days for the completion of all the biochemical tests. In recent years, it has become important to develop rapid, automated, and modern methods for species identification and strain differentiation in fungi [129–138]. In this context, the newer miniaturized fungal identification systems such as the API 20C Yeast Identification System (API Analytab Products, Plainview, NY), Biomerieux Vitek Yeast Biochemical Card (Hazelwood, MO), and Abbott Yeast Identification System (Abbott Laboratories, Irving, Texas) provide biochemical testing in 3–7 days with an acceptable degree of reliability [98, 139]. The biochemical tests designed in such automated systems are those routinely followed in conventional identification systems, except the test for assimilation of mixture of ρ-hydroxybenzoic acid and protocatechuic acid, as in the Abbott Yeast Identification System, which helps in the identification of *Candida parapsilosis* and certain other yeasts [140]. These three fungal identification systems are based on modifications of the classic auxanographic technique of carbohydrate assimilation [141, 142]. They are easy to use, require less preparation of reagents, and offer a significant saving of time compared to conventional tube tests and other currently available biochemical approaches. These commercially available kits have been widely used to identify filamentous fungi and yeasts [143–145].

API 20C Yeast Identification System

The commercially available API 20C Yeast Identification System is easy to use, requires less preparation of reagents, and has been widely used to identify both yeast and filamentous fungi. The kit consists of 20 microtubes containing dehydrated substrates in which 19 assimilation tests are performed. After inoculation and incubation, the reactions are interpreted by comparison to growth controls and use of a reference identification table that is provided with each kit. The system is based on modifications of the classic auxanographic technique of carbohydrate

assimilation. When the yeast is able to assimilate a particular carbohydrate accompanied with a color change, the system must also be supplemented with morphological studies. Germ tube tests should be done in conjunction with them as a means of obtaining a more complete profile of the yeast cells being identified (see Note 17) [143–145].

1. Melt the basal medium in the ampoules by placing them in an autoclave for 2 min.
2. Place the ampules in a water bath at 50 °C and allow to cool.
3. Dispense 20 mL of water into the incubation tray.
4. Place the strips into incubation tray.
5. Open the ampules according to manufacturer's instructions and inoculate the molten medium with an applicator stick that has touched one or two colonies of >2 mm diameter.
6. Inoculate the strip (20 cupules of approximately 200 µl each) using a Pasteur Pipette and then place the lid on the tray.
7. Incubate the trays at 28–30 °C for 24–72 h and check for growth.
8. Record and compare the results with the identification table to identify yeast (see Note 18).

Biomerieux Vitek Yeast Biochemical Card

The disposable plastic Vitek 30-well Yeast Biochemical Card contains 26 conventional biochemical tests and 4 negative controls. The card is one of several testing packages used with the Vitek System, which includes a programmed computer, a reader-incubator unit, a filling module, a sealer module, and a printer. Identification is generally based on conversion of biochemical test results into nine-digit biocodes that will be analyzed by the Vitek computer. This automated system has been used in the identification of clinically significant yeasts, particularly *Candida albicans* [146–151].

1. Use 1–3 colonies to prepare yeast suspension in 1.8 mL saline tubes (see Note 19).
2. Adjust the suspension to a McFarland no. 2 standard using colorimeter (46–56% transmittance, 450 nm).

3. Label the yeast cards with a marker and place in the filling stand with a transfer tube kept in the yeast suspension.
4. Inoculate the cards via the filling module.
5. Seal the cards via the sealer module and incubate at 30 °C for 24–48 h, depending on the readings provided by the instrument.
6. Vitek computer converts biochemical test results into nine-digit biocodes in printed form.
7. Most of the results are obtained after 24 h and a few isolates may require additional incubation.
8. Identification can be accepted if the printed result had a reliability of more than 85%.

    If the reliability is less than 85%, an API 20C Yeast Identification System or any manual biochemical tests in addition to morphological features can be used to identify the yeast (see Note 18).

Abbott Yeast Identification System

The Abbott Yeast Identification System is an instrumental method based on matrix analysis of 19 biochemical reactions in addition to the germ tube test for identifying yeast. It is a disposable plastic unit of 20 chambers containing lyophilized biochemical substrates such as arabinose, cellobiose, dulcitol, erythritol, galactose, glucose, inositol, lactose, maltose, melibiose, melizitose, nitrate, ρ-hydroxybenzoic acid and protocatechuic acid, raffinose, rhamnose, sucrose, trehalose, urea, xylose, and negative control. This rapid and automated system has now been widely in use for the identification of yeasts within 20–24 h after test initiation and offers significant saving in time compared to conventional biochemical and other currently available manual systems [152–157].

1. Isolate yeast pure colonies from Sabouraud agar plates.
2. Prepare the inocula and incubate for 24–48 h at 30 °C.
3. Select colonies of a test organism and suspend in 0.05% Noble agar using a sterile inoculating loop or cotton-tipped applicator (see Note 20).
4. Mix the suspension on a vortex mixer and deliver 0.2 mL into each chamber of the Abbott IDS cartridge.
5. Seal the cartridge with a polyester film (see Note 21).
6. Take the initial optical density of each reaction chamber with the MS-2, which will be automatically stored in the computer memory.
7. Incubate the cartridges off-line at 30 ° C for 22–25 h.
8. Reinsert the cartridge into the MS-2 for final reading.
9. Enter the results of several other morphological and additional tests (see Note 22).
10. Take the MS-2 data printout with a record of positive and negative biochemical tests.
11. Identify the yeast with up to five species listed in the printout in descending order of likelihood based on the percent likelihood value for each species.

Notes

1. Wear gloves, goggles, and lab coat. Nitrile gloves are suggested when working with solvents and acids. Avoid contact and inhalation of dyes and chemicals. Reagents should be prepared in a fume hood [72].
2. Addition of glycerol to KOH solution will prevent crystallization from occurring in the solution, thus enhancing the shelf life of this reagent. It will also help in preserving KOH preparation for a couple of days.
3. The components have to be mixed in a brown bottle and allowed to stand for 1 month before use. The stain must be stored at 4 °C; otherwise the components may degrade.
4. Under fume hood, dissolve pararosaniline with distilled water and slowly bubble sulfur dioxide gas through the solution until solution color begins to change. Stopper well and store in the dark for one or two days. Add activated charcoal and shake for about a minute if solution is not clear; filter and store at 0–5 °C.
5. Chromic acid is corrosive to skin and mucous membranes, highly toxic, and carcinogen. Aniline and xylene are moderate skin and severe eye irritants, possible carcinogens, and combustible liquids.

6. Mix stain in test tube and heat it in a water bath for 2 min. Liquid becomes almost black and syrupy. Dilute with 100 mL of 50% alcohol and let stand for 24 h and filter. Again dilute to 1:4 with tap water for use.
7. Periodic acid solution and the stock of periodic acid (a white powder) should be kept in dark bottles.
8. Add HCl to distilled water in a brown bottle before adding sodium metabisulphite, cool at 5 °C, and filter. Then add 20 mL diluted HCl (83 mL concentrated HCl/1,000 mL distilled water) and cool to 25 °C. Add 1 g sodium bisulphite and store in screw-top bottle in dark for 2 days. Add 0.5 g activated charcoal, shake intermittently for 1 h, and filter. Store in dark-colored, tightly closed bottle in refrigerator (expiration 5 years); pour into a Coplin jar for further use. Solution may be reused until it turns pink, at which time it must be discarded.
9. Silver nitrate is toxic, and skin contact should be avoided; methenamine is a flammable solid and an irritant. Any spill should be mopped up immediately with water.
10. Neutralize the ammonium silver solution immediately after use, as it can be explosive when allowed to dry.
11. To avoid splashing while mixing sulphuric acid with glacial acetic acid, place the jar in a plastic tub filled with cold water.
12. For more precision, stains like methylene blue or Parker blue-black fountain ink or chlorazol may be used along with KOH.
13. If background fluorescence is too bright for fungi to be distinguished, it may be quenched with alum hematoxylin for 1 min or potassium permanganate for 1 min. Quench immediately before the final dehydration step. This should be done with caution because it may reduce fungal fluorescence.
14. An epifluorescent microscope equipped with a mercury vapor lamp and either an ultraviolet (UV) or a blue-violet (BV) excitation filter to achieve radiation on the slide below 412 nm should be used, because the maximum absorbance of CFW is 347 nm. A microscope with selective filters that will prevent radiation below 490 nm should not be used for CFW [56].
15. Glucose, sucrose, lactose, arabinose, galactose, xylose, mannose, dulcitol, ethanol, etc.
16. Potassium nitrate, sodium nitrate, amino acids, urea, glycine, ammonium sulfate, asparagine, peptone, aliphatic amines, etc.
17. Germ tube tests and morphological studies should be included, as API 20C Yeast Identification System does not include rhamnose and urea. The API yeast profiles sometimes give many different yeast identifications for an individual isolate; this warrants supplemental tests.
18. For quality control, include known isolates of *Torulopsis glabrata*, *Candida albicans,* and *Cryptococcus laurentii.*
19. Heavy encapsulated yeasts and isolates with extensive mycelial growth are difficult to suspend [146].
20. Prepare a slightly turbid suspension that matches with the turbidity of 0.5 McFarland standard.
21. To eliminate the adverse effects of volatile metabolic end products of adjacent reactions, if any.
22. To get a comprehensive record of all the results, including morphological observations of hyphae, chlamydospores, arthroconidia, germ tube formation, capsules, and phenol oxidase activity [156].

References

1. Jimenez L (2007) Microbial diversity in pharmaceutical product recalls and environments. PDA J Pharm Sci Tech 61:383–399
2. Meyer SAM, Payne RW, Yarrow D (1988) Candida. In: Kurtzman CP, Fell JW (eds) The yeasts: a taxonomic study. Elsevier Science Publishers, Amsterdam, The Netherlands, pp 476–477
3. Borman AM, Linton CJ, Miles SJ, Johnson EM (2008) Molecular identification of pathogenic fungi. J Antimicrob Chemother 61:i7–i12
4. Williamson PR (1994) Biochemical and molecular characterization of the diphenol oxidase of *Cryptococcus neoformans*: identification as a laccase. J Bacteriol 176:656–664
5. Guo LD, Huang GR, Wang Y, He WH, Zheng WH, Hyde KD (2003) Molecular identification of white

morphotype strains of endophytic fungi from *Pinus tabulaeformis*. Mycol Res 107:680–688

6. Hinrikson HP, Hurst SF, Lott TJ, Warnock DW, Morrison CJ (2005) Assessment of ribosomal large-subunit D1-D2, internal transcribed spacer 1, and internal transcribed spacer 2 regions as targets for molecular identification of medically important *Aspergillus* species. J Clin Microbiol 43:2092–2103

7. Flórez AB, Álvarez-Martín P, López-Díaz TM, Mayo B (2007) Morphotypic and molecular identification of filamentous fungi from Spanish blue-veined Cabrales cheese, and typing of *Penicillium roqueforti* and *Geotrichum candidum* isolates. Int Dairy J 17:350–357

8. Nilsson RH, Kristiansson E, Ryberg M, Hallenberg N, Larsson KH (2008) Intraspecific ITS variability in the kingdom Fungi as expressed in the International Sequence Databases and its implications for molecular species identification. Evol Bioinform Online 4:193–201

9. Rodriguez-Tudela JL, Diaz-Guerra TM, Mellado E, Cano V, Tapia C, Perkins A et al (2005) Susceptibility patterns and molecular identification of *Trichosporon* species. Antimicrob Agents Chemother 49:4026–4034

10. Cooke DEL, Drenth A, Duncan JM, Wagels G, Brasier CM (2000) A molecular phylogeny of *Phytophthora* and related Oomycetes. Fungal Genet Biol 30:17–32

11. Bertini L, Agostini D, Potenza L, Rossi I, Zeppa S, Zambonelli A et al (1998) Molecular markers for the identification of the ectomycorrhizal fungus, *Tuber borchii*. New Phytol 139:565–570

12. Galagan JE, Calvo SE, Borkovich KA, Selker EU, Read ND, Jaffe D et al (2003) The genome sequence of the filamentous fungus *Neurospora crassa*. Nature 422:859–868

13. Brookman JL, Mennim G, Trinci APJ, Theodorou MK, Tuckwell DS (2000) Identification and characterization of anaerobic gut fungi using molecular methodologies based on ribosomal ITS1 and 18S rRNA. Microbiology 146:393–403

14. Ciardo DE, Schär G, Böttger EC, Altwegg M, Bosshard PP (2006) Internal transcribed spacer sequencing versus biochemical profiling for identification of medically important yeasts. J Clin Microbiol 44:77–84

15. Hussain Z, Martin A, Youngberg GA (2001) *Blastmyces dermatitides* with large yeast forms. Arch Pathol Lab Med 125:663–664

16. Chandler FW, Watts JC (1987) Pathologic diagnosis of fungal infections. ACSP Press, Chicago, pp 193–263

17. Haque AK, McGinnis MR (2008) Dail and Hammer's Pulmonary Pathology. Springer, New York

18. Woods GL, Walker DH (1996) Detection of infection or infectious agents by use of cytologic and histologic stains. Clin Microbiol Rev 9:382–404

19. Ryan KJ, Ray CG (2004) Sherris medical microbiology. McGraw-Hill, New York

20. Kim SO (2003) Molds identification. Kor J Med Mycol 8:97–102

21. Nag DR, Chatterjee S, Chatterjee S, Khan M (2002) Role of potassium hydroxide mount in rapid diagnosis of fungal corneal ulcers. J Ind Med Assoc 100:18–20

22. Shearer CA, Langsam DM, Longcore JE (2004) Fungi in freshwater habitats. In: Mueller GM, Bills GF, Foster MS (eds) Biodiversity of fungi, inventory monitoring methods. Elsevier Science Publishers, New York, pp 520–521

23. Parija SC, Prabhakar PK (1995) Evaluation of lacto-phenol cotton blue for wet mount preparation of feces. J Clin Microbiol 33:1019–1021

24. Larone DH (2002) Medically important fungi: a guide to identification. ASM Press, Washington, D.C

25. Chapin KC (2007) Principles of stains and media. In: Murray PR, Baron EJ, Landry ML, Jorgensen JH, Pfaller MA (eds) Manual of clinical microbiology. ASM Press, Washington, D.C., pp 182–191

26. Kern ME, Blevins KS (1997) Laboratory procedures for fungal culture and isolation in medical mycology. F.A. Davis, Philadelphia

27. Gurr E (1966) Rational use of dyes in biology. Williams & Wilkins, Baltimore, pp 276–277

28. Humason GL (1967) Animal tissue techniques. W.H. Freeman, San Francisco

29. Murray PR, Baron EJ, Pfaller MA, Tenover FC, Yolken RH (1999) Manual of clinical microbiology. ASM Press, Washington, D.C

30. Brucker MC (1986) Gram staining a useful laboratory technique. J Nurse Midwifery 31:156–158

31. Clark G (1981) Staining procedure. Williams & Wilkins, Baltimore

32. Clarridge JE, Mullins JM (1987) Microscopy and staining. In: Howard BJ, Klass J, Weissfeld AS, Tilton RC (eds) Clinical and pathogenic microbiology. C.V. Mosby, St. Louis, pp 87–103

33. Macher AM, Shelhamer J, MacLowery JD, Parker M, Masur H (1983) *Pneumocystis carinii* identified by Gram stain of lung imprints. Ann Intern Med 99:484–485

34. Bottone EJ (1980) *Cryptococcus neoformans*: pitfalls in diagnosis through evaluation of Gram-stained smears of purulent exudates. J Clin Microbiol 12:790–791

35. Felegie TP, Pasculle AW, Dekker A (1984) Recognition of *Pneumocystis carinii* by Gram stain in impression smears of lung tissue. J Clin Microbiol 2:1190–1191

36. Mahan CT, Sale GE (1978) Rapid methenamine silver stain for *Pneumocystis* and fungi. Arch Pathol Lab Med 102:351–352

37. Ward EWB, Ciurysek KW (1961) Somatic mitosis in a basidiomycete. Can J Bot 39:1497–1503

38. Wilson AD (1992) A versatile Giemsa protocol for permanent nuclear staining of fungi. Mycologia 84:585–588

39. Lillie RD (1965) Histopathologic techniques and practical histochemistry. McGraw-Hill, New York

40. Lillie RD (1977) H. J. Conn's biological stains. Williams & Wilkins, Baltimore, pp 289–291

41. Domingo J, Waksal HW (1984) Wright's stain in rapid diagnosis of *Pneumocystis carinii*. Am J Clin Pathol 81:511–514

42. Pattengale P (2005) Task for the veterinary assistants. Lippincott Williams & Wilkins, Baltimore, pp 258–260

43. Kiernan JA (2008) Histological and histochemical methods: theory and practice. Scion, Bloxham, UK

44. Lillie RD, Pizzolato P, Donaldson PT (1976) Nuclear stains with soluble metachrome mordant lake dyes - the effect of chemical endgroup blocking reactions and the artificial introduction of acid groups into tissues. Histochemistry 49:23–35

45. Llewellyn BD (2009) Nuclear staining with alum-hematoxylin. Biotech Histochem 84:159–177

46. Puchtler H, Meloan SN, Waldrop FS (1986) Application of current chemical concepts to metal-hamatein and -brazilein stains. Histochemistry 85:353–364

47. Bancroft JD, Stevens A (1982) Theory and practice of histological techniques. Churchill Livingstone, London, UK, pp 188–201

48. Kronvall G, Myhre E (1977) Differential staining of bacteria in clinical specimens using acridine orange buffered at low pH. Acta Pathol Microbiol Scand Sect B 85:249–254

49. De Brauwer E, Jacobs J, Nieman F, Bruggeman C, Drent M (1999) Test characteristics of acridine orange, Gram and May-Grunwald-Giemsa stains for enumeration of intracellular organisms in bronchoalveolar lavage fluid. J Clin Microbiol 37:427–429

50. Lauer BA, Reller LB, Mirret S (1981) Comparison of acridine orange and Gram stains for detection of microorganisms in cerebrospinal fluid and other clinical specimens. J Clin Microbiol 14:201–205

51. Culling CFA (1974) Handbook of histopathological and histochemical techniques. Butterworth, London, UK

52. Mote RF, Muhm RL, Gigstad DC (1975) A staining method using acridine orange and auramine O for fungi and mycobacteria in bovine tissue. Stain Technol 50:5–9

53. Pickett JP, Bishop CM, Chick EW, Baker RD (1960) A simple fluorescent stain for fungi: selective staining of fungi by means of a fluorescent method for mucin. Am J Clin Pathol 34:197–202

54. Drury RAB, Wallington EA (1980) Carleton's histological technique. Oxford University Press, Oxford, UK

55. Luna L (1980) Manual of histologic staining methods. Armed Forces Institute of Pathology, Washington, D.C., pp 228–229

56. Hageage GJ, Harrington BJ (1984) Use of calcofluor white in clinical mycology. Lab Med 15:109–112

57. Monheit JE, Cowan DF, More DG (1984) Rapid detection of fungi in tissue using calcofluor white and fluorescence microscopy. Arch Pathol Lab Med 108:616–618

58. McGowna KL (1987) Practical approaches to diagnosing fungal infections in immunocompromised patients. Clin Microbiol Newsletter 9:33–36

59. Baselski VS, Robinson MK, Pifer LW, Woods DR (1990) Rapid detection of *Pneumocystis carinii* in bronchoalveolar lavage samples by using Cellufluor staining. J Clin Microbiol 28:393–394

60. Totty BA (2002) Mucins. In: Bancroft JD, Gamble M (eds) Theory and practice of histological techniques. Churchill Livingstone, Edinburgh, pp 163–200

61. Prophet EB, Mills B, Arrington JB, Sobin LH (1992) Laboratory methods in histotechnology. Armed Forces Institute of Pathology, Washington, D.C

62. Mikel UV (1994) Advanced laboratory methods in histology and pathology. Armed Forces Institute of Pathology, Washington, D.C

63. Mallory FB (1961) Pathological techniques. Hafner Publishing Co., New York

64. Sheehan D, Hrapchak B (1980) Theory and practice of histopathology. Battelle Press, Columbus, OH

65. Jackson JA, Kaplan W, Kaufman L (1983) Development of fluorescent-antibody reagents for demonstration of *Pseudallescheria boydii* in tissue. J Clin Microbiol 18:668–673

66. Grocott RG (1955) A stain for fungi in tissue sections and smears. Am J Clin Pathol 25:975–979

67. Schumann GB, Swensen JJ (1991) Comparison of Papanicolaou's stain with the Gomori methenamine silver (GMS) stain for the cytodiagnosis of *Pneumocyctis catrinii* in bronchoalveolar lavage (BAL) fluid. Am J Clin Pathol 95:583–586

68. Hayashi I, Tome Y, Shimosato Y (1989) Thiosemicarbazide used after periodic acid makes methenamine silver staining of renal glomerular basement membranes faster and cleaner. Stain Technol 64:185–190

69. Lazcano O, Speights VO Jr, Stickler JG, Bilbao JE, Becker J, Diaz J (1993) Combined histochemical stains in the differential diagnosis of *Cryptococcus neoformans*. Mod Pathol 6:80–84

70. Witebsky FG, Andrews JWB, Gill VJ, MacLowry JD (1988) Modified toluidine blue O stain for *Pnemocystis corinii*: further evaluation of some technical factors. J Clin Microbiol 26:774–775

71. Gosey LL, Howard RM, Witebsky FG, Ognibene FP, Wu TC, Gill VJ et al (1985) Advantages of a modified toluidine blue O stain and bronchoalveolar lavage for the diagnosis of *Pneumocystis carinii* pneumonia. J Clin Microbiol 22:803–807

72. Crookham JN, Dapson RW (1991) Hazardous chemicals in the histopathology laboratory: regulations, risks, handling and disposal. Anatech Ltd., Battle Creek, MI

73. Furtado JS (1970) Alcoholic toluidine blue: a rapid method for staining nuclei in unfixed mycetozoa and fungi. Mycologia 62:406–407

74. Paterson RRM, Bridge PD (1994) Biochemical techniques for filamentous fungi. CAB International, Wallingford, UK

75. Wickerham LJ (1943) A simple technique for the detection of melibiose-fermenting yeasts. J Bacteriol 46:501

76. Wickerham LJ (1946) A critical evaluation of the nitrogen assimilation tests commonly used in the classification of yeasts. J Bacteriol 52:293

77. Wickerham LJ, Burton KA (1948) Carbon assimilation tests for the classification of yeasts. J Bacteriol 56:363

78. Schwarz P, Lortholary O, Dromer F, Dannaoui E (2007) Carbon assimilation profiles as a tool for identification of zygomycetes. J Clin Microbiol 45:1433–1439

79. Hobbie EA, Watrud LS, Maggard S, Shiroyama T, Rygiewicz PT (2003) Carbohydrate use and assimilation by litter and soil fungi assessed by carbon isotopes and BIOLOG® assays. Soil Biol Biochem 35:303–311

80. Steadham JE, Geis PA, Simmank JL (1986) Use of carbohydrate and nitrate assimilations in the identification of dematiaceous fungi. Diagn Microbiol Infect Dis 5:71–75

81. Middelhoven WJ, Hoog GS, Notermans S (1989) Carbon assimilation and extracellular antigens of some yeast-like fungi. Antonie Van Leeuwenhoek 55:165–175

82. Murray IG (1968) Some aspects of the biochemical differentiation of pathogenic fungi: a review. J Gen Microbiol 52:213–221

83. Sangtiean T, Schmidt S (2002) Growth of subtropical ECM fungi with different nitrogen sources using a new floating culture technique. Mycol Res 106:74–85

84. Watkinson S, Bebber D, Darrah P, Fricker M, Tlalka M (2006) The role of wood decay fungi in the carbon and nitrogen dynamics of the forest floor. In: Gadd GM (ed) Fungi in biogeochemical cycles. Cambridge University Press, Cambridge, UK, pp 151–181

85. Crawford NM, Arst HN (1993) The molecular genetics of nitrate assimilation in fungi and plants. Annu Rev Genet 27:115–146

86. Krappmann S, Braus GH (2005) Nitrogen metabolism of Aspergillus and its role in pathogenicity. Med Mycol 43:S31–S40

87. Gorfer M, Blumhoff M, Klaubauf S, Urban A, Inselsbacher E, Bandian D, et al. (2011) Community profiling and gene expression of fungal assimilatory nitrate reductases in agricultural soil. ISME J. Accessed 12 May 2011; doi: 10.1038/ismej.2011.53

88. Keller G (1996) Utilization of inorganic and organic nitrogen sources by high-subalpine ectomycorrhizal fungi of Pinus cembra in pure culture. Mycol Res 100:989–998

89. Ye ZH, Garrad RC, Winston MK, Bhattacharjee JK (1991) Use of α-aminoadipate and lysine as sole nitrogen source by Schizosaccharomyces pombe and selected pathogenic fungi. J Basic Microbiol 31:149–156

90. Bhavan PS, Rajkumar R, Radhakrishnan S, Seenivasan C, Kannan S (2010) Culture and identification of Candida albicans from vaginal ulcer and separation of enolase on SDS-PAGE. Int J Biol 2:84–93

91. Freydiere AM, Robert R, Ploton C, Marot-Leblond A, Monerau F, Vandenesch F (2003) Rapid identification of Candida glabrata with a new commercial test, GLABRATA RTT. J Clin Microbiol 41:3861–3863

92. Felek S, Asci Z, Kilic SS, Yilman M, Kokcamm I (1989) Yeasts and yeast like fungi as causative agents in diarrhea. J Islam Acad Sci 2:182–184

93. Fontes CO, Carvalho MAR, Nicoli JR, Hamdan JS, Mayrink W, Genaro O et al (2005) Identification and antimicrobial susceptibility of micro-organisms recovered from cutaneous lesions of human American tegumentary leishmaniasis in Minas Gerais, Brazil. J Med Microbiol 54:1071–1076

94. Bonfante R, Barroeta S (1966) Development and evaluation of a rapid identification test for Candida albicans. Mycopathologia 34:33–39

95. Sandven P (1990) Laboratory identification and sensitivity testing of yeast isolates. Acta Odontol Scand 48:27–36

96. Waltimo TMT, Sirén EK, Torkko HLK, Olsen I, Haapasalo MPP (2003) Fungi in therapy-resistant apical periodontitis. Int Endodont J 30:96–101

97. Land GA, Salkin IF, Zaatari ME, McGinnis MR, Hashem G (1991) Evaluation of the Baxter-MicroScan 4-hour enzyme-based yeast identification system. J Clin Microbiol 29:718–722

98. Huppert M, Harper G, Sun SH, Delanerolle V (1975) Rapid methods for identification of yeasts. J Clin Microbiol 2:21–34

99. Merheb CW, Cabral H, Gomes E, Da-Silva R (2007) Partial characterization of protease from a thermophillic fungus, Thermoascus auranthacus, and its hydrolytic activity on bovine casein. Food Chem 104:127–131

100. Rodarte MP, Dias DR, Vilela DM, Schwan RF (2011) Proteolytic activities of bacteria, yeast and filamentous fungi isolated form coffee fruit (Coffea arabica L.). Acta Scientiarum Agron 33:457–464

101. Eriksson K (1978) Enzyme mechanisms involved in cellulose hydrolysis by the rot fungus Sporotrichum pulverulentum. Biotechnol Bioengineer 20:317–332

102. Hankin L, Anagnostakis SL (1977) Solid media containing carboxymethyl cellulose to detect Cx cellulose activity of micro-organisms. J Gen Microbiol 98:109–115

103. Tomme P, Warren RA, Gilkes NR (1995) Cellulose hydrolysis by bacteria and fungi. Adv Microb Physiol 37:1–81

104. Ali-Shtayeh MS, Jamous RMF, Abu-Ghdeib SI (1999) Ecology of cycloheximide-resistant fungi in field soils receiving raw city wastewater or normal irrigation water. Mycopathologia 144:39–55

105. Crouzet M, Begueret J (1980) A new mutant form of the ribosomal protein L21 in the fungus Podospora anserina: identification of the structural gene for this protein. Mol Gen Genet 180:177–183

106. Bagy MM, El-Shanawany AA, Abdel-Mallek AY (1998) Saprophytic and cycloheximide resistant fungi isolated from golden hamster. Acta Microbiol Immunol Hung 45:195–207

107. Caldwell BA, Castellano MA, Griffiths RP (1991) Fatty acid esterase production by ectomycorrhizal fungi. Mycology 83:233–236

108. Brunke S, Hube B (2006) MfLIP1, a gene encoding an extracellular lipase of the lipid-dependent fungus *Malassezia furfur*. Microbiology 152:547–554

109. Byrde RJW, Fielding AH (1955) Studies on the acetylesterase of *Sclerotinia laxa*. Biochemistry 61:337–341

110. Rapp P, Backhaus S (1992) Formation of extracellular lipases by filamentous fungi, yeasts and bacteria. Enzyme Microbiol Technol 14:938–943

111. Shishiyama J, Araki F, Akai S (1970) Studies on cutin-esterase II. Characeteristics of cutin-esterase from *Botrytis cinerea* and its activity on tomato cutin. Plant Cell Physiol 11:937–945

112. Singh R, Gupta N, Goswami VK, Gupta R (2006) A simple activity staining protocol for lipases and esterases. Appl Microbiol Biotechnol 70:679–682

113. Valiente C, Quesada E (1991) Morphologic and physiologic characteristics of Costa Rica pathogenic fungi (Dermatiaceae). Rev Biol Trop 39:103–106

114. Kitancharoen N, Hatai K (1998) Some biochemical characteristics of fungi isolated from salmonid eggs. Mycoscience 39:249–255

115. Reddy NG, Ramakrishna DPN, Gopal SVR (2011) A morphological, physiological and biochemical studies of marine *Streptomyces rochei* (MTCC 10109) showing antagonistic activity against selective human pathogenic microorganisms. Asian J Biol Sci 4:1–14

116. Abrusci C, Martín-González A, Amo AD, Catalina F, Collado J, Platas G (2005) Isolation and identification of bacteria and fungi from cinematographic films. Int Biodeterior Biodegrad 56:58–68

117. Pointin SB (1999) Qualitative methods for the determination of lignocellulolytic enzyme production by tropical fungi. Fungal Divers 2:17–33

118. Hayanko K, Tubaki K (1985) Origin and properties of β-glucosidase activity of tomato-field soil. Soil Biol Biochem 17:553–557

119. Zanoelo FF, Moraes MLT, Terenzi HF, Jorge JA (2004) β-glucosidase activity from the thermophilin fungus *Scytalidium thermophilium* is stimulated by glucose and xylose. FEMS Microbiol Letters 240:137–143

120. Breuil C, Mayers P, Saddler JN (1986) Substrate conditions that influence the assays used for determining the β-glucosidase activity of cellulolytic microorganisms. Biotechnol Bioengineer 28:1653–1656

121. Kakde RB, Chavan AM (2011) Effect of carbon, nitrogen, sulfur, phosphorus, antibiotic and vitamin sources on hydrolytic enzyme production by storage fungi. Recent Res Sci Technol 3:20–28

122. Iftikhar T, Niaz M, Hussain Y, Abbas SQ, Ashraf I, Zia MA (2010) Improvement of selected strains through gamma irradiation for enhanced lipolytic potential. Pak J Bot 42:2257–2267

123. Kakde RB, Chavan AM (2011) Extracellular lipase enzyme production by seed-borne fungi under the influence of physical factors. Int J Biol 3:94–100

124. Mateos JC, Ruiz K, Rodriguez JA, Cordova J, Baratti J (2007) Mapping substrate selectivity of lipases from thermophilic fungi. J Mol Catal B: Enzymatics 49:104–112

125. Philpot C (1967) The differentiation of *Trichophyton mentagrophytes* from *T. rubrum* by a simple urease test. Sabouraudia 5:189

126. Shepard MC, Lunceford CD (1970) Urease color test medium U-9 for the detection and identification of "T" mycoplasmas in clinical material. Appl Microbiol 20:539–543

127. Rosenthal SA, Sokolsky H (1965) Enzymatic studies with pathogenic fungi. Int J Dermatol 4:72–79

128. Mahmoud ALE, El-Shanawany AA, Omar SA (1996) Factors affecting growth and urease production by *Trichophyton* spp. Mycopathologia 135:109–113

129. Wise MG, Healy M, Reece K, Smith R, Walton D, Dutch W et al (2007) Species identification and strain differentiation of clinical *Candida* isolates using the DiversiLab system of automated repetitive sequence-based PCR. J Med Microbiol 56:778–787

130. Healy M, Reece K, Walton D, Huong J, Frye S, Raad II et al (2005) Use of the DiversiLab system for species and strain differentiation of *Fusarium* species isolates. J Clin Microbiol 43:5278–5280

131. Jackson CJ, Barton RC, Evans GV (1999) Species identification and strain differentiation of dermatophyte fungi by analysis of ribosomal-DNA intergenic spacer regions. J Clin Microbiol 37:931–936

132. Shin JH, Sung JH, Park SJ, Kim JA, Lee JH, Lee DY et al (2003) Species identification and strain differentiation of dermatophyte fungi using polymerase chain reaction amplification and restriction enzyme analysis. J Am Acad Dermatol 48:857–865

133. Gherbawy Y, Voigt K (2010) Molecular identification of fungi. Springer, Heidelberg

134. Bhardwaj S, Sutar R, Bachhawat AK, Singhi S, Chakrabarti A (2007) PCR-based identification and strain typing of *Pichia anomala* using the ribosomal intergenic spacer region IGS1. J Med Microbiol 56:185–189

135. Lopes MB, Soden A, Martens AL, Henschke PA, Langridge P (1998) Differentiation and species identification of yeasts using PCR. Int J Syst Evol Microbiol 48:279–286

136. Gupta AK, Kohli Y, Summerbell RC (2000) Molecular differentiation of seven *Malassezia* species. J Clin Microbiol 38:1869–1875

137. Hopfer RL, Walden P, Setterquist S, Highsmith WE (1993) Detection and differentiation of fungi in clinical specimens using polymerase chain reaction

(PCR) amplification and restriction enzyme analysis. Med Mycol 31:65–75

138. Odds FC, Abbott AB (1980) A simple system for the presumptive identification of *Candida albicans* and differentiation of strains within the species. Med Mycol 18:301–317

139. Land GA, Harrison BA, Hulme KL, Cooper BH, Byrd JC (1979) Evaluation of the New API 20C strip for yeast identification against a conventional method. J Clin Microbiol 10:357–364

140. McGinnis MR (1980) Laboratory handbook of medical mycology. Academic, New York

141. Wadlin JK, Hanko G, Stewart R, Pape J, Nachamkin I (1999) Comparison of three commercial systems for identification of yeasts commonly isolated in the clinical microbiology laboratory. J Clin Microbiol 37:1967–1970

142. Sand C, Rennie RP (1999) Comparison of three commercial systems for the identification of germ-tube negative yeast species isolated from clinical specimens. Diagn Microbiol Infect Dis 33:223–229

143. Espinel-Ingroff A, McGinnis MR, Pincus DH, Goldson PR, Kerkering TM (1989) Evaluation of the API 20C Yeast Identification System for the differentiation of some dematiaceous fungi. J Clin Microbiol 27:2565–2569

144. Rath AC, Carr CJ, Graham BR (1995) Characterization of *Metarrhizium anisopliae* strains by carbohydrate utilization (API 50 CH). J Invertebr Pathol 65:152–161

145. Wasfy EH, Bridge PD, Brayford D (1987) Preliminary studies on the use of biochemical and physiological tests for the characterization of *Fusarium* isolates. Mycopathologia 99:9–13

146. Heelan JS, Sotomayor E, Coon K, D'Arezzo JB (1998) Comparison of the rapid yeast plus panel with the API20C Yeast System for identification of clinically significant isolates of *Candida* species. J Clin Microbiol 36:1443–1445

147. Aubertine CL, Rivera M, Rohan SM, Larone DH (2006) Comparative study of the new colorimetric VITEK 2 yeast identification card versus the older fluorometric card and of CHROM agar *Candida* as a source medium with the new card. J Clin Microbiol 44:227–228

148. Fricker-Hidalgo H, Vandapel O, Duchesne MA, Mazoyer MA, Monget D, Lardy B et al (1996) Comparison of the new API Candida System to the ID 32C system for identification of clinically important yeast species. J Clin Microbiol 34:1846–1848

149. Ahmad S, Khan Z, Mustafa AS, Khan ZU (2002) Seminested PCR for diagnosis of Candidemia: comparison with culture, antigen detection, and biochemical methods for species identification. J Clin Microbiol 40:2483–2489

150. Moghaddas J, Truant AL, Jordan C, Buckley HR (1999) Evaluation of the RapID Yeast Plus System for the identification of yeast. Diagn Microbiol Infect Dis 35:271–273

151. Costa AR, Silva F, Henriques M, Azeredo J, Oliveira R, Faustino A (2010) *Candida* clinical species identification: molecular and biochemical methods. Ann Microbiol 60:105–112

152. Rohm H, Lechner F, Lehner M (1990) Evaluation of the API ATB 32C System for the rapid identification of foodborne yeasts. Int J Food Microbiol 11: 215–223

153. Sekhon AS, Padhye AA, Garg AK, Pruitt WR (1987) Evaluation of the Abbott Quantum II yeast identification system. Mycoses 30:408–411

154. Qadri SMH, Flournoy DJ, Qadri SGM, Ramirez EG (1986) Rapid identification of yeasts by semi-automated and conventional methods. Med Microbiol Immunol 175:307–316

155. Kiehn TE, Edwards FF, Tom D, Lieberman G, Bernard EM, Armstrong D (1985) Evaluation of the Quantum II yeast identification system. J Clin Microbiol 22:216–219

156. Salkin IF, Schadow KH, Bankaitis LA, McGinnis MR, Kemna ME (1985) Evaluation of Abbott Quantum II yeast identification system. J Clin Microbiol 22:442–444

157. Cooper BH, Prowant S, Alexander B, Brunson DH (1984) Collaborative evaluation of the Abbott yeast identification system. J Clin Microbiol 19: 853–856

Protocol for the In Vivo Quantification of Superoxide Radical in Fungi

20

Konstantinos Grintzalis, Ioannis Papapostolou, and Christos Georgiou

Abstract

The presented protocol for superoxide radical in vivo quantification in fungal tissues is based on the quantification of 2-hydroxyethidium (2-OH-E^+) after its isolation and fluorometric quantitation. The protocol is ultrasensitive (i.e., <1 pmol) and applicable to any kind of fungal tissue; it can also be used for in vitro studies.

Keywords

Superoxide radical • In vivo • Fungi • 2-Hydroxyethidium (2-OH-E^+)

Introduction

Superoxide radical has been implicated in many physiological conditions, such as aging, differentiation, development, reproduction, cell cycle, and apoptosis, as well as in pathological conditions, such as cancer, atherosclerosis, hypertension, diabetes, ischemia, and epilepsia [1]. Several methods have been employed for its detection (e.g., superoxide dismutase-inhibited cytochrome *c* reduction, reduction of nitroblue tetrazolium, and chemiluminescence, or even much more cumbersome assays such as EPR). The presented protocol has been used for the in vivo quantification of superoxide radical in fungi [2]. It is based on the quantification of 2-hydroxyethidium (2-OH-E^+), resulting from the reaction of superoxide radical with hydroethidine (HE). 2-OH-E^+ is isolated by alkaline-acetone extraction, microcolumn cation exchange and hydrophobic chromatographies, and quantified by fluorescence (after its enzymatic destruction by horseradish peroxidase/H_2O_2 system). It is ultrasensitive (i.e., <1 pmol) and applicable to any kind of fungal tissue for the in vivo detection of superoxide radical. It can also be used for in vitro studies (e.g., superoxide radical production in culturing/growth mediums) [3].

Materials

1. Acetone
2. Acetonitrile (ACN)
3. Albumin from bovine serum (BSA)
4. Balance (Kern, 770/65/6J)

K. Grintzalis • I. Papapostolou • C. Georgiou (✉)
Department of Biology, University of Patras,
University Campus, 25600 Patras, Greece
e-mail: c.georgiou@upatras.gr

V.K. Gupta et al. (eds.), *Laboratory Protocols in Fungal Biology: Current Methods in Fungal Biology*,
Fungal Biology, DOI 10.1007/978-1-4614-2356-0_20, © Springer Science+Business Media, LLC 2013

5. Cation exchanger Dowex 50WX-8 (400) resin
6. Centrifugal vacuum concentrator (Savant, model SPD111V), connected to a vacuum pump (N 820.3 FT.18, KNF)
7. Centrifuge tubes, 15 mL (ISC BioExpress, cat. no. C-3394-1)
8. Chloroform ($CHCl_3$)
9. Coomassie Brilliant Blue G-250 (CBB G-250)
10. Cuvette for absorbance measurements (quartz $45 \times 10 \times 10$ mm, 3.5 mL) (Starna, England)
11. Dimethyl sulfoxide (DMSO)
12. DNA from salmon testes
13. Glass tubes, 15 mL
14. Glass Pasteur pipette (internal diameter 0.5 cm, 22 cm long, by Hirschmann Laborgeräte GmbH & Co, Germany)
15. HLB hydrophobic cartridges 30 μM, 30 mg (Waters, Oasis cat. no. WAT 094225)
16. Horseradish peroxidase (HRP)
17. Hydrochloric acid, HCl, ≥37%
18. Hydroethidine (HE or dihydroethidium)
19. Hydrogen peroxide (H_2O_2)
20. Methanol (MeOH)
21. Microcentrifuge clear tubes, 2 mL (VWR, cat. no. 89000-028)
22. Micropipettes (adjustable volume pipettes) 2.5 μL, 10 μL, 20 μL, 100 μL, 200 μL, 1 mL, and tips (Eppendorf Research)
23. Microcuvette for fluorescence measurements (quartz $45 \times 4 \times 4$ mm, 0.5 mL) with its FCA4 adaptor (Starna, England)
24. Nitrogen gas, 99.999%
25. pH meter (Metrohm, 827 pHlab)
26. Phosphate buffer (Na_2HPO_4)
27. Potassium nitrosodisulfonate (or Fremy's salt)
28. Refrigerated microcentrifuge (Eppendorf, model 5417R)
29. Sodium chloride (NaCl)
30. Sodium hydroxide (NaOH)
31. Spectrofluorometer (Shimadzu, model RF-1501)
32. Spectrophotometer (Shimadzu, model UV-1200)
33. SPE cartridges-Extract Clean SPE Prevail C18 1000 (Alltech, cat. no. 605430)
34. Tris
35. Trifluoroacetic acid (TFA)
36. Water (ddH_2O) was purified by a Milli-Q system (Millipore, Billerica, MA)

Methods

Solutions and Standard Curves

For applying the present protocol for the accurate quantification of 2-OH-E$^+$, the following stock solutions and standard curves need to be prepared.

Preparation of 20 mM HE Stock Solution

Prepare fresh by weighing 7.9 mg HE in a 1.5-mL microcentrifuge tube and dissolving with 1 mL of DMSO. Keep the solution protected from light and use within 1 h of its preparation.

Quantification of the HE Stock Solution by Fluorescence

For measuring superoxide radical accurately via the quantification of 2-OH-E$^+$, it is crucial to ensure that the superoxide radical trap HE is in excess in the biological sample. This can be done by isolating HE from the analyzed tissues (see step 14 in the section "2-OH-E$^+$ and HE Purification by Microcolumn Hydrophobic Chromatography") and converting it into HE concentration via an HE standard curve. The HE-standard curve is made from a series of stock solutions of HE (0–0.5 μM, in 60% ACN) against their fluorescence at ex/em 370/420 nm.

Synthesis and Fluorescence Extinction Coefficient of 2-OH-E$^+$

2-OH-E$^+$ is not commercially available; therefore, a standard for 2-OH-E$^+$ is synthesized by the reaction of HE with nitrosodisulfonate radical dianion (NDS, Fremy's salt) and purified by an Alltech Prevail SPE C18 cartridge [4] as follows:

Nds Stock Solution Preparation

1. Prepare fresh an aqueous solution of 1 mM NDS in 50 mM phosphate buffer, pH 7.4, and 100 μM DTPA by dissolving 3.6 mg NDS in

10 mL of 50 mM phosphate buffer, pH 7.4, containing 100 µM DTPA.

Note: Excess of NDS should be avoided because it reacts with 2-OH-E$^+$ decreasing its yield.

Note: NDS should be stored at 4 °C and used within an hour of its preparation.

2. Using a quartz microcuvette (1.4 mL), adjust baseline of the UV–Vis absorption spectrum (from 200 to 950 nm) with 50 mM aqueous solution of phosphate buffer, pH 7.4, containing 100 µM DTPA and run the spectrum of the prepared NDS solution. Record the absorbance values of NDS solution at 248, 545, and 900 nm.

3. Subtract the absorbance value measured at 900 nm from the values measured at 248 and 545 nm, and with the corrected absorbance values at 248 and 545 nm, calculate the concentration of NDS using the corresponding extinction coefficients 1.6×10^3 and 20.8 M/cm at 248 and 545 nm. Then, the average concentration of the NDS solution is calculated and adjusted to 1 mM.

Synthesis of 2-OH-E$^+$

Mix 24 mL ddH$_2$O with 4 mL 0.5 M phosphate buffer, pH 7.4, 4 mL 1 mM aqueous solution of DTPA, and 200 µL 20 mM solution of HE in DMSO. Add slowly 8.0 mL 1.0 mM NDS while slowly stirring the solution. Incubate the solution at room temperature for 2 h and purify the synthesized 2-OH-E$^+$ by an Alltech Prevail SPE C18 cartridge as follows:

1. Activate the cartridge by passing 6 mL ddH$_2$O, followed by 3 mL ddH$_2$O/MeOH (50/50) mixture, 3 mL of pure MeOH, and 6 mL ddH$_2$O.

2. Load the reaction mixture for 2-OH-E$^+$ synthesis onto the cartridge.

3. Wash the cartridge with 4×3 mL ddH$_2$O, then with 2×3 mL ddH$_2$O/MeOH (50/50) mixture, and finally with 3×3 mL ddH$_2$O/MeOH (20/80) mixture. 2-OH-E$^+$ should elute in this last wash step. If the eluate is still orange (indicating incomplete elution) use more ddH$_2$O/MeOH (20/80).

4. Wash the cartridge with 2×3 mL of pure MeOH and continue washing the cartridge until the second band (containing E$^+$) is eluted.

5. Split the fraction containing pure 2-OH-E$^+$ in 1.5 mL-microcentrifuge clear tubes and place them in a centrifugal vacuum concentrator, dissolve the red pellets in 0.05 mL 0.1 M HCl, combine them and quantify the stock solution of 2-OH-E$^+$ (ca. 4 mM).

Determination of 2-OH-E$^+$ Standard Solution Concentration

The concentration of the standard 2-OH-E$^+$ stock solution is determined as follows:

1. Mix 0.5 mL of aqueous solution of 100 mM phosphate buffer pH 7.4 containing 0.2 mM DTPA with 0.5 mL from a 100-fold dilution (ca. 40 µM) of the standard stock solution and transfer mixture to in a quartz microcuvette (1.4 mL)

2. Record the UV–Vis spectrum in the range 200–800 nm (subtracting the spectrum with the appropriate blank using 0.5 mL 1 mM HCl instead of standard stock solution).

3. Read the absorbance values at 470 and 800 nm.

4. Correct the measured absorbance value for the background by subtraction of the absorbance value measured at 800 nm, calculate the concentration of standard in the cuvette using the corrected absorbance values and the corresponding extinction coefficient 1.2×10^4 M/cm, and then calculate the concentration of the standard in stock solution.

Note: 2-OH-E$^+$ can be prepared and stored for months at 4 °C in 1 mM HCl.

Fluorescence Extinction Coefficient of 2-OH-E$^+$

For quantifying the 2-OH-E$^+$ isolated from tissues, a fluorescent standard curve is performed (using the 2-OH-E$^+$ standard stock solution), in the absence and presence (for higher sensitivity) of DNA. Specifically, to determine the fluorescence extinction coefficient (FEC) in the absence of DNA, prepare a series of dilutions (0–8 µM) of 2-OH-E$^+$ in 50 mM phosphate buffer, pH 7.4 (in 1% DMSO) and measure fluorescence at ex/em 480/583 nm in a quartz cuvette. For estimating the FEC in the presence of DNA prepare a series of dilutions (0–100 nM) of 2-OH-E$^+$ in 50 mM phosphate buffer, pH 7.4 (in 1% DMSO) to which 55 μL/mL 2 mg/mL DNA stock solution is added, and measure their fluorescence at ex/em 515/567 nm in a quartz cuvette. In this case, where the FEC of 2-OH-E$^+$ in the absence and presence of DNA is calculated for the Shimadzu RF-1501 spectrofluorometer (with 10 nm excitation/emission slit width, at high sensitivity, and with a quartz microcuvette of internal dimensions 45×4×4 mm, placed in its appropriate holder), the following steps are needed:

1. *FEC without DNA*: Prepare various dilutions (0–8 μM) of the 2-OH-E$^+$ standard stock solution in 50 mM phosphate buffer, pH 7.4 (in 1% DMSO) and place 0.3 mL of each in a quartz cuvette and measure their fluorescence units (FU) at ex/em 480/583 nm.

   For the Shimadzu, model RF-1501, the FEC (-DNA) = 115 FU/1 μM 2-OH-E$^+$ [5].

2. *FEC with DNA*: Prepare various dilutions (0–100 nM final concentration) of the 2-OH-E$^+$ standard stock solution in 50 mM phosphate buffer, pH 7.4 (in 1% DMSO) and add to each one 20 μL from 2 mg/mL DNA stock solution (prepared in ddH$_2$O after dissolving it over night at 0–4 °C in an ice-water bath) (DNA final concentration 0.15 mg/mL) and place in a quartz cuvette and measure their fluorescence units (FU) at ex/em 515/567 nm. For example, with the Shimadzu, model RF-1501, FEC (+DNA) = 2,265 FU/1 μM 2-OH-E$^+$ [5].

   Note: FEC either with or without DNA needs to be recalculated each time a different fluorometer is used.

Protocol for 2-OH-E$^+$ Quantification

The following steps describe the in vivo quantification of 2-OH-E$^+$ in fungal tissues:

Sample Treatment

1. Place 0.2 g (wet weight) fungal tissue sample in 2 mL of its culture medium and incubate for 30–60 min with 20 μL 5 mM HE stock (final 50 μM). During incubation, HE will enter the fungal tissue and will react with superoxide radical in vivo, forming and 2-OH-E$^+$.

2. After incubation, remove any bound HE and 2-OH-E$^+$ (either externally formed in the growth medium or contaminating the commercial HE) on the fungal tissue, by washing it with an equal (1 g = 1 mL) to tissue wet weight volume 10 N HCl, followed by three times washes with equal volume ddH$_2$O. Discard the supernatant after centrifugation at 3,000 × g for 5 min.

   Note: For yeast cells collect cells by centrifugation and wash them with 1 mL culture medium. Discard the supernatant after centrifugation at 3,000 × g for 5 min and resuspend cells in 0.2 mL 50 mM phosphate buffer, pH 7.4.

3. Homogenize the above washed tissue (e.g., with liquid nitrogen) and 50 mM phosphate buffer, pH 7.4 containing 5 mM KCN or 5 units catalase/mL (to inhibit destruction 2-OH-E$^+$ of nonspecific peroxidases and H$_2$O$_2$), keeping the homogenate volume as minimum as possible (e.g., in a final volume 0.3 mL).

   Note: For yeast and cell pellets resuspended in 50 mM phosphate buffer, pH 7.4 homogenization is not required.

   Note: Samples should be analyzed immediately. However, the homogenates can be frozen and stored at −80 °C for 2 weeks at most.

4. Collect a small portion of the homogenate for protein determination.

Sample Protein Determination

5. Quantify total protein amount in homogenate by an ultrasensitive modification of the Bradford assay.[6] Prepare a CBB-HCl reagent by dissolving 60 mg CBB G-250 in 100 mL 1 N HCl, stirring for 30 min and filtering through a Whatman no.1 filter paper by water pump aspiration. This solution is stable for months (kept light protected at 4 °C). Prepare fresh the hydrophobic CBB-TCA reagent by bringing 20 mL of the above CBB-HCl reagent to 1% ethanol and 2% TCA (0.4 g solid TCA) and adjusting its pH to 0.4 by Na$_3$PO$_4$. Remove any particulates by centrifugation and use within 1 h of its preparation. Prepare a series of 50 μL BSA standards (100–3,000 ng BSA made in ddH$_2$O) or sample appropriately diluted with ddH$_2$O and add to it 0.95 mL

hydrophobic CBB-TCA reagent. Measure absorbance at 610 nm against a reagent blank containing 50 μL ddH$_2$O in place of BSA standard. From the BSA standard curve, calculate the protein concentration in the homogenates using appropriate homogenate dilutions.

Alkaline Acetone Extraction of 2-OH-E$^+$ from Homogenates

6. To 1 volume homogenate add 9 volumes 100% acetone and 1/100 volume 10 N NaOH (final concentration 0.1 N NaOH). Vortex mixtures vigorously and centrifuge for 5 min at 15,000 × g and collect supernatant.

7. Dilute the supernatant-acetone mixture to 60% acetone with ddH$_2$O and add 1/50 volume 2.5 M Tris–HCl, pH 7.0. Adjust the pH of the diluted mixtures to 7.0 and remove cloudiness (due to tissue material) by incubating it on ice for 5 min and centrifuging it at 25,000 × g and 4 °C for 5 min. Collect the clear supernatant.

2-OH-E$^+$ Isolation by Microcolumn Cation Exchange Chromatography

8. Activate Dowex resin cation exchange microcolumn by washing resin with regenerating it with 1 mL 0.1 N HCl, 5 mL ddH$_2$O, 1 mL 0.1 N NaOH, and 5 mL ddH$_2$O.

   Note: In order to prepare the Dowex microcolumn wash (by swirling) 4 g cation exchanger Dowex 50WX-8 (400) resin with 40 mL 100% ACN and 40 mL ddH$_2$O. Equilibrate (by stirring) the resin with 15 mL 0.1 N HCl for 30 min, discard the supernatant and wash (by stirring) the resin with 15 mL ddH$_2$O for 30 min. Discard the supernatant and equilibrate (by stirring) resin with 15 mL 0.1 N NaOH for 30 min

   Prepare microcolumns by packing 0.25 mL bed volume (approx. 0.25 g activated resin) in a glass Pasteur pipette plugged with glass wool. After first use, the microcolumn can be stored at RT for at least 3 months and reused for up to ten times after regeneration.

9. Pass the 2-OH-E$^+$ clear alkaline acetone extract through the Dowex microcolumn at a free flow rate. 2-OH-E$^+$ (as well as HE) is bound to the Dowex resin.

10. Wash the microcolumn in sequence with 1 mL 4 M NaCl, 1 mL ddH$_2$O, 2 mL 100% ACN, and 2 mL ddH$_2$O and elute 2-OH-E$^+$ from the microcolumn with 1 mL 10 N HCl and dilute to 3 N HCl with ddH$_2$O.

2-OH-E$^+$ and HE Purification by Microcolumn Hydrophobic Chromatography

11. Activate and equilibrate HLB hydrophobic cartridge by passing 1 mL MetOH and 1 mL ddH$_2$O at a flow rate 2 mL/min.

12. Pass the eluted 2-OH-E$^+$ through the activated HLB microcolumn at a flow rate 2 mL/min. Wash off impurities by passing through 1 mL 17% ACN-phosphate and elute 2-OH-E$^+$ (as well as HE and other oxidation products of HE) by passing through 1.5 mL 25% ACN-phosphate.

13. Extract 2-OH-E$^+$ by adding 1.5 mL chloroform, vortex and centrifuge to collect organic bottom layer and evaporate it (e.g., using a vacuum centrifuge evaporator).

   Note: The red dry residue of 2-OH-E$^+$ is contaminated with other oxidation products of HE such as ethidium (E$^+$).

   Note: Samples should be analyzed immediately. However, the evaporated residues can be frozen and stored at −80 °C for 2 weeks at most.

14. Elute unreacted excess HE by passing through 1 mL 60% ACN-phosphate and collect the eluate to estimate the amount of HE present in sample by measuring its fluorescence at ex/em 370/420 nm. Convert the fluorescence to HE moles using the fluorescent standard curve for HE in 60% ACN (see Reagent setup).

   Note: For accurate determination of 2-OH-E$^+$ quantification, HE excess is required and should be established by testing various incubation time and HE concentration conditions.

15. Clear the HLB column from any other hydrophobic interferences by washing with 1 mL 100% ACN (containing 0.1% trifluoroacetic acid) and 1 mL ddH$_2$O with a flow rate 1 mL/min, and then equilibrate it by passing through 1 mL MetOH and 1 mL ddH$_2$O at a flow rate 2 mL/min.

Note: After first use, the HLB microcolumn can be stored at RT for at least 3 months and reused for up to ten times after regeneration.

Fluorometric Quantification of 2-OH-E⁺

16. The fluorometric quantification of 2-OH-E⁺ is based on measuring the total fluorescence of concentrated 2-OH-E⁺ in the presence of other contaminating oxidation products of HE (e.g., E⁺) and measuring fluorescence again after the enzymic destruction (by HRP/ hydrogen peroxide treatment) of 2-OH-E⁺. Resuspend the red 2-OH-E⁺ residue from the sample by dissolving it in 50 μL 50 mM phosphate buffer, pH 7.4 containing 6% DMSO. Add 250 μL 50 mM phosphate buffer, pH 7.4 (final DMSO 1%).

17. Measure fluorescence at ex/em 480/583 nm (of 2-OH-E⁺ and other oxidation products, designated as Total Fluorescent Units or TFU) of the 0.3 mL 2-OH-E⁺ solution.
Note: If fluorescence is limiting, it can be enhanced 25-fold at ex/em 515/567 nm by addition of 20 μL from 2 mg/mL DNA stock solution (DNA final concentration 0.15 mg/mL).

Record the FU in the absence/presence of DNA which are designated as total FU (TFU$_{\pm DNA)}$ since they result from 2-OH-E⁺ possibly mixed with other HE oxidation products like E⁺.

18. To the above mixture (±DNA) add 25 μL 0.003% H_2O_2 and 10 μL 100 units/mL HRP and incubate for 5 min at RT in order to destroy 2-OH-E⁺. Measure fluorescence

resulting from all other oxidation products except of 2-OH-E⁺ and correct it by multiplying with the factor 0.335/0.300 = 1.11 (to account for the dilution of the 0.30 mL solution from HRP and H_2O_2).
Note: If DNA was added the correction factor is the same (0.350/0.320 = 1.11). This fluorescence is designated as FU.

19. Calculate the fluorescence of 2-OH-E⁺ which is equal to TFU$_{\pm DNA}$ − FU$_{\pm DNA}$, convert it to 2-OH-E⁺ moles using the fluorescent standard curve for 2-OH-E⁺ (see Reagent setup), and express it per mg of protein.

References

1. Halliwell B, Gutteridge CMJ (1999) Free radicals in biology and medicine, 3rd edn. Oxford University Press, Oxford, UK
2. Papapostolou I, Georgiou DC (2010) Superoxide radical induces sclerotial differentiation in filamentous phytopathogenic fungi: a SOD mimetics study. Microbiology 156:960–966
3. Georgiou CD, Papapostolou I, Grintzalis K (2008) Superoxide radical detection in cells, tissues, organisms (animals, plants, insects, microorganisms) and soils. Nat Protoc 3:1679–1692
4. Zielonka J, Vasquez-Vivar J, Kalyanaraman B (2008) Detection of 2-hydroxyethidium in cellular systems: a unique marker product of superoxide and hydroethidine. Nat Protoc 3:8–21
5. Georgiou DC, Papapostolou I, Patsoukis N, Tsegenidis T, Sideris T (2005) An ultrasensitive fluorescent assay for the in vivo quantification of superoxide radical in organisms. Anal Biochem 347:144–151
6. Georgiou CD, Grintzalis K, Zervoudaki G, Papapostolou I (2008) Mechanism of Coomassie brilliant blue G-250 binding to proteins: a hydrophobic assay for ng quantities of proteins. Anal Bioanal Chem 391:391–403

Isolation of Intact RNA from Sorted S. cerevisiae Cells for Differential Gene Expression Analysis

21

Jeannette Vogt, Frank Stahl, Thomas Scheper, and Susann Müller

Abstract

Individuals within natural populations are highly diverse. They can vary regarding their physiological states and thus contribute differently to the performance of the whole community [1]. Hence, the characterization of microbial physiology on the single cell level is crucial for optimization and understanding of biotechnological processes.

Here a protocol is presented for detecting and monitoring the functional heterogeneity in a given yeast population, on both the cellular and molecular levels, using flow cytometry and microarray analysis (first published in Nature Protocols by Achilles et al. [2]). The protocol includes staining of living S. cerevisiae cells with three different fluorescent dyes, flow cytometric analysis, sorting of live cells and stabilization of their status, the subsequent isolation of RNA from small amounts of separated cells, and finally the quantification and integrity check of this RNA. Owing to the high quality and quantity of the isolated RNA from sorted cells, gene expression analysis can be performed by using microarrays, for example.

Keywords

Multiparametric flow cytometry • Cell sorting • Population dynamics • RNA isolation • Saccharomyces cerevisiae

Introduction

Microorganisms are widely used for biotechnological production processes. For example, yeasts are employed not only to produce beer, wine, and bread, which has been done for thousands of years, but also for the production of vaccines [3–5]. pharmaceuticals [6, 7], and chemicals [8–10].

The optimization of such biotechnological processes is focused on the receipt of high

J. Vogt • S. Müller (✉)
Department of Environmental Microbiology, Helmholtz Centre for Environmental Research—UFZ, Permoserstr. 15, Leipzig, Saxony 04318, Germany
e-mail: susann.mueller@ufz.de

F. Stahl • T. Scheper
Institute of Technical Chemistry, Callinstr. 5, 30559 Hanover, Lower Saxony, Germany

amounts of desired target products with minimal consumption of raw substrates. Commonly, the analyses of biomass, temperature, pH value, consumption of oxygen, and formation of carbon dioxide, for example, are used for monitoring of the processes. However, these bulk-scale measurement methods do not afford information about individual variations—for example, regarding size, shape, age, vitality, enzyme activity, content of storage material, and proliferation activity. Since the performance of a whole population is a sum of the individuals' performances, knowledge of specific characteristics of the individuals within the population is a valuable tool for optimizing and understanding of biotechnological processes.

Flow cytometry is a technique that allows analysis of the heterogeneity of cells within microbial populations or communities quantitatively [11–14]. Other techniques, like image analysis, fluorescence microscopy, confocal laser scanning microscopy (CLSM), confocal RAMAN-microscopy (CRM), or slide-based laser scanning cytometry (LSC), afford single cells analyses, too, but not in similar quantity and/or rapidness. By using flow cytometry thousands of cells within a suspension are analyzed within a few seconds and can be characterized regarding their functional and structural parameters, such as size, shape, granularity, proliferation activity, enzyme activity, membrane potential, and amount of storage material like neutral lipids and hydroxysterols (for review see Müller and Nebe-von-Caron [14]). Hence, this technique is suitable for studying microbial communities and has often been used to get insights into the physiology and morphology of single cells of *S. cerevisiae* populations [15–26]. In addition, flow cytometers can be equipped with cell sorters, which separate subpopulations from each other, showing differences in their characteristics. The sorted cells can be used for further cultivation and/or analysis, for example, regarding their gene expression profile [27–30], their proteome [31–33], or phylogenetic composition [34–37]. Consequently, flow cytometry is a useful tool to understand biotechnological processes not only on the single-cell level, but also on the level of

transcriptome and proteome by using cell sorting and further analysis.

The yield of biotechnological production processes is dependent on the viability, growth, proliferation activity, and metabolic activity of each single cell within the population influenced by the availability and utilization of carbon and energy sources. Therefore, there is a great interest in analyzing physiological states of single cells. Of particular importance is knowledge about the consumption of the substrate, including the uptake and metabolization and the regulation of these processes on the cellular and molecular levels, even under different microenvironmental conditions. For investigation of *S. cerevisiae* cell populations well-established flow cytometric methods are already available for certain cell parameters [38, 39] and were used for development of control strategies in industrial processes. However, these techniques relied mainly on measurement of proliferation activity or storage product accumulation.

In order to get an idea about the physiology of cells and their potential to synthesize the desired target and other products independent of the kind and amount of the carbon source, it is advantageous to analyze the carbon consumption of the cells. By using batch cultivation it is possible to measure the heterogeneous substrate consumption pattern in detail and to find the best conditions for product synthesis in a simple, fast, and low-cost way. Here, cellular glucose consumption can be followed in vital cells using fluorescent 2-NBDglucose [40]. In Fig. 21.1 a batch cultivation of *S. cerevisiae* H155 using a defined yeast nitrogen base medium (2 % glucose, Difco) is shown [40]. A number of physiological parameters like proliferation activity and cell size were flow-cytometrically analyzed in addition to the measurement of glucose and ethanol within the medium as well as the optical density (see Fig. 21.1a). As expected, during the first hours of the batch cultivation the glucose concentration decreased, whereas the concentration of ethanol in the medium and the biomass measured by optical density increased. The batch cultivation started with a high number of cells being in the G_2 phase of the cell cycle. This was caused by

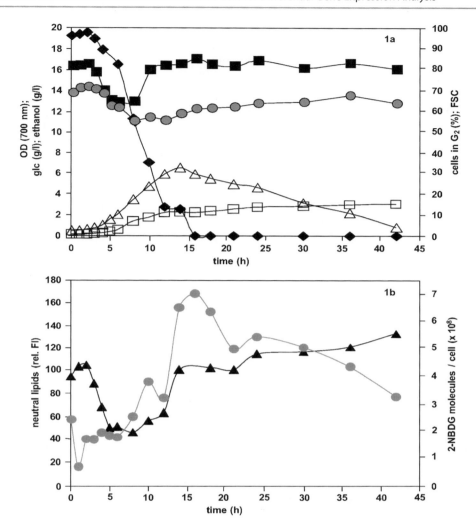

Fig. 21.1 Batch cultivation of *S. cerevisiae* H155. **(a)** Concentration of glucose in the medium (g/L; *filled diamond*), concentration of ethanol in the medium (g/L; *open triangle*), optical density (700 nm; *open square*), cells in the G_2 phase of the cell cycle (%; *filled square*), and mean values of cell size distributions (FSC; *shaded circle*).

(b) Cellular neutral lipid content (rel. FI; *filled triangle*) and cells' affinity to 2-NBD-glucose (molecules / cell (x 10^8); *shaded circle*). Reprinted with permission from Achilles J, Müller S, Bley T, Babel W. Affinity of single *S. cerevisiae* cells to 2-NBDglucose under changing substrate concentrations. Cytometry A. 2004; 61: 88–98 [40]

using cells from a stationary phase, as the new inoculum. *S. cerevisiae* cells taken from this phase commonly arrest the process of cell division owing to limited substrate concentration. Within the first 8 hours of the batch cultivation, the cells started to divide. Consequently, the amount of cells containing the double chromosome content as well as the size of the cells decreased. At the ninth hour the number of G_2 phase cells increased again, as a result of

replication, until the beginning of glucose limitation. The cell size, analyzed by forward scatter (FSC) signal of the cells, increased at the same time owing to bud formation and growth. The results of flow-cytometrical measurements of the cells' content of neutral lipids stained by nile red as well as affinity of the cells to glucose using the fluorescent glucose analogue 2-NBDglucose is shown in Fig. 21.1b. At the beginning of the batch cultivation the neutral lipid content within the cells declined rapidly, obviously caused by cell division activity and the associated high carbon turnover. Afterward the amount of neutral lipids increased with glucose consumption and finally glucose limitation. Yeast cells are known for synthesizing neutral lipids under limiting conditions to ensure their survival [20]. These results of neutral lipid analyses were confirmed by the side scatter (SSC) signal, which is a measure for the granularity of the cells (not shown). The analyses of the cells' affinity to glucose were performed by measuring the 2-NBDglucose fluorescence intensities of the cells. This fluorescent glucose analogue was previously used by Natarajan and Srienc [41, 42] to analyze the glucose uptake by single *Escherichia coli* cells. The higher the 2-NBDglucose fluorescence, the higher is the cells' affinity to glucose. At the beginning of the batch cultivation, characterized by high extracellular glucose concentration, the cells' affinity to 2-NBDglucose was very low. The cells' affinity to glucose was maximal at about 15 hours, when glucose was almost exhausted, probably caused by expression of high affinity glucose transport systems. Indeed, during batch cultivation, subpopulations could be determined differing in their affinities to the substrate and not correlated with the size and/or the budding status of the cells. During batch cultivation, however, there is usually a very short time period in which the rate of product synthesis reaches nearly its maximum. Additionally, batch cultivations are generally characterized by permanent concentration changes of the provided substrates, formed metabolites and end products, usually accompanied by changes of the pH-value, the redox potential and the number of cells within subpopulations differing in their age, vitality, proliferation activity or answer to stress conditions.

To study the behavior and physiological states of the cells as well as product formation under more constant and defined conditions, cells must be cultivated in a chemostat under steady state or transient state conditions. During chemostat cultivation cellular regulations can be detected (e.g., regulations depending on the extracellular substrate concentration). The concentration of sugar, especially glucose as the preferred carbon

and energy source of *S. cerevisiae,* mainly determines the used metabolic pathways and therefore the product formation [43]. Under oxic conditions and extracellular glucose concentrations below 2 gl^{-1} glucose is exclusively metabolized by respiration via the tricarboxylic acid cycle. Here, two different substrate flux conditions called "subcritical" and "critical" can be distinguished according to the limited respiratory capacity hypothesis of Sonnleitner and Käppli [44]. Under "subcritical" substrate flux conditions the respiratory capacity is not utilized; thus, ethanol can be used as second substrate in addition to glucose. If the oxygen demand equates to the respiratory capacity the substrate flux is "critical." Higher substrate concentrations result in "supracritical" substrate flux. Under these conditions the substrate cannot be completely oxidized and thus ethanol is produced as an overflow product by reductive metabolism. As long as the respiratory capacity is exceeded, the overflow products accumulate in the medium. However, during continuous cultivation of *S. cerevisiae* the yield of biomass per gram of glucose is at least fivefold lower under respiro-fermentative conditions in comparison to complete respirative conditions [43–46]. Hence, in biotechnological processes, yeasts should be incubated at first under respirative conditions to generate sufficient biomass for producing the desired target product under respiro-fermentative conditions in the second step of the process.

In Fig. 21.2 results of flow-cytometrical analyses of transient state cultivated live cells of *S. cerevisiae* stained with three different colors are shown [47]. The three-color technique enables the simultaneous analyses of the cell affinity to glucose by 2-NBDglucose, the cell proliferation activity by Hoechst 33342, as well as the dead cell amount within the population by using propidium iodide. The physiology of *S. cerevisiae* H155 cells independent of changing substrate concentrations using transient state cultivation with different and increasing glucose concentration is presented elsewhere [40]. In Fig. 21.2 it is shown that during the transient state cultivation, the affinities of cells to 2-NBDglucose were high at low substrate concentration and low when the

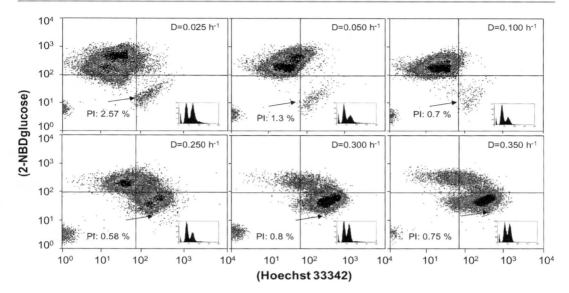

Fig. 21.2 Transient state cultivated live cells of S. *cerevisiae*, stained with the three-color assay. The dead cell amount is given in percent within the left lower quadrant of each dot plot. For comparison, the DNA of the identical but fixed samples were presented within the insets, analyzed at linear scale. Data of at least 20,000 events are displayed. Reprinted with permission from Achilles J, Harms H, Müller S. Analysis of living S. *cerevisiae* cell states—a three color approach. Cytometry A. 2006; 69: 173–7 [47]

glucose concentration started to rise in the bioreactor (from a dilution rate of D=0.25 h^{-1} onward). In addition, it was observed that the cells' affinities to 2-NBDglucose varied at identical extracellular glucose concentrations. This phenomenon was observed at different substrate concentrations. Thus, the number of cells in the different subpopulations showing differences in their affinities to the substrate varied at different extracellular glucose concentrations. The results of the flow cytometric analysis suggest that the cell affinities to 2-NBDglucose varied to a high degree independent of the budding status of the cell. For comparison, the insets present the DNA distributions of fixed cells of the same samples. The dead cell amount within the population was also determined over the course of the transient state cultivation and is given in percent within each dot plot. The knowledge of the dead cell amount is important to exclude these cells from further analysis, like gene expression analysis by microarrays. In order to understand the phenomena of different cell affinities at identical substrate concentrations at the molecular level, a

method had to be established allowing the isolation of intact RNA out of small subpopulations of living *S. cerevisiae* cells, stained and then separated by a cell sorter for gene expression analysis. The general flow chart of the established protocol for the isolation of intact RNA from cytometrically sorted 5×10^7 cells of *S. cerevisiae* for the analysis of intrapopulation diversity of gene expression is shown in Fig. 21.3. This protocol includes the fluorescence staining, the flow cytometric analysis, the sorting of live yeast cells, the stabilization of the sorted yeast cells, subsequently the isolation of RNA from the resulting subpopulations, and finally the quantification and the integrity check of the isolated RNA. This protocol is a slightly modified version of the protocol previously published in Nature Protocols [2].

The use of this protocol enables one to compare gene expression profiles of small numbers of live microbial cells showing different properties. Consequently, the protocol is useful for the examination of functional heterogeneities within microbial populations on both cellular and molecular levels.

Isolation of intact RNA from cytometrically sorted *S. cerevisiae* for the analysis of intra-population diversity of gene expression

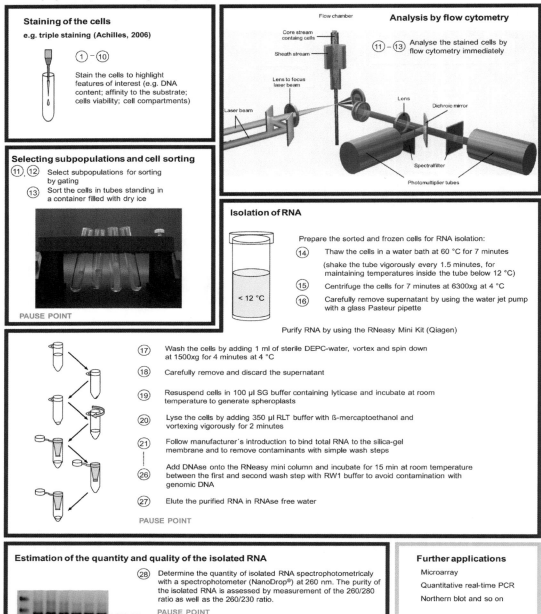

Fig. 21.3 Work flow of the isolation of intact RNA from cytometrically sorted *S. cerevisiae* for the analysis of intra-population diversity of gene expression. Reprinted with permission from Achilles J, Stahl F, Harms H, Müller S.

Isolation of intact RNA from cytometrically sorted *Saccharomyces cerevisiae* for the analysis of intrapopulation diversity of gene expression. Nat Protoc. 2007; 2: 2203–11 [2]

Materials

1. Schatzmann medium
2. 2-L Biostat®MD laboratory stirring bioreactor (Braun) equipped with a pH-electrode (model 405-DPAS-SC-K8S/200; Mettler-Toledo GmbH) and an oxygen electrode (model InPro 6000; Mettler-Toledo GmbH)
3. Sterile HEPES buffer (2-(4-(2-hydroxyethyl)-1-piperazinyl)-ethanesulfonic acid; Merck; 10 mM; pH 7.2)
4. 2-NBDglucose (2-(N-(7-nitrobenz-2-oxa-1,3-diazol-4yl)amino)-2-deoxyglucose; Invitrogen; 20 mM in double distilled water)
5. Hoechst 33342 (2,5'-Bi-1 H-benzimidazole; Invitrogen; 0.325 mM in HEPES buffer)
6. PI (propidium iodide; Sigma-Aldrich; 1 mM in phosphate buffered saline pH 7.0)
7. Verapamil hydrochloride (Sigma-Aldrich; 1 mM in double distilled water)
8. Sodium azide (Merck; 10 % in distilled water (wt/vol))
9. Glass tubes (Duran, 10 mL; treated with Hellmanex®II (Hellma GmbH Co. KG))
10. UV-visible spectrophotometer
11. Water bath
12. Water jet pump
13. Centrifuge
14. Vortex
15. RNAseZap® (Ambion)
16. Flow-Check™ Fluorospheres (Beckman Coulter)
17. Dry ice (dry ice to be used in direct contact with cells must be of highest purity)
18. Sheath fluid
    The sheath fluid contained 1.9 mM KH_2PO_4, 3.8 mM KCl, 16.6 mM Na_2HPO_4, and 139 mM NaCl, pH 7.0. A 10× stock solution of sheath fluid was used that was autoclaved 20 min at 121°C and afterward diluted with bidistilled water filtered using a glass frit (0.2 μm). Alternatively, a ready-to-use sheath fluid can be purchased from Beckman Coulter.
19. Sterile 50-mL plastic tubes
20. MoFlo Cell sorter (Beckman Coulter)
    The flow cytometer was equipped with two lasers for excitations at 488 nm and multi-line UV (333–365 nm; Innova 90C and Innova 70C from Coherent). FSC and SSC were analyzed after excitation of 60 mW at 488 nm. For 2-NBDglucose and PI fluorescence measurements, 530/40 and 630/30 band-pass filters were used, respectively. Detection of Hoechst 33342 fluorescence involved excitation by 40 mW multi-line UV and a 450/65 band pass filter. The trigger signal was FSC. The flow cytometer was equipped with a sort unit, allowing separating subpopulations. The size of the flow tip was 70 μm. The Summit®V.3.1 software (Dako) or updated versions (Beckman Coulter) were used to evaluate the data. Before analyzing and sorting the target population, the flow cytometer was calibrated with calibration beads (Flow-Check Fluorospheres) and a *S. cerevisiae* standard with known subpopulations (e.g., ethanol fixed and Hoechst 33342 or DAPI-stained cells after 11 h batch cultivation showing distinct subpopulations in the G_1 and G_2 phase of the cell cycle, respectively). In addition, the flow cytometer was prepared for stable sorting by the appropriate settings (for example the determination of the right drop delay).
21. RNeasy Mini Kit (Qiagen)
22. Rnase-free DNAse Set (Qiagen)
23. Lyticase (from *Arthrobacter luteus*; Sigma-Aldrich)
24. β-Mercaptoethanol (Sigma-Aldrich)
25. Ethanol (≥99.9 %)
26. Sterile RNAse-free water (diethyl pyrocarbonate (DEPC)-treated; from Sigma-Aldrich)
    Sterile RNAse-free water was prepared by adding 1 mL DEPC to 1 liter distilled water and shaken until DEPC had completely dissolved. The DEPC-water was incubated 12 h at 37 °C and then autoclaved 20 min at 121 °C to eliminate DEPC.

27. SG buffer (1 M sorbitol; 0.1 M ethylenedi-aminetetraacetate (EDTA) pH 7.4; in DEPC-water)
28. 2-mL microcentrifuge tubes
29. Microcentrifuge
30. Agarose
31. $0.5 \times$ TAE (40 mM Tris-acetate, 1 mM EDTA, pH 8.0)
32. Ethidium bromide (13 μM in $0.5 \times$ TAE-buffer)
33. NanoDrop®ND-1000 spectrophotometer V.3.1.0 (NanoDrop Technologies)
34. Gel electrophoresis system (Biometra®)
35. Gel image analysis system (software GeneSnap V.6.08; Synoptics Ltd.).

Cultivation Conditions

Saccharomyces cerevisiae H155 was continuously cultivated in a bioreactor (e.g., 2-L BiostatMD laboratory stirring bioreactor [Braun]) in 1 L Schatzmann medium (see later) at 30 °C. The pH was constantly maintained at 5.4 by adding 1 M NaOH as required controlled by a pH-electrode (model 405-DPAS-SC-K8S/200; Mettler-Toledo GmbH). The aeration rate was constantly kept at 3 L of air/L of medium/min controlled by an oxygen electrode (model InPro 6000; Mettler-Toledo GmbH) and the stirrer velocity at 600 rpm.

Schatzmann medium was made as described by Schatzmann [48] containing 167 mM glucose, 22 mM sodium citrate, 34 mM $(NH_4)_2SO_4$, 14 mM $(NH_4)_2HPO_4$, 12 mM KCl, 1.4 mM $MgSO_4$, 2.9 mM $CaCl_2$, 56 μM $FeCl_3$, 31 μM $ZnSO_4$, 56 μM $MnSO_4$, 10 μM $CuSO_4$, 333 μM myo-inositol, 126 μM Ca-pantothenate, 18 μM thiamine × HCl, 7 μM pyridoxine × HCl, 12 μM biotin; pH 5.4.

Method

Staining of the Cells

1. The cells were harvested, washed in HEPES buffer, and diluted to 5×10^7 cells using a calibration curve adjusted at a

spectrophotometer (OD of 700 nm). The staining was carried out in glass tubes since staining in plastic tubes (e.g., microcentrifuge tubes) may influence the staining process because of dye and/or cell adherences to the plastic surface.

2. 5×10^7 cells were centrifuged for 5 min at $3,200 \times g$ and 5 °C. Afterward the supernatant was carefully discarded.
3. 25 μL of Hoechst 33342 solution (stock solution: 0.325 mM in HEPES buffer) was added to the cell pellet for DNA staining.
4. 25 μL of verapamil (stock solution: 1 mM in double distilled water) was added to stabilize the Hoechst 33342 staining, as verapamil prevents the efflux of the Hoechst dye by multidrug membrane transporters.
5. 20 μL of 2-NBDglucose (stock solution: 20 mM in double distilled water) was added to visualize the cell's affinity to glucose.
6. The assay was gently mixed by pipetting up and down three times.
7. The assay was incubated in a water bath at 30 °C for 20 min.
8. 50 μL PI solution (stock solution: 1 mM in phosphate buffered saline) was added followed by vortexing to stain dead cells.
9. 100 μL of sodium azide (stock solution: 10 % in distilled water (wt/vol)) was added followed by vortexing to stop the staining reaction.
10. 780 μL of HEPES buffer was added and the cells were analyzed immediately by flow cytometry. (see Note 1)

Cell Sorting

1. A ribonuclease-free working place was prepared by using for example RNAseZap. Particular attention was paid to the sample unit and the nozzle of the flow cytometer. Additionally, gloves were worn to avoid ribonuclease contamination.
2. Subpopulations for sorting were selected using the Summit V.3.1. software (or updated versions) by appropriate gating (e.g., regarding the DNA content or the substrate affinity).

3. The cells were sorted into 50-mL plastic tubes standing in a container filled with dry ice. The sorting was realized in the sort mode "purify one" with a sample pressure resulting in a sorting rate of approximately 30,000 cells/s (coincidence rate: 4,700–5,000). This procedure was repeated until 5×10^7 cells per subpopulation were sorted. Always freshly harvested cells were used. (see Note 2)

Isolation of RNA

For the extraction of RNA, the RNeasy Mini Kit (Qiagen) and the protocol "Yeast II" were used with some modifications.

1. The cells were thawed in a water bath at 60 °C for 7 min. The tube was vigorously shaken every 1.5 min to maintain the temperature inside the tube below 12 °C.
2. The cells were centrifuged for 7 min at $6,300 \times g$ at 5 °C.
3. The supernatant was carefully removed by using a water jet pump with a glass Pasteur pipette submerged in the supernatant.
4. The cell pellet was resuspended in 1 mL of sterile DEPC-water and spun down at $1,500 \times g$ for 4 min at 4 °C.
5. The supernatant was carefully removed and discarded. As much liquid as possible should be removed from the yeast pellet.
6. The cells were resuspended in 100 μL SG buffer containing 100 Units (U) lyticase and incubated 20 min at room temperature. The cell suspension was gently swirled every 5 min to generate spheroplasts. (see Note 3)
7. 350 μL RLT buffer with β-mercaptoethanol was added and followed by vigorously vortexing for 2 min to lyse the cells.
8. 250 μL of 100 % ethanol was added to provide appropriate binding conditions.
9. The sample was briefly vortexed and applied to an RNeasy mini column. The tube was gently closed and centrifuged for 15 s at $8,000 \times g$ at 21 °C. The flow through tube was discarded and the RNeasy mini column was placed in a new tube.

10. The RNeasy mini column was washed by adding 350 μL RW1 buffer provided by the manufacturer and centrifuged for 15 s at $8,000 \times g$ at 21 °C.
11. A DNase treatment was used to avoid contamination with genomic DNA. A DNaseI incubation mix (10 μL DNaseI stock solution plus 70 μL Buffer RDD provided by the manufacturer [Qiagen]) was added onto the RNeasy mini column and incubated for 15 min at room temperature. Afterward, the column was spun for 15 s at $8,000 \times g$ at 21 °C.
12. The RNeasy mini column was washed again by adding 350 μL RW1 buffer and centrifuged for 15 s at $8,000 \times g$ at 21 °C.
13. The RNeasy mini column was washed two times by adding 500 μL of RPE buffer provided by the manufacturer and centrifuged for 15 s at $8,000 \times g$ at 21 °C, as described in the manufacturer's protocol.
14. To recover the RNA, the RNeasy mini column was transferred to a new RNAse-free microcentrifuge tube. 30 μL of RNAse-free water was added directly onto the center of the silica gel membrane and allowed to sit for 1 min. Afterward, the RNeasy mini column was centrifuged for 1 min at $8,000 \times g$ at 21 °C to elute the RNA. The elution step was repeated by adding 20 μL RNAse free water onto the silica-gel membrane and the RNA was eluted into the same collection tube. RNA samples can be stored at −20 °C up to 2 months, or at −80 °C for more than 2 months (see Note 4).

Estimation of the Quantity and Integrity of the Isolated RNA

1. The isolated RNA was spectrophotometrically quantified by using the NanoDrop spectrophotometer at 260 nm. The RNA purity was assessed by measuring the 260/280 ratio and the 260/230 ratio. A 260/280 ratio of 2.1 ± 0.15 indicated that the samples were free of protein. If the 260/230 ratio was at least 0.15 higher than the 260/280 ratio, the samples were free of salts (see Note 5).

2. To perform the RNA integrity check using a standard agarose gel, the RNA was loaded onto a standard 1 % agarose/ 0.5×TAE gel and the gel was run for 1.5 h at 70 V. Subsequently, the gel was stained with ethidium bromide for 20 min. The stained gel was analyzed by a gel image analysis system (see Note 6).

To check the integrity of the RNA using a Bioanalyzer 2100 (Agilent Technologies, Santa Clara, CA, USA) the manufacturer's instructions were followed.

Notes

1. The staining protocol was optimized for sort rates of 30,000 cells/s to obtain RNA of high quantity and integrity. Lower sort rates required more time to obtain the 5×10^7 cells necessary for RNA isolation resulting in lower RNA quantitity and integrity.

   The use of ethanol, methanol, formaldehyde, or formalin for fixation should be avoided, because these substances are known to hamper the analysis of mRNA [49, 50].

   All staining solutions and equipment need to be RNAse free.

2. The cells were always sorted in plastic tubes. Using glass would have been resulted in tube rupture when subjected to large temperature changes. Sorted cells were kept frozen during the whole sorting and collection procedure. It was advantageous to add some splits of pure dry ice directly into the collection tubes every 2–3 min. The tubes were changed when nearly half-filled, to avoid splashing onto deflection plates. Thinner collection tubes assure more efficient freezing but will not hold the required sample volume. Additionally, during centrifugation of multiple tubes, more cells can be lost. The stable run of the flow cytometer was checked by analyzing the yeast standard and the calibration beads at regular intervals. After every cytometer check with standards or other interruption of sorting it was necessary to repeat RNase activity removing step. Frozen samples for RNA isolation can be stored at −20 °C up to 2 months, or at −80 °C for longer than 2 months.

3. The cell wall composition can strongly depend on microenvironmental conditions [51] and/or the growth phase. Concentration of lyticase and exposure time might thus influence the success of cell wall lysis. Therefore, the exposure time and the concentration of lyticase need an appropriate adjustment.

4. Low yield of RNA can be caused by RNA damage and degradation in the preceding steps. Therefore, the use of RNAse-contaminated solutions and materials should be avoided. Additionally, the sheath fluid for flow cytometry should be prepared with DEPC- water and RNAse should be removed from the flow cytometer by extensively rinsing with DEPC-treated distilled and filtered water. The solution used to elute RNA from the silica-gel membrane (e.g., RNAse-free water) needs a pH of approximately 7. Lower pH could cause incomplete elution of RNA. The manufacturer's (Qiagen) instructions provide further help.

5. As RNA dissolved in unbuffered water might have a high variance of the 260/280 and 260/230 ratios, RNA should be dissolved in 10 mM Tris–HCl, pH 7.5 for determining these ratios. For low RNA concentrations, the NanoDrop spectrophotometer can display erroneous 260/280 and 260/230 ratios, due to the detection limit of the spectrophotometer. The detection limit regarding the RNA concentration is 2 ng/µL according to the manufacturer.

   As all nucleic acids absorb at 260 nm, a high value does not guarantee a high quantity and quality of the isolated RNA. The integrity of the isolated RNA was checked by agarose gel electrophoresis or by using a Bioanalyzer 2100. Excellent RNA integrity was indicated by clear bands of 26S rRNA and 18S rRNA with a 26S rRNA/18S rRNA band intensity ratio of around 2.

6. In order to avoid RNA degradation during gel electrophoresis, the loading dye solution needs to be RNAse free. It was helpful to prepare several tubes containing small volumes of this

solution in order to avoid contamination during the handling. Additionally, RNAse contamination of the gel electrophoresis system should be avoided by exhaustive cleaning and rinsing the gel chamber and the comb with H_2O_2 (1 %) and RNAseZap. Fresh 0.5xTAE buffer should always be used.

Analyzing the gene expression of life cells is highly difficult, especially when the cells have to be stained, flow cytometrically analyzed, and sorted beforehand. On the one hand, this is due to the small RNA amounts of cells from the separated subpopulations, making any RNA extraction very difficult. Additionally, mRNA is permanently and rapidly degraded, thus precluding long time periods for cell sorting. Hence, a fast and well-designed RNA isolation procedure was required and several control steps were essential. It was recommended to check the influence of various preparation steps on the RNA yield: cell staining, fixation, storing in buffer, cell sorting, freezing, and recovery by thawing.

The cellular amounts of RNA varied a lot independent of the physiological states of the cells. When *S. cerevisiae* was grown in a chemostat, the RNA content was proportional to the growth rate: the faster the growth, the higher the RNA content [52–54]. In addition, the yield of isolated RNA was affected by the cell wall composition, which varied depending on both microenvironmental conditions [51] and the growth phase. Consequently, the concentration of lyticase, an enzyme that breaks the yeast cell wall, and the exposition time of cells to lyticase was optimized for cells with different cultivation histories.

Notably, samples must be quickly prepared, optimally preserved, and stored to guarantee a successful application of the protocol. The described protocol enables one to isolate total-RNA from *S. cerevisiae,* which can be used for additional purposes—for example, microarray-based analyses of gene expression profiles of individual *S. cerevisiae* cells independent of the different cells' affinities to 2-NBDglucose. Consequently, functional heterogeneities within yeast populations can be examined on both the cellular and the molecular level. With regard to bioprocess optimization this is useful, assuming that glucose—as an often-used substrate in biotechnological processes with yeast as whole cell biocatalyst can be metabolized by different pathways, resulting in heterogeneities within a pure yeast cell population and therefore different cell abilities for product synthesis.

Acknowledgments We thank H. Engewald and C. Süring for technical assistance. We thank T. Hübschmann for technical assistance as well as for helpful discussions. Additionally, we want to thank M. Pähler from the working group "Chip Technology" at the Institute of Technical Chemistry of the University of Hannover. This work was supported by the Deutsche Forschungsgemeinschaft (MU 1089/5-3).

References

1. Müller S, Harms H, Bley T (2010) Origin and analysis of microbial population heterogeneity in bioprocesses. Curr Opin Biotechnol 21:100–113
2. Achilles J, Stahl F, Harms H, Müller S (2007) Isolation of intact RNA from cytometrically sorted *Saccharomyces cerevisiae* for the analysis of intrapopulation diversity of gene expression. Nat Protoc 2:2203–11
3. Ro DK, Paradise EM, Ouellet M, Fisher KJ, Newman KL, Ndungu JM et al (2006) Production of the antimalarial drug precursor artemisinic acid in engineered yeast. Nature 440:940–943
4. Kim SN, Jeong HS, Park SN, Kim HJ (2007) Purification and immunogenicity study of human papillomavirus type 16 L1 protein in *Saccharomyces cerevisiae*. J Virol Methods 139:24–30
5. Kundi M (2007) New hepatitis B vaccine formulated with an improved adjuvant system. Expert Rev Vaccines 6:133–140
6. Pemberton PA, Bird PI (2004) Production of serpins using yeast expression systems. Methods 32:185–190
7. Chang MC, Keasling JD (2006) Production of isoprenoid pharmaceuticals by engineered microbes. Nat Chem Biol 2:674–681
8. Ostergaard S, Olsson L, Nielsen J (2000) Metabolic engineering of *Saccharomyces cerevisiae*. Microbiol Mol Biol Rev 64:34–50
9. ChemLer JA, Yan Y, Koffas MA (2006) Biosynthesis of isoprenoids, polyunsaturated fatty acids and flavonoids in *Saccharomyces cerevisiae*. Microb Cell Fact 5:e20
10. Granström TB, Izumori K, Leisola M (2007) A rare sugar xylitol. Part II: biotechnological production and future applications of xylitol. Appl Microbiol Biotechnol 74:273–276

11. Davey HM, Kell DB (1996) Flow cytometry and cell sorting of heterogeneous microbial populations: the importance of single-cell analyses. Microbiol Rev 60:641–696

12. Gasol JM, Zweifel UL, Peters F, Fuhrman JA, Hagström A (1999) Significance of size and nucleic acid content heterogeneity as measured by flow cytometry in natural planktonic bacteria. Appl Environ Microbiol 65:4475–4483

13. Shapiro HM (2000) Microbial analysis at the single-cell level: tasks and techniques. J Microbiol Methods 42:3–16

14. Müller S, Nebe-von-Caron G (2010) Functional single-cell analyses—flow cytometry and cell sorting of microbial populations and communities. FEMS Microbiol Rev 34:554–587

15. Strässle C, Sonnleitner B, Fiechter A (1989) A predictive model for the spontaneous synchronization of Saccharomyces cerevisiae grown in continuous culture II. Experimental verification. J Biotechnol 9:191–208

16. Münch T, Sonnleitner B, Fiechter A (1992) The decisive role of the Saccharomyces cerevisiae cell cycle behaviour for dynamic growth characterization. J Biotechnol 22:329–352

17. Müller S, Hutter KJ, Bley T, Petzold L, Babel W (1997) Dynamics of yeast cell states during proliferation and non-proliferation periods in a brewing reactor monitored by multidimensional flow cytometry. Bioprocess Eng 17:287–293

18. Alberghina L, Smeraldi C, Ranzi BM, Porro D (1998) Control by nutrients of growth and cell cycle progression in budding yeast, analyzed by double-tag flow cytometry. J Bacteriol 180:3864–3872

19. Aon MA, Cortassa S (1998) Catabolite repression mutants of Saccharomyces cerevisiae show altered fermentative metabolism as well as cell cycle behavior in glucose-limited chemostat cultures. Biotechnol Bioeng 59:203–213

20. Müller S, Lösche A (2004) Population profiles of a commercial yeast strain in the course of brewing. J Food Eng 63:375–381

21. Klevecz RR, Bolen J, Forrest G, Murray DB (2004) A genomewide oscillation in transcription gates DNA replication and cell cycle. Proc Natl Acad Sci U S A 101:1200–1205

22. Tu BP, Kudlicki A, Rowicka M, McKnight SL (2005) Logic of the yeast metabolic cycle: temporal compartmentalization of cellular processes. Science 310:1152–1158

23. Valli M, Sauer M, Branduardi P, Borth N, Porro D, Mattanovich D (2005) Intracellular pH distribution in Saccharomyces cerevisiae cell populations, analyzed by flow cytometry. Appl Environ Microbiol 71:1515–1521

24. Herrero M, Quiros C, Garcia LA, Diaz M (2007) Use of flow cytometry to follow the physiological states of microorganisms in cider fermentation processes. Appl Environ Microbiol 72:6725–6733

25. Cipollina C, Vai M, Porro D, Hatzis C (2007) Towards understanding of the complex structure of growing yeast populations. J Biotechnol 128:393–402

26. Davey HM, Hexley P (2011) Red but not dead? Membranes of stressed Saccharomyces cerevisiae are permeable to propidium iodide. Environ Microbiol 13:163–71

27. Bryant Z, Subrahmanyan L, Tworoger M, LaTray L, Liu CR, Li MJ et al (1999) Characterization of differentially expressed genes in purified Drosophila follicle cells: toward a general strategy for cell type-specific developmental analysis. Proc Natl Acad Sci U S A 96:5559–5564

28. Juan G, Hernando E, Cordon-Cardo C (2002) Separation of live cells in different phases of the cell cycle for gene expression analysis. Cytometry 49:170–175

29. Szaniszlo P, Wang N, Sinha M, Reece LM, van Hook JW, Luxon BA et al (2004) Getting the right cells to the array: Gene expression microarray analysis of cell mixtures and sorted cells. Cytometry A 59:191–202

30. Fox RM, Von Stetina SE, Barlow SJ, Shaffer C, Olszewski KL, Moore JH et al (2005) A gene expression fingerprint of C. elegans embryonic motor neurons. BMC Genomics 6:42

31. Wiacek C, Müller S, Benndorf D (2006) A cytomic approach reveals population heterogeneity of Cupriavidus necator in response to harmful phenol concentrations. Proteomics 6:5983–5994

32. Becker D, Selbach M, Rollenhagen C, Ballmaier M, Meyer TF, Mann M et al (2006) Robust Salmonella metabolism limits possibilities for new antimicrobials. Nature 440:303–7

33. JehmLich N, Hübschmann T, Gesell SM, Völker U, Benndorf D, Müller S et al (2010) Advanced tool for characterization of microbial cultures by combining cytomics and proteomics. Appl Microbiol Biotechnol 88:575–584

34. Ben-Amor K, Heilig H, Smidt H, Vaughan EE, Abee T, de Vos WM (2005) Genetic diversity of viable, injured, and dead fecal bacteria assessed by fluorescence-activated cell sorting and 16S rRNA gene analysis. Appl Environ Microbiol 71:4679–4689

35. Kleinsteuber S, Riis V, Fetzer I, Harms H, Müller S (2006) Population dynamics within a microbial consortium during growth on diesel fuel in saline environments. Appl Environ Microbiol 72:3531–3542

36. Müller S, Vogt C, Laube M, Harms H, Kleinsteuber S (2009) Community dynamics within a bacterial consortium during growth on toluene under sulfate-reducing conditions. FEMS Microbiol Ecol 70:586–596

37. Bombach P, Hübschmann T, Fetzer I, Kleinsteuber S, Geyer R, Harms H et al (2011) Resolution of natural microbial community dynamics by community fingerprinting, flow cytometry, and trend interpretation analysis. Adv Biochem Eng Biotechnol 124:151–181

38. Hutter KJ, Eipel HE (1978) Protein content distribution in populations of baker's yeast. Eur J Appl Microbiol 5:203–206

39. Müller S, Lösche A, Schmidt M, Babel W (2001)
Optimisation of high gravity and diet beer production
in a German brewery by flow cytometry. J Inst Brew
107:373–382
40. Achilles J, Müller S, Bley T, Babel W (2004) Affinity
of single *S. cerevisiae* cells to 2-NBDglucose under
changing substrate concentrations. Cytometry A 61:
88–98
41. Natarajan A, Srienc F (1999) Dynamics of glucose
uptake by single *Escherichia coli* cells. Metab Eng
1:320–33
42. Natarajan A, Srienc F (2000) Glucose uptake rates of
single E. coli cells grown in glucose-limited chemostat cultures. J Microbiol Methods 42:87–96
43. Fiechter A, Seghezzi W (1992) Regulation of glucose
metabolism in growing yeast cells. J Biotechnol 27:
27–45
44. Sonnleitner B, Käppeli O (1986) Growth of
Saccharomyces cerevisiae is controlled by its limited
respiratory capacity: formulation and verification of a
hypotheses. Biotechnol Bioeng 28:927–937
45. Diderich JA, Schepper M, van Hoek P, Luttik MA,
van Dijken JP, Pronk J et al (1999) Glucose uptake
kinetics and transcription of HXT genes in chemostat
cultures of *Saccharomyces cerevisiae*. J Biol Chem
274:15350–15359
46. de Kock SH, du Preez JC, Kilian SG (2000) Anomalies
in the growth kinetics of *Sacharomyces cerevisiae*
strains in aerobic chemostat cultures. J Ind Microbiol
Biotechnol 24:231–236
47. Achilles J, Harms H, Müller S (2006) Analysis of
living S. cerevisiae cell states—a three color approach.
Cytometry A 69:173–7
48. Schatzmann H (1975) Anaerobes wachstum von
Saccharomyces cerevisiae. Regulatorische aspekte
des glycolytischen und respirativen stoffwechsels.
PhD thesis No. 5504, ETH Zürich
49. Esser C, Göttlinger C, Kremer J, Hundeiker C,
Radbruch A (1995) Isolation of full-size mRNA from
ethanol-fixed cells after cellular immunofluorescence
staining and fluoresccence-activated cell sorting
(FACS). Cytometry A 21:282–386
50. Diez C, Bertsch G, Simm A (1999) Isolation of
full-size mRNA from cells sorted by flow cytometry. J
Biochem Biophys Methods 40:69–80
51. Klis FM, Mol P, Hellingwerf K, Brul S (2002)
Dynamics of cell wall structure in *Saccharomyces
cerevisiae*. FEMS Microbiol Rev 26:239–256
52. Sebastian J, Mian F, Halvorson HO (1973) Effect of
the growth rate on the level of the DNA-dependent
RNA Polymerases in *Saccharomyces cerevisiae*.
FEBS Lett 34:159–162
53. Ertugay N, Hamamci H (1997) Continuous cultivation of bakers' yeast: change in cell composition at
different dilution rates and effect of heat stress on trehalose level. Folia Microbiol (Praha) 42:463–467
54. Nissen T, Schulze U, Nielsen J, Villadse J (1997) Flux
distributions in anaerobic, glucose-limited continuous
cultures of *Saccharomyces cerevisiae*. Microbiology
143:203–218

Quantitative PCR Analysis of Double-Stranded RNA-Mediated Gene Silencing in Fungi

José J. de Vega-Bartol, Vega Tello, Jonathan Niño,
Virginia Casado, and José M. Díaz-Mínguez

Abstract

Gene silencing in fungi produces a range of phenotypes based on the different amounts of target mRNA that are degraded by the RNAi machinery in each transformed strain. Detection of this range of variation when analyzing groups of transformants requires a fast and sensitive method. Quantitative or real-time PCR of reverse-transcribed target mRNA is particularly well suited for this analysis.

Keywords

Silencing • Transformation • Quantitative polymerase chain reaction • Reverse transcription • PCR amplification

Introduction

Since its discovery in *Caenorhabditis elegans* [1], dsRNA-mediated or RNAi silencing has been widely used in many organisms, including fungi, for gene functional analysis [2]. A main feature of RNAi in fungi is the range of phenotypes that can be observed in silenced or *knock-down* transformants [3–5] as a consequence of the variable reduction in the amount of target RNA. The transcript abundance has to be quantified to demonstrate that gene silencing is the underlying mechanism leading to the phenotypic effect. Real-time quantitative polymerase chain reaction (RT-qPCR) is the method of choice for the expression analysis of a limited number of different samples. Among its advantages are the low template input required, due to the high sensitivity of the method, and the high resolution, as small differences in expression between different transformants and the control can be measured. In addition, it is less time-consuming and cumbersome than other methods, such as Northern analysis, and the cost per sample is relatively low.

RT-qPCR is a combination of two steps: (1) reverse transcription from RNA to cDNA, followed by (2) PCR amplification of the cDNA and quantification of the amplification products in real time. There are different commercially available procedures to obtain a fluorescent signal from the synthesis of product that could be measured by real-time PCR instruments [6].

J.J. de Vega-Bartol • V. Tello • J. Niño • V. Casado •
J.M. Díaz-Mínguez (✉)
Department of Microbiologia y Genetica-CIALE,
Universidad de Salamanca, C/Río Duero, 12, Campus de
Villamayor, Villamayor, Salamanca, 37185, Spain
e-mail: josediaz@usal.es

V.K. Gupta et al. (eds.), *Laboratory Protocols in Fungal Biology: Current Methods in Fungal Biology*,
Fungal Biology, DOI 10.1007/978-1-4614-2356-0_22, © Springer Science+Business Media, LLC 2013

Quantitative detection of cDNA transcribed from the RNA template involved in gene silencing can be obtained using a fluorescent nucleic acid dye as *SYBR Green* or *EvaGreen*, which undergo a conformational change after binding to double-stranded DNA that results in an increase in fluorescence.

Individual qPCR reactions are characterized by the PCR cycle at which fluorescence, which is proportional to the amount of DNA produced in each PCR cycle, rises above a defined threshold, a parameter known as the threshold cycle (C_t) or crossing point (C_p). The more the target there is in the starting material, the lower the C_t. Measured variation is caused by both true biological variation and technical factors resulting in non-specific variation [6, 7]. Therefore, C_t values must be normalized against the initial concentration in each sample to correct for variability associated with the various steps of the experimental procedure, such as differences in the template input quantity and quality, yields of the extraction process and enzymatic reactions, and differences in the overall transcriptional activity of the cells analyzed. Once the C_t is measured there are two methods to quantify the amount of target DNA: the absolute quantification method calculates the amount of target DNA in the reaction by interpolation in a calibration curve that relates C_t to known amounts of template DNA; the relative quantification method compares the C_t of the target DNA with that of an endogenous control, which should be cDNA obtained from a steadily transcribed gene. Absolute quantification is required when a precise determination of the amount of amplicon is desired, for example for the calculation of fungal biomass in a host, but relative quantification is simpler and informative enough to characterize expression of silenced genes in fungi. Among the several normalization methods proposed [8], the use of expression of reference genes is currently preferred, because they are internal controls that are affected by all sources of variation during the experimental workflow in the same way as the genes of interest. However, a major problem is that silencing may produce unexpected alterations of important pathways involving down or upregulation of commonly used endogenous

genes. Therefore, the assessment of the expression stability of the gene(s) to be used as internal controls, under the experimental conditions employed, prior to its use for normalization is of paramount importance [9]. Some authors strongly recommend using several endogenous genes in parallel [10] to avoid the problems generated by RNAi side effects in gene expression.

One of the relevant aspects to take into account when designing RT-qPCR analysis of putatively silenced genes is primer design. During RNAi the RNA-induced silencing complex (RISC) cleaves the target mRNA sequence in the region complementary to the dsRNA [11]. Complete nucleolytic degradation of the resulting fragments is not always guaranteed, which might result in variations of the measured expression depending on the primer binding positions [12]. Thus, quantification of target mRNA may lead to different results depending on the pair of primers selected.

Software tools[1] and mathematical models have been developed to improve the accuracy and precision of RT-qPCR. A *de facto* standard is the $\Delta\Delta C_t$ method [13, 14] based on a mathematical method [15] dependent on cycle threshold (C_t) values and amplification efficiencies, and lately improved to include multiple reference genes [10]. Efforts have focused on the improvement of the determination of amplification efficiency because it is a known source of errors [16], and considering a fixed efficiency value is not acceptable. PCR efficiency could be achieved by standard curves, but novel quicker methods based on regression analysis of the PCR reaction kinetics after qPCR [17] lead to reproducible efficiency values [18].

Each step of the experimental workflow should be meticulously standardized to avoid introducing undesirable variation in the results that cannot be eliminated by applying the final normalization. Reverse transcription is awkward. Comparative results demonstrate that different RNA quantification methods produce different data and it is prudent to measure all samples

---

[1]Check www.gene-quantification.info for examples and download links.

using the same technique [19]. Absence of proteins, DNA contamination and inhibitors, and RNA integrity have to be determined. Moderately degraded RNA samples can be reliably analyzed and quantitated, as long as amplicons are kept short (<250 bp) and expression is normalized against a reference gene [20, 21]. DNA contamination within the RNA and cDNA samples can be respectively checked by the absence of amplification and the right size product after PCR amplification of a known gene containing an intron. Reverse transcription yields depend on the target, the reverse transcriptase enzyme, priming strategy, and experimental conditions. Also, the use of random primers, oligo(dT) or gene-specific primers has to be studied in each case [6]. Each approach has advantages and no strategy always works better [22, 23].

The PCR amplification step is remarkably reproducible under optimal conditions. As stated before, primer design is quite important but recent trends toward high throughput have resulted in a reduction of the need to optimize primer concentration [6]. However, optimization of primer concentration can significantly improve sensitivity. Melting curve analysis is a single step after amplification, and consists in a slow decrease of temperature that causes the melting of amplicons. The melting temperature (T_m) is characteristic of the size and nucleotide composition of the PCR product: those longer and richer in G/C content melt at higher temperatures. Melting causes a loss of fluorescence that is quantified and represented as a melting peak by calculating the first negative derivate of the fluorescence. These peaks provide the same information as DNA band visualization in an electrophoresis gel, such as the number of different amplicons obtained by reaction. However, routine visualization of RT-qPCR products in agarose gels is still recommended.

Finally, we advise reading and adhering to the recommendations proposed in the Minimum Information for Publication of Quantitative Real-Time PCR Experiments (MIQE) [24, 25], which is a set of guidelines that describes the minimum information necessary for evaluating RT-qPCR experiments and ensuring the integrity of the scientific literature.

Materials

Isolation of RNA

1. RNase-free 1.5-mL tubes and barrier tips.
2. Liquid nitrogen.
3. Mortar and pestle.
4. Commercial tri-reagent, such as Invitrogen Trizol Reagent.
5. DEPC-treated water.
6. Chloroform (trichloromethane).
7. Isopropanol (2-propanol).
8. Absolute ethanol.
9. 3 M Sodium acetate (NaOAc or CH_3COONa).
10. 1.5-mL tubes centrifuge.
11. *Speed vac* or desiccator connected to vacuum.

Determination of RNA Concentration and Quality

1. Nanodrop spectrophotometer.
2. DEPC-treated water and distilled water.
3. Agarose gel electrophoresis reactives (agarose, TAE buffer, etc.) and equipment. Check specific protocols.

DNase Treatment

1. RNase-free DNase I.
2. DNase kit buffer.
3. RNase-free water.
4. DNase inactivator.
5. 0.2-mL RNase-free tubes.
6. Thermocycler or water bath.

cDNA Synthesis

1. Random hexamers or oligo(dT)n primers.
2. RNase-free water.
3. cDNA synthesis kit buffer.
4. DTT.
5. dNTP mix.

6. Reverse transcriptase.
7. RNase activity inhibitor (*Invitrogen RNaseOUT* or *Roche Protector RNase Inhibitor*).
8. Thermocycler.

qPCR Quantification and Analysis

1. Gene-specific oligonucleotides.
2. SYBR Green reagent.
3. Molecular biology grade water.
4. RT-qPCR thermocycler.
5. Programs LinRegPCR, geNorm, and Microsoft Excel.

Methods

Reverse Transcription from RNA to cDNA

Isolation of RNA

In fungi, Tri-reagent[2]-based protocols generally provide a better yield than column-based methods. All the handling steps have to be done placing the tubes in ice. Use RNase-free consumables and barrier tips.

1. Grow the untransformed wild-type strain and several transformed strains in appropriate liquid medium. Previously, verify plasmid insertion into the transformant genome by PCR amplification of the promoter and a part of the target gene. Centrifuge the culture, wash the tissue (mycelia, spores, etc.) with distilled water, and centrifuge to eliminate the remaining liquid before freezing the samples under in liquid nitrogen. Samples may be stored at −80 °C until processing.

2. Remove fungal samples from the −80 °C freezer and grind 100 mg of mycelia with mortar and pestle. Add liquid nitrogen and grind to obtain a fine dust.

3. Place the ground material in a 1.5-mL tube and add 1 mL of Tri-reagent.

4. Repeatedly pipette to obtain a homogeneous suspension. Place the tubes on ice.

5. Add 0.2 mL of chloroform, shake gently several times and incubate for 5–10 min.

6. Centrifuge at 4 °C for 15 min at 12,000×g, and transfer the supernatant to a fresh tube.

7. Add 0.5 mL of cold isopropanol and precipitate RNA for 10 min at room temperature.

8. Centrifuge at 4 °C for 8 min at 12,000×g.

9. Remove the liquid phase with a micropipette, being careful not to disturb the pellet. Allow the pellet to dry for 5–10 min in a fume hood.

10. Dissolve the pellet in 0.4 mL of DEPC-treated water pipetting up and down.

11. Add 0.8 mL of cold phenol to precipitate remaining contaminations, shake by hand several times, and incubate 5–10 min in ice.

12. Centrifuge at 4 °C for 15 min at 8,000×g and transfer the supernatant to a fresh 1.5-mL tube.

13. Add 2.2 volumes of cold absolute ethanol and 0.1 vol of 3 M NaOAc to precipitate the nucleic acids. Incubate at −20 °C (the longer the better). This is a good stopping point.

14. Centrifuge at 4 °C for 20 min at 12,000×g.

15. Remove the remaining liquid with a micropipette, being careful not to disturb the pellet It is not necessary to remove all of the supernatants.

16. Wash the pellet with 70% ethanol.

17. Centrifuge being careful not to disturb the pellet for 5 min at 8,000×g.

18. Remove the remaining liquid with a micropipette. Dry the pellet in *speed vac* for 5–10 min until dry.

19. Dissolve the RNA pellet in 50 μL of DEPC-treated water.

Determination of RNA Concentration and Quality

It is very important to assess the quality of the RNA samples and accurately measure the RNA concentration. For both purposes convenient absorbance readings may be obtained from very

[2]For example Invitrogen Trizol Reagent. Cat. Number: 15596-026. http://products.invitrogen.com/ivgn/product/15596026.

small volumes by using a Nanodrop spectrophotometer.

1. For an accurate measurement heat the RNA for 5–10 min at 55 °C to completely resuspend it.

2. Set the Nanodrop to zero with 1 µL of distilled water. Configure the system for RNA measurements and measure 1 µL of DEPC-treated water as blank. Use 1–2 µL of each undiluted RNA sample. Clean the pedestal after measuring each sample. Measurement of blank samples between RNA samples is not needed. In nanodrop measurements, a ratio of absorbances at 260/280 nm around 2 is accepted as pure RNA. Lower values indicate the presence of proteins, phenol, or other contaminants. The ratio at 260/230 nm has to be higher than the previous 260/280 nm one for pure RNA. Values over 2.2 or less than 1.8 indicate the presence of contaminants. In agarose gels, intact RNA is observed as two bands around 2 and 4 kb corresponding to 18S and 28S ribosomal RNA, respectively. The degraded RNA appears as a lower molecular weight smear.

3. Visualize your RNA sample by agarose gel electrophoresis in 1% gels. Dilute 1 g of agarose in 100 mL of TAE or TBE buffer.[3] Heat at 70 °C for 10–15 min a sample containing1 µg of RNA (Optional: mix it with 0.5 vol of formaldehyde loading dye before loading on the gel well).

DNase Treatment

DNA traces that may be present in the RNA samples should be removed by means of a DNase treatment. It is best to use a procedure that does not require phenol/chloroform extraction, heating or addition of EDTA. The quantities given allow the amplification of ten target/endogenous genes with three replicates per run (30rx).

1. Transfer a volume of sample containing 2 µg of RNA to a fresh 0.2-mL PCR tube. Add water for a final volume of 44 µL. Use the same amount of total RNA in each reaction.

2. Prepare a master mix with 5 µL of 10× Buffer and 1 µL of DNase per sample.

3. Add 6 µL from the master mix to each sample tube to sum up 50 µL.

4. Incubate at 37 °C in a thermocycler for 30 min.

5. Add 5 µL of DNase inactivator reagent to each tube and incubate during 5 min at room temperature.

6. Centrifuge for 1.5 min at 10,000×g and room temperature. Proceed immediately with the cDNA synthesis.

cDNA Synthesis

The following information is for *Roche Transcription High Fidelity cDNA Synthesis kit*[4] (kit A) and Random hexamers. Differences when using *Oligo(dT)n* or other common kits such as *Invitrogen Supercript III Reverse Transcriptase*[5] (kit B) are indicated in each step.

1. Transfer 37.6 µL from the aqueous phase obtained in step 1.3.6 to a fresh 0.2-mL PCR tube.

2. Add 8 µL of Random hexamers and denature the mixture by heating the tube for 10 min at 65 °C in a thermocycler. Also, set a program in the thermocycler that immediately cools the tubes to 4 °C.

- Random hexamers primered RT showed a smaller yield than other options in controlled experiments. However, for silenced gene expression analysis, oligo(dT)n should only be used with intact RNA or experiments that require examination of splice variants. In that case, use 4–8 µL of oligo(dT)n (complete with water until 8 µL).

---

[3] Optional: the addition of formaldehyde denatures the high secondary structure of the RNA molecule for a clear visualization.

[4] Roche cat. Numbers: 05081955001/05091284001/05081 963001. http://www.roche-applied-science.com/proddata/gpip/3_6_8_39_1_3.html.

[5] Invitrogen cat. Numbers: 18080-093/18080-044/18080-085. http://www.invitrogen.com/site/us/en/home/Products-and-Services/Applications/Nucleic-Acid-Amplification-and-Expression-Profiling/Reverse-Transcription-and-cDNA-Synthesis/RT___cDNA_Synthesis-Misc/SuperScript.html.

Table 22.1 Quantities per tube for preparing a master mix for cDNA synthesis. Kit A: Roche Transcription High Fidelity cDNA Synthesis kit. Kit B: Invitrogen Supercript III Reverse Transcriptase

	Kit A	Kit B
Buffer	16	16
DTT	4	4
dNTP mix	8	Previously added
Reverse transcriptase	4.4	4
Inhibitor	2 (Protector RNase Inhibitor)	4 (RNaseOUT)
	34.4	28

Table 22.2 Conditions for cDNA synthesis. Kit A: Roche Transcription High Fidelity cDNA Synthesis kit. Kit B: Invitrogen Supercript III Reverse Transcriptase

	Random hexamers		Oligo(dT)	
	Kit A	Kit B	Kit A	Kit B
Preincubation	29°C for 10 min	25°C for 5 min	-	-
Elongation	48°C for 60 min	50–55°C for 60 min	45–55°C for 30 min	50–55°C for 60 min
Inactivation	85°C for 5 min	70°C for 15 min	85°C for 5 min	70°C for 15 min
Cooling	4°C less than 2 h. Store it at −20°C until 6 months			

- For kit B use the same quantities of primers and also add 4 µL from the dNTP mix and 6.4 µL of water.

3. Prepare a master mix with the quantities per tube that are shown in Table 22.1. Add 34.4 µL (or 28 µmL) to each tube for a final volume of 50 µL.

4. Set up the conditions shown in Table 22.2 in your thermocycler of choice according to the kit used.

5. PCR amplification of synthesized cDNA and quantification of the amplification products in real time.

Quantitative Polymerase Chain Reaction

In this example, we expect the simultaneous silencing of two highly similar transcription factors [26–28]. The wild-type strain and five transformed strains will be analyzed by using three sets of primers for the 5′, central and 3′ regions of each of the target genes, and 4 endogenous genes. Also, controls for primer contamination and RNA contamination will be included. Our qPCR equipment accepts 96rx plates, so three plates will be needed. Also, a calibrator (a common sample) for each gene must be included in all the plates to allow later normalizing for interplate variation.

Place the tubes on ice during the whole process. Keep SYBR Green that contains reagents in darkness.

1. Mix 1,344 µL of 2X SYBR Green Buffer with 591.36 µL of water in a 2-mL tube and vortex. Mix 5 µL of 2× Buffer with 3.2 µL of water for each reaction and add a 5% more to the final volume to prevent high volumes pipetting deviation.

2. Place six 1.5 mL tubes on ice. Mark them 1 to 6. Transfer 648 µL (90rx) to the first tube (1) and 216 µL (30rx) to the other (2 to 6). Set two tubes for the negative controls and mark them. Transfer 72 µL (10rx) and 43.2 µL (6rx) to each of them.

3. Add 180 µL of wild-type cDNA (RT product) to the first tube (1) and 60 µL of each of the five transformants to the other tubes (2 to 6). Add 20 µl and 4.8 µL of water to the respective negative controls. Always add cDNA before primers even if the number of samples is higher than the number of primers. Use 2 µL of RT product per reaction.

4. Place 30 tubes (3 columns and 10 rows) in a rack on ice and mark them 1 to 30. Add 27.6 µL from tubes 1, 2 and 3 to each tube of the first, second and third rack columns, respectively.

5. In ten new tubes mix 12 µL of forward and reverse 10 µM primers of each set. Add 2.4 µL of the primers mix to the three tubes of each row. Three tubes in the first row from the first primer tube, second row from the second primer, etc.

6. Centrifuge the tubes for 1 min at 12,000×g. Place all the tubes on ice in darkness.

7. Mix by pipetting. Pipette 9 μL in three contiguous wells and discard the rest. If you followed the numerical order, the first 3 wells will containing the tube 1 mix which corresponds with the first sample and first primer, the next 3 wells will contain the tube 2 which corresponds with the second sample and the second primer, and so on, as indicated in the previous schema. Pipetting 9 μL instead of 10 μL will prevent a smaller volume in the last well by previous-steps pipetting errors.

8. In the last plate, pipette 9.2 μL from the first negative control tube (10rx) in 10 contiguous wells. Add 0.8 μL form the first primer mixture in the first well and so on. Pipette 8 μL from the second negative control tube (6rx) in 6 contiguous wells. Add 2 μL of RNA extraction of the first sample (wild-type) in the first well and so on.

9. Briefly centrifuge the qPCR plate and load it in the thermocycler. Set up the appropriate program.

Data Analysis

Determine PCR Efficiency from RAW Data

1. Export the fluorescence raw data (not the C_t values) without baseline correction. Import the data in an Excel workbook. In most cases[6] raw data can be exported as a text or cvs file that can be imported in Excel using the import wizard. Set a sample per row and the fluorescence values for each cycle in columns. Keep the first row for the cycle numeration

(1–40/45) and the first two columns for the gene and sample information.

2. With the Excel program running, start the LinRegPCR (http://www.hartfaalcentrum.nl/index.php?main=files&sub=0) program [17]. Select the Excel workbook with the fluorescence raw data and appropriate range of rows and columns.

3. Determine the baselines. Select *Amplicon group* for group-based window-of-linearity and open the *Amplicon Groups* tab. Make a group for each amplicon according to the information in the first two columns. If you place the gene information in the first column, select "base groups on 1st part of the sample name from the front" and press *Group Samples* and *Set W-o-L* (Window of linearity) *per group*.

4. Save the results in the Excel workbook. You should have an average PCR efficiency value (column N) for each gene (group).

Data Normalization

1. Prepare a new Excel workbook with the average C_t and standard deviation for each combination of sample and gene. Values for each three replicas should be homogeneous, otherwise discard the value that clearly deviates from the other two. Place genes in the columns and strains (samples) in the rows. In our example that would make for 6 rows and 10 columns (plus headers). Also, insert the previous efficiency value of each gene into the new Excel workbook.

2. Calculate the ratio (R) between two plates of the average C_ts for each gene in the calibrator (wild-type sample). You can accept variations lower than 3% ($1.03 > R_n > 0.97$).

$$R = (\text{average } C_{t_{\text{wild - type , gene m}}})^{\text{Reference plate}} \div (\text{average } C_{t_{\text{t - type , gene m}}})^{\text{Other plate}}$$

3. Calculate the ratio (R) between two plates of the average C_ts for each gene in the calibrator (wild-type sample). You can accept variations lower than 3% ($1.03 > R_n > 0.97$).

4. Calculate for each gene the difference (ΔC_t) between the average C_t of a transformant and the wild-type strain, repeat it with the other transformants. Calculate also the ΔC_t of the wild-type strain, which must be 1. For each ΔC_t also calculate the error (E) of the two respective deviations.

[6]For *Roche Lightcycler 480* there is an import tool available in http://www.hartfaalcentrum.nl/index.php?main=files&sub=0

$$\Delta C_{t_{\text{sample n, gene m}}} = \text{average } C_{t_{\text{sample n, gene m}}} - \text{average } C_{t_{\text{wild-type, gene m}}}$$

$$E_{\text{sample n, gene m}} = \left((SD_{\text{sample n, gene m}})^2 + (SD_{\text{wild-type, gene m}})^2\right)^{0.5}$$

Validate Reference Genes

1. Prepare a new Excel workbook with the ΔC_t for each combination of sample and (only) endogenous gene. Place genes in the columns and strains in the rows.
2. Open geNorm plugin (http://medgen.ugent. be/~jvdesomp/genorm/) in a new Excel. Load the previous table with ΔC_t values. Press *Calculate*. Check the result matrix and discard

any endogenous gene with a stability value (*M*) higher than 1.5.

Quantify Relative Expression Values

1. Check that your Excel workbook includes the efficiency values of all the genes, the ΔC_t and prolongated error of all the gene and sample combinations, and that unstable endogenous genes have been discarded.
2. Calculate the relative expression of a target gene with respect to the calculated mean of the endogenous genes for each of the samples. For each relative expression value, calculate the expanded error (EE) of the ΔC_t (target gene and *n* endogenous genes).

$$RQ_{\text{sample n, target gene i}} = \text{Efficiency}_{\text{target gene i}}(\Delta C_{t_{\text{sample n, target gene i}}}) \div \prod^j (\text{Efficiency}_{\text{endogenous gene j}}(\Delta C_{t\text{sample, endogenous gene j}}))$$

$$EE_{\text{sample n, target gene i}} = \left((E_{\text{sample n, target gene i}})^2 + (E_{\text{samplen, endogenous gene 1}})^2 + \cdots + (E_{\text{sample n, endogenous gene j}})^2\right)^{0.5}$$

3. Calculate the minimum and maximum relative expression considering the relative expression error (EE).

$$RQ_{\text{max sample n, target gene i}} = \text{Efficiency}_{\text{target gene i}}\left(\Delta C_{t_{\text{sample n, target gene i}}} + EE_{\text{sample n, target gene i}}\right) \div \prod^j \left(\text{Efficiency}_{\text{endogenous gene j}}(\Delta C_{t_{\text{sample, endogenous gene j}}})\right)$$

$$RQ_{\text{max sample n, target gene i}} = \text{Efficiency}_{\text{target gene i}}\left(\Delta C_{t_{\text{sample n, target gene i}}} - EE_{\text{sample n, target gene i}}\right) \div \prod^j \left(\text{Efficiency}_{\text{endogenous gene j}}(\Delta C_{t_{\text{sample, endogenous gene j}}})\right)$$

4. Express the result as a value and a range in the form $RQ_{\text{sample, target gene}}$ ($RQ \min_{\text{sample, target gene}}$ $RQ \min_{\text{sample, target gene}}$). Notice that though wild-type relative expression is 1, its EE is not 0, so its expression can also show a range of values.
5. dsRNA-mediated silenced transformants should show a clear reduction of the relative expression of the analyzed gene with respect to the wild-type strain. Analyze the results obtained for each gene expression separately, as it is likely that results vary depending on the primer binding place in the target mRNA. If the results obtained when using a certain set of primers show a higher expression in the putative silenced strains than in the wild-type strain discard them, as it can be the result of incomplete degradation of the flanking mRNA regions.

References

1. Fire A, Xu S, Montgomery MK, Kostas SA, Driver SE, Mello CC (1998) Potent and specific genetic interference by double-stranded RNA in *Caenorhabditis elegans*. Nature 391:806–811
2. Salame TM, Ziv C, Hadar Y, Yarden O (2011) RNAi as a potential tool for biotechnological applications in fungi. Appl Microbiol Biotechnol 89:501–512
3. Liu H, Cottrell T, Pierini L, Goldman W, Doering T (2002) RNA interference in the pathogenic fungus *Cryptococcus neoformans*. Genetics 160:463–470
4. Ngô H, Tschudi C, Gull K, Ullu E (1998) Double-stranded RNA induces mRNA degradation in *Trypanosoma brucei*. Proc Natl Acad Sci U S A 95:14687–14692
5. Janus D, Hoff B, Hofmann E, Kück U (2007) An efficient fungal RNA-silencing system using the DsRed reporter gene. Appl Environ Microbiol 73:962–970

6. Nolan T, Hands RE, Bustin SA (2006) Quantification of mRNA using real-time RT-PCR. Nat Protoc 1:1559–1582
7. Bustin SA (2002) Quantification of mRNA using real-time reverse transcription PCR (RT-PCR): trends and problems. J Mol Endocrinol 29:23–39
8. Thellin O, ElMoualij B, Heinen E, Zorzi W (2009) A decade of improvements in quantification of gene expression and internal standard selection. Biotechnol Adv 27:323–333
9. Vandesompele J, Kubista M, Pfaffl M (2009) Reference gene validation software for improved normalization. In: Logan J, Edwards K, Saunders N (eds) Real-time PCR: current technology and applications. Caister Academic, Norwich, pp 47–64
10. Vandesompele J, De Preter K, Pattyn F, Poppe B, Van Roy N, De Paepe A et al (2002) Accurate normalization of real-time quantitative RT-PCR data by geometric averaging of multiple internal control genes. Genome Biol 3(7):research0034.1–research0034.11
11. Elbashir SM, Harborth J, Lendeckel W, Yalcin A, Weber K, Tuschl T (2001) Duplexes of 21-nucleotide RNAs mediate RNA interference in cultured mammalian cells. Nature 411:494–498
12. Van Maerken T, Mestdagh P, De Clercq S, Pattyn F, Yigit N, De Paepe A et al (2009) Using real-time qPCR to evaluate RNAi-mediated gene silencing. In: BioTechniques protocol guide 2009. p. 47. Biotechniques, NY, USA. DOI: 10.2144/000113006
13. Schmittgen T, Livak K (2008) Analyzing real-time PCR data by the comparative CT method. Nat Protoc 3:1101–1108
14. Livak K, Schmittgen T (2001) Analysis of relative gene expression data using real-time quantitative PCR and the 2ΔΔCT method. Methods 25:402–408
15. Pfaffl MW (2001) A new mathematical model for relative quantification in real-time RT-PCR. Nucleic Acids Res 29:e45
16. Ramakers C, Ruijter J, Deprez R, Moorman A (2003) Assumption-free analysis of quantitative real-time polymerase chain reaction (PCR) data. Neurosci Lett 339:62–66
17. Ruijter J, Ramakers C, Hoogaars W, Karlen Y, Bakker O, van den Hoff M et al (2009) Amplification efficiency: linking baseline and bias in the analysis of quantitative PCR data. Nucleic Acids Res 37(6):e45
18. Nordgård O, Kvaløy JT, Farmen RK, Heikkilä R (2006) Error propagation in relative real-time reverse transcription polymerase chain reaction quantification models: the balance between accuracy and precision. Anal Biochem 356:182–193
19. Huggett J, Dheda K, Bustin S, ZumLa A (2005) Real-time RT-PCR normalisation; strategies and considerations. Genes Immun 6:279–284
20. Fleige S, Pfaffl MW (2006) RNA integrity and the effect on the real-time qRT-PCR performance. Mol Aspects Med 27:126–139
21. Antonov J, Goldstein DR, Oberli A, Baltzer A, Pirotta M, Fleischmann AR et al (2005) Reliable gene expression measurements from degraded RNA by quantitative real-time PCR depend on short amplicons and a proper normalization. Lab Invest 85:1040–1050
22. Nolan T, Bustin SA (2004) Pitfalls of Quantitative Real-Time Reverse-Transcription Polymerase Chain Reaction. J Biomol Tech 15:155
23. Stahlberg A (2004) Properties of the Reverse Transcription Reaction in mRNA Quantification. Clin Chem 50:509–515
24. Bustin S, Benes V, Garson J, Hellemans J, Huggett J, Kubista M et al (2009) The MIQE guidelines: minimum information for publication of quantitative real-time PCR experiments. Clin Chem 55:611
25. Bustin S, Beaulieu J, Huggett J, Jaggi R, Kibenge F, Olsvik P et al (2010) MIQE precis: Practical implementation of minimum standard guidelines for fluorescence-based quantitative real-time PCR experiments. BMC Mol Biol 11:74
26. Ramos B, Alves-Santos FM, García-Sánchez MA, Martín-Rodrigues N, Eslava AP, Díaz-Mínguez JM (2007) The gene coding for a new transcription factor (ftf1) of Fusarium oxysporum is only expressed during infection of common bean. Fungal Genet Biol 44:864–876
27. de Vega-Bartol JJ, Martín-Domínguez R, Ramos B, García-Sánchez MA, Díaz-Mínguez JM (2011) New virulence groups in Fusarium oxysporum f. sp. phaseoli: the expression of the gene coding for the transcription factor ftf1 correlates with virulence. Phytopathology 101:470–479
28. Alves-Santos F, Ramos B, García-Sánchez MA, Eslava AP, Díaz-Mínguez JMA (2002) DNA-based procedure for in-planta detection of Fusarium oxysporum f. sp. phaseoli. Phytopathology 92(3):237–244

Semi-Nested PCR Approach to Amplify Large 18S rRNA Gene Fragments for PCR-DGGE Analysis of Soil Fungal Communities

23

Miruna Oros-Sichler and Kornelia Smalla

Abstract

Denaturing gradient gel electrophoresis (DGGE) of 18S rRNA gene fragments PCR-amplified from total community DNA is a powerful tool for the parallel comparative analysis of environmental fungal communities. The 18S rRNA gene has the advantages of universality, high phylogenetic information content, and a large number of sequences in the data banks. The comparative analysis of soil fungal communities from large numbers of samples by PCR-DGGE requires consistent amplification and separation efficiency, as achieved by the following semi-nested PCR-DGGE protocol based on two-step PCR of 1,650 bp rRNA gene fragments from bulk soil DNA and their separation in DGGE.

Keywords

Denaturing gradient gel electrophoresis (DGGE) • PCR-DGGE • 18S rRNA gene fragments • Soil fungal communities

Introduction

Along with bacteria, fungi are involved in soil functionality, comprising physical, chemical, and biological aspects [1]. As components of the complex interactive soil food web, fungal assemblages in soil respond to changes of the different soil trophic levels with community changes, which, in turn, affect the soil properties. Therefore, fungal community shifts may serve as indicators for soil food web modifications and their analysis is of crucial importance for the understanding of soil ecosystems [2].

For a long time the soil fungal community composition was studied only by methods based

M. Oros-Sichler
Julius Kühn Institut, Institute for Epidemiology and Pathogen Diagnostics, Zimmerstrasse 26, Braunschweig, Lower Saxony 38106, Germany

K. Smalla (✉)
Julius Kühn Institut, Federal Research Centre for Cultivated Plants, Institute for Epidemiology and Pathogen Diagnostics, Messeweg 11-12, Braunschweig, Lower Saxony 38104, Germany
e-mail: kornelia.smalla@jki.bund.de

on isolation of fungi directly from environmental samples plated onto nutrient media. However, the isolation techniques are very fastidious and confined by the boundaries set by unculturability of many fungi. The analysis of the community structure and dynamics of fungal communities from soils achieved important advances in the past two decades thanks to the molecular techniques [3]. Analysis methods based on PCR amplification of marker gene fragments from total DNA extracted from environmental samples brought forth suitable approaches to analyze comparatively high numbers of samples in a rapid and efficient manner.

The genes of the ribosomal gene complex, consisting of the small subunit (SSU) 18S rRNA gene, the large subunit (LSU) 28S rRNA gene, the internal transcribed spacer (ITS), and the intergenic spacer (IGS), are frequently used in fungal community profiling [4]. These marker genes comprise both highly conserved domains and variable regions [5, 6], allowing the design of suitable primer systems and the high resolution analysis of fungal communities at taxonomical levels ranging from phylum to strain [7, 8]. The fungal sequence data bases are considerably informative, especially for intensively studied taxonomic groups (e.g., arbuscular mycorrhizal fungi) [9].

Several primer systems, group specific or fungal universal, were designed to amplify either SSU, LSU rRNA gene fragments or the ITS/IGS regions and used for the characterization of fungal diversity in soils [10]. Theoretical and practical evaluation of primers targeting the 18S rRNA gene or the ITS region revealed that some of them amplify also nonfungal sequences, or other primers exclude major fungal taxa. Four different primer pairs were tested for their specificity toward fungal rRNA genes and their suitability for assessing the diversity of fungal communities in grassland soils [11]. Based on cloning and sequencing of amplicons obtained with each primer system from soil DNA, the authors concluded that primer biases might be less significant than previously expected.

Subsequent to the PCR amplification, the amplicon pools can be identified taxonomically

by sequence analysis or separated by means of molecular profiling methods that exploit the differences in their DNA sequence or conformation and result in taxonomically anonymous fingerprints that allow comparisons between different sample types.

Primer systems targeting fungal rRNA genes, coupled with molecular fingerprinting techniques such as denaturing gradient gel electrophoresis (DGGE) to analyze PCR products obtained from total community DNA, provide appropriate strategies for descriptive and comparative analysis of soil fungal community structure [12]. Due to the fact that the fungal rRNA genes are more conserved than bacterial 16S rRNA genes, the molecular fingerprints obtained for 18S rRNA gene fragments are less complex and, thus, easier to evaluate than the bacterial profiles. If different taxa contribute to the same band in the DGGE fingerprints [13, 14], the analysis resolution can be improved by using taxon-specific primers (e.g., for *Trichoderma* community composition) [15].

The specificity of the primer system and the phylogenetic information contained within the amplified fragments are decisive factors for the degree of resolution with which the community structure is revealed. The fungal universal primer NS1 [7], combined with the fungus-specific reverse primer FR1 [8], amplifies about 1,650 bp of the 18S fungal rRNA gene and thus allows the use of the most phylogenetic information contained by this gene. Initially designed and used for the study of wood-inhabiting fungi [8], this primer system was used later in only a few studies—for example, to investigate the fungal communities associated with the bulk and rhizosphere soil of maize from tropical climate [14] and to compare the fungal community structure under different agricultural practices in soil mini ecosystems [16] or in the endorhiza of different potato lines at two different sites [17]. These studies asserted the reproducibility of this system and suitability even for a soil with high contents of humics [16]. However, when analyzing different soils originating from 36 sites with known properties, we encountered difficulties in obtaining PCR amplicons from total community

DNA of a broad sample range using directly the primer combination NS1-FR1-GC [18].

The literature mentioned several different factors affecting the efficiency of DNA amplification from soil samples. For example, insufficient amount or low priming availability of the template DNA is known as a limitation for successful amplification [19, 20]. Also, co-extracted humic substances were often reported to inhibit the yield of amplicons from soil samples [19, 21]. Both the proportion in which the fungal DNA contributes to the total community DNA extracted, and the amount of co-extracted contaminants (e.g., humic acids), cannot be assessed by agarose gel electrophoresis and might reduce the amplification efficiency [19]. In addition to that, it was already speculated in the literature that the GC-clamp necessary for the DGGE analysis [22] might bias direct PCR amplification [23, 24].

Nested PCR approaches have the advantage of enhanced sensitivity, allowing the detection of problematic DNA (e.g., low target amount, high contaminant amount) and the reduction of nonspecific amplification [25]. Moreover, the GC-clamp necessary for DGGE analysis can be included without difficulties in the second PCR step, after the specific templates reached a sufficient amount in the first PCR step.

Therefore, we designed the semi-nested PCR protocol presented in this chapter, which consists of a first amplification with the novel primer combination NS1-EF3, followed by a second amplification with the primers NS-FR1-GC [8, 18]. EF3 was designed and used formerly in other primer combinations [26–28]. The semi-nested PCR protocol presented in this chapter was used successfully for the comparative analysis of soil fungal communities from 36 different sites [18], as well as for the study on the impact of the site, the sugar beet cultivar, the seasonal dynamics, and the rhizosphere effect on the fungal community structure at three different sites planted with sugar beet [29].

However, the semi-nested procedure presented in this chapter is far from free of limitations.

For example, inconsistent PCR amplification was reported when using the protocol presented in this chapter, resulting in a high variability of the replicate DGGE patterns when investigating fungi of the rhizosphere of strawberry and oil seed rape at different sites [30]. The problem was partially solved by these authors by switching to a nested PCR version, replacing in the first PCR step the primer NS1 with NS0, which is located upstream from NS1.

Furthermore, two different studies reported on the retrieval of nonfungal sequences when working with the PCR protocol presented in this chapter. Firstly, sequences of ubiquitous soil flagellates were retrieved in the analysis of fungal communities from bulk soils of three different sites [14]. Secondly, it was impossible to compare fungal communities from the gut of *Diabrotica virgifera* feeding on maize roots in different soils with the protocol in this chapter, as the DGGE patterns generated were dominated by a band of *D. virgifera* [31].

Last but not least, because of the relatively high conservation of the 18S rRNA gene within the fungi, some of the fragments might not contain enough variation to allow DGGE separation and thus migrate with similar electrophoretic mobility, as previously observed [13, 14].

Therefore, the ITS should be mentioned here as a valuable alternative marker for the analysis of soil fungal communities. The use of ITS fragments, complementary or independently to the analysis of 18S rRNA gene fragments, might allow DGGE separation up to an intraspecific level [32]. Thus, additional insights might be gained for comparative studies (e.g., when analyzing potential effects of transgenic crops on the microbial communities in comparison with nontransgenic lines) [33]. The number of ITS entries in the GenBank (32,050) exceeded already by June 2005 the number of submitted 18S rRNA gene fragment sequences (30,651) [34] and attained 50,956 fully identified and 27,364 insufficiently identified ITS sequences as of February 2008 [35]. Several PCR-DGGE protocols based on ITS fragments are available for the analysis of soil fungal communities, of which a semi-nested protocol with the primers ITS1 and ITS4 in the first PCR and the primers ITS1-GC and ITS2 in the second PCR was described recently in detail [36].

Equipment and Materials

Equipment

1. FastPrep® Instrument FP 120 for bead beating (BIO101, Carlsbad, California).
2. Microcentrifuge.
3. Vortex.
4. Magnetic stirrer.
5. Electrophoresis chamber with power supply and accessories.
6. Thermocycler.
7. DGGE DCode™ Universal Mutation Detection System (Biorad, München, Germany) and accessories.
8. Gradient maker with peristaltic pump.
9. Gel documentation system with UV transilluminator and camera (e.g., UV System, INTAS®, Mitsubushi Electric Corporation).
10. Fast DNA®Spin®Kit for Soil (BIO101, Carlsbad, California).
11. GENECLEAN®SPIN® Kit (BIO101, Carlsbad, California).
12. 1 kb molecular weight DNA marker (e.g., Invitrogen).
13. Ethidium bromide.
14. Agarose
15. AmpliTaq DNA Polymerase Stoffel Fragment (Applied Biosystems, Foster City, California).
16. Deoxynucleotide Triphosphate Set (Roche Diagnostics, Germany).
17. Primers (Table 23.1).
18. 2% dimethyl sulfoxide (DMSO).
19. 0.5 M EDTA pH 8.
20. Deionized formamide (Stir slowly for about 30 min 10 g Serdolit MB-1 and 1 L formamide. Filter through Whatman paper to remove ionic exchange resin. Store at -20 ° C as 50 mL Falcon tube aliquots).
21. 18% denaturant 7.5% acrylamide stock solution (see Notes 1, 2).
22. 38% denaturant 7.5% acrylamide stock solution (see Notes 1, 3).
23. 10% Ammonium peroxodisulfate (APS) (w/v) in MilliQ water (stored as aliquots at -20 ° C).
24. Tetramethylethylendiamine (TEMED).

25. MilliQ (deionized) water.
26. 5× TBE Buffer (27.5 g boric acid, 54 g Tris base, 20 mL 0.5 M EDTA pH 8.0 in 1 L distilled water).
27. 50× TAE Buffer: (242.2 g Tris base, 18.6 g EDTA, 57.1 mL acetic acid in 1 L distilled water; diluted 1:50 for DGGE run).
28. 6× DGGE loading buffer (25 mg bromophenol blue, 25 mg xylene cyanole, and 3 mL glycerol in 10 mL distilled water, stored at 4 ° C).
29. DGGE standard (see Note 4).
30. GelBond film (Lonza, Switzerland).
31. Reaction vials (1.5 and 2.0 mL).
32. Pipette tips (10, 20, 100, 200, 1,000 µL) and capillary pipette tips.
33. Syringe needles.
34. 15 mL polypropylene Falcon tubes.
35. DGGE fixative solution (10 mL acetic acid and 200 mL ethanol in 1,790 mL MilliQ water).
36. DGGE staining solution (0.2 g silver nitrate in 100 mL MilliQ water, made freshly for each gel).
37. DGGE developing solution (400 µL 37% formaldehyde in 100 mL 1.5% sodium hydroxide, made freshly for each gel).
38. DGGE stopping solution (7.5 g of sodium carbonate in 1 L MilliQ water).
39. DGGE conservation solution (250 mL ethanol and 100 mL glycerol in 650 mL MilliQ water).

Methods

Initial Material

Total community DNA extracted directly from replicate composite bulk soil samples taken from random plots at different sites (see Notes 5–8).

Semi-Nested PCR

The First Amplification Step

The primer combination NS1-EF3 (see Table 23.1) is used, amplifying almost the entire

Table 23.1 Primers used for the semi-nested PCR amplification of 18S rRNA gene fragments from bulk soil total community DNA (GC-clamp underlined)

Primer	Sequence	References
NS1	5′-GTA GTC ATA TGC TTG TCT C-3′	White et al. [7]
EF3	5′-TCC TCT AAA TGA CCA AGT TTG-3′	Smit et al. [26]
FR1-GC	5′-<u>CCC CCG CCG CGC GCGGCG GGCGGG GCG GGG GCA CGG GCC G</u>-AIC CAT TCA ATC GGT AIT-3′	Vainio and Hantula [8]

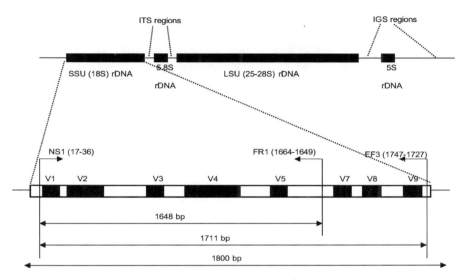

Fig. 23.1 Position of the annealing sites of the primers used in this chapter (NS1, FR1, EF3) on the 18S rRNA gene of *Saccharomyces cerevisiae* and length of the amplicons generated with different primer combinations

18S rRNA gene (Fig. 23.1). Perform the reaction with ca. 15–20 ng DNA extract in 25 μl volume containing: Stoffel buffer (10 mM KCl, 10 mM Tris–HCl pH 8.3), 0.2 mM deoxynucleoside triphosphates, 3.75 mM MgCl$_2$, 2% (w/v) dimethyl sulfoxide (see Note 9), 0.2 μM of each primer, and 2 U/μl of *Taq* DNA polymerase Stoffel fragment. PCR cycling program: 5 min denaturation at 94 ° C, followed by 25 thermal cycles of 30 s at 94 °C, 45 s at 47 °C, 3 min at 72 °C, and final extension at 72 °C for 10 min.

The Second Amplification Step

The primer combination NS1-FR1-GC (see Table 23.1) is used, amplifying 1,650 bp of the 18S rRNA gene (see Fig. 23.1). Perform the reaction with optimized dilutions of the amplicons from the first PCR step in 25 μL volume containing Stoffel buffer (10 mM KCl, 10 mM Tris–HCl pH 8.3),

0.2 mM deoxynucleoside triphosphates, 3.75 mM MgCl$_2$, 2% (w/v) dimethyl sulfoxide, 0.2 μM of each primer and 2 U/μl of *Taq* DNA polymerase Stoffel fragment). PCR cycling program: 5 min denaturation at 94 ° C, followed by 20 thermal cycles of 30 s at 94 ° C, 45 s at 48 ° C, 3 min at 72 ° C, and final extension at 72 °C for 10 min.

DGGE Fingerprinting of Soil Fungal Communities

See Figs. 23.2a,b and Note 10.

Gel Casting

Assembly of the gel sandwich:

1. Place the glass plates on a plane table. Carefully clean the surface of the glass plates with 97% ethanol.

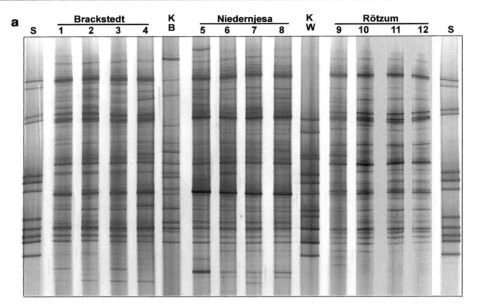

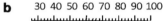

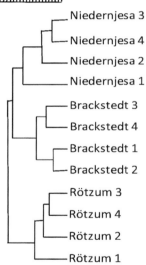

Fig. 23.2 (a) DGGE fungal community fingerprints of 18S rRNA gene fragments amplified from bulk soil DNA from three sites with different soil types: Lanes *1–4*, replicates from Brackstedt (sandy soil); lanes *5–8*, replicates from Niedernjesa (alluvial silt); lanes *9–12*, replicates Rötzum (clay-rich blacksoil) loess loam; lane *S*, standard mixture of PCR-amplified 18S rRNA gene fragments of fungal isolates; lanes *KB* and *KW*, standard mixtures of PCR-amplified 18S rRNA gene fragments cloned from total soil DNA of similar soils. (b) Dendrogram based on the Pearson correlation indices and UPGMA cluster analysis of the fungal community fingerprints of 18S rRNA gene fragments amplified from bulk soil DNA from Brackstedt, Niedernjesa, and Rötzum. The differences between the profiles are indicated by percentage of similarity. Patterns of soil samples originating from different site clusters separately

2. Spread a couple of tap-water drops on the large glass plate.
3. Place the GelBond film with the hydrophobic side in direct contact with the large glass plate, ensuring the perfect alignment of the film to the bottom of the glass.
4. Fix the film to the glass free of air bubbles (e.g., with a Drigalski spatula). Remove excess tap water.
5. Place two spacers at the outer sides of the large glass plate and the small glass on the top.
6. Insert the glass plate assembly within sandwich clamps ensuring that the bottoms correspond perfectly.
7. Place the sandwich assembly in a casting stand with a rubber strip at the bottom to prevent leakage. Ensure tight contact of the sandwich assembly bottom with the rubber strip and a stable position of the casting assembly. Close both clamps with equal pressure using the alignment card. Do not over-tighten clamps to avoid usage of plates.
8. Insert the comb in the glass plate sandwich.

Casting of denaturing gradient gels and polymerization:

1. Thaw and keep denaturing stock solutions on ice.
2. Add 45 µL 10% APS and 26 µl TEMED to each 15 mL of 18% respectively 38% denaturing solution and mix by inverting the vials. Work on ice to prevent premature polymerization.
3. Place the gradient maker on a stir plate at speed 300 rounds/min with a magnet stirrer in the outlet port chamber.
4. Connect the gradient maker to the peristaltic pump (the pump is off and the gradient maker channel is closed). Provide the pump tube with a syringe needle. Insert the needle centrally between comb and small glass plate.
5. Pour the 38% denaturing solution into the outlet port chamber of the gradient maker. Briefly open and close the valve to remove air between chambers. Pour the 18% denaturing solution in the remaining chamber.
6. Turn on the peristaltic pump and open simultaneously the valve between chambers.

An optimal flow of 5 mL/min is recommended. Ensure the solutions flow without leaking from the sandwich until air bubbles reach the syringe needle.

7. After gel casting, remove the needle and water-flush the gradient maker and tubing to discard solution rests.
8. Let the gel polymerize unmoved for at least 1 h. Use the same day or keep at 4 ° C wrapped in wet towels.

Pre-Run

1. Insert two gel sandwiches into the electrophoresis core. If only one gel is used, replace second gel with a glass plate sandwich without spacers.
2. Place the core assembly into the buffer tank filled with 1× TAE buffer. Renew 50% of the buffer between runs. Check buffer level, set up the temperature for 58 °C and start the pump.
3. When the buffer reaches the run temperature, turn off the system and remove the comb from the gel.

Sample Loading and Electrophoresis

1. Adjust the volume of PCR products to load at similar DNA concentrations. The different concentrations of the samples can interfere with software analysis of the gel.
2. Mix PCR products 1:1 with DGGE loading buffer and load with microcapillary pipette tips. Ensure that a maximum of 20 µL of sample is loaded to prevent well overflowing.
3. Load a DGGE standard in the outside lanes to assess the gradient formation and the band positions and to normalize the gel in further software analysis.
4. Close the system; check buffer, temperature, and pump; and start electrophoresis.

Gel Staining, Drying, and Scanning

1. Work on a switch rocker.
2. Transfer the gel in a recipient with 100 mL fixative solution for 10 min. or unmoved overnight.
3. Discard the fixative solution and add 100 mL 0.2% silver nitrate fresh solution for 15 min.

4. Discard the silver nitrate solution in a specific waste. Wash the gel at least twice for 1 min with MilliQ water.

5. Change the gel in a new recipient and add 100 mL fresh developing solution.

6. Discard the developing solution as pale bands appear. Add 100 mL stopping solution for ca. 10 min.

7. Discard the stopping solution. Add 100 mL conservation solution for at least 7 min.

8. Transfer the gel on a rigid frame. Cover the gel without air bubbles with a cellophane film soaked with conservation solution. Fix with clamps. Air-dry at room temperature for 2 days.

9. Transform the gel image in a digital picture using any transparency scanning system available.

10. Analyze the digitalized gel by means of software—e.g., GelCompar 4.0 (Applied Maths, Ghent, Belgium)—with unweighted pair group method using arithmetic averages (UPGMA) cluster analysis.

11. Apply any statistical method available to ensure statistical significance of results (see reference [37]).

Notes

1. Consider that 100% denaturant solution contains 40% deionized formamide and 7 M urea.

2. 18% denaturant 7.5% acrylamide stock solution: Dissolve 18.93 g urea in 100 mL MilliQ water. Add 5 mL 50× TAE, 18 mL deionized formamide, and 62.5 mL acrylamide Rotiphorese Gel 30 (37.5:1) (Roth, Germany). Adjust the volume to 250 mL in a volumetric flask and filter. Aliquot 15 mL solution in Falcon tubes. Store at -20 °C.

3. 38% denaturant 7.5% acrylamide stock solution: Dissolve 39.94 g urea in 100 mL MilliQ water. Add 5 mL of 50× TAE, 38 mL deionized formamide and 62.5 mL acrylamide Rotiphorese Gel 30 (37.5:1) (Roth, Germany). Adjust the volume to 250 mL in a volumetric flask and filter. Aliquot 15 mL solution in Falcon tubes. Store at -20 ° C.

4. The DGGE standard is an artificial mixture of 18S rRNA gene fragments PCR-amplified from single fungal isolates or clones known to migrate with different electrophoretic mobilities in DGGE protocol used.

5. Ensure at least four replicate samples per site for representative statistical results.

6. Dig a number of bulk soil cores per plot representative for the plot dimensions; e.g., 8–10 cores (15–20 cm of top soil) per 10 m². Avoid root material if bulk soil analysis is intended. Mix well by sieving.

7. Ensure an amount of 0.3–1 g dry weight of soil per replicate for DNA extraction to minimize eventual heterogeneous distribution of fungal DNA targets and to ensuring representative results.

8. For total DNA extraction, use one of the commercial kits for soil, preferably BIO101 Fast DNA®Spin®Kit for Soil, combined with a harsh cell lysis to break fungal cell walls, e.g., with the FastPrep® Instrument for 1 min at 4,000 rpm. Purify the crude DNA from eventual humic contaminants, e.g., with the GENECLEAN®SPIN® Kit. Store DNA extracts at -20 °C until further procedures.

9. Dimethyl sulfoxide in the reaction mixture is known to enhance PCR by eliminating nonspecific amplification and to improve the primer annealing efficiency by destabilizing secondary structures within the template.

10. All DGGE materials, gel casting procedures, and running conditions presented in this chapter are strictly referred to the DGGE DCode™ Universal Mutation Detection System (Biorad, München, Germany). Use an 18–38% denaturing gradient.

References

1. Ritz K, Young IM (2004) Interactions between soil structure and fungi. Mycologist 18:52–59

2. Schwarzenbach KA (2008) Monitoring soil fungal communities structures and specific fungal biocontrol strains for ecological effects and fate studies used for risk assessment. Dissertation. Swiss Federal Institute of Technology, Switzerland

3. Anderson IC, Cairney JGW (2004) Diversity and ecology of soil fungal communities: increased under-

standing through the application of molecular techniques. Environ Microbiol 6(6):769–779

4. Bidartondo MI, Gardes M (2005) Fungal diversity in molecular terms: profiling, identification, and quantification in the environment. In: Dighton J, White JF, Oudemans P (eds) The fungal community: its organisation and role in the ecosystem, 3rd edn. CRC Press, New York, pp 215–239

5. Woese CR (2000) Interpreting the universal phylogenetic tree. Proc Natl Acad Sci USA 97:8392–8396

6. Tehler A, Little DP, Farris JS (2003) The full-length phylogenetic tree from 1551 ribosomal sequences of chitinous fungi, Fungi. Mycol Res 107(8):901–916

7. White TJ, Bruns TD, Lee S, Taylor JW (1990) Amplification and direct sequencing of fungal ribosomal RNA genes for phylogenetics. In: Innis MA, Gelfand DH, Sninsky JJ, White TJ (eds) PCR protocols: a guide to methods and applications. Academic, New York, pp 315–322

8. Vainio EJI, Hantula J (2000) Direct analysis of wood-inhabiting fungi using denaturing gradient gel electrophoresis of amplified ribosomal DNA. Mycol Res 19:6823–6831

9. Bridge PD, Spooner BM, Roberts PJ (2005) The impact of molecular data in fungal systematics. Adv Bot Res 42:33–67

10. Hagn A, Geue H, Pritsch K, Schloter M (2002) Assessment of fungal diversity and community structure in agricultural used soils. Recent Res Dev Microbiol 6:551–569

11. Anderson IC, Campbell CD, Prosser JI (2003) Diversity of fungi in organic soils under a moorland Scots pine (*Pinus sylvestris* L.) gradient. Environ Microbiol 5(11):1121–1132

12. Kowalchuk GA (1999) New perspectives towards analysing fungal communities in terrestrial environments. Curr Opin Biotechnol 10:247–251

13. Gomes NCM, Fagbola O, Costa R, Rumjanek NG, Buchner A, Mendonça-Hagler L et al (2003) Dynamics of fungal communities in bulk and maize rhizosphere soil in the tropics. Appl Environ Microbiol 69: 3758–3766

14. Oros-Sichler M, König M, Smalla K. Polyphasic approach reveals comparable results obtained with different methods and limitations of using large 18S rRNA gene fragments in the analysis of soil fungal communities (in preparation)

15. Meincke R, Weinert N, Radl V, Schloter M, Smalla K, Berg G (2010) Development of a molecular approach to describe the composition of *Trichoderma* communities. J Microbiol Methods 80:63–69

16. Pennanen T, Paavolainen L, Hantula J (2001) Rapid PCR-based method for the direct analysis of fungal communities in complex environmental samples. Soil Biol Biochem 33:697–699

17. Götz M, Nirenberg H, Krause S, Wolters H, Draeger S, Buchner A et al (2006) Fungal endophytes in potato roots studied by traditional isolation and cultivation-independent DNA-based methods. FEMS Microbiol Ecol 58:404–413

18. Oros-Sichler M, Gomes NCM, Neuber G, Smalla K (2006) A new semi-nested PCR protocol to amplify large 18S rDNA fragments for the PCR-DGGE analysis of soil fungal communities. J Microbiol Methods 65:63–75

19. Wilson IG (1997) Inhibition and facilitation of nucleic acid amplification. Appl Environ Microbiol 63:3741–3751

20. Polz MF, Cavanaugh CM (1998) Bias in template-to-product ratios in multitemplate PCR. Appl Environ Microbiol 64:3724–3730

21. Tebbe CC, Vahjen W (1993) Interference of humic acids and DNA extracted directly from soil in detection and transformation of recombinant DNA from bacteria and a yeast. Appl Environ Microbiol 59: 2657–2665

22. Muyzer G, Smalla K (1998) Application of denaturing gradient gel electrophoresis (DGGE) and temperature gradient gel electrophoresis (TGGE) in microbial ecology. Antonie Leeuwenhoek 73:127–141

23. Bourne DG, McDonald IR, Murrell JC (2001) Comparison of *pmo*A PCR primer sets as tools for investigating methanotroph diversity in three Danish soils. Appl Environ Microbiol 67:3802–3809

24. Castle D, Kirchman DL (2004) Composition of estuarine bacterial communities assessed by denaturing gradient gel electrophoresis and fluorescence in situ hybridisation. Limnol Oceanogr Methods 2:303–314

25. Jacquot E, van Tuinen D, Gianinazzi S, Gianinazzi-Pearson V (2000) Monitoring species of arbuscular mycorrhizal fungi in planta and in soil by nested PCR: application to the study of the impact of sewage sludge. Plant Soil 226(2):179–188

26. Smit E, Leeflang P, Glandorf B, van Elsas JD, Wernars K (1999) Analysis of fungal diversity in the wheat rhizosphere by sequencing of cloned PCR-amplified genes encoding 18S rRNA and temperature gradient gel electrophoresis. Appl Environ Microbiol 65:2614–2621

27. Van Elsas JD, Frois Duarte G, Keijzer-Wolters A, Smit E (2000) Analysis of the dynamics of fungal communities in soil via fungal-specific PCR of soil DNA followed by denaturing gradient gel electrophoresis. J Microbiol Methods 43:133–151

28. Glandorf DCM, Verheggen P, Jansen T, Jorritsma J-W, Smit E, Leeflang P et al (2001) Effect of genetically modified *Pseudomonas putida* WCS358r on the fungal rhizosphere microflora of field-grown wheat. Appl Environ Microbiol 67:3371–3378

29. Oros-Sichler M, König M, Smalla K. Variability of fungal communities associated with field-grown sugar beet as revealed by molecular and statistical methods (in preparation).

30. Costa R, Götz M, Mrotzek N, Lottmann J, Berg G, Smalla K (2006) Effects of site and plant species on rhizosphere community structure as revealed by molecular analysis of microbial guilds. FEMS Microbiol Ecol 56:236–249

31. Dematheis F, Smalla K. Dominant gut associated microorganisms in *Diabrotica virgifera virgifera* LeConte are not influences by the soil type (in preparation)

32. Nilsson RH, Kristiansson E, Ryberg M, Hallenberg N, Larsson K-H (2008) Intraspecific ITS variability in the Kingdom *Fungi* as expressed in the international sequence databases and its implication for molecular species identification. Evol Bioinform 4: 193–201

33. Weinert N, Meincke R, Gottwald C, Heuer H, Gomes NCM, Schloter M et al (2009) Rhizosphere communities of genetically modified zeaxanthin-accumulating potato plants and their parent cultivar differ less than those of different potato cultivars. Appl Environ Microbiol 75(12):3859–3865

34. Neubert K, Mendgen K, Brinkmann H, Wirsel SGR (2006) Only a few fungal species dominate highly diverse mycofloras associated with the common reed. Appl Environ Microbiol 72(2):1118–1128

35. Ryberg M, Kristiansson E, Sjökvist E, Nilsson RH (2009) An outlook on the fungal internal transcribed spacer sequences in GenBank and the introduction of a web-based tool for the exploration of fungal diversity. New Phytol 181:471–477

36. Dematheis F, Smalla K. (2011) Study of fungal endophytes in plant roots versus rhizosphere and soil fungal communities. In: Pirttilä AM, Sorvari S (eds) Prospects and applications for plant associated microbes laboratory manual. Part B: Fungi. BioBien Innovations, Finland

37. Oros-Sichler M, Costa R, Heuer H, Smalla K (2007) Molecular fingerprinting techniques to analyse soil microbial communities. In: Van Elsas JD, Jansson J, Trevors J (eds) Modern soil microbiology, 2nd edn. CRC Press, Boca Raton, FL, pp 355–386

Proteomic Protocols for the Study of Filamentous Fungi

24

Raquel González Fernández
and Jesús V. Jorrín Novo

Abstract

In the last few years, proteomics has experienced rapid improvement in technologies and applications. Gel-based strategies have become the method of choice for both identification and quantification of proteins in most studies. The workflow of a standard gel-based proteomic experiment includes experimental design, sampling, protein extraction, protein separation, mass spectrometry analysis, protein identification, data statistical analysis, validation of the identification, quantification, and data analysis. The appropriate protocol to be used depends on and must be optimized for the biological system (i.e., fungal species, plant species, organ, tissue, cells). Preliminary steps are relevant. The choice of a good extraction protocol in a proteomic experiment is crucial because only if you can extract and solubilize a protein you have a chance of detecting and identifying it. This is more important in the case of filamentous fungi, which, owing to their particular cellular characteristics, can be considered recalcitrant biological material, making it difficult to obtain quality protein samples to proteomic analysis.

Fungi have an exceptionally robust cell wall, consisting largely of chitin, which makes up the majority of the cell mass. Because of its rigidity, cell lysis is an important element in fungal proteomics. For protein extraction, various buffer- and precipitation-based protocols are available. In most of these protocols, trichloroacetic acid (TCA) and/or acetone are used for protein precipitation, or a phenol extraction is made, where proteins are solubilized in the phenolic phase and then are precipitated with methanol and ammonium sulfate. These methods also eliminate some

R.G. Fernández (✉) • J.V.J. Novo
Department of Biochemistry and Molecular Biology,
University of Córdoba, Agro-forestry and Plant
Biochemistry and Proteomics Research Group,
Ed. Severo Ochoa, planta baja. Campus de Rabanales,
Córdoba 14071, Spain
e-mail: q42gofer@uco.es

contaminants abundant in fungal material (such as polysaccharides, lipids, nucleic acids, or phenolic compounds) that affect the protein isoelectrofocusing and electrophoresis processes.

Key words

fungal proteomics • fungal secretome • cell lysis • protein precipitation • protein isoelectrofocusing

Introduction

In a post-genomic era, proteomic technologies have become a powerful tool to study the proteome and to examine alterations in protein profiles [1]. Similar to genomics and transcriptomics, proteomics has evolved to incorporate high-throughput processes, which allow faster analysis of a larger number of proteins [2, 3]. Proteomics involves the combined applications of advanced separation gel-based, namely mono- and two-dimensional electrophoresis (1-DE and 2-DE), and gel-free, such as liquid chromatography (LC) techniques, identification techniques such as mass spectrometry (MS) analysis and bioinformatics tools to characterize the proteins in complex biological mixtures [4].

Plant pathogenic fungi cause significant yield losses in crops. Molecular studies of the fungal biological cycle and their interaction with their hosts are necessary, in order to develop efficient and environment-friendly crop protection strategies [5, 6]. Proteomics, in combination with other techniques, constitutes a successful tool for providing important information about pathogenicity and virulence factors. Moreover, proteomics also allows location-specific analysis (i.e., subproteomes at the level of organelles, cell membranes, cell wall, secretory proteins, etc.), the study of post-translational modifications [7] and interactions of host-pathogen, as well as host-pathogen-biocontrol agents [8, 9]. As a consequence, proteomics is opening up new possibilities for crop disease diagnosis and crop protection.

Several areas can be defined in proteomics, including descriptive and differential expression proteomics. In the case of fungi, a new area can also be defined as secretomics (the secretome is defined as being the combination of native proteins and cell machinery involved in their secretion), since many fungi secrete an arsenal of proteins to accommodate their saprotrophic lifestyle, namely proteins implicated in the adhesion to the plant surface, host-tissue penetration, and invasion effectors, together with other virulence factors [10]. Fungal proteomics research has experienced great advances over the last years, because of the availability of powerful proteomics technologies and the increasing number of fungal genome sequencing projects. Currently, more than 50 pathogenic fungal genomes have been sequenced.[1] Excellent reviews on fungal proteomic methodologies have been recently published [4, 11]. The workflow of a fungal gel-based proteomics experiment includes, among others, the following steps: experimental design, fungal growth, sampling, sample preparation, protein extraction, separation, MS analysis, protein identification, statistical analysis of data, quantification, and data analysis, management, and storage.

Most of plant pathogenic fungi are filamentous fungi. This type of fungi can be considered, similarly to plants, recalcitrant biological material, so the preparation of protein samples is a critical step. Cell breakdown and protein extraction are difficult because of the presence of a cell wall that makes up the majority of the cell mass [12]. To overcome this challenge, early studies were performed using mechanical lysis via glass beads [13–15], a cell mill [16], or sonication [17–19], because these methods are more efficient than

---

[1] Broad Institute Database, http://broadinstitute.org/science/project/fungal-genomeinitiative

those based on chemical or enzyme extraction [20]. Shimizu and Wariishi [21] employed an alternative approach to avoid the difficulty of lysing the fungal cell wall by generating protoplasts of *Tyromyces palustris*. A better 2-DE pattern was obtained from protoplast than from intact cells. The most widely used method for cell disruption consists of pulverizing the mycelium in liquid nitrogen using a mortar and pestle [17, 22–33]. The production of high-quality protein samples is also a crucial step for proteomic analysis. The protocol most widely employed for fungal proteins uses protein precipitation media containing organic solvents, such as trichloroacetic acid (TCA), followed by solubilization of the precipitate in an appropriate buffer. This method minimizes protein degradation/modification. Furthermore, it removes interfering compounds such as polysaccharides, polyphenols, pigment and lipids, which may be a problem during IEF [34], and prevents protease activities [35]. TCA-treatment complicates subsequent protein solubilization for IEF, especially with hydrophobic proteins. These problems have been partly overcome by the use of chaotropes (urea and thiourea) [36], new zwitterionic detergents [37–41], and by a brief treatment with sodium hydroxide [35], which led to an increase in resolution and capacity of 2-DE gels. Other protein extraction methods have reported an improvement when using acidic extraction solution to reduce streaking of fungal samples caused by their cell wall [42], as well as using a phosphate buffer solubilization before the precipitation [23, 24]. Finally, the combined use of TCA precipitation and phenol extraction provides a better spot definition, due to the fact that it reduces streaking and leads to a higher number of detected spots [22, 43]. Alternative protocols for protein extraction from spores of *Aspergillus ssp.* have been optimized, since they use acidic conditions, step organic gradient, and variable sonication treatment (ultrasonic homogenizer and sonic water bath) [19].

Special protocols are required for secreted proteins, due to the fact that there may be problems such as a very low protein concentration, sometimes below the detection limit of colorimetric methods (Bradford, Lowry, or BCA), or the presence of polysaccharides, mucilaginous material, salts, and secreted metabolites (low-molecular organic acids, fatty acids, phenols, quinones, and other aromatic compounds). The presence of these extracellular compounds may impair standard methods for protein quantification and may result in a strong overestimation of total protein number [44]. This determination can also be affected by the high concentration of reagents from the solubilization buffer (such as urea, thiourea, or DTT) that may interfere in the spectrophotometric measurement, producing an overestimation of the total amount of protein in which, depending on the method, the differences varied in the order of two magnitudes [45]. Comparison of different standard methods for protein precipitation has demonstrated their limited applicability to analyzing the whole fungal secretome [45–54].

Electrophoresis is almost the only protein separation technique employed in fungal research. Despite its simplicity, 1-DE remains as quite a valid technique providing relevant information, especially in the case of comparative proteomics with large numbers of samples to analyze. Thus, it is possible by using this technique to distinguish between phenotypes of different wild-type strains of *Botrytis cinerea* and to identify proteins involved in the pathogenicity mechanisms (Gonzalez-Fernandez et al., personal communication). With appropriate software, 1-DE is a simple, reliable technique for finger-printing crude extracts, and it is especially useful in the case of hydrophobic and low-molecular-weight proteins [53]. Therefore, the 1-DE is a good approach to obtain preliminary results before carrying on 2-DE analysis [54–58].

Two-DE is the dominant platform in fungal proteomics. Briefly, the 2-DE consists of a tandem pair of electrophoretic separations: in the first dimension, proteins are resolved according to their isoelectric points (pIs), normally using IEF; while in the second dimension, proteins are separated according to their approximate molecular weight using SDS-PAGE. Excellent reviews describing and discussing the features and protocols of electrophoretic separations in proteomics strategies have been published [34, 59]. The main advantages of 2-DE are its high protein separation capacity and the possibility of making large-scale protein-profiling experiments. Nevertheless, reproducibility and resolution of this technique are still remaining challenges. This method was reported to under-represent proteins with extreme physicochemical properties (size, isoelectric point, transmembrane domains), as well as those with a low abundance [60].

After separating proteins, they can then be detected by a variety of staining techniques [34, 59] namely: (1) organic dyes, such as colloidal Coomassie blue staining; (2) zinc–imidazole staining; (3) silver staining; and (4) fluorescence-based detection, such as Sypro Ruby. The criteria to choose the staining method are the level of sensibility and its compatibility with MS. Gels are digitalized, and bands or spots are studied by specific software of image analysis (i.e., Quantity-One, PD-Quest, BioRad). Bands or spots are excised from gels and prepared for MS analysis.

The limitations of gel-based analysis have led to the more recent development of techniques based on LC separation of proteins or peptides, including two-dimensional liquid-phase chromatography 2-D LC-MS/MS (based on a high performance chromatofocusing in the first dimension followed by high-resolution reversed-phase chromatography in the second) [61], and one-dimensional electrophoresis(1-DE)-nanoscale capillary LC-MS/MS, namely GeLC-MS/MS (this technique combines a size-based protein separation with an in-gel digestion of the resulting fractions) [62]. This GeLC-MS/MS strategy paves the way toward the analysis on a large-scale fungal response environmental cues on the basis of quantitative shotgun protein- profiling experiments. The case of Multidimensional Protein Identification Technology (MudPIT), which allows the identification of a much larger number of proteins compared to gel-based methods, is drawback being the lack of quantitative data [2, 63]. MudPIT was used to analyze the mechanisms of germling growth in *Uromyces appendiculatus* by comparing germinating asexual uredospores with inactive spores [64].

MS is the basic technique for global proteomic analysis due to its accuracy, resolution, and sensitivity (in the femtomole to attomole concentration range), and due to the fact that is has the capacity for a high throughput. Not only does it allow one to profile a proteome, but more importantly, it allows one to identify the protein species and characterize post-translational modifications and interactions. Proteins are identified from mass spectra of intact proteins (top-down proteomics), or peptide fragments obtained after enzymatic (mostly digested with trypsin) or chemical treatment (bottom-up proteomics). Protein species are identified by comparison of the experimental spectra, while the theoretical ones were obtained *in silico* from protein, genomic, ESTs sequence, or MS spectra databases. For that purpose, different instrumentation, algorithms, databases, and repositories are available [65, 66].

Although methods for proteomic analysis of limited fungal species have been published [4, 11, 67–69], procedures for protein extraction and 2-DE gel analysis conditions are progressively evolving according to individual characteristics of fungal species.

Materials

(See Note 1)
  1. Distilled water.
  2. Liquid nitrogen.
  3. Freeze-dryer.
  4. Mortar and pestle.
  5. Cell strainer, 100 μm Nylon (BD Falcon).
  6. Vortexer.
  7. Micropestles.
  8. Ultrasonic homogenizer.
  9. Microcentrifuge and centrifuge.
 10. Disposable microcentrifuge tubes: 1.5 mL and 2 mL.
 11. Centrifuge tubes: 50 mL.
 12. Trichloroacetic acid (TCA) (10% w/v)/acetone (80% v/v) solution.
 13. 0.1 M Ammonium acetate/methanol (100% and 80% v/v) solution.
 14. Acetone (80% v/v) solution.

15. Phenol solution equilibrated with 10 mM Tris–HCl pH 8 (Sigma-Aldrich).
16. SDS buffer: 30% (w/v) sucrose, 2% (w/v) SDS, 5% (v/v) β-mercaptoethanol, 0.1 M Tris–HCl pH 8.
17. Solubilization solution: 9 M urea, 2 M thiourea, 4% (w/v) CHAPS, 0.5% (v/v) Tritón-X100, 20 mM DTT.
18. Microtube mixer.
19. Bradford solution (Sigma-Aldrich).
20. Extraction buffer: 8 M urea, 1% (w/v) SDS, 1 mM EDTA, 100 mM DTT, 50 mM Tris–HCl pH 8.
21. TE buffer for secreted proteins: 50 mM EDTA, 2% (v/v) β-mercaptoethanol, 1 mM PMSF, 10 μL/mL buffer of protease inhibitor cocktail for fungi (Sigma-Aldrich), Tris–HCl pH 8.
22. Running buffer: 192 mM Glycine, 1% (w/v) SDS, 50 mM Tris–HCl pH 8.
23. Vertical electrophoresis equipment; for example, Criterion System (BioRad).
24. Precast free stain gels (Criterion System, BioRad): 4–20% Tris–HCl multi-wells for 1-DE and 8–16% Tris–HCl IPG+1 for 2-DE.
25. IPG strips, 11 cm, pH 5–8 (BioRad).
26. IPG strips rehydration solution: 7 M urea, 2 M thiourea, 4% (v/v) CHAPS, 2% (v/v) ampholytes (BioRad), 20 mM DTT.
27. Equilibration buffer: 6 M urea, 20% (v/v) glycerol, 2% (w/v) SDS, 1.5 M Tris–HCl pH 8.8.
28. Shaker.
29. Densitometer; for example, GS-800 (BioRad).

Methods

The methods below have been optimized to mycelium, secreted proteins in liquid media, and conidia from *B. cinerea*, although these procedures can be applied in proteomic analysis of filamentous fungi in general.

Sample Collection

For in vitro cultures, conidia are produced using rich-media plates at 22 °C under constant black light (UV) for 3–4 weeks. Mycelium and secreted proteins can be obtained from liquid cultures inoculated with conidia or nonsporulating mycelia (see Note 2). Mycelia and media can be separated by centrifugation and filtration, frozen in liquid nitrogen, and lyophilized.

Protein Extraction by TCA/ Acetone-Phenol/Methanol Method

Protein extraction is carried on by using the TCA/ acetone–phenol/methanol method [70, 71] with some modifications [4] and adapted to started material (conidia, mycelium or secreted proteins).

Mycelium

The lyophilized mycelium is ground to a fine powder in liquid nitrogen using a cooled mortar and pestle, and stored at −80 °C for later analysis (see Note 3). For protein extraction, the following protocol is applied:

1. Transfer 50–100 mg of mycelial powder into a 2-mL tube.
2. Add 1 mL of 10% (w/v) TCA/acetone and mix well using a micropestle and then by vortexing.
3. Sonicate 3 × 10 s (50 W, amplitude 60) at 4 °C, breaking on ice at 1 min.
4. Fill the tube with 10% (w/v) TCA/acetone. Mix well by vortexing.
5. Centrifuge at 16,000 × g for 5 min (4 °C) and remove the supernatant by decanting (see Note 4).
6. Fill the tube with 0.1 M ammonium acetate in 80% (v/v) methanol. Mix well by vortexing.
7. Centrifuge at 16,000 × g for 5 min (4 °C) and discard the supernatant.
8. Fill the tube with 80% (v/v) acetone. Mix well by vortexing.
9. Centrifuge at 16,000 × g for 5 min (4 °C) and discard the supernatant.
10. Air-dry at room temperature to remove residual acetone.
11. Add 1.2 mL of 1:1 phenol (pH 8, SIGMA)/ SDS buffer. Mix well using a pipette and by vortexing. Incubate for 5 min in ice.

12. Centrifuge at $16,000 \times g$ for 5 min. Transfer the upper phenol fase into a new 1.5-mL tube (see Note 5).

13. Fill the tube with 0.1 M ammonium acetate in 100% (v/v) methanol, mix well, and allow the precipitation overnight at −20 °C.

14. Centrifuge at $16,000 \times g$ for 5 min (4 °C) and discard the supernatant (a white pellet should be visible).

15. Wash the pellet with 100% methanol and mix by vortexing.

16. Centrifuge at $16,000 \times g$ for 5 min (4 °C) and discard the supernatant.

17. Wash the pellet with 80% (v/v) acetone and mix by vortexing.

18. Centrifuge at $16,000 \times g$ for 5 min (4 °C) and discard the supernatant.

19. Dry the pellet at room temperature.

20. Dissolve the proteins in solubilization solution for 2 h, shaking in a microtube mixer at 4 °C (see Note 6).

21. Quantify proteins using Bradford method [72].

22. Store the protein extracts at −20 °C for further analysis.

Secreted Proteins

Lyophilized media are re-suspended in 5 mL of TE buffer and proteins are precipitated according to the following protocol:

1. Transfer the medium resolubilized into a 50-mL tube and add 2/1 (v/v) (10 mL) of 20% (w/v) TCA/acetone. Mix well by vortexing and allow protein precipitation overnight at 4 °C.

2. Centrifuge at $16,000 \times g$ for 10 min (4 °C) and remove the supernatant by decanting (see Note 7).

3. Add a volume 4/1 (v/v) (20 mL) of 0.1 M ammonium acetate in 80% (v/v) methanol. Mix well by vortexing.

4. Centrifuge at $16,000 \times g$ for 10 min (4 °C) and discard the supernatant.

5. Add a volume 4/1 (v/v) (20 mL) of 80% (v/v) acetone. Mix well by vortexing.

6. Centrifuge at $16,000 \times g$ for 10 min (4 °C) and discard the supernatant.

7. Air-dry at room temperature to remove residual acetone.

8. Add 4 mL of 1/1 (v/v) phenol (pH 8, SIGMA)/SDS buffer. Mix well by vortexing and transfer the 4-mL to two 2-mL eppendorf. Incubate for 5 min in ice.

9. Centrifuge at $16,000 \times g$ for 10 min. Transfer the upper phenol phase into a new 2-mL tube (1 mL per 2-mL tube).

10. Fill the tube with 0.1 M ammonium acetate in 100% (v/v) methanol, mix well, and allow to precipitate overnight al −20 °C.

11. Centrifuge one 2-mL tube at $16,000 \times g$ for 5 min (4 °C) and discard the supernatant (a slight pellet should be visible). Fill the same 2-mL tube with the other eppendorf (mix well before changing). Centrifuge at $16,000 \times g$ for 5 min (4 °C) and discard the supernatant.

12. Follow the steps in Mycelium section, starting with step 15.

Conidia

Conidia can be harvested with H_2Od with 0,01% Tween-80 scraping on the surface of agar plate. The conidia suspension is filtered through a cell strainer, concentred in 1.5-mL tubes, centrifuged at $16,000 \times g$ for 5 min (4 °C), lyophilized and stored at −80 °C for further analysis. For protein extraction, the TCA/acetone-phenol/methanol [70, 71] method was used, with some modifications [4, 19].

1. Add 300 µL of extraction buffer to conidia. Mix well using a micropestle and by vortexing.

2. Sonicate 3×10 s (50 W, amplitude 60), breaking on ice at 1 min. Mix well using a micropestle and by vortexing.

3. Centrifuge at $16,000 \times g$ for 5 min (4 °C).

4. Fill the tube with 10% (w/v) TCA/acetone. Mix well using by vortexing.

5. Centrifuge at $16,000 \times g$ for 5 min (4 °C) and discard the supernatant.

6. Fill the tube with 0.1 M ammonium acetate in 80% (v/v) methanol. Mix well using a micropestle and by vortexing.

7. Follow the steps in Mycelium, starting with step 5.

Protein Separation

One-Dimensional Electrophoresis

Proteins can be separated by SDS-PAGE according to Laemmly electrophoresis system, [73] for example, using Criterion System (BioRad) with precast Criterion Stain Free Gels, Tris–HCl, 4–20% linear gradient (BioRad). The 1-DE is visualized using the Image Lab System (BioRad), and stained by CBB (Coomassie Blue Brilliant) method [74] (see Note 8). After the staining of proteins, bands can be analyzed using the Quantity-One software (BioRad).

Two-Dimensional Electrophoresis
Isoelectrofocusing

Focusing conditions will vary with sample composition, sample complexity, and strip pH range. In our conditions, the 11 cm IPG strips, pH 5–8 (BioRad), are rehydrated with 50 µg of protein extract in 185 µL rehydration solution applying 50 V for 16 h (active rehydratation) at 20 °C. Before the focusing a wet wick is inserted under each end of the strip (catode). The conditions for IEF have been adapted to our system from reference [45]: 150 V for 1 h, 1 h at 200 V, 1 h at 500 V, 1,000 V·h at 1,000 V, followed by 2.5 h gradient from 1,000 to 8,000 V, and finally focused for 30,000 V·h at 8,000 V, with a cell temperature of 20 °C (see Note 9). After IEF, IPG strips are stored at −20 °C.

Before the second dimension, IPG strips are equilibrated in two steps. Firstly, it is carried on with 2% (w/v) DTT in equilibration buffer for 10 min in agitation at room temperature; secondly, it is done with 2.5% (w/v) iodoacetamide in equilibration buffer for 10 min in agitation at room temperature.

The second dimension is performed in the same way as SDS-PAGE, but using precast Criterion Stain Free Gels, Tris–HCl, 8–16% linear gradient for IPG strips (BioRad). After the staining of proteins, spots can be analyzed using the PD-Quest software (BioRad).

Protein Identification

The bands or spots are cut out and digested with trypsin. Tryptic peptides are analyzed in a mass spectrometer; for example, a 4,800 Proteomics Analyzer MALDI–TOF/TOF (Applied Biosystems). In this case, the most abundant peptide ions are subjected to MS/MS analysis. A PMF search and a combined search (+MS/MS) are performed in nrNCBI database of proteins using the MASCOT algorithm (see Note 10).

Notes

1. Gloves and lab coat should be used in these procedures, and particular care should be taken when handling TCA and phenol (consult safety data sheets) because they are corrosive products. Steps involving phenol and β-mercaptoethanol should be performed in a fume hood.
2. Examples of rich-media are PDAB (potato, dextrose, agar + bean leaves), solid synthetic complete medium (CM) [75], or solid malt extract medium (1.5% w/v).
3. Be careful to work with liquid nitrogen because its cool temperature (−195.8 °C) could cause severe frostbite. The nitrogen evaporated reduces the concentration of oxygen in the air and can act as an asphyxiant, especially in confined spaces, so it may be dangerous because nitrogen is odorless, colorless, and tasteless, and could cause suffocation without any sensation or warning.
4. Be careful to not throw out the pellet.
5. Three phases appear, namely: the upper phase (which is the phenolic phase where are the proteins), a white interphase, and a lower aqueous phase. Try to not to get part of the white interphase.
6. The volume of solubilization solution added will depend on quantity of precipitated proteins. It is advisable that samples are well concentrated.
7. In this case, maybe that the precipitated pellet is faint because the proteins secreted to medium are at very low concentration.
8. More details about 1-DE and 2-DE separation methods are described in two excellent reviews [34, 59].
9. The condition of protein focusing must be optimized for each system of study. In our case, we use the PROTEAN IEF cell by

BioRad. The conditionating phase involves the application of previous steps at low voltage that allow to remove ions and other contaminants containing the sample, and that interfere on protein focusing. The current should not exceed 50 μA per strips. For more information see the 2-D Electrophoresis for Proteomics Manual by BioRad.

10. More details about MS analysis are described in references [65, 68, 76].

Acknowledgments This work was supported by the Spanish Ministry of Science and Innovation (BotBank Project, EUI2008-03686), the Regional Government of Andalusia (Junta de Andalucía) and the University of Córdoba (AGR-0164: Agricultural and Plant Biochemistry and Proteomics Research Group).

References

1. Weiss W, Gorg A (2009) High-resolution two-dimensional electrophoresis. Methods Mol Biol 564:13–32
2. Washburn MP, Wolters D, Yates JR 3rd (2001) Large-scale analysis of the yeast proteome by multidimensional protein identification technology. Nat Biotechnol 19:242–247
3. Wolters DA, Washburn MP, Yates JR 3rd (2001) An automated multidimensional protein identification technology for shotgun proteomics. Anal Chem 73:5683–5690
4. Gonzalez-Fernandez R, Prats P, Jorrin-Novo JV (2010) Proteomics of plant pathogenic fungi. J Biomed Biotechnol Volume 2010 (2010), Article ID 932527, 36 pages, doi:10.1155/2010/932527
5. Choquer M, Fournier E, Kunz C, Levis C, Pradier J-M, Simon A et al (2007) *Botrytis cinerea* virulence factors: new insights into a necrotrophic and polyphageous pathogen. FEMS Microbiol Lett 277:1–10
6. Egan MJ, Talbot NJ (2008) Genomes, free radicals and plant cell invasion: recent developments in plant pathogenic fungi. Curr Opin Plant Biol 11:367–372
7. Kim Y, Nandakumar MP, Marten MR (2007) Proteomics of filamentous fungi. Trends Biotechnol 25:395–400
8. Marra R, Ambrosino P, Carbone V, Vinale F, Woo SL, Ruocco M et al (2006) Study of the three-way interaction between *Trichoderma atroviride*, plant and fungal pathogens by using a proteomic approach. Curr Genet 50:307–321
9. Rampitsch C, Bykova NV, McCallum B, Beimcik E, Ens W (2006) Analysis of the wheat and *Puccinia triticina* (leaf rust) proteomes during a susceptible host-pathogen interaction. Proteomics 6:1897–1907
10. Deising HB, Kamoun S (2009) The secretome of plant-associated fungi and oomycetes. Springer, Berlin, Heidelberg, pp 173–180
11. González-Fernández R and Jorrin-Novo JV (2010) Proteomics of fungal plant pathogens: the case of *Botrytis cinerea*. Vilas AM (ed.). Formatex Research Center, p. 205–217
12. Ruiz-Herrera J (1992) Fungal cell wall: structure, synthesis and assembly. CRC Press, Boca Raton, FL
13. Ebstrup T, Saalbach G, Egsgaard H (2005) A proteomics study of in vitro cyst germination and appressoria formation in *Phytophthora infestans*. Proteomics 5:2839–2848
14. Grinyer J, Kautto L, Traini M, Willows RD, Te'o J, Bergquist P et al (2007) Proteome mapping of the *Trichoderma reesei* 20S proteasome. Curr Genet 51:79–88
15. Melin P, Schnurer J, Wagner EG (2002) Proteome analysis of Aspergillus nidulans reveals proteins associated with the response to the antibiotic concanamycin A, produced by *Streptomyces* species. Mol Genet Genomics 267:695–702
16. Bohmer M, Colby T, Bohmer C, Brautigam A, Schmidt J, Bolker M (2007) Proteomic analysis of dimorphic transition in the phytopathogenic fungus Ustilago maydis. Proteomics 7:675–685
17. Grinyer J, Hunt S, McKay M, Herbert BR, Nevalainen H (2005) Proteomic response of the biological control fungus *Trichoderma atroviride* to growth on the cell walls of *Rhizoctonia solani*. Curr Genet 47:381–388
18. Grinyer J, McKay M, Nevalainen H, Herbert BR (2004) Fungal proteomics: initial mapping of biological control strain *Trichoderma harzianum*. Curr Genet 45:163–169
19. Sulc M, Peslova K, Zabka M, Hajduch M, Havlicek V (2009) Biomarkers of *Aspergillus* spores: strain typing and protein identification. Int J Mass Spectrom 280:162–168
20. Nandakumar MP, Marten MR (2002) Comparison of lysis methods and preparation protocols for one- and two-dimensional electrophoresis of *Aspergillus oryzae* intracellular proteins. Electrophoresis 23:2216–2222
21. Shimizu M, Wariishi H (2005) Development of a sample preparation method for fungal proteomics. FEMS Microbiol Lett 247:17–22
22. Fernandez-Acero FJ, Colby T, Harzen A, Cantoral JM, Schmidt J (2009) Proteomic analysis of the phytopathogenic fungus *Botrytis cinerea* during cellulose degradation. Proteomics 9:2892–2902
23. Fernandez-Acero FJ, Jorge I, Calvo E, Vallejo I, Carbu M, Camafeita E et al (2007) Proteomic analysis of phytopathogenic fungus *Botrytis cinerea* as a potential tool for identifying pathogenicity factors, therapeutic targets and for basic research. Arch Microbiol 187:207–215
24. Fernandez-Acero FJ, Jorge I, Calvo E, Vallejo I, Carbu M, Camafeita E et al (2006) Two-dimensional electrophoresis protein profile of the phytopathogenic fungus *Botrytis cinerea*. Proteomics 6(Suppl 1): S88–S96
25. Hernandez-Macedo ML, Ferraz A, Rodriguez J, Ottoboni LM, De Mello MP (2002) Iron-regulated proteins in *Phanerochaete chrysosporium* and *Lentinula edodes*: differential analysis by sodium

dodecyl sulfate polyacrylamide gel electrophoresis
and two-dimensional polyacrylamide gel electrophoresis profiles. Electrophoresis 23:655–661

26. Shimizu M, Yuda N, Nakamura T, Tanaka H, Wariishi
    H (2005) Metabolic regulation at the tricarboxylic
    acid and glyoxylate cycles of the lignin-degrading
    basidiomycete *Phanerochaete chrysosporium* against
    exogenous addition of vanillin. Proteomics
    5:3919–3931

27. Yajima W, Kav NN (2006) The proteome of the phy-
    topathogenic fungus *Sclerotinia sclerotiorum*.
    Proteomics 6:5995–6007

28. Bringans S, Hane JK, Casey T, Tan KC, Lipscombe R,
    Solomon PS et al (2009) Deep proteogenomics; high
    throughput gene validation by multidimensional liq-
    uid chromatography and mass spectrometry of pro-
    teins from the fungal wheat pathogen *Stagonospora
    nodorum*. BMC Bioinformatics 10:301

29. Cao T, Kim YM, Kav NN, Strelkov SE (2009) A pro-
    teomic evaluation of *Pyrenophora tritici-repentis*,
    causal agent of tan spot of wheat, reveals major differ-
    ences between virulent and avirulent isolates.
    Proteomics 9:1177–1196

30. El-Bebany AF, Rampitsch C, Daayf F (2010)
    Proteomic analysis of the phytopathogenic soilborne
    fungus *Verticillium dahliae* reveals differential pro-
    tein expression in isolates that differ in aggressive-
    ness. Proteomics 10:289–303

31. Lakshman DK, Natarajan SS, Lakshman S, Garrett
    WM, Dhar AK (2008) Optimized protein extraction
    methods for proteomic analysis of Rhizoctonia solani.
    Mycologia 100:867–875

32. Tan KC, Heazlewood JL, Millar AH, Thomson G,
    Oliver RP, Solomon PS (2008) A signaling-regulated,
    short-chain dehydrogenase of *Stagonospora nodorum*
    regulates asexual development. Eukaryot Cell
    7:1916–1929

33. Taylor RD, Saparno A, Blackwell B, Anoop V, Gleddie
    S, Tinker NA et al (2008) Proteomic analyses of
    Fusarium graminearum grown under mycotoxin-
    inducing conditions. Proteomics 8:2256–2265

34. Gorg A, Weiss W, Dunn MJ (2004) Current two-
    dimensional electrophoresis technology for proteom-
    ics. Proteomics 4:3665–3685

35. Nandakumar MP, Shen J, Raman B, Marten MR
    (2003) Solubilization of trichloroacetic acid (TCA)
    precipitated microbial proteins via naOH for two-
    dimensional electrophoresis. J Proteome Res
    2:89–93

36. Rabilloud T (1998) Use of thiourea to increase the
    solubility of membrane proteins in two-dimensional
    electrophoresis. Electrophoresis 19:758–760

37. Everberg H, Gustavasson N, Tjerned F (2008)
    Enrichment of membrane proteins by partitioning in
    detergent/polymer aqueous two-phase systems.
    Methods Mol Biol 424:403–412

38. Kniemeyer O, Lessing F, Scheibner O, Hertweck C,
    Brakhage AA (2006) Optimisation of a 2-D gel elec-
    trophoresis protocol for the human-pathogenic fungus
    Aspergillus fumigatus. Curr Genet 49:178–189

39. Luche S, Santoni V, Rabilloud T (2003) Evaluation of
    nonionic and zwitterionic detergents as membrane
    protein solubilizers in two-dimensional electrophore-
    sis. Proteomics 3:249–253

40. Rabilloud T (1996) Solubilization of proteins for elec-
    trophoretic analyses. Electrophoresis 17:813–829

41. Rabilloud T, Adessi C, Giraudel A, Lunardi J (1997)
    Improvement of the solubilization of proteins in two-
    dimensional electrophoresis with immobilized pH
    gradients. Electrophoresis 18:307–316

42. Herbert BR, Grinyer J, McCarthy JT, Isaacs M, Harry
    EJ, Nevalainen H et al (2006) Improved 2-DE of
    microorganisms after acidic extraction. Electrophoresis
    27:1630–1640

43. Guais O, Borderies G, Pichereaux C, Maestracci M,
    Neugnot V, Rossignol M et al (2008) Proteomics anal-
    ysis of "Rovabiot Excel," a secreted protein cocktail
    from the filamentous fungus *Penicillium funiculosum*
    grown under industrial process fermentation. J Ind
    Microbiol Biotechnol 35:1659–1668

44. Kao SH, Wong HK, Chiang CY, Chen HM (2008)
    Evaluating the compatibility of three colorimetric
    protein assays for two-dimensional electrophoresis
    experiments. Proteomics 8:2178–2184

45. Fragner D, Zomorrodi M, Kues U, Majcherczyk A
    (2009) Optimized protocol for the 2-DE of extracel-
    lular proteins from higher basidiomycetes inhabiting
    lignocellulose. Electrophoresis 30:2431–2441

46. Abbas A, Koc H, Liu F, Tien M (2005) Fungal degra-
    dation of wood: initial proteomic analysis of extracel-
    lular proteins of *Phanerochaete chrysosporium* grown
    on oak substrate. Curr Genet 47:49–56

47. Medina ML and Francisco WA (2008) Isolation and
    enrichment of secreted proteins from filamentous
    fungi. SpringerLink (ed.). Humana Press, p. 275–285

48. Ravalason H, Jan G, Molle D, Pasco M, Coutinho
    PM, Lapierre C et al (2008) Secretome analysis of
    Phanerochaete chrysosporium strain CIRM-BRFM41
    grown on softwood. Appl Microbiol Biotechnol
    80:719–733

49. Vincent D, Balesdent MH, Gibon J, Claverol S,
    Lapaillerie D, Lomenech A et al (2009) Hunting down
    fungal secretomes using liquid-phase IEF prior to
    high resolution 2-DE. Electrophoresis 30:4118–4136

50. Zorn H, Peters T, Nimtz M, Berger RG (2005) The
    secretome of *Pleurotus sapidus*. Proteomics
    5:4832–4838

51. Fernandez-Acero FJ, Colby T, Harzen A, Carbu M,
    Wiencke U, Cantoral JM et al (2010) 2-DE proteomic
    approach to the *Botrytis cinerea* secretome induced
    with different carbon sources and plant-based elici-
    tors. Proteomics 10(12):2270–80

52. Espino JJ, Gutierrez-Sanchez G, Brito N, Shah P,
    Orlando R, Gonzalez C (2010) The *Botrytis cinerea*
    early secretome. Proteomics 10:3020–3034

53. Supek F, Peharec P, Krsnik-Rasol M, Smuc T (2008)
    Enhanced analytical power of SDS-PAGE using
    machine learning algorithms. Proteomics 8:28–31

54. Fryksdale BG, Jedrzejewski PT, Wong DL, Gaertner
    AL, Miller BS (2002) Impact of deglycosylation

methods on two-dimensional gel electrophoresis and matrix assisted laser desorption/ionization-time of flight-mass spectrometry for proteomic analysis. Electrophoresis 23:2184–2193

55. Medina ML, Haynes PA, Breci L, Francisco WA (2005) Analysis of secreted proteins from *Aspergillus flavus*. Proteomics 5:3153–3161
56. Medina ML, Kiernan UA, Francisco WA (2004) Proteomic analysis of rutin-induced secreted proteins from *Aspergillus flavus*. Fungal Genet Biol 41:327–335
57. Matis M, Zakelj-Mavric M, Peter-Katalinic J (2005) Mass spectrometry and database search in the analysis of proteins from the fungus *Pleurotus ostreatus*. Proteomics 5:67–75
58. Vanden Wymelenberg A, Sabat G, Mozuch M, Kersten PJ, Cullen D, Blanchette RA (2006) Structure, organization, and transcriptional regulation of a family of copper radical oxidase genes in the lignin-degrading basidiomycete *Phanerochaete chrysosporium*. Appl Environ Microbiol 72:4871–4877
59. Rabilloud T, Vaezzadeh AR, Potier N, Lelong C, Leize-Wagner E, Chevallet M (2009) Power and limitations of electrophoretic separations in proteomics strategies. Mass Spectrom Rev 28:816–843
60. Haynes PA, Roberts TH (2007) Subcellular shotgun proteomics in plants: looking beyond the usual suspects. Proteomics 7:2963–2975
61. Pirondini A, Visioli G, Malcevschi A, Marmiroli N (2006) A 2-D liquid-phase chromatography for proteomic analysis in plant tissues. J Chromatogr B Analyt Technol Biomed Life Sci 833:91–100
62. Recorbet G, Rogniaux H, Gianinazzi-Pearson V, Dumasgaudot E (2009) Fungal proteins in the extraradical phase of arbuscular mycorrhiza: a shotgun proteomic picture. New Phytol 181:248–260
63. Ye M, Jiang X, Feng S, Tian R, Zou H (2007) Advances in chromatographic techniques and methods in shotgun proteome analysis. Trends Anal Chem 26:80–84
64. Cooper B, Neelam A, Campbell KB, Lee J, Liu G, Garrett WM et al (2007) Protein accumulation in the germinating *Uromyces appendiculatus* uredospore. Mol Plant Microbe Interact 20:857–866

65. Domon B, Aebersold R (2006) Mass spectrometry and protein analysis. Science 312:212–217
66. Nesvizhskii AI, Vitek O, Aebersold R (2007) Analysis and validation of proteomic data generated by tandem mass spectrometry. Nat Methods 4:787–797
67. Harder A (2008) Sample preparation procedure for cellular fungi. Methods Mol Biol 425:265–273
68. Gusakov A, Semenova M, Sinitsyn A (2010) Mass spectrometry in the study of extracellular enzymes produced by filamentous fungi. J Anal Chem 65:1446–1461
69. de Oliveira JM, de Graaff LH (2011) Proteomics of industrial fungi: trends and insights for biotechnology. Appl Microbiol Biotechnol 89:225–237
70. Maldonado AM, Echevarria-Zomeño S, Jean-Baptiste S, Hernandez M, Jorrin-Novo JV (2008) Evaluation of three different protocols of protein extraction for *Arabidopsis thaliana* leaf proteome analysis by two-dimensional electrophoresis. J Proteomics 71:461–472
71. Wang W, Vignani R, Scali M, Cresti M (2006) A universal and rapid protocol for protein extraction from recalcitrant plant tissues for proteomic analysis. Electrophoresis 27:2782–2786
72. Bradford MM (1976) A rapid and sensitive method for the quantitation of microgram quantities of protein utilizing the principle of protein-dye binding. Anal Biochem 72:248–254
73. Laemmli UK (1970) Cleavage of structural proteins during the assembly of the head of bacteriophage T4. Nature 227:680–685
74. Neuhoff V, Arold N, Taube D, Ehrhardt W (1988) Improved staining of proteins in polyacrylamide gels including isoelectric focusing gels with clear background at nanogram sensitivity using Coomassie Brilliant Blue G-250 and R-250. Electrophoresis 9:255–262
75. Pontecorvo G, Roper JA, Forbes E (1953) Genetic recombination without sexual reproduction in *Aspergillus niger*. J Gen Microbiol 8:198–210
76. Han X, Aslanian A, Yates JR 3rd (2008) Mass spectrometry for proteomics. Curr Opin Chem Biol 12:483–490

Detection and Quantification of Endoprotease Activity Using a Coomassie Dye-Binding Assay

Anthony J. O'Donoghue and Cathal S. Mahon

Abstract

Traditional methods for detecting proteases in fungi require the separation of product from substrate. These methods are time-consuming, laborious, and not amenable to high-throughput analysis. A simple alternative method is described here that utilizes Coomassie dye reagent to follow the time-dependent proteolytic loss of a macromolecular protein substrate.

Keywords

Protease assay • Peptidase • Fungi • Colorimetric • Coomassie dye

Introduction

Fungi possess multiple proteases with highly diverse functions that are central to their physiology, metabolism, and development. These functions range from simple digestion of protein for food to complex regulation of signal transduction pathways. There are six classes of protease found in fungi that are defined based on the nature of the functional group in the active site and are therefore termed aspartic, glutamic, serine, threonine, cysteine, and metallo proteases. Some enzymes in these families exhibit exopeptidase activity, catalyzing the release of mono, di-, or tripeptides from the amino or carboxy termini of a polypeptide. However, most of the proteolytic enzymes isolated and characterized from fungi possess endoprotease activity and cleave polypeptides distal from the termini [1].

There are several methods to detect and quantify endoproteolytic activity in fungus. One technique utilizes synthetic peptide substrates with colorimetric or fluorescent labels that produce a signal upon protease cleavage. While these substrates are highly specific and sensitive to proteolysis, knowledge of the protease specificity is a prerequisite in choosing a substrate. Where substrate specificity is unknown, there are several general endoprotease assays for detection of proteolytic activity in fungal extracts and spent media. Traditional techniques employ common macromolecular substrates such as casein, albumin, and hemoglobin. Proteolytic cleavage can be detected by SDS PAGE followed by Coomassie staining or trichloroacetic acid precipitation

A.J. O'Donoghue (✉) • C.S. Mahon
Department of Pharmaceutical Chemistry,
University of California—San Francisco,
600 16th Street, San Francisco, CA 94158, USA
e-mail: aodonoghue@picasso.ucsf.edu

V.K. Gupta et al. (eds.), *Laboratory Protocols in Fungal Biology: Current Methods in Fungal Biology*,
Fungal Biology, DOI 10.1007/978-1-4614-2356-0_25, © Springer Science+Business Media, LLC 2013

followed by spectrophotometry [2]. These techniques are time-consuming, laborious, and not compatible with microplate assay formats. Fluorescently labeled proteins such as casein, elastin, and gelatin are useful for high-throughput protease assays [3]. These macromolecular substrates are heavily conjugated with dye such that fluorescence is efficiently quenched. Proteolytic release of dye-labeled peptides results in an increase in fluorescence that correlates with enzyme activity.

A simple, rapid, and quantitative colorimetric assay is described here that is a modification of assays developed by Saleemuddin and co-workers [4] and Buroker-Kilgore and Wang [5]. The protocol utilizes Coomassie Brilliant Blue G-250 dye to follow the decrease of protein substrate following endoprotease digestion and functions on the basis that the dye does not bind to peptides less than ~3,000 Da [6]. This assay can be used to investigate endoprotease activity using any number of proteins as substrate and digestion produces a quantitative and time-dependent loss of substrate. Variations of this technique have been used successfully to detect and characterize endoproteases from bacteria [7], protozoa [8], and multiple fungi [9–12]. We used the dye-binding endoprotease assay described here to detect and subsequently purify two acid-acting proteases from the fungus *Talaromyces emersonii* [13]. In our experimental strategy we utilized bovine serum albumin (BSA) as the protein substrate but other groups have successfully used casein [11], gelatin [7], IgG, hemoglobin, and mucin [8]. Furthermore, proteolytic activity in the presence of inhibitors can be readily performed [13]. This assay is pH- and temperature-independent, amenable to microplate assay format and requires only dye, a source of protein substrate and a spectrophotometer.

Materials

1. 50 mL of sterile culture media in 250-mL shake flask.
2. Cheese cloth.
3. Liquid nitrogen.
4. Mortar and pestle.

5. Phosphate buffered saline (PBS).
6. Microcentrifuge with 1.5-mL microcentrifuge tubes.
7. Coomassie Brilliant Blue G-250 in phosphoric acid and methanol. Available from Thermo Scientific (Coomassie Plus Bradford Assay Reagent), Sigma-Aldrich (Coomassie Protein Assay Reagent) or BioRad (Quick Start Bradford 1X Dye Reagent).
8. Protease sample; cell-free culture supernatant, or cell lysate diluted to <0.1 mg/mL in assay buffer.
9. Protease assay buffer; choice of buffer will depend on pH preference of protease. Suggested buffers include 25 mM citrate–phosphate buffer pH 2.5–6.0, 25 mM Phosphate buffer pH 6.0–8.0, 25 mM Tris–HCl buffer pH 7.5–9.0 or 20 mM Glycine–NaOH pH 9.0–10.5. All buffers can be supplemented with salts, reducing reagents, and detergents at concentrations compatible with Coomassie dye-binding reagent (see supplier's compatibility chart).
10. 2 mg/mL of BSA or any other protein that produces a linear response curve to Coomassie dye binding.
11. Microplate spectrophotometer with 580- to 610-nm filter (595 nm is optimal).
12. 96-well microtiter assay plates.
13. PCR tubes.
14. Thermal cycler.
15. Protease inhibitors (1,10-Phenanthroline, Pepstatin A, PMSF, and E-64; all available from Sigma-Aldrich).
16. Aspergillopepsin (Sigma-Aldrich, Catalogue# P2143).
17. Trypsin (Sigma-Aldrich, Catalogue # T8003).

Method

Fungal Culture and Enzyme Sample Preparation

1. Inoculate a flask of sterile liquid media with fungal cells and grow for 2–5 days under optimal conditions.

2. Separate cells from media using several layers of cheese cloth (or centrifugation).

3. Transfer 1.5 mL of filtered media to a micro-centrifuge tube and spin at $12,000 \times g$ for 20 min at 4 °C. Remove supernatant and store at 4 °C. This is the cell-free media sample.

4. "Squeeze dry" filtered cells and freeze with liquid nitrogen. Grind to a fine powder using mortar and pestle.

5. Transfer powder to a microcentrifuge tube and dissolve in PBS. Spin at $12,000 \times g$ for 20 min at 4 °C.

6. Remove supernatant and determine protein concentration. Dilute to <0.1 mg/mL in PBS.

Generation of Standard Curve for Protein Substrate

1. Remove Coomassie dye reagent from refrigerator and equilibrate to room temperature.

2. Make a 2-mg/mL stock solution of BSA (or other protein substrate) in assay buffer.

3. Make a dilution series of substrate in assay buffer starting at 1.5 mg/mL.

4. Pipette 180 μL of each dilution in triplicate into wells of a microplate, known as the protease assay (PA) plate.

5. Add 20 μL of control protease sample (fresh media or PBS) to each well, mix by pipetting, and remove 20 μL into wells of a second plate, known as the dye-binding (DB) plate.

6. Add 180 μL of Coomassie dye reagent, shake for 10 s, and incubate at room temperature for 5 min.

7. Measure the absorbance between 580 and 610 nm (595 nm is optimal) and generate a standard curve. (see Note 1)

Endoprotease Assay

1. Choose a substrate concentration that produces the highest absorbance on the linear portion of the standard curve (Fig. 25.1a). Make a 20-mL stock in assay buffer.

2. Add 180 μL of this protein stock into a PA well or PCR tube. A PCR tube and a thermocycler are important for assays performed at elevated temperature (>30 °C).

3. Combine 20 μL of cell-free media or cell lysate with substrate. Mix by pipetting and immediately remove 20 μL into triplicate wells on the DB plate.

4. Add 180 μL of Coomassie dye reagent and measure absorbance as outlined previously. This sample is the zero minute time point (T_0). (see Note 2)

5. Remove 20 μL in triplicate from the reaction after 5, 10, 15, and 20 mins and repeat treatment with Coomassie dye reagent in DB plate.

6. Use the standard curve to calculate the concentration of substrate remaining at each time point and plot versus time (see example in Fig. 25.1b). Note: If no significant loss of substrate is observed after 20 min, the assay can be run for several hours to days. Use a PCR tube to minimize evaporation.

7. Optional: Set up a control reaction with either 0.2 μg of Aspergillopepsin I at pH 1.5–4.5 or Trypsin at pH 7–9 to monitor time-dependent cleavage of protease substrate.

Inhibition Assay

1. Combine an equal volume of fungal sample containing protease with 20× Protease Inhibitor. Mix briefly and incubate for 15 min at room temperature. Set up a control reaction containing no inhibitor.

2. Pipette 20 μL of this reaction into a well in the PA plate or PCR tube containing 180 μL of substrate, incubate, and repeat dye-binding assay as outlined above. (see Note 3).

Calculation of Protease Cleavage

Loss of substrate at each time point can be calculated using the standard curve and the following formula:

$$\Delta (\text{delta})S = [S_0] - [S_x]$$

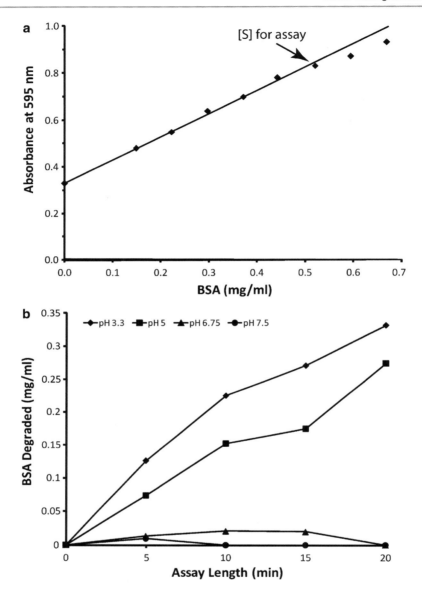

Fig. 25.1 (a) Standard curve of BSA absorbance at 595 nm following Coomassie dye binding. The highest substrate concentration in the linear portion of the graph was chosen as the substrate concentration for the protease assay. (b) Time-dependent cleavage of BSA measured by Coomassie dye-binding assay. Spent media from shake-flask cultures of *Talaromyces emersonii* were assayed for proteolytic activity as described in Methods section. Assays were carried out in citrate-phosphate pH 3.3, ammonium acetate pH 5.0, and sodium phosphate pH 6.75 and 7.5.

where S_0 and S_X are substrate concentrations at T_0 and T_X minutes, respectively. It is important to always take a T_0 sample because the addition of enzyme to substrate may result in an overall increase in absorbance, particularly when using a cell lysate as the source of protease.

Notes

1. To increase sensitivity of the assay, a higher ratio of protein substrate to Coomassie dye reagent may be used. Some protocols use an equal volume of protein sample to dye.

2. For most assays, addition of Coomassie dye will instantly quench reaction due to the low pH of the reagent. However, for acid-acting enzymes quenching can be performed by heat denaturation in a thermocycler or by the addition of inhibitor to dye reagent.

3. The total enzyme concentration is 50 % relative to the previous assay so extended incubation may be required to observe cleavage.

References

1. Yike I (2011) Fungal proteases and their pathophysiological effects. Mycopathologia 171(5):299–323
2. Waxman L (1981) Calcium-activated proteases in mammalian tissues. Methods Enzymol 80:664–680
3. Jones LJ, Upson RH, Haugland RP, Panchuk-Voloshina N, Zhou M (1997) Quenched BODIPY dye-labeled casein substrates for the assay of protease activity by direct fluorescence measurement. Anal Biochem 251(2):144–152
4. Saleemuddin M, Ahmad H, Husain A (1980) A simple, rapid, and sensitive procedure for the assay of endoproteases using coomassie brilliant blue G-250. Anal Biochem 206:202–206
5. Buroker-Kilgore M, Wang K (1993) A Coomassie brilliant blue G-250-based colorimetric assay for measuring activity of calpain and other proteases. Anal Biochem 208:387–392
6. Sedmak JJ, Grossberg SE (1977) A rapid, sensitive, and versatile assay for protein using Coomassie brilliant blue G250. Anal Biochem 79(1–2):544–552
7. Alvarez VM, von der Weid I, Seldin L, Santos a L S (2006) Influence of growth conditions on the production of extracellular proteolytic enzymes in *Paenibacillus peoriae* NRRL BD-62 and *Paenibacillus polymyxa* SCE2. Lett Appl Microbiol 43(6):625–630
8. Elias CGR, Aor AC, Valle RS, d'Avila-Levy CM, Branquinha MH, Santos ALS (2009) Cysteine peptidases from *Phytomonas serpens*: biochemical and immunological approaches. FEMS Immunol Med Microbiol 57(3):247–256
9. dos Santos ALS, de Carvalho IM, da Silva BA, Portela MB, Alviano CS, de Araújo Soares RM (2006) Secretion of serine peptidase by a clinical strain of *Candida albicans*: influence of growth conditions and cleavage of human serum proteins and extracellular matrix components. FEMS Immunol Med Microbiol 46(2):209–220
10. Silva BA, Pinto MR, Soares RM, Barreto-Bergter E, Santos AL (2006) *Pseudallescheria boydii* releases metallopeptidases capable of cleaving several proteinaceous compounds. Res Microbiol 157(5):425–32
11. Lavens SE, Rovira-Graells N, Birch M, Tuckwell D (2005) ADAMs are present in fungi: identification of two novel ADAM genes in *Aspergillus fumigatus*. FEMS Microbiol Lett 248(1):23–30
12. Marchal R, Warchol M, Cilindre C, Jeandet P (2006) Evidence for protein degradation by *Botrytis cinerea* and relationships with alteration of synthetic wine foaming properties. J Agric Food Chem 54(14):5157–5165
13. O'Donoghue AJ, Mahon CS, Goetz DH, O'Malley JM, Gallagher DM, Zhou M et al (2008) Inhibition of a secreted glutamic peptidase prevents growth of the fungus *Talaromyces emersonii*. J Biol Chem 283(43):29186–29195

Protocol of a LightCycler™ PCR Assay for Detection and Quantification of *Aspergillus fumigatus* DNA in Clinical Samples of Neutropenic Patients

Birgit Spiess and Dieter Buchheidt

Abstract

The increasing incidence of life-threatening systemic fungal infections emphasizes the need to improve molecular diagnostic tools. Polymerase chain reaction method (PCR) was used to establish sensitive and rapid molecular detection assays of pathogens not detectable in cultures. Using the LightCycler™ technology that combines amplification of DNA with an immediate fluorescence detection of the amplicon, a real-time PCR assay was established to achieve an improved, specific, sensitive, and rapid method for quantification of the *Aspergillus fumigatus* fungal load in clinical samples of hematological patients in order to improve antifungal treatment monitoring.

Keywords

Aspergillus fumigatus • Polymerase chain reaction (PCR) • Clinical samples • Fungal load • Neutropenic patients

Introduction

The increasing incidence of life-threatening systemic fungal infections, especially invasive aspergillosis (IA), correlates with an increased number of immunocompromised patients [1]. Patients at highest risk are those who received intensive cytotoxic chemotherapy for acute leukemia or bone marrow or allogeneic hematopoietic stem cell transplantation, which can lead to prolonged periods of neutropenia [2–7].

Due to the limited prognosis of patients with invasive *Aspergillus* infections, all diagnostic approaches primarily aim at an early confirmation of an infection, to optimize antifungal treatment [8].

It remains difficult to diagnose IA at all, since the current diagnostic tools either lack specificity or sensitivity or both, at worst. At present, only positive results from conventional cultures or histological examination provide the definitive proof of invasive aspergillosis. However, establishing cultures from blood, bronchoalveolar lavage

B. Spiess (✉) • D. Buchheidt
3rd Department of Internal Medicine,
Hematology and Oncology, Scientific Laboratory,
Mannheim University Hospital, Pettenkoferstr. 22,
Mannheim 68169, Germany
e-mail: birgit.spiess@medma.uni-heidelberg.de

V.K. Gupta et al. (eds.), *Laboratory Protocols in Fungal Biology: Current Methods in Fungal Biology*,
Fungal Biology, DOI 10.1007/978-1-4614-2356-0_26, © Springer Science+Business Media, LLC 2013

(BAL), or other clinical samples is often unsuccessful because of the low yields of colony-forming units of the pathogen [9].

Polymerase chain reaction method (PCR) was used to establish sensitive and rapid molecular detection assays of pathogens not detectable in cultures.

Against this background we established and evaluated a highly sensitive and *Aspergillus* specific two-step PCR assay for the analysis of clinical samples (blood, BAL, CSF, tissue samples) [3–7,10,11].

In order to estimate the fungal burden and to monitor and evaluate the response to antifungal drugs, the quantification of the fungal burden is of great clinical relevance.

Using the LightCycler™ technology that combines amplification of DNA with an immediate fluorescence detection of the amplicon, we established an assay for clinical samples and achieved an improved, specific, sensitive, and rapid method for quantification of the *A. fumigatus* fungal load [12].

Materials

1. Sterile distilled water.
2. Puffer 10× RCLB: 1.55 M NH_4Cl, 0.1 M NH_4HCO_3, 1 mM EDTA pH 7.4.
3. Puffer 1× PBS.
4. Lyticase.
5. Proteinase K.
6. 10% SDS.
7. 2× APEX: 400 mM Tris/Cl, 20 mM EDTA, 2% SDS.
8. Phenol:chloroform (1:1).
9. 70% isopropanol.
10. 70% ethanol.
11. LightCycler Fast Start DNA Master Hybridization Probes kit (Roche Applied Science, Mannheim, Germany).
12. Oligonucleotide primers and labeled probes (custom-made TIB MOLBIOL, Berlin, Germany).
13. Microcentrifuge (Heraeus, Biofuge, Frankfurt, Germany).
14. LightCycler (Roche Applied Science, Mannheim, Germany).
15. LightCycler glass capillaries.
16. Agarose, molecular biology grade (BioRad, Munich, Germany).
17. Puffer 10× TBE.
18. Ethidium bromide.
19. Horizontal electrophoresis equipment (e.g., BioRad, Munich, Germany).
20. U.V. transilluminator and camera suitable for imaging agarose gels (e.g., ChemiDoc XRS+, BioRad, Munich, Germany).

Methods

Clinical Samples from Immunocompromised Patients

Blood samples were obtained under sterile conditions by venipuncture, in a sterile vessel containing potassium EDTA to a final concentration of 1.6 mg EDTA per milliliter of blood. The sample volume was 1–5 mL.

Bronchoalveolar lavage samples: bronchoscopy was performed by experienced physicians according to guidelines [9] and BAL samples were obtained in a sterile vessel without conservation media. The mean sample volume was 8–10 mL.

CSF and tissue samples were obtained in a sterile vessel without conservation media by puncture under sterile conditions, according to specific clinical guidelines.

All clinical samples were drawn after informed consent of the individual patient.

DNA Preparation

1. 3–5 mL peripheral blood were mixed with 5 volumes of RCLB (10× RCLB; red cell lysis buffer: 1.55 M NH_4Cl, 0.1 M NH_4HCO_3, 1 mM EDTA pH 7.4) and incubated on ice for 10 min for lysis of the erythrocytes.
2. After incubation, the sample was centrifuged for 10 min at 300×g. Supernatant was discarded, the leukocytes were washed once with 1× PBS (10× PBS; phosphate buffered saline: 1.4 M NaCl, 50 mM KCl, 90 mM $Na_2PO_4 \cdot 2H_2O$, 20 mM KH_2PO_4, pH 7.4) and recentrifuged.

3. BAL samples were transferred into 1.5 mL tubes and centrifuged for 5 min at 300×*g*. The sedimented leukocyte pellet was resuspended in 300 µl 1× PBS and incubated with 10–125 U of lyticase (50,000 U, Sigma-Aldrich, Deisenhofen, Germany) for 30 min at 37 °C to achieve degradation of the fungal cells.

4. 500–1000 µg proteinase K (Roche Molecular Biochemicals, Mannheim, Germany) and 0.5% SDS (Sigma-Aldrich) were added and the suspension was incubated at 55 °C for 1 h.

5. By treatment with additional 100 µl 2× APEX (2× APEX; *Aspergillus* extraction buffer: 400 mM Tris/Cl, 1 M NaCl, 20 mM EDTA, 2% SDS) for 30 min at 65 °C, residual cell material was lysed. DNA isolation was performed under a laminar flow.

6. The purification of the fungal and human DNA mixture was performed by conventional phenol–chloroform extraction [13].

7. The DNA was precipated with 70% (v/v) of isopropanol and the DNA pellet was washed once with 70% ethanol and air dried.

8. DNA concentration of human DNA was measured by spectrophotometry at 260 nm/280 nm.

Primers and Hybridization Probes for the LightCycler™-based PCR Assay

PCR primers and probes were derived from fungal mitochondrial Cytochrome B genes.

Several regions of the mitochondrial Cytochrome B gene of minor homology between four *Aspergillus*, four *Candida* species, and human DNA were the presupposition for the design of the primers and hybridization probes (*A. fumigatus* (GenBank accession no. AB025434), *A. flavus* (GenBank accession no. AB000596), *A. terreus* (GenBank accession no. AB000603) and *A. niger* (GenBank accession no. AB000597), *C. albicans* (GenBank accession no. AB0044919), *C. parapsilosis* (GenBank accession no. AB044929), *C. glabrata* (GenBank accession no. AB044922), *C. tropicalis* (GenBank accession no. AB044930), and human (GenBank accession no. M28016) (Fig. 26.1).

Alignment of the DNA sequences was performed using the program Geneworks (Intelligenetics, Inc.) with standard algorithms.

After testing of 16 theoretically wise combinations of primers and hybridization probes, the optimum pairs were chosen for all subsequent PCR assays.

Out of the mitochondrial Cytochrome B gene sequence of *Aspergillus fumigatus* (GenBank accession no. AB025434) the sequence for the forward primer was 5′-AATGCACGATACTGTA GGATCTG-3′ (AfLC2s), and for the reverse primer 5′-TGCATTGGATTAGCCATAACA-3′ (AfLC2as). The length of the amplified fragment was 194 bp. Hybridization probes were selected from the region between forward and reverse primers of the primer pair.

For labeling of one probe at the 5′ end, the LightCycler Red 640 fluorophore was used: 5′-TAATCTATCATAATTACCAGAAATACCT AAAGGA-3′ (Cyt3A). The other probe was labeled at the 3′ end with fluorescein: 5′-AATCTTTAAATACAAAGTAAGGAGCG AAAG-3′ (Cyt3B) [12]. Primers and hybridization probes were obtained from TIB MOLBIOL, Berlin, Germany.

Quantification of *A. fumigatus* DNA

Amplification and quantification of *A. fumigatus* DNA was performed using the LightCycler™ PCR and detection system (Roche Applied Science, Mannheim, Germany). The Hot Start PCR reaction was performed in glass capillaries using the LightCycler Fast Start DNA Master Hybridization Probes kit (Roche Applied Science) as described by the manufacturer. PCR mixture contained 1× Fast Start reaction mix including the Fast Start Taq DNA polymerase, reaction buffer, dNTPs and 10 mM $MgCl_2$, all together 3.5 mM $MgCl_2$, 20 pmol of each primer, and 60 nmol of hybridization probes.

In a volume of 20 µl PCR was performed under following conditions: initial denaturing for 8 min at 95 °C, 45 cycles with 4 s at 95 °C, annealing for 8 s at 58 °C, and enzymatic chain extension for 20 s at 72 °C. Each analysis done by PCR included a H_2O negative control without any template

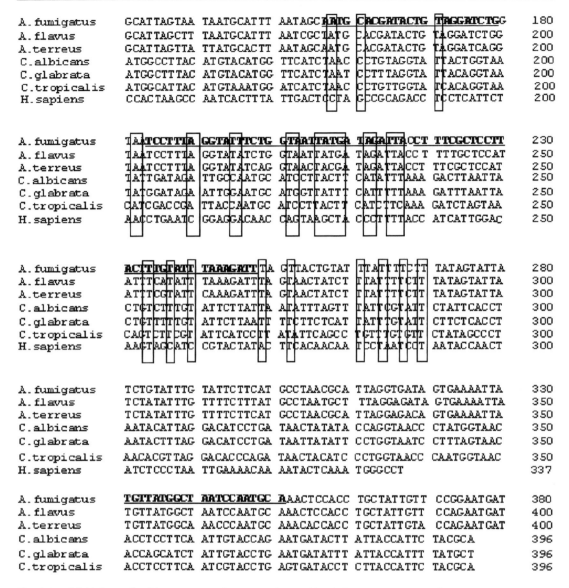

```
A.fumigatus   GCATTAGTAA TAATGCATTT AATAGCARTG CRCGATACTG TAGGATCTGG   180
A.flavus      GCATTAGCTT TAATGCATTT AATCGCTATG CACGATACTG TAGGATCTGG   200
A.terreus     GCATTAGTTA TTATGCACTT AATAGCAATG CACGATACTG TAGGATCAGG   200
C.albicans    ATGGCCTTAC ATGTACATGG TTCATCTAAC CCTGTAGGTA TTACTGGTAA   200
C.glabrata    ATGGCTTTAC ATGTACATGG TTCATCTAAT CCTTTAGGTA TTACAGGTAA   200
C.tropicalis  ATGGCATTAC ATGTAAATGG ATCATCTAAC CCTGTTGGTA TCACAGGTAA   200
H.sapiens     CCACTAAGCC AATCACTTTA TTGACTGCTA GCCGCAGACC TCCTCATTCT   200
```

```
A.fumigatus   TAATCCTTTA GGTATTTCTG GTAATTATGA TAGATTACCT TTCGCTCCTT   230
A.flavus      TAATCCTTTA GGTATATCTG GTAATTATGA TAGATTACC T TTTGCTCCAT   250
A.terreus     TAATCCTTTA GGTATATCAG GTAACTACGA TAGATTACCT TTCGCTCCAT   250
C.albicans    TATTGATAGA TTGCCAATGC ATCCTTACTT CATATTTAAA GACTTAATTA   250
C.glabrata    TATGGATAGA ATTGGAATGC ATGGTTATTT CATTTTTAAA GATTTAATTA   250
C.tropicalis  CATCGACCGA TTACCAATGC ATCCTTACTT CATCTTCAAA GATCTAGTAA   250
H.sapiens     AACCTGAATC GGAGGACAAC CAGTAAGCTA CCCTTTTACC ATCATTGGAC   250
```

```
A.fumigatus   ACTTTGTATT TAAAGATTTA GTTACTGTAT TTATTTTCTT TATAGTATTA   280
A.flavus      ATTTCATATT TAAAGATTTA GTAACTATCT TTATTTTCTT TATAGTATTA   300
A.terreus     ATTTCGTATT CAAAGATTTA GTAACTATCT TTATTTTCTT TATAGTATTA   300
C.albicans    CTGTCTTTGT ATTCTTATTA ATATTTAGTT TATTCGTATT CTATTCACCT   300
C.glabrata    CTGTTTTTGT ATTCTTAATT TTCTTCTCAT TATTTGTATT CTTCTCACCT   300
C.tropicalis  CAGTCTTCGT ATTCATCCTT ATATTCAGCC TGTTTGTGTT CTATAGCCCT   300
H.sapiens     AAGTAGCATC CGTACTATAC TTCACAACAA TCCTAATCCT AATACCAACT   300
```

```
A.fumigatus   TCTGTATTTG TATTCTTCAT GCCTAACGCA TTAGGTGATA GTGAAAATTA   330
A.flavus      TCTATATTTG TTTTCTTTAT GCCTAATGCT TTAGGAGATA GTGAAAATTA   350
A.terreus     TCTATATTTG TTTTCTTCAT GCCTAACGCA TTAGGAGACA GTGAAAATTA   350
C.albicans    AATACATTAG GACATCCTGA TAACTATATA CCAGGTAACC CTATGGTAAC   350
C.glabrata    AATACTTTAG GACATCCTGA TAATTATATT CCTGGTAATC CTTTAGTAAC   350
C.tropicalis  AACACGTTAG GACACCCAGA TAACTACATC CCTGGTAACC CAATGGTAAC   350
H.sapiens     ATCTCCCTAA TTGAAAACAA AATACTCAAA TGGGCCT               337
```

```
A.fumigatus   TGTTATGGCT AATCCAATGC AAACTCCACC TGCTATTGTT CCGGAATGAT   380
A.flavus      TGTTATGGCT AATCCAATGC AAACTCCACC TGCTATTGTT CCAGAATGAT   400
A.terreus     TGTTATGGCA AACCCAATGC AAACACCACC TGCTATTGTA CCAGAATGAT   400
C.albicans    ACCTCCTTCA ATTGTACCAG AATGATACTT ATTACCATTC TACGCA      396
C.glabrata    ACCAGCATCT ATTGTACCTG AATGATATTT ATTACCATTT TATGCT      396
C.tropicalis  ACCTCCTTCA ATCGTACCTG AGTGATACCT CTTACCATTC TACGCA      396
```

Fig. 26.1 Multiple nucleotide sequence alignment of mitochondrial Cytochrome B genes of *A. fumigatus* (GenBank accession no. AB025434), *A. flavus* (GenBank accession no. AB000596), *A. terreus* (GenBank accession no. AB000603), *C. albicans* (GenBank accession no. AB0044919), *C. glabrata* (GenBank accession no. AB044922), *C. tropicalis* (GenBank accession no. AB044930), and human (GenBank accession no. M28016). Locations of primers and hybridization probes are underlined. Homologous regions are in boxes

DNA to monitor for possible contamination. Aliquots of DNA from healthy control persons were prepared in parallel to clinical sample DNA specimens and analyzed as negative controls.

A serially diluted standard of genomic *A. fumigatus* DNA was used. 1×10^6 copies of the mitochondrial Cytochrome B gene were corresponding to 1.32 ng and 1×10 copies to 13.2 fg of genomic *A. fumigatus* DNA or 1–5 CFU per mL blood.

The logarithmic linear phase was distinguished from the background by online monitoring

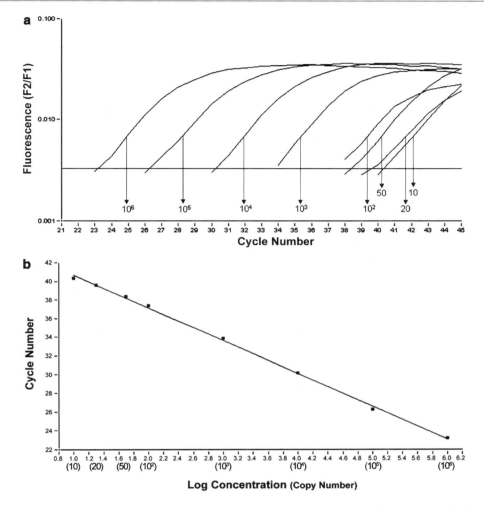

Fig. 26.2 Determination and quantification of serially diluted standard of *Aspergillus fumigatus* DNA (**a**) using the LightCycler PCR technique. (**b**) LightCycler standard curve report for serially diluted *Aspergillus fumigatus* DNA. A representative evaluation is shown

(Figs. 26.2a, b). The amounts of *Aspergillus* DNA in unknown samples were calculated by comparing the Cytochrome B gene copy numbers of the logarithmic linear phase of the sample with the copy numbers of the standards (Fig. 26.3) [12].

Specificity of the LightCycler PCR Assay

DNA from several fungal and bacterial strains was subjected to the LightCycler PCR to determine the specificity of the assay. Cross reactivity of the primers and hybridization probes with human DNA was excluded by testing of DNA of

10 healthy control persons in the LightCycler PCR assay. Only DNA from *A. fumigatus* (DSM 819 CS) was detectable in the LightCycler PCR assay. All PCR assays with other fungal and bacterial strains were negative.

PCR has been shown to be a highly sensitive and specific diagnostic tool for the detection of *Aspergillus* species in clinical samples. We aimed to extend the diagnostic value of our nested PCR assay [10] to a quantification of the pathogen load in order to improve the antifungal treatment monitoring.

For clinical evaluation we investigated clinical samples from neutropenic patients suffering

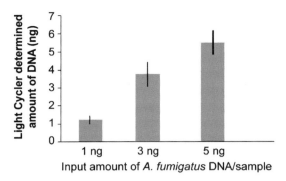

Fig. 26.3 Correlation between LightCycler quantification of defined amounts of *Aspergillus fumigatus* DNA and sample amounts (1–5 ng) in vitro. Data represent means ± SD from five separate experiments

from malignant hematological diseases. BAL and blood samples gave positive results in the LightCycler PCR assay and were also tested with our previously described [10] nested PCR assay. The PCR-mediated quantification of the fungal burden showed 15–269,018 CFU per mL of BAL and 298–104,114 CFU per mL of blood sample. BAL and blood samples from subjects without evidence for invasive pulmonary aspergillosis were PCR-negative [12].

In studies, up to now, we screen clinical samples of high-risk patients first with our nested PCR assay providing highest sensitivity and general specificity for *Aspergillus* species. Samples tested positive in the nested PCR assay are subsequently quantified with the LightCycler PCR assay [4–6].

Notes

In summary, our highly specific and sensitive LightCycler-based real-time PCR assay is applicable for the rapid and early detection of *Aspergillus* species and the quantification of the fungal load from clinical samples of high-risk patients.

References

1. Buchheidt D (2008) Molecular diagnosis of invasive aspergillosis in patients with hematologic malignancies—new answers to a diagnostic challenge? Expert Opin Med Diagn 2:753–761
2. Lehrnbecher T, Frank C, Engels K, Kriener S, Groll AH, Schwabe D (2010) Trends in the postmortem epidemiology of invasive fungal infections at a university hospital. J Infect 61:259–265
3. Buchheidt D, Baust C, Skladny H, Baldus M, Brauninger S, Hehlmann R (2002) Clinical evaluation of a polymerase chain reaction assay to detect *Aspergillus* species in bronchoalveolar lavage samples of neutropenic patients. Br J Haematol 116:803–811
4. Buchheidt D, Hummel M, Schleiermacher D, Spiess B, Schwerdtfeger R, Cornely OA et al (2004) Prospective clinical evaluation of a LightCycler-mediated polymerase chain reaction assay, a nested-PCR assay and a galactomannan enzyme-linked immunosorbent assay for detection of invasive aspergillosis in neutropenic cancer patients and haematological stem cell transplant recipients. Br J Haematol 125:196–202
5. Buchheidt D, Baust C, Skladny H, Ritter J, Suedhoff T, Baldus M et al (2001) Detection of *Aspergillus* species in blood and bronchoalveolar lavage samples from immunocompromised patients by means of 2-step polymerase chain reaction: clinical results. Clin Infect Dis 33:428–435
6. Hummel M, Spiess B, Kentouche K, Niggemann S, Bohm C, Reuter S et al (2006) Detection of *Aspergillus* DNA in cerebrospinal fluid from patients with cerebral aspergillosis by a nested PCR assay. J Clin Microbiol 44:3989–3993
7. Hummel M, Spiess B, Roder J, von Komorowski G, Durken M, Kentouche K et al (2009) Detection of *Aspergillus* DNA by a nested PCR assay is able to improve the diagnosis of invasive aspergillosis in paediatric patients. J Med Microbiol 58:1291–1297
8. Verweij PE, Maertens J (2009) The changing face of febrile neutropenia-from monotherapy to moulds to mucositis. Moulds: diagnosis and treatment. J Antimicrob Chemother 63:i31–i35
9. Maschmeyer G, Beinert T, Buchheidt D, Cornely OA, Einsele H, Heinz W et al (2009) Diagnosis and antimicrobial therapy of lung infiltrates in febrile neutropenic patients: guidelines of the infectious diseases working party of the German Society of Haematology and Oncology. Eur J Cancer 45:2462–2472
10. Skladny H, Buchheidt D, Baust C, Krieg-Schneider F, Seifarth W, Leib-Mösch C et al (1999) Specific detection of *Aspergillus* species in blood and bronchoalveolar lavage samples of immunocompromised patients by two-step PCR. J Clin Microbiol 37:3865–3871
11. Hummel M, Spiess B, Cornely OA, Dittmer M, Mörz H, Buchheidt D (2010) *Aspergillus* PCR testing: results from a prospective PCR study within the AmBiLoad trial. Eur J Haematol 85:164–169
12. Spiess B, Buchheidt D, Baust C, Skladny H, Seifarth W, Zeilfelder U et al (2003) Development of a LightCycler PCR assay for detection and quantification of *Aspergillus fumigatus* DNA in clinical samples from neutropenic patients. J Clin Microbiol 41:1811–1818
13. Sambrook J, Fritsch E, Maniatis T (1989) Molecular cloning: a laboratory manual. Cold Spring Harbor Laboratory Press, Cold Spring Harbor, NY

Application of Polymerase Chain Reaction and PCR-Based Methods Targeting Internal Transcribed Spacer Region for Detection and Species-Level Identification of Fungi

K. Lily Therese, R. Bagyalakshmi, and H.N. Madhavan

Abstract

This chapter focuses on the application of molecular technique based on polymerase chain reaction (PCR) targeting internal transcribed spacer (ITS) region for detection and identification of fungi from normally sterile body fluids and tissue biopsies; detection of nucleotide polymorphisms in *Aspergillus flavus*; PCR-based DNA sequencing for identification of non-sporulating molds (NSM); and detection and identification of dermatophytes from dermatological specimens by PCR-based restriction fragment length polymorphism (PCR-RFLP).

Keywords

ITS region • PCR • PCR-based DNA sequencing • Nonsporulating molds (NSM) • PCR-based RFLP • Dermatophytes

Introduction

Conventional methods for the detection of fungal infections are less sensitive because the microbial threshold is low and the techniques are laborious and time-consuming. Rapid diagnosis by molecular methods aid in the institution of specific anti-fungal drug and management [1]. Molecular methods are rapid, extremely sensitive, and specific. Fungi have a ribosomal DNA (rDNA) complex that includes a sequence coding for the 18S rDNA gene, an internal transcribed spacer region 1(ITS1), the 5.8S rDNA gene coding region, another ITS region called ITS2, and the sequence coding for 28S rDNA gene [1, 2]. Polymerase chain reaction (PCR) assays have been developed and applied targeting the 28S rDNA [3, 4] and 18S rDNA [5]. The ITS region is a multicopy gene and consists of considerable variation to differentiate the fungal species.

Several studies that target the ITS region to detect and identify fungal genome using PCR and PCR-based methods from clinical specimens have been carried out. Target genes used in molecular diagnosis of fungal infections include

K.L. Therese (✉) • R. Bagyalakshmi
H.N. Madhavan
Larsen and Toubro Microbiology Research Centre,
Kamal Nayan Bajaj Research Centre,
Vision Research Foundation,
41 (Old No. 18) College Road, Chennai,
Tamil Nadu 600006, India
e-mail: drklt@snmail.org

V.K. Gupta et al. (eds.), *Laboratory Protocols in Fungal Biology: Current Methods in Fungal Biology*,
Fungal Biology, DOI 10.1007/978-1-4614-2356-0_27, © Springer Science+Business Media, LLC 2013

single and multicopy nuclear and mitochondrial genes. Multicopy genes provide a better detection threshold than single-copy genes and are employed in different molecular assays [6–16]. Molecular diagnostic assays involve the use of PCR followed by restriction fragment length polymorphisms (RFLP) [17, 18] to identify the species and to analyze the strain variations and DNA sequencing for identification, detection of nucleotide polymorphisms, and mutations existing in the fungal genome [19–21].

Application of a Semi-Nested Polymerase Chain Reaction Targeting the Internal Transcribed Spacer Region

Protocol for Detection of Panfungal Genome by Application of snPCR Targeting ITS Region

Materials
Standard Strains of Fungi
1. *C. albicans* ATCC 24433
2. *C. albicans* ATCC 90028
3. *C. parapsilosis* ATCC 90018
4. *C. parapsilosis* ATCC 22019
5. *C. krusei* ATCC 6258
6. *C. tropicalis* ATCC 750
7. *A. fumigatus* ATCC 10894

Reagents (Commercially Available QIAGEN Kit, Germany)
1. Proteinase K
2. Lysis buffer
3. Absolute ethanol
4. Washing buffer-1 and 2
5. Elution buffer
6. Microfuge (Eppendorf, Germany)
7. New sterile disposable polypropylene microfuge tubes, 1.5 mL
8. Waterbath

Methods
DNA extraction is carried out using QIAamp DNA extraction kit (Qiagen, Germany) according to the manufacturer's instruction.

1. Pipette 20 µl Qiagen Proteinase K into the bottom of a 1.5-mL microfuge tube.
2. Add 200 µl of the sample to the microfuge tube. Use up to 200 µl whole blood, plasma, serum, buffy coat or body fluids or up to 5×10^6 lymphocytes in 200 µl PBS.
3. Add 200 µl of lysis buffer (AL buffer) to the sample. Mix by pulse vortexing for 15 s.
4. Incubate at 56 °C in a water bath for 10 min. (see Note 1)
5. Add 200 µl of ethanol (96–100%) to the sample, mix by gentle pipetting for 15 s.
6. Carefully apply the mixture from step 5 to the QIAamp mini-spin column (in a 2-mL collecting tube) without wetting the rim. Close the cap, and centrifuge at 8,000 rpm for 1 min. Place the QIAamp mini-spin column in a clean 2-mL collecting tube and discard the tube containing the filtrate.
7. Carefully open the QIAamp mini-spin column and add 500 µl buffer AW1 without wetting the rim. Close the cap and centrifuge at 8,000 rpm for 1 min. Place the QIAamp mini-spin column in a clean 2-mL collecting tube and discard the tube containing filtrate.
8. Carefully open the QIAamp mini-spin column and add 500 µl buffer AW2 without wetting the rim. Close the cap and centrifuge at 14,000 rpm for 3 min, followed by an empty spin at 14,000 rpm for 1 min.
9. Place the QIAamp mini-spin column in a new 1.5-mL microfuge tube and discard the tube containing filtrate. Carefully open the QIAamp mini-spin column and add 200 µl buffer AE. Incubate at room temperature for 1 min, and then centrifuge at 8,000 rpm for 1 min. Discard the column and store the DNA at −20 ° C.

Semi-Nested PCR Targeting the Internal Transcribed Spacer Region
The snPCR targeting ITS region involves the use of primers designed by Ferrer et al. [22].

Materials
1. Deoxyribonucleotide triphosphates (dNTPs) — dATP, dGTP, dCTP, dTTP (10 mM each, Bangalore Genei, India) stored at −20 °C.

2. Thermostable *Taq* DNA polymerase 3 units/μl (Bangalore Genei, India) supplied with a10× PCR buffer (Tris–Cl buffer pH 8.8) which contains 500 mM KCl, 15 mM Magnesium chloride.

3. 25 mM Magnesium chloride (Bangalore Genei, India).

4. Oligonucleotide primers (custom made, Bangalore Genei, India).

5. PCR tubes thin walled (0.5 mL, 0.2 mL, Axygen, USA).

6. Thermal cycler (Eppendorf, Germany PerkinElmer 2700, USA).

Procedure

1. A 50-μl reaction, consists of 8 μl of dNTPs (200 μmol), 5 μl of 1× PCR buffer (1.5 mM MgCl$_2$,50 mM KCl, 10 mM Tris–Cl,0.001% gelatin), 6 μl of 25 mM MgCl$_2$ (1 in 10 diluted to get a final concentration of 3 mM), 1 microlitre of forward primer (10 picomoles) of, ITS1—5′-TCC GTA GGT GAA CCT GCG G-3′and 1 microlitre of reverse primer ITS4—5′-TCC TCC GCT TAT TAT GC-3′ targeting ITS region, 1 unit of Taq polymerase, 18.7 microlitre of milli-Q water and 10 μl of template DNA [22].

2. Amplification is allowed to occur in a PCR machine (Perkin Elmer Model 2700). The first round of amplification yields 520–611 bp product depending on the fungal species present in the clinical specimen.

3. The thermal profile includes initial denaturation at 95 °C for 5 min followed by 35 cycles of denaturation at 95 °C for 30 s, annealing at 55 °C for 60 s and extension at 72 °C for 60 s followed by final extension at 72 °C for 6 min.

4. This is then subjected to semi-nested amplification using same PCR conditions as that of the first round with forward primer ITS86—5′-GTG AAT CAT CGA ATC TTT GAA C-3′ and reverse primer ITS 4 as indicated above.

5. 5 μl of amplified product is transferred from the first round to the second round and subjected to amplification.

6. The thermal profile consists of initial denaturation at 95 °C for 5 min followed by 35 cycles

of denaturation at 95 °C for 30 s, annealing at 55 °C for 30 s, and extension at 72 °C for 30 s followed by final extension at 72 °C for 5 min.

Detection of Amplified Products

1. Agarose molecular grade (SRL, India).

2. Tris–borate EDTA buffer (TBE buffer) consisting of 50 mM Tris, 50 mM boric acid, and 1 mM EDTA. A 1-in-10 dilution is made and used to dissolve agarose.

3. Ethidium bromide −0.5 mg/mL stock.

4. Tracking dye: Bromophenol blue (Hi Media, India) consisting of 0.1% Bromophenol blue, 40% (w/v) sucrose, 0.1 M EDTA. 1-in-10 dilution is made with electrophoresis buffer and used.

5. Parafilm for aliquoting the amplified products.

6. Horizontal electrophoresis equipment with power pack (Sri Balaji Scientific Supplies, India) along with gel-casting accessories.

7. Molecular weight marker *Hinf* I digest of ϕX174 Bacteriophage DNA.

8. Gel documentation system (Vilber Lourmat, France).

Preparation of 2% Agarose Gel

1. Clean the gel trough with ethanol and seal the ends with cellophane tape with the combs placed within groove to form wells.

2. Prepare 2% agarose gel by dissolving 1 g of agarose in 50 mL of 1× TBE buffer. The agarose is dissolved by boiling it in a microwave oven and add 8 μl of ethidium bromide. (Final concentration in the gel 0.5 μg/mL).

3. Mix thoroughly and pour onto the trough.

4. The gel is allowed to solidify for 30 min.

5. After the gel gets solidified, the combs are carefully removed.

6. Place the gel in the submarine electrophoresis tank and remove the combs.

7. Mix 10 microlitre of PCR amplified product using 2 microlitre of Bromophenol blue in a parafilm strip.

8. Load the negative control, samples, positive control, and then the molecular weight marker.

9. Carry out the electrophoresis at 100 V for half an hour.

Table 27.1 Amplicon size in base pairs of fungi

Fungi	I Round	II Round
Candida albicans	586	282
Candida parapsilosis	524	270
Candida glabrata	820	360
Candida krusei	510	294
Aspergillus flavus	595	300
Aspergillus fumigatus	596	299
Aspergillus niger	599	300
Aspergillus terreus	608	308
Fusarium solani	569	286
Fusarium oxysporum	544	283
Cryptococcus neoformans	556	320
Scedosporium apiospermum	611	329
Alternaria alternata	570	292

10. Carry out the gel documentation using the gel documentation system Vilber Lourmat, France and analyze using BioID software.
11. The amplified product size varies according to the species of fungi and the respective amplicon size of representative fungi is given in Table 27.1.

Sensitivity of snPCR Targeting ITS Region for Detection of Fungal Genome

1. 1 μl of the extracted DNA is dissolved in 999 μl of water and quantified spectrophotometrically at 260 and 280 nm.
2. The reading at 260 nm gives the nucleic acid concentration. The ratio of readings at OD 260/280 nm gives the purity of the nucleic acid.
3. The sensitivity of snPCR is determined using serial tenfold dilutions of standard strain of C. albicans ATCC 24433.
4. 10 μg of C. albicans and A. fumigatus are used as the template DNA.

Specificity of snPCR

The specificity of snPCR is determined using DNA extracts of microorganisms; C. albicans (ATCC 90028), C. tropicalis (ATCC 750),

C. parapsilosis (ATCC 90018), C. krusei (ATCC 6258), A. flavus (ATCC 204304), A. fumigatus (ATCC 10894), A. niger (ATCC16404), F. solani (ATCC 36031), laboratory isolates of A. terreus, Curvularia and Alternaria species, S. aureus (ATCC 12228), Pseudomonas aeruginosa (ATCC 27853), M. tuberculosis (H37Rv), M. fortuitum (ATCC 1529), M. chelonae (ATCC 1524) and laboratory isolates of Nocardia asteroides, Actinomyces species, Herpes Simplex virus (ATCC 733 VR), Acanthamoeba polyphaga (ATCC 30461), and human leukocyte DNA.

Standardization of PCR: Sensitivity of snPCR Targeting ITS Region

1. The expected sensitivity of ITS primers is 1–10 fg of C. albicans DNA (single cell of C. albicans) and 10 fg of A. fumigatus and F. lichenicola DNA.
2. The ITS PCR is specific amplifying only the fungal DNA. The application of semi-nested PCR targeting ITS region on clinical specimens is shown in Fig. 27.1.

PCR targeting ITS region of fungi is a rapid and sensitive diagnostic test used for species-level identification of fungi from clinical specimens [23].

Application of Semi-Nested PCR Targeting ITS Region to Determine the Nucleotide Polymorphisms Associated with Ocular Isolates of *Aspergillus flavus*

Strain typing of medically important fungi (i.e., the ability to identify them to the species level and to discriminate among individuals within species) has been galvanized by new methods of tapping the tremendous variation found in fungal DNA. There are several methods like multilocus enzyme electrophoresis, electrophoretic karyotype analysis, RFLP, randomly amplified polymorphic DNA, sequence confirmed amplified region analysis, DNA fingerprinting, PCR-based DNA sequencing with repetitive sequences

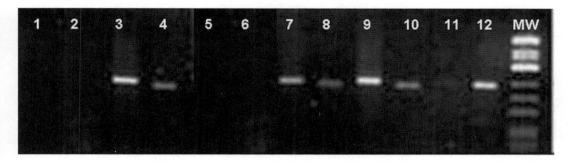

Fig. 27.1 Agarose gel electrophoretogram showing application of ITS PCR on ocular specimens. Lane *1*: Negative control II round; Lane *2*: Negative control I round; Lane *3*: AH-positive (*Aspergillus flavus*) (300 bp); Lane *4*: VF-positive *Fusarium solani* (286 bp); Lane *5*: VF-negative; Lane *6*: AH-negative; Lane *7*: AH-positive *Aspergillus terreus* (308 bp); Lane *8*: VF positive *Candida tropicalis* (293 bp); Lane *9*: VF positive (*Aspergillus flavus*) (300 bp); Lane *10*: AH-positive *Candida albicans* (282 bp); Lane *11*: AH-positive *A. niger* 300 bp; Lane *12*: Positive control : *Candida albicans* ATCC 24433 (282 bp); Lane *MW*: Molecular weight marker Hinf I digest of Phi X174 bacteriophage DNA. *Note*: The species identity detected by ITS PCR was further confirmed by PCR-based DNA sequencing

available for studying the strain variation existing among the species of fungi rRNA genes. The use of DNA sequence diversity in the ribosomal regions as an aid to species identification has been exploited using PCR amplification of targets followed by either fragment length analysis [19], DNA probe hybridization [20], or DNA sequence analysis.

Study on Strain Variations of *Aspergillus flavus*

In a study conducted in the authors' center, seven ocular isolates of *A. flavus* from aqueous humor (AH, 2), vitreous fluid (VF, 2), corneal scraping (1), eviscerated material (1), corneal button (1) were isolated from four patients clinically suspected to have fungal endophthalmitis.

Molecular Microbiological Investigations: DNA Extraction from Isolates

DNA extraction from clinical specimens and fungal isolates is done using QIAamp kit as described in "Methods".

PCR Targeting ITS Region

PCR targeting ITS region is carried out on culture isolates of *Aspergillus flavus* according to the method provided in Semi-Nested PCR (snPCR) Targeting the ITS Region.

PCR-Based Restriction Fragment Length Polymorphism

1. PCR-RFLP is carried out on ITS amplicons (first round products) using restriction enzyme *Hae* - III.
2. For a 25 µl reaction, 2.5 µl of Buffer C, 0.5 µl of Hae – III enzyme and 10 µl of amplified product are added and incubated at 37 ° C for 3 h.
3. The digested products are loaded on 4% agarose gel incorporated with ethidium bromide and the electrophoresis is carried out at 100 V for 45 min.
4. The digested pattern (385,188,82 bp) is visualized and documented using gel documentation system (Vilber Lourmat, France).

DNA Sequencing of ITS Amplicons

DNA sequencing of ITS amplicons is carried out after purification of amplified products.

In our study, all *A. flavus* isolates showed similar pattern of digestion with *Hae*-III. In an earlier study Henry et al. [24]. have reported a variation of 1% in epidemiological analysis of *A. flavus* strains. Alignment of contiguous fungal sequences demonstrated that both single nucleotide differences and short lengths of sequence diversity due to insertion or deletion existed in the ITS regions among the pathogenic *A. flavus* strains. *A. flavus* isolates in the study had a BLAST score of 97.7% identity with the standard strain of *A. flavus* (ATCC 16883)—GenBank accession no. AB008415. The inspection of BLAST alignments generated with *A. flavus* ITS1 and ITS 2 data from GenBank revealed that many *A. flavus* sequences in the data base had truncated ends and/or heterogeneities at positions found to be conserved at the subgeneric level among reference sequence of type and authenticated culture collection strains. The study by Bagyalakshmi et al. [25] on *A. flavus* isolates revealed a variation of 2.3% as compared with the standard strain of *A. flavus*. The sequences have been deposited in GenBank and the accession nos. DQ683118, DQ683119, DQ683120, DQ683121, DQ683122, DQ683123, DQ683124 have been assigned for the isolates. The nucleotide polymorphisms existing among *A. flavus* strain in the study was novel and the first to be reported in literature [25].

Application of PCR-Based DNA Sequencing Targeting ITS Region for Genus- and/or Species-Level Identification of Nonsporulating Molds

In the clinical laboratory, isolates that cannot be identified by reproductive structures are described as Mycelia sterilia or nonsporulating molds (NSM), a name indicating a filamentous fungus that displays no distinguishing phenotypes recognized by routine clinical laboratory analysis. Such fungi can be identified using PCR-based DNA sequencing.

Identification of NSM by PCR-based DNA Sequencing

1. DNA extraction from fungal isolates of NSM should be done using Qiagen kit according to the manufacturer's instructions (see Methods).
2. PCR is carried out as per the procedure given in "Semi-Nested PCR Targeting the ITS Region".

Table 27.2 Details of DNA sequencing technique (the four nucleotide bases with the respective acceptor dyes and color emission)

Terminator	Acceptor dye	Color of raw data on ABI PRISM310 electrophoretogram
A	dR6G	Green
C	dROX	Red
G	dR110	Blue
T	dTAMRA	Black

3. The amplified products are subjected to cycle sequencing, purified, and then subjected to DNA sequencing.

Sequencing in *ABI Prism 310/3100 AVANT* Genetic Analyzer

1. The sequence of the PCR amplified DNA is deduced with the help of the ABI Prism.
2. 310/3100 AVANT genetic analyzer that works based on the principle of Sanger dideoxy sequencing.
3. The fluroscent-based detection by automated sequencer adopts the Sangers method and incorporates the fluorescent dyes into DNA extension products using 5′-dye labeled primers or 3′-dye labeled ddNTPs (dye terminators called commercially as RR MIX).
4. Each dye emits light at a different wavelength when excited by an argon ion laser (Table 27.2). All four colors and, therefore, all four bases can be detected and distinguished in a single gel lane or capillary injection.
5. The protocol and thermal profile for cycle sequencing are provided in Tables 27.3 and 27.4, respectively
6. The amplified products with the dye at the terminated 3′end are subjected to capillary electrophoresis by an automated sample injection.
7. The emitted fluorescence from the dye labels on crossing the laser area is collected in the rate of one per second by cooled, charge-coupled device camera at particular wavelength bands (virtual filters) and stored as digital signals on the computer for processing that are analyzed by software called as the

Table 27.3 Protocol for cycle sequencing

Components	Volume (μl) 28S rRNA amplicons	Volume (μl) ITS amplicons
Amplfied products	1.0	1.0
Sequence buffer	3.0	2.5
Primer (2 pmol/μl)	2.0	2.0
RRMIX	1.0	1.5
Water	3.0	3.0

Table 27.4 PCR conditions for cycle sequencing

PCR step	Temperature (°C)	Time
Initial denaturation	96	1 min
Denaturation	96	10 s
Annealing	50	5 s
Extension	60	4 min

The reaction to be carried out for 25 cycles.

sequence analysis softwares (Sequence Navigator in ABI 310 and seqscape manager in ABI 31000 AVANT machine).

8. The ABI AVANT genetic analyzer can be upgraded from 4 capillary to 16 capillary so that facilitates the electrophoresis of 16 samples at a given time.

Purification of Extension Products

The products are purified to remove the unincorporated dye terminators before sujecting the samples to capillary electrophoresis.

Procedure

1. 2 μl of 125 mM EDTA and 2 μl of 3 M sodium acetate (pH 4.8) 10 microlitre of Milli Q water is mixed with 10 microlitre of cycle sequenced prodcuts followed by the addition of 50 μl of absolute ethanol and incubated at room temperture for 15 min follwed by centrifugation at 12,000 rpm for 20 min to precipitate the amplified product and remove the unutilized ddNTPs, primer (short-length molecules), etc.

2. The pellet is washed twice with 70% ethanol at 12,000 rpm for 10 minutes followed by air drying. The purified samples is suspended in formamide and subjected for capillary electro-

phoresis in ABI PRISM 310/3100 genetic analyzer.

3. The sequence is then analyzed in Sequence Navigator software (version 1.0.1; ABI Prism 310) or seqscape manager (version 2.1; ABI Prism 3100 AVANT).

The use of PCR-based DNA sequencing has several advantages over the conventional methods in terms of rapidity, accuracy, and definite identification. Out of the 50 NSM fungal isolates, 27 were found to be emerging pathogens involving 7 genera (*Botryosphaeria* species, *Lasiodiplodia* species, *Thielavia tortuosa*, *Glomerulla singulata*, *Macrophomina phaseolina*, *Rhizoctonia bataticola*, *Podospora* species) and 23 as established pathogens involving 8 genera (*Aspergillus, Fusarium, Bipolaris, Pythium, Cochliobolus, Exserohilum, Pseudallescheria* and *Scedosporium* species) (Table 27.5) [26].

Application of Polymerase Chain Reaction-Based Restriction Fragment Length Polymorphism for Species-Level Identification of Dermatophytic Fungi

Molecular Microbiological Investigations

DNA Extraction from Isolates

DNA extraction from clinical specimens and fungal isolates is done using QIAamp kit as described in "Methods".

PCR Targeting ITS Region

PCR targeting ITS region is carried out on dermatological specimens according to the method provided in "Semi-Nested PCR Targeting the ITS Region".

PCR-Based Restriction Fragment Length Polymorphism

1. PCR-RFLP is carried out on ITS amplicons (first round products) using restriction enzyme *Hae*-III.

Table 27.5 Nonsporulating molds identified by PCR-based DNA sequencing targeting ITS region

Newer emerging pathogens identified by PCR-based DNA sequencing targeting ITS region	Ocular specimens	GenBank accession number
Emerging pathogens 27		
Botryosphaeria species 10		
Botryosphaeria rhodina 8	Corneal scraping 4[a]	
	Conjunctival scraping 1[a]	
	Corneal necrotic tissue 1[a]	
	Corneal button 2[a]	EF446281
Botryosphaeria dothidea 1	Corneal button 1	
Botryosphaeria species 1	Corneal scraping 1	EF446291
Lasiodiplodia theobromae 2	Corneal scraping 1	
	Eviscerated material 1	
Rhizoctonia bataticola 5	Corneal scraping 4	EF446282
	Eviscerated material 1	
Glomerulla singulata 3	Corneal scraping 1	EF446289
	Corneal button 1	
	Eviscerated material 1	
Cochliobolus species 3		
Cochliobolus species 2	Corneal scraping 2	
Cochliobolus heterophrynus 1	Corneal button 1	
Macrophomina phaseolina 2	Corneal scraping 2	EF446288
Podospora species 1	Corneal scraping 1	
Thielavia tortuosa 1	Corneal button 1	EF446287
Established pathogens: 27		
Pythium insidiosum 9	Corneal scraping 6, Corneal button 2, Vitreous aspirate 1	
Fusarium species 7		
Fusarium solani 2	Corneal scraping 2	
F. solani 1, *F. falciforme* 2[b] 3	Corneal scraping 3	
Fusarium proliferatum 2	Corneal scraping 1, Corneal button 1	
Exserohilum species 1	Corneal scraping 1	
Aspergillus terreus 1	Corneal scraping 1	EF446283
Aspergillus fumigatus 1	Corneal button 1	
Scedosporium species 1	Infected suture 1	
Pseudallescheria species 1	Corneal button 1	
Bipolaris species 2	Corneal scraping 1	EF446284
	Donor corneal rim 1	

[a]The five isolates of *Botryosphaeria rhodina* were obtained from the same patient (corneal scraping 2, conjunctival scraping, corneal button, necrotic tissue)
[b]The three *Fusarium* species were identified by species specific PCR as *F. solani* (1) and *F. falciforme* (2) by PCR-based DNA sequencing on 28S rRNA gene

2. For a 25 μl reaction, 2.5 μl of buffer C, 0.5 μl of *Hae*-III enzyme (Bangalore Genei, India), 12 microlitre of milli-Q water and 10 μl of amplified product are added and incubated at 37 °C for 3 h.

3. The digested products are loaded on 4% agarose gel incorporated with ethidium bromide and the electrophoresis is carried out at 100 V for 45 min.

4. The digested pattern (*Microsporum gypseum* 420, 95 bp, *Trichophyton rubrum* (300, 200, 95 bp), *Epidermophyton floccosum* (350, 95 bp) [27] is visualized and documented using gel documentation system (Vilber Lourmat, France).

PCR-based RFLP is a rapid method for detection and identification of dermatophytes (24 h) as compared to conventional culture (which may require up to 21 days for growth) and always may not result in successful isolation.

Notes

1. Depending on the purulence of body fluids and thickness of tissue biopsies incubation can be extended up to 1 h at 56 °C. Before proceeding with the next step, it is absolutely essential that complete digestion of the specimen is attained. When complete digestion takes place, the specimen becomes clear. While extracting DNA from tissue specimens, tissue lysis buffer provided in the kit should be used.

2. Wear laboratory coat, gloves, safety goggles while handling clinical specimens.

3. Extraction of DNA and preparation of PCR premix should be carried out in a Laminar flow hood.

4. While extracting DNA from standard strains it is strictly recommended to handle the fungal strains in a level III biosafety cabinet.

5. For all molecular mycological procedures, autoclaved deionized water should be used or molecular grade water available commercially can be used.

6. New pre-sterilized vials to be used for PCR.

7. The pre-amplification, amplification, and post-amplification areas need to be physically separated and dedicated pipettes should be used for DNA extraction, preparation of PCR cocktail, and post-amplification analysis.

8. Filter-guarded tips should be used to avoid cross contamination.

9. Wear face mask, cap, and gloves while handling ethidium bromide since it is carcinogenic.

10. Gel incorporated with ethidium bromide and pipette tips should be wrapped in a foil, discarded in a labeled container and disposed off by incineration.

11. While transferring the first round products to the second round, always the negative control should be transferred first followed by the specimens and finally the positive control to avoid amplicon carryover.

12. The place of transfer of negative controls and positive controls should be physically separated.

Acknowledgments The financial support in the form of a Research Grant by Department of Science and Technology, (DST) Government of India, and the infrastructure facility provided by Vision Research Foundation, Chennai are gratefully acknowledged.

References

1. White TJ, Bruns T, Lee S (1990) Amplification and direct sequencing of fungal ribosomal RNA genes for phylogenetics. In: Taylor J, Innis MA, Gefland DH, Sninsky JJ, White TJ (eds) PCR protocols: a guide to methods and applications. Academic Press, Inc., New York, pp 315–322

2. Iwen PC, Hinriche H, Rupp ME (2001) Utilization of ITS regions as molecular targets to detect and identify fungal pathogens. Med Mycol 40:87–109

3. Sandhu GS, Kline BC, Stockman L, Roberts GD (1995) Molecular probes for diagnosis of fungal infections. J Clin Microbiol 33:2913–2919

4. Anand AR, Madhavan HN, Sudha NV, Therese KL (2001) PCR in the diagnosis of *Aspergillus endophthalmitis*. Ind J Med Res 114:133–140

5. Jaeger EEM, Carroll NM, Choudhury S, Dunlop AA, Towler HM, Matheson MM et al (2000) Rapid detection and identification of *Candida, Aspergillus,* and *Fusarium* species in ocular samples using nested PCR. J Clin Microbiol 38:2902–2908

6. Hendolin PH, Paulin L, Koukila-Kähkölä P, Anttila V, Malmberg H, Richardson M et al (2000) Panfungal PCR and multiplex liquid hybridization for detection of fungi in tissue specimens. J Clin Microbiol 38:4186–4192

7. Vollmer T, Stormer M, Kleesiek K, Dreier J (2008) Evaluation of novel broad range Real Time PCR assay for rapid detection of human pathogenic fungi in various clinical specimens. J Clin Microbiol 46: 1919–1926

8. Kumar M, Shukla PK (2005) Use of PCR targeting of Internal Transcribed spacer regions and Single-Stranded Conformation Polymorphism analysis of

sequence variation in different regions of rRNA genes in fungi for rapid diagnosis of mycotic keratitis. J Clin Microbiol 43:662–668

9. Sato T, Takayanagi A, Nagao K, Tomatsu N, Fukui N, Kawaguchi M et al (2010) Simple PCR-Based DNA microarray system to identify human pathogenic fungi in skin. J Clin Microbiol 48:2357–2364

10. Nakamura A, Sugimoto Y, Ohishi K, Sugawara Y, Fujieda A, Monma F et al (2010) Diagnostic value of PCR analysis of bacteria and fungi from blood in empiric-therapy-resistant febrile neutropenia. J Clin Microbiol 48:2036

11. Bouchara J, Yi Hsieh H, Croquefer S, Barton R, Marchais V, Pihet M et al (2009) Development of an oligonucleotide array for direct detection of fungi in sputum samples from patients with cystic fibrosis. J Clin Microbiol 47:142–152

12. Soeta N, Terashima M, Gotoh M, Mori S, Kyoko (2009) An improved rapid quantitative detection and identification method for a wide range of fungi. J Med Microbiol 58:1037–1044

13. Martin C, Roberts D, Weide M, Rossau R, Jannes G, Smith T et al (2000) Development of a PCR-based line probe assay for identification of fungal pathogens. J Clin Microbiol 38:3735–3742

14. Narutaki S, Takatori K, Nishimura H, Terashima H, Sasaki T (2002) Identification of fungi based on the nucleotide sequence homology of their internal transcribed spacer 1 (ITS1) region. PDA J Pharm Sci Technol 56:90–98

15. Mancini N, Perotti M, Ossi CM, Cavallero A, Matuska S, Paganoni G et al (2006) Rapid molecular identification of fungal pathogens in corneal samples from suspected keratomycosis cases. J Med Microbiol 55:1505–1509

16. Hsiao CR, Huang L, Bouchara J, Barton R, Li HC, Chang TC (2005) Identification of medically important molds by an oligonucleotide array. J Clin Microbiol 43:3760–3768

17. Carter DA, Burt A, Taylor JW, Koenig JL, White DB (1997) A set of electrophoretic molecular markers for strain typing and population genetic studies of *Histoplasma capsulatum*. Electrophoresis 18: 1047–1053

18. De Baere T, Claeys G, Swinne D, Verschraegen G, Muylaert A, Massonet C et al (2002) Identification of cultured isolates of clinically important yeast species using fluorescent fragment length analysis of the amplified internally transcribed rRNA spacer 2 region (ITS2). BMC Microbiol 2:21

19. Diaz-Guerra TM, Martinez-Suarez JV, Laguna F, Rodriguez-Tudela JL (1997) Comparison of four molecular typing methods for evaluating genetic diversity among *C. albicans* isolates from human immunodeficiency virus-positive patients with oral candidiasis. J Clin Microbiol 35:856–861

20. Chen YC, Eisner JD, Kattar MM, Rassoulian-Barrett SL, LaFe K, Yarfitz SL et al (2000) Identification of medically important yeasts using PCR-based detection of DNA sequence polymorphisms in the internal transcribed spacer 2 region of the rRNA genes. J Clin Microbiol 38:2302–2310

21. Chen YC, Eisner JD, Kattar MM, Rassoulian-Barrett SL, LaFe K, Yarfitz SL et al (2001) Polymorphic internal transcribed spacer region 1 DNA sequences identify medically important yeasts. J Clin Microbiol 39:4042–4051

22. Ferrer C, Colom F, Frases S, Mulet E, Abad JL, Alio JL (2001) Detection and identification of fungal pathogens by PCR and by ITS2 and 5.8S ribosomal DNA typing in ocular infections. J Clin Microbiol 39:2873–2879

23. Bagyalakshmi R, Therese KL, Madhavan HN (2007) Application of a semi nested polymerase chain reaction targeting Internal Transcribed Spacer (ITS) region for detection of panfungal genome in ocular specimens. Ind J Ophthalmol 55:261–265

24. Henry T, Iwen PC, Hinriche H (2000) Identification of *Aspergillus* species using ITS regions 1 and 2. J Clin Microbiol 38:1510–1515

25. Bagyalakshmi R, Therese KL, Madhavan HN (2007) Nucleotide polymorphisms associated with Internal Transcribed Spacer (ITS) regions of ocular isolates of *Aspergillus flavus*. J Microbiol Methods 68:1–10

26. Bagyalakshmi R, Therese KL, Madhavan HN, Prasanna S (2008) Newer emerging pathogens of ocular non-sporulating molds (NSM) identified by polymerase chain reaction (PCR)-based DNA sequencing technique targeting internal transcribed spacer (ITS) region. Curr Eye Res 33:139–147

27. Bagyalakshmi R, Senthilvelan B, Therese KL, Murugusundram S, Madhavan HN (2008) Application of polymerase chain reaction (PCR) and PCR-based restriction fragment length polymorphism for detection and identification of dermatophytes from dermatological specimens. Ind J Dermatol 53:15–20

Naomichi Yamamoto

Abstract

Traditional growth-based methods for characterizing environmental fungi are biased by selection of culture media and incapable of detecting non-cultivable fungi, which still retain allergenicity and/or pathogenicity. Meanwhile, real-time quantitative polymerase chain reaction (qPCR)-based methods have been recently developed and used to characterize fungal concentrations in environmental samples such as air and house dust. As qPCR-based methods are independent of fungal cultivability or viability, they are expected to be toxicologically more relevant than conventional growth-based methods for assessing health effects caused by environmental fungi. This chapter presents protocols for the collection of environmental fungal samples and subsequent qPCR analysis.

Keywords

Fungi • qPCR • Allergens • Spores • DNA • 18S rRNA

Introduction

Growth-based methods have been traditionally and widely used to detect environmental fungi [1, 2]. However, the results of these methods are often biased by selection of culture media [1]. Furthermore, identification of fungal species by growth-based methods is often based on observation of their microscopic and macroscopic morphologies, and accuracy of the identification relies on operator's experience and level of expertise. Most importantly, growth-based methods are incapable of detecting non-cultivable microorganisms [3, 4], which are still potential allergens and pathogens [5]. For instance, research has indicated that growth-based methods underestimated concentrations of fungi in dust by two to three orders of magnitude compared to real-time quantitative polymerase chain reaction (qPCR) assays [3].

In recent years, several studies have utilized qPCR-based methods to characterize fungal concentrations in environmental samples such as air [6–9] and house dust [3, 4, 10–12]. qPCR-based methods are expected to be toxicologically more

N. Yamamoto (✉)
Department of Environmental Health, Graduate School of Public Health, Seoul National University,
1 Gwanak-ro, Gwanak-gu, Seoul 151-742, Korea
e-mail: naomichi.yamamoto@yale.edu

V.K. Gupta et al. (eds.), *Laboratory Protocols in Fungal Biology: Current Methods in Fungal Biology*,
Fungal Biology, DOI 10.1007/978-1-4614-2356-0_28, © Springer Science+Business Media, LLC 2013

relevant than conventional growth-based methods because they are not affected by cultivability or viability of the fungi. Although qPCR-based methods have become common in recent years, care must be taken in their implementation as the inadequate reporting of experimental detail and the use of flawed protocols lead to the technically inappropriate results [13]. Here I introduce a protocol for measurements of fungi in air by qPCR [9, 14]. Sections include creation of a DNA standard, Andersen impactor environmental sampling, personal sampling, DNA extraction, qPCR protocol, avoiding qPCR inhibition, and precision, accuracy, and method detection limit (MDL) of qPCR.

Materials

1. Sterile distilled water.
2. Pipettes: 0.1–10 μL, 10–100 μL, 100–1,000 μL.
3. Disposable filter pipette tips: 0.1–10 μL, 10–100 μL, 100–1,000 μL.
4. Petri dishes: 90 mm diameter.
5. Disposable polypropylene tubes: 15-mL conical.
6. Disposable polypropylene microcentrifuge tubes: 1.5-mL conical.
7. Sterile nylon-flocked swabs.
8. Ethanol (70%).
9. Vortexer.
10. Tween 20 solution (0.1% in deionized water).
11. Crystal violet.
12. Microscopic counting chamber (depth 20 μm).
13. Optical microscope (×400).
14. Tally counter.
15. Microcentrifuge.
16. DNA extraction kit; e.g., PowerSoil DNA Isolation Kit (Mobio Laboratories, Carlsbad, CA, USA).
17. Disposable polypropylene screw-capped microcentrifuge tubes: 2.0-mL conical.
18. Sterile glass beads: 0.1- and 0.5-mm diameters.
19. Bead beater; e.g., Model 607 (Bio Spec Products Inc., Bartlesville, OK, USA).

20. TE buffer (10 mM Tris–HCl, 1 mM EDTA, pH 8.0).
21. Sterile tweezers.
22. For airborne fungal sampling: eight-stage nonviable Andersen samplers or the Personal Environmental Monitor Model 200 (PEM) (New Star Environmental, Inc., Roswell, GA, USA). The Andersen sampler uses quartz glass fiber filters (81-mm diameter; New Star Environmental Inc., Roswell, GA, USA) while the PEM uses polycarbonate track etch (PCTE) membrane filters (37-mm diameter, 0.8-μm pore size; SKC Inc., Eighty Four, PA, USA).
23. Sterile scissors.
24. Molecular biology grade water.
25. TaqMan Universal PCR Master Mix; e.g., Applied Biosystems, Carlsbad, CA, USA.
26. Primers.
27. TaqMan probes; e.g., The BHQplus™ probe with FAM dye (Biosearch Technologies Inc., Novato, CA, USA).
28. Real-time PCR well reaction plate; e.g., Applied Biosystems MicroAmp® Optical 96-Well Reaction Plate.
29. Real-time PCR clear adhesive seal.
30. Centrifuge; e.g., 5804R (Eppendorf, Hamburg, Germany).
31. Real-time PCR system; e.g., Applied Biosystems ABI 7500.

Methods

Preparation of Standard Fungal DNA Samples

This method has been used to prepare standard fungal DNA samples containing known numbers of fungal spores for use in qPCR. Fungal species producing unicellular amerospores such as *Aspergillus fumigatus*, *Cladosporium cladosporioides*, and *Penicillium chrysogenum* are considered here.

1. Swab and re-suspend the spores in 10 mL of ethanol (70%) in a 15-mL tube. Harvest only the spores of the entire surface of fungal colonies grown on a 90-mm petri dish. Multiple dishes may be used if needed.

2. Aliquot the suspension into 1.5-mL microcentrifuge tubes to make several subsamples. Transfer 1-mL suspension into each microcentrifuge tube. Vortex the 15-mL conical tube each time before transfer to prevent gravitational settling of the spores in the suspension.

3. Centrifuge the microcentrifuge tubes at 10,000×g for 3 min to pellet the spores.

4. Remove ethanol supernatant.

5. Select 5 subsamples for cell enumeration by microscopy. The remaining subsamples are used for DNA extraction. The samples used for the DNA extraction are added with the Mobio power lysis solution (750 mL), and the suspensions are transferred to 2-mL screw-capped microcentrifuge tubes. For DNA extraction, follow the steps in "Extraction of Fungal DNA", starting from step 2.

6. For cell enumeration, add 1-mL Tween 20 solution (0.1%) and 10 μL crystal violet to each subsample tube.

7. Pipette 10 μL of the suspension into the microscopic counting chamber.

8. Enumerate the spores on an entire area of the counting chamber (=1 mm^2) using an optical microscope (×400) and a tally counter. Calculate the total number of fungal spores in each subsample. 10^8–10^9 spores per sample are desirable. Dilution of the suspension may be necessary. The optimum counts per 1/25 mm^2 are in the 5–15 cell range.

Collection of Airborne Spores by 8-Stage Nonviable Andersen Sampler

The 8-stage nonviable Andersen sampler allows for collection of size-fractionated airborne fungal cells on glass fiber filters [15]. This sampler is suitable to characterize particle size distributions of airborne fungi.

1. Load glass fiber filters on each stage of the Andersen sampler using sterile tweezers.

2. Operate the sampler with an air flow rate of 28.3 Lmin^{-1}. The sampling time will vary depending on the expected fungal concentrations in the air.

3. After sampling, recover the filters and place them in 90-mm Petri dishes.

4. Cut each filter to one-eighth piece and shred them into smaller pieces using sterile scissors and tweezers.

5. Place all the shredded pieces of the one-eighth filter to 2-mL screw-capped microcentrifuge tube and follow the steps for DNA extraction in "Extraction of Fungal DNA", starting from step 1.

Collection of Airborne Spores by Personal Environmental Monitor

The PEM is used to collect airborne fungi in the PM2.5 or PM10 size ranges on a polycarbonate filter. This sampler is suitable to assess personal exposures to airborne fungi in the PM2.5 or PM10 fractions, which are generally respirable or inhalable fractions, respectively.

1. Load a 37-mm PCTE filter on the PEM.

2. Operate the PEM with a specified air flow rate. The sampling time will vary depending on the expected fungal concentrations in the air.

3. Cut the PCTE filter to one-fourth piece using sterile scissors.

4. Place the cut filter to 2-ml screw-capped microcentrifuge tube, and follow the steps for DNA extraction in "Extraction of Fungal DNA", starting from step 1.

Extraction of Fungal DNA

This method has been used to extract DNA from fungal cells in environmental samples such as those collected on air filters from the above or other methods. A protocol by the PowerSoil DNA Isolation Kit is modified to use for the isolation of fungal DNA from air samples. This kit is suitable to remove PCR inhibitors from environmental samples.

1. Add 750 mL of Mobio lysis solution.

2. Add the Mobio power beads (1.0 g) supplemented with 0.1-mm diameter glass beads (300 mg) and 0.5-mm diameter glass beads (100 mg).

3. Disrupt fungal cells for 5 min at 3,450 rpm by a bead beater.

4. Follow the Mobio's protocol (see Note 1) and elute DNA into 50 μL of TE buffer.

5. For the standard DNA extracted from the samples containing known numbers of fungal spores prepared according to the steps in Preparation of Standard Fungal DNA Samples, make tenfold dilution series from 1/10 to $1/10^7$. Calculate the numbers of fungal spores in 1 μL of each diluted standard (see Note 2).

Quantitation of Fungal Spores Using Real-Time qPCR

The ribosomal DNA (rDNA) is the most commonly targeted region for fungal PCR [16]. Sensitive fungal detection is possible by targeting rDNA since there are large numbers of gene copies present in each fungal genome. For instance, *A. fumigatus* has an average of 54 copies (range from 38 to 91 depending on the strain) of rDNA per genome [17] (see Note 3). Selecting species-specific rDNA regions for PCR primer sites allows for species-specific fungal detection. For instance, the U.S. Environmental Protection Agency (EPA) has provided the DNA sequences that can be used for qPCR primers and probes to characterize more than 130 of the major indoor air fungal species.[1] [18] The following qPCR protocol can be used with primers and probes reported by the U.S. EPA for species-specific fungal quantification. This method uses TaqMan chemistry. Here the ABI 7500 is used as a real-time PCR system, but other instruments with equivalent functions may also be used.

1. Prepare a sufficient amount of TaqMan reaction mixture for the number of samples to be analyzed. Each 24 μL of the master mix contains 1× TaqMan Universal PCR Master Mix, 1 μM of each primer, and 0.08 μM of the probe. Vortex and centrifuge the mixture, and distribute 24 μL aliquots into each well of the reaction plate.

2. Add 1 μL of the DNA samples and standard DNA to the appropriate well of the reaction plate. The samples are in triplicate. The tenfold dilution series (from 1/10 to $1/10^7$) of

the standard DNA prepared according to the steps in Extraction of Fungal DNA and the no-template control; i.e., molecular biology grade water, are used for qPCR calibration. Each standard is in duplicate.

3. Use a clear adhesive seal to seal the PCR well reaction plate.

4. Centrifuge the plate briefly.

5. Turn on the real-time PCR machine.

6. Load the reaction plate in the real-time PCR machine.

7. Follow the procedure specified by the manufacturer of the real-time PCR system (e.g., ABI 7500 system).

8. Enter the information of each sample and standard including concentrations of each standard and the names of each sample using the ABI 7500 software. For the standard, enter the number of fungal spores in 1 μL calculated in "Extraction of Fungal DNA", starting from step 5.

9. Run the appropriate thermal condition: 50 °C for 2 min, 95 °C for 15 min of initial denaturation and 45 cycles of 95 °C for 15 s of dissociation, and 60 °C for 1 min of annealing and extension. For the ABI 7500 system, use the standard mode. The fast mode is not recommended.

10. When cycling is complete, create a calibration plot and calculate the concentrations of each sample. For the ABI 7500 system, this can be done by selecting "Auto Ct" as an analysis setting and clicking the "Analyze" button on the sheet of "Amplification Plot." Use the auto function to calculate cycle threshold (Ct) values.

11. Export the file containing the concentration results (e.g., csv format).

12. Process the data using spreadsheet software (e.g., MS-Excel) (see Note 4).

PCR Inhibition Assay

PCR can be inhibited by a wide range of materials including those originating from environmental compounds (e.g., phenolic compounds, humic acids, and heavy metals), constituents of microbial cells, non-target DNA, etc. [19]. PCR

---

[1] http://www.epa.gov/nerlcwww/moldtech.htm.

inhibitors affect amplification efficiencies and lead to inaccurate qPCR measurements [20]. To test for PCR inhibition, sample extracts are spiked with subsets of the standard fungal DNA, and standard curves are produced according to the steps described in PCR Inhibition Assay. The standard DNA used to spike the samples should have a significantly higher concentration than any fungal DNA present in the sample itself. The amplification efficiencies with and without spiking the sample extracts are compared to assess the presence of PCR inhibitors. PCR inhibition should be mitigated if a significant reduction in the amplification efficiencies is observed. Dilution of the sample extracts is a simple method that can increase amplification efficiencies as it dilutes inhibiting materials, although it also dilutes target DNA [6]. Alternatively, the samples may be re-analyzed using the remaining pieces of air filters with special care taken during the washing step of the DNA extraction (see Note 1).

Accuracy, Precision, and Method Detection Limits of Fungal Measurement by qPCR

Accuracy of fungal measurement by qPCR is limited by the efficiency of DNA extraction from cells and whole-cell recovery from sampling air filters while precision is separated into the total precision (reproducibility) and precision associated with the analytical instrument (instrument repeatability). For the MDL, a probability distribution is used to estimate the qPCR MDL as PCR detection is based on a logarithmic signal amplification and is binary, i.e., either positive or negative. Further details about methods to determine the accuracy, precision, and MDLs of fungal measurement by qPCR were reported in our previous work [12, 14].

ethanol-based washing buffer, is critical to eliminate all traces of residual ethanol.

2. For example, if there are 10^8 spores in the standard sample prepared according to the steps in Preparation of Standard Fungal DNA Samples, you can assume 2×10^6 spores in 1 μL of the extract as a final elution volume of the extract is 50 μL. The tenfold dilution series (from 1/10 to $1/10^7$) of this extract will contain 2×10^5 to 2×10^{-1} spores in 1 μL of each dilute, respectively.

3. Variations in the numbers of rDNA gene copies as well as those in the numbers of nuclei in a fungal cell [21, 22] can lead to the biases in the qPCR measurements. Regardless of these uncertainties, we use a cell count as a basis of qPCR calibration since it allows for more intuitive data presentation. Alternatively, you may calibrate qPCR based on the number of gene copies. In this case, synthesized DNA oligomers with known sequences containing the region to be amplified, including the sites for forward and reverse primers, can be used as standard DNA. The number of gene copies in the synthesized DNA oligomers can be estimated by the following [23]: gene copy number = DNA content (pg) × 0.978 × 10^9 (bp/pg)/ genome size (bp). The mass of synthesized DNA oligomers can be obtained by spectrophotometer or fluorescent methods.

4. The DNA quantities, as a basis of the number of fungal cells, obtained here are those in 1 μL of the extract. As fungal DNA is eluted in 50 μL of TE buffer in the extraction process, these values must be multiplied by a factor of 50 to obtain fungal quantities on the analyzed filters (i.e., one-eighth of glass fiber filter for the Andersen sampler and one-fourth of polycarbonate filter for the PEM).

Acknowledgment I thank Karen Dannemiller at Yale University for invaluable comments on the manuscript.

Notes

1. Care must be taken in the washing step of the DNA extraction as residual ethanol in the washing buffer can inhibit PCR. The second spin of the empty columns, after removing the

References

1. Takahashi T (1997) Airborne fungal colony-forming units in outdoor and indoor environments in Yokohama, Japan. Mycopathologia 139:23–33

2. Shelton BG, Kirkland KH, Flanders WD, Morris GK (2002) Profiles of airborne fungi in buildings and outdoor environments in the United States. Appl Environ Microbiol 68:1743–1753

3. Meklin T, Haugland RA, Reponen T, Varma M, Lummus Z, Bernstein D et al (2004) Quantitative PCR analysis of house dust can reveal abnormal mold conditions. J Environ Monitor 6:615–620

4. Lignell U, Meklin T, Rintala H, Hyvarinen A, Vepsalainen A, Pekkanen J et al (2008) Evaluation of quantitative PCR and culture methods for detection of house dust fungi and streptomycetes in relation to moisture damage of the house. Lett Appl Microbiol 47:303–308

5. Hirvonen MR, Ruotsalainen M, Savolainen K, Nevalainen A (1997) Effect of viability of actinomycete spores on their ability to stimulate production of nitric oxide and reactive oxygen species in RAW264.7 macrophages. Toxicology 124:105–114

6. Yamamoto N, Kimura M, Matsuki H, Yanagisawa Y (2010) Optimization of a real-time PCR assay to quantitate airborne fungi collected on a gelatin filter. J Biosci Bioeng 109:83–88

7. Goebes MD, Boehm AB, Hildemann LM (2011) Contributions of foot traffic and outdoor concentrations to indoor airborne *Aspergillus*. Aerosol Sci Technol 45:352–363

8. Kaarakainen P, Meklin T, Rintala H, Hyvaerinen A, Karkkainen P, Vepsalainen A et al (2008) Seasonal variation in airborne microbial concentrations and diversity at landfill, urban and rural sites. Clean Soil Air Water 36:556–563

9. Yamamoto N, Schmechel D, Chen BT, Lindsley WG, Peccia J (2011) Comparison of quantitative airborne fungi measurements by active and passive sampling methods. J Aerosol Sci 42:499–507

10. Pitkaranta M, Meklin T, Hyvarinen A, Paulin L, Auvinen P, Nevalainen A et al (2008) Analysis of fungal flora in indoor dust by ribosomal DNA sequence analysis, quantitative PCR, and culture. Appl Environ Microbiol 74:233–244

11. Cai GH, Broms K, Malarstig B, Zhao ZH, Kim JL, Svardsudd K et al (2009) Quantitative PCR analysis of fungal DNA in Swedish day care centers and comparison with building characteristics and allergen levels. Indoor Air 19:392–400

12. Yamamoto N, Shendell DG, Peccia J (2011) Assessing allergenic fungi in house dust by floor wipe sampling and quantitative PCR. Indoor Air 21:521–530

13. Bustin SA, Beaulieu JF, Huggett J, Jaggi R, Kibenge FSB, Olsvik PA et al (2010) MIQE precis: Practical implementation of minimum standard guidelines for fluorescence-based quantitative real-time PCR experiments. BMC Mol Biol 11:74

14. Hospodsky D, Yamamoto N, Peccia J (2010) Accuracy, precision, and detection limits of quantitative PCR for airborne bacteria and fungi measurement. Appl Environ Microbiol 76:7004–7012

15. Yamamoto N, Qian J, Hospodsky D, Peccia J (2010) Particle size distribution and seasonal concentrations of selected airborne fungi in the northeastern United States. In: The American Association for aerosol research. 29th Annual conference, Portland, Oregon

16. Borneman J, Hartin RJ (2000) PCR primers that amplify fungal rRNA genes from environmental samples. Appl Environ Microbiol 66:4356–4360

17. Herrera ML, Vallor AC, Gelfond JA, Patterson TF, Wickes BL (2009) Strain-dependent variation in 18S ribosomal DNA copy numbers in aspergillus fumigatus. J Clin Microbiol 47:1325–1332

18. Haugland R, Vesper S (2002) Method of identifying and quantifying specific fungi and bacteria. US Patent 6,387,652

19. Wilson IG (1997) Inhibition and facilitation of nucleic acid amplification. Appl Environ Microbiol 63:3741–3751

20. Kontanis EJ, Reed FA (2006) Evaluation of real-time PCR amplification efficiencies to detect PCR inhibitors. J Forensic Sci 51:795–804

21. Yuill E (1950) The numbers of nuclei in conidia of Aspergilli. Trans Br Mycol Soc 33:324–331

22. Campbell TH, Backus MP, Stauffer JF (1956) Cytological studies on Penicillium chrysogenum Thom. Bull Torrey Bot Club 83:93–106

23. Dolezel J, Bartos J, Voglmayr H, Greilhuber J (2003) Nuclear DNA content and genome size of trout and human. Cytometry 51A:127–128

Quantitative Sampling Methods for the Analysis of Fungi: Air Sampling

29

Mary C. O'Loughlin, Katherine D. Turner, and Kevin M. Turner

Abstract

Quantitative sampling of fungi can be carried out in a variety of ways for a large array of purposes. These include approaches such as tape sampling, settled dust sampling, bulk material sampling, and air sampling followed by macroscopic analysis, polymerase chain reaction (PCR), or immuno-chemical methods for quantitation. Air sampling is widely used in a variety of industries and settings as a means of isolating material for the identification and potential enumeration of fungal strains thus we have chosen to discuss this approach to sampling in detail in this chapter. PCR overcomes the limitations of traditional culturing and macroscopic methods as it is not dependant on culturability or viability of the microorganisms. This chapter discusses a general approach to carrying out such air sampling and PCR-based quantitation.

Keywords

Quantitative sampling • Air sampling • Polymerase chain reaction

M.C. O'Loughlin (✉)
Department of Life Sciences, University of Limerick, Castletroy, Limerick, Ireland
e-mail: maryc.oloughlin@gmail.com

K.D. Turner
Centre for Chromosome Biology, School of Natural Sciences, National University of Ireland Galway, Galway, Ireland

K.M. Turner
Manufacturing Sciences and Technology, Pfizer Ireland Pharmaceuticals, The Pfizer Biotech Campus at Grange Castle, Clondalkin, Dublin 22, Ireland

Introduction

In the nineteenth century Louis Pasteur disposed of the supposition of spontaneous generation by showing that airborne microscopic organisms account for biologic growth on previously sterile media. It has been estimated in recent times that up to 40% of homes in Northern Europe and Canada have mould contamination [1]. Various health effects, such as respiratory symptoms, allergic rhinitis, asthma, and hypersensitivity pneumonitis, are associated with mould exposure. Traditional methods for the isolation and identification of fungal spores can be time-consuming and laborious.

V.K. Gupta et al. (eds.), *Laboratory Protocols in Fungal Biology: Current Methods in Fungal Biology*,
Fungal Biology, DOI 10.1007/978-1-4614-2356-0_29, © Springer Science+Business Media, LLC 2013

Sampling should be performed by using validated methods and must be planned so that the smallest amount of sampling and interpretation is done to meet the information requirements. Air sampling is widely used in the detection of such moulds for eventual identification and if possible enumeration [2–5].

These air techniques are generally categorized as both passive (gravitational) and active (volumetric). Traditionally, passive air sampling (Settle plate) has been used and is still commonly in use to determine the types of microorganisms by exposing a Petri dish containing nutrient rich agar medium to the air [6, 7]. The method is somewhat criticized for being considered semiquantitative with a potential for bias towards microorganisms of larger spore size. This is primarily due to its reliance on gravity. However, it does have the potential to mimic the natural deposition of airborne spores on the surface of food products and some regard it as a dependable method to assess airborne microbial food contamination. The technique is straightforward to perform and does not involve additional investment on specialized equipment.

Active air sampling uses devices that draw a predetermined volume of air at a particular speed over a definite period of time for the assessment of viable airborne microorganisms. Though both are widely used, there has been criticism that the methodologies for sampling and analysis are neither consistent nor definitive [8]. Quantitation of sampled specimens is by and large carried out by direct microscopy, culture, or biochemical analysis but new DNA-based methods for fungal detection can now be used to enumerate the spores of fungi. Airborne spores can be collected and identified by PCR allowing identification of the species [9, 10]. However, the sample volume, short collection period, and artificial air disturbance produced by the devices may in fact affect the types and quantities of fungi captured [6]. This chapter outlines protocols for both passive and active air sampling in the context of quantitation of fungal strains from indoor areas.

Materials

All chemicals sourced from Sigma-Aldrich, St. Louis, MO, USA, unless specified.

1. SAS-super-180 sampler (PBI International, Milan, Italy).
2. Reuter Centrifugal Air Sampler (Biotest, Frankfurt, Germany) with high volume pump (Gast Inc., Benton Harbor, MI, USA).
3. Rotameter (Zefon International, Ocala, FL, USA).
4. DG18 media composed of glucose, 10 g/L; peptone, 5 g/L; NaH_2PO_4, 1 g/L; Mg_2SO_4, 0.5 g/L; dichloran, 0.002 g/L; agar, 15 g/L; glycerol, 220 g/L; and chloramphenicol, 0.5 g/L.
5. Rose Bengal agar medium, MEA (Malt extract agar), CYA (Czapaek yeast extract agar), YES (Yeast extract sucrose agar), CREA (Creatine sucrose agar), and NO^2 (Nitrite sucrose agar) (Difco; Becton Dickinson, Sparks, MD, USA).
6. 9 cm Petri dish (Fisher Scientific Company, Pittsburgh, PA).
7. Parafilm-M™ (Fisher Scientific Company, Pittsburgh, PA).
8. 70% Ethanol.
9. 0.05% Tween 80.
10. 13-mm mixed cellulose ester filter 0.8 μm pore size (Fisher Scientific Company, Pittsburgh, PA).
11. Acetone-vaporizing unit (Quickfix, Environmental Monitoring Systems, Charleston, SC).
12. Glycerin jelly (20 g gelatin, 2.4 g phenol crystals, 60 mL glycerol, and 70 mL water).
13. 0.1% IGEPAL CA-630®.
14. Acid-washed Ballotini beads (8.5 grade, 400–455 mm in diameter) and ball mill (Glen Creston, Stanmore, UK).
15. 2× Lee and Taylor lysis buffer (100 mM Tris–HCl pH 7.4, 100 mM EDTA, 6% SDS, 2% β-mercaptoethanol).
16. Phenol:chloroform (1:1).
17. 20 μg/μL Glycogen (Roche diagnostics Ltd., Lewes, UK).

18. 6 M ammonium acetate.
19. Isopropanol.
20. TE buffer: 10 mM Tris–HCl pH 7.5 (25°C), 0.1 mM EDTA.
21. RNaseA: 10 mg/mL in TE.
22. Phenol.
23. ABI 7000 Fast Real-time PCR System (Applied Biosystems, CA, USA).
24. Universal fungal primers NS5 (5′-AA CTTAAAGGAATTGACGGAAG-3′) and NS6(5′-GCATCACAGACCTGTTATTG CCTC-3′).
25. SYBR® Premix Ex Taq™ II (×2) and ROX Reference Dye (×50) (Takara Bio., Shiga, Japan).

Methods

Active Air Sampling Using a Portable SAS-Super-180 Sampler

1. Operate the SAS-super-180 sampler at a sampling rate of 180 L air/min.
2. Follow the instruction manual provided by the producer.
3. Charge the portable battery fully prior to use.
4. Sample 500 L of air from the center of the room at a height of 1 m above the floor.
5. Use a 9-cm Petri dish containing Dichloran 18% glycerol agar medium (DG 18) [11] to trap viable fungal particles.
6. Disinfect the sampler with 70% ethanol before each use.
7. Cover the Petri dish with the lid immediately following sampling and seal with Parafilm-M™.
8. Incubate the Petri dishes containing DG18 at 25°C for 1 week in the dark and examine every 24 h.

Active Air Sampling Using a Reuter Centrifugal Air Sampler

1. Operate the Reuter centrifugal air sampler (RCS) at a flow rate of 40 L air/min [8, 12–14].

2. Follow the instruction manual provided by the producer.
3. Calibrate the high volume pump to 28.3 L/min using a rotameter.
4. Collect 15 samples of 1–15 min duration in random order at a height of 1 m above the floor.
5. Use a 9-cm Petri dish containing Rose Bengal agar medium supplemented with 100 mg/L chloramphenicol to trap viable fungal particles.
6. Disinfect the sampler with 70% ethanol before each use.
7. Cover the Petri dish with the lid immediately following sampling and seal with Parafilm-M™.
8. Incubate the Petri dishes containing Rose Bengal agar at 25°C for 1 week and examine every 24 h.

Isolation and Enumeration of Mycological Samples

1. Count the numbers of fungal colonies on each Petri dish and then subculture on Petri dishes with suitable agar media for species identification.
2. Prepare all media as per manufacturers' instructions.
3. Plate moulds belonging to *Penicillium* on the following media; MEA, CYA, YES, CREA, and NO^2.
4. Plate other moulds and yeasts on MEA and PDA (Potato dextrose agar) [11].
5. Incubate MEA, CYA, YES, and PDA in the dark at 25°C and CREA and NO2 at 20°C for 7 days [15].
6. Convert the number of CFU per plate to the number of CFU/L of air and analyze data using an appropriate statistical package.

Identification of Mycological Samples

1. Suspend individual colonies in 20 mL 0.05% Tween® 80 prepared with sterile deionized water in a test tube [5].

2. Vortex for 10 s at 20,000×*g*.
3. Filter each sample through a 0.8 μm mixed cellulose ester filter and then place on a glass slide.
4. Allow slides to dry overnight.
5. Clear the slides using a modified instant acetone-vaporizing unit
6. Mount a 25×25-mm cover glass on the slide using glycerin jelly.
7. Observe the slides and identify the collected fungal spores at ×400 magnification using a light microscope.

Extraction of DNA from Mycological Samples

1. Suspend individual colonies in 0.5 mL 0.1% IGEPAL CA-630® prepared with sterile deionized water in a tube.
2. Adjust the spore suspensions were adjusted to 2–3×10⁴ spores/mL.
3. Transfer 0.4 mL of the spore suspension to a fresh tube and add 0.4 g of acid-washed Ballotini beads (8.5 grade, 400–455 mm in diameter).
4. Shake the mixture for 8 min in a ball mill.
5. Add 0.4 mL 2× Lee and Taylor lysis buffer [16].
6. Vortex the sample and incubate at 65°C for 1 h.
7. Add 0.8 mL phenol:chloroform (1:1), vortex briefly, and centrifuge at 20,000×*g* for 15 min
8. Transfer the top aqueous layer to a clean tube.
9. Add 1 μL glycogen (20 μg/μL), 40 μL 6 M ammonium acetate, and 600 μL isopropanol
10. Invert tube gently to mix and incubate at −20°C for 10 min.
11. Centrifuge at 20,000×*g* for 2 min and remove the supernatant.
12. Resuspend the pellet in 50 μL TE buffer containing RNase A (10 mg/mL) to the pellet and incubate at 37°C for 15 min.
13. Add 150 μm TE and then add 200 μL phenol
14. Vortex briefly and centrifuge at 20,000×*g* for 6 min.

15. Transfer the top aqueous layer to a clean tube.
16. Add 10 mL 6 M ammonium acetate and 600 mL isopropanol.
17. Invert tube gently to mix and incubate at −20°C for 10 min.
18. Centrifuge at 20,000×g for 10 min and remove the supernatant.
19. Add 800 mL ice cold 70% ethanol, centrifuge at 20,000×g for 2 min, and remove the supernatant.
20. Centrifuge at 20,000×g for 10 s and remove the remaining liquid.
21. Dry pellet for 20 min in a fume hood and resuspend in 50 mL TE buffer.

Quantitation of Mycological Samples Using RT-PCR

1. Use the universal fungal primer pair NS5 (5′-AACTTAAAGGAATTGACGGAAG-3′) and NS6 (5′-GCATCACAGACCTGTTATT GCCTC-3′) to amplify the 310 bp of 18S rDNA region [17].
2. Thaw the reagents and DNA preparations and keep on ice until required.
3. Dilute the extracted fungal DNA in sterile PCR-grade water to several ratios; i.e., 0.5/20, 1/20, 2/20, 4/20, and 8/20.
4. Prepare a 50-μL reaction mixture consisting of 25 μL of SYBR® Premix Ex Taq™ II (×2), 1 μL of ROX Reference Dye (×50), 2 μL of each primer (10 μM), and 20 μL of the diluted DNA extracts.
5. Program the ABI 7000 fast real-time PCR System, or equivalent system, with the following cycling conditions 95°C for 10 s, 4 cycles of 95°C for 30 s, 60°C for 31 s, and then 40 cycles of 95°C for 4 s and 60°C for 31 s.
6. Set a threshold level of 0.2 and use the auto-baseline function on the ABI 7000 software.
7. Carry out quantitative analysis of the sample using the ABI 7000 software.

References

1. Schlatter J (2004) Toxicity data relevant for hazard characterization. Toxicol Lett 153:83–89
2. Al Maghlouth A, Al Yousef Y, Al Bagieh N (2004) Qualitative and quantitative analysis of bacterial aerosols. J Contemp Dent Pract 5:91–100
3. Engelhart S, Glasmacher A, Simon A, Exner M (2007) Air sampling of Aspergillus fumigatus and other thermotolerant fungi: comparative performance of the Sartorius MD8 airport and the Merck MAS-100 portable bioaerosol sampler. Int J Hyg Environ Health 210:733–739
4. Grisoli P, Rodolfi M, Villani S, Grignani E, Cottica D, Berri A et al (2009) Assessment of airborne microorganism contamination in an industrial area characterized by an open composting facility and a wastewater treatment plant. Environ Res 109:135–142
5. Sivasubramani SK, Niemeier RT, Reponen T, Grinshpun SA (2004) Assessment of the aerosolization potential for fungal spores in moldy homes. Indoor Air 14:405–412
6. Asefa DT, Langsrud S, Gjerde RO, Kure CF, Sidhu MS, Nesbakken T et al (2009) The performance of SAS-super-180 air sampler and settle plates for assessing viable fungal particles in the air of dry-cured meat production facility. Food Control 20:997–1001
7. Scherwing C, Golin F, Guenec O, Pflanz K, Dalmaso G, Bini M et al (2007) Continuous microbiological air monitoring for aseptic filling lines. PDA J Pharm Sci Technol 61:102–109
8. Saldanha R, Manno M, Saleh M, Ewaze JO, Scott JA (2008) The influence of sampling duration on recovery of culturable fungi using the Andersen N6 and RCS bioaerosol samplers. Indoor Air 18:464–472
9. Bellanger AP, Reboux G, Murat JB, Bex V, Millon L (2009) Detection of Aspergillus fumigatus by quantitative polymerase chain reaction in air samples impacted on low-melt agar. Am J Infect Control 38:195–198
10. Lee SH, Lee HJ, Kim SJ, Lee HM, Kang H, Kim YP (2009) Identification of airborne bacterial and fungal community structures in an urban area by T-RFLP analysis and quantitative real-time PCR. Sci Total Environ 408:1349–1357
11. Pitt JI, Hocking AD (1999) Fungi and food spoilage. Aspen Publishers, Gaithersburg, MD
12. An HR, Mainelis G, Yao M (2004) Evaluation of a high-volume portable bioaerosol sampler in laboratory and field environments. Indoor Air 14:385–393
13. Yao M, Mainelis G (2007) Analysis of portable impactor performance for enumeration of viable bioaerosols. J Occup Environ Hyg 4:514–524
14. Yao M, Mainelis G (2007) Use of portable microbial samplers for estimating inhalation exposure to viable biological agents. J Expo Sci Environ Epidemiol 17:31–38
15. Amend AS, Seifert KA, Samson R, Bruns TD (2010) Indoor fungal composition is geographically patterned and more diverse in temperate zones than in the tropics. Proc Natl Acad Sci U S A 107:13748–13753
16. Lee SB, Taylor JW (1990) Isolation of DNA from fungal mycelia and single spores. In: Innis MA, Gelfand DH, Sninsky JJ, White TJ (eds) PCR protocols: a guide to methods and applications. Academic, San Diego, CA, pp 282–287
17. Wu Z, Wang X-R, Blomquist G (2002) Evaluation of PCR primers and PCR conditions for specific detection of common airborne fungi. J Environ Monit 4:377–382

Transformation of Filamentous Fungi in Microtiter Plate

30

Bianca Gielesen and Marco van den Berg

Abstract

The polyethylene glycol (PEG) protoplast method for transformation of filamentous fungi is developed for use in microtiter plates, enabling one to perform a multitude of transformations in parallel.

Keywords

Polyethylene glycol • PEG protoplast • Genetic transformation • Filamentous fungi • Microtiter plates

Introduction

Genetic transformation is one of the most commonly used tools in molecular biology. For filamentous fungi, various methods have been described, including *Agrobacterium*-mediated transformation [1], particle bombardment [2], and electroporation [3]. Still, the method mostly used is the classical polyethylene glycol (PEG)-mediated protoplast transformation [4]. Although all methods do work, yielding from a few dozen to well over 1,000 transformants per microgram DNA, none of them has been described so far as a high-throughput method. With the availability of many fungal genome sequences, such a method is needed to embark on genome-wide functional studies.

In this chapter we describe a protocol for protoplast transformation and subsequent regeneration of filamentous fungi in microtiter plate (MTP). The examples given are obtained with the ß-lactam antibiotic producing *Penicillium chrysogenum*, but similar results were obtained for *Aspergillus niger*; therefore, the protocol can be used as a basis for other fungal species as well.

Materials

1. Penicillium YGG medium: per liter: 8.0 g KC, 16.0 g glucose, 6.7 g Difco yeast nitrogen base, 1.5 g citric acid, 6.0 g K_2HPO_4, 2.0 g yeast extract, pH 6.2, supplemented with penicillin and streptomycin (Gibco).
2. KC buffer: per liter: 60 g KCl, 2 g citric acid, pH 6.2.

B. Gielesen • M. van den Berg (✉)
DSM Biotechnology Center, Alexander Fleminglaan 1, Delft, Zuid, Holland 2613 AX, The Netherlands
e-mail: marco.berg-van-den@dsm.com

3. Glucanex (GLX) stock solution. Dissolve 5 g GLX in 50 mL KC, cool overnight at 4 °C, filter sterilize, and store in small portions at −20 °C.

4. STC buffer: per liter: 145.6 g sorbitol, 2.8 g $CaCl_2$, 3.0 g Tris, pH 8.0; sterilize in autoclave.

5. Aurintricarboxylic acid (ATA) stock solution. Dissolve 53 mg ATA in 50 mL and store in small portions at −20 °C.

6. 20 % PEG solution. Dissolve 200 g PEG4000 in 1 L water; sterilize in autoclave.

7. 30 % PEG solution. Dissolve 300 g PEG4000 in 1 L water; sterilize in autoclave.

8. 10 % Acetamide stock solution in water; filter sterilize.

9. 1.5 M CsCl stock solution in water; filter sterilize.

10. Trace element solution: per liter (in g): citric acid (150), $FeSO_4 \cdot 7H_2O$ (15), $MgSO_4 \cdot 7H_2O$ (150), $H_3BO_3 \cdot (0.0075)$, $CuSO_4 \cdot 5H_2O$ (0.24), $CoSO_4 \cdot 7H_2O$ (0.375), $ZnSO_4 \cdot 7H_2O$ (5), $MnSO_4 \cdot H_2O$ (2.28), $CaCl_2 \cdot 2H_2O$ (0.99); filter sterilize.

11. Acetamide regeneration medium with sucrose: (1) 15.0 g agar, 342.0 g sucrose, 5.0 g glucose, 35.0 g lactose, 2.9 g Na_2SO_4, 5.2 g KH_2PO_4, 4.8 g K_2HPO_4 in 900 mL; (2) set pH 6.5 at before sterilization; (3) after sterilization in autoclave and cooling to ~50 °C add 10 mL acetamide stock, 10 mL CsCl stock and 10 mL of a trace element solution, and fill up to 1 L with sterilized water.

12. Acetamide regeneration medium with KCl: (1) 15.0 g agar, 22.4 g KCl, 5.0 g glucose, 35.0 g lactose, 2.9 g Na_2SO_4, 5.2 g KH_2PO_4, 4.8 g K_2HPO_4 in 900 mL; (2) set pH 6.5 at before sterilization; (3) after sterilization in autoclave and cooling to ~50 °C add 10 mL acetamide stock, 10 mL CsCl stock and 10 mL of a trace element solution, and fill up to 1 L with sterilized water.

13. 96-well MTP, standard 200 μL shallow-well, 200 μL wide-well, and 1 mL deep-well.

14. 24-well deep-well MTP.

15. Filters 14 × 14 mm; sterilize in autoclave.

Methods

Protoplastation

1. Inoculate 25 mL of YGG with 10^4–10^5 spores/mL of *P. chrysogenum*.

2. Incubate o/n at 25° C and 250 rpm.

3. Dilute the overnight culture 1:30 into 100 mL fresh YGG and incubate o/n at 25 °C and 250 rpm.

4. Spin down the cells for 5 min at 5,000 rpm.

5. Wash the cell pellet thoroughly with 10 volumes of KC.

6. Resuspend the cells in 15 mL of KC and add 4 mL of GLX.

7. Incubate at 25 °C and 100 rpm.

8. Monitor the protoplast formation under a light microscope every 30 min and cool on ice when >90 % mycelium is converted into protoplasts.

9. Add cold KC up to 50 mL and spin down protoplasts for 5 min at 1500 rpm.

10. Discard the supernatant and resuspend the protoplasts in cold KC up to half the volume of the tube and add another half of cold STC. Spin down for 5 min at 1,500 rpm.

11. Discard the supernatant. Wash the protoplasts in cold STC. Spin down for 5 min at 1,500 rpm.

12. Discard the supernatant and resuspend the protoplasts in cold STC.

13. Count the protoplasts in the supernatant and adjust towards 10^8 protoplasts/mL. The protoplasts are now ready for direct use or can be frozen at −80 °C for later use.

Transformation

In order to study the robustness towards varying types of MTP, transformations were performed in standard (i.e., shallow-well) 96-well MTP, wide-well 96-well MTP, and deep-well 96-well MTP. Transformants were obtained with all MTP types. For our experiments, we typically use either the standard (i.e., shallow-well) 96-well MTP or wide-well 96-well MTP.

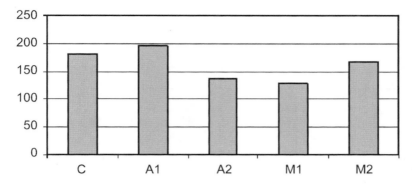

Fig. 30.1 Comparison of manual versus automated handling of *Penicillium chrysogenum* MTP transformation. *P. chrysogenum* strain DS17690 was transformed with plasmid pHELY-A1 according to the classical manual method [6] and the MTP method as described in this chapter. The transformation mixtures were plated out on acetamide-selective agar in petri dishes. *y*-axis, number of transformants per 50 μL of protoplasts (or 2 μg of DNA) as obtained on standard agar plates; *x*-axis, different handling protocols. *C* control, i.e., the classical glass tube protocol with manual pipetting and swirling; *A1* MTP method with automated pipetting and swirling; *A2* MTP method with automated pipetting and mixing; *M1* MTP method with manual pipetting and swirling; *M2* MTP method with manual pipetting and mixing

1. Pipette in each well of a standard 96-well MTP 4 μL of DNA (typically, 0.5–2 μg) or H$_2$O as control. In our experiments we used as DNA 0.25 μg/μL of the plasmid pHELY-A1 harboring an acetamidase expression cassette [5]. Upon uptake of the DNA, this expression cassette enables transformants to grow on media with acetamide as the sole nitrogen source.

2. Mix together the protoplasts, the nuclease inhibitor (ATA) and 20 % PEG in the ratio of 10:1:5.

3. Add 40 μL of the mixture from step 2 to the DNA-containing wells. This can be done either manually (pipetting using a repetitive or a multichannel pipette) or automated (e.g., using a Multidrop from Thermo Electron Corporation). Large differences between the number of transformants obtained with either the manual or the automated method (Fig. 30.1) were not detected.

4. If step 3 is done manually, the mixture is carefully mixed, either via swirling, or via pipetting up-and-down.

5. Subsequently, the MTP is incubated on ice for 30 min.

6. Add to each well 190 μL 30 % PEG. As in step 4, this can be done manually or automated. No additional mixing is required at this step.

7. Incubate for 15 min at 25 °C.

8. Add to each well 700 μL STC. No additional mixing is required at this step.

9. Spin down the protoplasts via centrifugation for 5 min at 1,500 rpm. Discard the supernatant and try not to disturb the pellet.

10. Resuspend the pellet by adding 100 μL STC per well. The mixtures are now ready for plating out on selective media and regeneration of the protoplasts.

Regeneration

In order to study the robustness towards varying conditions for regeneration, we looked at four parameters: (1) the type of osmotic stabilizer, i.e., sucrose or KCl; (2) the concentration of KCl as osmotic stabilizer; (3) the MTP type, i.e., 24- and 96-well (both shallow- and deep-well); (4) control of water surface tension/humidity by plating on a filter or not. As illustrated by Figs. 30.2 and 30.3, all variations do lead to transformants, with slight differences in the numbers obtained. Typically 24-well deep-well plates, filled with 3 mL of medium (acetamide selective medium with 0.3 M of KCl) and sterile filters placed on top were used.

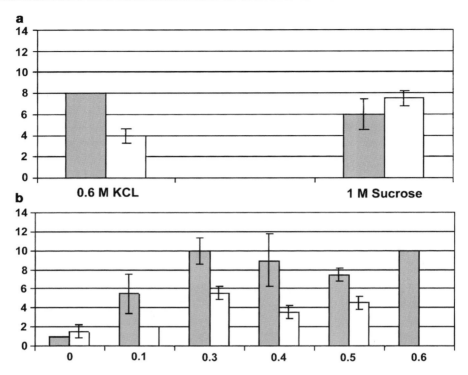

Fig. 30.2 Comparison of osmostabilizers during MTP transformation of *Penicillium chrysogenum. P. chrysogenum* strain DS17690 was transformed with plasmid pHELY-A1 according to MTP transformation method as described in this chapter. (**a**) Osmostabilizing either by 0.6 M KCl or 1 M sucrose does not lead to a drastic difference, either with (*gray bars*) or without (*white bars*) filters. (**b**) Increasing the concentration of KCl does have a clear effect on the number of transformants, whereas plating on filters (*gray bars*) leads to, on average, 50 % more transformants as compared to plating without filters (*white bars*). y-axis, average number of transformants per well, with standard deviation indicated (*n*=3-5); *x*-axis, osmostabilizer concentration

1. The protoplast suspensions are transferred at different dilutions to selective acetamide MTP, either manually (using a multichannel pipette) or automated (any pipetting station with sterility control will do)
2. Incubate the MTP for 5–7 days at 25 °C. For improved humidity control one might leave the lids off the MTP in a flow cabinet for 1 h before closing and transfer to incubator. Or, alternatively, after the first day in the incubator, the plates are put in a box inside the incubator.
3. Transfer obtained transformants to a second selective plate for purification.
4. If needed, apply further characterization techniques to selected transformants in order to verify the altered genotype and the subsequent altered phenotype. Typically this method results in several dozens of transformants per μg DNA.

Notes

1. Instead of the YGG medium, the standard yeast medium (YEPD, per liter: yeast extract 10 g, pepton 10 g, glucose 20 g) can be used, or any other medium most suitable to the fungus of choice.
2. It is advisable to determine the best effective MTP format, regenerative medium, and osmostabilizer (concentration) for each specific filamentous fungal species.
3. When selection markers other than acetamidase are used (e.g., phleomycin, nourseotricin, or hygromycin resistance), the regeneration medium needs to be adapted accordingly.
4. The MTP transformation method can be very well combined with the so-called NHEJ mutants of filamentous fungi, which have a significant improved frequency of homologous

Fig. 30.3 *Penicillium chrysogenum* transformants regenerated in MTP. *P. chrysogenum* strain DS17690 was transformed with plasmid pHELY-A1 according to the MTP transformation method as described in this chapter. (**a**) Regeneration in standard shallow-well 96-well MTP on acetamide medium with 1 M sucrose. (**b**) Regeneration in deep-well 24-well MTP on acetamide medium with 0.3 M KCl on a filter

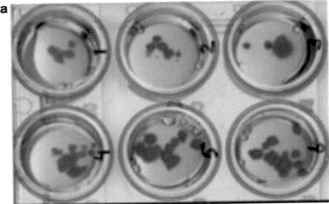

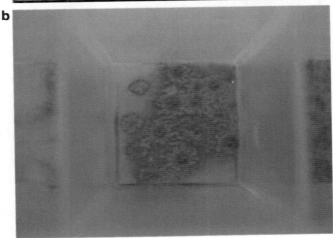

recombination (example as reported in reference [6]).

5. The transfer of transformants to the second selective plates preferably is done manually with a toothpick and re-streaking to single spore-derived colonies.

References

1. de Groot MJ, Bundock P, Hooykaas PJ, Beijersbergen AG (1998) *Agrobacterium tumefaciens*-mediated transformation of filamentous fungi. Nat Biotechnol 16:839–842

2. Aboul-Soud MA, Yun BW, Harrier LA, Loake GJ (2004) Transformation of *Fusarium oxysporum* by particle bombardment and characterisation of the resulting transformants expressing a GFP transgene. Mycopathologia 158:475–482

3. Chakraborty BN, Patterson NA, Kapoor M (1991) An electroporation-based system for high-efficiency transformation of germinated conidia of filamentous fungi. Can J Microbiol 37:858–863

4. Tilburn J, Scazzocchio C, Taylor GG, Zabicky-Zissman JH, Lockington RA, Davies RW (1983) Transformation by integration in *Aspergillus nidulans*. Gene 26:205–221

5. van den Berg MA, Bovenberg RAL, Raamsdonk LML, Sutherland JD, de Vroom E, Vollinga RCR (2007) Cephem compound. PCT/NL2004/000367

6. Snoek IS, van der Krogt ZA, Touw H, Kerkman R, Pronk JT, Bovenberg RA et al (2009) Construction of a *hdfA Penicillium chrysogenum* strain impaired in non-homologous end-joining and analysis of its potential for functional analysis studies. Fungal Genet Biol 46:418–426

Molecular Fingerprinting of Fungal Communities in Soil

31

Roberto A. Geremia and Lucie Zinger

Abstract

Study of fungal spatio-temporal dynamics requires analyzing a high number of samples with robust methods in a reasonable time. Fingerprinting DNA-based methods are better adapted than classical cultural methods for such purposes. Although molecular fingerprinting does not allow identification of phylotypes, they provide a snapshot of fungal communities, allowing comparison of a large number of samples.

Keywords

Fungal communities • Phylogenetic structure • Molecular fingerprinting • Molecular signature • Spatio-temporal dynamics • Soil

Introduction

The fungal biomass represents a large amount of soil biota, and sometimes it is the major biotic community of soil [1]. Soil fungal communities hold a large panel of functional properties that place them as the first step of the soil food web. Indeed, the fungal extracellular enzymes are involved in degradation of vegetal organic matter [2]. They also influence plant development and productivity by mutualistic nutrient relocation through mycorrhiza [2–5]. However, little is known about the factors that influence the dynamics and functional implications of fungal communities as well as the systematic study of different habitats. Indeed, the identification of situations that lead to changes in the genetic structure of fungal communities would allow better understanding of the ecological role of fungi in different habitats. For instance, studies in alpine tundra habitats show the prominent role of snow as well as plant cover in the distribution and dynamics of fungal communities [6–8]. Molecular tools for characterization of fungal communities play a prominent role in this kind of study.

The phylogenetic structure of fungal communities is most commonly achieved using molecular methods. Actually, cultural methods are time-consuming and only allow one to cultivate less than 20% of the fungal strains [9]. The simplest molecular methods to assess the phylogenetic structure of fungal communities are the so-called molecular fingerprinting or molecular signature. Although these methods do not provide

R.A. Geremia (✉) • L. Zinger
Laboratoire d'Ecologie Alpine, CNRS/UJF,
Université Joseph Fourier, BP 53 Bat D Biologie,
Grenoble 38041, France
e-mail: roberto.geremia@ujf-grenoble.fr

taxonomic identification, they display a snapshot of fungal diversity. The emergence or loss of fungal phylotypes would be detected as changes in the electrophoretic pattern of the community. Total soil DNA is used as a template for the PCR-amplification of a marker gene, the most common marker being the internal transcribed spacer 1 (ITS1) [10]. The PCR amplicons, actually a mixture of different fungal phylotypes, are separated through three kinds of methods relying on electrophoresis. First, the marker gene is separated according its length (size polymorphism); the most common are fragment length analysis (FLA) and automated ribosomal intergenic spacer analysis (ARISA). Second, the marker gene is separated according to its conformation (sequence polymorphism) either by denaturant (temperature gradient gel electrophoresis, TGGE; denaturing gradient gel electrophoresis, DGGE), or native electrophoresis (single-stranded conformation polymorphism, SSCP). Finally, the third technique relies on both previous methods: PCR amplicons are digested according to their sequence with restriction enzyme, and then separated by their length (restriction fragment length polymorphism (RFLP) and terminal-RFLP (T-RFLP)).

The robustness of final results requires a high sampling density and PCR replicates, which increases the number of samples to be processed. To cope with such constraints, a high-throughput and robust strategy was set up in our laboratory. We first adapted a protocol of the commercial Power-Soil kit for extraction. Second, to reduce the experimental steps, we chose a straightforward method like SSCP. Finally, we used capillary electrophoresis (CE), which provides the possibility of analyzing a large number of samples in a few hours. Although CE has already been coupled to T-RFLP for soil fungal community studies [11, 12], the use of CE-SSCP remains limited. Additionally, and owing to different fluorophores, CE allows the analysis of at least three different primer pairs. We have used CE-SSCP to study fungal, bacterial, and chrenoarcheal communities. Here, we present a protocol allowing the study of fungal and bacterial communities.

Materials

1. Manual drill.
2. Plastic gloves.
3. Sieves <5 and <2 mm.
4. Disposable polypropylene tubes: 50 mL conical.
5. Ultra-high-quality (UHQ) water.
6. PowerSoil® DNA Isolation Kit catalog no. 12888-100 (Mobio Lab. Inc, Carlsbad, CA).
7. FastPrep®-24 Instrument (MP Biomedicals, Inc., Illkirch, France).
8. Vortexer.
9. Centrifuge.
10. PCR hood.
11. Disposable polypropylene microcentrifuge tubes: 1.5 mL conical.
12. Spectrophotomer UV.
13. PCR machine.
14. Automatic micropipettes (i.e., Pipetman).
15. Filter tips, 0.01–1 mL.
16. 96-well PCR Plate, rigid semi-skirted, ABI.
17. Fungal primers (12): ITS5 (5′-GGAAG-TAAAAGTCGTAACAACG-3′), ITS2-FAM (5′-fluorescein phosphoramidite GCTGCG-TTCTTCATCGATGC-3′).
18. AmpliTaqGold (Applied Biosystems, Stockholm, Sweden).
19. Agarose.
20. Ethidium bromide: 0.5 mg/mL stock.
21. Gel loading mixture: 40% (w/v) sucrose, 0.1 M EDTA, 0.15 mg/mL bromophenol blue.
22. Horizontal electrophoresis equipment.
23. UV transilluminator and camera suitable for photographing agarose gels, e.g., Syngene Gene Genius Bioimaging system.
24. TBE buffer.
25. Mark XIV gene ladder.
26. Hi-Di™ Formamide (Applied Biosystems).
27. NaOH 0.3 M.
28. HD-Rox-400 markers (Applied Biosystems).
29. Multi-well plate for sequencer.

Methods

The methods described as follows were designed to obtain a fungal molecular profile specific to a given habitat. It allows one to withdraw a soil sample, extract its DNA, amplify the ITS1 from this DNA, obtain the SCCP profile, and generate distance trees. Depending on the content and quality of soil organic matter, relative amount of fungal DNA, some modifications should be introduced at the PCR level. Our suggestions are indicated in the Notes.

Soil Sampling

The DNA extraction requires 0.25 g of soil. However, in order to minimize the bias due to spatial heterogeneity in fungal distribution, larger samples from different points of the sampling site should be obtained. As previously suggested [12], at least five samples per habitat replicate should be obtained. Usually, we obtain 50 g of soil per sampling point.

1. Select the points to be sampled based on the principles of the experiment.
2. Drill the first 15 cm of the soil; wearing gloves place them in a plastic bag or directly sieve them (see Note 1).
3. To perform the sieving wear gloves and keep the sample out of the refridgerator for as little time as possible.
4. Manually desegregate the soil.
5. Sieve the soil first at 5 mm.
6. Recover the material passing through the sieve.
7. Sieve the soil at 2 mm.
8. Recover the material passing through the sieve; transfer to a polypropylene 50-mL tube.
9. Store the tube at −20°C.

DNA Extraction

This step is performed in a dedicated laboratory, to prevent contamination with DNA of other organisms or PCR products. The protocol used is the one provided by the manufacturer for individual tubes (available at www.mobio.com/images/custom/file/protocol/12888.pdf) (see Note 2), with only one modification in step 5: instead of the flat vortex, we used the Fastprep 24 instrument at maximum power two times 20 s. DNA integrity is estimated by electrophoresis in a 1.6% W/V

agarose gel; DNA is considered suitable when only a high molecular weight (MW) with no or low smear. DNA concentration and purity are estimated by absorbance at 260/280 nm. DNA could be stored a few days at 4°C, but for longer time it should be stored at −20°C.

Amplification of ITS1 Fragments by PCR

The protocol below uses the primers ITS5 and ITS2, which work nicely in our laboratory [13–15]. This protocol was also used to obtain amplicons for pyrosequencing, which allowed to test their specificity; only 1.3% of the obtained sequences were assigned to Virideplantae, 0.01% to Metazoa, and 0.02% to Alveolata, Rhizaria, and Rhodophyta [16]. Fungal ITS1 was amplified with the primers ITS5 and ITS2-FAM labeled [17]; however, new primers have recently been proposed that are even more specific [18], but we do not have experience on SSCP.

1. The setup of the PCR reaction is performed in a dedicated hood, located in the DNA extraction laboratory. Wear gloves through all the preparation.
2. For each sample, make three independent PCR reactions and prepare at least three negative controls.
3. The PCR reactions (25 μL) consisted of 2.5 mM of $MgCl_2$, 1 U of AmpliTaq GoldTM buffer, 20 g/L of bovine serum albumin (BSA), 0.1 mM of each dNTP, 0.26 mM of each primer, 0.2 U of AmpliTaqGold DNA polymerase (Applied Biosystems), and around 10 ng of DNA template in 1 μL.
4. Thaw the reagents and DNA preparations. Reactions are carried out in 25 μL. Prepare a reaction mixture sufficient for the number of samples to be tested.
5. To prepare the reaction mixture, use a set of dedicated micropipettes.
6. Calculate the amount of the reaction mixture to be prepared using 24 μL/sample. Each 96 μL (for 4 samples) of the master mix, contains 62.4 μL of UHQ water, 10 μL of 25 mM $MgCl_2$, 4 μL of a mixture containing 2.5 mM of each dNTP, 4 μL of each 5 mM ITS5 and 5 mM ITS2-FAM, 0.8 μL of L 20 mg/mL

BSA, and 0.8 μL of AmpliTaq Gold polymerase (1 U/μL) (see Note 2.) Gently vortex, centrifuge, and add 24 μL aliquots into the wells of a suitable multi-well PCR plate.

7. With a micropipette equipped with filter tips, add 1 μL of soil DNA (10–20 μg/μL) to the PCR reaction well; mix briefly with the pipet tip. To the negative controls add 1 μL of UHQ water. Cover the plate with a suitable adhesive film and centrifuge. Carry out the PCR reaction as follows: an initial step at 95°C (10 min), followed by 30 cycles at 95°C (30 s), 56°C (15 s), and 72°C (15 s), and final step at 72°C (7 min).

8. The PCR products were checked on a 1.6% agarose gel. A positive reaction shows one or several bands between 150 and 500 bp (see Note 3).

9. The samples where amplification was successful are further submitted to CE-SSCP.

10. Calculate the amount of denaturing mixture considering 10 μL of Hi-Di™ Formamide (Applied Biosystems) (see Note 4), 0.5 μL of 0.3 M NaOH, and 0.2 μL of internal DNA molecular weight (MW) standard Genescan-400HD ROX (Applied Biosystem) per sample. Aliquot 10 μL of the mixture in an ABI sample plate, add 1 μL of the sample, and place the plate cover (see Note 5).

11. Denature for 3 min at 95°C in a PCR machine. Cool on ice to prevent re-annealing of double strands. Keep at 4°C until running.

12. Perform the SSCP run on an ABI Prism 3130xl Genetic Analyzer™ (Applied Biosystem) using a capillary 36 cm in length. The nondenaturing polymer consists of 5% Genscan polymer, 10% glycerol, and 16Tris–borate–EDTA (TBE) (Euromedex, France). Running buffer consists of 10% glycerol and 1× TBE Injection time and voltage should be set to 22 s and 6 kV. Electrophoreses are performed at 32°C; data were collected for 25 min.

Data Analysis

In order to analyze the data by statistical methods, the electrophoregram should be recovered as a table. We have developed software to produce this kind of table, standardizing with the molecular weight markers (Supplementary Material 1, see Note 6). The fluorescence units might vary a lot according to PCR or electrophoresis' capillaries. To make these profiles suitable to use for abundance studies, it is necessary to normalize peak height. A Scilab script is available under request (see Note 6) that also selects the size of fragments (150–400 bp) to include in the final analysis. The resulting tables are then suitable to construct distance-based trees or multivariate analysis. It worth mentioning that peaks often shows "shoulder" precluding the use of SSCP data for study of alpha-diversity. On the other hand, these data are suitable for beta-diversity analysis. We currently use the R packages ade4 or vegan for these purposes.

Notes

1. It is highly recommended to sieve the soil at the sampling site. However, it is possible that field conditions do not allow one to do so, in which case the soils should be stored at 4°C and sieved as soon as possible. Sieving 96 samples by three persons can be achieved in 1 day of work.

2. A kit based on multi-well plates is also available at MoBio. However, the possibility of cross contamination between wells makes the use of individual tubes preferable.

3. In certain cases, no PCR products are found. This can be due to the presence of inhibitors or a tiny amount of fungal DNA. In both cases the best solution to get amplicons is to increase the number of cycles to 32–35. In the second case, up to 50 μg of DNA can be added.

4. Work under a chemical hood.

5. It is possible that the fluorescence in the region of interest is saturated, in which case a 1/10 or 1/50 dilution should be done.

6. The software "Lucieanalyzer" to read the *.fsa or *.ab1 files from the ABI Prism Genetic Analyzer™ is available under request to roberto.geremia@ujf-grenoble.fr.

SSCP Data Analysis

1. A. File import

Goal: Prepare a table.txt Fluorescence *vs* size
Software "LucieAnalyzer" (in French).
1.Transfer in the working folder the *.fsa or *.ab1 electrophoregrams files.
2. In the same folder save a ROX400.txt file with the sizes of the Rox marker

2.

Fichier->Nouveau Projet…

1.

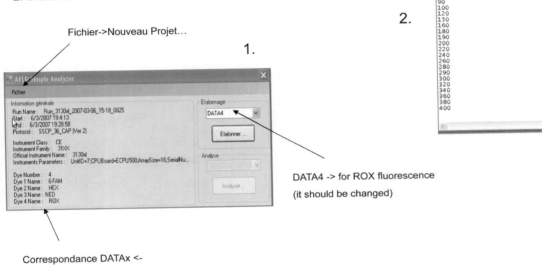

DATA4 -> for ROX fluorescence

(it should be changed)

Correspondance DATAx <-
>fluorophore

1. B. Standarisation using MW Standards

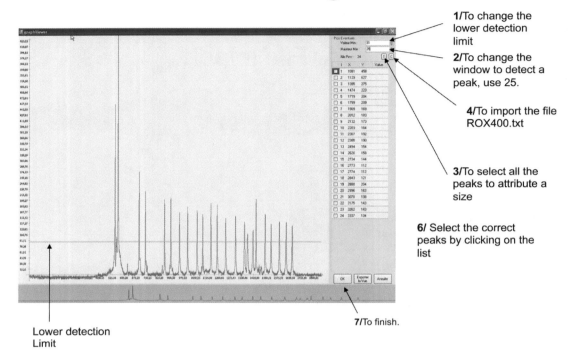

1/To change the lower detection limit

2/To change the window to detect a peak, use 25.

4/To import the file ROX400.txt

3/To select all the peaks to attribute a size

6/ Select the correct peaks by clicking on the list

7/To finish.

Lower detection Limit

1. C. Analyse a profile

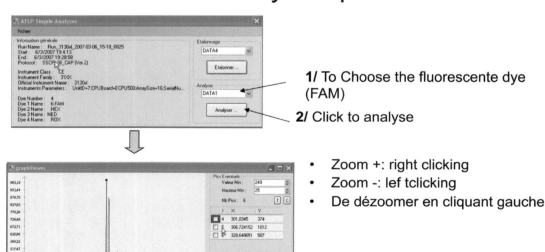

1/ To Choose the fluorescente dye (FAM)

2/ Click to analyse

- Zoom +: right clicking
- Zoom -: lef tclicking
- De dézoomer en cliquant gauche

1. D. Table export

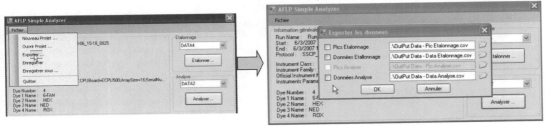

To export all the points of the profile in
a .csv file

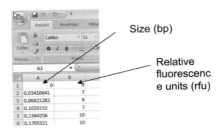

Size (bp)

Relative
fluorescenc
e units (rfu)

Do for each electrophoresis file

2. Profile normalisation

- Prepare a file with all the samples

- Replace "," by "."
- Complete the void boxes with 0 (zero)

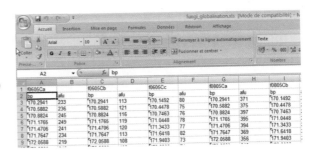

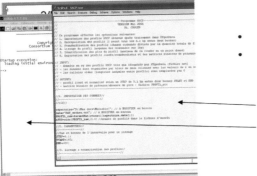

- Download Scilab
- Normalize the file using The
SSCP_normalisation.txt using Scilab.

References

1. Ananyeva ND, Susyan EA, Chernova OV, Chernov IY, Makarova OL (2006) The ratio of fungi and bacteria in the biomass of different types of soil determined by selective inhibition. Microbiology 75:702–707
2. De Boer W, Folman LB, Summerbell RC, Boddy L (2005) Living in a fungal world: impact of fungi on soil bacterial niche development. FEMS Microbiol Rev 29:795–811
3. Torsvik V, Overas L (2002) Microbial diversity and function in soil: from genes to ecosystems. Curr Opin Microbiol 5:240–245
4. Wardle DA, Bardgett RD, Klironomos JN, Setälä H, van der Putten WH, Wall DH (2004) Ecological linkages between aboveground and belowground biota. Science 304:1629–1633
5. Wardle DA (2006) The influence of biotic interactions on soil biodiversity. Ecol Lett 9:870–886
6. Schadt CW, Martin AP, Lipson DA, Schmidt SK (2003) Seasonal dynamics of previously unknown fungal lineages in tundra soils. Science 301:1359–1361
7. Zinger L, Shahnavaz B, Baptist F, Geremia RA, Choler P (2009) Microbial diversity in alpine tundra soils correlates with snow cover dynamics. ISME J 3:850–859
8. Zinger L, Lejon DPH, Baptist F, Bouasria A, Aubert S, Geremia RA et al (2011) Contrasting diversity patterns of crenarchaeal, bacterial and fungal soil communities in an alpine landscape. PLoS One 6:e19950
9. Hawksworth DL, Rossman AY (1997) Where are all the undescribed fungi? Phytopathology 87:889–891
10. Nilsson RH, Ryberg M, Abarenkov K, Sjokvist E, Kristiansson E (2009) The ITS region as a target for characterization of fungal communities using emerging sequencing technologies. FEMS Microbiol Lett 296:97–101
11. Klamer M, Roberts MS, Levine LH, Drake BG, Garland JL (2002) Influence of elevated CO_2 on the fungal community in a coastal scrub oak forest soil investigated with terminal-restriction fragment length polymorphism analysis. Appl Environ Microbiol 68:4370–4376
12. Schwarzenbach K, Enkerli J, Widmer F (2007) Objective criteria to assess representativity of soil fungal community profiles. J Microbiol Methods 68:358–366
13. Zinger L, Gury J, Alibeu A, Rioux D, Gielly L, Sage L et al (2008) CE-SSCP and CE-FLA, simple and high throughput alternatives for fungal diversity studies. J Microbiol Methods 72:42–53
14. Gury J, Zinger L, Gielly L, Tarberlet P, Geremia RA (2008) Exonuclease activity of proofreading DNA polymerases is at the origin of artifacts in molecular profiling studies. Electrophoresis 29:2437–2444
15. Baptist F, Zinger L, Clement JC, Gallet C, Guillemin R, Martins JMF et al (2008) Tannin impacts on microbial diversity and the functioning of alpine soil. Environ Microbiol 10:799–809
16. Lentendu G, Zinger L, Manel S, Coissac E, Choler P, Geremia RA et al (2011) Assessment of soil fungal diversity in different alpine tundra habitats by means of pyrosequencing. Fungal Div 49:113–123
17. White TJ, Bruns T, Lee S, Taylor J (1990) Amplification and direct sequencing of fungal ribosomal RNA genes for phylogenetics. In: Innis MA, Gelfand DH, Shinsky JJ, White TJ (eds) PCR protocols: a guide to methods and applications. Academic, New York, pp 315–322
18. Bellemain E, Carlsen T, Brochmann C, Coissac E, Taberlet P, Kauserud H (2010) ITS as an environmental DNA barcode for fungi: an in silico approach reveals potential PCR biases. BMC Microbiol 10:189

Development of Microsatellite Markers from Fungal DNA Based on Shotgun Pyrosequencing

Shaobin Zhong

Abstract

Traditional methods for the isolation and identification of fungal microsatellite markers mostly rely on construction of microsatellite-enriched DNA libraries and Sanger sequencing of clones from these libraries. These methods are time-consuming, labor-intensive, and relatively expensive. In this chapter, a quick and cost-effective approach is described for the discovery of microsatellites in fungi based on direct shotgun pyrosequencing. With this approach, high molecular weight DNA is extracted from the fungus of interest and subjected to 454 genome sequencing. The sequence reads are assembled into contigs or unique sequences in a fasta format. Then, free softwares such as MSATCOMMANDER and QDD are used for microsatellite search and primer design. The designed primer pairs are tested for PCR amplification and polymorphism using DNA samples from a diverse collection of fungal isolates.

Keywords

Microsatellite • Single sequence repeat • Fungi • DNA • Shotgun pyrosequencing • 454 Genome Sequencer FLX

Introduction

Microsatellites or single sequence repeats (SSRs) are short, tandemly repeated motifs of 1–6 bases, which are found in genomes of all eukaryotes, including plants, animals, and fungi [1, 2]. Due to their high level of polymorphism, co-dominance, ease to score, and reproducibility, SSR markers have been widely used in many fields of biology, such as genome mapping and population genetics [3, 4]. However, development of SSR markers is still a big challenge, especially for the non-model organisms, which have no or limited genomic sequences available and/or relatively low frequency of microsatellites in their genomes [5, 6]. Traditional methods for the development of SSR markers mostly rely on construction of SSR-enriched libraries and then sequencing of the clones from these libraries by the Sanger sequencing technology [2]. Although this approach has

S. Zhong (✉)
Department of Plant Pathology, North Dakota State University, Walster Hall 306, Fargo, ND 58102, USA
e-mail: shaobin.zhong@ndsu.edu

V.K. Gupta et al. (eds.), *Laboratory Protocols in Fungal Biology: Current Methods in Fungal Biology*,
Fungal Biology, DOI 10.1007/978-1-4614-2356-0_32, © Springer Science+Business Media, LLC 2013

been used for developing SSR markers in a large number of organisms, the labor-intensive and time-consuming procedure as well as the relatively high Sanger sequencing costs prevent it from becoming a high-throughput method for large-scale SSR discovery. With the development of next-generation sequencing (NGS) technology [7], rapid and cost-effective discovery of microsatellite markers in species without prior genome sequence information has become feasible [6, 8–10]. In general, two approaches have been used for microsatellite discovery based on NGS. One approach is to generate random genomic DNA sequences from the organism of interest by direct shotgun pyrosequencing and then search for those sequences or contigs containing microsatellites [8–12]. The other approach uses the shotgun pyrosequencing technology to sequence microsatellites-enriched DNA libraries [13–15]. For those organisms with low microsatellite abundance or with large genome size, microsatellite enrichment increases the amount of target microsatellites in the sequences generated [13–15], but this can lead to systematic biases in the type of microsatellite detected [6]. Due to the decreasing costs of shotgun pyrosequencing and the relatively small sizes of fungal genomes, the development of SSR markers based on the direct shotgun pyrosequencing approach has become one of the most favorable choices. In this chapter, I will describe procedures for the development and characterization of SSR markers from sequence reads generated by 454 Genome Sequencer FLX.

Materials

Equipment and Consumables

1. Computers (see Note 1).
2. LI-COR 4300 DNA Analyzer (Li-Cor Inc., Lincoln, NE, USA).
3. Bench-top microcentrifuge, e.g., Eppendorf 5415D.
4. Refrigerated centrifuge, e.g., Sorvall RC-5 Refrigerated Centrifuge.

5. Vortexer.
6. Lyophilizer.
7. Mortars and pestles.
8. Rubber policeman.
9. Flasks (1,000 mL).
10. Beakers (250 mL).
11. Miracloth (CalBiochem, EMD Chemicals, Inc., San Diego, CA, USA).
12. Cheesecloth.
13. Environmental incubator shaker.
14. pH meter.
15. Horizontal electrophoresis equipment.
16. UV transilluminator and camera suitable for photographing agarose gels.
17. Water bath.
18. Petri plates.
19. Microcentrifuge tubes.
20. Centrifuge tubes (50 mL).
21. PCR tubes.
22. Thermal cycler, e.g., Master 100.
23. Compound microscope.
24. Hemocytometer.

Media and Reagents

1. Sterile distilled water.
2. V8-PDA (Add 150 mL V8 juice, 10 g PDA, 3 g $CaCO_3$, 10 g Agar, and 850 mL H_2O to make a total volume of 1,000 mL).
3. Potato dextrose broth (PDB).
4. DNA extraction buffer (50 mM Tris–HCl pH 8.0, 150 mM EDTA, 1% Sarkosyl [n-lauroyl sarcosine], and 300 μg/mL proteinase K [add fresh]).
5. Electrophoresis-grade agarose.
6. TE (10 mM Tris–HCl pH 8.0, 1 mM EDTA).
7. Proteinase K.
8. Isopropanol.
9. Ethanol (70%).
10. Phenol solution equilibrated with 10 mM Tris–HCl pH 8, 1 mM EDTA (EMD Chemicals, Inc., San Diego, CA, USA).
11. Phenol:chloroform:isoamyl alcohol (25:24:1) (USB Corporation, Cleveland, OH, USA).
12. Chloroform:isoamyl alcohol (24:1).

13. RNaseA (10 mg/mL in TE, heat treated by standing in a boiling water bath for 15 min).
14. 20% PEG8000/2.5 M NaCl (w/v).
15. 3 M NaOAc.
16. Lamda DNA.
17. dNTPs [a mixture of dATP, dCTP, dGTP, and dTTP, 10 mM each from Promega, Madison, WI, USA, stored at −20°C].
18. Thermostable DNA polymerase (e.g., Taq, and reaction buffer supplied by manufacturer) (see Note 2).
19. Oligonucleotide primers (custom-made from Eurofins MWG Operon, Huntsville, Alabama, USA, resuspended to a concentration of 100 µM using sterile distilled water and stored at −20°C).
20. M13 primer labeled by IRD700 or IRD800 at the 5′ end (custom-made from Eurofins MWG Operon, Huntsville, Alabama, USA, resuspended to a concentration of 100 µM using sterile distilled water and stored at −20°C).
21. TBE buffer (50 mM Tris, 50 mM boric acid, 1 mM EDTA).
22. Ethidium bromide (0.5 mg/mL in stock).
23. Gel loading mixture (40% [w/v] sucrose, 0.1 M EDTA, 0.15 mg/mL bromophenol blue).
24. RapidGel-XL-6% Liquid Acrylamide (6% modified acrylamide, 7 M urea, 89 mM Tris, 89 mM boric acid, and 2 mM EDTA. USB Corporation, Cleveland, OH, USA) (see Note 3).

Methods

Extraction of High Molecular Weight DNA from Fungi for Shotgun Pyrosequencing

See Note 4.

1. Grow the fungus on V8-PDA plates for 7–10 days (see Note 5).
2. Harvest the spores by adding 10 mL of ddH$_2$O to each plate, scraping the agar surface using a rubber policeman, and pouring the spore suspension through two layers of cheesecloth into a 250-mL beaker.
3. Measure the spore concentration using a hemocytometer.
4. Inoculate 10^6–10^7 spores (see Note 6) into 250 mL of liquid medium (PDB) in a 1-L flask. Incubate 1–2 days with shaking (150–200 rpm at RT).
5. Harvest the mycelia by pouring the fungal culture through one layer of miracloth into a 1-L beaker.
6. Collect the mycelia mat retained on the miracloth and place in a 50 mL Falcon tube, and store at −20°C overnight or until needed.
7. Lyophilize for 25 h or until the mycelia is dry and brittle.
8. Grind the lyophilized mycelia in a mortar with liquid nitrogen and a pestle to a fine powder.
9. Suspend the mycelial powder in disposable 50-mL polypropylene tubes each containing 20 mL DNA extraction buffer. Approximately 1.6 g dry weight is loaded to each tube. Vortex vigorously for 30–60 s.
10. Add an equal volume of Tris-saturated phenol and gently mix well. Centrifuge for 10 min, at 5 K rpm and at 4°C in a rotor. Transfer the upper (aqueous) phase to a clean tube.
11. Add an equal volume of 25:24:1 phenol:chloroform:isoamyl alcohol. Centrifuge for 10 min, at 5 K rpm and at 4°C in a rotor. Transfer the upper (aqueous) phase to a clean tube.
12. Repeat step 10 one more time.
13. Repeat step 11 one more time.
14. Add 1/10 volume of 3 M NaOAc to the aqueous solution and add two volumes of cold absolute ethanol slowly into the aqueous solution to precipitate the DNA. Invert to mix and leave on ice or−20°C for >10 min.
15. Spool out the floating DNA from the aqueous solution with a clean glass rod (or 1-mL pipette tip) and transfer to a 1.5-mL tube, wash with 70% ethanol twice, and dry the DNA for 5–10 min or until dry (see Note 7).
16. Dissolve the DNA in 200 µL TE (see Note 8).
17. Add 2 µL of RNase A to each tube and incubate at 37°C for 1 h.

18. Run 1.5 μL on 0.6% gel to check if the RNA is still present.

19. Extract the DNA once with 25:24:1 phenol:chloroform:isoamyl alcohol and once with 24:1 chloroform:isoamyl alcohol.

20. Add 1/10 volume of 3 M NaOAc to the aqueous solution and add two volumes of cold absolute ethanol slowly into the aqueous solution to precipitate the DNA. Invert to mix and leave on ice or−20°C for >10 min.

21. Spool out the floating DNA from the aqueous solution with a clean glass rod (or 1-mL pipette tip) and transfer to a 1.5-mL tube, wash with 70% ethanol twice, and dry the DNA for 5–10 min or until dry (see Note 7).

22. Dissolve the DNA in 200 μL TE (see Note 8).

23. Combine two tubes of the DNA solution in one 2-mL tube. Bring the volume to 1.2 mL with H$_2$O.

24. Add 800 μL of 20% (w/v) PEG 8000/2.5 M NaCl. Mix well and incubate on ice for 60 min to precipitate DNA.

25. Spool the DNA out and wash with 70% ethanol (see Note 9). Drain and dry the DNA pellet.

26. Resuspend the DNA pellet in TE and leave it in a water bath at 60°C for 10–20 min to dissolve the DNA completely (see Note 10).

27. Check the DNA quality and quantity by electrophoresis using lamda DNA as standards.

Shotgun Pyrosequencing

1. Send the high molecular weight DNA samples to a DNA facility for library construction and sequencing using a 454 Genome Sequencer FLX (see Note 11).

Microsatellite Discovery and Primer Design

1. Assemble the DNA sequences into contigs and single reads using the Newbler (see Note 12).

2. Run the software MSATCOMMANDER [16] to search for all microsatellite loci in the contigs and single reads (see Note 13).

3. Design primers from the flanking regions of each microsatellite locus using Primer3[1] (see Note 14) with the default settings except for length of expected PCR product set between 100 and 400 bp, optimal primer pair annealing temperature of 60°C (range 58–65°C), 50% GC content (range 40–60%), and optimal primer length of 24 bp (range 21–30 bp).

4. Add an M13 tag (5-CACGACGTTGTAAA-ACGAC) to the 5′ end of each forward primer during primer synthesis so that the fluorescent-labeled M13 primer can be incorporated in PCR to generate fluorescent-labeled amplicons to be detected with an LI-COR 4300 DNA Analyzer (see Note 15).

Screening of Primer Pairs by PCR Amplification

1. The designed primers should be tested for PCR amplification using the DNA sample extracted from the fungal strain originally used for the shotgun pyrosequencing.

2. Each PCR amplification contains 1 × PCR buffer (10 mm of Tris–HCl, 50 mm of KCl), 200 μm of dCTP, dGTP, dTTP, and dATP, 1.5 mm MgCl$_2$, 1 pmol M13 primer labeled by IRD700 or IRD800 at the 5′ end, 0.5 pmol 5′-tagged forward primer, 0.5 pmol reverse primer, 1 U of Taq polymerase, and 20 ng of genomic DNA in a final volume of 10 μl.

3. PCR is performed in an MJ Research PTC-100 thermal cycler with the following profile: 95°C for 2 min, 3 cycles at 95°C for 30 s, 56°C for 30 s, 72°C for 60 s, 25 cycles at 94°C for 30 s, 52°C for 30 s, and 72°C for 45 s, and 1 cycle at 72°C for 5 min followed by a 4°C holding step.

4. The PCR products are diluted 10- to 20-fold and analyzed on an LI-COR 4300 DNA Analyzer using a 6% polyacrylamide gel (see Note 3).

5. Primer pairs that give clear amplicons are further used for polymorphism evaluation (see Note 16).

---

[1]http://frodo.wi.mit.edu/primer3/

Analysis of Polymorphism of Microsatellite Markers

1. The polymorphism for each microsatellite marker can be evaluated by PCR using DNA samples isolated from at least 40 isolates for haploid fungi (Ascomycota) or at least 20 isolates for diploid or dikaryotic fungi (Basidiomycota).
2. Each PCR amplification contains $1 \times$ PCR buffer (10 mm of Tris–HCl, 50 mm of KCl), 200 μm of dCTP, dGTP, dTTP, and dATP, 1.5 mm $MgCl_2$, 1 pmol M13 primer labeled by IRD700 or IRD800 at the 5′ end, 0.5 pmol 5′-tagged forward primer, 0.5 pmol reverse primer, 1 U of Taq polymerase, and 20 ng of genomic DNA in a final volume of 10 μl.
3. PCR is performed in an MJ Research PTC-100 thermal cycler with the following profile: 95°C for 2 min, 3 cycles at 95°C for 30 s, 56°C for 30 s, 72°C for 60 s, 25 cycles at 94°C for 30 s, 52°C for 30 s, and 72°C for 45 s, and 1 cycle at 72°C for 5 min followed by a 4°C holding step.
4. The PCR products are diluted 10- to 20-fold and analyzed on an LI-COR 4300 DNA sequence using a 6% polyacrylamide gel (see Note 3).
5. Primers that show polymorphism among the isolates are chosen for future applications.

Notes

1. A Windows PC that has XP SP2 or Window 7 as its operating system is needed for running MSATCOMMANDER. A Mac computer with OSX 10.6 and greater also works for MSATCOMMANDER.
2. We typically use Taq from New England Biolabs (Ipswich, MA, USA), which is supplied with a $10 \times$ reaction buffer containing 20 mM $MgSO_4$. All of these reagents are stored at −20°C.
3. To make a 6% polyacrylamide gel, add 600 μL 10% ammonium per sulfate and 60 μL TEMED to 100 μL of the RapidGel-XL-6% Liquid Acrylamide.

4. The protocol described here is modified based on the protocol provided by Drs. Dongliang Wu and Gillian Turgeon at Department of Plant Pathology and Plant-Microbe Biology, Cornell University. It has been used to extract high molecular weight DNA from Cochliobolus species for 454 sequencing. Other protocols may be suitable for isolation of high molecular weight DNA from fungi for shotgun pyrosequencing.
5. We use V8-PDA for culturing Cochliobolus sativus. Other fungi may need a different medium for spore production.
6. For fungi that do not produce spores, mycelia can be collected from the agar plates and blended with a blender before they are used to inoculate the liquid medium.
7. The DNA pellet should not be too dry. Otherwise, it is hard to dissolve.
8. Incubating the tube in a water bath at 65°C can help dissolve the DNA pellet. Check frequently and do not incubate longer than 30 min.
9. The DNA can be recovered by centrifugation at 10 K rpm and 4°C for 20 min in a microcentrifuge.
10. If the DNA is still not dissolved, keep the tube at 4°C for overnight to dissolve the DNA pellet completely.
11. The 454 GS FLX Titanium system can generate more than ~400 million nucleotides (bases) per run with average read length of 400 bp. For a fungal genome with size of 40 million bases, a full 454 run can generate sequences with 10× genome coverage, which is enough to develop hundreds or even thousands of microsatellite markers. We identified hundreds of microsatellite loci from C. sativus sequences generated by a ½ 454 run at the Advanced Studies in Genomics, Proteomics and Bioinformatics (ASGPB) of University of Hawaii. The sequencing costs were only $5,250.[2]
12. Newbler is a software package for de novo DNA sequence assembly. It is designed specifically for assembling sequence data generated by the 454 GS-series of pyrosequencing platforms.

[2] http://asgpb.mhpcc.hawaii.edu/sequence/

13. We use MSATCOMMANDER [16] because it is a free and user-friendly software for finding microsatellites and for designing primers. There are other free softwares such as QDD [17] that can be used for the same purpose.

14. MSATCOMMANDER has a function for primer design and for automatically adding tags to the primers, but it only designs the primer pairs for the first 196 microsatellite loci.

15. If an LI-COR DNA analyzer is not available, other methods such as vertical polyacrylamide gel electrophoresis with silver staining [18] can also be used to detect the microsatellite markers.

16. For haploid fungi, primer pairs that generate one allele or amplicon are usually chosen for further characterization. For diploid or dikaryotic fungi, primer pairs that generate one or two alleles can be chosen for further evaluation. It depends on if the fungal isolate used is homozygous or heterozygous.

Acknowledgements The author thanks Drs. Dongliang Wu and Gillian Turgeon for sharing the protocol for DNA extraction from *Cochliobolus* species and Yueqiang Leng, Rui Wang, and Krishna D. Puri for technical assistance.

References

1. Toth G, Gaspari Z, Jurka J (2000) Microsatellites in different Eukaryotic genomes: survey and analysis. Genome Res 10:967–981
2. Zane L, Bargelloni L, Patarnello T (2002) Strategies for microsatellite isolation: a review. Mol Ecol 11:1–16
3. Jarne P, Lagoda PJL (1996) Microsatellites, from molecules to populations and back. Trends Ecol Evol 11:424–429
4. Luikart G, England PR, Tallmon D, Jordan S, Taberlet P (2003) The power and promise of population genomics: from genotyping to genome typing. Nat Rev Genet 4:981–994
5. Dutech C, Enjalbert J, Fournier E et al (2007) Challenges of microsatellite isolation in fungi. Fungal Genet Biol 44:933–949
6. Castoe TA, Poole AW, Gu W et al (2010) Rapid identification of thousands of copperhead snake (Agkistrodon contortrix) microsatellite loci from modest amounts of 454 shotgun genome sequence. Mol Ecol Resour 10:341–347
7. Metzker ML (2010) Sequencing technologies – the next generation. Nat Rev Genet 11:31–46
8. Abdelkrim J, Robertson BC, Stanton JAL, Gemmell NJ (2009) Fast, cost effective development of species-specific microsatellite markers by genomic sequencing. Biotechniques 46:185–192
9. Allentoft ME, Schuster SC, Holdaway RN et al (2009) Identification of microsatellites from an extinct moa species using high-throughput (454) sequence data. Biotechniques 46:195–200
10. Perry JC, Rowe L (2010) Rapid microsatellite development for water striders by next-generation sequencing. J Hered 102:125–129
11. Rasmussen DA, Noor MAF (2009) What can you do with 0.1x genome coverage? A case study based on a genome survey of the scuttle fly *Megaselia scalaris* (Phoridae). BMC Genomics 10:382
12. Csencsics D, Brodbeck S, Holderegger R (2010) Cost-effective, species specific microsatellite development for the endangered dwarf bulrush (Typha minima) using next-generation sequencing technology. J Hered 101:789–793
13. Santana QC, Coetzee MPA, Steenkamp ET et al (2009) Microsatellite discovery by deep sequencing of enriched genomic libraries. Biotechniques 46:217–223
14. Martin JF, Pech N, Megle'cz E et al (2010) Representativeness of microsatellite distributions in genomes, as revealed by 454 GS-FLX Titanium pyrosequencing. BMC Genomics 11:560
15. Malausa T, Gilles A, Megle'cz E et al (2011) High-throughput microsatellite isolation through 454 GS-FLX Titanium pyrosequencing of enriched DNA libraries. Mol Ecol Resour 11(4):638–644. doi:10.1111/j.1755-0998.2011.02992.x
16. Faircloth BC (2008) MSATCOMMANDER: detection of microsatellite repeat arrays and automated, locus-specific primer design. Mol Ecol Resour 8:92–94
17. Megle'cz E, Costedoat C, Dubut V et al (2010) QDD: a user-friendly program to select microsatellite markers and design primers from large sequencing projects. Bioinformatics 26:403–404
18. Zhong S, Steffenson BJ (2000) A simple and sensitive silver-staining method for detecting AFLP markers in fungi. Fungal Genet Newsletter 47:101–102

Multiplex and Quantifiable Detection of Infectious Fungi Using Padlock Probes, General qPCR, and Suspension Microarray Readout

Magnus Jobs, Ronnie Eriksson, and Jonas Blomberg

Abstract

By combining the multiplexing qualities of padlock probes and Luminex™ (Luminex Corporation, Austin, Texas, USA) technology, together with the well-established quantitative feature of qPCR, a ten-plex fungal detection protocol that quantitatively reveals ten different fungal species in a single experiment has been devised. Padlock probes are oligonucleotides designed to form circular DNA when hybridizing to specific target DNA. The 5′ and 3′ regions of the probes meet and ligate only when a specific target sequence is present in the examined sample. The region of the padlock probes that separates the target specific 5′ and 3′ ends contains general primer sequences for amplification of circularized probes by means of rolling circle amplification and qPCR. The interspersed region also contains specific tag sequences for subsequent Luminex recognition.

Keywords

Multiplex • Padlock probes • Rolling circle amplification • Fungal detection • Suspension microarray

Introduction

A protocol for simultaneous quantitative detection of ten different fungal species is described [1]. The protocol involves padlock probes, qPCR, and Luminex™ (Luminex Corporation, Austin, Texas, USA) technology. A padlock probe is a long oligonucleotide designed to hybridize to a specific target sequence so that the 5′ and 3′ ends of the probe meet. The DNA nick between the two ends can be closed via enzymatic ligation, resulting in a circularized probe [2]. Circularized probes can be detected via rolling circle amplification (RCA) and/or PCR followed by amplification

M. Jobs (✉)
Research Scientist, Dalarna University,
School of Health and Social Studies,
Högskolegatan 2, Falun 79188, Sweden
e-mail: mjb@du.se

R. Eriksson
Livsmedelsverket Sweden,
Box 662, Uppsala 75126, Sweden

J. Blomberg
Department of Medical Sciences Uppsala University,
Uppsala Academic Hospital, Dag Hammarskjölds v. 17,
Uppsala 75237, Sweden

V.K. Gupta et al. (eds.), *Laboratory Protocols in Fungal Biology: Current Methods in Fungal Biology*,
Fungal Biology, DOI 10.1007/978-1-4614-2356-0_33, © Springer Science+Business Media, LLC 2013

product detection [3]. The amplification of circularized probes is made possible by including targets for PCR primers in the padlock probe region separating the 5′ and 3′ regions. Detection of amplification products can be done in several ways, but a neat way is to introduce a specific address tag sequence in the interspersed region that amplifies along with the rest of the circularized probe [4]. If one or two fluorescent primers are used, amplification products can then hybridize to solid phase-bound anti-tag sequences (i.e., microarray spots or Luminex beads) and be detectable [5].

Padlock probes have successfully been used in various multiplex detection systems. The advantage of padlock probes in multiplexing is both the specificity of the ligation reaction that forms the circularized probes and the possibility to use a single PCR amplification primer-pair common for all padlock probes included in the multiplex assay when amplifying the ligation products [6–8]. The protocol described here is for a ten-plex fungal detection panel that quantitatively detects ten clinically important fungi. The technique is built on the padlock probe concept, but two readouts are used: (1) a SybrGreen™ real-time PCR for quantification of circularized padlock probes, and (2) suspension array technology (i.e., Luminex) for identification of amplified sequences [1]. In Fig. 33.1 the padlock probe concept has been outlined, illustrating the different sections of the padlock probe and how the probe can hybridize and become circularized when a target sequence is present (A-B). The RCA and subsequent real-time qPCR using a single universal primer-pair (one labeled primer) are also illustrated in Fig. 33.1 (C-E). The Luminex suspension array consists of fluorescence-coded microspheres with coupled oligonucleotide anti-tag sequences for capturing of amplimers containing the sequence tag. Thus, a bead is analogous to a microarray spot [5]. The combination of two readouts of padlock probe ligation, first via SybrGreen qPCR and then by suspension microarray technology, allows the well-established quantitative aspects of qPCR with a wide dynamic range to be combined with specific identification via the high multiplex capacity of padlock probes. Included in the protocol is a strategy for improved hybridization of PCR amplimers to the bead bound anti-tag. By letting labeled oligo-

nucleotides hybridize to the 3′ and 5′ end regions of the amplimers, these regions become blocked for rehybridization to the complementary PCR amplimer strand. This reduces competition between the anti-tag and the competing strand. Moreover, the fluorophores on the end covering oligonucleotides contribute to an improved detection signal.

Materials

Solutions should be prepared using ultrapure water. Pre-PCR reagents should be handled in a clean room well separated from post-PCR activities. We use "PCR-grade water" when diluting stocks and working solutions of oligonucleotides. Shelf products in use in this protocol should be handled and stored according to the manufacturer's instructions. Oligonucleotides and PCR reagents should be stored at −20°C and buffers and microspheres at 4°C.

Materials for DNA Isolation

1. Use 0.5-mm-diameter zirconium-silica beads (Techtum Lab AB, Umeå, Sweden) for bead beating in a Tissuelyser (Qiagen AB, Stockholm, Sweden). Use 2-mL Eppendorf safe-lock tubes (VWR International, Stockholm, Sweden) in the bead beating procedure.
2. Use NucliSENS® easyMAG™ lysis buffer (BioMerieux Sverige AB, Askim, Sweden) to lyse clinical samples. For sputum use 2× Sputolysin® (Calbiochem, San Diego, CA, USA).
3. Extract DNA on a NucliSENS® easyMAG™ according to manufacturer's instructions.

Materials for Dilution Series of Standards

Prepare stock solutions of synthetic targets 100 μM (100 pmol/μL) according to instructions from the oligonucleotide manufacturer (Note 1). Prepare two pools of synthetic targets, one pool

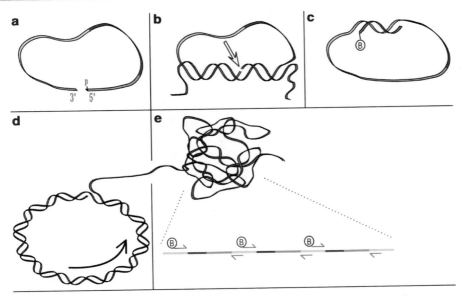

Fig. 33.1 Outline of the padlock probe concept. (**a**) The 3′ and 5′ regions of the padlock probe (indicated in *green*) are complementary to a specific target sequence. The internal region contains a sequence designed for a general primer-pair (indicated in *yellow*) and an address sequence (indicated in *blue*). (**b**) When the target specific regions hybridize to the target (*gray*), the 3′ and 5′ ends meet, and via a ligation reaction the padlock probe becomes circularized. (**c**) A primer, complementary to one part of the general primer-pair region, is then hybridized. (**d**) A rolling circle amplification (RCA) is performed generating a long repetitive sequence. (**e**) Finally, using the general primer-pair (one of the primers is the same as that used for the RCA), PCR is performed on the RCA product. The PCR product contains the address sequence and a biotin moiety for subsequent detection in the Luminex instrument. Reprinted with permission from Eriksson [1]

of targets for low efficiency padlock probes (marked with down arrow in Table 33.1) and one pool of targets for high efficiency padlock probes (marked with up arrow in Table 33.1). Prepare the two pools so that each of the five targets included in a pool makes up one-fifth of the total copy number in the pool. Do this by mixing 10 μL from each stock of synthetic targets belonging to the same category. Add 251 μL PCR-grade water to the 50 μL pool stocks so that the pools contain 10^{13} molecules/μL (all five molecule species combined). From here, use 0.05× Denhardt's solution (Sigma-Aldrich, St. Louis, Missouri, USA, art. # D2532) instead of pure water when preparing the standard dilution series. Dilute 1,000 times (1 μL target pool plus 999 μL 0.05× Denhardt's solution) and from that solution dilute five times (50 μL target pool plus 200 μL 0.05 Denhardt's solution) and from that solution serially dilute with a factor 10 (10 μL target pool plus 90 μL 0.05× Denhardt's solution) until a series ranging from 0.2×10^2 molecules/μL to $0.2 * 10^7$ molecules/μL has been established. Make sure to vortex every new dilution thoroughly before preparing the next dilution in the series.

Materials for Padlock Probe Ligation and RCA

1. Use padlock probes for *Candida albicans, Candida glabrata, Candida tropicalis, Candida parapsilosis, Cryptococcus neoformans, Aspergillus fumigatus, Aspergillus flavus, Aspergillus niger, Aspergillus nidulans,* and *Pneumocystis jirovecii* (see Table 33.1), and general forward primer (Biot-5′-AAGATATCGTAAGGAT-3′) and general reverse primer (5′-TTGGATAAGTGGG-ATA-3′) synthesized by Thermo Electron

Table 33.1 Padlock probe sequences for the ten targeted fungi marked with a star (*)

Fungal species	*Padlock probe sequence (5'–3')	**Synthetic target sequence (5'–3')
↑ high eff. probe		
↓ low eff. probe		
A. flavus ↑	*PO$_4$-TTGCGTTCGGCAAGCGCCATTGGTAAATTGGTAAATGAATTGATCCTTACGATA TCTTGGATAAGTGGGATATCCAAGGTCAACCTGAAAAAGATTGAT	**TCCGGCATCGATGAAGAACGCAGCCCCGGCCGGCGCTTGCCGAACGCAAAATCAATCTTTTTCCAGGTTGACCTTGGATCAGGTAG CATATCAATAAGCGGAGGA
A. fumigatus ↑	*PO$_4$-GGGTGTCGGCTGGCGCTTAGATGAATGAATTGTGAAGTATTTAGATCCTTACGATATCT TGGATAAGTGGGATAATCCGAGGTCAACCTTAGAAAAAATAAAGTT	**TCCGGCATCGATGAAGAACGCAGCCCCGGCCGGCGCCAGCCGACACCCIAACTTTATTTTTCTAAGGTTGACCTCGGATCAGGTAGC ATATCAATAAGCGGAGGA
A. nidulans ↓	*PO$_4$-TCGAGCGGGTGACAAAGCCCTGAAAATGAATGAATGATGAAATTGATCCTTACGA TATCTTGGATAAGTGGGATAGCCCGGCCGGCCTAA	**TCCGGCATCGATGAAGAACGCAGCCGAGCGTATGGGGCTTTGTCACCCGCTCGAITAGGGCGCCGGCCGGCGCCAGCCGACGCATA TCAATAAGCGGAGGA
A. niger ↑	*PO$_4$-AGGCGCCGGCCAATCCTACGTAAAAAGAAAGGTATAAAGGTAAATCCTTACGAT ATCTTGGATAAGTGGGATAGAAAAGAATGGTTGGAAAACGTCGGC	**TCCGGCATCGATGAAGAACGCAGCTGCTCTGTAGGAITGGCCGGCGCCTIGCCGACGTTTTCCAACCATTCTTTCCAGGTTGCATAT CAATAAGCGGAGGA
C. albicans ↑	*PO$_4$-CGCTACCGCCGCAAGCAATGATGATTTGAAGATTATTGGTAATGTAAATCCTTACGAT ATCTTGGATAAGTGGGATAAAGGTCAAAGTTTGAAGATATACGTGGTAGA	**TCTCGCATCGATGAAGAACGCAGCAACAITGCTTGCGGCGGTAGCGITCTACCACGTATATCTTCAAACTTTGACCTAAGCATATCAA TAAGCGGAGGA
C. glabrata ↑	*PO$_4$-GTTGGTAAAACCTAATACAGTATTAACCCCGATTGATTATTGTGATTTGAATTG ATCCTTACGATATCTTGGATAAGTGGGATACTTATCCCTCCCTAGATCAACACCGA	**TCTCGCATCGATGAAGAACGCAGCCGGGGGTTAATACTGTATTAGGTTTTACCAACITCGGTGTTGATCTAGGGAGGGATAAGTGG CATATCAATAAGCGGAGGA
C. tropicalis ↓	*PO$_4$-CCACTAGCAAAATAAGCGTTTTTGGATAAAATGATATGAATTGGATTATTGGTAT ATCCTTACGATATCTTGGATAAGTGGGATAAGGTCAAAGTTATGAAATAAAITGTGGTGG	**TCTCGCATCGATGAAGAACGCAGCTITATCCAAAACGCTTATTTGCTAGTGGICACCACCAAITTATTTCATAACTTTGACCTGC ATATCAATAAGCGGAGGA

C. parapsilosis	*PO$_4$-GGAGTTTGTACCAATGAGTGGAAAAAAACGTTAGTTAGAATTATTGTTAGTTAGA
→	TCCTTACGATATCTTGGATAAGTGGGATATGATTGAGGTCGAAATTGGAAGAAGTTTT
	**TCTCGCATCGATGAAGAAACGCAGCGTTTTTTCCACTCATTGGTACAAACTTCTTCCAAATTCGACCTCAAATCAGCAT
	ATCAATAAGCGGAGA
Cr. neoformans	*PO$_4$-GCCGAAGACTACCCCATAGGCCGTAAGATGTTGATATAGAAGATTAATCCTTA
→	CGATATCTTGGATAAGTGGGATAAAACAAAAAGAGATGGTTGTTATCAGCAA
	**TTCCACATCGATGAAGAAACGCAGCTGGCCTATGGGGTAGTCTTCGGCTTGCTGATAACAACCATCTCTTTTTGTTGAGCATAT
	CAATAAGCGGAGGA
P. Jiroveci	*PO$_4$-GAATTTCAGACTAGCATGCATATAATTATTTAATGTTGTGAATAATGTAGAAAG
→	ATCCTTACGATATCTTGGATAAGTGGGATAGACACTAGGCAAAGAAAAAAGTACTTTT
	**TCTCGCGTCGATGAAGAAACGTGGCAATAATTATATGCATGCTAGTCTGAAATTCAAAAGTAGCTTTTTTTCTTTGCCTAGTGTCGCA
	TATCAATAAGCGGAGGA

The target matching sections are underlined and the address-tags are shown in bold. The remaining unmodified section is the general primer-pair sequence, identical for all padlock probes. The synthetic target sequences are marked with a double star (**). Here the underlined section represents the padlock probe matching section. The padlock probe ligation position has been pointed out with a bar (|). Extra sequences flanking the target region have been included and represent interspecies conserved sequences (in vivo these conserved regions may not be immediately adjacent to the padlock probe matching sections)

Table 33.2 Outline of amino-modified carbon 12-linked anti-tag sequences coupled to different color-coded microspheres[a]

Anti-tag for	Anti-tag sequence	Microplex™ xTAG
A. flavus	N-C12-ATTGGTAAATTGGTAAATGAATTG	LUA-7
A. fumigatus	N-C12-TTAGATGAATTGTGAAGTATTTAG	LUA-90
A. nidulans	N-C12-TGAAATGAATGAATGATGAAATTG	LUA-35
A. niger	N-C12-GTAAAAAGAAAGGTATAAAGGTAA	LUA-30
C. albicans	N-C12-GATTTGAAGATTATTGGTAATGTA	LUA-4
C. glabrata	N-C12-GATTGATTATTGTGATTTGAATTG	LUA-5
C. tropicalis	N-C12-TGATATGAATTGGATTATTGGTAT	LUA-70
C. parapsilosis	N-C12-GTTAGTTAGATTATTGTTAGTTAG	LUA-80
Cr. Neoformans	N-C12-GTAAGATGTTGATATAGAAGATTA	LUA-9
P. jiroveci	N-C12-TAATGTTGTGAATAATGTAGAAAG	LUA-40

[a]The anti-tags match tags included in padlock probes targeting the different fungal species indicated. Pre-coupled microspheres, with the same anti-tags, can be ordered directly from the Luminex, and the table points out the corresponding Microplex microspheres

GmbH, Ulm, Germany. All padlock probes must be synthesized 5′ phosphorylated and the general forward primer must have a 5′ biotin moiety. Dilute padlock probes and primers in PCR-grade water to 100 μM (100 pmol/μL) stock solutions (the manufacturer provide information on what volume to use). Prepare 1 μM work solutions of each padlock probe by diluting the main stock solutions a hundred times. Then prepare a mixture of all ten padlock probes with a 10 nM concentration of each padlock probe (for a 500-μL mixture take 5 μL of each padlock probe work solution and add 450 μL PCR-grade water). Prepare two different general forward primer work solutions, one 1 μM solution for use when setting up the RCA reaction, and one 10 μM solution for use when setting up the PCR. Dilute the general reverse primer to a work solution of 10 μM.

2. Ligation reagents: 10× Ampligase® reaction buffer, Ampligase thermostable DNA ligase (Epicentre Biotechnologies, WI, USA), PCR-grade water (Applied Biosystems, Stockholm, Sweden), and padlock probe mix with 10 nM of each padlock probe (for padlock probe mix preparation, see previous).

3. RCA reagents: 10× Phi29 buffer (Fermentas, Vilnius, Lithuania) dNTP 10 mM, BSA 2 μg/μL, Phi29 DNA polymerase 10 U/μL, and general forward primer 1 μM.

Materials for PCR

1. PCR reagents: 10× SYBR® Green PCR Buffer, dNTPs including dUTP 12.5 mM, MgCl$_2$ 25 mM, AmpliTaq Gold® DNA Polymerase 5 U/μL, AmpErase® UNG 1 U/μL (Applied Biosystems, Stockholm, Sweden), and general forward and reverse primers 10 μM of each.

2. PCR equipment: Carry out PCR on a Rotor-Gene 3000 (Corbett Life Science, Concorde, New South Wales, Australia) in thin walled 0.2 mL PCR tubes (Qiagen, Stockholm Sweden). Instrument software: Rotor-gene 6.1.

Materials for Luminex (Coupling of Anti-Tag Oligonucleotides)

The following material is needed for coupling of anti-tag oligonucleotides to FlexMap™ microspheres: 5′C-12 amino-modified oligonucleotide anti-tags 100 μM (Table 33.2) (Biomers.net GMbH, Ulm, Germany), ten sets ("microsphere regions") of carboxylated polystyrene FlexMap™ microspheres (Luminex Corporation, Austin, Texas, USA), 0.1 M 2-morpholinoethane sulfonic acid (MES, pH 4.5) MES buffer, desiccated 1-ethyl-3-3-dimethylaminopropyl carbodiimide (EDC) (Pierce Thermo Fisher Scientific), 0.02% Tween-20 and 0.1% SDS, Tris–EDTA (TE, pH 8.0) buffer.

Coupling of anti-tag oligonucleotides is performed as follows:

1. Bring two 10 mg aliquots of desiccated EDC powder from the freezer and the stocks of the ten different FlexMap™ microspheres from the fridge. Let the reagents adjust to room temperature (Note 2).

2. Resuspend the microspheres by first vortexing and then sonication for 20 s.

3. Transfer 2.5 million (200 μL from stock) of each set of microspheres to ten pre-labeled 1.5-mL Eppendorf tubes (Note 3).

4. Pellet the microspheres by microcentrifugation at $\geq 8,000 \times g$ for 1–2 min.

5. Remove the supernatants and resuspend the microspheres in 25 μL MES buffer (vortex and sonicate for 20 s).

6. Add 2 μL of the ten anti-tag oligonucleotides to the corresponding microsphere sets and vortex.

7. Prepare a fresh solution of 10 mg/mL EDC by adding 1 mL water to one of the aliquots of desiccated EDC.

8. Quickly add 2.5 μL of the fresh EDC solution to all ten microsphere sets and vortex.

9. Incubate at room temperature for 30 min in darkness.

10. Prepare another fresh solution of EDC by adding 1 mL water to the second EDC aliquot.

11. Again quickly add 2.5 μL of the fresh EDC solution to all ten microsphere sets and vortex.

12. Incubate for another 30 min at room temperature in darkness.

13. Add 0.5 mL 0.02% Tween-20 to each set.

14. Pellet the sets by microcentrifugation at $\geq 8000 \times g$ for 1–2 min.

15. Remove the supernatant and resuspend the sets in 0.5 mL 1% SDS (vortex).

16. Pellet the sets by microcentrifugation at $\geq 8,000 \times g$ for 1–2 min.

17. Remove the supernatant and resuspend the sets in 50 μL TE buffer (pH 8.0).

18. Prepare a mixture of all ten sets by mixing equal volumes of each.

19. Store the sets and the mixture of coupled microspheres (50,000 spheres/μL) in the fridge (dark).

The coupling procedure is a modification of an original Luminex protocol (see Notes 4 and 5).

Materials for Luminex (Hybridization and Detection)

1. Tetramethyl ammonium chloride (TMAC) hybridization buffer (4.5 M TMAC, 0.15% sodium lauryl sarcosinate; "sarkosyl"(Sigma-Aldrich), 75 mM Tris–HCl pH 8 and 6 mM EDTA pH 8) and Tris–EDTA (TE, pH 8.0) buffer are used in the hybridization of tagged PCR products to microsphere-bound anti-tags. In the hybridization mixture 5′ Cy3 labeled end covering oligonucleotides (Cy3-5′-ATC-CTTACGATATCTT-3′ and Cy3-5′-TTGGAT-AAGTGGGATA-3′) 10 μM (Thermo Electron GmbH, Ulm, Germany) is included. Streptavidin-R-Phycoerythrin 0.15 μg/μL (Qiagen AB, Stockholm, Sweden) is used to label the biotinylated PCR products. Wells from non-skirted thin-wall 96×0.2 mL plates (Bioplastics, Landgraaf, The Netherlands) are suitable for the hybridization solution (the plates fit the heater block part# 67-50066-00-001).

2. The Luminex 200 (Luminex Corporation, Austin, Texas, USA) system together with STarStation software (Applied Cytometry System, Sheffield, UK) is used to analyze the amount of hybridized products.

Method

DNA Isolation

Clinical samples may be in the form of charcoal swabs, urine, vaginal swab culture broth, bronchoalveolar lavage, or sputum. Procedures for initiating DNA isolation vary for the different types. Steps 1–3 represent the start of the DNA isolation procedure for the different types of clinical samples.

1. Wash charcoal swabs in 600 μL NucliSENS® easyMAG™ lysis buffer by inserting the swab in a 1.5-mL Eppendorf tube containing the

buffer and rotate the swab thoroughly. Then proceed to step 4.

2. Transfer 1 mL urine, vaginal swab broth, or bronchoalveolar lavage to 1.5 mL Eppendorf tube and centrifuge the sample at 13,000 × g for 20 min. Discard the supernatant and dissolve the pellet in 600 μL NucliSENS® easyMAG™ lysis buffer. Then proceed to step 4.

3. Dilute sputum samples in an equal volume of 2× Sputolysin® (1× Sputolysin® final concentration), and vortex thoroughly. Incubate for 20 min and then proceed to step 4.

4. Perform cell lysis by bead beating. First transfer the suspension to 2-mL Eppendorf safe-lock tubes containing 600 μL 0.5-mm-diameter zirconium-silica beads. Then shake the samples in a Tissuelyser for 10 min at 30 Hz.

5. Transfer the lysates to clean 15-mL Falcon tubes (leave the beads in the original tube). Rinse the beads two times in 1 mL NucliSENS® easyMAG™ lysis buffer and add the rinse solution to the collected lysate in the Falcon tube. The final volume of cell lysate will be approximately 2.5 mL (it is impossible to recover the entire volume from the bead slur).

6. Add the full sample volume to a sample vessel for NucliSENS® easyMAG™ extraction robot and run the extraction according to manufacturer's instructions. Set the elution volume to 60 μL. Store extracts at −20°C.

Ligation and RCA

The reaction mixtures of both the ligation reaction and the RCA reaction must be prepared in a clean room well separated from samples, synthetic targets, and PCR products from previous experiments. The following protocol is based on a single reaction and needs to be multiplied to match the number of samples analyzed. In fact, preparing a reaction mixture for a single reaction is not recommended, because it is difficult to pipette such small volumes accurately. Including the two standard dilution series, counting five concentrations each (described above), means

that at least a volume for ten reactions must be prepared only to cover the dilution series.

1. Bring out and thaw the material for the ligation reaction (described above). Prepare the ligation reaction mixture by mixing 1 μL of 10× Ampligase® reaction buffer, 0.2 μL Ampligase thermostable DNA ligase, 1 μL Padlock probe mixture, and 4.8 μL PCR-grade water. Add 3 μL of DNA extract to the reaction mixture, pipette-mix, and place the reaction on a thermal cycler at 95°C for 2 min, 55°C 30 min, and keep at 4°C until the next step.

2. Bring out and thaw the RCA components (described previously). Prepare the RCA reaction by mixing 2 μL 10× Phi29 reaction buffer, 2 μL BSA, 0.5 μL forward primer, 0.25 μL dNTP and 0.3 μL Phi29 DNA polymerase. Add 4.95 μL PCR-grade water (to reach a volume of 10 μL). Finally, mix the RCA reaction mixture with the fully incubated ligation reaction (20 μL final volume), and incubate at 37°C for 30 min, 85°C for 4 min, and keep at 4°C until the next step.

PCR

The PCR reaction mixture must be prepared in a clean room well separated from samples and PCR products from previous experiments.

1. Prepare a 25 μL PCR mixture by mixing 2.5 μL 10× SYBR® Green PCR Buffer, 0.4 μL dNTP (including dUTP), 1.5 μL MgCl$_2$, 0.25 μL AmpliTaq Gold® DNA Polymerase, 0.25 μL AmpErase® UNG, and 0.75 μL of each general padlock probe primer. Add 16.1 μL PCR-grade water. Finally add 2.5 μL of the RCA product.

2. Program your real-time PCR instrument (we use a Rotor-Gene 3000 from Corbett Life Science) as follows: 50°C for 2 min, 95°C for 10 min, followed by 30 cycles of 95°C for 15 s, 52°C for 15 s and 68°C for 30 s. Monitor fluorescence during the 68°C step in each temperature cycle. Monitor in the FAM/Sybr channel (source: 470 nm, detector: 510 nm).

3. The real-time PCR will reveal any ligation event in the previous steps in the procedure so that at this point it is appropriate to review the real-time PCR data to decide whether to proceed in the protocol or not.

Measurements in the Luminex 200 Instrument

1. Bring all the Luminex components (see previous) to room temperature. Prepare the hybridization reaction by vortexing and sonicating (for 20 s) the anti-tag coupled microsphere mixture (containing all ten anti-tag coupled microspheres). Dilute the microsphere mixture in TMAC hybridization solution by mixing 1.5 μL microsphere mixture with 33 μL TMAC (use wells from low non-skirted thin-wall 96 well plates). Then add 1 μL of each end covering oligonucleotide, 8.5 μL TE buffer, and finally 5 μL PCR product.
2. Pipette-mix the hybridization mixture and denature it for 2 min at 95°C (preferably in a 96-well PCR instrument). Place the wells in a heater block on a shaking table heated to 50°C and allow hybridization for 30 min. Then add 2 μL streptavidin-R-Phycoerythrin, pipette-mix, and let the incubation continue for another 15 min.
3. Prepare the Luminex instrument by adjusting the analysis probe (the suction needle) to the current sample plate and run the start up scripts. Program the instrument to analyze median fluorescence intensity (MFI) from the ten microsphere regions to which anti-tags have been coupled. Initiate the assay by setting the instrument to calculate MFI based on 100 measurements from each microsphere set and allow the Luminex XYP reach 50°C. Place the heater block with the samples in the instrument and run.

Analysis

PCR amplification data are preferably reviewed in the dedicated instrument software. Instruct the PCR software to automatically identify cut off (Ct) values. MFI from all ten microsphere regions are presented as numbers by the Luminex instrument and can easily be exported to an Excel file (or be reviewed directly in the instrument software). Figures 33.2a and 33.2b show data from a titration series of a synthetic *Candida glabrata* sequence. The MFIs have been exported to Excel and converted into a diagram.

When running clinical samples, the use of standard titration series with known starting copy numbers run in parallel makes it possible to estimate sample copy numbers. In a ten-plex assay ten such dilution series of the different targets would be very impractical and expensive to include. Instead all ten targets could be pooled so that each target makes up one-tenth of the total copy number in the pool. The SybrGreen-based qPCR would then detect the pool as a single target species and the copy number of the sample could be calculated based on the amplification of the pool (given that the sample contains only one fungal species). However, we have noticed that the padlock probes matching the different fungal species amplify with different efficiencies resulting in different Ct-values despite the same target copy numbers. Therefore, we have categorized the padlock probes in two categories and made two pools of targets: one pool containing targets for the more efficient padlock probes and one for the less efficient. Both pools contain five target species. The total copy numbers of the five targets in a pool together make up the standard for each dilution step. In this way two standard curves with slightly different slopes will be produced. The two standard curves encompass most of the differences between padlock probes, and make it possible to approximate sample copy numbers once the Luminex analysis has revealed what fungal species was present in the sample.

Notes

1. Take special care to avoid contamination when handling concentrated synthetic targets. Prepare and store the high concentration targets in a room separated from where the main laboratory work is conducted.
2. Prepare several Eppendorf tubes (1.5 mL) with 10 mg EDC aliquots in each in advance

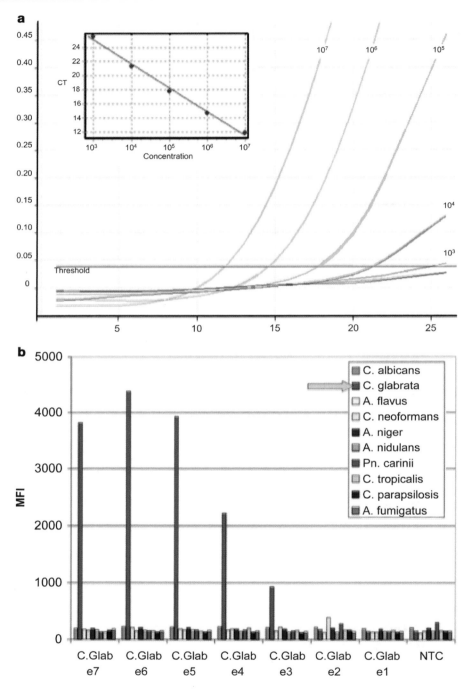

Fig. 33.2 The figure shows the result of padlock probe-based analyses of a titration series of a synthetic fungal DNA sequence (*Candida glabrata*). (**a**) Results from SybrGreen-based real-time PCR data. (**b**) Luminex median fluorescence intensity (MFI) data. The figure demonstrates the methods' ability to quantify a sample in ten-plex mode. Reprinted with permission from Eriksson [1]

and store them in a sealed container together with desiccant.

3. Label the tubes clearly with information on microsphere region and what fungus the corresponding anti-tag will detect.

4. Coupled microspheres for 500 reactions are produced by following this coupling protocol so that it is not necessary to do this every time you run an assay.

5. Microspheres with pre-coupled anti-tags (microplex xTAG microspheres) can be ordered from Luminex avoiding the coupling procedure all together. In Table 33.2 the corresponding microplex xTAG microspheres are indicated.

References

1. Eriksson R, Jobs M, Ekstrand C, Ullberg M, Herrmann B, Landegren U et al (2009) Multiplex and quantifiable detection of nucleic acid from pathogenic fungi using padlock probes, generic real time PCR and specific suspension array readout. J Microbiol Methods 78:95–202

2. Nilsson M, Malmgren H, Samiotaki M, Kwiatkowski M, Chowdhary BP, Landegren U (1994) Padlock probes: circularizing oligonucleotides for localized DNA detection. Science 265(5181):85–2088

3. Baner J, Nilsson M, Mendel-Hartvig M, Landegren U (1998) Signal amplification of padlock probes by rolling circle replication. Nucleic Acids Res 26:5073–5078

4. Baner J, Gyarmati P, Yacoub A, Hakhverdyan M, Stenberg J, Ericsson O et al (2007) Microarray-based molecular detection of foot-and-mouth disease, vesicular stomatitis and swine vesicular disease viruses, using padlock probes. J Virol Methods 143:200–206

5. Fulton RJ, McDade RL, Smith PL, Kienker LJ, Kettman JR Jr (1997) Advanced multiplexed analysis with the FlowMetrix system. Clin Chem 43:1749–1756

6. Nilsson M, Landegren U, Antson DO (2002) Single-nucleotide sequence discrimination in situ using padlock probes. Curr Protoc Hum Genet. Chapter 4:Unit 4: 11

7. Baner J, Isaksson A, Waldenstrom E, Jarvius J, Landegren U, Nilsson M (2003) Parallel gene analysis with allele-specific padlock probes and tag microarrays. Nucleic Acids Res 31:e103

8. Edwards KJ, Reid AL, Coghill JA, Berry ST, Barker GL (2009) Multiplex single nucleotide polymorphism (SNP)-based genotyping in allohexaploid wheat using padlock probes. Plant Biotechnol J 7:375–390

Rapid Deletion Plasmid Construction Methods for Protoplast and *Agrobacterium*-based Fungal Transformation Systems

34

María D. García-Pedrajas, Zahi Paz,
David L. Andrews, Lourdes Baeza-Montañez,
and Scott E. Gold

Abstract

The increasing availability of genomic data and the sophistication of analytical methodology in fungi have increased the need for functional genomics tools in these organisms. Gene deletion is a critical tool for functional analysis. The targeted deletion of genes requires both a suitable method for the transfer of foreign DNA to fungal cells and the generation of deletion constructs. The deletion constructs should contain the regions flanking the gene of interest, while the ORF is replaced by a DNA fragment harboring a marker that allows selection of cells transformed with this foreign DNA. Deletion mutants are produced upon transformation by integration of this construct into the

M.D. García-Pedrajas (✉) • L. Baeza-Montañez
Instituto de Hortofruticultura Subtropical y Mediterránea
"La Mayora", Consejo Superior de Investigaciones
Científicas (IHSM-UMA-CSIC), Estación Experimental
"La Mayora", Algarrobo-Costa, Málaga, E-29750, Spain
e-mail: mariola@eelm.csic.es; lourdes@eelm.csic.es

Z. Paz • D.L. Andrews
Department of Plant Pathology, Miller Plant Science
Building, University of Georgia, Athens, Georgia
30602-7274, USA
e-mail: zpaz@uga.edu; dandrew@uga.edu

S.E. Gold
Research Plant Pathologist, United States Department of
Agriculture - Agricultural Research Unit (USDA-ARS),
Toxicology & Mycotoxin Research Unit, 950 College
Station Road, Athens, Georgia 30605, USA
e-mail: scott.gold@ars.usda.gov

V.K. Gupta et al. (eds.), *Laboratory Protocols in Fungal Biology: Current Methods in Fungal Biology*,
Fungal Biology, DOI 10.1007/978-1-4614-2356-0_34, © Springer Science+Business Media, LLC 2013

fungal genome by homologous recombination. Protoplasts have been widely used as starting material for genetic transformation in fungal species. However, a number of fungi have proven to be recalcitrant to protoplast-mediated transformation (PMT). Among the alternative methodologies developed for those species, *Agrobacterium tumefaciens*-mediated transformation (ATMT) has been particularly successful, becoming the preferred genetic transformation method for an increasing number of fungi. Here we describe two methods to rapidly generate plasmid-based gene deletion constructs, namely, DelsGate and OSCAR, which are compatible with PMT and ATMT, respectively. Both procedures are based on PCR of the target gene flanks and Gateway cloning technology, allowing generation of deletion constructs in a very simple and robust manner in as little as 2 days. Gateway vectors have been modified so that a single Gateway cloning step generates the deletion construct itself. The PCR and transformation steps these methodologies involve should be well suited for high-throughput approaches to gene deletion construction in fungal species in which either of the two major DNA transformation methods, PMT or ATMT, is used. We describe here the entire process, from the generation of the deletion constructs to the analysis of the fungal transformants for gene replacement confirmation, with the Basidiomycete fungus *Ustilago maydis* for DelsGate and PMT, and with the Ascomycete fungus *Verticillium dahliae* for OSCAR and ATMT.

Keywords

Gene deletion • Plasmid construction • Protoplast • *Agrobacterium* • Fungal transformation • Deletions via Gateway (DelsGate) • One-step construction of *Agrobacterium*-recombination-ready plasmids

Introduction

The genomic era is generating large sets of candidate genes with potential roles that are worth functional exploration, through either the sequencing and annotation of fungal genomes or other large-scale approaches such as transcriptome analysis. Therefore, there is a great demand for fast and simple methods to generate deletion constructs suitable for high-throughput approaches to gene deletion. Here we describe two methods to rapidly generate gene deletion constructs, namely DelsGate (Deletions via Gateway) [1] and OSCAR (one-step construction of *Agrobacterium*-recombination-ready plasmids) [2], which are compatible with PMT and ATMT, respectively. Both approaches to gene

deletion construction combine PCR with Gateway cloning technology [3]. Gene flanking regions are amplified by PCR, allowing for precise deletion of the gene of interest. Although the frequency of integration by homologous recombination can vary among fungal species, and even within species among genes, we regularly use 1 kb of the 5′ and 3′ gene flanks in the deletion construct to promote homologous recombination. That flank length works well to promote homologous recombination for both *U. maydis* and PMT and for homologous integration of the T-DNA in the ATMT of *V. dahliae*, and should be sufficient in most species. As the Gateway system developed by Invitrogen is used for very efficient cloning of the amplified gene flanks, appropriate recombination sites are introduced via the primers into the PCR products during amplification. To speed up the process, appropriate vectors to be used with Gateway cloning technology have been developed separately for DelsGate and OSCAR, so that after PCR of the gene flanks, a single cloning step generates the deletion construct itself.

In DelsGate, during PCR the *attB1* and *attB2* recombination sites are introduced in the 5′ flank and 3′ flank PCR products, respectively, to promote in vitro recombination with a donor vector containing the *attP1* and *attP2* sites. In the donor vector the *attP1* and *attP2* sites flank the *ccdB* gene and the in vitro recombination by the BP clonase replaces this gene with the PCR products generating an entry clone. The *ccdB* gene is lethal for most *Escherichia coli* strains including the commonly used strain employed in this study, DH5α, thus transformants harboring the PCR products are highly favored to produce colonies. The Invitrogen pDONR201 vector was modified to be used with DelsGate by introduction of suitable selectable markers for fungal transformation. We also greatly simplified the process by taking advantage of a report by Suzuki et al. [4] showing that two independent PCR fragments, each carrying an *attB1* or *attB2* site on one end, can simultaneously recombine with a single pDONR vector. This generates a linear construct that is then circularized in vivo via *E. coli* transformation, provided that there are homologous sequences at the free ends of the PCR fragments to promote recombination. In the DelsGate method the sequence added to the 5′ and 3′ flanks to promote homologous recombination in vivo is the 18 bp recognition site for the homing endonuclease I-*Sce*I, absent in most fungal genomes. Thus, in addition to promoting homologous recombination, this site is then universally used to generate the linear DNA for fungal transformation without concern for inadvertent digestion of the gene flanks.

The OSCAR method is in turn an adaptation of the available MultiSite Gateway system; however, as with DelsGate a single BP clonase reaction generates the deletion construct. ATMT requires the generation of deletion constructs in a binary vector. When using the approach for targeted gene deletion, constructs are designed to contain, between the T-DNA borders, the flanking regions of the gene of interest with the intervening ORF replaced by a selectable marker. As above stated, the OSCAR approach has also been designed so that a single cloning step is required to produce the deletion construct in the binary vector ready for ATMT. With that purpose, two new vectors were developed. The first vector developed is a binary vector suitable for ATMT in fungi that harbors the toxic gene *ccdB* flanked by the recombination sites *attP2r* and *attP3*, between the left and right T-DNA borders. The second vector contains a *hygR* marker suitable for selection of transformants in Ascomycete fungi, flanked by the recombination sites *attP1r* and *attP4* in pBluescript II KS(+). During PCR, recombination sites *attB2r* and *attB1r* and recombination sites *attB4* and *attB3* are introduced at the ends of the 5′ and 3′ gene flanks, respectively. When these PCR products are incubated with the two vectors in the presence of BP clonase, each *attB* site should recombine only with its single compatible *attP* site generating the deletion construct. Again, transformation of *E. coli* DH5α, or similar strain, prevents selection of the original binary vector because it contains the *ccdB* gene, and selection for spectinomycin resistance prevents selection of the marker vector since this confers ampicillin and not spectinomycin resistance. Deletion constructs have then only to be transformed into *Agrobacterium tumefaciens* in preparation for fungal transformation.

For user convenience we describe here the entire DelsGate process from the production of the deletion construct to transformation of the fungus and confirmation of gene deletion for the Basidiomycete *U. maydis*. Similarly we describe the entire process of generation of OSCAR constructs and their use in ATMT with the Ascomycete fungus *V. dahliae*.

Materials

Culture Media

1. Potato dextrose agar supplemented to 2% agar (2PDA): 39 g PDA powder (Difco, Franklin Lakes, NJ), 5 g supplemental agar, 1 L dH$_2$O.
2. Potato dextrose both (PDB): 24 g PDB powder (Difco), 1 L dH$_2$O. After autoclaving, store at room temperature (RT) (see Note 1).

3. Low Na LB antibiotic plates: 1% bactotryptone, 0.5% yeast extract, 0.1% NaCl, antibiotic of choice. To prepare 1 L: 10 g tryptone, 5 g yeast extract, 1 g NaCl, 20 g agar, dH$_2$O to 1 L. After autoclaving add 0.5 mL of a 100 mg/mL solution of kanamycin sulfate (Sigma-Aldrich, Saint Louis, MO), or 0.5 mL of a 100 mg/mL solution of ampicillin (Sigma-Aldrich), or 0.5 mL of 100 mg/mL solution of spectinomycin (Sigma-Aldrich), according to the plasmid to be selected.

4. YEPS medium: 1% yeast extract, 2% bactopeptone, and 2% sucrose. For 500 mL: dissolve 5 g of yeast extract, 10 g of bactopeptone, and 10 g sucrose in a final volume of 500 mL dH$_2$O, dispense in aliquots of 100 mL in 500-mL flasks. After autoclaving, store at 4°C.

PCR Amplification of Gene Flanks and Clean Up of PCR Products

1. Primers: gene-specific primers 1 and 2 to amplify 5′ flanks and 3 and 4 to amplify 3′ flanks. For DelsGate primers 1 and 2 contain at their 5′ end the I-*Sce*I recognition sequence (in the forward orientation) and the *attB1* sequence, respectively, and primers 3 and 4 contain at their 5′ ends the *attB2* sequence and the I-*Sce*I recognition sequence (in the reverse orientation), respectively (Table 34.1). For OSCAR, primers 1 and 2 contain at their 5′ ends the *attB2r* and *attB1r* sequences, respectively, and primers 3 and 4 contain at their 5′ ends the *attB4* and *attB3* sequences, respectively (see Table 34.1).

2. *Taq*-polymerases and reaction buffers. Because the bands to be amplified are only 1 kb long any *taq*-polymerase of general use works well for this step. We regularly use homemade *taq*-polymerase. For commercial *taq*-polymerases the reaction buffer provided with them is used. For homemade *taq*-polymerase the 10× buffer we prepare contains: 0.5 M KCl, 100 mM Tris–HCl pH 8.3, 0.1% gelatin, 1% Triton X-100.

3. 30% PEG 8000/30 mM MgCl$_2$. To prepare 100 mL dissolve 30 g of polyethylene glycol (PEG) 8000 (Sigma-Aldrich) in approximately 70 mL dH$_2$O, bring volume to 90 mL, after autoclaving add 10 mL of sterile 0.3 M MgCl$_2$.

Store at 4°C or dispense in aliquots and freeze at −20°C.

4. QIAquick PCR purification kit (QIAGEN, Valencia, CA).

5. QIAquick gel extraction kit (QIAGEN).

BP Clonase Reaction

1. Gateway BP clonase II enzyme mix (Invitrogen, Carlsbad, CA).

2. Modified Gateway donor vectors for DelsGate: For PMT of *U. maydis*: pDONR-Cbx and pDONR-Hyg; for PMT of Ascomycete fungi: pDONR-A-Hyg (Fig. 34.1a). Note that these vectors are freely available from the Fungal Genetics Stock Center.

3. Modified Gateway binary and marker vectors for OSCAR: pOSCAR (binary vector) and pA-Hyg-OSCAR (marker vector) (see Fig. 34.1b). Note that these vectors are also freely available upon request from the Fungal Genetics Stock Center.

4. Donor vectors and the binary vector are maintained in *E. coli* strain DB3.1 (Invitrogen) since they contain the *ccdB* gene which is toxic to most other *E. coli* strains used in molecular biology.

Transformation of Bacterial Strains

1. *E. coli* strain DH5α (Bethesda Research Laboratories).

2. To increase transformation frequencies, commercial One Shot® MAX Efficiency™ DH5α-T1^R, One Shot® Mach1™ T1^R, or One Shot® OmniMAX™ 2-T1^R *E. coli* competent cells (Invitrogen) can be used.

3. *A. tumefaciens* strain AGL-1 [5].

Verification of Deletion Constructs

1. Primers to verify DelsGate constructs: SceI-F and SceI-R (see Table 34.1) combined with gene-specific primers from the sect. PCR Amplification of Gene Flanks and Clean Up of PCR Products or alternatively with vector

Table 34.1 Primers used for DelsGate and OSCAR deletion construct generation, verification of deletion constructs, and testing of deletion mutants

Primer	Method	Use	Sequence
Primer 1-(I-SceIF)	DelsGate	Amplification of 5′ flank, primer forward	5′- TAGGGATAACAGGGTAAT-(gene-specific sequence, N_{20-25})-3′
Primer 2-(attB1)	DelsGate	Amplification of 5′ flank, primer reverse	5′-GGGGACAAGTTTGTACAAAAAAGCAGGC TAA-(gene-specific sequence N_{20-25})-3′
Primer 3-(attB2)	DelsGate	Amplification of 3′ flank, primer forward	5′-GGGGACCACTTTGTACAAGAAAGCTGGG TA-(gene-specific sequence, N_{20-25})-3′
Primer 4-(I-SceIR)	DelsGate	Amplification of 3′ flank, primer reverse	5′-ATTACCCTGTTATCCCTA-(gene-specific sequence, N_{20-25})-3′
Primer 1-(attB2r)	OSCAR	Amplification of 5′ flank, primer forward	5′-GGGGACAGCTTTCTTGTACAAAGTGGAA- (gene-specific sequence, N_{20-25})-3′
Primer 2-(attB1r)	OSCAR	Amplification of 5′ flank, primer reverse	5′-GGGGACTGCTTTTTTGTACAAACTTGT- (gene-specific sequence, N_{20-25})-3′
Primer 3-(attB4)	OSCAR	Amplification of 3′ flank, primer forward	5′-GGGGACAACTTTGTATAGAAAAGTTGTT- (gene-specific sequence, N_{20-25})-3′
Primer 4-(attB3)	OSCAR	Amplification of 3′ flank, primer reverse	5′-GGGGACAACTTTGTATAATAAAGTTGT- (gene-specific sequence, N_{20-25})-3′
SceI-F	DelsGate	Verification of deletion construct	5′- TAGGGATAACAGGGTAAT-3′
SceI-R	DelsGate	Verification of deletion construct	5′-ATTACCCTGTTATCCCTA-3′
DonrF-C	DelsGate	Verification of deletion construct when using pDONR-Cbx	5′-TCGCGTTAACGCTAGCATGGATCTC-3′
DonrF-H	DelsGate	Verification of deletion construct when using pDONR-A-Hyg and pDONR-Hyg	5′-ATCAGTTAACGCTAGCATGGATCTC-3′
DonrR	DelsGate	Verification of deletion construct for all vectors	5′-GTAACATCAGATTTTGAGACAC-3′
OSC-F	OSCAR	Verification of presence of 5′ flank in deletion construct	5′-CTAGAGGCGCGCCGATATCCT-3′
Hyg-R(210)	OSCAR	Verification of presence of 5′ flank in deletion construct and verification of gene deletion	5′-GCCGATGCAAAGTGCCGATAAACA-3′
Hyg-F(850)	OSCAR	Verification of presence of 3′ flank in deletion construct	5′-AGAGCTTGGTTGACGGCAATTTCG-3′
OSC-R	OSCAR	Verification of presence of 3′ flank in deletion construct	5′-CGCCAATATATCCTGTCAAACACT-3′
CbxF-DG	DelsGate	Verification of gene deletion when using pDONR-Cbx	5′-GACAGCCTATTGTGGCAGCC- 3′
Hyg-DG	DelsGate	Verification of gene deletion when using pDONR-Hyg	5′-AGAGCTTGGTTGACGGCAATTTCG-3′

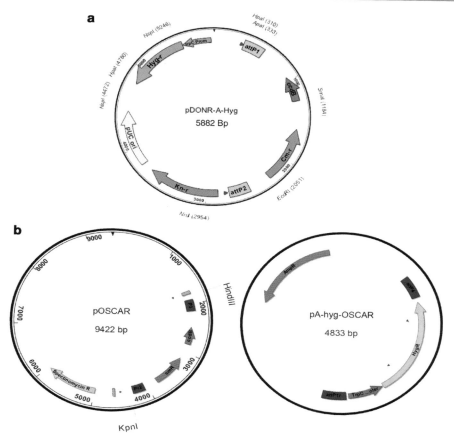

Fig. 34.1 Maps of selected plasmids for transformation of Ascomycota. (**a**) Map of DelsGate vector pDONR-A-Hyg. (**b**) Map of OSCAR binary (pOSCAR) and selection (pA-Hyg-OSCAR) plasmids

primers DonrF-C or DonrF-H (for donor vectors harboring carboxin and hygromycin as selectable markers, respectively) and DonrR (see Table 34.1).

2. Primers to verify OSCAR constructs: OSC-F and Hyg-R(210), and Hyg-F(850) and OSC-R (see Table 34.1).

3. Restriction enzymes *Kpn*I and *Hind*III (New England Biolabs, Ipswich, MA).

Preparation of DelsGate Deletion Constructs for Fungal Protoplast-Mediated Transformation

1. QIAprep spin miniprep kit (QIAGEN).
2. Restriction enzyme I-*Sce*I (New England Biolabs).

Protoplast-Mediated Fungal Transformation

1. SCS buffer: 20 mM sodium citrate pH 5.8, 1 M sorbitol. For 200 mL: dissolve 1.18 g Na$_3$-citrate and 36.44 g sorbitol (Sigma-Aldrich) in approximately 180 mL dH$_2$O, bring volume to 200 mL and autoclave. Store at 4°C.

2. STC buffer: 10 mM Tris–HCl pH 7.5, 100 mM CaCl$_2$, 1 M sorbitol. For 200 mL: dissolve 36.44 g sorbitol in approximately 160 mL dH$_2$O, bring volume to 178 mL and autoclave, then add 2 mL sterile 1 M Tris–HCl (pH 7.5) and 20 mL sterile 1 M CaCl$_2$. Store at 4°C.

3. Buffer II: 5 mM Tris–HCl, pH 7.5, 25 mM CaCl$_2$, 1 M sorbitol. For 100 mL: dissolve 18.22 g sorbitol in approximately 80 mL dH$_2$O,

bring volume to 95 mL and autoclave. Then add 2.5 mL sterile 1 M Tris–HCl (pH 7.5) and 2.5 mL sterile 1 M CaCl$_2$. Store at 4°C.

4. Lallzyme MMX solution: 500 mg/mL in Buffer II. For 10 mL: dissolve 5 g of Lallzyme MMX (standard activities: 1840 poly-galacturonase units/g, 24 pectin lyase units/g and 545 pectin esterase units/g) (Lallemand) in a final volume of 10 mL Buffer II by gently pipetting up and down, use fresh or dispense in aliquots, and store at −80°C. Thaw at RT upon use.

5. Alternatively Vinoflow® FCE (Novozyme) can be used instead of Lallzyme MMX to digest cell walls. Vinoflow solution: 384 mg/mL in Buffer II. For 10 mL: dissolve 3.84 g of Vinoflow in a final volume of 10 mL Buffer II, use fresh or dispense in aliquots, and store at −80°C. Thaw at RT upon use.

6. 40% PEG in STC. For 10 mL: autoclave 4 g PEG 4000 (Sigma-Aldrich) and 1.82 g sorbitol with 3 mL dH$_2$O (see Note 2), then add 0.1 mL sterile 1 M Tris–HCl (pH 7.5), 1 mL sterile 1 M CaCl$_2$, and sterile dH$_2$O to 10 mL. Store at 4°C and keep on ice when in use.

7. YEPS with sorbitol (YEPS-S). For 1 L: dissolve 10 g yeast extract, 20 g bactopeptone, 20 g sucrose, and 182.2 g sorbitol in approximately 800 mL of dH$_2$O, bring volume to 1 L, add 20 g of agar, and autoclave. After autoclaving add 3 µg/mL carboxin (cbx) or 150 µg/mL hygromycin (hyg) depending on the vector used.

Agrobacterium tumefaciens-mediated Fungal Transformation

1. Potassium buffer (K-buffer) pH 7.0. For 100 mL: dissolve 20 g K$_2$HPO$_4$ and 14.5 g KH$_2$PO$_4$ in 80 mL of dH$_2$0, bring volume to 100 mL with dH$_2$O. Adjust pH with 10 N NaOH. Filter sterilize and store at 4°C.

2. M-N solution: 3% (w/v) MgSO$_4$·7H$_2$O, 1.5% (w/v) NaCl. For 100 mL: dissolve 3 g MgSO$_4$·7H$_2$O and 1.5 g NaCl in 80 mL dH$_2$O, bring volume to 100 mL with dH$_2$O. Filter sterilize and store at 4°C.

3. Spore elements. For 500 mL: dissolve 50 mg ZnSO$_4$·7H$_2$O, 50 mg CuSO$_4$·5H$_2$O, 50 mg H$_3$BO$_3$, 50 mg MnSO$_4$·H$_2$O, and 50 mg Na$_2$MoO$_4$·2H$_2$O in 500 mL dH$_2$O. Filter sterilize, dispense in aliquots, and store at −20°C.

4. 1 M 2-[N-Morpholino]ethanesulfonic acid (MES) pH 5.3. For 100 mL: dissolve 21.33 g of MES in 80 mL dH$_2$O, adjust pH with 10 N NaOH (see Note 3), bring volume to 100 mL. Sterilize by autoclaving and store at −20°C.

5. 2 M glucose. For 100 mL: dissolve 36.03 g of glucose in 80 mL dH$_2$O, bring volume to 100 mL. Filter sterilize and store at RT.

6. 50% glycerol. Mix 50 mL of dH$_2$O with 50 mL glycerol. Autoclave and store at RT.

7. 20 mM 3′,5′-Dimethoxy-3′-hydroxyacetophenone acetosyringone. For 10 mL: dissolve 0.039 g of acetosyringone in 10 mL ethanol. Store at −20°C.

8. 200 mM Cefotaxime. Dissolve 0.191 g of cefotaxime in 2 mL of dH$_2$O. Filter sterilize.

9. 100 mg/mL Moxalactam. Dissolve 0.1 g of moxalactam in 1 mL of dH$_2$0. Filter sterilize.

10. Minimal Medium (MM). For 100 mL: to 94.15 mL of sterilized dH$_2$O add 1 mL K-buffer (pH 7.0), 2 mL of M-N solution, 0.1 mL of 1% CaCl$_2$H$_2$O (w/v), 1 mL of 0.01% FeSO$_4$ (w/v), 1 mL of 20% glucose (w/v), 0.5 mL of spore elements, and 0.25 mL of 20% NH$_4$NO$_2$ (w/v). Prepare from stock solutions upon use.

11. Induction Medium (IM). For 100 mL: to 89.87 mL of sterilized dH$_2$O add 1 mL K-buffer (pH 7.0), 2 mL M-N solution, 0.1 mL of 1% CaCl$_2$H$_2$O (w/v), 1 mL of 0.01% FeSO$_4$ (w/v), 0.5 mL of spore elements and 0.25 mL of 20% NH$_4$NO$_2$ (w/v), 1 mL 50% glycerol, 4 mL 1 M MES, and 0.5 mL 2 M glucose. Prepare from stock solutions upon use.

12. Cocultivation medium plates. This medium is the same as MM but with 100 µM final concentration of acetosyringone and 1.5% agar. For 100 mL: mix 89.87 mL of dH2O with 1.5 g of agar, autoclave. Then add the same components as for MM plus 1 mL of 20 mM acetosyringone. Store plates at 4°C.

13. Selection medium. For 500 mL: mix 19.5 g Potato Dextrose Agar (PDA, Difco) and 2.5 g of agar with 500 mL of dH$_2$O and autoclave.

After autoclaving add 50 µg/mL of hygromycin B, 0.5 mL of 200 mM cefotaxime and 0.5 mL of 100 mg/mL moxalactam. Store plates at 4°C.

Analysis of Fungal Transformants to Confirm Gene Deletion

1. For DelsGate, gene-specific primer 5 and CbxF-DG or Hyg-DG (see Table 34.1 and Fig. 34.2c) for cbx and hyg vectors, respectively. For OSCAR, gene-specific primer 5 and Hyg-R(210) (see Table 34.1 and Fig. 34.2d). Additionally, it is useful to use ORF-specific primers to confirm deletion.
2. Alternatively use gene-specific primers 2-O and 3-O (Fig. 34.2c, d).

Methods

Figures 34.2a and 34.2b outlines the DelsGate and OSCAR construction steps, respectively. Figure 34.2c shows a schematic representation of the entire procedure to generate deletion mutants using DelsGate: production of deletion constructs by DelsGate, manipulation of the deletion construct for fungal transformation, and finally analysis of transformants to test for gene replacement. Figure 34.2d shows a schematic representation of the entire procedure to generate deletion mutants using OSCAR: production of deletion constructs by OSCAR, introduction of the deletion construct into A. tumefaciens, and analysis of fungal transformants for gene replacement.

DelsGate deletion construction involves the following three primary steps: (1) simultaneous independent PCRs of the 5′ and 3′ ORF flanks; (2) Gateway BP cloning; (3) E. coli transformation. A number of features have been included in the design of our DelsGate method to accelerate generation of deletion constructs. For the amplification of the 5′ flank the recognition sequence for the homing endonuclease I-SceI is included at the 5′ end of the forward primer (primer 1), while an attB1 sequence is included at the 5′ end of the reverse primer (primer 2).

For the 3′ flank, an attB2 sequence is included at the 5′ end of the forward primer (primer 3) and the I-SceI recognition sequence in the reverse orientation is included at the 5′ end of the reverse primer (primer 4). PCR products are then inserted into a deletion plasmid vector via the Invitrogen Gateway BP clonase system. During the BP clonase reaction the co-purified 5′ and 3′ gene flank PCR products recombine with the attP1 and attP2 sequences of the donor vector, respectively. This reaction generates a linear molecule harboring 18 bp of homologous sequences in opposite orientation at the free ends (the I-SceI recognition site). After E. coli transformation, this homologous sequence recombines in vivo, as reported by Suzuki et al. [4], generating a circular construct. Selection of E. coli transformants containing the original donor vector is prevented by the presence on these plasmids of the ccdB gene, which is lethal to DH5α and most other E. coli strains used for cloning. DelsGate modified donor vectors pDONR-Cbx and pDONR-Hyg for use with U. maydis and pDONR-A-Hyg for use in Ascomycete fungi have been produced by addition of appropriate selectable markers. Therefore, the entry clone resulting from the BP clonase reaction and in vivo recombination is the final deletion construct itself, which has the gene precisely replaced by the vector containing the selectable marker for fungal transformation and both flanks separated by the I-SceI recognition site. Thus, in addition to promoting homologous recombination to generate the circular molecule in vivo, this 18 bp sequence is then used to generate the linear DNA for fungal transformation. The I-SceI site is extremely rare; it does not exist in the U. maydis genome and is likely completely absent from most fungal genomes. The deletion constructs generated with this method are compatible with PMT; we describe here the formation and transformation of protoplasts of the basidiomycete fungus U. maydis with DelsGate deletion constructs, and finally the analysis of fungal transformants for gene deletion. However, in addition to its extensive use with U. maydis DelsGate has to our knowledge been successfully applied to Cryptococcus

a

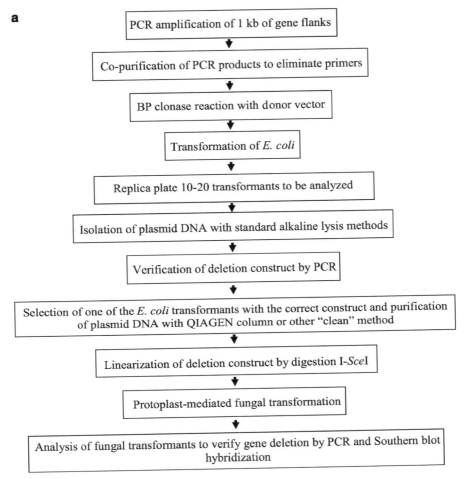

PCR amplification of 1 kb of gene flanks

↓

Co-purification of PCR products to eliminate primers

↓

BP clonase reaction with donor vector

↓

Transformation of *E. coli*

↓

Replica plate 10-20 transformants to be analyzed

↓

Isolation of plasmid DNA with standard alkaline lysis methods

↓

Verification of deletion construct by PCR

↓

Selection of one of the *E. coli* transformants with the correct construct and purification of plasmid DNA with QIAGEN column or other "clean" method

↓

Linearization of deletion construct by digestion I-*Sce*I

↓

Protoplast-mediated fungal transformation

↓

Analysis of fungal transformants to verify gene deletion by PCR and Southern blot hybridization

Fig. 34.2 DelsGate and OSCAR deletion methodologies. (**a**) Flowchart of steps in the DelsGate method. (**b**) Flowchart of steps in the OSCAR method. (**c**) Schematic representation of DelsGate deletion construction method, and generation of deletion mutants using DelsGate constructs and PMT. (**d**) Schematic representation of OSCAR deletion construction method, and generation of deletion mutants using OSCAR constructs and *A. tumefaciens*-mediated transformation

neoformans [6], *Fusarium verticillioides*,[1] and *Colletotrichum graminicola*.[2]

OSCAR deletion construction is also based on Gateway technology and similarly to DelsGate involves three primary steps: (1) simultaneous independent PCRs of the 5′ and 3′ ORF flanks; (2) a single Gateway BP cloning step; (3) *E. coli*

transformation. For OSCAR we have specifically adopted the *attB* sites as described in the MultiSite Gateway system. Thus, during PCR, recombination sites *attB2r* and *attB1r* and recombination sites *attB4* and *attB3* are introduced at the ends of the 5′ and 3′ gene flanks, respectively. To simplify the method so that only a single cloning step is required to generate a structure in which the 5′ and 3′ target gene flanks are separated by the *hygR* marker all placed between the right and left T-DNA borders of the

[1] A. Glenn, personal communication.

[2] S. Sukno, personal communication.

b

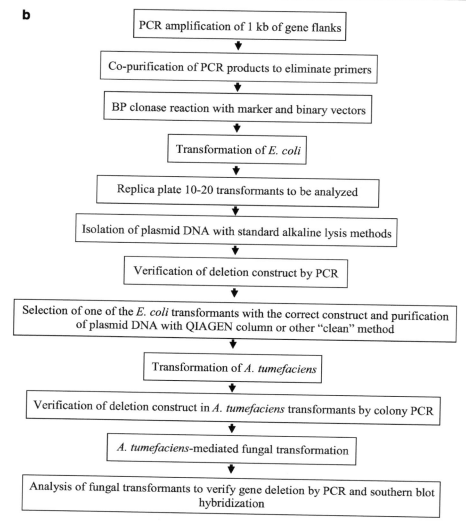

PCR amplification of 1 kb of gene flanks

↓

Co-purification of PCR products to eliminate primers

↓

BP clonase reaction with marker and binary vectors

↓

Transformation of *E. coli*

↓

Replica plate 10-20 transformants to be analyzed

↓

Isolation of plasmid DNA with standard alkaline lysis methods

↓

Verification of deletion construct by PCR

↓

Selection of one of the *E. coli* transformants with the correct construct and purification of plasmid DNA with QIAGEN column or other "clean" method

↓

Transformation of *A. tumefaciens*

↓

Verification of deletion construct in *A. tumefaciens* transformants by colony PCR

↓

A. tumefaciens-mediated fungal transformation

↓

Analysis of fungal transformants to verify gene deletion by PCR and southern blot hybridization

Fig. 34.2 (continued)

binary plasmids, two new vectors were developed. The first vector developed is a binary vector suitable for ATMT in fungi that harbors the toxic gene *ccd*B flanked by the recombination sites *attP2r* and *attP3*, between the left and right T-DNA borders. We named this vector pOSCAR. The second vector contains a *hygR* marker driven by the *Aspergillus nidulans trpC* promoter, suitable for selection of transformants in Ascomycete fungi, flanked by the recombination sites *attP1r* and *attP4* in pBluescript II KS(+). This is the marker vector, here named pA-Hyg-OSCAR. When the PCR products are incubated with

pOSCAR and pA-Hyg-OSCAR in the presence of BP clonase, each *attB* site should recombine only with its single compatible *attP* site (e.g., *attB2r* with *attP2r*) such that the favored plasmid structure generated is the deletion construct itself. Transformation of *E. coli* DH5α, or similar strain, and selection for spectinomycin resistance prevent selection of the original pOSCAR and pA-Hyg-OSCAR vectors since the former contains the *ccd*B gene and the latter confers ampicillin and not spectinomycin resistance. Deletion constructs produced are then transformed into *A. tumefaciens* in preparation

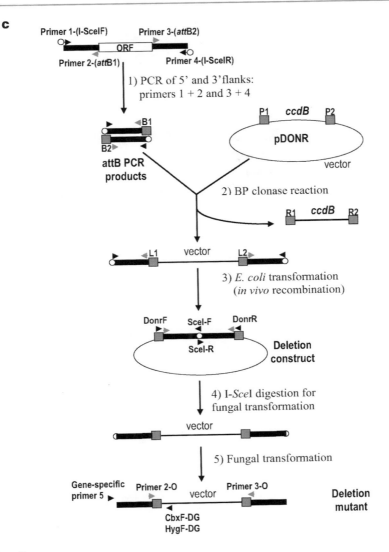

Fig. 34.2 (continued)

for fungal transformation. We describe here the ATMT of *Verticillium dahliae* and analysis of fungal transformants for gene deletion.

Primer Design for the Amplification of Gene Flanks

1. Primers are designed to separately amplify 1 kb of the 5′ and 3′ sequences flanking the ORF of the gene of interest. Specific sequences are added to the 5′ end of each primer.

2. Primers 1 and 2 are designed to amplify the 5′ flank and for DelsGate their 5′ ends contain the I-*Sce*I recognition sequence (in the forward orientation) and the *attB1* sequence, respectively (see Table 34.1). For OSCAR, on the other hand, the 5′ end of primer 1 contains the *attB2r* sequence while the 5′ end of primer 2 contains the *attB1r* (see Table 34.1). The gene-specific sequence of primer 1 (forward primer) is designed for the primer to anneal approximately 1 kb before the start codon while primer 2 (reverse primer) is gen-

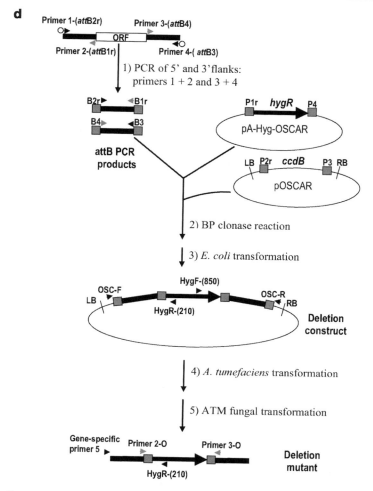

Fig. 34.2 (continued)

erally designed to anneal as close as practical before the start codon (see Fig. 34.2c, d).

3. Primers 3 and 4 are designed to amplify the 3′ flank. For DelsGate their 5′ ends contain the *attB2* sequence and the I-*Sce*I recognition sequence (in the reverse orientation), respectively (see Table 34.1). For OSCAR the 5′ ends of primers 3 and 4 contain the *attB4* and *attB3* sequences, respectively (see Table 34.1). The gene-specific sequence of primer 3 (primer forward) is designed for the primer to anneal as close as practical after the termination codon, while the gene-specific

sequence of primer 4 (primer reverse) is generally designed for the primer to anneal approximately 1 kb distal to primer 3 (see Fig. 34.2c, d).

4. Gene-specific sequences between 20 and 25 nucleotides work well for these primers.

PCR Amplification of 1 kb Gene Flanks and Clean Up of PCR Products

1. The 1 kb 5′ and 3′ gene flanks are amplified separately. PCRs are performed in a total volume

of 50 μL. For each of the two reactions combine in a PCR tube: 32.0 μL PCR-grade ddH₂O, 5.0 μL 10× *taq*-polymerase reaction buffer, 5.0 μL MgCl₂ 25 mM (see Note 4), 1.0 μL 50× dNTP mix (10 mM each), 2.5 μL forward primer 20 μM, 2.5 μL reverse primer 20 μM, 1.0 μL *taq*-polymerase, 1.0 μL (10–100 ng) fungal genomic DNA.

2. Amplification is performed under the following conditions: an initial denaturation of 1 min at 94 °C, 30 cycles of 30 s at 94 °C, 30 s at 60 °C, and 1 min at 72 °C, and completed with a final extension of 5 min at 72 °C (see Note 5).

3. After amplification, load 5 μL of each PCR reaction on a 0.8% agarose gel to confirm amplification of the desired bands.

4. To eliminate primers from the sample prior to BP clonase reaction, the 5′ and 3′ PCR products are co-purified either by PEG/MgCl₂ precipitation or by using the QIAquick PCR purification system (QIAGEN) (see Notes 6 and 7).

5. For PEG precipitation 90 μL of the combined PCR products (see Note 8) are mixed with 270 μL of TE buffer (pH 7.5–8) and 180 μL of 30% PEG 8000/30 mM MgCl₂, vortexed thoroughly and centrifuged 15 min at 14,000 rpm. Finally, the pellet is resuspended in 15 μL of TE buffer or ddH₂O.

6. Purification of the combined 5′ and 3′ flanks with QIAquick columns is performed according to the manufacturer's instructions (see Note 8). In the final step DNA is eluted in 30 μL of buffer EB (10 mM Tris-Cl, pH 8.5) or ddH₂O.

7. After purification of the combined 5′ and 3′ flanks by either method DNA concentration is determined by A_{260}. One μL of the purified DNA can be directly used for this measurement when using a NanoDrop spectrophotometer. When using a standard spectrophotometer measure concentration of co-purified PCR products by mixing 2 μL DNA with 58 μL TE buffer or ddH₂O (Dilution Factor (DF) = 30) and measuring A_{260}. To calculate concentration: $A \times DF \times 50/1,000 = $ concentration (μg/μL).

BP Clonase Reaction

1. For each DelsGate deletion construct set up a 5 μL BP clonase reaction, adding the following components in a 0.2-mL microcentrifuge tube at RT: 3 μL (75–125 ng) of combined purified 5′ and 3′ gene flanks, 1 μL (75 ng) of pDONR vector (see Note 9). Proceed to step 3.

2. For each OSCAR deletion construct set up a 5 μL BP clonase reaction, adding the following components in a 0.2-mL microcentrifuge tube at RT: 2 μL (15–20 ng) of combined purified 5′ and 3′ gene flanks, 1 μL (60 ng) of pA-Hyg-OSCAR vector, 1 μL (60 ng) of pOSCAR (see Note 9). Proceed to step 3.

3. Thaw BP Clonase II enzyme mix on ice for about 2 min and vortex briefly twice (2 s each time).

4. Add 1 μL of BP Clonase II enzyme mix to the reaction and vortex briefly twice. Spin down briefly.

5. Incubate overnight (o/n) at 25°C in a PCR thermocycler with a heated lid (see Note 10).

6. To terminate reaction add 0.5 μL of proteinase K, vortex briefly, and incubate sample for 10 min at 37°C.

Transformation of *E. coli*

1. 5 μL of BP clonase DelsGate or OSCAR reaction is used to transform *E. coli* competent cells. Generally, transforming homemade DH5α competent cells with standard heat shock methods is sufficient to produce a number of colonies to analyze. However, the use of commercial One Shot® MAX Efficiency™ DH5α-T1ᴿ, One Shot® Mach1™ T1ᴿ, or One Shot® OmniMAX™ 2-T1ᴿ *E. coli* competent cells (Invitrogen) can be an option to increase transformation frequencies (see Note 11).

2. After transformation plate aliquots of 200 μL on low Na kanamycin plates for DelsGate and low Na spectinomycin plates for OSCAR (see Note 12), incubate plates o/n at 37 °C.

Verification of Deletion Constructs

1. Prepare a replica LB kanamycin or LB spectinomycin plate with the transformants to be analyzed for the presence of the correct construct. Between 10 and 20 transformants are routinely analyzed.
2. Plasmid DNA is then purified to confirm construct structure by PCR and/or restriction enzyme digestion.
3. Inoculate cells from colonies in the replica plate in 5 mL of LB amended with 50 μg/mL kanamycin or 50 μg/mL spectinomycin and grow o/n at 37 °C and 200 rpm.
4. Purify plasmid DNA using standard plasmid miniprep alkaline lysis methods (see Note 13).
5. For OSCAR constructs the standardized test to confirm correct deletion structure initially involves a double restriction digestion with *Kpn*I and *Hind*III.
6. Each reaction is performed in a total volume of 20 μL that contains: 1–2 μg of plasmid DNA (see Note 14), 2 μL of reaction buffer, 1 μL (10 units) of *Kpn*I and 1 μL (10 units) of *Hind*III, ddH₂O to 20 μL. Incubate reactions for 2 h at 37 °C.
7. After digestion load 20 μL of each reaction on a 0.8% agarose gel to test for the presence of a band of the expected size.
8. A PCR analysis standardized to test for the presence of the 5' and 3' gene flanks is performed to confirm construct structure with both DelsGate and OSCAR methodology. For DelsGate the two primer combinations used are: SceI-F and gene-specific primer 2, and SceI-R and gene-specific primer 3 (see Fig. 34.2c). Primers SceI-F and SceI-R are designed from the I-*Sce*I recognition site in the forward and reverse orientation, respectively (see Table 34.1). Alternatively, this analysis can be performed using primers from the vector in substitution for the gene-specific primers. In this case the primer combinations used are DonrF-C or DonrF-H (for donor vectors harboring carboxin and hygromycin and selectable markers, respectively) and DonrR (see Table 34.1) in combination with the SceI-F and SceI-R primers, respectively.

For OSCAR the two primer combinations are: OSC-F and Hyg-R(210), and Hyg-F(850) and OSC-R (see Table 34.1 and Fig. 34.2d).

9. Each reaction is performed in a total volume of 25 μL that contains: 2.5 μL 10× *taq*-polymerase reaction buffer, 2.5 μL MgCl₂ 25 mM (see Note 4), 0.5 μL 50× dNTP mix (10 mM each), 1.25 μL forward primer 20 μM, 1.25 μL reverse primer 20 μM, 0.5 μL *taq*-polymerase, approximately 10 ng of plasmid DNA, PCR-grade ddH₂O to 25 μL.
10. For each of the two primer combinations prepare a master mix containing all the components except the DNA, for all the reactions you need to perform plus one. For each reaction mix 24 μL of the master mix with 1 μL of plasmid DNA sample.
11. Amplification is then performed under the following conditions: an initial denaturation of 1 min at 94 °C, 30 cycles of 30 s at 94 °C, 30 s at 60 °C, and 1 min at 72 °C, and completed with a final extension of 5 min at 72 °C.
12. After PCR load 10 μL of each PCR product on a 0.8% agarose gel to test for the presence of a band of the expected size.

Preparation of DelsGate Deletion Construct for Fungal Transformation

1. After confirmation of proper deletion construct structure, plasmids are digested with I-*Sce*I for fungal transformation. This generates a linear vector molecule containing the selectable marker flanked by the 5' and 3' flanks of the ORF to be deleted.
2. Select one of the colonies confirmed to have the desirable construct and grow again in 5 mL of LB amended with 50 μg/mL kanamycin.
3. Purify plasmid DNA using a QIAGEN column following the manufacturer's manual. Measure DNA concentration as above described.
4. For I-*Sce*I digestion, combine in a microcentrifuge tube, 3–5 μg of plasmid DNA, 5 μL of 10× I-*Sce*I buffer, 1 μL (5 units) of I-*Sce*I and ddH₂O to 50 μL. Incubate 2 h at 37 °C.

5. Generally, this incubation is sufficient to fully digest the vector DNA; however, it is a good practice to test if the digestion is completed by loading 5 μL of the digestion reaction on a 0.8% agarose gel. A single band of about between 6.8 and 5.5 kb should be visible depending on the pDONR vector that was used to make the deletion construct.

6. After digestion is complete precipitate DNA by adding 5 μL of sodium acetate 3 M (pH 5.2) and 100 μL 95% ethanol. Vortex briefly and incubate on ice for 15 min. Centrifuge 15 min at 14,000 rpm. Discard supernatant and wash pellet with 70% ethanol. Dry pellet and resuspend in 12 μL of ddH$_2$O.

7. Concentration of transforming DNA can then be measured by A$_{260}$ as described previously (see Note 15).

Transformation of *A. tumefaciens* with OSCAR Deletion Constructs

1. After confirming correct OSCAR deletion constructs, these are transformed into *A. tumefaciens* AGL-1, following the same standard heat shock methods used to transform *E. coli*. Deletion construct plasmid DNA purified as described in the sect. Verification of Deletion Constructs by standard plasmid miniprep alkaline lysis methods can be used directly for *A. tumefaciens* transformation. Alternatively, plasmid DNA can be purified with a QIAGEN column for this purpose.

2. Transformants are selected on low Na LB spectinomycin plates.

3. The presence of the deletion construct in the strain resistant to spectinomycin selected to be used in the fungal transformation may be confirmed by colony PCR with primer combinations OSC-F and Hyg-R(210), and/or Hyg-F(850) and OSC-R (see Table 34.1), used to confirm deletion constructs as described in the sect. Verification of Deletion Constructs (see Note 16).

4. For colony PCR, strains to be tested are replicated on an LB spectinomycin plate and

grown o/n to have enough bacterial biomass.

5. Take approximately half of the bacterial biomass from each replicated colony with a toothpick or a yellow pipette tip and mix it with 20 μL of water by vortexing in a microcentrifuge tube.

6. Incubate 10 min at 100 °C to lyze cells and then place samples on ice for 2 min.

7. Spin cells down for a few seconds at maximum speed and take 5 μL of the supernatant for PCR.

8. Each reaction is performed in a total volume of 25 μL which contains: 5 μL of plasmid DNA solution, 2.5 μL 10× *taq*-polymerase reaction buffer, 2.5 μL MgCl$_2$ 25 mM (see Note 4), 0.5 μL 50× dNTP mix (10 mM each), 1.25 μL forward primer 20 μM, 1.25 μL reverse primer 20 μM, 0.5 μL *taq*-polymerase, and PCR-grade ddH$_2$O to 25 μL.

9. Amplification is performed under the following conditions: an initial denaturation of 1 min at 94 °C, 30 cycles of 30 s at 94 °C, 30 s at 60 °C, and 1 min at 72 °C, and completed with a final extension of 5 min at 72 °C.

10. After PCR, load 15 μL of each PCR reaction on a 0.8% agarose gel to test for the presence of a band of the expected size.

Protoplast-Mediated Fungal Transformation

1. For PMT of *U. maydis* we use a variation of the method described by Tsukuda et al. [7] (see Note 17).

2. Inoculate 5 mL of PDB with one isolated colony of *U. maydis* grown on 2PDA, incubate o/n at 28 °C and 250 rpm.

3. Use approximately 100 μL of the o/n culture to inoculate 100 mL YEPS medium in 500-mL flasks.

4. Incubate o/n at 28 °C and 250 rpm until an OD$_{600}$ between 0.6 and 0.8.

5. Spin cells down in two 50-mL conical tubes, fairly gently approximately 1,100 × g, discard supernatant (s/n).

6. Add 10 mL of SCS to each tube and resuspend with gentle vortexing. Combine the two tubes from each strain into one. Bring the volume to 30 mL with SCS, and spin as in step 5.

7. Resuspend cells in 1 mL SCS and add 200 μL Lallzyme MMX solution (500 mg/mL in Buffer II). Alternatively add 200 μL Vinoflow solution (384 mg/mL in Buffer II).

8. Incubate with gentle mixing at RT checking protoplast formation on a microscope every 15 min. It takes around 45 min for a majority of cells to generate protoplasts when using Lallzyme solution. With Vinoflow incubation times are shorter, ranging from 10 to 30 min (see Note 18).

9. When a majority of cells have formed protoplasts spin them down for 10 min at $1,000 \times g$, discard s/n, resuspend pellet in 1 mL SCS, and transfer this suspension to a 1.5-mL microfuge tube.

10. Spin protoplasts down in a microcentrifuge for 5 min at $1,000 \times g$ at RT (see Note 19), carefully remove s/n, and resuspend protoplasts in 1 mL SCS.

11. Spin down as above, resuspend in 1 mL STC, and spin down in the same conditions.

12. Resuspend protoplasts, on ice, in 1 mL of ice cold STC. Protoplasts are now ready for transformation; they can be either used fresh or stored at −80 °C for future transformations. For freezing protoplasts add filter sterilized DMSO to 7% (70 μL for 1 mL of protoplast suspension) and dispense in aliquots.

13. For transformation, in a 1.5-mL microcentrifuge tube combine approximately 1 μg of transforming linear DNA in a maximum volume of 5 μL (from step 4, in the sect. Preparation of DelsGate Deletion Construct for Fungal Transformation), 1 μL heparin (15 mg/mL in STC; stored at 4 °C), and 50 μL of protoplast suspension, mix gently, and incubate on ice for 20 min.

14. Add 500 μL of PEG 4000 solution, mix by inverting the tube several times or by very gentle pipetting, and incubate on ice for 30 min.

15. Add 500 μL STC, mix by inverting the tubes several times and pellet protoplasts by centrifuging at $1,000 \times g$ for 5 min (see Note 20).

16. Aspirate off the s/n carefully and resuspend protoplasts in 200 μL of STC.

17. Plate cells on YEPS-S plates containing the appropriate selection (carboxin or hygromycin) and incubate plates at 30 °C, checking them every 24 h. Usually transformants start appearing after 4–5 days.

18. When transformants start to appear transfer them to 2PDA with the appropriate selection.

Agrobacterium tumefaciens-Mediated Fungal Transformation

1. For *A. tumefaciens*-mediated transformation of *V. dahliae* we use the method described by Mullins et al. [8] adapted for *V. dahliae* by Dobinson et al. [9] with minor variations.

2. On day 1, streak the *A. tumefaciens* strain harboring the deletion construct on LB supplemented with 50 μg/mL of spectinomycin, and grow for two days at 28 °C.

3. On day 2, inoculate 5 mL of YEPS (see Note 21) with fragments of *V. dahliae* mycelium from the edge of an actively growing colony, and incubate culture for 3–4 days at 24 °C and 250 rpm, to produce conidia.

4. On day 3, inoculate 5 mL of MM supplemented with 50 μg/mL spectinomycin with a colony of the *A. tumefaciens* strain containing the deletion construct. Grow at 28 °C and 250 rpm for 2 days.

5. On day 5, measure O.D. of the *A. tumefaciens* culture at 600 nm, harvest cells by centrifugation at $5,500 \times g$ for 10 min, discard supernatant, and resuspend cells in 500 μL of IM.

6. Dilute the *A. tumefaciens* suspension in 5 mL IM supplemented with 50 μg/mL spectinomycin, to a final O.D. of 0.15, and grow for 6–7 h to an optical density, at 600 nm, of 0.6–0.8 (see Note 22).

7. On day 6, harvest *V. dahliae* conidia by filtration through Miracloth (Calbiochem)

and then centrifugation at $5,500 \times g$ for 10 min. Finally, resuspend conidia in sterilized dH$_2$O to a final concentration of 10^6–10^7 cells/mL.

8. In a 1.5-mL microcentrifuge tube mix 100 μL of spore suspension and 100 μL of *A. tumefaciens* culture.

9. Spread mixture on a plate of cocultivation medium overlaid with a sterile 0.45 μm pore nitrocellulose membrane. Incubate for 2 days at RT.

10. Cut nitrocellulose membrane in strips and distribute these strips on 2 plates of selection medium.

11. When fungal transformants start to appear (see Note 23) transfer them individually to 2PDA supplemented with 50 μg/mL hygromycin.

12. Prepare single spore cultures of each transformant in preparation for analysis to determine absence of the gene of interest.

Analysis of Fungal Transformants to Confirm Gene Deletion

1. Analysis of fungal transformants for gene deletion has been standardized by performing PCR with a gene-specific primer located approximately 1.1 kb upstream of the start codon (i.e., outside the deletion construct) (gene-specific primer 5), combined with a primer that anneals to the selectable marker (see Fig. 34.2c, d). The marker primers used are: CbxF-DG when using the *U. maydis* pDONR-Cbx vector, Hyg-DG when using DelsGate vectors containing *hygR* as selectable marker, and Hyg-R(210) for OSCAR (see Table 34.1).

2. The reaction for each transformant to be analyzed is performed in a total volume of 25 μL that contains: 2.5 μL 10× *taq*-polymerase reaction buffer, 2.5 μL 25 mM MgCl$_2$ (see Note 3), 0.5 μL 50× dNTP mix (10 mM each), 1.25 μL 20 μM gene-specific primer, 1.25 μL selectable marker primer 20 μM, 0.5 μL *taq*-polymerase, 1.0 μL genomic DNA (10–100 ng), PCR-grade ddH$_2$O to 25 μL.

3. Prepare a master mix containing all the components except the DNA, for all the reactions you need to perform plus one. For each reaction mix 24 μL of the master mix with 1 μL of DNA.

4. Amplification is then performed under the following conditions: an initial denaturation of 1 min at 94°C, 30 cycles of 30 s at 94°C, 30 s at 60°C, and 90 s at 72°C, and completed with a final extension of 5 min at 72°C.

5. After PCR, load 10 μL of each PCR product on a 0.8% agarose gel; only transformants in which the gene has been replaced by the deletion construct through homologous recombination are expected to produce a band with this primer combination.

6. When using this approach to test for gene deletion it is good practice to confirm absence of the gene of interest in those transformants that generated a band in the above PCR by designing primers that anneal within the ORF. Perform this PCR in the manner described in steps 2, 3, and 4. Use a wild type strain as positive control in the amplification.

7. Replacement of the gene ORF by the deletion construct through homologous recombination can also be tested using primers with the same sequence as the gene-specific primers 2 and 3 but in the opposite orientation, we name these primers 2-O and 3-O (see Fig. 34.2c, d) (see Note 24).

8. Prepare reactions like in step 2 but substituting gene-specific and marker primers by primers 2-O and 3-O. Proceed as in step 3.

9. Amplification is then performed under the following conditions: an initial denaturation of 1 min at 94°C, 40 cycles of 30 s at 94°C, 30 s at 60°C, and 2–5 min at 72°C (see Note 25), and completed with a final extension of 7 min at 72°C.

10. After PCR load 18 μL of each PCR product on a 0.8% agarose gel. In transformants in which gene replacement has taken place you should observe absence of the wild type band and presence only of the deletion construct band, which ranges in size from about

3.4–4.4 kb in DelsGate, depending of which pDONR vector was used, and approximately 1.5 kb in OSCAR.

11. Further confirmation of gene of interest deletion is carried out by Southern blot hybridization.

Notes

1. While the rest of media are generally stored at 4 °C, PDB tends to precipitate in cold and it is stored at RT.

2. This is a very dense solution; to prepare it mix the PEG 4000, sorbitol, and dH_2O and directly autoclave. After autoclaving a clear viscous solution is obtained to which the rest of components are added.

3. Addition of drops of NaOH to increase the pH helps bring the MES into solution.

4. If the 10× *taq*-polymerase reaction buffer used already contains the $MgCl_2$, do not add any extra to the reaction and increase volume of ddH_2O accordingly.

5. As with any PCR, the annealing temperature may have to be adjusted for your particular primer combination. However, 60 °C generally works well for the primer lengths proposed.

6. Although PCR products can be used directly for BP clonase reactions, if not eliminated from the sample, primers can also recombine with vectors carrying the *attP* sites and greatly increase the background of colonies that do not harbor the desired construct. We therefore strongly recommend performance of this clean up step.

7. The PEG/$MgCl_2$ precipitation removes primer-dimers and DNA molecules smaller than 300 bp. Purification through QIAquick columns eliminates primers and DNA molecules smaller than 100 bp. We favor the use of PEG/$MgCl_2$ precipitation since it is an inexpensive method and produces very clean DNA.

8. It is important that both flanks are roughly at the same concentration during the BP clonase reaction; therefore if one of the flanks was amplified more efficiently it is advised to mix appropriate volumes of each PCR product to result in similar molar ratios in the combined purified sample.

9. Vector DNA suitable for the BP clonase reaction can be purified using standard methods. This includes alkaline lysis; however this DNA cannot be quantified by A_{260} due to contamination with RNA and we therefore prefer to use the QIAprep miniprep method from QIAGEN or any other "clean" plasmid purification method.

10. This incubation time can be reduced to as short as 1 h. However, we favor overnight incubations which increase the number of colonies by five- to tenfold.

11. Transformation can also be performed by electroporation; however in our experience this method increases the background of colonies that do not have the right construct.

12. We have consistently found that selection works much better on low Na LB than on standard LB. In standard LB sometimes colonies appear that contain no plasmid.

13. Any method for purification of plasmid DNA compatible with digestion with restriction enzymes can be used. At this stage we do not use QIAGEN columns since of the plasmids analyzed, only one with the correct structure will be selected as the deletion construct for further work.

14. Standard plasmid miniprep alkaline lysis methods do not allow quantification of plasmid DNA by measuring O.D. as the samples contain RNA. In that case, after extracting DNA, load an aliquot on an agarose gel and quantify approximate concentration visually.

15. We start with 3–5 μg of plasmid DNA to make sure that after the precipitation step we have enough DNA for transformation. We regularly use ≥1 μg of linear DNA for each transformation.

16. Protocols to extract plasmid DNA do not work well with *A. tumefaciens*; thus we have found that the best way to confirm the presence of deletion construct is by colony PCR.

17. The DelsGate method can be used with any organism with an efficient recombination system. We have used deletion constructs

generated by DelsGate for PMT to produce deletion mutants in *U. maydis* and this is the methodology we describe here. For other fungi, aspects such as formation of protoplasts or amount of linear DNA that give efficient rates of fungal transformation may vary.

18. Since the sale of Novozyme, previously used for the formation of protoplasts, was discontinued we have tried several commercial enzyme mixtures; we have found that Vinoflow generates protoplasts very efficiently and that Lallzyme MMX, although requiring longer incubation times, also generates protoplasts with high transformation frequencies.

19. Be certain to use a variable speed centrifuge and that you have it set at the appropriate setting since centrifugation at higher speed at this stage will burst protoplasts.

20. STC is added to decrease the density of the solution to allow protoplasts to pellet.

21. Other authors use Complete Medium (CM) [9]. We have found that both media, YEPS and CM, are suitable to produce conidia for ATMT.

22. Sometimes longer incubations, such as to grow o/n, are required to reach this O.D.

23. Transformants are visible growing from the edges of the strips of nitrocellulose membrane into the agar.

24. This approach can only be used for DelsGate when the length of the deletion is different from that of the vector in the deletion construct. Therefore it is not suitable for genes of about 4.4 kb when using pDONR-Cbx to generate the deletion construct, for genes of about 4.8 kb when using pDONR-Hyg, and for genes of about 3.5 kb when using pDONR-A-Hyg. For OSCAR this approach can only be used when the length of the ORF is different from the *hygR* marker, approximately 1.5 kb. However, when it can be applied we consider this approach desirable because whether there was homologous or ectopic integration a positive PCR result is expected, allowing these events to be distinguished by amplicon length. We therefore find this approach very accurate in testing for

gene deletion by PCR. Its disadvantage is that larger bands need to be amplified; however, except for large ORFs the expected size of bands is still within the range that can be amplified with standard *taq*-polymerase-based PCR reactions.

25. Because larger bands need to be amplified with this approach, longer extension times are used.

References

1. García-Pedrajas MD, Nadal M, Kapa LB, Perlin MH, Andrews DL, Gold SE (2008) DelsGate, a robust and rapid gene deletion construction method. Fungal Gen Biol 45:379–388

2. Paz Z, García-Pedrajas MD, Andrews DL, Klosterman SJ, Baeza-Montañez L, Gold SE (2011) One Step Construction of Agrobacterium-Recombination-ready plasmids (OSCAR), an efficient and robust tool for ATMT based gene deletion construction in fungi. Fungal Gen Biol 48:677–684

3. Walhout AJ, Temple GF, Brasch MA, Hartley JL, Lorson MA, van den Heuvel S et al (2000) GATEWAY recombination cloning: application to the cloning of large numbers of open reading frames or ORFeomes. Methods Enzymol 328:575–592

4. Suzuki Y, Kagawa N, Fujino T, Sumiya T, Andoh T, Ishikawa K et al (2005) A novel high-throughput (HTP) cloning strategy for site-directed designed chimeragenesis and mutation using the Gateway cloning system. Nucleic Acids Res 33:e109

5. Hellens R, Mullineaux P, Klee H (2000) Technical focus: a guide to Agrobacterium binary Ti vectors. Trends Plant Sci 5:446–451

6. Kmetzsch L, Staats CC, Simon E, Fonseca FL, Oliveira DL, Joffe LS et al (2011) The GATA-type transcriptional activator Gat1 regulates nitrogen uptake and metabolism in the human pathogen *Cryptococcus neoformans*. Fungal Gen Biol 48:192–199

7. Tsukuda T, Carleton S, Fotheringham S, Holloman WK (1988) Isolation and characterization of an autonomously replicating sequence from *Ustilago maydis*. Mol Cell Biol 8:3703–3709

8. Mullins ED, Chen X, Romaine P, Raina R, Geiser DM, Kang S (2001) *Agrobacterium*-mediated transformation of *Fusarium* oxysporum: an efficient tool for insertional mutagenesis and gene transfer. Phytopathology 91:173–180

9. Dobinson KF, Grant SJ, Kang S (2004) Cloning and targeted disruption, via *Agrobacterium tumefaciens*-mediated transformation, of a trypsin protease gene from the vascular wilt fungus *Verticillium dahliae*. Curr Genet 45:104–110

Improved Transformation Method for *Alternaria Brassicicola* and Its Applications

Yangrae Cho, Akhil Srivastava, and Christopher Nguyen

Abstract

Alternaria brassicicola is a filamentous fungus that causes black spot disease on most plants in the Brassicaceae, including cultivated *Brassica* species and weedy Arabidopsis. Since the concept of transformation constructs of linear minimal elements was developed [1], we have optimized protoplast production by reducing melanin accumulation in fungal mycelium. We have used this method either to delete targeted genes or to insert exogenous genetic information in almost any location of research interest.

Key Words

Alternaria brassicicola • Transformation • Melanin • Protoplast production

Introduction

The imperfect filamentous fungus, *Alternaria brassicicola* (Schwein, Wiltshire), causes black spot disease on a broad range of cultivated and weedy members of the Brassicaceae. Of note, *A. brassicicola* has been used as an example of a true necrotrophic fungus in studies with Arabidopsis. Since genome sequences and functional methodologies have been developed for both the plant[1] and its pathogen,[2] this has become an attractive system for the discovering events that occur at the host–pathogen interface. These events ultimately determine the outcome of the interaction.

Knocking out gene functions either by targeted gene disruption or gene replacement [1–5] has been very helpful for targeted mutational analysis. The targeted gene approach will be especially powerful when combined with large numbers of expressed sequence tags using rapidly evolving technologies. They include serial analysis of gene expression (SAGE) [6], cap analysis of gene expression (CAGE) [7], massively parallel signature sequencing (MPSS) [8],

Y. Cho (✉) • A. Srivastava • C. Nguyen
Department of Plant and Environmental Protection Sciences, University of Hawaii at Manoa, 3190 Maile Way, Honolulu, HI 96822, USA
e-mail: yangrae@hawaii.edu

[1] http://www.arabidopsis.org/
[2] http://genome.jgi-psf.org/Altbr1/Altbr1.home.html

V.K. Gupta et al. (eds.), *Laboratory Protocols in Fungal Biology: Current Methods in Fungal Biology*,
Fungal Biology, DOI 10.1007/978-1-4614-2356-0_35, © Springer Science+Business Media, LLC 2013

and RNA-seq [9]. Targeted gene knockout methods have been widely used within the fungal research community. Investigators have adopted a similar approach for other filamentous fungi, including plant pathogens such as *Magnaporthe grisea*, *Fusarium oxysporum*, and *Colletotrichum lagenarium* [10–12]. Although there were previous reports on the transformation of Alternaria species, including *A. brassicicola* [13–15], studying gene functions by targeted gene knockout has remained difficult owing to the low efficiency of both transformation and targeted integration. To improve targeted gene disruption efficiency and expedite production of gene disruption constructs, a short linear construct with minimal elements that contains an antibiotic resistance selectable marker gene and a 250–600 bp-long partial target gene, were developed and successfully used with an almost 100 % targeted gene efficiency [1]. The production of protoplasts, however, continued to be a limiting factor in high throughput transformation.

During the last several years we have optimized a method of producing large numbers of protoplasts. We used the protoplasts to detect pathogenicity associated genes among 200 transcription factor-coding genes. We modified protoplast production by inoculating an artificial medium with hyphae instead of conidia and then growing the mycelium under conditions that reduced melanin production.

Melanin is a ubiquitous pigment that plays an important role in protecting fungi from the damaging effects of environmental stress. It is produced by and accumulates in the cell walls of hyphae and conidia during the late stationary phase of mycelial growth. Melanin also forms under stressful conditions, including ultraviolet irradiation, a hyperosmotic environment, nutrient deficiency, or an accumulation of toxic wastes during in vitro culture. Melanin increases the tolerance of fungi to UV irradiation [16–18], enzymatic lysis [19], and extreme temperatures [17–20]. Previously, we grew mycelia for 7 days to acquire an adequate biomass. After 7 days the mycelia were moderately melanized and digesting them for 4 h produced about ten million protoplasts. With our new method, we inoculate a nutrient broth with hyphae instead of conidia.

After only 2 days in culture and one change of medium at 24 h, melanin accumulation in the *A. brassicicola* mycelium is dramatically reduced. We then digest the mycelium with the same amount of enzyme for 1–1.5 h. With this method, we can produce up to 200 million protoplasts and use them to transform up to 20 constructs.

We have observed an increase in transformation and targeted gene efficiency using protoplasts from mycelia with reduced melanin. We speculate that reducing melanin accumulation during mycelial growth was an important factor in this increased efficiency. We use this method of protoplast production not only for the transformation of linear minimal element constructs for gene disruption, but also for linear gene deletion constructs with a selectable marker gene flanked on both sides by a targeted gene sequence. We also use the protoplasts to transform various constructs that are designed to tag targeted promoters or targeted coding sequences. They can be used to investigate the spatial and temporal expression patterns of the gene or the localization of the tagged proteins, or to purify tagged proteins (Fig. 35.1).

Materials

The following solutions need to be prepared several days before the transformation procedures.

1. STC buffer: 1 M Sorbitol (54.615 g), 50 mM Tris (15 mL of 1 M Tris, pH 8.0), 50 mM $CaCl_2$ (1.66485 g) in 300 mL distilled water. Stored at 4 °C, it can be used for more than 1 year.

2. 40 % PEG in STC: Dissolve 40 g of PEG 3500 in STC with a final volume of 100 mL. Stored at 4 °C, it can be used for up to 1 year.

3. Molten regeneration medium: 0.5 % yeast extract (1.5 g), 0.5 % casamino acid (1.5 g), 1 M sucrose (102.6 g), 0.8 % agar (2.4 g) in 300 mL distilled water. Autoclave to liquefy the medium. Cool to about 40–45 °C to pour. Medium stored at room temperature can be used for several months.

4. 0.7 M NaCl: 140 mL of 5 M NaCl + 860 mL distilled water. Autoclave. Stored at 4 °C, it can be used for more than 1 year.

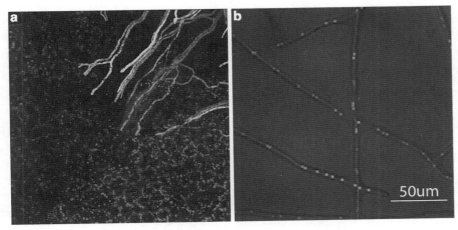

Fig. 35.1 Confocal microscopy showing expression of a targeted gene tagged with green fluorescent protein (GFP). The green fluorescence appears white in this image. (**a**) Hyphae (*upper right* corner) invading a host plant leaf. The GFP gene was inserted at the end of a targeted promoter by replacing the targeted gene. The expression of the GFP gene is controlled by the targeted gene promoter. (**b**) Bright dots represent the nuclei in hyphal cells. The GFP protein gene was fused to the end of the targeted gene coding sequence. The targeted gene was expressed as a fusion protein along with the GFP protein. The targeted gene expression and its protein localization were examined in real time

5. PDA with 30 ng/mL Hygromycin B: Add 19.5 g of Potato Dextrose Agar powder to 500 mL distilled water and autoclave.
6. 1 % Glucose and 0.5 % yeast extract broth in 250 mL flasks.
7. DNA constructs to transform. 5–10 μg of DNA constructs in 10 μL water in each 1.5 mL aliquot. Stored constructs at −20 °C and put on ice before the procedure.

The following equipment and materials are needed:
1. Kitalase (Wako cat #114-00373)
2. Shaker/incubator
3. Centrifuge (Eppendorf 5180R or equivalent)
4. Water bath
5. An incubator or a second water bath
6. Hemocytometer
7. Sterile 50-mL tubes and 12 or 15 mL tubes with screw caps
8. Tube holder(s) for 1.5, 12, and 50 mL-sized tubes
9. Miracloth (Calbiochem cat #475855)
10. 0.2 μm Millex-FG syringe-driven filter (Millipore, Carritowhill, Ireland)
11. 10 mL Syringe
12. Funnel
13. Sterilized 250-mL flasks with aluminum foil covers
14. Scalpel
15. Ice box and ice
16. Petri plates

Methods

The following method is written for people with little laboratory experience. It may be used to train new students, but it may be too detailed for experienced researchers. This method can be used by the latter, however, as a guideline for making simplified protocol. This method was also tested with *Alternaria alternata* and *Maganporthe grisea* to increase the efficiency of protoplast production and transformation. Note: Contamination of the environment should be avoided. Do all work with conidia in a biological safety cabinet. Always perform experiments using proper sterilization techniques. For example, avoid touching the openings of flasks, tubes, and caps with your fingers; flame the rims of flask or tube openings before opening and closing them.

Fungal Culture

1. Sterilize the scalpel blade by dipping it in ethanol and then igniting it with a flame. Put the scalpel to the side and allow it to cool.
2. Remove the lid of a Petri plate containing sporulating *A. brassicicola*. Before actually touching the fungus, insert the scalpel blade into the agar near the edge of the plate. This will further cool the blade.
3. Gently scrape the conidia toward the edges of the plate, being careful not to disturb the agar. You want the hyphae growing in the agar for protoplast production, not the conidia.
4. After clearing conidia to the side, collect only hyphae by gently scraping, not slicing the hyphae-containing agar from the surface. Note: Each piece of agar will serve as the core of a mycelia ball. If the agar pieces are small, there will be many small mycelia balls; and if the agar pieces are large, there will be only a few large mycelia balls. Harvest enough mycelium-impregnated agar to produce 3–5 mL of fungal biomass after washing with NaCl (see Transformation, step 8).
5. Transfer the agar pieces containing mycelium into a 250-mL flask containing 50 mL of 1 % glucose and 0.5 % yeast extract broth (GYEB). Shake the scalpel to release the agar into the broth. Flame the scalpel before each transfer of agar and hyphae to the flask.
6. Lightly swirl the flask, flame the rim of the opening, and then seal it with foil. Flaming the rim of the flask opening and aluminum cover will help prevent contamination.
7. Label the foil cover with the sample name, type of medium, date, and time.
8. Incubate the flasks in the shaker/incubator at 25 °C and 100 rpm in the dark.
9. Let the fungal mycelium grow for about 36–48 h.
10. Optional: the day before the transformation procedure, replace the old GYEB with fresh GYEB and continue to incubate the culture.
11. After incubation the broth should contain separate, spiky, milky white mycelia balls. A grey color to the mycelia or conidia is a sign of melanin accumulation, which inhibits enzyme digestion, as described in the following section.

Transformation

1. Turn on the water bath and set it at 42 °C. Gather the necessary materials and turn on the biosafety cabinet.
2. Place the molten regeneration medium in the 40–45 °C water bath or incubator.
3. Put the prepared 0.7 M NaCl in the ice box to keep it cold.
4. Remove the incubating mycelium from the shaker/incubator and pour it into a sterilized 50 mL falcon tube. Centrifuge at 3,600 rpm for 5 min.
5. While the mycelium is being centrifuged, prepare 10 mg/mL Kitalase. Measure 100 mg of the Kitalase powder and pour it into a sterilized 12 mL polypropylene tube; add 0.7 M NaCl to make 10 mL. Place the tube in the shaker/incubator until the contents are completely dissolved.
6. For the mycelium harvested from the 50 mL falcon tube, use 10 mL of Kitalase solution.
7. Pour off the GYEB medium and fill the tube with 0.7 M NaCl. Centrifuge at 3,600 rpm for 5 min and then decant the NaCl solution.
8. Continue washing the mycelium by repeating step 6. We prefer to have 3–5 mL of fungal biomass when the two washings with NaCl are finished.
9. Get a sterile 250-mL flask, 20-μm pore size filter, and a syringe kit and prepare to filter-sterilize the Kitalase solution as it is added to the culture sample. Do not touch the filter with your hands or allow anything else to touch it. Grip it through the plastic wrap.
10. After washing the culture sample with NaCl (step 8), open the syringe and pull out the plunger, attach the filter to the front of the syringe, and pour the Kitalase into the syringe. Reassemble the syringe, then apply pressure to the plunger to force the Kitalase through the filter and into the washed mycelium.

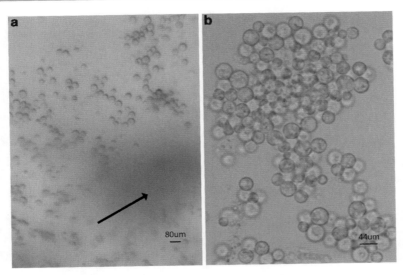

Fig. 35.2 Protoplasts produced by enzyme digestion. (a) Protoplasts before purification. Digestion is complete and the protoplasts in the digestion solution are clean enough to be used for transformation after precipitation by centrifugation. Floating cell wall debris (*arrow*) does not inhibit transformation. (b) Protoplasts after purification. The size of protoplasts varies from less than 10 to almost 50 μm in diameter

11. Pour the washed mycelium and Kitalase into a sterile 250-mL flask. Flame the mouth of the flask and the foil, then cover. Place the flask in the shaker/incubator and set the temperature at 28 °C. Digestion of the cell walls of the mycelium will be complete within 90 min. Check the progress of the digestion after 1 h by removing a 10-μL sample from the flask and viewing it under a compound microscope. Undigested hyphae look like long, smooth hairs. Partially digested hyphae look like strings of sausages. Completely digested hyphae will normally float like cotton balls and the protoplasts are well separated from the floating cell-wall debris (Fig. 35.2).

12. During the digestion process, pour 50 mL of molten regeneration medium into each 50 mL tube and incubate in the 40–45 °C water bath or incubator. The number of 50-mL tubes needed depends on the number of samples being prepared for transformation.

13. Once the fungal mycelium has been properly digested, filter the protoplasts through a miracloth-lined funnel and collect them in a sterile 12-mL polypropylene tube. Gently swirl the miracloth funnel while filtering. Wash the miracloth funnel with a small volume of 0.7 M NaCl to recover uncollected protoplasts. Add 0.7 NaCl until the 10 mL line of the polypropylene tube is reached.

14. Centrifuge the polypropylene tube at 4 °C and 700 rpm for 10 min. Decant the liquid, add 3 mL of cold 0.7 M NaCl and resuspend the protoplasts by gently inverting the tube back and forth. Ideally, the pellet of protoplasts will remain attached to the bottom of the tube during the inverting. If the pellet breaks loose, it may be difficult to completely resuspend the protoplasts. Gentle inversion will slowly release the pellet into the liquid. Add the remaining 0.7 M NaCl until the volume reaches 10 mL.

15. Centrifuge again at 4 °C and 700 rpm for 10 min. Decant the liquid and add 3 mL of STC buffer and gently invert to mix. When the pellet is completely resuspended, add STC to make 10 mL and mix well. Collect 10 μL of the sample and place it in a hemocytometer. Centrifuge again at the same settings to precipitate the protoplasts.

16. Count the number of protoplasts with the hemocytometer. Five to ten million protoplasts will be needed for each transformation construct; more is better than less.

17. When this centrifugation is done, decant the liquid and add STC to make a concentration of ~10 million protoplasts in 70 μL.

18. Add the 70 μL of protoplasts using 1-mL pipette to the DNA transformation constructs in 1.5-mL aliquots. Pipetting with a smaller pipette may damage protoplasts. Smaller pipettes can be used after cutting off the pipette tips to make the opening wider. Place the aliquots on ice for 10 min. Move the ice box next to the water bath so aliquots can be quickly transferred from the ice to the water bath. Place the aliquot holder in the ice box and insert the aliquots as they are filled. Move the aliquot holder and aliquots to the 42 °C water bath and incubate for 2–10 min.

19. Remove the aliquots from the water bath and quickly put them back on ice for 10 min. Collect the Petri plates that will be needed.

20. Add 40 % PEG in STC at room temperature to the aliquots. Mix by gently inverting each tube. Incubate for 15 min at room temperature.

21. Transfer 400 μL of the sample to a 50-mL tube of molten regeneration medium. Thoroughly mix the protoplasts and PEG into the molten regeneration medium.

22. Pour each 50-mL tube of molten regeneration medium equally into two 100×15 mm Petri plates.

23. Incubate these plates in an incubator at 25 °C for 16–24 h.

24. Overlay the plates with 25 mL of PDA with Hygromycin B (30 ng/mL).

25. Transformants will begin to appear in 5–7 days and continue to emerge for about 2 weeks (Fig. 35.3). They are ready for single-spore purifications as they emerge from the selective medium. Cross-contamination of the transformation plates is rare as long as the plates are kept undisturbed.

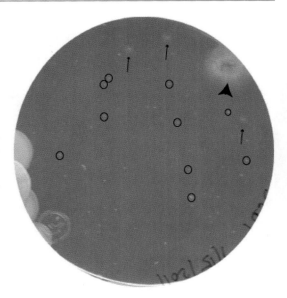

Fig. 35.3 Transformation plate 4 days after overlaying it with the selective medium. The *arrowhead*, *arrows*, and *circles* mark emerged transformants, growing transformants in the selective medium, and putative transformants, respectively, that may grow within a few days

Acknowledgements We thank Lindsay Oxalis for assisting with the research and Dr. Fred Brooks for his critical review of the manuscript. This research was supported by USDA-TSTAR 2009-34135-20197 and HATCH funds to Yangrae Cho, administered by the College of Tropical Agriculture and Human Resources, University of Hawaii at Manoa, Honolulu, HI.

References

1. Cho Y, Davis JW, Kim KH, Wang J, Sun QH, Cramer RAJ et al (2006) A high throughput targeted gene disruption method for Alternaria brassicicola functional genomics using linear minimal element (LME) constructs. Mol Plant Microbe Interact 19:7–15

2. Adachi K, Nelson GH, Peoples KA, Frank SA, Montenegro-Chamorro MV, DeZwaan TM et al (2002) Efficient gene identification and targeted gene disruption in the wheat blotch fungus Mycosphaerella graminicola using TAGKO. Curr Genet 42:123–127

3. Hamer L, Adachi K, Montenegro-Chamorro MV, Tanzer MM, Mahanty SK, Lo C et al (2001) Gene discovery and gene function assignment in filamentous fungi. Proc Natl Acad Sci U S A 98:5110–5115

4. Lo C, Adachi K, Shuster JR, Hamer JE, Hamer L (2003) The bacterial transposon Tn7 causes premature polyadenylation of mRNA in eukaryotic organisms: TAGKO mutagenesis in filamentous fungi. Nucleic Acids Res 31:4822–4827

5. Yang L, Ukil L, Osmani A, Nahm F, Davies J, De Souza CP et al (2004) Rapid production of gene replacement constructs and generation of a green fluorescent protein-tagged centromeric marker in Aspergillus nidulans. Eukaryot Cell 3:1359–1362

6. Velculescu VE, Zhang L, Vogelstein B, Kinzler KW (1995) Serial analysis of gene expression. Science 270:484–487

7. Shiraki T, Kondo S, Katayama S, Waki K, Kasukawa T, Kawaji H et al (2003) Cap analysis gene expression for high-throughput analysis of transcriptional starting point and identification of promoter usage. Proc Natl Acad Sci U S A 100:15776–15781

8. Brenner S, Johnson M, Bridgham J, Golda G, Lloyd DH, Johnson D et al (2000) Gene expression analysis by massively parallel signature sequencing (MPSS) on microbead arrays. Nat Biotechnol 18:630–634

9. Wang Z, Gerstein M, Snyder M (2009) RNA-Seq: a revolutionary tool for transcriptomics. Nat Rev Genet 10:57–63

10. Parsons KA, Chumley FG, Valent B (1987) Genetic transformation of the fungal pathogen responsible for rice blast disease. Proc Natl Acad Sci U S A 84:4161–4165

11. Malardier L, Daboussi MJ, Julien J, Roussel F, Scazzocchio C, Brygoo Y (1989) Cloning of the nitrate reductase gene (niaD) of Aspergillus nidulans and its use for transformation of Fusarium oxysporum. Gene 78:147–156

12. Perpetua NS, Kubo Y, Okuno T, Furusawa I (1994) Restoration of pathogenicity of a penetration-deficient mutant of Collectotrichum lagenarium by DNA complementation. Curr Genet 25:41–46

13. Tsuge T, Nishimura S, Kobayashi H (1990) Efficient integrative transformation of the phytopathogenic fungus Alternaria alternata mediated by the repetitive rDNA sequences. Gene 90:207–214

14. Shiotani H, Tsuge T (1995) Efficient gene targeting in the filamentous fungus Alternaria alternata. Mol Gen Genet 248:142–150

15. Yao C, Köller W (1995) Diversity of cutinases from plant pathogenic fungi: different cutinases are expressed during saprophytic and pathogenic stages of Alternaria brassicicola. Mol Plant Microbe Interact 8:122–130

16. Kawamura C, Tsujimoto T, Tsuge T (1999) Targeted disruption of a melanin biosynthesis gene affects conidial development and UV tolerance in the Japanese pear pathotype of Alternaria alternata. Mol Plant Microbe Interact 12:59–63

17. Rehnstrom AL, Free SJ (1996) The isolation and characterization of melanin-deficient mutants of Monilinia fructicola. Physiol Mol Plant Pathol 49:321–330

18. Wang Y, Casadevall A (1994) Decreased susceptibility of melanized Cryptococcus neoformans to UV light. Appl Environ Microbiol 60:3864–3866

19. Hyakumachi M, Yokoyma K, Ui T (1987) Role of melanin in susceptibility and resistance of Rhizoctonia solani to microbial lysis. Trans Br Mycol Soc 89:27–33

20. Rosas AL, Casadevall A (1997) Melanization affects susceptibility of Cryptococcus neoformans to heat and cold. FEMS Microbiol Lett 153:265–272

Vijai Kumar Gupta, Maria G. Tuohy, and Rajeeva Gaur

Abstract

Because of breakage of rigid fungal cell walls, the major challenge for DNA isolation from fungi is to obtain samples of good quality and quantity. We developed a fast and efficient DNA isolation method from fungi that was successfully applied to *Fusarium* spp., *Colletotrichum* spp., *Trichoderma* spp., *Gliocladium roseum*, and *Lasiodiplodia theobromae*.

Keywords

DNA isolation • Fungi • *Fusarium* spp. • *Colletotrichum* spp. • *Trichoderma* spp. • *Gliocladium roseum* • *Lasiodiplodia theobromae*

Introduction

Fungal biology has become important in relation to the study of evolutionary biology and in the industrial process. Consequently, there is a need to investigate the molecular genetics of this fungus. A wide range of molecular manipulation techniques of fungi are currently available, including gene disruption, various PCR applications (random amplified polymorphic DNA analysis, microsatellite typing, etc.), and DNA-based epidemiological studies (restriction fragment length polymorphism analysis, fingerprinting, etc.). [1–3] Each of these techniques requires the recovery of good-quality genomic DNA. Most DNA extraction protocols for fungi are based on mechanical isolation methods that employ grinding mycelia after freezing them in liquid nitrogen or glass bead disruption, followed

V.K. Gupta (✉)
Molecular Glycobiotechnology Group,
Department of Biochemistry,
School of Natural Sciences,
National University of Ireland Galway,
University Road, Galway, Ireland

Assistant Professor of Biotechnology, Department
of Science, Faculty of Arts, Science & Commerce,
MITS University, Rajasthan, India
e-mail: vijai.gupta@nuigalway.ie; vijaifzd@gmail.com

M.G. Tuohy
Molecular Glycobiotechnology Group, Department
of Biochemistry, School of Natural Sciences,
National University of Ireland Galway,
University Road, Galway, Ireland

R. Gaur
Department of Microbiology, Dr. R.M.L. Avadh
University, Faizabad, Uttar Pradesh, 224001, India

V.K. Gupta et al. (eds.), *Laboratory Protocols in Fungal Biology: Current Methods in Fungal Biology*,
Fungal Biology, DOI 10.1007/978-1-4614-2356-0_36, © Springer Science+Business Media, LLC 2013

by additional purification steps. Although the quantity and quality of DNA obtained by these methods are generally satisfactory, the techniques are time-consuming and, therefore, not suitable for analysis of a large number of samples. Consequently, new protocols have been developed in order to reduce the time required for DNA isolation. [4, 5] However, most of these protocols involve extractions using hazardous chemicals such as phenol, chloroform, and isoamyl alcohol and are, therefore, not a preferred option.

In this chapter, we investigate a simple method for isolation of good-quality DNA from fungi. This method is convenient and easy to use, does not require physical methods, and works directly with mycelia.

Materials

1. 1 M Tris HCl
2. 0.5 M Na2 EDTA (pH-8.0)
3. 1 M NaCl
4. 0.5 % SDS
5. 25 mM EDTA
6. 200 mM Tris HCl
7. 2.5 M Sodium acetate (pH 5.2)
8. 10x TAE buffer
9. Extraction buffer: Prepared using 200 mM Tris HCl, 20 mM Nacl, 25 mM ethylenediaminetetraacetic acid (EDTA) and 1 % sodium dodecyl sulfate (SDS) (pH 8.5)
10. Saturated isopropanol
11. 70 % alcohol
12. Tris-EDTA (TE): Prepared by using 10 mM Tris and 1 mM EDTA, pH 8.0
13. 6× gel loading dye: Prepared using 25 mg bromophenol blue, 25 mg xylene cyanol, 3 mL glycerol in 6.75 mL DW (for 10 mL of 6× loading dye). The dye was then autoclaved and stored at 4 °C
14. Ethidium bromide: 10 mg/mL stock

Methods

The methods given as follows describe general procedures for obtaining DNA from different fungal sources. The volumes and number of tubes

used per sample may need to be varied, depending on the type of sample and size. This method was employed to prepare DNA from *Fusarium* spp., including those of *Colletotrichum* spp., *Trichoderma* spp., *Gliocladium roseum,* and *Lasiodiplodia theobromae.* [6–8]

1. Grow fungal cultures in suitable broth for 72 h at 28±2 °C on a shaker at 120 rpm.
2. Decant the sample containing growth media including fungal growth in small centrifugal tubes as per size of the experiment. Pellet the sample by centrifugation at high speed (13,000 rpm) for 5 min. Wash with sterile distilled water for 5 min for two times at same speed (see Note 1).
3. Overdry the pelleted sample at 65 °C for 30 min to remove excess liquid (see Note 2).
4. Homogenize the required amount of the pellet in an Eppendorf tube/sterile tube for 5 min in a suitable ratio of fungal mycelia sample and extraction buffer (1:6), respectively (i.e., for our experiments we used 50 mg of mycelia with 300 μL of extraction buffer in an Eppendorf tube) (see Note 3).
5. Add 200 μL of 2.5 M sodium acetate, vortex the mixture, and cool the mixture at −20 °C for 10–15 min.
6. Centrifuge for 5 min at 13,000 rpm and transfer the supernatant in fresh Eppendorf tubes and then add equal volume of chilled saturated isopropanol. If the DNA is in good quantity, spool out the DNA into a fresh tube and follow step 8 (see Notes 4 and 5).
7. Centrifuge for 5 min at 13,000 rpm to pellet the DNA.
8. Supernatant was discarded without interfering with the DNA pellet.
9. Rinse the DNA pellet twice with 500 μL of 70 % ethanol for 5 min at 13,000 rpm (see Note 5).
10. DNA pellet was air dried in an aseptic condition for 1.5 h to remove traces of ethanol and was subsequently dissolved in 100 μL TE buffer.
11. The Eppendorf tubes were kept in water bath at 55 °C for 4 h to inhibit DNAse activity and to resuspend the pellet completely in TE buffer or allow the pellets to resuspend overnight at 4 °C.

12. After 4 h of resuspension of DNA in TE buffer at 55 °C in water bath or overnight resuspension at 4 °C, the TE-dissolved DNA was cooled and stored at −20 °C for further use.

13. Purify the DNA using DNA purification kits. We found PowerClean® DNA Clean-Up Kit (Mo Bio Laboratories, USA) to be very effective for purifying DNA (see Note 6).

14. To determine the approximate concentration of the DNA, take 1 μL of DNA solution and dilute it by adding 999 μL of Milli-Q water. Mix well and measure absorbance values at 260 nm and 280 nm in a UV spectrophotometer. For good-quality DNA the ratio is 1–1.8, but OD is not useful in the presence of any contaminants. Therefore, run 10 μL genomic (uncut) DNA using 6× gel loading dye on a 0.8 % agarose gel with ethidium bromide. The amount of fluorescence compared to the λ(lambda) DNA markers will provide a rough estimate of the DNA concentration (see Notes 7 and 8).

Notes

1. Washing fungal mycelia with sterile distilled water is necessary to remove excess metabolites attached with mycelia.

2. DNA pellets should not be overdried because this may cause a problem in dissolving the DNA in TE buffer. Undissolved residues should be removed using sterile pipette tips.

3. Completely mixing the pelleted mycelia sample with extraction buffer into a homogenate increases the efficiency of DNA extraction.

4. If the DNA does not spool out, do not discard; centrifugation helps to recover the DNA.

5. 70 % alcohol should be freshly prepared. Both saturated isopropanol and 70 % alcohol should be used after being chilled.

6. If the extracted DNA is not sufficiently pure (i.e., not suitable to restriction enzyme digestion and/or PCR amplification), repeat the saturated isopropanol extraction. Alternatively, use any standard DNA purification kit to purify DNA, following the manufacturer's instructions.

7. The ethidium bromide acts as an intercalating agent. It has four cyclic rings similar to purine and pyrimidine nitrogen bases of the nucleotides. After the addition of ethidium bromide to the gel and loading of the gel was over, the ethidium bromide slipped in between the adjacent base pairs of the DNA double helix. This intercalated ethidium bromide showed fluorescence in the presence of the UV light. The density of ethidium bromide is directly proportional to the concentration of DNA present in the given volume of sample loaded. Concentration of ethidium bromide prepared is 10 mg/mL, and it is autoclaved after preparation. Ethidium bromide helps us to analyze the position and concentration of DNA in the gel).

8. Note: Because λ(lambda) markers are the known molecular weight strands from λ(lambda) phage DNA, by using these markers one can quantitate the molecular weight of the DNA to be quantified. This is a very sophisticated and accurate method for quantitation of a genomic DNA as compared to UV spectroscopy.

Acknowledgements The authors are very thankful to Head, Department of Microbiology, R. M. L. Avadh University, Faizabad, UP, India, and Prof. Shakti Baijal, Ex-Dean, FASC, MITS University, Rajasthan, for providing the necessary research grants.

References

1. Raeder U, Broda P (1985) Rapid preparation of DNA from filamentous fungi. Appl Microbiol 1:17–20

2. Cubeta MA, Echandi E, Abernethy T, Vilgalys R (1991) Characterization of anastomosis groups of binucleate Rhizoctonia species using restriction analysis of an amplified ribosomal RNA. Phytopathology 81:1395–1400

3. Clarke DL, Woodlee GL, McClelland CM, Seymour TS, Wickes BL (2001) The Cryptococcus neoformans STE11alpha gene is similar to other fungal mitogen-activated protein kinase kinase kinase (MAPKKK) genes but is mating type specific. Mol Microbiol 40:200–213

4. Graham GC, Mayers P, Henry RJ (1994) A simplified method for the preparation of fungal genomic DNA for PCR and RAPD analysis. Biotechniques 16:48–50

5. Min J, Arganoza MT, Ohrenberger J, Xu C, Akins RA (1995) Alternative methods of preparing whole-cell

DNA from fungi for dot-blot, restriction analysis, and colony filter hybridization. Anal Biochem 225:94–100

6. Gupta VK (2009) Molecular characterization of Fusarium wilt pathogens of guava (*Psidium guajava* L.) using RAPD and microsatellite maker. Ph.D. Thesis. Dr. R.M.L. Avadh University, Faizabad, Uttar Pradesh, India

7. Gupta VK (2012) Genetic Diversity of *Fusarium* wilt pathogens of Guava in India. J Environ Sci Health [B] 47:315–325

8. Gupta VK (2012) Characterization of virulence gene loci in the genome of *Fusarium* spp. associated with wilt disease of guava in India. Arch Phytopathol Plant Protect (accepted), APPP ID: 45(2):244–259

Production of Recombinant Proteins from *Pichia pastoris*: Interfacing Fermentation and Immobilized Metal Ion Affinity Chromatography

37

Berend Tolner, Gaurav Bhavsar, Bride Foster, Kim Vigor, and Kerry Chester

Abstract

The methods describe a *Pichia pastoris* fermentation system for generation and purification of recombinant proteins. The proteins are secreted with hexahistidine tags and purified from feedstock by immobilized metal ion affinity chromatography (IMAC) using either radial flow or expanded bed adsorption. IMAC allows for an initial fast capture and isolation step that omits the need for filtration or centrifugation as primary procedures. The methods are applicable to production of recombinant protein in the laboratory and can be adapted to good manufacturing practice (GMP) compliant processes.

Keywords

P. pastoris • Recombinant • Protein • Purification • His-tag • IMAC • Radial flow bed • Expanded bed

Introduction

The use of *Pichia pastoris* as a production platform for recombinant proteins has been highly successful on both the laboratory and good manufacturing practice (GMP) bioprocess scale. The system can produce heterologous recombinant protein at yields of grams per liter [1–8]. To date some 600 genes have been cloned and expressed in *P. pastoris* [9], and the overall production of proteins from this yeast is likely to increase [10, 11].

Depending on application, yeast production can be an attractive alternative to mammalian cells or the commonly used *Escherichia coli* platform. Relative to production of recombinant proteins in mammalian cell lines, yeast expression has a shorter fermentation time, shows fast progression from gene synthesis to first-time protein production, makes use of simple chemically defined media without animal-derived products (exclusion of viral contaminants), and has an overall reduced operation cost [12]. *P. pastoris* performs posttranslational modifications such as disulphide bonds and glycosylation [3, 4, 5]. However, *Pichia* glycoforms are of the high mannose type and therefore differ from mammalian structures. This

B. Tolner (✉) • G. Bhavsar • B. Foster
K. Vigor • K. Chester
Department of Oncology, University College London
Cancer Institute, Paul O'Gorman Building, 72 Huntley
Street, London, WC1E 6BTUK
e-mail: b.tolner@ucl.ac.uk

V.K. Gupta et al. (eds.), *Laboratory Protocols in Fungal Biology: Current Methods in Fungal Biology*,
Fungal Biology, DOI 10.1007/978-1-4614-2356-0_37, © Springer Science+Business Media, LLC 2013

can be problematic for some in vivo therapeutic purposes because mannosylation facilitates accelerated clearance from circulation; in some cases, such accelerated clearance is a desirable design feature [13, 14]. *Pichia* glycoforms may also be disadvantageous if the protein required depends on specific glycosylation for optimal activity. For example, antibody functions such as antibody-dependent cell cytotoxicity (ADCC) depend on glycosylation [15, 16], and differences relative to human glycosylation may cause wild-type *P. pastoris* recombinant proteins not to engage effector functions. However, advances in development of glyco-engineered strains [17–20] allow production of recombinant proteins with mammalian-like glycosylation. The availability of the complete sequence of the *P. pastoris* genome [21, 22], as well as public access to this dataset [23–25], support further strain development and are likely to result in a toolbox of strains for future applications with designer glycoforms.

In *E. coli*, the expression constructs usually have an episomal location and therefore must be maintained under selective pressure to ensure stable plasmid copy number and counteract segregational plasmid instability. Furthermore, structural plasmid stability is critical. In *P. pastoris* an expression cassette can be stably integrated into the chromosome, although this could lead to clonal variation.

However, relative to *E. coli* the major advantage of yeast is its capability to perform post-translational modifications and secretion of product directly into the media. The latter simplifies downstream processing because *P. pastoris* secretes only very low concentrations of host cell proteins into the media. Therefore, *P. pastoris* fermentation with secretion of recombinant His-tagged protein can be readily interfaced with expanded bed adsorption–immobilized metal ion affinity chromatography (EBA-IMAC) [26, 27] or radial flow bed adsorption IMAC (RBA-IMAC) and could be advantageous in bringing new products into the clinic [28]. Indeed, several His-tagged recombinant proteins are in clinical applications (Table 37.1). For a listing of other *P. pastoris* produced therapeutic proteins see, for example, Refs. [11, 12].

The methods described in this chapter have been used in our laboratory for production of a range of proteins including single chain Fv antibodies and diabodies [29–32] and N-A1 domains of carcinoembryonic antigen [33]. The yields ranged from 4 to 200 mg/L of supernatant. The quality of proteins generated with this protocol is sufficient for applications where purity is critical—for example, affinity studies using surface plasmon resonance [33], biodistribution [13, 14, 34], flow cytometry analysis [13], crystallization (Fig. 37.1), and nanoparticle functionalization [28]. Furthermore, the protocol can be readily adapted to GMP-compliant manufacture [13, 14, 28, 35].

Materials

See Notes 1–3.

Chemicals

1. 70% Ethanol
2. $CuSO_4 \cdot (H_2O)_5$
3. H_2O: sterile deionized or water for irrigation (Baxter, UKF7114)
4. Imidazole
5. Na-EDTA
6. NaCl
7. NaOH
8. PBS
9. Virkon
10. pH indicator strips pH 1–14 (VWR 315082P)
11. Solutions: Prepare as per Table 37.2.

Equipment

1. Radial flow column: Proxcys, model AXCIS MD 62 MKIII; depth 6 cm and volume 125 mL
2. IDA chelating matrix (Sterogen, Cellthru BigBead resin; 300 mL, catalog no. 37311)
3. Peristaltic pump, capacity up to 300 mL/min (e.g., Watson Marlow 520S)

Table 37.1 His-tagged proteins in clinical trials

Product names	Company/University	Description/trial status	Reference
NY-ESO-1	Cornell University	Early trial	http://www.news.cornell.edu/stories/Aug09/cancerVaccine.htmL http://www.ncbi.nlm.nih.gov/pubmed/21365782
OFA/iLRP & Quantum Immunologics	University of South Alabama	Phase I/II	http://quantumimmunologics.us/About-Quantum-Immunologics.htmL
rCR2	Medical Research Council UK (MRC)	Preclinical testing	http://www.ncbi.nlm.nih.gov/pmc/articles/PMC3071809/
Vicinium	Viventia Biotechnology	Phase I/II	http://www.viventia.com/
Proxinium	Viventia Biotechnology	Phase I/II/III	http://www.viventia.com/
MFECP1	Royal Free Hospital/University College London	Phase I	http://www.ucl.ac.uk/cancer/research-groups/recombinant-therapeutics/index.htm http://clincancerres.aacrjournals.org/content/12/21/6509.full
OprF-OprI	University of Hannover	At least 4 Phase I/II studies	http://www.ncbi.nlm.nih.gov/pubmed/17683588 http://www.old-herborn-university.de/literature/books/OHUni_book_17_article_7.pdf
IdioVax	CellGenix	Phase II & has Orphan Drug status in EU granted 2004	http://www.cellgenix.com/en/news/EN-news-press-2004-Orphan.htmL
rMAGE-A3	ASCI/GSK	Phase II/III	http://www.ascitrials.com/
Endostar (or Endu)	Simcere Pharmaceutical Group	Phase I, II, III, IV	http://www.simcere.com/english2009/news/News_show.asp?gongs_id=44

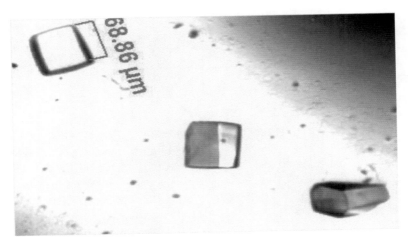

Fig. 37.1 Crystals of a single chain Fv (scfv). Protein was isolated using *P. pastoris* fermentation interfaced with EBA-IMAC. Crystals were isolated and images prepared by Dr. Noelia Sainz-Pastor

4. Pump head tubing, ID 9.5 mm (Cole-Parmer HV-06508-73)
5. Tee connectors, PFA, ½-inch sanitary clamp fitting (Cole-Parmer, KH-31530-10)
6. Four sanitary to hose barb connection (Cole-Parmer WZ-30560-10)
7. Four sanitary clamps (Cole-Parmer WZ-30562-01)

Table 37.2 Buffers and solutions required

Buffer/solution	Composition	Per liter (g)	This protocol (g)
Packing buffer[a]	0.5 M NaCl	29.22	58.44
	0.5× PBS	4.8	9.6
			2 L H_2O
Storage solution	0.1 M NaOH	4	4
	0.5 M NaCl	29.22	29.22
			1 L H_2O
Copper sulphate solution[a]	0.1 M $CuSO_4 \cdot (H_2O)_5$	25	50
			2 L H_2O
Equilibration buffer[a]	0.5 M NaCl	29.22	146.1
	0.5× PBS	4.8	24
	10 mM Imidazole	0.68	3.4
			5 L H_2O
Biomass dilution buffer[a,b]	2 M NaCl	116.88	467.52
	2× PBS	19.2	76.8
	40 mM Imidazole	2.7	10.8
			4 L H_2O
Matrix wash buffer[a]	0.5 M NaCl	29.22	292.2
	0.5× PBS	4.8	48
	40 mM Imidazole	2.7	27
			10 L H_2O
Elution buffer[a]	0.5 M NaCl	29.22	58.44
	0.5× PBS	4.8	9.6
	200 mM Imidazole	13.6	27.2
			2 L H_2O
EDTA solution[a]	50 mM Na-EDTA	18.75	37.5
			2 L H_2O
Strip solution	1 M NaCl	58.44	58.44
	1 M NaOH	40	40
			1 L H_2O

[a]pH of this buffer is not adjusted
[b]Assume harvesting a 10-L bioreactor

8. Sanitary gaskets 10/pk (Cole-Parmer EK-30548-00)
9. Sanitary pressure gauge 0–30 psi (Cole-Parmer WZ-68056-72)
10. Clear tubing, 1/4″ ID×3/8″ OD (0.6 cm ID×1 cm OD), Tygon ACG00017-CP (Cole-Parmer RZ-95635-53)
11. Male-pipe-thread-CPC nonvalved (Proxcys APC24B)
12. Female-in line-CPC barbed fitting nonvalved (Proxcys APC17)
13. Male-pipe-thread-CPC valved (Proxcys APCD24B)
14. Female-in line-CPC barbed fitting valved (Proxcys APCD17)

Spares:

15. 40-μm inlet and outlet frits (Proxcys; PX-FS1-AXCIS-MD62-MKII)
16. Inlet and outlet O-rings (Proxcys; PX-UP1-AXCIS- MD62-MKII)
17. Five autoclaveable waste containers (e.g., Nalgene carboy, 10 L)
18. Five Duran 1-L bottles
19. Five Duran 2-L bottles
20. Stirrer platform plus stirrer bar (suitable for 1- to 2-L bottles)
21. One mixer vessel: 2-L aspirator bottle with bottom sidearm (e.g., Corning PYREX® Product #1220-2L)

Methods

In this protocol we describe our methods for methanol inducible recombinant protein production in *P. pastoris* with secretion of the target protein into the media, followed by IMAC as primary capture step. Subsequent steps comprise concentration and dialysis (using a tangential flow filtration (TFF) system) and size exclusion chromatography. We describe the complete process flow (Table 37.3). However, to be concise, we refer to our published procedures where applicable but outline in detail our recently developed RBA-IMAC procedure. Suggestions for alternatives and optimization of the protocol are given or referred to (see also Refs. [26, 27]).

Seed Lot Production and Fermentation

Genetic engineering of *P. pastoris* is straightforward using standard molecular biology techniques [36], with a range of vectors and strains available [37, 38]. The vector used in this method is pPICZα(alpha) [37], which carries the methanol inducible alcohol oxidase I (AOXI, [39]) promoter. As an alternative the constitutive glyceraldehyde-3-phosphate dehydrogenase promoter (GAP) [9, 37, 40–42] or formaldehyde dehydrogenase promoter (FLD1) [37, 43] from *P. pastoris* can be used. Considerations to choose either promoter can be related to, for example, (1) the need to produce instable proteins or those that are toxic to the host cell (both require inducible expression); or (2) health and safety risks and cost of bulk amounts of highly flammable methanol on site. Vector pPICZα(alpha) carries the α(alpha)-factor secretion signal, which directs expression of the recombinant protein into the broth. This simplifies downstream processing because *P. pastoris* secretes only a small amount of host cell proteins (Fig. 37.2); for example, on glucose media there are only some 20 proteins secreted at detectable levels [22]. Upon initial construction of the expression vector in *E. coli*, the construct is stably integrated in the *P. pastoris* chromosome, and following expression testing a clone is taken forward

to produce a seed lot. Subsequently, this clone is tested in the bioreactor.

Vector Construction and Seed Lot Preparation

For details on plasmid construction and *P. pastoris* transformation, see Ref. [37].

See Notes 4 through 6 for additional considerations regarding vector and seed lot preparation.

Fermentation

For details on recombinant protein production using *P. pastoris* by fermentation, see Ref. [26].

Primary Capture: Expanded and Radial Flow Bed Adsorption IMAC

Packed bed columns can be divided in axial and radial flow devices (Fig. 37.3; Table 37.4). Axial flow columns have a long bed length and consequently contain a large number of theoretical plates, which gives the column a potential for high resolution. Only clarified feed can be used and therefore the major application of axial columns will be for secondary purification and polishing steps in downstream processing. Preconditioning of the feed can be performed by (continuous) centrifugation or (cross-flow or dead-end) filtration. For larger volumes this is a challenge due to the high biomass of *P. pastoris* at harvest time.

In contrast, an axial column where the bed is allowed to expand (see Fig. 37.3B1, B2) can be fed with feedstock almost "as is" directly from the bioreactor. The only preconditioning would involve dilution of the biomass, buffer additions, (e.g., NaCl), and/or pH changing (acid or base addition). The extent to which the bed expands is related to particle size of the matrix, flow rate, and viscosity of the applied sample (and buffer). Unclarified feed may cause deposition of biomass onto the matrix that limits product binding; furthermore, cell-to-cell aggregation can restrict the flow rate. Both potentially can be counteracted by the addition of chemical additives [44]. Amongst others, suppliers for laboratory and process scale

Table 37.3 Process flow scheme and timeline

Process stage	Timing
Clone selection and seed lot production	1–4 months
Primary seed	Day 1
Secondary seed	Day 2
Start bioreactor AOXI: generate biomass (glycerol) GAP: generate product (e.g., glucose)	Day 3
Product generation AOXI: generate product (methanol) GAP: generate product (e.g., glucose) *Set up downstream processing* *(EBA- or RBA-IMAC; TFF; FPLC)*	Day 4–6
Protein isolation	Day 7

The flow process and timelines of the production processes using either AOXI or GAP promoter to drive generation of recombinant protein are indicated. *TFF* tangential flow filtration; *FPLC* fast protein liquid chromatography

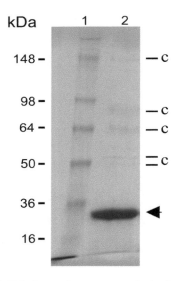

Fig. 37.2 Relative to the target protein (a single chain Fv), the abundancy of secreted native *P. pastoris* proteins is low at 40 h post methanol feed start (preharvest sample). The bioreactor conditions were as outlined in Ref. [26]. Gel: 12% poly acryl amide, tris–glycine buffer; staining with Coomassie brilliant blue R250. Lane *1*, molecular weight markers (see Blue® Plus2 Pre-Stained Standard, Invitrogen); lane *2*, preharvest sample. *c1–5*, contaminating proteins; *arrow*, target protein

expanded bed equipment and matrices are GE Healthcare [45] and Upfront [46]. We have used EBA-IMAC (GE Healthcare StreamLine 50 column with chelating matrix) successfully for laboratory- and clinical-grade recombinant protein production [13, 14, 28–33, 35].

The disadvantages of EBA-IMAC, such as high buffer consumption, slow flow, and overall long process time, can be overcome by implementation of a radial flow column. Radial flow columns are available from small laboratory to full process scale (for review, see Ref. [47]). The two main radial flow column suppliers are Sepragen [48] and Proxcys [49]. A radial flow column is cylindrical at process scale, and the chromatography step takes place by passing the feedstock through the matrix from the outside to the inside of the column (see Fig. 37.3C). Consequently the feed enters through a large surface area and exits through a much smaller area. The distance between entry and exit is the bed height or column length and is short relative to axial columns. This short length combined with large matrix beads makes it very suitable for primary capture and gradient elution. Also, the relatively low height of the radial column relative to the axial column allows for a much lower pressure differential between the highest and lowest point of the column. This pressure differential is prohibitive to build large tall axial columns; consequently, large axial columns tend to be pancake-shaped. (For a performance comparison between similar-sized axial and radial columns, see Ref. [50].)

In this protocol we use radial flow adsorption in IMAC mode (RBA-IMAC) with a chelating matrix consisting of large-size beads (Sterogene

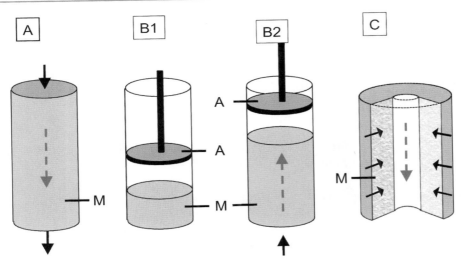

Fig. 37.3 Axial and radial flow chromatography. (**a**) Axial flow column (packed bed). (**b**) Axial flow expanded bed column, collapsed (*B1*), and expanded in up flow (*B2*). (**c**) Radial flow column (packed bed). *A*, adapter; *M*, matrix; *arrows*, flow direction. Dimensions are not to scale: smaller axial flow columns are thin and long, and larger axial flow columns tend to be pancake-shaped

Table 37.4 Characteristics of radial and axial flow columns

	Axial flow bed	Expanded bed	Radial flow bed
Bed type	Fixed size	Expanding	Fixed size
Relative flow rate	Slow-medium	High	High
Feed	Preconditioned	Bioreactor content as is; or crude part purified	Bioreactor content as is; or crude part purified
Usage	Secondary and polishing steps	Primary capture step (secondary steps)	Primary capture step (secondary steps)
Buffer consumption	Medium	High	Medium
Total process time (including setup and cleaning)	Medium	Slow	Fast
Resolution[a]	High	Low	low
Application surface	Small	Small	Large
Scalable	Low	Medium	High

[a]This is application-dependent; see Refs. [47,50]

Cellthru BigBead) that allow capture of His-tagged proteins directly from an unclarified feed. The column used is a small process/laboratory scale wedge-shaped column. In essence this wedge is a "slice" from the larger cylindrical columns with the same characteristics and therefore is inherently scaleable. Relative to EBA [27], RBA has several advantages: the buffer consumption is lower, the flow rate is faster, and the overall process time is reduced (see also Table 37.4). However, RBA-IMAC is more susceptible to clumping of cells, which, depending on severity, could restrict flow and demand a further dilution of the feed.

The following section presents guidelines for installing and implementing a Proxcys radial flow column (AXCIS MD 62 MKIII) in the process flow (see Note 7).

Installation in the Process Flow and Packing of the Column

Use the Proxcys AXCIS MD 62 MKIII manual to assemble the column and use the drawings to familiarize yourself with all connections.

1. Verify the flow rate of the Watson Marlow pump: Place the outlet line into the measuring cylinder and measure the actual flow rate at a specific pump setting. Determine the pump setting for flow rates of 10, 50, 100, and 200 mL/min (see Note 8).

2. Prepare the following radial flow column setup (Fig. 37.4a):
   • Connect the feed line from the feed container to the Watson Marlow pump head tubing.
   • Use connecting tubing, to connect pump head tubing and pressure gauge.
   • Use connecting tubing, to connect pressure gauge to the column inlet port using the push fitting.
   • Connect the column via the outlet line to the waste container using the push fitting on the outlet port. (see Note 9).

3. Fill the feed container with packing buffer (0.5 M NaCl/0.5× PBS) and flush the empty column with packing buffer to remove any debris or dust particles attached to the inside walls and frits.

4. Mix the IDA chelating Cellthru BigBead resin bottle and transfer approximately 150 mL resin into a 1-L Duran bottle; add 500 mL of packing buffer and place on a stirrer platform (see Note 10).

5. Disconnect the column outlet line from the outlet port; this closes the outlet port (see Fig. 37.4b; see Note 9).

6. Place the feed line into a feed bottle with packing buffer.

7. Remove the tubing located between the gauge and the inlet port; put a new section of tubing between the gauge and the MPC connector on the packing port; and open the column packing port.

8. Place the column upside down to facilitate entry of the beads into the column while packing (see Note 11).

9. Connect tubing to the column inlet port using a push fitting, and place the other end of the tubing into the feed bottle to allow recirculation.

10. Start the Watson Marlow pump slowly at 10 mL/min to ensure that flow into the column is unrestricted and that there is no leakage.

11. Increase the flow rate to 50 mL/min, again to ensure that flow into the column is unrestricted and that there is no leakage.

12. Stop the pump. The system is now ready for packing.

13. Transfer the feed line into the resin bottle (leave the stirrer on).

14. Transfer the column inlet line into the resin bottle to allow recirculation while packing.

15. Start the Watson Marlow pump slowly at 10 mL/min to ensure that flow of the beads into the column is unrestricted.

16. Increase the speed to 50 mL/min. While packing is in progress the gauge will show that there is (virtually) no pressure (reading about 0) in the system.

17. When the bed is almost full (15 min), the pressure will increase sharply to about 20 psi. Decrease the flow rate to 10–20 mL/min and continue packing under low pressure.

18. The bed is packed when the pressure again increases sharply at a slow flow rate.

19. Close the column packing port and remove the tubing from this port by disconnecting the MPC connector and replace it with blank MPC connector. Disconnect the line on the inlet port using the push fitting. Reconnect the tubing section between inlet port and gauge.

20. Connect the outlet line back to the outlet port with the push fitting.

At this point the process can be paused and the system prepared for storage by alkali treatment (see "Treating the Column with NaOH"). Upon storage the column has to be neutralized (see "Neutralizing the Column with Deionized Water"). Alternatively, the column can be primed with copper (see "Copper Priming and Equilibrating with Equilibration Buffer").

Treating the Column with NaOH

1. Transfer the feed line into storage solution (0.1 M NaOH and 0.5 M NaCl)

2. Transfer the outlet line into the waste container

3. Feed five column volumes of storage solution at 50 mL/min through the column; subsequently disconnect the inlet and outlet lines and store the column at 2–8 °C until further use.

Fig. 37.4 Configuration of a radial flow column in the process flow. (**a**) Configuration for system flushing and loading of feed stock. (**b**) Configuration for packing of the column

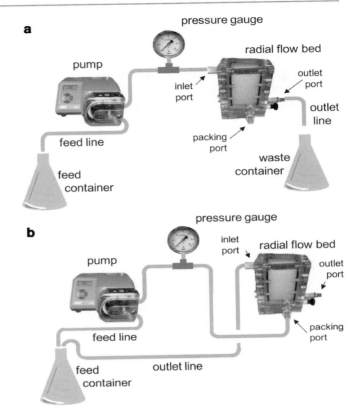

Neutralizing the Column with Deionized Water

1. Transfer the feed line into the feed container and fill this with sterile deionized water (or water for irrigation).
2. Transfer the outlet line into the waste container
3. Start the pump at 200 mL/min and flush the column with sterile water until the pH of the waste matches the pH of the applied water (see Note 12).

Copper Priming and Equilibrating with Equilibration Buffer

1. Transfer the feed line into 2 L 0.1 M copper sulphate solution.
2. Start the pump and apply all copper sulphate to the column at 200 mL/min.
3. Stop the pump once the copper has been applied.
4. Transfer the feed line into sterile water.
5. Wash the column with sterile water to remove any unbound copper from the column. Check the OD600 of the waste outlet and continue washing the column until it matches the OD of the feed.
6. Transfer the feed line into 2 L sterile equilibration buffer (0.5 M NaCl/0.5× PBS/10 mM Imidazole).
7. Equilibrate the column with minimum of ten column volume of equilibration buffer.

Adding the Biomass to the Column

1. Leave the feed line in the equilibration buffer container.
2. Transfer the outlet line into the equilibration buffer container.
3. Start the Watson Marlow pump slowly at 10 mL/min and allow the air to be displaced in both lines.
4. Apply a short pulse of the feed pump to shock remove air from the frits and the column.
5. Perform the same procedure in reverse direction to remove the air from the column.

6. Repeat steps 3–5, if needed (see Note 13).
7. Connect the feed line to the side arm of a 2-L "side arm aspirator" bottle (mixer vessel).
8. Place a sterile magnetic stirrer bar inside the mixer vessel.
9. Place the mixer vessel on magnetic stirrer platform.
10. Connect the bioreactor harvest line to the inlet of the mixer vessel.
11. Connect the dilution buffer container outlet line to the inlet of the mixer vessel.
12. Add 500 mL of biomass dilution buffer (2 M NaCl/2× PBS/40 mM Imidazole) into the mixer vessel and transfer 1,500 mL of bioreactor culture into the mixer vessel.
13. Allow to mix for 1 min and subsequently apply to the radial flow column at 200 mL/min.
14. After each liter, collect a 1-mL aliquot of the flow through, spin down the cells, and store the supernatant of each fraction for later analysis (column breakthrough analysis).
15. Repeat steps 12–14 until the complete bioreactor content has been applied (see Note 14).
16. Stop the pump and transfer the feed line into a container with wash buffer.
17. Restart the pump to wash any remaining cells and cell debris from the matrix.
18. After each liter, collect a 1-mL aliquot of the wash flow through, spin down the cells, and store the supernatant of each fraction for later analysis (leakage from matrix).
19. Apply wash buffer until the OD280 is steady.

Eluting of Nonspecifically Bound Contaminants from the Matrix

1. Apply 2 L matrix wash buffer (0.5 M NaCl/0.5× PBS/40 mM Imidazole) to the column, at 200 mL/min; collect 50-mL fractions.
2. Measure OD280 of each fraction and store an aliquot of each fraction for later analysis. (see Note 15) (Fig. 37.5).

Eluting Target Protein from the Matrix

1. Apply 2 L elution buffer (0.5 M NaCl/0.5× PBS/200 mM Imidazole) to the column, at 200 mL/min; collect 50-mL fractions.
2. Measure OD280 of each fraction and store an aliquot of each fraction for later analysis.
3. See Fig. 37.5.

4. Pool the fractions with the highest OD280 for further downstream processing
5. Continue the elution buffer feed to remove any remaining cells and cell debris from the matrix by applying equilibration buffer until the OD280 is steady.

Regeneration of the Column: Removing Copper and Washing with Sodium Hydroxide

1. Apply 2 L EDTA solution to the column, at 200 mL/min; collect the first 10 fractions of 50 mL. Store an aliquot of each fraction for later analysis (detection of protein still bound to the column).
2. Apply all EDTA to the column.
3. Transfer the feed line into the strip solution.
4. Apply sodium hydroxide until the A280 of the waste is the same as that of the applied solution.
5. Neutralize the column as per instructions in "Neutralizing the Column with Deionized Water".

The column can now be prepared for storage (see "Treating the Column with NaOH") or recharged with copper for the next round of purification (see "Copper Priming and Equilibrating with Equilibration Buffer").

Downstream Processing

At this stage many options are available for further downstream processing, depending on the nature of the protein and/or final application in question. These are beyond the scope of this chapter (see Note 16).

Notes

1. Health and safety considerations: before starting practical work, it should be verified that the procedures given in this method are in alignment with local rules and regulations. Risk assessments, waste disposal, and decontamination procedures should be in place. Also, any hazard sheets (Material Safety Data Sheet, MSDS) should be read and first-aid

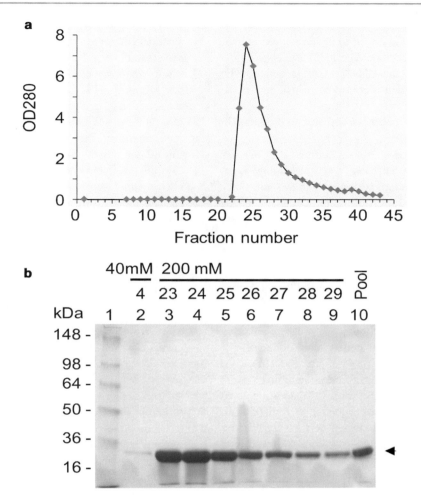

Fig. 37.5 Anticipated purification profile during radial flow bed elution, illustrated with a single chain Fv (scfv). (**a**) OD280 profile of eluted fractions. Fractions 1–20, 40 mM elution; fractions 21–43, 200 mM elution. Each fraction was 50 mL. (**b**) 12% polyacrylamide Coomassie (brilliant blue R250) stained gel of eluted fractions. Lane *1*, molecular weight marker; lane *2*, fraction 4 of 40 mM elution; lanes *3–9*, fraction 23–29 of 200 mM elution; lane *10*, pool of fractions 23–30. *Arrow*, target protein

measures should be known and available. Where applicable, personal protection equipment (PPE) should be used.

- Ensure acid and base are stored separately in suitable storage cupboards. Special care should be taken with ammonia, which develops pressure on standing, particularly when warm. Release pressure in an appropriate ventilated hood!
- Methanol is highly flammable, and suitable containers should be used.

2. The equipment, chemicals, disposables, and so on described in this method are currently in use in the authors' laboratory; however, these should be easily exchangeable with equipment/materials of equivalent specification.

3. *Pichia pastoris* belongs to the group of microorganisms that are "generally regarded as safe" (GRAS).

- Decontamination: Spray work area and other surfaces with 70% (v/v) ethanol and leave for 5 min.
- Spills: Decontaminate the area by sprinkling Virkon granules.[1] For larger spills adsorbent containment booms[2] can be used;

---

[1] e.g., http://www2.dupont.com/DAHS_EMEA/en_GB/products/disinfectants/virkon_s/index.htmL.

[2] e.g., http://www.newpig.com.

these effectively adsorb the liquid waste and subsequently need to be autoclaved (center of the load should be 121 °C for 30 min).

4. Regarding the DNA/protein sequence to be expressed and vector to be used, consider the following:
   • Codon optimization for expression in *Pichia*.
   • Presence of amber codons, e.g., *E. coli* TG1 (often used in phage display) is an amber suppressor and *Pichia* is not.
   • Restriction sites *Pme*I, *Sac*I, or *Bst*XI; these enzymes are used to linearize plasmid pPICZα(alpha) before insertion into *Pichia* and consequently should not be located in the gene.
   • Presence of LysArg encoding sequences; this site can be prone to cleavage by the native *Pichia* Kex2 protease and cause protein breakdown.
   • Presence of consensus sequence for N-glycosylation (Asn-Xaa-Ser/Thr).
   • Addition of a stop codon: ligation in frame at the 3′ prime causes the addition from the vector of a sequence that encodes an Myc and His-tag. Alternatively, a stop codon or any tag of choice followed by a stop coden could be placed directly following the inserted gene.
   • 5′ modifications to the vector sequence encoding KREAEA: this sequence is cleaved by the *Pichia* enzymes Kex2 (KR) and Ste13. However, Ste13 cleavage can be partial (e.g., Refs. [51–53]) and removal of the Ste13 recognition sequence could be considered.

5. Expression of clones can vary substantially; therefore screening should be performed on at least 10 or more clones; of each clone the supernatant and cell pellet should be examined for the target protein.

6. The seed lot can be very valuable; therefore, it should be stored in two locations at −80 °C and should be subjected to minimal characterization like viability, production level sterility (monosepsis), and stability.

7. Before implementing the RBA-IMAC process, pilot studies can be performed to establish binding conditions for the target protein. At least optimal conditions for ionic strength,

imidazole concentration, and pH should be established. For a detailed practical approach, see appendix 3 in Ref. [27].

8. Pump performance and tubing can deteriorate in time and "as good practice" should be measured at a regular interval.

9. Disconnecting the push fitting will close the port and reconnecting will open the port.

10. Stir moderately to avoid damaging the resin.

11. Turning the column upside down at this stage allows a more even spread of the beads and reduces the pressure while filling column.

12. Follow neutralization using pH strips or a pH meter.

13. Before adding the biomass it is important to ensure that all air is removed from the column; ensure even application of biomass to the column; and maximize access of the protein to the column matrix. Also, at this point each connection should be verified and checked for leakage.

14. During the biomass application the pressure should not or only minimally rise. A rise could suggest clocking up of frits and or matrix and should be investigated immediately.

15. It is important to establish that the protein remains bound during this washing step. If an excessive amount of protein elutes, then other concentrations of imidazole should be explored (e.g., 20 or 30 mM in the wash buffer).

16. A simple procedure which can be easily adapted and is in use in the author's laboratory is the following.
    • Sterile filter the protein fraction on a 0.2 μm filter, for example 0.5 L Nalgene,[3] especially PES (polyethersulfone) filters are suitable since they show very low protein binding.
    • Concentrate the protein fraction on a TFF system, e.g., a Labscale system from Millipore with Pellicon XL membrane[4] or a Pall system.[5]
    • Using the TFF system, dialyze the protein fraction into a suitable buffer, e.g., 1× PBS.

---

[3] http://www.nalgenelabware.com/techdata/technical/FiltrationIndex.asp.

[4] http://www.millipore.com/catalogue/item/xx42lss11.

[5] http://www.pall.com/laboratory_53238.asp.

- Apply the sample to an FPLC, fitted with a size exclusion column (e.g., a sample can be loaded using a superloop to a 500-mL Superdex [GEHealthcare]) and collect the fractions of interest.

- If required, remove endotoxins by successive passaging over an endotoxin removal gel (Thermo Scientific). Progress can be monitored using a Limulus Amebocyte Lysate (LAL) test (Pyrogent-Plus 64 Gel Clot Assay, Lonza Biologics).

Acknowledgements This work was supported by Cancer Research UK; Department of Health (ECMC, Experimental Cancer Medicine Network Centre); Engineering and Physical Sciences Research Council (EPSRC); The Breast Cancer Campaign; and UCL Cancer Institute Research Trust.

References

1. Sreekrishna K, Tschopp JF, Thill GP, Brierley RA, Barr KA (1998) Expression of human serum albumin in *Pichia pastoris*. US Patent 5707828
2. Kobayashi K, Kuwae S, Ohya T, Ohda T, Ohyama M, Ohi H et al (2000) High-level expression of recombinant human serum albumin from the methylotrophic yeast *Pichia pastoris* with minimal protease production and activation. J Biosci Bioeng 89:55–61
3. Cregg JM, Cereghino JL, Shi J, Higgins DR (2000) Recombinant protein expression in Pichia pastoris. Mol Biotechnol 16:23–52
4. Daly R, Hearn MT (2005) Expression of heterologous proteins in *Pichia pastoris*: a useful experimental tool in protein engineering and production. J Mol Recognit 18:119–138
5. Macauley-Patrick S, Fazenda ML, McNeil B, Harvey LM (2005) Heterologous protein production using the *Pichia pastoris* expression system. Yeast 22:249–270
6. Baez J, Olsen D, Polarek JW (2005) Recombinant microbial systems for the production of human collagen and gelatin. Appl Microbiol Biotechnol 69:245–252
7. Xie T, Liu Q, Xie F, Liu H, Zhang Y (2008) Secretory expressin of insulin precursor in *Pichia pastoris* and simple procedure for producing recombinant human insulin. Prep Biochem Biotechnol 38:308–317
8. Murasugi A (2010) Review. Secretory expression of human protein in the yeast Pichia pastoris by controlled fermentor culture. Recent Pat Biotechnol 4:153–166
9. Zhang AL, Luo JX, Zhang TY, Pan YW, Tan YH, Fu CY et al (2008) Recent advances on the GAP promoter derived expression system of *Pichia pastoris*. Mol Biol Rep 36:1611–1619
10. Kovar K, Looser V, Hyka P, Merseburger T, Meier C (2010) Recombinant yeast technology at the cutting edge: robust tools for both designed catalysts and new biologicals. Chimia (Aarau) 64:813–818
11. Meyer HP, Brass J, Jungo C, Klein J, Wenger J, Mommers R (2008) An emerging star for therapeutic and catalytic protein production. BioProcess Int 6(No. S6):10–21
12. Gerngross TU (2004) Review. Advances in the production of human therapeutic proteins in yeasts and filamentous fungi. Nat Biotechnol 11:1409–1414
13. Kogelberg H, Tolner B, Sharma SK, Lowdell MW, Qureshi U, Robson M et al (2007) Clearance mechanism of a mannosylated antibody-enzyme fusion protein used in experimental cancer therapy. Glycobiology 17:36–45
14. Sharma SK, Pedley RB, Bhatia J, Boxer GM, El-Emir E, Qureshi U et al (2005) Sustained tumor regression of human colorectal cancer xenografts using a multifunctional mannosylated fusion protein in antibody-directed enzyme prodrug therapy. Clin Cancer Res 11:814–825
15. Jefferis R (2005) Glycosylation of recombinant antibody therapeutics. Biotechnol Prog 2:11–16
16. Rothman RJ, Perussia B, Herlyn D, Warren L (1989) Antibody-dependent cytotoxicity mediated by natural killer cells is enhanced by castanospermine-induced alterations of IgG glycosylation. Mol Immunol 26:1113–1123
17. Li H, Sethuraman N, Stadheim TA, Zha D, Prinz B, Ballew N et al (2006) Optimization of humanized IgGs in glycoengineered *Pichia pastoris*. Nat Biotechnol 24:210–215
18. Potgieter TI, Cukan M, Drummond JE, Houston-Cummings NR, Jiang Y, Li F et al (2009) Production of monoclonal antibodies by glycoengineered *Pichia pastoris*. J Biotechnol 139:318–325
19. De Pourcq K, De Schutter K, Callewaert N (2010) Review. Engineering of glycosylation in yeast and other fungi: current state and perspectives. Appl Microbiol Biotechnol 87:1617–1631
20. Jacobs PP, Geysens S, Vervecken W, Contreras R, Callewaert N (2009) Engineering complex-type N-glycosylation in *Pichia pastoris* using GlycoSwitch technology. Nat Protoc 4:58–70
21. De Schutter K, Lin YC, Tiels P, Van Hecke A, Glinka S, Weber-Lehmann J et al (2009) Genome sequence of the recombinant protein production host *Pichia pastoris*. Nat Biotechnol 27:561–566
22. Mattanovich D, Graf A, Stadlmann J, Dragosits M, Redl A, Maurer M et al (2009) Genome, secretome and glucose transport highlight unique features of the protein production host *Pichia pastoris*. Microb Cell Fact 8:29
23. Mattanovich D, Callewaert N, Rouzé P, Lin YC, Graf A, Redl A et al (2009) Open access to sequence: browsing the *Pichia pastoris* genome. Microb Cell Fact 16:53
24. http://pichiagenome.org. Accessed 16 Sep 2011
25. http://bioinformatics.psb.ugent.be/webtools/bogas. Accessed 16 Sep 2011
26. Tolner B, Smith L, Begent RH, Chester KA (2006) Production of recombinant protein in Pichia pastoris by fermentation. Nat Protoc 1:1006–1021

27. Tolner B, Smith L, Begent RH, Chester KA (2006) Expanded-bed adsorption immobilized-metal affinity chromatography. Nat Protoc 1:1213–1222

28. Tolner B, Smith L, Hillyer T, Bhatia J, Beckett P, Robson L et al (2007) Review. From laboratory to Phase I/II cancer trials with recombinant biotherapeutics. Eur J Cancer 43:2515–2522

29. Vigor KL, Kyrtatos PG, Minogue S, Al-Jamal KT, Kogelberg H, Tolner B et al (2010) Nanoparticles functionalized with recombinant single chain Fv antibody fragments (scFv) for the magnetic resonance imaging of cancer cells. Biomaterials 31:1307–1315

30. Kogelberg H, Tolner B, Thomas GJ, Di Cara D, Minogue S, Ramesh B et al (2008) Engineering a single-chain Fv antibody to alpha v beta 6 integrin using the specificity-determining loop of a foot-and-mouth disease virus. J Mol Biol 382:385–401

31. Tolner B, Vigor K, Mather S, Robinson M, Adams G, Plueckthun A et al (2010) Anti-HER2 imaging agents for breast cancer imaging. Breast Cancer Research 2010 Conference. Breast Cancer Res 12(Suppl 1):P42

32. Tolner B, Hodgson D, Smith L, Begent RHJ, Robinson M, Adams G et al (2008) Anti-HER2 diabody for breast cancer imaging: GMP-compliant production and preclinical analysis. In: (Proceedings) NCRI Cancer Conference Proceedings (http://www.ncri.org.uk/ncriconference)

33. Sainz-Pastor N, Tolner B, Huhalov A, Kogelberg H, Lee YC, Zhu D et al (2006) Deglycosylation to obtain stable and homogeneous *Pichia pastoris*-expressed N-A1 domains of carcinoembryonic antigen. Int J Biol Macromol 39:141–150

34. Francis RJ, Mather SJ, Chester K, Sharma SK, Bhatia J, Pedley RB et al (2004) Radiolabelling of glycosylated MFE-23::CPG2 fusion protein (MFECP1) with 99mTc for quantitation of tumour antibody-enzyme localisation in antibody-directed enzyme pro-drug therapy (ADEPT). Eur J Nucl Med Mol Imaging 31:1090–1096

35. Mayer A, Francis RJ, Sharma SK, Tolner B, Springer CJ, Martin J et al (2006) A phase I study of single administration of antibody-directed enzyme prodrug therapy with the recombinant anti-carcinoembryonic antigen antibody-enzyme fusion protein MFECP1 and a bis-iodo phenol mustard prodrug. Clin Cancer Res 12:6509–6516

36. Sambrook J, MacCallum P (2001) Molecular cloning: a laboratory manual, 3rd edn. Cold Spring Harbor Laboratory Press, Cold Spring Harbor, NY

37. http://invitrogen.com. Accessed 16 Sep 2011

38. http://faculty.kgi.edu/cregg/index.htm. Accessed 16 Sep 2011

39. Tschopp JF, Brust PF, Cregg JM, Stillman CA, Gingeras TR (1987) Expression of the lacZ gene from two methanol-regulated promoters in *Pichia pastoris*. Nucleic Acids Res 15:3859–3876

40. Waterham HR, Digan ME, Koutz PJ, Lair SV, Cregg JM (1997) Isolation of the *Pichia pastoris* glyceraldehyde-3-phosphate dehydrogenase gene and regulation and use of its promoter. Gene 186:37–44

41. Zhao W, Wang J, Deng R, Wang X (2008) Scale-up fermentation of recombinant *Candida rugosa* lipase expressed in *Pichia pastoris* using the GAP promoter. J Ind Microbiol Biotechnol 35:189–195

42. Heyland J, Fu J, Blank LM, Schmid A (2010) Quantitative physiology of *Pichia pastoris* during glucose-limited high-cell density fed-batch cultivation for recombinant protein production. Biotechnol Bioeng 107:357–368

43. Shen S, Sulter G, Jeffries TW, Cregg JM (1998) A strong nitrogen source-regulated promoter for controlled expression of foreign genes in the yeast *Pichia pastoris*. Gene 216:93–102

44. Vennapusa RR, Fernandez-Lahore M (2010) Effect of chemical additives on biomass deposition onto beaded adsorbents. J Biosci Bioeng 110:564–571

45. http://gelifesciences.com. Accessed 16 Sep 16, 2011

46. http://upfront-dk.com/. Accessed 16 Sep 2011

47. Gu J (2009) Chromatography, radial flow. In: Flickinger MC (ed) Encyclopedia of industrial biotechnology: bioprocess, bioseparation, and cell technology. John Wiley & Sons, New York

48. http://sepragen.com. Accessed 16 Sep 2011

49. http://proxcys.nl. Accessed 16 Sep 2011

50. Cabanne C, Raedts M, Zavadzky E, Santarelli X (2007) Evaluation of radial chromatography versus axial chromatography, practical approach. J Chromatogr B Analyt Technol Biomed Life Sci 845:191–199

51. Emberson LM, Trivett AJ, Blower PJ, Nicholls PJ (2005) Expression of an anti-CD33 single-chain antibody by *Pichia pastoris*. J Immunol Methods 305:135–151

52. Chen Z, Wang D, Cong Y, Wang J, Zhu J, Yang J et al (2011) Recombinant antimicrobial peptide hPAB-β expressed in *Pichia pastoris*, a potential agent active against methicillin-resistant *Staphylococcus aureus*. Appl Microbiol Biotechnol 89:281–291

53. Buensanteai N, Mukherjee PK, Horwitz BA, Cheng C, Dangott LJ, Kenerley CM (2010) Expression and purification of biologically active Trichoderma virens proteinaceous elicitor Sm1 in *Pichia pastoris*. Protein Expr Purif 72:131–138

Télesphore Sime-Ngando and Marlène Jobard

Abstract

Recently, molecular environmental surveys of the eukaryotic microbial community in lakes have revealed a high diversity of sequences belonging to uncultured zoosporic fungi. Although they are known as saprobes and algal parasites in freshwater systems, zoosporic fungi have been neglected in microbial food web studies. Recently, it has been suggested that zoosporic fungi, via the consumption of their zoospores by zooplankters, could transfer energy from large inedible algae and particulate organic material to higher trophic levels. However, because of their small size and their lack of distinctive morphological features, traditional microscopy does not allow the detection of fungal zoospores in the field. Hence, quantitative data on fungal zoospores in natural environments are missing. We provide a simplified step-by-step real-time quantitative PCR laboratory protocol, for the assessment of uncultured zoosporic fungi and other zoosporic microbial eukaryotes in natural samples.

Keywords

Fungi • Spores • DNA extraction • Real-time qPCR • Environmental samples

T. Sime-Ngando (✉)
UMR CNRS 6023, Université Blaise Pascal, Clermont
II, 24 Avenue des Landais, BP 80026, Aubière 63171,
Cedex, France
e-mail: telesphore.sime-ngando@univ-bpclermont.fr

M. Jobard
LMGE UMR CNRS, U.F.R. Sciences et Technologies,
24 avenue des Landais, BP 80026, Aubière 63171,
Cedex, France

V.K. Gupta et al. (eds.), *Laboratory Protocols in Fungal Biology: Current Methods in Fungal Biology,*
Fungal Biology, DOI 10.1007/978-1-4614-2356-0_38, © Springer Science+Business Media, LLC 2013

Introduction

Recent molecular surveys of microbial eukaryotes have revealed overlooked uncultured environmental fungi with novel putative functions [1–3], among which zoosporic forms (i.e., chytrids) are the most important in terms of diversity, abundance, and functional roles, primarily as infective parasites of phytoplankton [4, 5] and as valuable food sources for zooplankton via massive zoospore production, particularly in freshwater lakes [6–8]. However, due to their small size (2–5 µm), their lack of distinctive morphological features, and their phylogenetic position, traditional microscopic methods are not sensitive enough to detect fungal zoospores among a mixed assemblage of microorganisms. Most chytrids occupy the most basal branch of the kingdom Fungi, a finding consistent with choanoflagellate-like ancestors [9]. These reasons may help explain why both infective (i.e., sporanges) and disseminating (i.e., zoospores) life stages of chytrids have been misidentified in previous studies as phagotrophic sessile flagellates (e.g., choanoflagellates) and "small undetermined" cells, respectively. These cells often dominate the abundance of free-living heterotrophic nanoflagellates (HNFs), and are considered the main bacterivores in aquatic microbial foodwebs [2, 10]. Their contribution ranges from 10 to 90% of the total abundance of HNFs in pelagic systems (see review in reference [11]). Preliminary data have provided that up to 60% of these unidentified HNFs can correspond to fungal zoospores [12], establishing HNF compartment as a black box in the context of microbial food web dynamics [4]. A recent simulation analysis based on Lake Biwa (Japan) inverse model indicated that the presence of zoosporic fungi leads to (1) an enhancement of the trophic efficiency index, (2) a decrease of the ratio detritivory/herbivory, (3) a decrease of the percentage of carbon flowing in cyclic pathways, and (4) an increase in the relative ascendancy (indicates trophic pathways more specialized and less redundant) of the system [13]. Unfortunately, because specific methodology for their detection

is not available, quantitative data on zoosporic fungi are missing.

Molecular approaches have profoundly changed our view of eukaryotic microbial diversity, providing new perspectives for future ecological studies [3]. Among these perspectives, linking cell identity to abundance and biomass estimates is highly important for studies on carbon flows and the related biogeochemical cycles in natural ecosystems [11]. Historically, taxonomic identification and estimation of *in situ* abundances of small microorganisms have been difficult. In this context, our inability to identify and count many of these small species in the natural environment limits our understanding of their ecological significance. Thus, new tools that combine both identification and quantification need to be developed. Fluorescent *in situ* hybridization (FISH) has been an assay of choice for simultaneous identification and quantification of specific microbial populations in natural environments [14, 15]. However, this technique is limited because of (1) the relatively low number of samples that can be processed at a time, and (2) its relatively low sensitivity due to background noise and the potentially low number of target rRNA molecules per cell in natural environments [16]. In contrast, real-time quantitative PCR (qPCR), which has been widely used to estimate prokaryotic and eukaryotic population abundances in natural ecosystems, allows the simultaneous analysis of a high number of samples with a high degree of sensitivity [15].

The main objective of this chapter is to provide, in a simplified step-by-step format, a qPCR assay for the quantitative assessment of uncultured zoosporic fungi and other zoosporic microbial eukaryotes in natural environments [15], together with practical advice on how to apply the method. QPCR was recently used to estimate fungal biomass in a stream during leaf decomposition [17] and in biological soil crusts [18]. The interpretation of the semi-quantitative data obtained in these studies was relatively difficult because the whole fungal community was targeted (including unicellular, multicellular, and multinuclear fungal species). Thus, an estimation of fungal density or even fungal biomass was not possible. In the following protocol, the primary targets are zoospores in liquid suspensions. Because zoospores are unicellular, qPCR data could be directly converted into cell density estimates (i.e., by multiplying semi-quantitative data by a number of rDNA copies per cells). Moreover, we designed primers targeting Rhizophidiales taxon to limit quantification bias generated by the variability in the number of rDNA copies within eukaryotic ribosomal operon.

Materials

1. Gloves (should be worn when manipulating most of the following materials).
2. Sterile distilled water.
3. 0.6 μm pore size polycarbonate filters.
4. Filtration columns.
5. Sodium dodecyl sulfate (SDS).
6. Proteinase K.
7. TE buffer—10 mM Tris–HCl pH 7.5 (25 °C), 1 mM EDTA.
8. NucleoSpin Plant kit® (Macherey-Nagel, Bethlehem, PA) with silica-membrane columns and the materials for running the provided protocol from the manufacturer.
9. Molecular-biology-grade agarose.
10. Ethidium bromide—because suspected as a mutagen, particular care should be taken when handling (consult safety data sheet).
11. Calf thymus DNA (Sigma-Aldrich, St. Louis, MO).
12. Oligonucleotidic primers resuspended in sterile distilled water and stored at −20 °C (see Note 1).
13. SYBR Green (Sigma-Aldrich).
14. dNTPs—a mixture of dATP, dCTP, dGTP, and dTTP (10 mM of each), stored at −20 °C.
15. Thermostable DNA polymerase and reaction buffer supplied by manufacturer. To avoid nonspecific amplicon, use hot-start (e.g., HotStarTaq, Qiagen, Valencia, CA).
16. Vortexer.
17. Centrifuge.
18. Water bath.
19. Horizontal electrophoresis machine.

20. TBE buffer: 50 mM Tris, 50 mM boric acid, 1 mM EDTA, diluted when needed from a 50× stock solution.
21. Spectrophotometer—the authors use NanoDrop (NanoDrop Technologies, Inc, Wilmington, DE).
22. Disposable conical tubes (1.5 mL); PCR tube strips or plate with adhesive film and cap adapted for real-time quantitative PCR assay.
23. Thermal cycler—we use Mastercycler ep realplex detection system (Eppendorf).
24. UV transilluminator equipped with a camera suitable for photographing agarose gels

Methods

DNA Extraction and Purification

Collect zoosporic organisms onto 0.6 μm pore size polycarbonate filters (after removal of the algal host by prefiltrations when only zoospores are targeted) (see Note 2).

1. For cell disruption, incubate the filters in 560 μL of a buffer containing 1% SDS and 1 mg/mL proteinase K in TE buffer for 1 h at 37 °C in a water bath (see Notes 3 and 4).
2. For DNA purification, use the silica-membrane columns provided with the NucleoSpin Plant kit® (Macherey-Nagel), following the instructions from the manufacturer (see Note 4).
3. Visualize the integrity and yield of the extracted genomic DNA in a 1% agarose gel stained with 0.3 μg/mL of ethidium bromide solution (Sigma-Aldrich), using UV transilluminator and photograph. For this (1) heat (45 s using a microwave oven) a mixture of agarose in 1× TBE buffer; (2) leave it to cool on the bench for 5 min down to about 60 °C before adding ethidium bromide (i.e., to avoid vapor formation); (3) mix and pour into suitable gel gray with comb and leave to set for at least 30 min; (4) remove the comb and submerge the gel to 2–5 mm depth in electrophoresis tank containing 1× TBE buffer; (5) transfer DNA sample aliquots (i.e., 2 volumes of sample and 1 volume of loading buffer), marker,

and the serial dilution of 5–10 ng of calf thymus DNA (Sigma-Aldrich-Aldrich) to the wells of the agarose gel; and (6) start the electrophoresis migration for about 30 min at 100 V.

4. Calculate DNA extract concentrations from dilutions of calf thymus DNA (Sigma-Aldrich), using a standard curve of calf thymus DNA vs. band intensity.

Real-Time qPCR Assays

1. PCR mix contained SYBR Green (Sigma-Aldrich), 200 μm of each dNTPs, 10 pM of each primer, 2.5 units of *Taq* DNA polymerase, the PCR buffer supplied with the enzyme, and 1.5 mM MgCl$_2$. Vortex briefly (less than 10 s) and centrifuge the mix before distributing aliquots in suitable PCR tubes (strips or plates) and place on ice.

2. Add variable quantity of DNA (we used 5 ng for our environmental freshwater samples, and 10 ng for DNA from appropriate PCR negative control strains) used as template in a final volume of 25 μL (see Note 5).

3. Standard curve of C_t (see Note 4) vs. DNA copy number required to calculate target copy numbers (see Note 6) in each reaction is generated using triplicates of PCR reactions of tenfold dilutions of linear plasmid (containing Rhizophidiales 18S rDNA insert; PFB11AU2004) ranging from 100 to 1×10^8 copy/μL (see Note 7). This number of copies was calculated using the equation: molecules/μL $= a/(b \times 660) \times 6.022 \times 10^{23}$, where a is the plasmid DNA concentration (g/μL), b the plasmid length in bp, including the vector and the inserted 18S rDNA fragment, 660 the average molecular weight of one base pair, and 6.022×10^{23} the Avogadro constant [15, 19].

4. Place all tubes (i.e., samples, controls, and standards) in the real-time qPCR cycler and run the appropriate cycling program: initial HotStarTaq activation at 95 °C for 15 min, 35 cycles with denaturation at 95 °C for 1 min, annealing at 63.3 °C for 30 s with Fchyt/Rchyt primers pair (see Note 1), elongation at 72 °C for 1 min, and a final additional elongation step at 72 °C for 10 min.

5. Using SYBR Green molecule, melting curves analysis can be performed immediately following each qPCR assay to check specificity of amplification products (to confirm the absence of primer dimers or unspecific PCR products) by increasing the incubation temperature from 50 to 95 °C for 20 min.

6. Analyze the real-time PCR result with the suitable software. Check to see if there is any bimodal dissociation curve or abnormal amplification plot (see Note 5) before calculating the initial concentration of the targeted uncultured fungal 18S rDNA (copies/mL) in the environmental genomic DNA (see Note 6).

Notes

1. Consensus (universal) primers can be used to amplify regions of fungal ribosomal RNA gene. For natural waters, we have designed primers specific to chytrids using a database containing about a hundred 18S rDNA environmental sequences recovered from surveys conducted in different lakes and sequences belonging to described fungi [15]. Sequences were aligned using BioEdit software (http://www.mbio.ncsu.edu/BioEdit/bioedit.html) and the resulting alignment was corrected manually. A great proportion of the environmental chytrid sequences recovered from lakes was closely affiliated to the Rhizophidiales. Thus, Rhizophidiales-specific primers F-Chyt (sequence 5′ > 3′: GCAGGCTT ACGCTTGAATAC) and R-Chyt (sequence 5′ > 3′: CATAAGGTGCCGAACAAGTC) were designed in order to fulfill three requirements: (1) a GC content between 40 and 70%, (2) a melting temperature (T_m) similar for both primers and close to 60 °C, and (3) PCR products below 500 bp (i.e., between 304 and 313 bp depending on the species considered). The absence of potential complementarities (hairpins and dimers) was checked using Netprimer (http://www.premierbiosoft.com/

netprimer/netprlaunch/netprlaunch.html), and confirmed by inspection of the melting curve following the qPCR assay.

2. For targeting uncultured zoosporic fungi, zoospores are discarded from other environmental microorganisms by successive prefiltrations through 150, 80, 50, 25, 10, and 5 μm filters before collected them onto 0.6-μm polycarbonate filters. Filters can be conserved at −80 °C until DNA extraction in appropriate tubes (2 mL).

3. Other enzymes such as lyticase can be used for cell disruption, with no significant difference compared to proteinase K. However, the one-step proteinase K yields higher amount of genomic DNA than the lyticase method and has a better reproducibility. Physical disruption procedures such as sonication or thermal shocks (i.e., freezing in liquid nitrogen and thawing) are to be avoided because of the possible degradation of DNA.

4. A standard phenol–chloroform purification procedure can also be used but when the genomic DNA extracts are used as template in PCR reactions, the DNA purification method using the commercial kit gave significantly better results (based on C_t, the threshold cycle during PCR when the level of fluorescence gives signal over the background and is in the linear portion of the amplified curve) than the phenol–chloroform method. Consequently, the DNA extraction method using Proteinase K and the commercial kit was selected and considered the best overall.

5. In the case of novel designed primers (see Note 1) for uncultured fungi, DNA from both positive and negative plasmids and different mixtures (e.g., 5, 10, 25, and 50% of the positive plasmids) will be used for the optimization of the conditions (annealing temperature, cycling), cross-reactivity, the detection limit (using serial tenfold dilutions of the positive plasmids; see Note 7), and the amplification efficiency of the qPCR essays, which should be at least 90%. Poor primer quality is the leading cause for poor PCR efficiency. In this case, the PCR amplification curve usually reaches plateau early and the final fluorescence intensity is significantly lower than that of most other PCRs. This problem may be solved with re-synthesized primers.

6. The initial concentration of target 18S rDNA (copies/mL) in environmental samples can be calculated using the formula $[(a/b) \times c]/d$, where a is the 18S rDNA copy number estimated by qPCR, b is the volume of environmental genomic DNA added in the qPCR reaction, c is the volume into which the environmental genomic DNA was resuspended at the end of the DNA extraction, and d is the volume of sample filtered from which environmental DNA was extracted.

7. In the absence of cultures, plasmids used in qPCR to construct standard curves and to optimize qPCR reactions come from genetic libraries constructed during previous environmental surveys [2]. Briefly, the complete 18S rRNA gene was amplified from environmental genomic DNA extracts using the universal eukaryote primers 1f and 1520r. An aliquot of PCR products was cloned using the TOPO-TA cloning kit (Invitrogen, Carlsbad, CA) following the manufacturer's recommendations. Plasmid containing the insert of interest was extracted with NucleoSpin® plasmid extraction kit (Macherey-Nagel) following the manufacturer's recommendations. The 18S rRNA gene was sequenced from plasmid products by the MWG Biotech services using M13 universal primers (M13rev (−29) and M13uni (−21)). Phylogenetic affiliation of sequences acquired was established using Neighbor-Joining and Bayesian methods. In our case, positive plasmids contain insert affiliated to target chytrid (i.e. Rhizophidiales species) and displaying less than two mismatches with primers F-Chyt and R-Chyt sequences (see Note 1). The plasmid PFB11AU2004 (Genbank accession number DQ244014) was selected to construct the standard curve required for qPCR. Linearized plasmids were produced from supercoiled plasmids by digestion with restriction endonuclease one-time cutting into the vector sequence. Linear plasmid DNA concentration can be determined by measuring the absorbance at 260 nm (A260) in a spectrophotometer.

Acknowledgements MJ was supported by a PhD Fellowship from the Grand Duché du Luxembourg (Ministry of Culture, High School, and Research). This study receives grant-aided support from the French ANR Programme Blanc #ANR 07 BLAN 0370 titled DREP: Diversity and Roles of Eumyctes in the Pelagos.

References

1. Jobard M, Rasconi S, Sime-Ngando T (2010) Diversity and functions of microscopic fungi: a missing component in pelagic food webs. Aquat Sci 72:255–268
2. Lefèvre E, Bardot C, Noël C, Carrias JF, Viscogliosi E, Amblard C et al (2007) Unveiling fungal zooflagellates as members of freshwater picoeukaryotes: evidence from a molecular diversity study in a deep meromictic lake. Environ Microbiol 9:61–71
3. Monchy S, Jobard M, Sanciu G, Rasconi S, Gerphagnon M, Chabe M et al (2011) Exploring and quantifying fungal diversity in freshwater lake ecosystems using rDNA cloning/sequencing and SSU tag pyrosequencing. Environ Microbiol 13(6):1433–1453. doi:10.1111/j.1462-2920.2011.02444.x
4. Gachon C, Sime-Ngando T, Strittmatter M, Chambouvet A, Hoon KG (2010) Algal diseases: spotlight on a black box. Trends Plant Sci 15:633–640
5. Rasconi S, Jobard M, Sime-Ngando T (2011) Parasitic fungi of phytoplankton: Ecological roles and implications for microbial food webs. Aquat Microb Ecol 62:123–137
6. Gleason FH, Kagami M, Marano AV, Sime-Ngando T (2009) Fungal zoospores are valuable food resources in aquatic ecosystems. Mycologia 60:1–3
7. Kagami M, Von Elert R, Ibelings BW, de Bruin A, Van Donk E (2007) The parasitic chytrid, *Zygorhizidium*, facilitates the growth of the cladoceran zooksplankter, *Daphnia*, in cultures of the inedible alga. Asterionella Proc R Soc B 274:1561–1566
8. Kagami M, Helmsing NR, Van Donk E (2011) Parasitic chytrids could promote copepod survival by mediating material transfer from inedible diatoms. In: Sime-Ngando T, Niquil N (eds) Disregarded microbial diversity and ecological potentials in aquatic systems. Springer, Heidelberg, pp 49–54
9. James TY, Letcher PM, Longcore JE, Mozley-Standridge SE, Porter D, Powell MJ et al (2006) A molecular phylogeny of the flagellated fungi (Chytridiomycota) and description of a new phylum (Blastocladiomycota). Mycologia 98:860–871
10. Lefèvre E, Roussel B, Amblard C, Sime-Ngando T (2008) The molecular diversity of freshwater picoeukaryotes reveals high occurrence of putative parasitoids in the plankton. PlosOne 3:2324
11. Sime-Ngando T, Lefèvre E, Gleason FH (2011) Hidden diversity among aquatic heterotrophic flagellates: ecological potentials of zoosporic fungi. In: Sime-Ngando T, Niquil N (eds) Disregarded microbial diversity and ecological potentials in aquatic systems. Springer, Heidelberg, pp 5–22
12. Jobard M, Rasconi S, Sime-Ngando T (2010) Fluorescence *in situ* hybridization of uncultured zoosporic fungi: testing with clone-FISH and application to freshwater samples using CARD-FISH. J Microbiol Methods 83:236–243
13. Niquil N, Kagami M, Urabe J, Christaki U, Viscogliosi E, Sime-Ngando T (2011) Potential role of fungi in plankton food web functioning and stability: a simulation analysis based on Lake Biwa inverse model. In: Sime-Ngando T, Niquil N (eds) Disregarded microbial diversity and ecological potentials in aquatic systems. Springer, Heidelberg, pp 65–79
14. Lefèvre E, Carrias J-F, Bardot C, Amblard C, Sime-Ngando T (2005) A preliminary study of heterotrophic picoflagellates using oligonucleotidic probes in Lake Pavin. Hydrobiologia 55:61–67
15. Lefèvre E, Jobard M, Venisse JS, Bec A, Kagami M, Amblard C et al (2010) Development of a real-time PCR essay for quantitative assessment of uncultured freshwater zoosporic fungi. J Microbiol Methods 81:69–76
16. Moter A, Göbel UB (2000) Fluoresence *in situ* hybridization (FISH) for direct visualization of microorganisms. J Microbiol Methods 41:85–112
17. Mayura AM, Seena S, Barlocher F (2008) Q-RT-PCR for assessing *Archaea*, *Bacteria*, and *Fungi* during leaf decomposition in a stream. Microb Ecol 56:467–473
18. Bates ST, Garcia-Pichel F (2009) A culture-independant study of free-living fungi in biological soil crusts of the Colorado plateau: their diversity and relative contribution to microbial biomass. Environ Microbiol 11:56–67
19. Zhu F, Massana R, Not F, Marie D, Vaulot D (2005) Mapping of picoeukaryotes in marine ecosystems with quantitative PCR of the 18S rRNA gene. FEMS Microbiol Ecol 52:79–92

Jochen Schmid, Dirk Mueller-Hagen, Volker Sieber, and Vera Meyer

Abstract

Exopolysaccharides (EPS) produced by filamentous fungi often interfere with common methods for the extraction of nucleic acids or proteins. Similar precipitation behavior (e.g., by formation of hetero-triple helixes between the EPS and nucleic acid strands) severely minimizes yields of DNA or RNA extraction. In fungal strains that produce very high amounts of EPS, the fraction of mycelium per volume of fermentation broth is low, which requires removal of EPS to generate sufficient concentration of proteins or nucleic acids for downstream applications.

In this chapter we provide adapted methods for the extraction of nucleic acids and proteins of strong EPS-producing fungal strains using *Sclerotium rolfsii* as an example. The concentration and purity achieved are suitable for applications such as sequencing, RT-PCR, or 2D-PAGE.

Keywords

Exopolysaccharides • DNA • RNA • Protein • Extraction • Fungi • Scleroglucan • *Sclerotium rolfsii*

Introduction

Traditional methods for RNA, DNA, and protein extraction do not account for the presence of exopolysaccharides (EPS) at higher concentrations. A similar precipitation behavior of polysaccharides and other biological polymers such as RNA, DNA, and proteins, as well as an adverse volumetric ratio between mycelia and secreted EPS, lead to a low yield, if any at all. Alcohol precipitation is commonly used for concentrating, desalting, and recovering nucleic acids. Using alcohols such as ethanol or isopropanol as

J. Schmid (✉) • V. Sieber
Chair of Chemistry of Biogenic Resources,
Technische Universität München, Schulgasse 16,
Straubing, 94315, Germany
e-mail: J.schmid@tum.de

D. Mueller-Hagen • V. Meyer
Department of Applied and Molecular Microbiology,
Berlin University of Technology, Gustav-Meyer-
Allee 25, Berlin 13355, Germany

V.K. Gupta et al. (eds.), *Laboratory Protocols in Fungal Biology: Current Methods in Fungal Biology*,
Fungal Biology, DOI 10.1007/978-1-4614-2356-0_39, © Springer Science+Business Media, LLC 2013

a precipitant, DNA or RNA as well as polysaccharides lose their hydration hull and are thereby co-extracted. Precipitation of DNA using isopropanol at 4 °C is very effective; however, it is accompanied by the co-precipitation of salts. This negative effect is further increased in the presence of polysaccharides. As a result, with organisms that secrete large amounts of extracellular polysaccharides, these methods result in less concentrated DNA that is salt-enriched and highly contaminated with polysaccharides and therefore not suitable for downstream applications such as PCR or restriction analysis. One solution to counter this problem involves CTAB, which either precipitates the DNA or the contaminants [1]. At concentrations of 0.7–0.8 M NaCl, polysaccharides precipitate while nucleic acids remain in solution. This method is also suitable for purification of DNA and was also modified for high EPS-producing organisms [2] and for other highly EPS-contaminated products such as banana [3]. As with nucleic acids, proteins may interact with EPS, which results in a loss of certain proteins for downstream applications [4, 5].

Product yield is also hampered by an unfavorable ratio between the amount of mycelia and EPS, which are intensively swollen due to incorporation of water. This has to be opposed either by removal of EPS or the incorporated water.

Commercially available DNA and RNA extraction kits focus on samples such as plant leaves and blood, which show only a low contamination with polysaccharides. However, in our lab, extensive tests of a spectrum of these kits showed that they are not applicable to high EPS levels, which are achieved in cultivation of *Sclerotium rolfsii,* for example, or the bacterial EPS producer *Xanthomonas campestris.*

Another approach to get a grip on nucleic acid and protein extraction from high EPS-producing fungi is the development of cultivation conditions that impede EPS production. However, this approach alone is not suitable for sophisticated methods such as comparative transcriptome analysis [6].

In this chapter, methods optimized for the extraction of DNA, RNA, and proteins from high EPS-producing fungal strains are presented. We developed these methods for transcriptomic and proteomic analyses of *S. rolfsii,* an industrially important EPS-producing filamentous fungus of the genus basidiomycota. DNA, RNA, and proteins extracted were of high purity and yield and are applicable for many methods, such as microarray analysis, cDNA synthesis, RT-PCR, cloning, and 2D-PAGE. These protocolls can also act as a starting point for other high-level polysaccharide-producing organisms, such as bacteria, microalgae, and yeast, and also for metagenomic approaches, which are very often limited by the presence of polymeric substances such as polysaccharides.

Materials

See Note 1.
1. Cultivation media (see Note 2).
2. Filter gauze (for *S. rolfsii* with a pore size of 70 μm, see Note 3).
3. Sterile double distilled water
4. Liquid nitrogen (see Note 4)
5. Dismembrator (e.g., FastPrep FP220A Instrument and Braun Mikro Dismembrator 2) (see Notes 5 and 6).
6. DNA extraction buffer (10 mM Tris/HCl pH 8.0, 0.1 M EDTA pH 8.0, 0.5% SDS) (see Note 7).
7. RNA extraction buffer (4 M GuSCN, 0.1 M Tris/HCl pH 7.5, 1% β-Mercaptoethanol (0.14 M). Solve 50 g of GuSCN in 10 mL of 1 M Tris/HCl pH 7.5. Add to 100 mL with H_2O. Solution is filtered sterile and stored at RT. Directly before usage, sterilized and stored 1% of Mecaptoethanol is added).
8. RNAse (10 μg/μL).
9. Proteinase K (20 μg/μL).
10. DEPC-H_2O (see Note 8).
11. Chloroform/isoamylalcohol (24:1).
12. Isopropanol (see Note 9).
13. Phenol (Te-buffered) (see Note 1).
14. Ice.
15. Vortexer.
16. Magnetic stirrer.
17. Water bath.
18. Microcentrifuge (e.g., Hettich Lab Technology, Tuttlingen Germany).

19. Disposable polypropylene microcentrifuge tubes, 1.5 and 2 mL conical (Greiner, Germany).
20. Ultracentrifuge (Sorvall, Dupont, Bad Homburg).
21. Single-use ultracentrifugation tubes (suitable to the rotor).
22. Centrifuge tubes (suitable for 45,000×g).
23. Rotation incubator.
24. 10% Na-Laurylsarcosine.
25. AN-Prot-Ex buffer (25 mM Hepes pH 7.5, 50 mM KCl, 5 mM MgCl$_2$, 0.1 mM EDTA pH 8.0, 10% glycerol, 0.7 μL mercaptoethanol/1 mL, 35 μg/mL PMSF, 0.7 μg/mL Pepstatin A, 0.5 μg/mL Leupeptin).
26. Membrane-lysis buffer (5 mM Tris/HCl pH 7.4, 2 mM EDTA, 0.7 μg/m. Pepstatin A, 35 μg/mL PMSF, 0.5 μg/mL Leupeptin).
27. Membrane-suspension buffer (75 mM Tris/HCl pH 7.4, 12.5 mM MgCl$_2$, 5 mM EDTA).
28. Impact-resistant 2-mL tubes pre-filled with Lysing Matrix C particles.
29. PE-buffer (0.1 M Tris/HCl pH 8.8, 10 mM EDTA, 0.4% 2-mercapto-ethanol, 0.9 M sucrose).
30. IEF-buffer (7 M urea, 2 M thiourea, 4% (w/v) CHAPS, 20 mM DTT, 20 mM Tris, 1% Zwittergent pH 3–10, 0.5% Pharmalyte pH 3–10, 0.002% bromophenol blue).

Methods

Here, we present methods suitable for the extraction of DNA, RNA, and proteins from fungi that produce high amounts of exopolysaccharides. Obtained nucleic acids or proteins can be used for many applications, such as transcriptomic or proteomic analyses. With minor modifications all methods are applicable to other fungi that produce low amounts of EPS or no EPS at all as well as to bacteria.

Small-Scale Preparation of Genomic DNA from *S. rolfsii*

The method describes a fast procedure for obtaining DNA from fungal exopolysaccharide

producers with sufficient purity for subsequent PCR assays and is based on a method described earlier [7]. The number of tubes can be adapted if needed.

1. Inoculate 1 mL of cultivation media in a 2-mL reaction tube with a piece of mycelia (1 or 2 mm in diameter of a freshly overgrown plate) and incubate overnight at 28 °C with shaking.
2. Harvest mycelia in a tabletop centrifuge for 5 min with 12,000 rpm at RT.
3. Wash once with 1 mL ddH$_2$O.
4. Resuspend the pellet subsequently in 600 μL ddH$_2$O.
5. Disrupt cells by three times repeated shock freezing in liquid nitrogen following thawing for 5 min at 75 °C.
6. Purify DNA by adding 500 μL phenol and thoroughly shaking the tube for 2 min.
7. Separate phases in a tabletop centrifuge for 10 min with 12,000 rpm at 4 °C.
8. Transfer the top phase to a new tube.
9. Add 400 μL of chloroform and homogenize thoroughly by shaking the tube for 2 min.
10. Separate phases in a tabletop centrifuge for 10 min with 12,000 rpm at 4 °C.
11. Transfer the top phase to a new tube.
12. Add two volumes of ethanol (96%) and 1/25 volume 3 M Na-acetate.
13. Precipitate DNA for 1 h at −20 °C or overnight (see Note 10).
14. Centrifuge for 15 min with 12,000 rpm at 4 °C.
15. Wash the pellet once with 70% ethanol.
16. Air-dry the pellet for 15 min (see Note 11).
17. Resuspend DNA pellet in 20 μL ddH$_2$O.
18. Use 1 μL for PCR.

Large-Scale Preparation of Genomic DNA from *S. rolfsii*

This method describes a general procedure for obtaining high concentrations of high-quality DNA from fungal exopolysaccharide producers.

1. Inoculate 100 mL of respective medium with mycelium of *S. rolfsii* (1/8th of a freshly overgrown Petri dish cut into stripes with a scalpel) and add a sterile magnetic stirrer bar.

2. Incubate at 28 °C at 250 rpm on a magnetic stirrer (see Note 12).

3. Harvest mycelia after 48 h of cultivation by filtration through a piece of gauze.

4. Wash mycelia twice with hot water (85 °C) to remove scleroglucan.

5. Shock-freeze mycelium in liquid nitrogen (mycelia can be stored for several weeks at −80 °C).

6. Transfer 500 mg of mycelium to a pre-frozen Teflon cup (see Note 5).

7. Add 700 μL DNA extraction buffer.

8. Disrupt mycelia with a dismembrator (e.g., Braun Mikro-Dismembrator 2) for 2 min (see Note 6). Use of FastPrep FP220A will result in heavily sheared DNA.

9. Thaw the mixture and transfer to a 2-mL tube.

10. Add DNA extraction buffer to final volume of 1.8 mL.

11. Add 10 μL of RNAse (10 μg/μL).

12. Incubate for 10 min at 37 °C.

13. Add 10 μL of proteinase K (20 μg/μL).

14. Incubate for 30 min at 50 °C.

15. Divide the mixture in two equal parts.

16. Mix each with 900 μL of phenol.

17. Homogenize the two resulting phases by gentle shaking for 2 min.

18. Centrifuge for 10 min with 12,000 rpm at RT in a tabletop centrifuge.

19. Carefully transfer the top phase to a new 2-mL tube (see Note 13).

20. Add again 900 μL of phenol.

21. Repeat steps 17 and 18.

22. Transfer the top phase carefully to a new 2-mL tube (see Note 13).

23. Add 900 μL of chloroform/isoamylalcohol (24:1).

24. Repeat steps 17 and 18.

25. Combine both top phases and add an equal amount of isopropanol.

26. Do not mix for the first 5 min (see Note 14).

27. After 5 min mix by inverting the tube carefully.

28. Precipitate DNA for 30 min at −20 °C or overnight (see Note 10).

29. Centrifuge the sample for 5 min with 8,000 rpm at 4 °C in a tabletop centrifuge.

30. Rinse the DNA pellet once with 70% ethanol.

31. Centrifuge for 2 min with 8,000 rpm at 4 °C.

32. Air-dry the pellet for 15 min (see Note 11).

33. Re-dissolve the DNA pellet in 50 μL ddH$_2$O and incubate for 10 min at 65 °C.

34. DNA is now ready to use for PCR, restriction analysis, or Southern blot etc. (see Note 15).

Isolation of RNA from *S. rolfsii* using CsCl-Pads

The method given is based on a method by Chirgwin et al [8] and describes an efficient procedure for obtaining RNA from fungal exopolysaccharide producers. In our hands, RNA purification based on CsCl is the method of choice for EPS-producing microorganisms.

1. Inoculate 100 mL of respective medium with mycelium of *S. rolfsii* (1/8th of a freshly overgrown Petri dish cut into stripes with a scalpel) and add a sterile magnetic stirrer bar.

2. Incubate the flask at 28 °C with 250 rpm on a magnetic stirrer.

3. Harvest mycelia after 48 h of cultivation by filtration through a piece of gauze (70 μM) (see notes 3, 16, and 17).

4. Wash mycelia with sterile ddH$_2$O.

5. Transfer approximately 1 g mycelium to a pre-chilled (liquid N$_2$) Teflon cup, filled with two steel balls.

6. Disintegrate the mycelium for 2.5 min with a dismembrator (see Note 6).

7. Freeze again in liquid N$_2$.

8. Add 5 mL of RNA-extraction buffer to the Teflon cup.

9. Resuspend the pulverized mycelium and transfer to an SS-34 centrifuge tube, seal with parafilm and agitate.

10. Add 250 μL of 10% Na-Laurylsarcosine and gently turn the tube.

11. Centrifuge for 10 min with 9,000 rpm at RT.

12. Pre-fill an ultracentrifuge tube with 5 mL of the CsCl-pad (3.5 mL 5.7 M CsCl+0.01 M EDTA).

13. Carefully overlay the CsCl-pad with 1.5 mL of the supernatant (keep ultracentrifuge tube angular for adding the supernatant gently at the boundary).

14. Centrifuge for 19 h with $125,000 \times g$ at 4 °C using a swing bucket rotor (e.g., AH650).

15. Carefully remove supernatant leaving approximately 1 cm of supernatant above bottom level of the ultracentrifuge tube.

16. Remove bottom of the ultracentrifugation tube (containing the RNA pellet) over the surface of the resting supernatant with a hot scalpel (see Note 18).

17. Carefully remove the last supernatant.

18. Dissolve the pellet in 300 μL TE-SDS buffer and 1/10 Vol of 8 M LiCl containing 2.2 Vol ice-cold EtOH (96%).

19. Transfer the dissolved pellet into 1.5-mL reaction tube.

20. Centrifuge for 15 min with 10,000 rpm at 4 °C.

21. Wash the pellet once in 500 μL ice-cold EtOH (80%).

22. Centrifuge for 15 min with 10,000 rpm at 4 °C.

23. Discard the supernatant and air-dry for 10 min.

24. Dissolve the pellet in 100 μL of DEPC-H_2O.

Methods for Protein Isolation from EPS Producers

Extraction of Soluble Proteins

The following method is an easy-to-use protocol for the extraction of cytosolic, soluble proteins from a fungal exopolysaccharide producer. The number of tubes can be adapted to the amount needed. Raising the volume is not advised.

1. Inoculate 100 mL of respective medium with mycelium of *S. rolfsii* (one-eight of a freshly overgrown Petri dish is cut into stripes with a scalpel) and add a sterile magnetic stirrer bar.

2. Incubate the flasks at 28 °C with 250 rpm on a magnetic stirrer.

3. Harvest the mycelia after 48 h of cultivation by filtration through a piece of gauze (70 μM) (see Note 3).

4. Wash mycelia with sterile ddH$_2$O.

5. To remove excess water, squeeze the mycelium in the gauze (see Note 17).

6. Immediately shock-freeze mycelium in liquid nitrogen (mycelia can be stored for several weeks at −80 °C).

7. Transfer 750–1,500 mg of mycelium into N$_2$-pre-frozen Teflon tubes with two steel balls.

8. Add one volume of An-Prot-Ex buffer.

9. Disrupt mycelia in a dismembrator (e.g., Braun Mikro-Dismembrator 2) for 2 min (see Note 6).

10. Refreeze in liquid N$_2$.

11. Thaw the mixture on ice and transfer to a 2-mL tube.

12. Remove cell debris by centrifugation for 15 min with 14,000 rpm at 4 °C (see Note 19).

13. Transfer the supernatant into a fresh 1.5-mL tube and centrifuge for an additional 60 min with 14,000 rpm at 4 °C to remove the exopolysaccharide.

14. Transfer the resulting supernatant into a fresh 1.5-mL tube and store on ice or use immediately.

Extraction of Membrane Proteins

The following method is based on a Web-protocol (http://www.westernblotting.org/protocol%20 membrane%20extraction.htm), and was optimized to meet the requirements of high-EPS-producing fungi.

1. Inoculate 100 mL of respective medium with mycelium of *S. rolfsii* (1/8th of a freshly overgrown Petri dish is cut into stripes with a scalpel) and add a magnetic stirrer bar.

2. Incubate the flask at 28 °C with 250 rpm on a magnetic stirrer.

3. Harvest the mycelia after 48 h of cultivation by filtration through a piece of gauze (70 μM) (see Note 3).

4. Wash mycelia with sterile ddH$_2$O.

5. To remove excess water, squeeze the mycelium in the gauze (see Note 17).

6. Immediately shock-freeze mycelium in liquid nitrogen (mycelia can be stored for several weeks at −80 °C).

7. Transfer 750–1,500 mg of mycelium into Teflon tubes with 2 steel balls pre-frozen in liquid nitrogen.

8. Disrupt mycelia in a dismembrator (e.g., Braun Mikro-Dismembrator 2) for 2 min (see Note 6).

9. Transfer the powdered mycelium into a tube (e.g., SS-34) pre-filled with 10 mL membrane-lysis buffer.

10. Vortex for 20 s to homogenize the pellet.

11. Centrifuge with 500×g for 15 min at 4 °C.

12. Transfer supernatant into a fresh SS-34 tube.

13. Add 5 mL lyses buffer to the pellet and resuspend by vortexing for 20 s.

14. Centrifuge with 500×g for 15 min at 4 °C.

15. Combine the supernatants.

16. Centrifuge with 45,000×g for 15 min at 4 °C.

17. Discard the supernatant and wash the pellet twice in 5 mL lysis buffer by centrifugation with 45,000×g for 15 min at 4 °C.

18. Dissolve the pellet in 750 µL of membrane-suspension buffer and transfer into a 2-mL reaction tube (see Note 20).

19. The protein solution can be stored at −80 °C.

Extraction of Soluble and Insoluble Proteins

The set of proteins isolated depends on the method used [9, and our own published observations]. We therefore recommend applying different protein extraction methods in order to isolate as many as possible proteins. In our experience, a method that allows extracting both proteins of soluble and insoluble origin in one run is helpful to isolate proteins that cannot be captured by the two previous methods (and *vice versa*). For example, (soluble or insoluble) proteins with an alkaline isoelectric point can efficiently be isolated with the following method, which was adapted from Hurkman and Tanaka [10].

1. Inoculate 100 mL of respective medium with mycelium of *S. rolfsii* (1/8th of a freshly overgrown Petri dish is cut into stripes with a scalpel) and add a sterile magnetic stirrer bar.

2. Incubate the flask at 28 °C with 250 rpm on a magnetic stirrer.

3. Harvest the mycelia after 48 h of cultivation by filtration through a piece of gauze (70 µM) (see Note 3).

4. Wash mycelia with sterile ddH$_2$O.

5. To remove excess water, squeeze the mycelium in the gauze (see Note 17).

6. Immediately shock-freeze mycelium in liquid nitrogen (mycelia can be stored for several weeks at −80 °C).

7. Transfer 1.5–2 g of mycelium into a pre-frozen Teflon tube with 2 steel balls (see Note 5).

8. Disrupt mycelia in a dismembrator for 2 min (see Note 6).

9. Carefully open the Teflon tube and place it into a ice pan filled to 2 cm with liquid nitrogen, then add 2.5 mL of Tris pH 8.8 buffered phenol and 2.5 mL of PE-buffer.

10. Continue disruption for 1 min.

11. Transfer the suspension into a 15-mL tube and homogenize for 30 min at 4 °C.

12. Centrifuge for 10 min with 5,000×g at 4 °C.

13. Transfer the phenol phase (top phase) into a fresh tube and store on ice.

14. Mix the aqueous phase with 2.5 mL of Tris pH 8.8 buffered phenol and 2.5 mL of PE-buffer by vortexing.

15. Centrifuge for 10 min with 5,000 × g at 4 °C, then combine the phenol phase with the phenol phase from the first extraction in a centrifuge tube (e.g., SS-34).

16. Precipitate proteins by adding 5 volumes of 0.1 M ammonium acetate in 100% methanol (stored at −20 °C).

17. Vortex and incubate at −20 °C for at least 1 h or overnight.

18. Collect the precipitate by centrifugation for 20 min with 20,000 × g at 4 °C.

19. Wash the pellet twice with 0.1 M ammonium acetate in methanol, then twice with ice-cold 80% acetone (see Note 21).

20. Wash the pellet with cold 70% ethanol.

21. Store the pellet in 70% EtOH at −20 °C, or dissolve it in 0.5–1 mL of IEF-buffer and incubate for 1 h at RT for direct use.

Notes

1. Gloves (Nitrile) should be worn throughout these procedures. At some point, it is advisable to sterilize gloves using a disinfectant. Particular care should be taken when handling phenol and chloroform (consult safety data sheets). Phenol is a corrosive agent and a potential mutagen. All steps involving phenol and chloroform should be performed in a fume hood.

2. We tested several cultivation media for the extraction of pure fungal DNA with the different methods. No interference of media components, extraction buffer, etc., was observed. When possible and available, a medium composition that leads to a low level of EPS should be used.

3. A stable gauze should be used that can be sterilized in 80% ethanol. Pore size of the gauze is crucial for removal of the EPS and retention of the fungal biomass. For different fungi, different pore sizes might be tested. Steps including gauze can be substituted by dilution and centrifugation of the broth.

4. Liquid nitrogen is hazardous; work should only be performed with extra gloves suitable for low temperatures and protective glasses.

5. When using the Teflon cups for biomass homogenization, the sterile Teflon cups (sterilized with 80% of EtOH) should be prechilled for at least 5 min in liquid nitrogen, to guarantee that the mycelium stays frozen during homogenization.

6. Using a Braun dismembrator, 2–3 periods of 2 min at maximum shaking frequency (2,000 min-1), followed by a 2-min resting time on ice, have been found to be most efficient for disruption of most samples tested. However, optimization may be needed when testing other fungal species. Bacteria might be homogenized by the typical SDS method [2].

7. DNA extraction buffer can be stored for few weeks at 4 °C; however, using fresh extraction buffer improves yields.

8. DEPC-water is prepared as follows: Add 1 mL of 0.1% Diethylpyrocarbonate (DEPC) to 1,000 mL distilled water. Mix well and let set at room temperature for 1 h. Autoclave and let cool to room temperature prior to use.

9. Work with the solvent isopropanol should be performed in a fume hood.

10. Precipitation overnight results in higher yields.

11. Drying the pellet for longer than 30 min will result in only low solubility of the DNA pellet.

12. Stirrer speed has an important influence on mycelial morphology. Stirring at 500 rpm results in small loosely attached mycelia with less EPS production than stirring at 250 rpm.

13. Cutting the top of the pipette tip results in lower shear forces within the cup, facilitating more careful removal of the upper phase.

14. The resting time of 5 min is crucial for DNA purification and should be kept in detail. Gently mixing is also essential, as vortexing will shear genomic DNA.

15. In case the DNA is still contaminated with EPS, the following modifications can be applied. See step 14 of the section: Isolation of RNA from *S. rolfsii* using CsCl-Pads, and insert these additional steps: Step 14.1: Divide the mixture into three equal parts. Step 14.2: Add 100 µL of 5 M NaCl. Step 14.3: Add 75 µL CTAB/NaCl solution to the extraction buffer (10% CTAB, 0.7 M NaCl) and mix thoroughly. For preparation of the CTAB/NaCl solution, first weigh in the NaCl and add CTAB slowly while heating the solution. Step 14.4: Incubate the 3 samples at 65 °C for 30 min. Continue the protocol of Isolation of RNA from *S. rolfsii* Using CsCl-Pads with step 16.

16. Steps including harvesting and washing of mycelia should be performed as fast as possible to prevent alteration of the transcriptional status of the fungus, which is crucial (e.g., for a reliable transcriptome analyses).

17. The exopolysaccharide may swell intensively due to incorporation of water. Thereby, the amount of mycelium per volume is reduced significantly. To remove the excess water after washing the mycelium, squeeze out the mycelium in the gauze wearing sterile gloves. Keep the time for washing and

squeezing as short as possible and immediately transfer the mycelium into liquid N_2.

18. Cutting the bottom of the tube proved to be better for pipetting the sample.

19. After centrifugation the reaction tube contains ~50% clear supernatant and a fluffy pellet containing cell debris and exopolysaccharide.

20. At this point, the solution will be turbid, since the proteins are not soluble in the aqueous environment. For further applications such as 2D-PAGE use appropriate buffer systems.

21. Completely resuspend the pellet each time by pipetting up and down. Place the resuspended sample at −20 °C for at least 15 min between each wash.

References

1. Ausubel FH, Brent R, Kingston RE, Moore DD, Seidman JG, Smith JA et al (1994) Current protocols in molecular biology. Wiley, Hoboken, NJ

2. Jaufeerally-Fakim Y, Dookun A (2000) Extraction of high quality DNA from polysaccharides-secreting xanthomonads. Sci Technol 6:33–40

3. Shankar K, Chavan L, Shinde S, Patil B (2011) An improved DNA extraction protocol from four in vitro banana cultivars. Asian J Biotechnol 3:84–90

4. Sakurai K, Mizu M, Shinkai S (2001) Polysaccharide-polynucleotide complexes. 2. Complementary polynucleotide mimic behavior of the natural polysaccharide schizophyllan in the macromolecular complex with single-stranded RNA and DNA. Biomacromolecules 2:641–650

5. De Kruif CG, Tuinier R (2001) Polysaccharide protein interactions. Food Hydrocolloids 15:555–563

6. Schmid J, Müller-Hagen D, Bekel T, Funk L, Stahl U, Sieber V et al (2010) Transcriptome sequencing and comparative transcriptome analysis of the scleroglucan producer *Sclerotium rolfsii*. BMC Genomics 11:329

7. Meyer V, Wedde M, Stahl U (2002) Transcriptional regulation of the antifungal protein in *Aspergillus giganteus*. Mol Genet Genomics 266:747–57

8. Chirgwin JM, Przybyla AE, MacDonald RJ, Rutter WJ (1979) Isolation of biologically active ribonucleic acid from sources enriched in ribonuclease. Biochemistry 18:5294–9

9. Jacobs DI, Olsthoorn MM, Maillet I, Akeroyd M, Breestraat S, Donkers S et al (2009) Effective lead selection for improved protein production in Aspergillus niger based on integrated genomics. Fungal Genet Biol 46:S141–52

10. Hurkman WJ, Tanaka CK (1986) Solubilization of plant membrane proteins for analysis by two-dimensional gel electrophoresis. Plant Physiol 81:802–806

Directed Evolution of a Fungal Xylanase for Improvement of Thermal and Alkaline Stability

Dawn Elizabeth Stephens, Suren Singh, and Kugen Permaul

Abstract

Pre-treatment of paper pulps with xylanases has been shown to decrease the amounts of toxic chlorine dioxide used to bleach pulp. Natural xylanases are unable to tolerate the extremes of pH and temperature during the paper bleaching process and have to be genetically modified to be made more suitable for such industrial conditions. Such modification can be achieved using site-directed or random mutagenesis methods. Random mutagenesis methods are more attractive because detailed information regarding sequence or structure of the enzyme is not required. This chapter outlines how the thermal stability and alkaline stability of a glycosyl hydrolase family 11 cellulase-free xylanase from the fungus *Thermomyces lanuginosus* were improved using two random mutagenesis methods, error-prone PCR and a DNA shuffling method called the staggered extension process.

Keywords

Random mutagenesis • DNA shuffling • Xylanase • Thermal stability • Alkaline stability • Error-prone PCR • Staggered extension process

Introduction

Evolution is a process occurring over eons of time, in which the selection for specific traits is accomplished by applying environmental pressure. In nature, genetic diversity is obtained by protracted spontaneous mutations that occur during DNA replication or by recombination events. Through recursive cycles of mutation, selection and amplification, new traits accumulate in a population of organisms. Those that provide an advantage under prevailing environmental conditions are passed from one generation to the next [1]. Man has exploited natural evolution by using techniques such as cross-breeding in a specific manner to produce plants and animals with useful characteristics. This

D.E. Stephens (✉) • S. Singh • K. Permaul
Department of Biotechnology and Food Technology,
ML Sultan Campus, Durban University of Technology,
Durban, Kwa-Zulu-Natal, 4001, South Africa
e-mail: dawnestephens@yahoo.com

form of sexual recombination is one of the most powerful evolutionary strategies to generate new variants. From these crossings, progeny with improved features are chosen for additional breeding cycles [2, 3].

Genetic and protein engineering (both rational and random) are modern laboratory techniques for increasing the robustness of proteins for improved stability to high temperatures, extremes of pH, oxidizing agents and organic solvents. Cloning and expression of suitable genes from a thermophile into a suitable and faster-growing mesophilic host have allowed enhanced production of the specific thermostable enzyme required for a particular biotransformation process. Rational engineering is labour-intensive and requires precise information about enzyme structure. Random mutagenesis or directed evolution offers the fastest and most effective means of improving biocatalysts, provided the screening method is sensitive enough to detect the altered properties being screened for. During recombination, the beneficial properties from different variant genes are combined into a single gene, generating a protein with superior properties to its wild-type parent [4–7].

Xylanases are the predominant enzymes responsible for the hydrolysis of plant xylans [8]. Their uses are well documented in literature [9–12], but their predominant application has been in the biobleaching of paper pulp [13–16]. For xylanases to be economically feasible for pulp application, they must be cellulase-free, and retain stability from 60 to 90 °C at pH 8–10 for 3–5 h [17]. *T. lanuginosus* DSM 5826 produces high amounts of a cellulase-free xylanase, exhibits stability over a wide pH and temperature range, has been cloned into *Escherichia coli* [18] and crystallized [19]. It is a potentially good candidate for protein engineering for pulp application. This chapter focuses on the use of random mutagenesis and DNA shuffling to genetically alter the xylanase from *T. lanuginosus* DSM 5826

to render it more temperature and alkaline stable for future pulp application.

Materials

Cloned Gene(s) of Interest and Strains

Gene(s) of interest should be cloned onto a plasmid vector and transformed into a bacterial host for easier mutagenesis. The vector and host (bacteria or yeast) are important issues for achieving maximal expression of cloned genes and valuable insight for such selection is provided [20]. The control strain was *E. coli* containing a pBSK vector without the xylanase gene (*xynA*). An alternative vector/host expression system for high-level expression in *E. coli* such as the Champion™ pET SUMO Protein Expression System is currently available from Invitrogen.

Ampicillin Preparation

1. Ampicillin-Na salt (Roche Applied Sciences).
2. Sterile double-distilled water.
3. Sterile 1.5 mL centrifuge tubes.
4. Vortex mixer.
5. 10 mL sterile syringes (Braun).
6. 0.22 μm filters (Millipore).

Luria–Bertani Medium Preparation

1. 1.10 g/L bactopeptone.
2. 5 g/L yeast extract.
3. 5 g/L sodium chloride.
4. 15 g/L technical agar (for broth, agar is omitted).
5. Distilled water.
6. Magnetic stirrer.
7. Sterile petri plates
8. Liquid-measuring dispenser for broth tubes or Erlenmeyer flasks.

Plasmid DNA Isolation

1. Sterile 1.5 mL centrifuge tubes.
2. Microcentrifuge.
3. Vortex mixer.
4. FastPlasmid Mini kit (Eppendorf).
5. Isopropanol (95–100%).
6. Micropipettes.
7. Sterile tips.
8. A suitable analytical instrument capable of accurately quantifying DNA.

Agarose Gel Electrophoresis

1. Agarose (0.8%).
2. Erlenmeyer flask.
3. Sterile 1× TAE buffer (50×, 242 g Tris, 57.1 mL acetic acid, 100 mL of 0.5 M EDTA, pH 8).
4. Microwave.
5. Gel casting tray with well-makers.
6. Gel loading buffer (0.0375 g bromophenol blue 4 g sucrose, 1.5 mL 10% SDS, 3 mL 0.5 M EDTA in a total volume of 15 mL).
7. Electrophoresis tank and power supply.
8. Ethidium bromide (0.5 μg/mL).
9. UV transilluminator or gel documentation system.

Random Mutagenesis

1. Sterile 200 μL PCR tubes.
2. Microcentrifuge.
3. Vortex mixer.
4. 10 ng plasmid DNA (pX4).
5. T3 and T7 primers (0.5 μM; Integrated DNA Technologies).
6. 1U *Taq* polymerase (Roche Applied Sciences).
7. Mg^{2+} (supplied with *Taq*), Mn^{2+}, dNTPs (Roche Applied Sciences).
8. Diversify random mutagenesis kit (Clontech).
9. 1× PCR buffer (supplied with *Taq*).
10. Microcentrifuge.
11. PCR thermocycler (Genius-Techne). See Table 40.1 for more details.

After PCR, the 786 bp PCR products were recovered using the GFX PCR DNA and Gel Band Purification Kit (Amersham).

Restriction Endonuclease Digestion

1. PCR products.
2. pBSK plasmid DNA.
3. *Eco*RI and *Xho*I restriction enzymes and buffers (Roche Applied Sciences).
4. Sterile 1.5 mL centrifuge tubes.
5. Sterile distilled water.
6. Microcentrifuge.
7. Vortex mixer.
8. 37 °C heating block.
9. GFX PCR DNA and Gel Band Purification Kit (Amersham).

Ligation

1. Purified restricted vector and PCR products.
2. Sterile 1.5 mL centrifuge tubes.
3. Rapid DNA Ligation Kit (Roche Applied Sciences).
4. Microcentrifuge.
5. Vortex mixer.

Preparation of Competent Cells

1. Sterile 1.5 mL centrifuge tubes.
2. Microcentrifuge.
3. Vortex mixer.
4. A single colony of *E. coli* XL1 Blue MRF' (Stratagene).
5. 5 mL sterile SOC medium (20 g/L tryptone, 5 g/L yeast extract, 40 mM glucose, 20 mM NaCl, 20 mM $MgCl_2$, 20 mM $MgSO_4$, 5 mM KCl) in a test tube.
6. 37 °C shaking incubator.
7. 29 mL SOC medium in a 50 or 100 mL Erlenmeyer flask.
8. Spectrophotometer to measure cell density.
9. Ice.
10. 35 mL of cold sterilized 100 mM $CaCl_2$ containing 10% glycerol.

Table 40.1 Mutagenic PCR conditions used for *xynA*

Concentration (mM)	Condition								
	A	B	C	DI[a]	DII	E[b]	F[b]	G[b]	H[b] (control)
Mg^{2+}	1.5	4.8	1.5	2	2	3.5	3.5	3.5	3.5
Mn^{2+}	–	0.5	–	0.04	–	–	0.64	0.64	–
dNTPs	0.1	0.2 AG 0.8 CT	0.04 AG 0.2 CT	0.02 AG 0.2 CT	0.02 AG 0.2 CT 0.04 dITP	0.04 G	0.04 G	0.2 G	0.2
PCR Program	94 °C–1 min	94 °C–1 min		94 °C–1 min 46 °C–1 min 72 °C–2 min	94 °C–1 min 46 °C–1 min 72 °C–2 min	94 °C–1 min 46 °C–30 s 68 °C–1 min	72 °C–2 min	72 °C–2 min	42 °C–1 min
No. of cycles	35			20	30	25			No. of cycles
Reference	Matsumura and Ellington [37]; Chen et al. [38]			Xu et al. [39]		Diversify kit manual			Reference

[a] 2 µL unpurified PCR product of DI was used as template for DII

[b] Diversify random mutagenesis kit conditions. Special Diversify dNTP mix with unspecified concentrations was used for conditions E, F and G

Preparation of RBB-Xylan-LB Plates for Transformation and Screening

1. 0.4% RBB-xylan.
2. 10 g/L bactopeptone.
3. 5 g/L yeast extract.
4. 5 g/L sodium chloride.
5. 15 g/L technical agar containing 100 μg/mL ampicillin.

Transformation

1. 2 μL of the ligated DNA solutions.
2. 150 μL SEM-competent cells.
3. Ice.
4. 42 °C heating block.
5. 800 μL of fresh SOC medium (no ampicillin) per transformation reaction.
6. 37 °C shaking incubator.
7. Hockey stick.
8. RBB-xylan-LB plates.
9. 37 °C incubator.

Growth of Mutant Library and Enzyme Extraction

1. Clones exhibiting xylanase activity.
2. Sterile toothpicks.
3. LB-amp plates.
4. 5 mL LB-amp broth.
5. 37 °C shaking incubator.
6. Sterile 1.5 mL centrifuge tubes.
7. 30% sterile glycerol.
8. Temperature-controlled centrifuge.
9. 500 μL Bugbuster Protein Extraction Reagent (Novagen).
10. Shaking incubator at room temperature.

Thermostability Screening Assay

1. Xylanase supernatant.
2. 80 °C water bath.
3. 1.5 mL centrifuge tubes.
4. Ice.
5. Temperature-controlled centrifuge.

6. 1% Birchwood xylan substrate (Roth) dissolved in 0.05 M citrate buffer (pH 6.5).
7. 50 °C water bath.
8. DNS (dinitrosalicylic acid) reagent (150 g potassium sodium tartrate, 8 g NaOH, 5 g DNS in 500 mL distilled water).

Alkaline Screening Assay

1. 0.1% Birchwood xylan (Roth).
2. 1% Agarose.
3. 0.05 M glycine-NaOH buffer (pH 10).
4. Sterile petri plates.
5. Ouchterlony well-maker.
6. 60 °C incubator.
7. 1% Congo Red.
8. 1 M NaCl.
9. 60 °C water bath.
10. 1% Birchwood xylan substrate (Roth) dissolved in 0.05 M citrate buffer (pH 6.5).
11. 50 °C water bath.
12. DNS reagent.

Long-Term Thermal and Alkaline Stabilities

Growth and Extraction

1. 5 mL LB-amp broth per selected mutant strain.
2. 37 °C shaking incubator.
3. 300 mL LB-amp medium.
4. Spectrophotometer for measuring cell density
5. 1 mM filter-sterilized IPTG (isopropyl-β-D-thiogalactoside pyranoside).
6. Centrifuge.
7. 50 mL centrifuge tubes.
8. 150 U DNase I (Roche Applied Sciences).
9. 2 g Lysozyme (Roche Applied Sciences).
10. 120 mL Breaking buffer solution (6.80 g/L KH_2PO_4, 0.61 g/L $MgCl_2 \cdot 6H_2O$, 0.77 g/L dithiothreitol, 0.37 g/L EDTA, 0.10 g/L phenylmethylsulfonylfluoride).

Stability

1. Xylanase-containing supernatant.
2. 80 °C water bath.

3. 1.5 mL centrifuge tubes.
4. Ice.
5. Temperature-controlled centrifuge.
6. 1% birchwood xylan substrate (Roth) dissolved in 0.05 M citrate buffer (pH 6.5).
7. 50 °C water bath.
8. DNS (dinitrosalicylic acid) reagent (150 g potassium sodium tartrate, 8 g NaOH, 5 g DNS in 500 mL distilled water).
9. 0.05 M glycine-NaOH buffer (pH 10).
10. 60 °C water bath.

DNA Recombination: Staggered Extension Process

1. Overnight cultures of G41 and G53 mutants.
2. FastPlasmid Mini Kit (Eppendorf).
3. 10 ng plasmid DNA.
4. 1.5 mM $MgCl_2$.
5. 0.5 μM each T3 and T7 primers.
6. 1× PCR buffer and 1 U *Taq*.
7. 640 μM $MnSO_4$, 0.2 mM dGTP.
8. Diversify Random Mutagenesis Kit (Clontech).

DNA Sequence Analysis

1. Automated DNA sequencer.
2. Chromas Lite software.
3. ExPASy protein translation tools.
4. CLUSTALW sequence alignment.

Methods

Cloning

The xylanase gene used for this research, *xynA* of *T. lanuginosus*, was cloned into a plasmid Bluescript SK vector (pBSK) using the λ-ZapII cloning system (Strategene) and transformed into *E. coli* [18]. This plasmid was called pX3. The earlier strategy for cloning the xylanase involved creating a cDNA library, so the *xynA* gene was present on a larger insert. The cloned insert was thus reduced in size to increase the mutational

efficiency of an ep-PCR reaction, and facilitate easier sequencing. The modified plasmid was renamed pX4 [21].

Ampicillin Preparation

A 5 mL stock solution of 100 mg/mL of ampicillin was aseptically made using sterile distilled water and filter sterilized with 0.22 μm filters. The solution was aliquoted in sterile 1.5 mL centrifuge tubes and stored at −20 °C. For use, the ampicillin was added to the growth media to a final concentration of 100 μg/mL. Note that ampicillin is never autoclaved.

LB Medium Preparation

LB medium supplemented with ampicillin (see previous) was used for plating of the *E. coli* harbouring the plasmid with the cloned gene of interest. This medium was used to maintain all recombinant strains. LB media components were combined, dissolved and sterilized. When the medium cooled, the prepared ampicillin was added and the medium was mixed gently (to avoid creation of air bubbles). Agar plates were then aseptically poured, allowed to set and stored at 4 °C.

LB medium (without agar) can be made up in bulk, then dispensed in tubes or Erlenmeyer flasks (depending on final use) and sterilized. They can be stored at 4 °C or room temperature and ampicillin can be added just prior to use.

Plasmid Isolation

E. coli cultures with plasmids were inoculated in 5 mL LB medium containing 100 μg/mL ampicillin and grown for 12–16 h at 37 °C in a shaking incubator at 150 rpm. Cells were harvested by centrifugation of broth cultures at 5,000×*g* for 5 min. For other details refer to the FastPlasmid Mini Kit (Eppendorf) user manual.

Measurement of DNA Concentration

This was estimated spectrophotometrically at 260 nm and calculated on the premise that an absorbance of 1 at $OD_{260\ nm}$ corresponds to 50 ng DNA/µL. Purity of the DNA sample was estimated using the absorbance ratio measured at both $OD_{260\ nm}$ and $OD_{280\ nm}$. A ratio greater than 1.7 is considered ideal purity for procedures like DNA sequencing. Currently, many instruments capable of accurately quantifying DNA are available.

Agarose Gel Electrophoresis

The desired amount of agarose was placed in an Erlenmeyer flask together with the required amount of 1× TAE buffer, which was diluted from the 50× TAE stock. The contents of the flask were then microwaved for 1 min and poured into a casting tray with well combs and allowed to set. Gel loading buffer was added to the DNA samples in a ratio of 1:5, which was then loaded into the agarose gel wells. Samples were run alongside a DNA molecular weight marker at 90 V for approximately 1 h. Gels were then stained in ethidium bromide for 20 min and destained in distilled water for a further 5–10 min. Care must be taken to strictly adhere to the recommended concentration for ethidium bromide staining as high concentrations can reduce ligation efficiency of target DNA. Stained gels were then viewed on a UV transilluminator and the band sizes compared to the DNA molecular weight marker. Ethidium bromide is a known mutagen and must be handled with extreme precaution. It is safer to purchase a ready-made stock to minimize handling [20, 22].

Random Mutagenesis

A broad scope of conditions (see Table 40.1) ranging from previously published research as well as the use of a commercial mutagenesis kit, Diversify Random Mutagenesis Kit (Clontech), were investigated for mutagenesis of *xynA*.

Commercial mutagenesis kits offer the advantage of being able to theoretically control the number of mutations per gene. Plasmid pX4 served as the template DNA for all mutagenic conditions tested. The final volume of each PCR reaction was always 50 µL. Five microlitres of each PCR product was analysed by agarose gel electrophoresis to determine if the target DNA was successfully amplified. Lighter bands generally imply a high mutagenic rate in ep-PCR since the *Taq* polymerase does not amplify optimally. Therefore, it cannot form as much product compared to PCR under normal conditions. This represents an early, visible indication of mutation in the target gene. The remaining PCR products were recovered after agarose gel elctrophoresis. The gel was stained for not more than 10 min in ethidium bromide and then viewed on the UV transilluminator whilst the PCR-gel bands were quickly excised. Ethidium bromide and UV light are mutagens and prolonged contact with the PCR products can cause irreversible DNA damage and also drastically reduce ligation efficiency. The GFX PCR DNA and Gel Band Purification Kit (Amersham) was used according to the manufacturer's instructions. Many such kits are available for commercial use.

Restriction

Standard protocols were followed for restriction analysis [20, 22]. Restriction analysis is necessary to create compatible sticky ends to facilitate ligation of the PCR product into the vector. Note that both the vector and PCR products must be restricted separately, and with the same combinations of restriction enzymes. If using more than one restriction enzyme, it is important to use buffers compatible with both restriction enzymes. Buffer compatibility charts are usually included with the restriction enzymes. For digestion, the following volumes were added to a sterile 1.5 mL centrifuge tube: 5 µL DNA, 2 µL sterile water, 1 µL buffer H, 1 µL *Eco*RI and 1 µL *Xho*I. The tubes were vortexed and spun down in a minicentrifuge for a few seconds and then incubated in a 37 °C heating block for 1 h. For notes on restriction digestion, refer to references [20, 22]. Each restriction reaction was purified from solution using the GFX PCR DNA and Gel Band Purification Kit (Amersham) according to the manufacturer's instructions.

Ligation

T4 DNA ligase was used to join the DNA between the 5′-phosphate and the 3′-hydroxyl-group of adjacent nucleotides of the vector and PCR inserts. For the creation of recombinant libraries, a molar vector, insert ratio of 1:3 was used to guarantee high ligation efficiency. Ligations were carried out using the Rapid DNA Ligation Kit (Roche Applied Sciences) according to the manufacturer's instructions and were directly transformed into competent *E. coli* cells.

Preparation of Competent Cells

Host cells were made "competent" or capable of taking up DNA from their surrounding environment, by exposing them to Ca^{2+}, which interacts with their cell envelopes. *E. coli* host cells were made competent prior to transformation [23].

E. coli XL1 Blue was cultured on LB medium without ampicillin. A single colony was used to inoculate 5 mL sterile SOC medium (20 g/L tryptone, 5 g/L yeast extract, 40 mM glucose, 20 mM NaCl, 20 mM $MgCl_2$, 20 mM $MgSO_4$, 5 mM KCl), incubated at 37 °C and shaken overnight. One millilitre of this culture was used to inoculate 29 mL of fresh SOC medium and was shaken at 150 rpm at 37 °C until it reached an $OD_{600\,nm}$ of 0.5. At this stage of growth, *E. coli* cells have maximum ability to uptake DNA. The culture was immediately placed on ice and kept cold for the duration of the procedure. Cold conditions suspend the continued growth of the cells and preserve their viability for subsequent transformation. The cells were pelleted at 5,000×*g* for 10 min and the supernatant was discarded. The cells were resuspended in 10 mL cold 100 mM $CaCl_2$, recentrifuged at the same speed and resuspended in 10 mL of cold 100 mM $CaCl_2$. The

entire mixture was incubated on ice for 20 min and then centrifuged. The competent cells were subsequently resuspended in 2 mL 100 mM $CaCl_2$ containing 10% glycerol. One hundred and fifty microlitres of the prepared competent cells were dispensed into eppendorfs, stored at 4 °C overnight, and then maintained at −70 °C, after snap-freezing in liquid nitrogen. According to the protocol followed [23], it is postulated that SEM-competent cells are most efficient when prepared 24 h prior to transformation. The preparations were therefore incubated at 4 °C overnight to enhance the effectiveness of the subsequent transformation procedure.

Preparation of RBB-Xylan-LB Plates for Transformation and Screening

Mutation sometimes produces clones that contain the gene of interest; however, it is often so heavily mutated that an active, fully functional protein is no longer produced [24, 25]. Also, mutations are more likely to destroy enzyme activity rather than enhance it. Thus, Remazol Brilliant Blue dye was linked to birchwood xylan for detection of xylanase producers during the transformation process [26, 27]. Only clones capable of producing a fully functional xylanase would be able to break the linkage between the xylan and the RBB dye to produce colourless xylooligosaccharides which was visible as a zone of hydrolysis around the colony. RBB-xylan was added to LB medium components, dissolved and sterilized. After cooling to just below 50 °C, filter-sterilized ampicillin was added to the medium and mixed gently (to avoid creation of air bubbles). Agar plates were then aseptically poured, allowed to set and stored at 4 °C.

Transformation

Two microlitres of the ligated DNA solutions were added to 150 μL of the SEM-competent cells and incubated on ice for 30 min and thereafter subjected to heat shock for 30 s at 42 °C. An increase in temperature allows for the creation of transient pores that allow for the uptake of DNA attached to the cell surfaces. Rapid addition of 800 μL of fresh SOC medium to each of the mixtures followed the heat shock procedure and they were then shaken at 37 °C for 1 h. SOC medium is rich in salts and ions that facilitate the rapid regeneration of the cell wall. No ampicillin is added to the SOC medium to prevent additional metabolic stress on the cells. The cells actively divide and the ligated products (plasmids) are also amplified. Any uncircularized (or unligated) DNA is degraded by cell DNases. The transformation mixtures were diluted 1:1 with fresh SOC medium and 100 μL was plated on RBB-xylan-LB containing ampicillin and incubated at 37 °C overnight.

Growth of Mutant Library and Enzyme Extraction

Generally, as the degree of mutagenesis in ep-PCR reactions increases, the number of transformants obtained also decreases because less DNA is amplified during the ep-PCR reactions. All positive transformants, that is, clones exhibiting xylanase activity, were picked with a sterile toothpick and initially streaked onto an LB plate containing 100 μg/mL ampicillin. Subsequently, a single colony from the overnight plate was inoculated into 5 mL LB broth containing 100 μg/mL ampicillin and incubated for 12–16 h at 37 °C in a shaking incubator at 150 rpm. One hundred and fifty microlitres of these overnight cultures were placed in sterile 1.5 mL centrifuge tubes containing 30% sterile glycerol. These tubes were stored at −20 °C and served as the master mutant library stock. Since the E. coli transformant host is a prokaryote and the xylanase is eukaryotic, most of the xylanase is located intracellularly and cannot be properly secreted. It must be lysed to release the functional xylanase protein. Mechanical lysis methods such as the use of a French press or sonicator are unsuitable since they result in significant heating of the sample which can denature the xylanase. A gentle chemical lysis agent such as the BugbusterProtein Extraction Reagent (Novagen) is much more effective. The use of Bugbuster, however, is uneconomical for large-scale cell lysis.

The rest of the culture was centrifuged at 5,000×g and the media discarded. The pellets were resuspended in 500 μL cold Bugbuster reagent. Pellet resuspension must be gentle to minimize foaming and heat generation. The suspensions were shaken gently at 40 rpm at room temperature for 20 min to gently lyse E. coli cells and release the mutant xylanases. The lysates were then centrifuged at 15,000×g at 4 °C to remove cell debris. The resulting pellets were discarded whilst the supernatants were stored at 4 °C for further analysis.

Thermostability Screening Assay

The clear lysate obtained after enzyme extraction contained the crude xylanase and was used to test the thermostability of the xylanase variants. The protocol followed was a combination of the methods used [28, 29]. A temperature of 80 °C was chosen for screening possible thermostable xylanase variants since XynA is documented to be stable up to 70 °C [18].

A short-term screen time of 40 min at 80 °C was used to decrease the number of mutants screened later. Prior to incubation at 80 °C in a water bath, 0 min (untreated) samples were removed from the clear cell lysates and placed on ice. The crude enzymes were subsequently heated for 40 min, chilled on ice for 15 min and incubated for 30 min at room temperature to prevent low-temperature denaturation of the enzymes. The samples were centrifuged and the supernatants assayed for residual xylanase activity [30]. Activities of the 0 min samples were considered as 100%, and activities of the 40 min incubation time were expressed as percentages of the untreated sample to determine the percentage of residual activity after heat treatment. The wild-type XynA served as the control.

Alkaline Screening Assay

Alkaline screening was performed in two parts, viz., a plate assay and a liquid assay that tested stability at pH 10. Wells were punched (using an Ouchterlony well-maker) into plates containing

0.1% birchwood xylan and 1% agarose made with 0.05 M glycine–NaOH buffer (pH 10). Ten microlitres of each crude enzyme was inoculated into the wells, with one well containing the control XynA on each screening plate. The plates were incubated at 60 °C for 2 h and then stained with 1% Congo Red for 25 min. Excess dye was flushed off with 1 M NaCl for up to 1 h, until zones of hydrolysis were clearly visible [31, 32]. The supernatants of mutants displaying larger or more distinct zones than XynA were diluted in 0.05 M glycine-NaOH buffer (pH 10) and incubated for 40 min in a 60 °C water bath and residual activities of the enzymes were determined. A temperature of 60 °C was used for detection of alkaline-stable mutants to effectively trim down the number of mutants screened during longer incubation periods.

Long-Term Thermal and Alkaline Stabilities

Growth and Extraction

Mutants displaying more than 60% residual activity after heat and alkaline treatment were selected for long-term stability testing. These mutants were inoculated into 5 mL LB-amp broth and incubated overnight at 37 °C at 150 rpm. One millilitre of this culture was used to inoculate 300 mL LB-amp medium and shaken at 37 °C until the $OD_{600\,nm}$ of all flasks reached 0.5 absorbance units. The cells were induced for xylanase production since the pBSK vector has a LacZ promoter. This means that lactose would induce the LacZ promoter, and thus leads to expression of the cloned *xynA*. IPTG is a lactose analogue, which constitutively induces the LacZ promoter, but is not metabolized. IPTG is added to the flasks and shaken overnight at 37 °C. Thereafter, the samples were centrifuged at 5,000×*g* with each 50 mL pellet resuspended in 2 mL of cold lysis solution. Bugbuster reagent (used previously for large-scale screening) is unstable for long-term incubation at high temperatures. Thus, a lysis solution was designed to eliminate this problem. First, a breaking buffer solution was made by dissolving 0.37 g/L EDTA in 200 mL distilled water,

followed by the addition of 6.80 g/L KH_2PO_4, 0.61 g/L $MgCl_2.6H_2O$ and 0.77 g/L dithiothreitol. The pH of the solution was brought to 6.8. After autoclaving and cooling of the breaking buffer solution, 0.10 g/L phenylmethylsulfonylfluoride was added. The lysis solution comprised 150 U DNase I, 2 g lysozyme in 120 mL breaking buffer solution. The suspensions were left at 4 °C overnight and then centrifuged at 15,000×*g*. The supernatant lysates containing the enzymes were stored at 4 °C until further use.

Stability

For determinations of long-term thermal stability, samples were incubated at 80 °C and samples were removed every 15 min for 90 min, incubated on ice and then assayed for residual xylanase activity as previously described. For determination of long-term alkaline stability, samples were diluted in 0.05 M glycine-NaOH buffer (pH 10), incubated at 60 °C and samples were removed every 15 min for 90 min, incubated on ice and then assayed for residual activity. Activities of each mutant were expressed as percentages of the 0 min sample.

DNA Recombination: Staggered Extension Process

In this study, highly thermostable xylanases displayed poor catalytic activities, whereas the opposite was found for alkali-stable xylanases. Thus, the two best thermostable and alkaline-stable *xynA* variants were good candidates for DNA recombination for incorporation of both properties to create a single, robust xylanase. The recombination method used in this study was a modification of the staggered extension process (StEP) reaction [33]. Advantages of the StEP method include the following: (1) it can be performed using a pair of flanking primers in a single PCR tube; (2) separation of parent templates from the recombined products is not necessary; and (3) higher recombination frequencies between highly homologous templates can be achieved. Disadvantages include that there is a smaller probability of generating enzymes with unique functions and the sometimes non-specific annealing and formation of undesirable recombinants. Other shuffling methods have been reviewed and there are benefits and disadvantages of each method [34].

The G41 and G53 plasmids were isolated using the FastPlasmid Mini Kit (Eppendorf). Thermal stability was considered more important than alkaline stability; thus, twice the amount of G41 plasmid was added to each reaction as compared with G53. The final concentration of DNA was maintained at 10 ng. StEP recombination PCR reactions contained 10 ng total DNA, 1.5 mM MgCl$_2$, 0.5 µM each T3 and T7 primers, 1× PCR buffer and 1 U *Taq*. Eighty cycles of denaturation for 30 s at 94 °C followed by annealing for 4 s at 42 °C were carried out on both the reactions. The recombination products were subsequently cloned and xylanase activity was tested as previously described. The study resulted in the creation of recombinant xylanases that exhibited properties of both parent, G41 and G53, xylanases [35].

DNA Sequence Analysis

DNA sequencing was carried out to determine the mutations that were responsible for the observed changes in stabilities of some xylanase variants. Many possible automated DNA sequencing methods are currently available. Raw DNA sequencing data were initially processed using the Chromas Lite software package (Technelysium, version 2.0) (www.technelysium.com.au) and both DNA strands were edited to yield complete gene sequences. The DNA sequences were then translated into their protein counterparts using the Translate tool from the ExPASy Website (www.expasy.org/tools/dna.htmL) and aligned to the wild-type parent using the CLUSTALW (version 1.81) alignment program on the GenomeNet server (www.clustalw.genome.ad.jp). The results showed that most substitutions occurred in the β-sheet of *xynA*. Studies of family 11 xylanases indicate that this long β-sheet is responsible for thermostability, because it stabilizes the overall xylanase structure [36]. Also, noteworthy was the observed increase in arginine content of the most stable xylanase variants.

Acknowledgements Ms. Siphi Dlungwane is duly acknowledged for providing technical assistance. The National Research Foundation of South Africa funded this research.

References

1. Taylor SV, Kast P, Hilvert D (2001) Investigating and engineering enzymes by genetic selection. Angew Chem Int Ed 40:3310–3335
2. Tobin MB, Gustafsson C, Huisman GW (2000) Directed evolution, the "rational" basis for "irrational" design. Curr Opin Struct Biol 10:421–427
3. Dalby PA (2003) Optimizing enzyme function by directed evolution. Curr Opin Struct Biol 13:500–505
4. Cedrone F, Ménez A, Quéméneur E (2000) Tailoring new enzyme functions by rational redesign. Curr Opin Struct Biol 10:405–410
5. Bacher JM, Reiss BR, Ellington AD (2002) Anticipatory evolution and DNA shuffling. Genome Biol 3:1021.1–1021.4
6. Haki GD, Rakshit SK (2003) Developments in industrially important thermostable enzymes, a review. Biores Technol 89:17–34
7. Tracewell CA, Arnold FH (2009) Directed enzyme evolution, climbing fitness peaks one amino acid at a time. Curr Opin Chem Biol 13:3–9
8. Biely P (1985) Microbial xylanolytic systems. Trends Biotechnol 3:286–290
9. Matsumura S, Sakiyama K, Toshima K (1999) Preparation of octyl-β-D-xylobioside and xyloside by xylanase-catalyzed direct transglycosylation reaction of xylan and octanol. Biotechnol Lett 21:17–22
10. Strauss MLA, Jolly NP, Lambrechts MG, van Rensburg P (2001) Screening for the production of extracellular hydrolytic enzymes by non-*Saccharomyces* wine yeasts. J Appl Microbiol 91:182–190
11. Kimura T, Suzuki H, Furuhashi H, Aburatani T, Morimoto K, Sakka K et al (2002) Molecular cloning, characterization and expression analysis of the *xynF3* gene from *Aspergillus oryzae*. Biosci Biotechnol Biochem 66:285–292
12. Colombatto D, Morgavi DP, Furtado AF, Beauchemin KA (2003) Screening of exogenous enzymes for ruminant diets, relationship between biochemical characteristics and *in vitro* ruminal degradation. J Anim Sci 81:2628–2638
13. Viikari L, Kantelinen A, Sunquist J, Linko M (1994) Xylanases in bleaching, from an idea to the industry. FEMS Microbiol Rev 13:335–350
14. Beg QK, Kapoor M, Mahajan L, Hoondal GS (2001) Microbial xylanases and their industrial applications, a review. Appl Microbiol Biotechnol 56:326–338
15. Salles BC, Medeiros RG, Bao SN, Silva FG, Filho EXF (2005) Effect of cellulase-free xylanases from *Acrophialophora nainiana* and *Humicola grisea* var. *thermoidea* on eucalyptus kraft pulp. Process Biochem 40:343–349

16. Valls C, Vidal T, Roncero MB (2010) The role of xylanases and laccases on hexenuronic acid and lignin removal. Process Biochem 45:425–430

17. Techapun C, Poosaran N, Watanabe M, Sasaki K (2003) Thermostable and alkaline-tolerant cellulase-free xylanases produced from agricultural wastes and the properties required for use in pulp bleaching processes, a review. Process Biochem 38:1327–1340

18. Schlacher A, Holzmann K, Hayn M, Steiner W, Schwab H (1996) Cloning and characterization of the gene for the thermostable xylanase XynA from *Thermomyces lanuginosus*. J Biotechnol 49:211–218

19. Gruber K, Klintschar G, Hayn M, Schlacher A, Steiner W, Kratky C (1998) Thermophilic xylanase from *Thermomyces lanuginosus*, high resolution x-ray structure and modelling studies. Biochemistry 37: 13475–13485

20. Sambrook J, Fritsch EF, Maniatis T (1989) Molecular cloning, a laboratory manual, 2nd edn. Cold Spring Harbour Laboratory, New York, NY

21. Stephens DE, Singh S, Permaul K (2009) Error-prone PCR of a fungal xylanase for improvement of its alkaline and thermal stability. FEMS Microbiol Lett 293:42–47

22. Sambrook J, Fritsch EF, Maniatis T (2001) Molecular cloning, a laboratory manual, 3rd edn. Cold Spring Harbor Laboratory, Cold Spring Harbor, NY

23. Ausubel FM, Brent R, Kingston RE, Moore DD, Siedman JG, Smith JA, Struhl K (1989) Current protocols in molecular biology, vol 2. Greene, New York, NY

24. Drummond DA, Iverson BL, Georgiou G, Arnold FH (2005) Why high error-rate random mutagenesis libraries are enriched in functional and improved proteins. J Mol Biol 350:806–816

25. Vanhercke T, Ampe C, Tirry L, Denolf P (2005) Reducing mutational bias in random protein libraries. Anal Biochem 339:9–14

26. Biely P, Mislovicova D, Toman R (1985) Soluble chromogenic substrates for the assay of endo-β-1,4-xylanases and endo-β-1,4-glucanases. Anal Biochem 144:142–146

27. Biely P, Mislovicova D, Toman R (1988) Remazol brilliant blue xylan, a soluble chromogenic substrate for xylanases. Methods Enzymol 160:536–541

28. Giver L, Gershenson A, Freskgard PO, Arnold FH (1998) Directed evolution of a thermostable esterase. Proc Natl Acad Sci U S A 95:12809–12813

29. Matsuura T, Miyai K, Trakulnaleamsai S, Yomo T, Shima Y, Miki S et al (1999) Evolutionary molecular engineering by random elongation mutagenesis. Nat Biotechnol 17:58–61

30. Bailey MJ, Biely P, Poutanen K (1992) Interlaboratory testing of methods for assay of xylanase activity. J Biotech 23:257–270

31. Teather RM, Wood PJ (1982) Use of Congo Red-polysaccharide interactions in enumeration and characterization of cellulolytic bacteria from the bovine rumen. Appl Environ Microbiol 43:777–780

32. Béguin P (1990) Molecular biology of cellulose degradation. Annu Rev Microbiol 44:219–248

33. Zhao H, Giver L, Shao Z, Affholter JA, Arnold FH (1998) Molecular evolution by staggered extension process (StEP) in vitro recombination. Nat Biotechnol 16:258–261

34. Gong J, Zheng H, Wu Z, Chen T, Zhao X (2009) Genome shuffling: progress and applications for phenotype improvement. Biotech Adv 27: 996–1005

35. Stephens, D.E. (2007). Protein engineering of a fungal xylanase. Doctoral Thesis. Durban University of Technology, South Africa.

36. Hakulinen N, Turunen O, Jänis J, Leisola M, Rouvinen J (2003) Three-dimensional structures of thermophilic β-1,4-xylanases from *Chaetomium thermophilum* and *Nonomuraea flexuosa*, comparison of twelve xylanases in relation to their thermal stability. Eur J Biochem 270:1399–1412

37. Matsumura I, Ellington AD (2001) Mutagenic PCR of protein-coding genes for *in vitro* evolution. In: Braman J (ed) In vitro mutagenesis. Humana, NJ, pp 261–269

38. Chen YL, Tang TY, Cheng KJ (2001) Directed evolution to produce an alkalophilic variant from a *Neocallimastix patriciarum* xylanase. Can J Microbiol 47:1088–1094

39. Xu H, Petersen EI, Petersen SB, El-Gewely MR (1999) Random mutagenesis libraries: optimization and simplification by PCR. Biotechniques 27: 1102–1108

Genome Shuffling Protocol for the Pentose-Fermenting Yeast *Scheffersomyces stipitis*

41

Paramjit K. Bajwa, Nicole K. Harner,
Terri L. Richardson, Sukhdeep Sidhu,
Marc B. Habash, Jack T. Trevors, and Hung Lee

Abstract

This chapter presents the protocol for genome shuffling based on recursive cross-mating in the pentose-fermenting yeast *Scheffersomyces* (*Pichia*) *stipitis*. Genome shuffling involves two stages. In the first stage, a pool of mutants with improved phenotypes is selected. Several rounds of random mutagenesis can be done using different mutagens, and mutant selection can be based on different criteria to generate different mutant cell lines. In the second stage, the genomes of mutants derived from different lines are mated recursively to allow for genetic recombination, followed by screening after each mating cycle to select for improved phenotypes in the recombinants. A number of reports have described genome shuffling based on recursive protoplast fusion in bacteria and yeasts. Recently, we developed mating-based genome shuffling in the pentose-fermenting yeast *S. stipitis*. We have used this approach to obtain genetically stable mutants of *S. stipitis* with considerably improved tolerance to hardwood spent sulphite liquor (HW SSL), a pulping waste liquor containing a complex mixture of inhibitory substances. This was achieved in the complete absence of knowledge as to the precise genetic modifications needed to confer HW SSL tolerance. Here we describe the protocols for recursive UV mutagenesis, cross-mating, sporulation and isolation of recombinants with improved phenotypic traits.

Keywords

Genome shuffling • Random mutagenesis • Recursive cross-mating • Pentose fermentation • *Scheffersomyces stipitis*

P.K. Bajwa • N.K. Harner • T.L. Richardson •
S. Sidhu • M.B. Habash • J.T. Trevors • H. Lee (✉)
School of Environmental Sciences, University of Guelph,
50 Stone Road East, Guelph, ON, Canada N1G 2W1
e-mail: hlee@uoguelph.ca

V.K. Gupta et al. (eds.), *Laboratory Protocols in Fungal Biology: Current Methods in Fungal Biology*,
Fungal Biology, DOI 10.1007/978-1-4614-2356-0_41, © Springer Science+Business Media, LLC 2013

Introduction

Genome shuffling is a microbial strain improvement technology that involves the generation of a pool of mutants with improved phenotypes, followed by iterative recombination between their genomes. This approach offers a number of advantages compared to the classical and molecular methods of microbial strain improvement. Classical strain improvement using mutagenesis and selection is time-consuming, laborious and based on a single starting strain. Molecular approaches of strain improvement are only applicable to genes that can be isolated, along with some knowledge of what genetic modifications are needed to produce the desired phenotypic effect. To improve a cell's complex phenotype, such as stress tolerance, likely requires modification in a number of known and unknown genes; hence, the molecular methods are inadequate. Genome shuffling is particularly suitable for the engineering of complex multi-genic phenotypic traits that are difficult to modify by either the classical or molecular strain improvement approaches, as it does not require a priori knowledge of the set of genes to be modified to confer beneficial changes.

Genome shuffling was originally developed for bacteria and then extended to yeasts. In bacteria, genome shuffling has been used successfully to increase the tylosin titer in *Streptomyces fradiae* [1], lactic acid tolerance by *Lactobacillus* [2], and pentachlorophenol degradation and tolerance by *Sphingobium chlorophenolicum* [3]. In yeasts, particularly *Saccharomyces cerevisiae*, genome shuffling has been used to improve thermotolerance, ethanol productivity, ethanol tolerance and acetic acid tolerance [4–7].

In genome shuffling, genome recombination is carried out using either recursive protoplast fusion or cross-mating. Most of the reports on genome shuffling are based on protoplast fusion [1–5]. We have developed a genome shuffling protocol based on recursive mating in the pentose-fermenting yeast *Scheffersomyces stipitis* and isolated mutant strains with improved tolerance to hardwood spent sulphite liquor (HW SSL) [8]. This approach led to a rapid improvement in tolerance to HW SSL in the selected strains, which retained their growth and fermenting ability. In our approach, the haploid cells from a pool of *S. stipitis* mutants are mated to form diploid zygotes. In malt extract agar, the diploid zygotes undergo meiosis to form two-spored asci [8, 9]. The ascus wall dissolves on its own to release the hat-shaped ascospores. When transferred to a rich medium, the ascospores germinate to produce vegetative haploid cells. The recombinants with improved phenotypes are identified and selected using suitable selection regimes (Fig. 41.1). The process of mating and sporulation is repeated several times to allow for recombination between multiple parents and the pooling of beneficial alleles from various genomes.

Materials

Random Mutagenesis

1. A physical (ultraviolet light) or a chemical (e.g., *N*-methyl-*N'*-nitro-*N*-nitrosoguanidine or ethylmethanesulphonate) mutagen.
2. *S. stipitis* culture grown for 48 h in minimal medium broth (0.67% [w/v] yeast nitrogen base (YNB) without amino acids and 2% [w/v] xylose).
3. Sterile Petri dishes, 15-mL centrifuge tubes and a tube rotator.
4. Selective agar (1.5% [w/v] medium).
5. Square plastic Petri dishes (120×120×17 mm).

Yeast Mating and Genome Shuffling

1. Two auxotrophic yeast strains with different nutritional requirements.
2. YEPD broth (1% [w/v] yeast extract, 2% [w/v] peptone, 2% [w/v] glucose).
3. Minimal medium plates (0.67% [w/v] YNB without amino acids, 2% [w/v] xylose, 1.5% [w/v] agar).
4. Malt extract (ME) agar plates (3% [w/v] ME, 1.5% [w/v] agar).
5. Sterile Petri dishes, 15-mL centrifuge tubes and a tube rotator.

Fig. 41.1 Schematic of
genome shuffling in yeasts

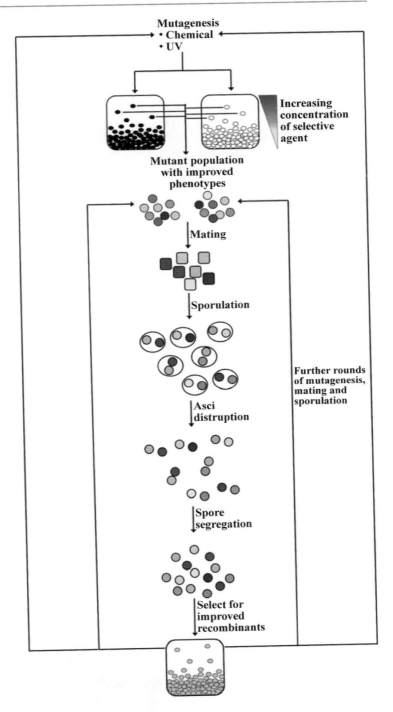

Methods

Random Mutagenesis Procedure

Random mutagenesis can be carried out using a physical or chemical mutagen. Freshly grown yeast cells are exposed to the mutagen for a predetermined time. The cells are then allowed to recover overnight in liquid or solid medium followed by screening and selection of the mutant colonies in a selective medium. In the following, we describe random mutagenesis using UV light.

1. Transfer a loopful of *S. stipitis* cells from an isolated colony on YEPD agar plate to 20 mL of broth containing 0.67% [w/v] YNB without amino acids and 2% [w/v] xylose in a 125-mL Erlenmeyer flask at 28 ± 1 °C with shaking at 180 rpm for 48 h.
2. Aseptically transfer 1 mL of the 48-h grown yeast culture (OD600 between 8 and 10) to an empty sterile Petri dish.
3. Place the Petri dish under the UV light source at a distance of about 40 cm.
4. Remove the Petri dish cover and turn on the UV light. Prepare a yeast survival curve based on length of UV exposure. For *S. stipitis* wild-type (WT) cells, the length of UV exposure tested ranged from 10 to 60 s and 50% survival was achieved at 20 s.
5. Transfer the UV-irradiated culture to a 15-mL sterile centrifuge tube covered with aluminium foil and incubate in a tube rotator (Fig. 41.2) for 24 h at 23 ± 1 °C. The tubes are rotated about their vertical axis at about 60° from horizontal at 90 cycles per min as previously described [10, 11].
6. Spread 100 μL of the UV-exposed culture on several selective medium plates and incubate at 28 ± 1 °C for 4–5 days.
7. Isolate the mutant colonies appearing on the selective medium plates and maintain them on YEPD agar plates.
8. Individual mutants are further tested in liquid broth. Mutants that retain good fermenting ability in liquid broth were selected for further rounds of mutagenesis and genome shuffling.

Fig. 41.2 Test tube rotator

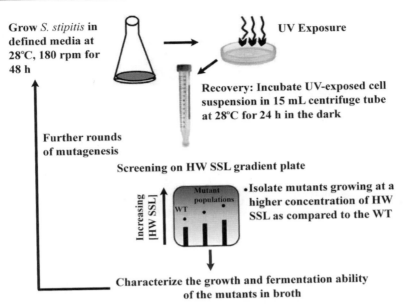

Fig. 41.3 Schematic of UV mutagenesis followed by screening on gradient plates to select for HW SSL tolerant mutants

In our earlier study [12], we used three consecutive rounds of UV mutagenesis followed by screening to obtain mutants of *S. stipitis* NRRL Y-7124 with improved tolerance to HW SSL. A schematic of the random mutagenesis procedure is presented in Fig. 41.3. HW SSL is a waste liquor from the Tembec pulp and paper mill in Témiscaming, Québec, Canada. It contained, in [w/v]: 0.076% arabinose, 0.25% galactose, 0.33% glucose, 0.55% mannose, 2.2% xylose, 1% acetic acid, 0.18% furfural and 0.11% hydroxymethyl furfural. Prior to use, the pH of HW SSL was increased from 2.5 to 5.5 with 10 M NaOH. The liquor was then boiled for 5 min in a microwave oven, followed by gradual cooling to room temperature.

For UV mutagenesis, the UV exposure time was optimized to achieve a survival rate of about 50%. We hypothesized that changes in many genes are needed to collectively confer tolerance to HW SSL in yeasts. Thus, we used a low UV dosage resulting in a high survival rate in order to maximize those surviving populations carrying a small number of mutations for screening. Higher UV doses resulting in a lower survival rate may result in many members of the surviving popula-

tions carrying multiple mutations, some of which may not be beneficial. The presence of nonbeneficial mutations will likely mask the effect of the beneficial mutations, resulting in these mutants not being selected in the screen.

An increasing concentration of HW SSL was used as the selective agent for initial mutant screening on agar plates after UV mutagenesis. This was done by spreading 100 µL of the UV-exposed culture on HW SSL (pH 5.5) gradient plates. The plates were incubated at 28 °C for 5–10 days. Colonies growing at a higher concentration of HW SSL as compared to the WT strain were isolated and maintained on YEPD plates. HW SSL gradient plates were prepared by successively pouring two layers of agar into a square plastic Petri dish ($120 \times 120 \times 17$ mm). The bottom layer consisted of 25 mL of HW SSL agar medium (Fig. 41.4a). The plate was allowed to solidify at a slightly inclined position. The Petri dish was then placed in a horizontal position and a second layer of plain agar was poured over the first layer (Fig. 41.4b). The Petri dish was incubated at 30 °C for 2 days before use. This allowed diffusion of the HW SSL components through the agar layer, thereby establishing a concentration gradient.

The first layer of agar with HW
SSL is poured into the Petri dish
and allowed to solidify while the
plate is held in an inclined
position

The second layer of plain agar
is poured into the Petri dish
and allowed to solidify while
the plate is held in a leveled
position

Fig. 41.4 Preparation of HW SSL gradient agar plate

Because performance of the mutants on the
plate screen may not translate to the broth, the
mutants were further tested for growth in liquid
HW SSL. Also, since our end goal is fermentation of the sugars in the waste liquor, the fermenting ability of the mutants needs to be verified by
screening in broth. Therefore, the mutants with
improved HW SSL tolerance were assessed for
the ability to ferment xylose, glucose, mannose,
galactose and arabinose in defined media as well
as HW SSL. Six improved mutants obtained after
several rounds of UV mutagenesis and screening
were used as the starting strains for genome
shuffling.

Optimization of Yeast Mating and Genome Shuffling

1. Grow each auxotrophic yeast strain in 5 mL of
   YEPD broth in a 15-mL centrifuge tube at
   28 °C with shaking at 180 rpm for 24 h.
   Auxotrophic yeast strains can be obtained by
   random mutagenesis [13].
2. Mix 2×10^8 cells of each auxotrophic strain
   together and spread the mixed cell suspension
   on ME plates. Also, spread 2×10^8 cells of

each auxotroph separately on ME plates as
controls.

3. Incubate the ME plates at 28 °C for 7–10 days
   to allow mating and sporulation to occur.
   Examine the cells periodically under the
   microscope to follow spore formation. *S. stipitis* forms two hat-shaped spores per ascus.
4. After sporulation, scoop out all the cells from
   each plate. Suspend them in sterile water, centrifuge and wash the cells several times with
   sterile water. The cells are then suspended in
   5 mL of sterile water.
5. Spread the cells on minimal media plates. The
   auxotrophic strains cannot grow on these
   plates. Only the recombinant cells resulting
   from mating between the two auxotrophic
   strains can grow on minimal media plates.
6. Incubate the plates at 28 °C for 4–5 days.
   Enumerate the colonies appearing on minimal
   media plates and determine the mating
   frequency.
7. To check for the stability of recombinants,
   repeatedly streak the colonies derived from
   the mating on minimal media plates.

In our lab, the mating and sporulation protocol
was optimized using two auxotrophic strains of
S. stipitis (FPL Y14—*ura3, ade2, met1* and FPL
Y18—*ura3, leu2 delta*). Each parent carried a
common auxotrophic requirement for uracil in
addition to the other different auxotrophic markers. If mating and successful recombination have
occurred, then the recombinants would require
only uracil. Therefore, the sporulated cells were
spread on uracil-containing minimal media plates
(0.67% w/v YNB w/o amino acids, 2% w/v xylose
and 0.005% w/v uracil). Only the recombinant
cells could grow on these plates (Fig. 41.5). The
mating efficiency of the haploid *S. stipitis* auxotrophs was estimated to be 0.05% after one
round of mating and sporulation. The estimate
was based on the number of recombinants
obtained on uracil-containing plates after mating
the two auxotrophic *S. stipitis* strains. The optimized protocol was used for genome shuffling of
UV-induced mutants of *S. stipitis*.

Six HW SSL-tolerant mutants obtained after
the third round of UV mutagenesis served as
the parental strains for genome shuffling. The

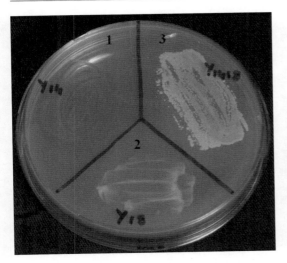

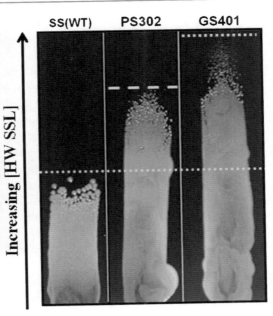

Fig. 41.5 Growth of the sporulated cell suspensions obtained from individual auxotrophs and mated populations on minimal media plates+uracil. *Sectors 1* and *2* represent the growth of individual auxotrophs FPL Y14—*ura3, ade2, met1* and *P. stipitis* FPL Y1—*ura3, leu2 delta*, respectively. *Sector 3* represents the growth of sporulated cells obtained after mating of the two auxotrophs

Fig. 41.6 Growth of *S. stipitis* WT (*left*), UV-induced mutant (PS302, *middle*) and genome shuffled strain (GS401, *right*) on HW SSL (pH 5.5) gradient plate

selected mutants were individually grown overnight in YEPD broth followed by mixing of 1×10^8 cells of each mutant on ME plates. The plates were incubated for 7 to 10 days to allow mating and sporulation to occur in the same way as described above for the auxotrophic strains. All the cells were scooped out from the ME plates and suspended in 5 mL of sterile water. After centrifugation and several washings with sterile water, the spore suspension was transferred to 20 mL of fresh YEPD broth in 125-mL Erlenmeyer flask and incubated at 28 °C with shaking (180 rpm). After 48 h of growth, 100 μl of cell suspension was spread on several HW SSL (pH 5.5) gradient plates. Colonies appearing at a higher concentration of HW SSL as compared to the best UV-induced mutant were isolated. Three consecutive rounds of genome shuffling involving mating, sporulation and selection on HW SSL plates were done. Improved mutants obtained from one round served as the starting strains for the next round of genome shuffling. Since we started with six UV-induced mutants, three rounds of genome shuffling were deemed sufficient to recombine all the beneficial mutations together.

A greater number of rounds would be desirable if we had started with a larger mutant pool. Figure 41.6 illustrates the growth of *S. stipitis* WT, UV-induced mutant and a genome shuffled mutant on the HW SSL gradient plate. The genome shuffled strain (GS401) was clearly more tolerant than the UV-induced mutant (PS302), which in turn was more tolerant than the WT, on HW SSL gradient plate. The isolated mutants were further tested for growth in liquid HW SSL to confirm their HW SSL tolerant character. The WT was unable to grow in HW SSL unless diluted to 65% (v/v) or lower. The UV-induced mutants grew in 75% (v/v) HW SSL. Two mutants obtained after three rounds of genome shuffling grew in 85% (v/v) HW SSL and one of these could survive in 90% (v/v) HW SSL, although no increase in cell number was seen [8]. The genome shuffled mutants consumed 4% [w/v] xylose or glucose in defined media more efficiently and produced more ethanol as compared to the *S. stipitis* WT strain. These mutants also utilized mannose and galactose and produced the same amount of ethanol as the WT.

Summary

The results from our research demonstrated the utility of genome shuffling via cross-mating as a means for industrial strain improvement in *S. stipitis*. The protocol is easy, inexpensive and convenient. The key requirement is the availability of a screening method that is specific and sensitive. The improved haploid recombinant(s) are genetically stable and amenable to be further improved (i.e., they are not dead-end strains). For example, the selected genome shuffled strains in our study can be subjected to further genome shuffling again to select for mutants with improved tolerance to other stresses such as acetic acid and other pretreatment-derived inhibitors. Beneficial mutations in different mutant lines can be easily recombined into one strain. The same method can be applied to other yeasts, including native pentose-fermenting yeasts such as *Pachysolen tannophilus*, for which a genetic mating system has been described [14].

Acknowledgments We thank Juraj Strmen and Frank Giust of Tembec Inc. (Témiscaming, Québec, Canada) for providing the HW SSL, and Tom Jeffries (USDA, Madison, Wisconsin, USA) for providing the auxotrophic *S. stipitis* strains.

References

1. Zhang Y, Perry K, Powell K, Stemmer PC, Del Cardayré SB (2002) Evolution of *Streptomyces fradiae* by whole genome shuffling. Nature 415:644–646
2. Patnaik R, Louie S, Gavrilovic V, Perry K, Stemmer WP, Ryan CM et al (2002) Genome shuffling of *Lactobacillus* for improved acid tolerance. Nat Biotechnol 20:707–712
3. Dai M, Copley SD (2004) Genome shuffling improves degradation of the anthropogenic pesticide pentachlorophenol by *Sphingobium chlorophenolicum* ATCC 39723. Appl Environ Microbiol 70:2391–2397
4. Wei PY, Li ZL, He P, Lin Y, Jiang N (2008) Genome shuffling in the ethanologenic yeast *Candida krusei* to improve acetic acid tolerance. Biotechnol Appl Biochem 49:113–120
5. Shi DJ, Wang CL, Wang KM (2009) Genome shuffling to improve thermotolerance, ethanol tolerance and ethanol productivity of *Saccharomyces cerevisiae*. J Ind Microbiol Biotechnol 36:139–147
6. Hou L (2010) Improved production of ethanol by novel genome shuffling in *Saccharomyces cerevisiae*. Appl Biochem Biotechnol 160:1084–1093
7. Zheng D-Q, Wu X-C, Wang P-M, Chi X-Q, Tao X-L, Li P, Jiang X-H (2010) Drug resistance marker-aided genome shuffling to improve acetic acid tolerance in *Saccharomyces cerevisiae*. J Ind Microbiol Biotechnol 38:415–422
8. Bajwa PK, Pinel D, Martin VJJ, Trevors JT, Lee H (2010) Strain improvement of the pentose-fermenting yeast *Pichia stipitis* by genome shuffling. J Microbiol Methods 81:179–186
9. Melake T, Passoth VV, Klinner U (1996) Characterization of the genetic system of the xylose fermenting yeast *Pichia stipitis*. Curr Microbiol 33:237–242
10. Schneider H, Wang PY, Chan YK, Maleszka R (1981) Conversion of D-xylose into ethanol by the yeast *Pachysolen tannophilus*. Biotechnol Lett 2:89–92
11. Barbosa Mde F, Lee H, Collins-Thompson DL (1990) Additive effects of alcohols, their acidic by-products, and temperature on the yeast *Pachysolen tannophilus*. Appl Environ Microbiol 56:545–550
12. Bajwa PK, Shireen T, D'Aoust F, Pinel D, Martin VJJ, Trevors JT et al (2009) Mutants of the pentose-fermenting yeast *Pichia stipitis* with improved tolerance to inhibitors in hardwood spent sulphite liquor. Biotechnol Bioeng 104:892–900
13. Hashimoto S, Ogura M, Aritomi K, Hoshida H, Nishizawa Y, Akada R (2005) Isolation of auxotrophic mutants of diploid industrial yeast strains after UV mutagenesis. Appl Environ Microbiol 71:312–319
14. James AP, Zahab DM (1983) The construction and genetic analysis of polyploids and aneuploids of the pentose-fermenting yeast, *Pachysolen tannophilus*. J Gen Microbiol 129:1489–2494

Detection and Identification of Fungal Microbial Volatile Organic Compounds by HS-SPME-GC–MS

42

Bernhard Kluger, Susanne Zeilinger,
Gerlinde Wiesenberger, Denise Schöfbeck,
and Rainer Schuhmacher

Abstract

A method based on solid phase microextraction coupled to gas chromatography–mass spectrometry (GC–MS) for the detection and identification of microbial volatile organic compounds (MVOCs) in the headspace of filamentous fungi is presented. MVOCs are identified by comparison of mass spectra and linear temperature programmed retention indices (LTPRIs) with database entries and LTPRIs published in literature. The presented method enables the monitoring of the formation of volatile metabolites for defined time intervals during cultivation of the investigated fungus. The experimental procedure is exemplified with *Fusarium graminearum* and *Trichoderma atroviride* but can also be used to detect, identify and profile MVOCs produced by other filamentous fungi.

Keywords

Solid phase microextraction • Gas chromatography–mass spectrometry • Microbial volatile organic compounds • Linear temperature programmed retention index

B. Kluger • D. Schöfbeck • R. Schuhmacher (✉)
Department for Agrobiotechnology (IFA-Tulln),
University of Natural Resources and Life Sciences
Vienna, Konrad Lorenz Str. 20, Tulln 3430, Austria
e-mail: rainer.schuhmacher@boku.ac.at

S. Zeilinger
Research Area Gene Technology and Applied
Biochemistry, Institute for Chemical Engineering,
Vienna University of Technology, Getreidemarkt 9,
Vienna 1060, Austria

G. Wiesenberger
Institute of Applied Genetics and Cell Biology,
University of Natural Resources
and Life Sciences Vienna, Konrad Lorenz Str. 24,
Tulln 3430, Austria

Introduction

Filamentous fungi produce and release microbial volatile organic compounds (MVOCs), which are frequently involved in self-signalling as well as the interaction with other organisms, such as plants and microbes. Because of their volatile nature, these compounds can easily be transported through air and therefore can act over relatively long distances [1]. Complex mixtures of MVOCs are being produced, individual metabolites of which belong to different structure classes, such as mono- and sesquiterpenes, alcohols, ketones, lactones and the so-called C_8-compounds [2–4].

V.K. Gupta et al. (eds.), *Laboratory Protocols in Fungal Biology: Current Methods in Fungal Biology*,
Fungal Biology, DOI 10.1007/978-1-4614-2356-0_42, © Springer Science+Business Media, LLC 2013

Nowadays, volatile components are usually detected and identified by gas chromatography–mass spectrometry (GC–MS) [5]. Gas chromatography (GC) is applied for the separation of complex mixtures of (at least partly) evaporable substances on capillary columns available with stationary phases of different polarity. GC can easily be coupled to mass spectrometry, which offers a powerful tool for sensitive and selective detection as well as the simultaneous identification and structure characterisation of natural products such as MVOCs.

Several sample preparation techniques have been used for the extraction of MVOCs from fungal cultures. For solid [6] or liquid growth media [7], solid phase extraction using C18 or silica gel columns or steam distillation extraction [8] have been described. While these methods require the destruction of the fungal culture, headspace (HS) techniques such as online gas enrichment on adsorption tubes [9], closed-loop stripping analysis [10], dynamic headspace (purge and trap) [11] or solid phase microextraction (SPME) [12–14] can be used to sample volatile metabolites during culture growth. Since these noninvasive HS methods allow the extraction, enrichment and detection of volatile metabolites from the living fungal culture, they provide a more representative picture of the MVOCs actually produced [15]. In addition, HS extraction helps to reduce the time-consuming manual preparation of samples. This holds particularly true for SPME because this technique allows quick sample extraction and can be fully automated and coupled online to GC–MS analysis.

The HS-SPME technique [16] has already been demonstrated to be very useful for the extraction of volatile compounds from a wide variety of samples, including MVOCs from plants [15] and different fungal genera (e.g. *Aspergillus, Penicillium, Mucor, Fusarium, Trichoderma*) [6, 12, 17].

Automated detection, identification and structure characterisation of MVOCs from GC–MS chromatograms require efficient software tools for comprehensive data evaluation, such as chromatographic peak detection, deconvolution (i.e., "purification") of mass spectra and calculation of retention indices. Suitable software programs are available and can freely be downloaded from the internet (e.g. AMDIS, MetaboliteDetector) [18, 19]. Assignment of chromatographic peaks and putative identification of metabolites require the comparison of measured mass spectra as well as experimentally derived retention indices (i.e., linear temperature programmed retention index, LTPRI) with reference spectra and values. For both parameters, (pre-)defined criteria have to be fulfilled for positive identification. In this protocol we present a comprehensive procedure that is based on a method by Stoppacher et al. [20] and has been further developed for the cultivation, HS-SPME extraction, GC–MS analysis, automated detection and identification of MVOCs produced by living cultures of filamentous fungi. The procedure allows high flexibility, as growth conditions can be controlled independently from the sampling and extraction steps.

Materials

1. Potato dextrose agar (PDA).
2. Distilled water.
3. Autoclave.
4. 50 °C water bath.
5. 85-mm plates (Petri dishes).
6. 20-mL headspace vials.
7. Cotton plugs and rubber straps.
8. Fungal culture, e.g. *Trichoderma atroviride* (ATCC 74058), *Fusarium graminearum* (PH-1, NRRL 31084).
9. Thermostat for incubator at 28 °C.
10. Cork borer for cutting 5-mm discs from agar plate
11. Sterile syringe needles.
12. Fusarium minimal medium (FMM) plates (1 g KH_2PO_4, 0.5 g $MgSO_4*7H_2O$, 0.5 g KCl, 2 g $NaNO_3$, 30 g sucrose, 20 g agar, 10 mg citric acid, 10 mg $ZnSO_4·6H_2O$, 2 mg $Fe(NH_4)_2(SO_4)_2·6H_2O$, 0.5 mg $CuSO_4*5 H_2O$, 0.1 mg $MnSO_4$, 0.1 mg H_3BO_4, 0.1 mg $Na_2MoO_4·2H_2O$ in 1 L deionised water).
13. Growth chamber 20 °C.
14. Mung beans substrate (MBS) medium (put 10 g of mung beans to 500 mL boiling water,

cook for 20 min, filtrate the extract through a paper filter, fill up to 1 L and autoclave for 20 min at 121 °C).

15. Sterile scalpels.
16. Orbital shaker.
17. Baffled flasks (250 mL).
18. Glass wool filter (remove the plunger of a 10-mL syringe and stuff it with glasswool to the 5-mL mark).
19. 50-mL falcon tubes.
20. Centrifuge with swing-out rotor for 50-mL falcon tubes (optional).
21. Light microscope and haemocytometer.
22. Laminar flow work bench; e.g. Heraeus Instruments LaminAir® HB2472 (Heraeus, Hanau, Germany).
23. Synthetic air (Messer, Gumpoldskirchen, Austria).
24. Sterile filter, e.g. 0.2 μm PTFE Midisart BV membrane filter (Sartorius, Göttingen, Germany).
25. Screw caps containing 1.3-mm gas-tight silicone/teflon septa (Gerstel, Mülheim, Germany).
26. Climate chamber: 28 °C, 70 % humidity.
27. SPME headspace autosampler, e.g. Gerstel MPS 2XL equipped with SPME fibre holder (Gerstel, Mülheim, Germany; see Note 1).
28. SPME fibre, e.g. with 65 μm PDMS/DVB coating (Supelco, Bellefonte, USA; see Note 2).
29. Headspace inlet glass liner, 1.5 mm i.d. (e.g. Supelco, Bellefonte, USA).
30. Gas chromatograph, e.g. Agilent 6890 N (Agilent, Waldbronn, Germany).
31. Helium (5.0) (Messer, Gumpoldskirchen, Austria).
32. GC capillary columns (see Note 3).
   • Column 1: HP-5MS 30 m × 0.25 mm × 0.25 μm (e.g. Agilent, Waldbronn, Germany).
   • Column 2: DB-Wax 30 m × 0.25 mm × 0.25 μm (e.g. Agilent, Waldbronn, Germany).
33. Mass Selective Detector, e.g. Agilent Mass Selective Detector 5975B (Agilent, Waldbronn, Germany).
34. Alkane standards (C_8–C_{20}), (C_{21}–C_{40}) (Sigma-Aldrich, Vienna, Austria) used for LTPRI calibration.

35. Pentane 99 % (Sigma-Aldrich, Vienna, Austria), hexane 99.5 % (Merck, Darmstadt, Germany), heptane 99.5 % (J.T. Baker, Deventer, Netherlands), octane 99 % (Sigma-Aldrich, Vienna, Austria), nonane 99 % (Sigma-Aldrich, Vienna, Austria), decane p.a. (Promochem, Wesen, Germany) also used for LTPRI calibration
36. Software AMDIS—automated mass spectral deconvolution and identification system (version 2.69) (see Note 4)

Method

The method presented below describes a detailed procedure for cultivation, sample preparation, HS-SPME-GC–MS analysis, identification and profiling of MVOCs produced by filamentous fungi. The protocol will be exemplified for the fungi *Trichoderma atroviride* and *Fusarium graminearum*. Depending on the fungus to be cultivated (using agar discs or alternatively spore suspension) and studied, different parts of the method may need modifications (e.g. cultivation conditions, SPME fibre coating or extraction temperature).

Preparation of PDA Plates and Slants in Headspace Vials

In addition to biological samples, a HS vial containing only PDA medium (medium blank) should be prepared and measured to exclude false identification of compounds released from the culture medium (see Note 5).

1. For preparing agar slants in headspace (HS)-vials, sterilise the vials sealed with cotton plugs and rubber straps.
2. Prepare 1 L of medium by suspending 39 g of PDA in 1 L of distilled water.
3. Sterilise at 120 °C for 15 min and afterwards cool the medium down to 50 °C in a waterbath.
4. For PDA plates, pour 25-mL aliquots of the medium into sterile 85-mm Petri dishes.

5. Transfer 5-mL aliquots of the autoclaved PDA medium equilibrated to 50 °C to the sterile HS vials.

6. Place HS vials in slanted position until the medium has solidified (see Note 6).

Cultivation of Fungi Using Agar Discs (e.g. *Trichoderma atroviride*)

1. Prepare a fungal pre-culture (e.g. *Trichoderma atroviride*) by cultivating the fungus on PDA plates for 7 days in a 28 °C incubator in the dark.

2. For cultivating the fungus in HS vials containing agar slants, cut a 5-mm disc from the pre-culture plate containing sporulated fungal mycelium using a cork borer.

3. Transfer the disc from the plate to the HS vial by placing it at a central position of the surface of the PDA slant using a sterile syringe needle.

4. Re-seal the HS vial with the sterile cotton plug and the rubber strap.

Cultivation of Fungi Using Spores (e.g. *Fusarium graminearum*)

As an alternative to cultivation using agar discs, a spore suspension at a defined concentration can be pipetted on top of agar surface and used for the cultivation in HS vials.

1. Grow fungi (e.g. *Fusarium graminearum*) on an FMM plate at 22 °C in the dark until the mycelium reaches the edge of the Petri dish.

2. For sporulation of the required strain, scrape five to ten small pieces of the mycelium (total area about 2 cm^2) with a sterile scalpel from the surface of the plate (see Note 7) and suspend them in 50 mL of MBS medium in a 250-mL baffled flask. Shake the flask on an orbital shaker at 120 rpm at 20 °C until spores are formed (2–4 days; see Note 8).

3. Separate the spores from the mycelium by filtration of the culture through a column containing tightly packed glass wool, collect the flow though in a 50-mL Falcon tube. If desired, the spores can be concentrated by gravity (overnight at 4 °C) or centrifugation (5 min at 4,000 rpm) and removal of the supernatant. Resuspend the spores in a small amount of the supernatant or in sterile water and store the suspension at 4 °C until further use (see Note 9).

4. Determine the spore count using a haemocytometer (e.g. Fuchs-Rosenthal).

5. Inoculate a HS vial with 8×10^3 spores by pipetting the required volume to the centre of the PDA surface.

6. Re-seal the HS vial with the sterile cotton plug and the rubber strap.

Sample Preparation of Fungal Cultures for Headspace Extraction

This method enables one to study the synthesis rate of volatiles in defined time periods at different time points during fungal cultivation. This step of the method is applied to purge culture vials with air and thereby remove all volatiles from the HS above the cultures (see Note 10).

1. Carry out the procedure at a defined time point (e.g. 6 h) prior to GC–MS measurement in a laminar flow work bench.

2. Remove the cotton plug from the HS vial.

3. Purge the headspace above culture with a gentle stream of air through a sterile filter for 30 s. Caution: Spores might be expelled from the HS vial to ambient air.

4. Tightly seal the HS vials with screw caps containing gas-tight silicone/teflon septa.

5. Incubate the cultures for a defined time period at controlled conditions in the tray of the SPME headspace autosampler (e.g. at 28 °C) until measurement (see Note 11) (Fig. 42.1).

HS-SPME-GC–MS Analysis of Filamentous Fungi

An SPME headspace autosampler is used to extract volatiles automatically from the headspace of fungal cultures and transfer the extracted MVOCs into the injector of the GC–MS system.

Fig. 42.1 HS vials with agar slants and culture agar discs before purging with synthetic air, sealed with cotton plug and rubber strap, (*left*) and after purging, sealed with screw cap (*right*)

Application of an autosampler increases precision and sample throughput.

1. Equilibrate the fungal cultures and blank sample at the extraction temperature of 30 °C for 5 min (see Note 12).

2. Insert the SPME fibre (e.g. 65 μm DVB/PDMS coating) into the sample vial and expose fibre coating to the HS above the cultivated fungus or blank sample (see Note 2).

3. Extract for 45 min at 30 °C without agitation (fibre penetration depth 21 mm).

4. Transfer the fibre to GC–MS injection port and desorb volatiles bound to the fibre for 2 min in a split/splitless injector (splitless mode, 250 °C, fibre penetration depth of 57 mm, headspace inlet glass liner, 1.5 mm i.d.).

5. After sample injection/desorption, clean the fibre for 3 min in a needle heater station (gentle nitrogen stream, 250 °C, fibre penetration depth of 57 mm).

6. Use helium (5.0) as carrier gas at a constant flow rate of 1 mL/min.

7. Analyse the compounds desorbed from the fibre on two columns for identification (see Note 3):
   - HP-5MS (30 m × 0.25 mm × 0.25 μm).
   - DB-Wax (30 m × 0.25 mm × 0.25 μm).

8. Temperature oven program: 40 °C (hold 2 min), 10 °C/min to 200 °C, 25 °C/min to 260 °C (hold 5 min) (see Note 13).

9. Mass selective detector settings: electron ionisation (EI) at 70 eV, source 230 °C, quadrupole 150 °C, solvent delay of 4 min, full scan (m/z 45–400) (Fig. 42.2).

GC–MS Measurements of Alkane Standards for Calibration of LTPRI Values

This step of the protocol describes the experimental details for the HS-SPME-GC–MS measurements of alkane standards (C_5–C_{40}). Alkane standards have to be measured with the same GC–MS method as the fungal samples. Their retention times are required to determine LTPRI values of fungal metabolites. LTPRI values were calculated according to the formula suggested by Van den Dool and Kratz in 1963 [21] (Fig. 42.3). Compared to the procedure described in the section "HS-SPME-GC–MS Analysis of Filamentous Fungi", different extraction parameters are necessary for alkane standards (see Note 14). The alkane standard mixture C_8–C_{20} can be measured with the protocol described in the section "HS-SPME-GC–MS Analysis of Filamentous Fungi".

1. Alkane standard mixtures C_8–C_{20} and C_{21}–C_{40} are commercially available as solutions in hexane with a concentration of 40 mg/L for each alkane.

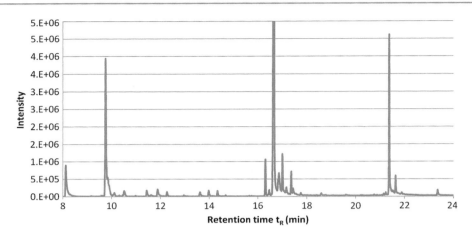

Fig. 42.2 Total ion chromatogram (displayed from 8 to 24 min) resulting from MVOCs of *Trichoderma atroviride* after 48 h of cultivation using the HS-SPME-GC–MS method

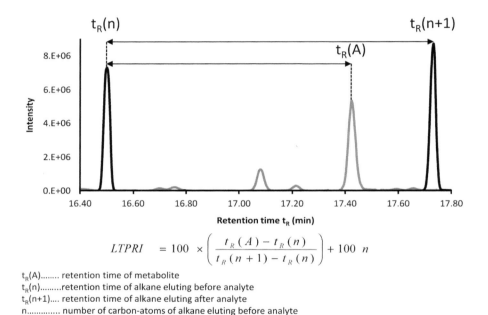

$$LTPRI = 100 \times \left(\frac{t_R(A) - t_R(n)}{t_R(n+1) - t_R(n)} \right) + 100\ n$$

$t_R(A)$........ retention time of metabolite
$t_R(n)$.........retention time of alkane eluting before analyte
$t_R(n+1)$.... retention time of alkane eluting after analyte
n.............. number of carbon-atoms of alkane eluting before analyte

Fig. 42.3 Overlay of two chromatograms (sample, alkane standard) and formula for the calculation of the linear temperature programmed retention index (LTPRI).

The LTPRI represents the relative retention time of an analyte, which is calculated based on the alkane standards, one eluting before and the other eluting after the analyte

2. Prepare C_5–C_{10} alkane standard mixture separately by using pure compounds (pentane:hexane:heptane:octane:nonane:decane 17:6:3:2:1:0.5 [v/v]).

3. Transfer each of the three alkane standard mixtures to a separate empty clean HS vial C_5–C_{10}: 1 μL, C_8–C_{20}: 10 μL, C_{21}–C_{40}: 40 μL

and immediately seal HS vials with screw caps containing gas-tight silicone/teflon septa.

4. Equilibrate the HS vials including the alkane standard mixture without agitation prior to SPME extraction (C_5–C_{10}: 0.01 min, 90 °C; C_8–C_{20}: 5 min, 30 °C; C_{21}–C_{40}: 30 min, 120 °C).

5. Extract without agitation using a fibre penetration depth of 21 mm (C_5–C_{10}: 0.01 min, 90 °C; C_8–C_{20}: 45 min, 30 °C; C_{21}–C_{40}: 60 min, 120 °C).

6. Follow the steps in the section HS-SPME-GC–MS Analysis of Filamentous Fungi, starting from step 4.

Data Evaluation

Putative identification of volatile metabolites is based on two parameters: (1) LTPRI values and (2) mass spectra. Consideration of both parameters requires the comparison with reference values. These reference LTPRI values or mass spectra can either be taken from databases, the scientific literature or from own measurements of authentic standard substances (see Note 15).

Independent on where the reference LTPRI values and mass spectra originate from, a database containing these reference values has to be created before measured chromatographic peaks can be assigned to certain metabolites. Moreover, specific criteria have to be defined, which must be fulfilled for putative identification of MVOCs, which can be used for the profiling of filamentous fungi.

Establishment of a Reference Database

A reference database has to be established that serves as a positive list for metabolite identification. Therefore, mass spectra, LTPRI values and a unique identifier (e.g. CAS number) have to be selected for each compound contained in the database. Comprehensive databases, such as the Wiley Registry 9th/NIST 2008 mass spectral library, may serve for initial identification of metabolites based on spectral similarity, because only metabolites that are contained in the reference database can be identified by this approach. For the determination of literature LTPRI values, different sources, such as the NIST Chemistry Webbook or the Adams library, are available (see Note 15). Exact description of the data format required for the reference database used by AMDIS is beyond the scope of this protocol.

A detailed description of the AMDIS software is given in the user manual, which can be accessed via the Internet.

1. Use reference mass spectra obtained from comprehensive databases (e.g. Wiley Registry 9th/NIST 2008, standard compounds).

2. Select appropriate reference spectra to be used with the reference database.

3. Convert selected mass spectra to a format that can be used by the program AMDIS to build a reference database.

4. Add LTPRI values obtained from the literature (e.g. NIST Chemistry Webbook) to reference database.

Calibration of LTPRI Values for Alkane Standards

In order to enable calibration of LTPRI values corresponding to chromatographic peaks of MVOCs (which follows in the steps of the section Automated Peak Detection, Deconvolution of Mass Spectra and Calculation of LTPRI Values), alkane standard chromatograms have to be evaluated by AMDIS.

1. Open AMDIS program and load an alkane standard data file.

2. Load settings for LTPRI calculation.

3. Use appropriate deconvolution settings (see Note 16).

4. Use an alkane standard database containing the spectra of C_5–C_{40}.

5. Analyse every alkane standard data file (C_5–C_{10}, C8–C_{20} and C_{21}–C_{40}). Alkane standard peaks are detected automatically, mass spectra are deconvoluted and retention times of alkanes are used for the calibration of LTPRI values according to the formula [21].

Automated Peak Detection, Deconvolution of Mass Spectra and Calculation of LTPRI Values

Metabolic profiling produces great quantities of data, and manual data evaluation is difficult to standardise and can be very labor-intensive and time-consuming [22]. Therefore, the application of tailor-made software programs (e.g. AMDIS, MetaboliteDetector) is highly recommended (see Note 4). The following protocol uses the program AMDIS for LTPRI determination, component detection and assignment of metabolites. The program is freely available and accepts a wide variety of manufacturers' raw data formats [23].

1. Load sample data file with AMDIS (blank or fungal culture chromatogram; see Note 5).
2. Load LTPRI calibration for the LTPRI calculation of detected peaks.
3. Use appropriate deconvolution settings (see Note 16).
4. Exclude *m/z* values corresponding to siloxanes that are released by the column or the fibre such as *m/z* 207 (see Note 17).
5. Analyse the opened data file for peak detection and deconvolution of mass spectra. LTPRI of every detected peak is calculated based on the alkane standard calibration made before.

Selection of Identification Criteria and Assignment of Metabolites

Several criteria have to be fulfilled to assure the identification of unknown MVOCs. Therefore, obtained gas chromatograms and mass spectra have to be compared to the reference database established in Establishment of a Reference Database. Moreover, when analysing fungal samples with the aim to identify MVOCs of this particular fungus for the first time, measurements should be carried out on two columns with stationary phases of different polarity. The following protocol uses the comparison of mass spectra and LTPRI values for the identification of MVOCs using the deconvolution program AMDIS.

1. Compare the mass spectra of detected peaks to the reference spectra contained in the reference database (e.g. Wiley Registry 9th/NIST 2008 mass spectral library) established in Establishment of a Reference Database.
2. Open the menu "Analyze—Search NIST library" in AMDIS.

3. Use following parameters for analysis:
   - Hits reported per search: "Min. match factor ≥90" (see Note 18).
   - Select from "all components".
   - Number of components searched: "All above threshold 0.0 % of total signal".
   - Libraries: Select the reference library to compare a component spectrum.
4. Put all components that fulfil the criteria mentioned in step 3 on a positive list of tentatively identified compounds.
5. Compare the experimentally defined LTPRI values of tentatively identified compounds with match factors ≥90 with published LTPRIs from literature with AMDIS (see Note 18).
6. Repeat the procedure for samples measured on both, an HP-5MS and DB-Wax column. For identification the compound has to fulfil the criteria on both columns.
7. Confirm the mass spectrum and LTPRI of a putatively identified substance with obtained data of an appropriate pure standard compound (see Note 19).
8. Store mass spectra and LTPRI values of all MVOCs identified according to the chosen criteria in a separate database (see Note 20).

Notes

1. Manual extraction and injection of extracted metabolites are also possible. For this purpose, a manual fibre holder is needed. It has to be noted that the use of a manual fibre holder is not only more labor-intensive, but also leads to less reproducible measurements results.
2. Choice of fibre coating material and fibre thickness depends on the MVOCs produced by the fungus and has to be optimised according to the analytical purpose. The selection of adequate fibre coating material for the extraction of metabolite classes is mainly associated with the polarity of the metabolites to be extracted and is crucial to achieve efficient extraction of MVOCs. Different

fibre coatings, such as polyacrylate (PA), polydimethylsiloxane (PDMS), PDMS/divinylbenzene (DVB), carboxene (CAR)/PDMS and DVB/CAR/PDMS, are available.

3. Reliable identification of metabolites generally requires the application of two stationary phases of different polarity. In general, retention behaviour of a selected analyte to be identified is investigated, on both a non-polar stationary phase (e.g. 100 % PDMS or 95 % dimethylpolysiloxane 5 % diphenylpolysiloxane), and a more polar stationary phase (e.g. polyethylene glycol) with different chromatographic selectivity. For the columns mentioned in this protocol, LTPRI values for various compounds are available (e.g. NIST Chemistry Webbook).

4. The software program AMDIS (http://chemdata.nist.gov/mass-spc/amdis/downloads/) [18] has been used for automated data processing in this protocol. The software program MetaboliteDetector (http://metabolitedetector.tu-bs.de/download.htmL) [19] can also be used alternatively. Both programs are freely available. They can be used for automated LTPRI calibration and evaluation of retention indices, automated deconvolution of raw mass spectra, peak detection and calculation of mass spectral match factors. Furthermore, both programs offer the option of automated compound identification and assignment of metabolites from a reference list, if predefined identification criteria are met. In contrast to the AMDIS program, MetaboliteDetector additionally offers batchwise processing and quantification of identified as well as "unknown" metabolites.

5. Always include a blank sample that contains only nutrition medium as a control to prevent false positives originating from culture medium, glassware, silicon septa, lab environment, for example. This sample should be treated the same as the fungal culture sample with regard to sample preparation, HS-SPME-GC–MS analysis and data evaluation. Identified compounds in the blank sample are not further considered for the database including identified MVOCs.

6. Alternatively, only 2-mL aliquots of the autoclaved PDA medium can be transferred to sterile HS vials and the vials placed in a horizontal position for medium solidification to prevent a nutrition gradient.

7. For a good yield of spores it is important to shred the mycelium into small pieces using a scalpel before inoculating the MBS.

8. The sporulation should be carried out for at least 2 days, but it should be stopped when the first germinating conida are visible.

9. It is preferable to use freshly made spore suspensions because the age of the spores can have a significant influence on germination onset and growth and, consequently, the HS-profile.

10. The purging step is recommended for two reasons. First, removal of all metabolites accumulated above the culture enables the monitoring of MVOCs synthesis starting from that particular time point. Second, replacement of cotton plug by tightly sealing screw caps usually leads to uncontrollable loss of metabolites from the culture vial and reduces the precision of the method.

11. Incubation temperature has to be selected as required for the fungal culture. Standardisation of incubation temperature is recommended in order to reduce biological variability.

12. Extraction temperature was set to 30 °C in order to minimise affection of the biological culture while simultaneously achieving standardised conditions by heating to a temperature above room temperature. If required, extraction can also be carried out without temperature control or under cooled conditions. However, it has to be considered that extraction yield will decrease at lower temperature.

13. Conditions have been chosen with the aim to achieve appropriate chromatographic separation of MVOCs. Improved chromatographic separation especially for terpenoids can be obtained by lowering the rate of temperature increase/time interval.

14. For calculation of correct LTPRI values, narrow and symmetric chromatographic peaks of Gaussian shape are required for each

alkane standard. Therefore, different SPME extraction methods have to be used for alkane standard mixtures C_5–C_{10} and C_{21}–C_{40}. Alkane standard mixtures injected at too high concentrations or after extraction with an inadequate SPME program will result in poor peak shapes and consequently lead to inaccurate LTPRI values. Hence, it is highly recommended to optimise the analysis of n-alkanes before the measurement of fungal cultures. It might happen that not all n-alkanes elute from the column until the end of a single chromatographic run. In particular, high-boiling alkanes might need higher temperatures in order to elute during the chromatographic run. Therefore, it is recommended to bake out the column at a temperature close to maximal tolerable temperature after measuring alkane standard mixtures.

15. Usually more than one LTPRI value can be found for a particular stationary GC-phase in the literature. However, the choice of the reference LTPRI should not be made arbitrarily. NIST chemistry Webbook (http://webbook.nist.gov/chemistry/) [24], for example, contains a collection of various LTPRI values published in the literature for every compound. LTPRI values from the literature should be as consistent as possible with measured LTPRI values regarding column coating, column dimensions and the gas chromatographic method to assure best compatibility. We recommend the use of the median of published LTPRI values.

16. Deconvolution settings have to be adjusted according to the peak intensity, width and shape in chromatograms. The following settings shall serve as an example and have been found suitable in our laboratory for the evaluation of raw chromatograms obtained with instrumentation used in the present study. These settings might serve as orientation and will have to be modified according to the samples to be evaluated:

- Component width: "32"
- Adjacent peak subtraction: "One"
- Resolution: "Medium"

- Sensitivity: "Medium"
- Shape requirements: "Medium"
- User manual and detailed program description tion can be downloaded from the NIST Webpage.

17. The following m/z values are well known to be caused by decomposition of polysiloxane material used as GC stationary phase or fibre coating and therefore should be excluded from further data evaluation: m/z 77, 207, 267, 281, 341, 355. These signals can also be recognised from typical isotopic distribution of silicium.

18. Identification criteria for compounds may vary according to the analytical study. For metabolic profiling of MVOCs the application of strict criteria is advised due to the fact that only few standards are commercially available for tentatively identified compounds. Strict criteria help to exclude false positives. Therefore, it is recommended to confirm the results for identification of a compound on two columns of different stationary phase. We set the tolerance for the maximum relative deviation to ±2 % of literature values and a minimum match factor for the similarity of mass spectra of ≥90 %. Additionally, the compound should be confirmed by a standard, if available.

19. Many MVOCs identified are not available as standard compounds; therefore, it is not possible to confirm every MVOC. Dilute available standard compounds in a concentration resulting in narrow and symmetric peak shapes. No general concentration for standards can be recommended due to the fact that every substance has a different binding behaviour on particular fibre coatings. Standards can be diluted in n-hexane or $ACN:H_2O$ (1:1) depending on solubility.

20. This database can be used for the further metabolic profiling of, for example, this particular fungal strain, species or genus.

Acknowledgement The financial support by the Austrian Science Fund (FWF projects F3702 and F3706) is gratefully acknowledged.

References

1. Wheatley RE (2002) The consequences of volatile organic compound mediated bacterial and fungal interactions. Antonie Van Leeuwenhoek 81:357–364
2. Schnürer J, Olsson J, Börjesson T (1999) Fungal volatiles as indicators of food and feeds spoilage. Fung Genet Biol 27:209–217
3. Korpi A, Järnberg J, Pasanen AL (2009) Microbial volatile organic compounds. Crit Rev Toxicol 39:139–193
4. Reino JL, Guerrero RF, Hernández-Galán R, Collado IG (2008) Secondary metabolites from species of the biocontrol agent Trichoderma. Phytochem Rev 7:89–123
5. Magan N, Evans P (2000) Volatiles as an indicator of fungal activity and differentiation between species, and the potential use of electronic nose technology for early detection of grain spoilage. J Stored Prod Res 36:319–340
6. Nemčovič M, Jakubíková L, Víden I, Vladimír F (2008) Induction of conidiation by endogenous volatile compounds in Trichoderma spp. FEMS Microbiol Lett 284:231–236
7. Keszler À, Forgács E, Kótai L (2000) Separation and identification of volatile components in the fermentation broth of Trichoderma atroviride by solid phase extraction and GC/MS. J Chromatogr Sci 38:421–424
8. Larsen TO, Frisvad CF (1995) Comparison of different methods for collection of volatile chemical markers from fungi. J Microbiol Methods 24:135–144
9. Wheatley R, Hackett C, Bruce A, Kundzewicz A (1997) Effect of substrate composition on production of volatile organic compounds from Trichoderma spp. Inhibitory to wood decay fungi. Int Biodeter Biodegr 39:199–205
10. Meruva NK, Penn JM, Farthing DE (2004) Rapid identification of microbial VOCs from tobacco molds using closed-loop stripping and gas chromatography/time-of-flight mass spectrometry. J Ind Microbiol Biotechnol 31:482–488
11. Deetae P, Bonnarme P, Spinnler HE, Helinck S (2007) Production of volatile aroma compounds by bacterial strains isolated from different surface-ripened French cheeses. Appl Microbiol Biotechnol 76:1161–1171
12. Fiedler K, Schütz E, Geh S (2001) Detection of microbial volatile organic compounds (MVOCs) produced by moulds on various materials. Int J Hyg Environ Health 204:111–121
13. Demyttenaere JCR, Moriña RM, De Kimpe N, Sandra P (2004) Use of headspace solid-phase microextraction and headspace sorptive extraction for the detection of the volatile metabolites produced by toxigenic Fusarium species. J Chromatogr A 1027:147–154
14. Van Lancker F, Adams A, Delmulle B, De Saeger S, Moretti A, Van Peteghem C et al (2008) Use of headspace SPME-GC–MS for the analysis of the volatiles produced by indoor molds grown on different substrates. J Environ Monit 10:1127–1133
15. Tholl D, Boland W, Hansel A, Loreto F, Röse USR, Schnitzler JP (2006) Practical approaches to plant volatile analysis. Plant J 45:540–560
16. Arthur CL, Pawliszyn J (1990) Solid phase microextraction with thermal desorption using fused silica optical fibers. Anal Chem 62:2145–2148
17. Jeleń HH (2003) Use of solid phase microextraction (SPME) for profiling fungal volatile metabolites. Lett Appl Microbiol 36:263–267
18. Stein SE (1999) An integrated method for spectrum extraction and compound identification from gas chromatography/mass spectrometry data. J Am Soc Mass Spectrom 10:770–781
19. Hiller K, Hangebrauk J, Jäger C, Spura J, Schreiber K, Schomburg D (2009) MetaboliteDetector: comprehensive analysis tool for targeted and nontargeted GC/MS based metabolome analysis. Anal Chem 81:3429–3439
20. Stoppacher N, Kluger B, Zeilinger S, Krska R, Schuhmacher R (2010) Identification and profiling of volatile metabolites of the biocontrol fungus Trichoderma atroviride by HS-SPME-GC–MS. J Microbiol Methods 81:187–193
21. Van Den Dool H, Kratz P (1963) A generalization of the retention index system including linear temperature programd gas–liquid partition chromatography. J Chromatogr A 11:463–471
22. Lu H, Dunn BD, Shen H, Kell DB, Liang Y (2008) Comparative evaluation of software for deconvolution of metabolomics data based on GC-TOF-MS. Trends Anal Chem 27:215–227
23. Styczynski MP, Moxley JF, Tong LV, Walther JL, Jensen KL, Stephanopoulos GN (2007) Systematic identification of conserved metabolites in GC/MS data for metabolomics and biomarker discovery. Anal Chem 79:966–973
24. NIST Mass Spec Data Center, Stein SE (2011) Retention indices. In: Linstrom PJ, Mallard WG (eds). NIST Chemistry WebBook. NIST Standard Reference Database Number 69. Gaithersburg, MD: National Institute of Standards and Technology; 2011: 20899, http://webbook.nist.gov (retrieved April 27, 2011)

Transformation Methods for Slow-Growing Fungi

43

Suman Mukherjee and Rebecca Creamer

Abstract

It is challenging to prepare protoplasts and establish a transformation system in any toxin that produces slow-growing fungus, like *Undifilum oxytropis*. This chapter describes a protocol to generate a nontransient transformation system with the help of introducing a foreign heterologous gene to a toxin producing slow-growing fungi. This system will be useful to study genes in metabolically important pathways.

Keywords

Slow-growing fungi • Protoplast • Transformation • *Undifilum oxytropis* • Green fluorescent protein

Introduction

The preparation and regeneration of protoplasts and the subsequent development of transformation systems are well established for several fungi. Several groups, such as Fierro et al. and Akamatsu et al., established transformation systems for ascomycete fungi, whereas others, such as Panaccione et al., developed transformation systems for the endophyte *Neotyphodium* [1–3]. All of these systems were developed for fungi that rapidly grow and reproduce.

Toxic locoweeds are found to be associated with a slow-growing endophytic fungus, *Undifilum oxytropis*. *U. oxytropis* is an *Embellisia*-like fungus that does not produce external mycelia on host [4]. When cultured in vitro, *U. oxytropis* grows up to 0.03–0.34 mm/day on potato dextrose agar (PDA) plates at room temperature [5]. *U. oxytropis* produces a polyhydroxy alkaloid, swainsonine (1, 2, 8-trihydroxyindolizidine), naturally and in culture [5, 6], which has been correlated with toxicity of locoweeds [5, 7]. This protocol details the generation of protoplasts and establishment of a stable transformation system from slow-growing toxin-producing fungi [8].

The current protocol was established for *U. oxytropis* but can be extended to several

S. Mukherjee (✉)
National Institutes of Health, Laboratory of Biochemistry and Genetics, NIDDK, 8 Center Drive, Room 326, Bethesda, MD 20892, USA
e-mail: suman.mukherjee@nih.gov; creamer@nmsu.edu

R. Creamer
Department of Entomology, Plant Pathology, and Weed Science, New Mexico State University, Box 30003, MSC 3BE, 945 College Ave, Las Cruces, NM 88003, USA

V.K. Gupta et al. (eds.), *Laboratory Protocols in Fungal Biology: Current Methods in Fungal Biology*, Fungal Biology, DOI 10.1007/978-1-4614-2356-0_43, © Springer Science+Business Media, LLC 2013

slow-growing toxin-producing fungi, including endophytes. Protoplasts of *U. oxytropis* were prepared and transformed with a fungal-specific vector that expresses green fluorescent protein (GFP). Regenerated fungal mycelia were screened for GFP expression. In this chapter we describe protoplast preparation, regeneration with selection, nontransient GFP expression, and a stable transformation system using a slow-growing fungus. This technology can provide tools to understand metabolically important pathways of toxin-producing fungi.

Materials

1. Potato dextrose agar plates (PDA)
2. Water agar media
3. Fungal-specific vector
4. Potato dextrose broth
5. Antibiotic hygromycin B (Hyg)
6. Platform shaker at 200 rpm
7. Miracloth
8. Buchner funnel
9. β-glucuronidase
10. Lysing enzyme
11. β-D-Glucanase G
12. Bovine serum albumin
13. $MgSO_4$
14. NaH_2PO_4
15. Sorbitol
16. Tris–HCl
17. Refrigerated centrifuge
18. $CaCl_2$
19. Hemocytometer
20. 4', 6-Diamidino-2-phenylindole
21. Deionized water
22. Fluorescence microscope
23. TE buffer
24. Sucrose
25. Yeast extract
26. Casein hydrolysate
27. Bacto Agar
28. Poly ethylene glycol, 4,000 molecular weight
29. NaCl
30. NaF
31. Phenylmethylsulphonyl floride
32. EDTA
33. Triton-X 100
34. Glycerol
35. SDS-polyacrylamide gel
36. Immobilon polyvinylidene fluoride (PVDF)
37. Tween-20
38. Tris buffered saline
39. Rocking platform

Methods

Fungal Strains and Culture Conditions

1. Fungal isolates can be isolated from plants or harvested from colonies grown on potato dextrose agar plates (PDA).
2. If collecting from plant tissues, surface sterilize for 30 s in 70 % ethanol, followed by 3 min in 20 % bleach, and then 30 s in sterile water.
3. Dry the plant tissue and place on water agar media; incubate media plates at room temperature (25 °C) for 2 weeks.
4. Transfer fungal hyphae from water agar media to PDA plates and grow at room temperature for at least 14 days [9].
5. Preserve fungal isolates and store at both 4 °C and −80 °C. Fungal plates can also be maintained in a sealed plastic box at room temperature for further experiments (Fig. 43.1).

Hygromycin B Selection

1. pPd-EGFP vector was used for *U. oxytropis* transformation; an ascomycete-specific vector can be used for transformation.
2. pPd-EGFP is a fungal-specific expression vector that encodes the antibiotic hygromycin B (Hyg) resistance gene driven by the *Aspergillus nidulans* trpC-promoter [10]. The vector also expresses GFP driven by the *Cryphonectria parasitica* glyceraldehydes-3-phosphate dehydrogenase (gpd) promoter.
3. Generate working stock of Hyg at a concentration of 1 mg/mL (dissolve in sterile water).

Fig. 43.1 *Unidifilum oxytropis* (25-1 isolate) grown on PDA plates for 4 weeks

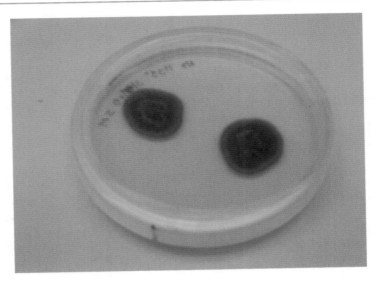

4. Add Hyg to a final concentration of 20 μg/mL for a 20 mL of PDA plate. For *U. oxytropis* 20 μg/mL of Hyg is the lethal dose, so any growth of mycelia on plate represents growth of positive transformants.

5. Lethal dose differs among fungal species, so a kill curve must be determined with the required antibiotic using serial dilutions of the antibiotic on PDA plates.

6. Measure fungal growth on the PDA plate in mm over a 14-day span or more depending on the growth rate of the respective fungus. Use the data to generate a kill curve for hygromycine sensitivity of the fungi.

Preparation of Protoplasts, Transformation, and Regeneration

This method was used to prepare fungal protoplasts.

1. Add 1 g of fungal mycelia into potato dextrose broth in a 250-mL conical flask.

2. Incubate with shaking for 2 weeks on a platform shaker at 200 rpm at room temperature.

3. Filter the fungi and media to collect the fungal mass using miracloth in a Buchner funnel.

4. Re-suspend mycelial mass in 100 mL of digestion buffer.

5. Digestion buffer composition: 1 mL of β-glucuronidase, 75 mg lysing enzyme, 800 mg β-D-Glucanase G, 600 mg bovine serum albumin dissolved in 100 mL osmotic medium (1.2 M MgSO$_4$ with NaH$_2$PO$_4$, pH 5.8).

6. Incubate at room temperature on an orbital shaker for 3 h.

7. The 3 h of incubation time after addition of digestion buffer provides optimum protoplasts yield for *U. oxytropis*. Optimum incubation time depends on fungal species and isolates.

8. After incubation, collect 8 mL of digested mycelial mass in 30-mL glass centrifuge tubes.

9. Combine mycelial mass with 10 mL of trapping buffer (0.4 M Sorbitol in 100 mM Tris–HCl, pH 7.0).

10. Centrifuge at 6,000 rpm at 4 °C for 15 min.

11. Collect protoplasts at the interface of the two layers.

12. Add 2 volumes of 1 M sorbitol to the collected protoplasts.

13. Centrifuge the protoplasts at 6,000 rpm at 4 °C for 5 min.

14. Collect pellets after decanting the supernatant.

15. Suspend the collected protoplasts (pellets) in 100 μL of STC buffer (STC buffer: 1 M sorbitol in 100 mM Tris–HCl, pH 8.0, 100 mM CaCl$_2$).

16. Count aliquots of protoplasts using a hemocytometer with a light microscope.

17. Keep suspended protoplasts on ice for transformation experiments (10^9 protoplasts per mL can be obtained using this method).

DAPI Staining

To check for quality/viability of protoplasts, DAPI (4′, 6-Diamidino-2- phenylindole) stain can be used.

1. DAPI stock solution is 5 mg/mL (10 mg in 2 mL of deionized water).

2. Add DAPI stock solution into the protoplasts to make a 100 ng/mL final concentration.

3. Incubate 10 μL of the protoplast /DAPI solution in a sterile microcentrifuge tube for 30 min at room temperature and observe in fluorescence microscope with excitation at 360 nm [8].

4. DAPI stock solution should be stored in a brown bottle, protected from light at −20 °C; working solution can be stored at 4 °C.

Transformation Procedure

1. Dilute 5 μg of vector in 10 μL of TE (0.5×, pH 8) buffer.

2. Add the vector solution (10 μL) to purified protoplasts.

3. Incubate on ice for 1 h.

4. GFP expression can be observed at this point using a fluorescence microscope [8].

5. Add 1 mL of PTC buffer (PTC buffer: 40 % poly ethylene glycol, 4000 molecular weight, 100 mM Tris–HCl, pH 8.0 and 100 mM CaCl$_2$) to each tube of protoplast.

6. Incubate 25 min at room temperature.

7. Plate the transformation mixture as 2, 20, and 200 μL droplets onto Petri dishes.

8. Add 12.5 mL regeneration medium (1 M of sucrose, 0.001 w/v of yeast extract, 0.001 w/v of casein hydrolysate and 0.016 % of Bacto Agar) to each plate.

9. Add further 12.5 mL regeneration media containing 40 μg/mL Hyg overlaid on each plate (double the amount of Hyg to be added to avoid dilution of the antibiotic).

10. Incubate plates for 4 days at room temperature (see Note 1).

Sporulation and Hyphal Tipping

1. Grow transformant on Hyg-containing water agar plates (complete separation of clonal hyphae can be observed in water agar plate).

2. Collect single hyphal tips under microscope to start a mononucleate culture on PDA plates again.

3. Observe GFP expression after 2 weeks of growth on PDA plates (see Note 2).

Microscopy

1. Take fungal mycelia on a slide, add 10 μl of water, and observe under fluorescence microscope.

2. Use fluorescence microscope to capture fluorescence images at a combined objective and eyepiece magnification of 40× and 10×, respectively. Mycelium demonstrated strong expression of the GFP gene, as observed by emission at 488 nm using an Axiovert 200 M microscope (any other fluorescence microscope can be used).

3. GFP expression of emerged mycelia was observed microscopically after 3 weeks of growth (Fig. 43.2). Expression was retained after several weeks and passages of mycelial growth on PDA plates as observed (Fig. 43.3) (see Note 3).

Detection of Transformed Gene or Marker by Immunoblot

1. Harvest fungal tissues from transformed and untransformed *U. oxytropis* for western blot experiment.

2. Homogenize filtered fungal tissue with liquid nitrogen in extraction buffer (Extraction

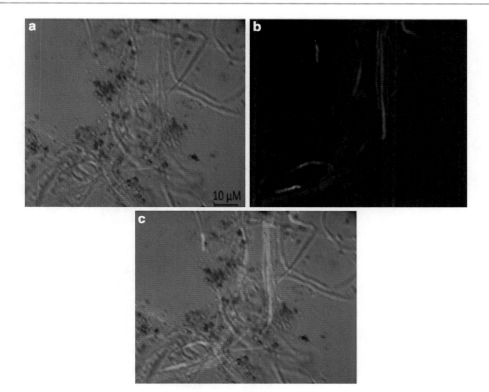

Fig. 43.2 (**a**) Mycelial mass of *U. oxytropis* 3 weeks post-transformation on hygromycin selection PDA plate. (**b**) Fluorescence image of mycelial growth of *U. oxytro-* *pis* 3 weeks post-transformation on hygromycin selection PDA plate expressing GFP. (**c**) Overlay of (**a**) and (**b**)

buffer: 50 mM Tris–HCl, pH 7.5, 100 mM NaCl, 50 mM NaF, 2 mM phenylmethylsulphonyl floride, 5 mM EDTA, 1 % Triton-X 100, 10 % glycerol).

3. Centrifuge at 4 °C for 10 min at 13,000 rpm.

4. Quantify protein concentration of lysates using a protein assay kit.

5. Load total protein (20 μg) on a 12 % SDS-polyacrylamide gel.

6. Transfer the gel to Immobilon polyvinylidene fluoride (PVDF) transfer membrane.

7. Block the membrane in 3 % (w/v) bovine serum albumin and dissolve in 1× Tris buffer saline with Tween-20 (TBST) for 1 h at room temperature (on rocking platform).

8. Dilute GFP or selective marker antibody as recommended by manufacturer.

9. Add primary antibody solution to the membrane and incubate for 1 h at room temperature (on rocking platform).

10. Wash the membrane by rocking in 10 mL of 1× TBST three times, 10 min each.

11. Add secondary antibody solution to the membrane and incubate for 1 h at room temperature (on rocking platform).

12. Wash the membrane by rocking in 10 mL of 1× TBST three times, 10 min each.

13. Visualize blots using the detection kit (Fig. 43.4) (see Note 4).

Notes

1. Although serial dilutions were performed on PDA plates, observation of individual colonies is difficult after 7 days of regenerated growth because the transformants can merge with each other. A count of individual colonies should be done at an early stage to measure transformation efficiency. Fourteen days after

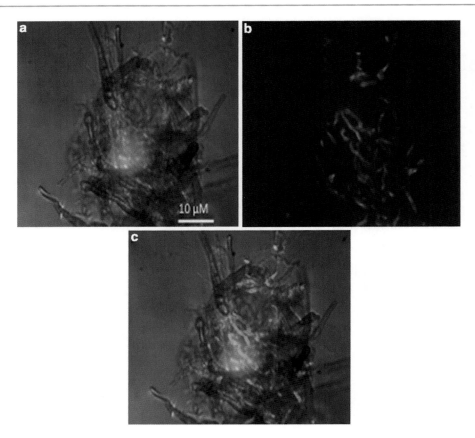

Fig. 43.3 (a) Mycelial mass of *U. oxytropis* 9 weeks post-transformation on hygromycin selection PDA plate. (b) Fluorescence image of mycelial growth of *U. oxytro-* *pis* 9 weeks post-transformation on hygromycin selection PDA plate expressing GFP. (c) Overlay of (a) and (b)

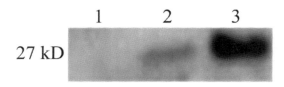

Fig. 43.4 Immunoblot of hyphal samples of *U. oxytropis* using anti-GFP antibody. 27 kDa band indicates GFP expression. *Lane 1*, non-transformed hyphae; *lane 2*, hyphae 3 weeks post-transformation; *lane 3*, hyphae 9 weeks post-transformation

transformation, complete mycelia formation was observed on Hyg-containing PDA plates. No hyphae were observed in untransformed protoplasts plated on Hyg-containing PDA plates [8].

Slow-growing fungi such as *U. oxytropis* may take 7 days to grow mycelia after regeneration. In other ascomycetes, mycelial growth can be observed within 24 or 48 h.

Perform a control transformation mixture without vector (containing antibiotic marker) following similar protocol as above. Plate on Hyg-containing PDA plate as control.

2. Sporulation is a rare event for *U. oxytropis*, and conidial germination rates are poor. Hyphal tipping was used to start new cultures. For other fungi, single spores can be used to start new cultures on PDA plates.

3. GFP expression was observed for several weeks and months in transformed fungal mycelia, which indicates the establishment of a stable transformation system. We used GFP

marker for fluorescence microscope study, but any marker can be used.

4. We used GFP as a marker to investigate transformation, but any marker can be detected with this method if suitable antibody is present. Blocking buffer can be made also with PBST with 2 % dry non-fat milk. Primary and secondary antibody solution was generated by diluting designated antibodies in blocking solution.

Acknowledgements The authors would like to thank Dr. Richard Richins, Dr. Soum Sanogo, Dr. Swati Mukherjee, and Deana Beacom of New Mexico State University for technical assistance and constructive discussion.

References

1. Akamatsu H, Itoh Y, Otani H, Kohmoto K (1997) AAL-toxin-deficient mutants of Alternaria alternata tomato pathotype by restriction enzyme-mediated-integration. Phytopathology 87:967–972

2. Fierro F, Gutierrez S, Diez B, Martin JF (1993) Resolution of four large chromosomes in penicillin-producing filamentous fungi: the penicillin gene cluster is located on chromosome II (9.6 Mb) in *Penicillium notatum* and chromosome I (10.4 Mb) in *Penicillium chrysogenum*. Mol Gen Genet 241:573–578

3. Panaccione DG, Johnson RD, Wang J, Young CA, Damrongkool P, Scott B et al (2001) Elimination of ergovaline from a grass-Neotyphodium endophyte symbiosis by genetic modification of the endophyte. Proc Natl Acad Sci U S A 98:12820–12825

4. Pryor B, Creamer R, Shoemaker R, McLain-Romero J, Hambleton S (2009) Undifilum, a new genus for endophytic Embellisia oxytropis and parasitic Helminthosporium bornmuelleri on legumes. Botany 87:178–194

5. Braun K, Romero J, Liddell C, Creamer R (2003) Production of swainsonine by fungal endophytes of locoweed. Mycol Res 107:980–988

6. Harris CM, Schneider MJ, Ungemach FS, Hill JE, Harris TM (1988) Biosynthesis of the toxic indolizidine alkaloids slaframine and swainsonine in Rhizoctonia leguminicola: metabolism of 1-hydroxy-indolizidines. J Am Chem Soc 110:940–949

7. Ralphs MH, Gardner DR, Turner DL, Pfister JA, Thacker E (2002) Predicting toxicity of tall larkspur (*Delphinium barbeyi*): measurement of the variation in alkaloid concentration among plants and among years. J Chem Ecol 28:2327–2341

8. Mukherjee S, Dawe AL, Creamer R (2010) Development of a transformation system in the swainsonine producing, slow growing endophytic fungus, *Undifilum oxytropis*. J Microbiol Methods 81:160–165

9. Ralphs MH, Creamer R, Baucom D, Gardner DR, Welsh SL, Graham JD et al (2008) Relationship between the endophyte Embellisia spp. and the toxic alkaloid swainsonine in major locoweed species (*Astragalus and Oxytropis*). J Chem Ecol 34:32–38

10. Suzuki N, Geletka LM, Nuss DL (2000) Essential and dispensable virus-encoded replication elements revealed by efforts To develop hypoviruses as gene expression vectors. J Virol 74:7568–7577

Enzymatic Saccharification of Lignocellulosic Biomass

Manimaran Ayyachamy, Vijai Kumar Gupta,
Finola E. Cliffe, and Maria G. Tuohy

Abstract

The conversion of polymers present in the lignocellulosic biomass into fermentable sugars can be achieved through physical/chemical and enzymatic pretreatments. The microbial conversion of biomass to bioenergy will be cost-effective only if all of the components in the biomass are converted into value-added products. The combination of appropriate chemical and enzymatic conversion methods is very important to develop an effective biomass to biofuels and biorefineries conversion technology.

Keywords

Lignocellulosic biomass • Pretreatment • Saccharification • Enzyme cocktails • Biofuels • Biorefineries

Introduction

Rapidly depleting stocks and increasing demand for petroleum-derived energy sources, in combination with exponentially increasing global warming, have forced the public and private sectors globally to look for alternative technologies to meet future energy needs. Biofuels are promising as future candidate fuels as they are largely produced from renewable plant resources or wastes. Biofuels have great potential not only as environmentally clean, renewable energy sources, but also they provide a mechanism to reduce reliance on imported energy sources [1, 2].

Plant biomass (or lignocellulosic biomass) is approximately composed of cellulose (20–50%) and hemicellulose (15–35%) and strongly intermeshed with the aromatic copolymer lignin

M. Ayyachamy (✉) • F.E. Cliffe • M.G. Tuohy
Molecular Glycobiotechnology Research Group,
Department of Biochemistry, School of Natural Sciences,
National University of Ireland,
University Road, Galway, Ireland
e-mail: manimaran.ayyachamy@nuigalway.ie

V.K. Gupta
Molecular Glycobiotechnology Research Group,
Department of Biochemistry, School of Natural Sciences,
National University of Ireland, University Road,
Galway, Ireland

Assistant Professor of Biotechnology, Department
of Science, Faculty of Arts, Science & Commerce,
MITS University, Rajasthan, India

V.K. Gupta et al. (eds.), *Laboratory Protocols in Fungal Biology: Current Methods in Fungal Biology*,
Fungal Biology, DOI 10.1007/978-1-4614-2356-0_44, © Springer Science+Business Media, LLC 2013

[3–8]. The conversion of polysaccharides present in the biomass into fermentable sugars can be achieved through physical/chemical and enzymatic methods [6, 9]. The microbial conversion of biomass to energy will be cost-effective only if all of the components in the biomass are converted into value-added products, which is the basis of the lignocellulosic biorefinery concept [10–13].

A primary target of a biomass to biorefinery strategy is to maximize the conversion of the significant reservoir of complex carbohydrates to feedstocks that are suitable for the production of biofuels and other commodity products through fermentation or thermochemical processing [12, 14]. To maximize recovery of fermentable sugars, lignocellulosic biomass is generally pretreated with chemicals and/or microbial enzymes to reduce the size of the feedstock and also to open up the plant structure [15, 16]. Cellulose and hemicellulose, which are the principal sugar components, are bound with lignin in a complex, highly ramified and cross-linked matrix that is designed to resist chemical and microbial attack. This structure limits bioconversion, either by microorganisms or the enzymes they produce; therefore, a pretreatment is generally required before lignocellulosic biomass can be subjected to microbial or enzymatic conversion.

Pretreatment technologies for lignocellulosic biomass can be assigned into four broad categories: physical, chemical, physicochemical, and biological methods. An ideal pretreatment should remove lignin and thus reduce the crystallinity of cellulose [17], increase the porosity and accessibility of the cellulose (and hemicellulose) to enzymatic hydrolysis, generate low levels of compounds inhibitory to both enzymes and fermentation microorganisms, be of low cost and have low energy requirements. Overall, the pretreatment should result in a reduction in the recalcitrance of lignocellulose and increase accessibility to enzymes. Pretreatments vary from hot-water extraction, steam pretreatments (often with an oxidant or other chemical), to weak and strong acid and alkali pretreatments [18, 19]. However, pretreatments may affect the composition and interactions between biomole-

cules in the substrate in a way that is not necessarily advantageous to downstream bioconversion [20, 21].

The ideal pretreatment should minimize the need for feedstock particle size reduction, limit the formation of sugar degradation compounds, minimizes energy demands, and reduces the overall costs. During the chemical pretreatment process, sugar-derived compounds such as furfural and 5-hydroxymethyl furfural are formed as co-products [22]. These compounds are very toxic to microorganisms and also inhibit subsequent enzymatic saccharification process. Therefore, these compounds must be removed or neutralized prior to the microbial fermentation and enzymatic hydrolysis. Biological pretreatments are based mainly on the use of white rot fungi to delignify biomass in a low-cost treatment approach (some loss of hemicellulose and minor amounts of cellulose occurs as a result of this type of microbial pretreatment). Microbial delignification, although gentle and effective, can remove upto 32% of the lignin from biomass materials, such as corn stover [23]. However, this pretreatment method generally does not give high sugar yields during subsequent hydrolysis. Furthermore, microbial pretreatment times are lengthy, typically requiring 18–35 days.

Enzymatic saccharification of lignocellulosic biomass is a very complex process and the hydrolysis of all of the polysaccharides requires a repertoire of several hydrolytic enzymes. In the biosphere, a variety of pathogenic and saprophytic microorganisms play an unparalleled role in bringing about the depolymerization and decomposition of plants and plant-derived residues and wastes. Fungi are key microbial players in the biological conversion of plants and plant-derived wastes. These eukaryotic microorganisms secrete a vast array of carbohydrate hydrolases (collectively termed glycosyl hydrolases) as well as a range of peptidases and lignin-modifying enzymes that reduce plant biomass to its simple building blocks. For decades, scientists have worked to identify the types and modes of action of cellulases and hemicellulases produced by fungi. Species of ascomycete fungi, namely *Trichoderma reesei* and *Aspergillus niger,* have been exploited as cell factories for production of commercial cellulase and hemicellulase enzyme products. More recently, additional fungal species have been investigated as sources of cellulases and hemicellulases with enhanced catalytic efficiency and thermostability [24, 25].

Thus, the combination of chemical and enzymatic methods is required for an efficient conversion of biomass into fermentable sugars, and selection of appropriate combinations of chemical and enzymatic conversion methods is very important to develop an effective biomass to biofuels and biorefinery conversion technology.

Materials

1. Feedstocks: lignocellulosic biomass collected from local sources (see Note 1).
2. Enzymes: cellulolytic and hemicellulolytic enzymes were obtained from Novozymes (Denmark) and Genencor (USA).
3. Novozymes: Celluclast, Cellic C Tech 2, Cellic H Tech 2.
4. Genencor: Accellerase 1500, Spezyme CP, Accellerase CY, Accellerase XC, Accellerase XY, Accellerase BG.
5. Buffers (100 mM): pH (5–6)—Sodium acetate buffer; pH (7–8)—Sodium phosphate buffer (pH 7-8); pH 9—Glycine-NaOH buffer (see Note 2).
6. Substrates: wheat arabino xylan (Megazyme International Ireland Ltd., Bray, Co. Wicklow), oat spelts xylan (Sigma-Aldrich, Dublin, Ireland), birch wood xylan (Sigma-Aldrich), β-glucan (from barley; Megazyme Intl. Ireland Ltd.), and carboxy methyl cellulose and 4-nitrophenyl α- or β-glycosides (Sigma-Aldrich) (see Note 3).
7. Dionex high-performance anion exchange chromatography (HPAEC) ICS-3000 ion chromatography system equipped with an autosampler, a UV detector, and ED40 pulsed amperometry detector, fitted with a gold electrode (Dionex Corporation, Sunnyvale, CA, USA). SD-10 and PA-100 analytical and guard columns (Dionex

Corporation, Sunnyvale, CA, USA) Chromeleon™ Version 6.70 software for data collection, processing, and analysis (Dionex Corp., Sunnyvale, CA, USA) (see Note 4).
8. Chemicals: Unless otherwise stated, all chemicals and reagents were purchased from Sigma-Aldrich (Dublin, Ireland).
9. Fine-grade muslin was purchased from a local supplier.
10. Grant water baths (Mason Technology, Ireland).
11. Thermostatically controlled New Brunswick Scientific Innova 44 platform, shaking incubator (New Brunswick Scientific, USA).
12. High-temperature oven (Fisher Scientific, Ireland).
13. BioTek Powerwave XS2 microplate reader with incubation facilities to 50 °C (BioTek, USA).
14. PDFE membrane (0.2 μm) syringe-less filters (Millipore, USA).
15. Laboratory plastics, i.e., disposable micropipette tips, microcentrifuge tubes, and flat-bottomed, high-optical-quality, low-protein-absorbing microtitre plates (Sarstedt, Sinnottstown, Co. Wexford, Ireland).
16. Safety apparel (Caulfields Ltd., Galway, Ireland).

Methods

Feedstock Collection and Characterization

1. Lignocellulosic biomass is collected from local areas and is washed thoroughly with tap water to remove contaminants.
2. Feedstock must be dried at 50 °C in a hot air oven for 3 days.
3. Feedstock is milled to a particle size of 2–3 mm before subjecting to either chemical or enzymatic pretreatments (see Note 5).
4. Total solids, ash, nitrogen, and carbohydrate contents in the biomass sample are analyzed using the standard methods [22, 26].

Commercial Enzymes

1. Cellulolytic and hemicellulolytic enzymes can be purchased from the leading commercial suppliers such as Novozymes (Denmark) and Genencor (USA).
2. Necessary enzyme activities have to be determined [27–32] using the standardized assay methods (see Note 6).
3. The enzyme solution should be prepared freshly every time from the stock prior to saccharification (see Note 7).
4. Enzymes should be kept at 4 °C and brought to room temperature prior to use.

Enzymatic Saccharification of Lignocellulosic Biomass

In order to achieve maximum biomass conversion, the reaction conditions for enzymatic hydrolysis should be optimum. The following experiments must be performed to determine the optimal reaction conditions for enzymatic saccharification.

pH Optimization

1. Prepare buffers at different pH values over the range 2.6–11. Sterilize the buffer solutions in an autoclave at 121 °C for 15 min. Allow the buffer solutions to cool to ambient temperature prior to use (see Note 8).
2. Weigh out 1.5 g of biomass and add 15 mL of buffer solution to obtain a 10% (w/v) solids ratio. Prepare all tests in triplicate.
3. Uniformity in the enzyme dosage has to be maintained since different types of commercial enzymes are used for saccharification. Therefore, a consistent enzyme dosage should be used for each test, when comparing different enzymes.
4. Controls in (triplicate) consisting of a denatured sample of the appropriate enzyme should be run in parallel with tests (see Note 9).
5. Sodium azide is added at 0.02% to prevent the microbial contamination. If using a thermostable enzyme preparation at higher

reaction temperature, preservative may not be required.
6. Reactions are carried out at 50 °C for 72 h under shaking conditions (150 rpm).
7. Sampling must be done every 1 h at the start of the reaction, and every 3–12 h thereafter. Samples should be centrifuged to remove residual biomass and then assayed for the release of reducing sugars [31] from the polysaccharides present in the biomass (see Note 10).

Optimization of Temperature

1. 1.5-g quantity of biomass is weighed and transferred to a 50-mL polypropylene centrifuge tube. A 15-mL volume of buffer solution (optimized pH) is added. All tests are prepared in triplicate.
2. The enzyme should be maintained at an identical dosage level in all tests comparing the commercial enzymes.
3. A set of controls (in triplicate) with denatured enzyme must be included in parallel with test.
4. Sodium azide is added at 0.02% to prevent the contamination.
5. The reaction is carried out at different temperatures (30–80 °C) for 72 h at 150 rpm.
6. Withdraw samples every 1 h at the start of the reaction, and every 3–12 h thereafter. Samples should be centrifuged to remove residual biomass and then assayed for the release of reducing sugars [31] (see Note 10).

Optimization of Enzyme Dosage

1. Add 1.5 g of biomass and transfer to 50-mL PP centrifuge tube. Add 15 mL of buffer (at the optimum pH). All tests are prepared in triplicate.
2. The enzyme should be added at different dosage levels, such as 1, 5, 10, 25, or 50 IU per gram of biomass (see Note 11).
3. Triplicate controls are prepared with denatured enzyme.
4. Sodium azide is added at 0.02% in all tests to prevent microbial contamination.
5. The reaction is carried out at the optimum temperature for 72 h at 150 rpm (see Note 12).

6. Sampling is done every 1 h at the start of the reaction, and every 3–12 h thereafter. Samples should be centrifuged to remove residual biomass and then assayed for the release of reducing sugars [31] (see Note 10).

Saccharification of Biomass with Mixtures of Enzyme Cocktails

1. The conversion of biomass to fermentable sugars can be further enhanced by mixing different enzyme preparations together at various proportions (see Note 13).
2. By quantifying the reducing sugars at regular time intervals, addition of selective accessory enzymes can be determined.
3. Saccharification should be carried out at optimized reaction conditions (see Note 14).
4. The total reaction volume should be identical in all tests.

Recovery of Reducing Sugars (Hydrolysates) from Enzyme Pretreated Biomass

1. The saccharification process is terminated by heating the enzyme pretreated biomass samples at 100 °C for 10 min (see Note 15).
2. The biomass sample is filtered through muslin or centrifuged at 4,000 rpm for 10 min.
3. A 5-mL quantity of buffer is then added to the residual biomass, and the components mixed using a vortex for 5 min to recover the sugars completely from biomass.
4. Store the hydrolysates at 4 °C until required for analysis.

Analysis of Lignin-Derived Compounds in the Hydrolysates

1. Test and control samples are centrifuged at 14,000×g for 15 min.
2. Lignin-derived compounds in the supernatant are observed by measuring the absorption spectrum between 200 and 465 nm using UV-Vis spectrophotometer (Varian Cary 100, USA), or in microassay format using a BioTek

Powerwave XS2 microplate reader in the spectral scan mode, against the control supernatant from biomass pretreated with denatured enzyme.
3. Compare the absorption values between control and test samples (see Note 16).

Quantification of Reducing and Total Sugars in the Hydrolysates

1. Centrifuge the hydrolysates from test and control samples at 10,000×g for 10 min and the supernatant is used for analyzing reducing sugars.
2. The appropriate dilution of hydrolysates has to be made.
3. Reducing sugars in the hydrolysates are quantified by DNS reagent method [31].
4. Total sugars in the hydrolysates is quantified by phenol-sulphuric acid method [24].
5. Calculate the saccharification (hydrolysis) efficiency of enzyme cocktails from the reducing sugars released in relation to the total sugars present in the lignocellulosic biomass [32].

Qualitative and Quantitative Analysis of Reducing Sugars in Hydrolysates

1. The concentrations of mono-, di- and oligosaccharides in the hydrolysates can be determined by HPAEC equipped with UV and pulsed amperometry ED40 gold electrode detectors (Dionex, France).
2. Hydrolysate samples at various time intervals are collected and centrifuged at 10,000×g for 5 min.
3. The supernatant is filtered through 0.2 μm PDFE membrane syringe-less filters (Millipore, USA).
4. Elute the samples through the column CarboPac SD-10 and CarboPac P100 for neutral monosaccharides and oligosaccharides, respectively, using 50–100 mM NaOH, or NaOH with increasing sodium acetate for oligosaccharide separation.

5. The flow rate is set at 1 mL per minute and the flow cell temperature is kept at 30 °C.
6. Samples used for HPAEC analysis should be diluted in such a way that the concentration lies within the detectable range of the column (see Note 17).

Notes

1. If specific plant biomass is used for any pretreatment, make sure it is not contaminated before processing the samples. Rheological factors may influence the nutrient contents of biomass and therefore replicate biomass samples should be collected from the same sources.
2. Buffers should be freshly prepared each week. Ionic strength and pH should be checked carefully, otherwise this will influence enzymatic hydrolysis.
3. Substrates used for enzyme assays should be prepared freshly. Make sure that they are not degraded, otherwise this will affect calculation of the exact activities.
4. Use HPLC-grade water only for sample preparation. Samples should be filtered through appropriate filters before analysis, otherwise this will affect the HPAEC column life and resolution.
5. If chemically pretreated sample is used for enzymatic saccharification, they must be washed properly until it reaches neutral pH. It is very important to use the same batch samples for all proximate analyses and saccharification experiments. This will minimize the variation in the results.
6. Most of the commercial enzyme suppliers will not give the details about the enzyme activity and types of enzymes present in their cocktails. Key enzyme activities should be determined using the standard assay methods.
7. It is possible that the diluted enzyme will lose its activity very quickly. Therefore, fresh enzyme dilutions should be made whenever required. Mix the contents properly before use.
8. The optimal pH may vary for each commercial cocktail. After reviewing the results from tests conducted at different pH values, further optimization studies using an enzyme cocktail can be conducted at the optimal pH.
9. In order to ensure that sugar released is not due to the potential reducing sugar content of an enzyme cocktail, a control run with denatured enzymes should be carried out.
10. Freeze the samples immediately or preferably heat at 100 °C for 15 min to terminate the enzymatic action.
11. Key enzyme activities in the cocktail should be determined to ensure uniform enzyme dosage in all cases. If necessary, appropriate dilutions have to be made using the appropriate buffers.
12. The optimal temperature and pH may vary between two enzymes. If a big difference does not exist in the conversion rate between two temperatures, it is better to perform the reaction at the lower temperature.
13. Some of the cocktails may not have the required accessory enzymes. In this case it is better to include accessory enzymes. This will not only increase the conversion efficiency but also potentially reduce the reaction time.
14. It is expected that the optimal conditions will not be same for two different enzymes. Therefore, a compromise may be required such that the best reaction conditions for one of the enzymes is used preferentially, or to compromise, reaction temperature is used that will work for both enzyme preparations.
15. It is difficult to recover all of the released sugars from the biomass. After filtering the hydrolysates, the residual biomass has to be washed with a minimum amount of buffer, generally one third of the total reaction volume.
16. Lignin-derived compounds in the hydrolysates can be compared with control samples. Generally, lignin is directly or indirectly removed due to enzymatic action. Increase in the absorbance values of the tests in the wavelength range within which lignans and phenolics absorb light is an indication that the enzyme has acted on lignocellulosic biomass.

17. Sufficient column washes should be carried out between HPAEC analyses. Internal sugar standards should be included in samples, as appropriate to ensure that separation and quantification are optimum.

References

1. Mielenz J (2001) Ethanol production from biomass: technology and commercialization status. Curr Opin Microbiol 4:324–329
2. Ragauskas AJ, Williams CK, Davison BH, Britovsek G, Cairney J, Eckert CA et al (2006) The path forward for biofuels and biomaterials. Science 311:484–489
3. Dekker RFH, Richards GN (1976) Hemicellulases, their occurrence, purification, properties and mode of action. Adv Carbohydr Chem Biochem 32:277–352
4. Balat M (2011) Production of bioethanol from lignocellulosic materials via the biochemical pathway: a review. Energy Conv Manag 52:858–887
5. Sanchez C (2009) Lignocellulosic residues: biodegradation and bioconversion by fungi. Biotechnol Adv 27:185–194
6. Boateng AA, Jung HG, Adler PR (2006) Pyrolysis of energy crops including alfalfa stems, reed canarygrass and eastern gamagrass. Fuel 85:2450–2457
7. Howard RL, Abotsi E, Jansen van Rensburg EL, Howard S (2003) Lignocellulose biotechnology: issues of bioconversion and enzyme production. Afr J Biotechnol 2:602–619
8. Malherbe S, Cloete TE (2002) Lignocellulose biodegradation: fundamentals and applications. Environ Sci Biotechnol 1:105–114
9. Sims REH, Mabee W, Saddler JN, Taylor M (2010) An overview of second generation biofuel technologies. Bioresour Technol 101:1570–1580
10. Cherubini F (2010) The biorefinery concept: using biomass instead of oil for producing energy and chemicals. Energy Conv Manag 51:1412–1421
11. FitzPatrick M, Champagne P, Cunningham MF, Whitney RA (2010) A biorefinery processing perspective: treatment of lignocellulosic materials for the production of value-added products. Bioresour Technol 101:8915–8922
12. Percival Zhang Y-H (2008) Reviving the carbohydrate economy via multi-product lignocellulose biorefineries. J Ind Microbiol Biotechnol 35:367–375
13. Taylor G (2008) Biofuels and the biorefinery concept. Energy Pol 36:4406–4409
14. Kamm B, Kamm M (2004) Principles of Biorefineries. Appl Microbiol Biotechnol 64:137–145
15. Lee J (1997) Biological conversion of lignocellulosic biomass to ethanol. Biotechnology 56:1–24
16. Gray KA, Zhao L, Emptage M (2006) Bioethanol. Curr Opin Chem Biol 10:141–146
17. Lynd LR, Weimer PJ, van Zyl WH, Pretorius IS (2002) Microbial cellulose utilization: fundamentals and biotechnology. Microbiol Mol Biol Rev 66:506–577
18. Hendriks ATWM, Zeeman G (2009) Pretreatments to enhance the digestibility of lignocellulosic biomass. Bioresour Technol 100:10–18
19. Sun Y, Cheng JJ (2002) Hydrolysis of lignocellulosic materials for ethanol production: a review. Bioresour Technol 96:1–11
20. Lynd LR, Wyman CE, Gerngross TU (1999) Biocommodity engineering. Biotechnol Prog 15:777–793
21. Hansen MAT, Kristensen JB, Felby C, Jorgensen H (2011) Pretreatment and enzymatic hydrolysis of wheat straw (Triticum aestivum L.) – The impact of lignin relocation and plant tissues on enzymatic accessibility. Bioresour Technol 102:2804–2811
22. APHA (1998) APHA, Standard methods for the examination of water and wastewater, 20th edn. American Public Health Association, Washington, DC.
23. Wan C, Li Y (2010) Microbial pretreatment of corn stover with Ceriporiopsis subvermispora for enzymatic hydrolysis and ethanol production. Bioresour Technol 101:6398–6403
24. Tuohy MG, Murray PG, Gilleran CT, Collins CM, Reen FJ, McLoughlin L et al (2007). Talaromyces emersonii enzyme systems. Patent WO/2007/091231
25. Voutilainen SP, Murray PG, Tuohy MG, Koivula A (2010) Expression of Talaromyces emersonii cellobiohydrolase Cel7A in Saccharomyces cerevisiae and rational mutagenesis to improve its thermostability and activity. Protein Eng Des Sel 23:69–79
26. Dubois M, Gilles KA, Hamilton JK, Rebers PA, Smith F (1956) Colorimetric method for determination of sugars and related substances. Anal Chem 28:350–356
27. Bailey MJ, Biely P, Poutanen K (1992) Inter laboratory testing of methods for assay of xylanase activity. J Biotechnol 23:257–270
28. Ghose TK (1987) Measurement of cellulase activities. Pure Appl Chem 59(2):257–268
29. Murray PG, Collins C, Grassick A, Penttila M, Saloheimo M, Tuohy MG (2004) Expression in Trichoderma reesei and characterization of a thermostable family 3 β-glucosidase from the moderately thermophilic fungus Talaromyces emersonii. Protein Expr Purif 38:248–257
30. Tuohy MG, Puls J, Claeyssens M, Vrsanska M, Coughlan MP (1993) The xylan-degrading enzyme system of Talaromyces emersonii: novel enzymes with activity against aryl β-D-xylosides and unsubstituted xylans. Biochem J 290:515–523
31. Miller GL (1959) Use of dinitrosalicylic acid reagent for determination of reducing sugar. Anal Chem 31:426–428
32. Gilleran CT, Hernon AT, Murray PG, Tuohy MG (2010) Induction of enzyme cocktails by low cost carbon sources for production of monosaccharide-rich syrups from plant materials. BioRes 5:634–649

Protoplast Fusion Techniques in Fungi

Annie Juliet Gnanam

Abstract

Traditionally, genetic manipulations of fungi were mostly dependent on the conventional mutation method, which has severe limitations. In comparison, protoplast fusion technique offers great potential as a biotechnological tool for improvement in industrial strains, establishing genetic systems, and overcoming incompatibility barriers in relative and non-relative fungi. This technique has been successfully used to create recombinant strains with desired properties by fusion and also allowed for transformation of desired gene/genes and viruses.

Keywords

Protoplast • Regeneration • Reversion • Fusion • KCl • Polyethylene glycol

Introduction

The protoplast fusion technique and recombinant DNA technology have opened doors for applied genetics in fungi. Protoplast fusion is an important tool for gene manipulation because it can break down the barriers to genetic exchange imposed by conventional mating systems [1]. Removing the wall and exposing the protoplast membrane allow for manipulation involving fusion, or uptake of nucleic acids, processes that are less achievable or impossible with intact cells [2]. Through protoplast fusion it is possible to transfer desired traits such as disease resistance, nitrogen fixation, rapid growth, frost hardiness, drought resistance, herbicide resistance and heat and cold resistance from one species to another [3].

The isolation of protoplasts from microbial cells involves the total digestion or localized puncturing of the walls by enzymes, allowing the cell contents enclosed by the plasma membrane to escape. For survival as intact structures, the protoplasts must be released into a hypertonic solution to provide osmolarity [2]. Several commercial lysing enzymes are available for use in the release of protoplasts in fungi. Lytic enzymes singly or in combinations could be used for effective lysis of cell walls. Interestingly, several lytic

A.J. Gnanam (✉)
University of Texas at Austin, Institute for Cellular and Molecular Biology, College of Natural Science, 2400 Inner Campus Drive, Austin, TX 78712, USA
e-mail: anniegnanam@yahoo.com

V.K. Gupta et al. (eds.), *Laboratory Protocols in Fungal Biology: Current Methods in Fungal Biology*,
Fungal Biology, DOI 10.1007/978-1-4614-2356-0_45, © Springer Science+Business Media, LLC 2013

enzymes originally used for lysis of cell walls in filamentous fungi were made from culture supernatants of several fungal species [4–7], so making lytic enzymes in the laboratory for small-scale use is a possibility.

The age of the mycelia markedly influences the protoplast yield, as the susceptibility of the cell walls to lytic enzyme/lenzymes is dependent on it. Most frequently, cultures of the early to mid-exponential phase have been used for good protoplast yields [8–14]. However, stationary phase cultures, spores and germ tubes have also been used for protoplast preparations. The culture media used also has some effect on protoplast release [15]. Pretreatment with compounds such as thiols has been shown to render the cell walls susceptible to lysis [16–18].

The lysed protoplasts have to be released into an osmotic buffer to keep them from bursting and also to maintain their structural integrity. Several inorganic salts and sugars in varying concentrations and pH have been used for this purpose. The most commonly and successfully used inorganic salts, sugars and sugar alcohols include potassium chloride (KCl), sodium chloride (NaCl), magnesium sulphate, sorbitol, mannitol, sucrose and glucose [8, 9, 19–25].

Not all protoplasts have the capability for cell wall regeneration and subsequent reversion into normal mycelial forms. The regeneration and reversion of protoplasts are influenced by the osmotic stabilizer in the medium, the carbon source and the type of medium. In most cases, the regeneration frequencies are higher than reversion frequencies. The ideal conditions for reversion could be determined before the use of protoplasts in fusion experiments.

The protoplast fusion and transformation systems developed in the past several years have led to our current understanding of phenomenon such as genetic incompatibility and to exploit economically important fungi used in the food and pharmaceutical industries and also in agriculture. The technique has been widely used for increased antibiotic production, antagonistic potential and multipesticide resistance, to name a few applications [26–28]. The most important

development in fusion of protoplasts happened following the introduction of polyethylene glycol (PEG) as a fusogen, which leads to aggregation of the protoplasts. The concentration of PEG and the type used play a very important role in the successful fusion of protoplasts. The general recommended concentrations are between 20 and 50 %. Below 20 % the stabilizing effect is lost, whereas above the recommended concentration the protoplast starts shrinking and results in lower fusion frequencies. The addition of the cation, Ca^{++}, to PEG stimulates the fusion of protoplasts and has been widely used in the fusion of protoplasts of several fungi [9, 19, 20, 22–25, 28].

Development of fungal genetic transformation systems was used to bring the power of genetic analysis to species lacking sexual and parasexual cycles [29]. Protoplast-mediated transformation could be the optimal method when heterologous integrations of multiple copies of a gene of interest is envisaged to integrate at random sites in the genome. This technique has been used to increase protein yields considerably in several fungi [30–33]. Transmission of double-stranded RNA mycoviruses via protoplast fusion has been reported in plant-pathogenic fungi such as *Aspergillus nidulans*, *Rosellinia necatrix* and *Fussarium boothi* [25, 34, 35].

Several selection procedures are employed to identify and screen for protoplast fusants. The most common methods of selection include identification based on the difference in colony morphology from the parent strains, using auxotrophic mutants, resistance/ susceptibility to heavy metal, catabolite repressors and antifungal agents [23, 24, 28]. The regeneration of the fusants is generally carried out in solid or in liquid media under non-selective or selective conditions. The regeneration and reversion of the protoplast to mycelia state could take between several hours to days.

The literature on protoplast isolation, regeneration and fusion and the reasons for making them varies depending on the organisms. By this technique it is possible to manipulate genomes of the organisms to a very high level.

Materials

1. Inoculating loop.
2. Sterile disposable rod/glass rod.
3. Glass slide.
4. Cover slips.
5. Hemocytometer.
6. 1-mL micropipettor.
7. 1-mL sterile disposable plastic tips.
8. Incubator set at 28 °C.
9. Bunsen burner.
10. Orbital shaker set at 28 °C.
11. Vortexer.
12. Sterile Erlenmeyer flasks, 25- and 500-mL capacity.
13. Microcentrifuge, e.g., Eppendorf, Centrifuge 5417C.
14. Sterile disposable polypropylene microcentrifuge tubes: 1.5-mL conical.
15. Sterile distilled water.
16. Potato dextrose yeast extract agar medium (PDYEA) (peeled potato, 200 g; dextrose, 20 g; yeast extract, 3 g; agar, 20 g/L of distilled water at pH 6.5).
17. PDYE agar plates.
18. PDYE broth.
19. PDYEA plates with 0.6 M KCl, pH adjusted to 5.5.
20. PDYEA plates with 0.6 M KCl, 0.5 μg/mL benomyl and 100 μg/mL griseofulvin, pH adjusted to 5.5.
21. Lytic enzymes—Novozym 234 (5 mg/mL), chitinase (3 mg/mL), pectinase (2 mg/mL), lysozyme (1 mg/mL) and cellulose (3 mg/mL).
22. 0.6 M KCl pH 5.5.
23. 1 M $CaCl_2$, autoclaved or filter sterilized.
24. PEG (polyethylene glycol, MW 3,500).
25. Sterile cheese cloth.
26. Phase contrast microscope, e.g., Carl Leitz photo microscope.

Methods

The methods given below describe general procedures for protoplast release, fusion, regeneration and reversion in filamentous fungi. Different enzyme combinations and concentrations may be needed for effective release of protoplasts. The age of the mycelium plays an important role in protoplast release and may have to be altered for different fungal species. The PEG used for fusion of protoplast has to be picked up from a range of 3,000–6,000 MW and the concentration used may have to be varied between 25 and 50 %. The regeneration media has to be made as per the requirement of the individual organisms used. In most cases the complete media with the osmotic stabilizers added is used. However, in some cases minimal media could also be used. Suitable markers have to be developed for screening of the fusant colonies. In case there is no defined marker available, colony morphology, estimation of the yield of desired product, etc., could be used for selection.

Protoplast Isolation from Young Hyphae (e.g., *Trichothecium roseum*)

The following method has been used to isolate high number of protoplasts from the filamentous fungi *Trichothecium roseum*, *Trichoderma harzianum*, *Trichoderma reesi*, *Trichoderma longibrachiatum*, *Trichoderma viride*, *Dreschlera oryzae* and *Venturia inaequalis* [8–13, 36].

1. The conidia were collected from 7-day-old plate cultures by adding 3–5 mL sterile water to the agar plate and gently scraping the surface with a sterile inoculation loop. The conidial suspension was drawn with a 1-mL micropipettor and transferred into a sterile 1.5-mL microcentrifuge tube.
2. The conidial suspension was then washed twice with sterile water (1.2 mL) by centrifugation at 500×g for 10 min at room temperature. The final pellet was resuspended in appropriate volume of water to bring the concentration of the conidia to 1×10^6/mL.
3. This conidial suspension was grown in 100 mL PDYE broth at 28 °C with shaking conditions for 24 h at 200 rpm.
4. The young mycelia (100 mg wet weight) were harvested by filtration through triple-layer sterile cheese cloth. It was then washed thrice with 5 mL sterile distilled water, followed by

3–4 washes with 5 mL sterile osmotic stabilizer (see Note 1).

5. The washed mycelia were scooped with a sterile loop and added to the lytic enzyme mix (see Note 2) and shaken at room temperature (28 ± 2 °C) for 4 h at 200 rpm (see Note 3).

6. After 4 h, the undigested hyphal material was removed by filtering the solution through 6 layers of sterile cheese cloth. To the resultant filtrate equal volume of 0.6 M KCl was added and the contents transferred into two sterile 1.5-mL microfuge tubes and centrifuged at $500 \times g$ for 5 min at room temperature. The supernatant was carefully removed with a micropipette and the resulting protoplast pellet was resuspended in 1 mL 0.6 M KCl.

PEG-Meditated Protoplast Fusion, Regeneration, Reversion and Selection (e.g., *Trichothecium roseum* Isolates AJ 102 and AJ 210)

The following method has been widely used for self-, inter- and intra-specific protoplast fusions in filamentous fungi and yeasts [9, 20, 22–24, 27, 28]. The success of this method lies in the ability of the protoplasts to regenerate and revert into normal vegetative forms after fusion.

1. Follow the steps in Protoplast Isolation from Young Hyphae (e.g., *Trichothecium roseum*), from steps 1–6, for both isolates. Adjust the concentrations of protoplasts to 1×10^6 with the osmotic stabilizer, 0.6 M KCl (see Note 4).

2. 1 mL of each of the protoplast suspension was mixed with 2 mL of 80 % PEG solution (see Note 5) and 0.2 mL of 1 M CaCl$_2$ and incubated at room temperature for 15 min (see Note 6).

3. The suspension was then diluted with 6 mL of 0.6 M KCl and incubated again for 15 min at room temperature and vortexed for 5–10 s (see Note 7).

4. The protoplast suspension was then transferred to 1.5-mL sterile microfuge tubes and harvested by centrifugation at $500 \times g$ for 5 min at room temperature. The supernatant was carefully removed with a micropipettor and the protoplast pellets were pooled into a 1.5-mL sterile microfuge tube.

5. The pooled protoplasts were washed twice with 1.2 mL of 0.6 M KCl and resuspended in the same.

6. The fused protoplasts were then spread plated on selective and non-selective agar plates at a concentration of 1×10^7 (see Notes 8–10).

7. The plates were incubated with the right side up at ± 28 °C for 24 h and then inverted the next day.

8. The fusant colonies appeared after 3 days. Single colonies were isolated from the selection media. The fusion frequency was calculated based on the ratio of the number of colonies on selective and non-selective agar. The colonies were subcultured several times on selective and non-selective agar before further studies.

Notes

1. The osmotic solution, 0.6 M KCl, pH 5.5, was autoclaved, cooled, and stored at room temperature. Care was taken to keep the concentration accurate, by ensuring that the liquid was not lost due to evaporation during autoclaving.

2. The lytic enzyme solution was made less than 20 min before use and filter sterilized with a 0.22-μm filter into a sterile 25-mL Erlenmeyer flask, the top of which was plugged with a cotton plug.

3. At every 30-min interval, samples were drawn and observed under the microscope to determine the maximum protoplast release. Ten microlitres of the sample were placed on a clean glass slide and a cover slip was gently placed over it for viewing under the microscope. A hemocytometer was used for counting the number of protoplasts.

4. The protoplasts were washed thoroughly to remove any traces of the lytic enzymes and the PEG, for effective regeneration and reversion of the fusants. The samples have to be handled carefully and gently, as the protoplasts are extremely fragile.

5. For the 80 % PEG solution, PEG 3,500 MW was dissolved in 0.6 M KCl solution, the pH adjusted to 5.5 and autoclaved. Before use

the solution was filter sterilized with a 0.22-μm filter. The solution containing the protoplasts and the PEG solution were mixed in equal volumes to bring the concentration of PEG to 40 %.

6. The incubation period with PEG could be variable for different organisms and protoplast preparations.

7. Step 3 was performed to dislodge the loosely bound protoplasts and the aggregates that did not fuse.

8. Selective agar plates were usually regular growth media, with the osmotic stabilizer and also markers such as heavy metals, fungicides or aminoacids (in the case of auxotrophic mutants) at appropriate concentrations.

9. Isolates *Trichothecium roseum* AJ102 was sensitive to benomyl and resistant to griseofulvin; AJ210 was resistant to benomyl and sensitive to griseofulvin. The concentrations required for selecting the fusants was determined as 0.5 μg/mL benomyl and 100 μg/mL griseofulvin.

10. Selective medium (PDYEA plates with 0.6 M KCl, 0.5 μg/mL benomyl and 100 μg/mL griseofulvin, pH adjusted to 5.5); non-selective medium (PDYEA plates with 0.6 M KCl, pH adjusted to 5.5). For the selective medium PDYEA medium was autoclaved and cooled down to 55 °C. The osmotic stabilizer (KCl) and the antifungal compounds benomyl and grisefulvin were added to it at 0.6 M, 0.5 μg and 100 μg/mL concentrations, respectively, mixed thoroughly and poured into Petri plates. The Petri plates were allowed to cool and left at room temperature for 24–48 h before use.

11. Plating usually involved using 100 μL protoplasts suspension per plate that was spread gently over the agar plate with a sterile disposable rod or glass rod. Extreme care was taken during plating as the protoplasts tend to be very fragile at this stage.

12. It might take several days for the fusant colonies to be visible on the selection plates. So it is better to leave the plates for several days in the incubator.

References

1. Murlidhar RV, Panda T (2000) Fungal protoplast fusion: A revisit. Bioprocess Biosyst Eng 22: 429–431
2. Peberdy JF (1979) Fungal protoplasts: isolation, reversion and fusion. Annu Rev Microbiol 33:21–39
3. Verma, N., Bansal, M. C. and Kumar, V. (2008) Protoplast fusion technology and its biotechnological applications. http://www.aidic.it/IBIC2008/webpapers/96Verma.pdf
4. Villanueva JR, Garcia Acha I (1971) In: Booth C (ed). Methods in microbiology, Vol. 4. Academic, London, pp. 665–718.
5. Siestma JH, Eveleigh DE, Haskins RH, Spencer JFT (1967) Protoplasts from *Pythium* sp. PRL 2142. Can J Microbiol 13:1701–1704
6. Gibson RK, Peberdy JF (1972) Fine structure of protoplasts of *Aspergillus nidulans*. J Gen Microbiol 72:529–538
7. De Vries OMH, Wessels JGH (1973) Effectiveness of a lytic enzyme preparation from *Trichoderma viride* in releasing spheroplasts from fungi, particularly basidiomycetes. Antonie van Leuwenhoek 39: 397–400
8. Balasubramanian N, Gnanam AJ, Srikalaivani P, Lalithakumari D (2003) Release and regeneration of protoplast from the fungus *Trichothecium roseum*. Can J Microbiol 49:263–268
9. Prabavathy VR, Mathivanan N, Sagadevan E, Murugesan K, Lalithakumari D (2006) Self-fusion of protoplast enhances chitinase production and biocontrol activity in *Trichoderma harzianum*. Bioresour Technol 97:2330–2334
10. Mrinalini C, Lalithakumari D (1996) Protoplast fusion: a biotechnological tool for strain improvement of *Trichoderma* sp. Curr Trend Life Sci 21:133–146
11. Annamalai P, Lalithakumari D (1991) Isolation and regeneration of protoplasts from mycelium of *Dreschlera oryzae*. J Plant Dis Prot 98:197–204
12. Karpagam, S. (1994) Variations in the protoplast regenerated isolates of *Trichoderma harzianum*. M. Phil. Thesis, University of Madras, Chennai, India.
13. Revathi R, Lalithakumari D (1993) *Venturia inaequalis*: a novel method for protoplast isolation and regeneration. J Plant Dis Prot 100:211–219
14. Buxton FP, Radford A (1983) Cloning of the structural gene for orotidine 5′-phosphate carboxylase of *Neurospora crassa* by expression in *Escherichia coli*. Mol Gen Genet 190:403–405
15. Musilkova M, Fencl Z (1968) Some factors affecting the formation of protoplasts in *Aspergillus niger*. Folia Microbiol 13:235–239
16. Fawcett PA, Loder PB, Duncan MJ, Beesley TJ, Abraham EP (1973) Formation and properties of protoplasts from antibiotic-producing strains of *Penicillium chrysogenum* and *Cephalosporium acremonium*. J Gen Microbiol 79:293–309

17. Berliner MD, Reca ME (1969) Protoplasts of systemic dimorphic fungal pathogens: *Histoplasma capsulatum* and *Blastomyces dermatitidis*. Mycopathol Mycol Appl 37:81–85

18. Dooijewaard-Kloosterziel AMP, Siestma JH, Wouters JTM (1973) Formation and Regeneration of *Geotrichum candidum* protoplasts. J Gen Microbiol 74:205–209

19. Szewczyk E, Nayak T, Oakley EC, Edgerton H, Xiong Y, Taheri-Talesh N et al (2006) Fusion PCR and gene targeting in *Aspergillus nidulans*. Nature Protoc 6:3111–3120

20. Reeves E, Kavanagh K, Whittaker PA (1992) Multiple transformation of *Saccharomyces cervisiae* by protoplast fusion. FEMS Microbiol Lett 99:193–198

21. May GS (1992) Fungal technology. In: Kinghorn JR, Turner G (eds) Applied molecular genetics of filamentous fungi. Blackie Academic and Professional, Glasgow, pp 1–27

22. van Diepeningen AD, Debets AJM, Hoekstra RF (1998) Intra- and inter species virus transfer in Aspergilli via protoplast fusion. Fungal Genet Biol 25:171–180

23. Hatvani L, Manczinger L, Kredics L, Szekeres A, Antal Z, Vagvolgyi C (2006) Production of *Trichoderma* strains with pesticide-polyresistance by mutagenesis and protoplast fusion. Antonie Van Leeuwenhoek 89:387–393

24. Ogawa K, Yoshida N, Gesnara W, Omumasaba AC, Chamuswarng C (2000) Hybridization and breeding of the Benomyl resistant mutant, *Trichoderma harzianum* antagonized to phytopathogenic fungi by protoplast fusion. Biosci Biotechnol Biochem 64:833–836

25. Lee K, Yu J, Son M, Lee Y, Kim H (2011) Transmission of *Fussarium boothi* mycovirus via protoplast fusion causes hypovirulence in other phytopathogenic fungi. PLoS One 6(6):e21629

26. Robinson M, Lewis E, Napier E (1981) Occurrence of reiterated DNA sequences in strains of *Streptomyces* produced by an interspecific protoplast fusion. Mol Gen Genet 182:336–340

27. Agbessi S, Beausejour J, Derry C, Beaulieu C (2003) Antagonistic properties of two recombinant strains of *Streptomyces melanosporofaciens* obtained by intraspecific protoplast fusion. Appl Microbiol Biotechnol 62:233–238

28. EL-Bondkyl AM (2006) Gene transfer between different *Trichoderma* species and *Aspergillus niger* through intergeneric protoplast fusion to convert ground rice straw to citric acid and cellulases. Appl Biochem Biotechnol 135:117–132

29. Esser K (1997) Fungal genetics: fundamental research to biotechnology. Prog Bot 58:3–38

30. Askolin S, Nakari-Setala T, Tenkanen M (2001) Overproduction, purification, and characterization of the *Trichoderma reesi* hydrophobin HFBI. Appl Microbiol Biotechnol 57:124–130

31. Lee DG, Nishimura-Masuda I, Nakamura A, Hidaka A, Masaki H, Uozumi T (1998) Over production of alpha-glucosidase in *Aspergillus niger* transformed with the cloned gene *agl*A. J Gen Appl Microbiol 44:177–181

32. Verdoes JC, Punt PJ, van den Hondel CA (1995) Molecular genetic strain improvement for the overproduction of fungal proteins by filamentous fungi. Appl Microbiol Biotechnol 43:195–205

33. Kinghorn JR, Lucena N (1994) Biotechnology of filamentous fungi. In: Iberghin L, Frontali L, Sensi P (eds) ECB6: Proceedings of the 6th European Congress on Biotechnology. Elsevier, Amsterdam, pp 277–286

34. van Diepeningen AD, Debets AJM, Slakhors SM, Fekete C, Hornok L, Hoekstra RF (2000) Interspecies virus transfer via protoplast fusion between *Fussarium poae* and black *Aspergillus* strain. Fungal Genet Newslett 47:99–100

35. Kanematsu S, Sasaki A, Onoue M, Oikawa Y, Ito T (2010) Extending the fungal host range of a partivirus and a mycoreovirus from *Rosellinia necatrix* by inoculation of protoplasts with virus particles. Physiology 100:922–930

36. Lalithakumari D (2001) Fungal protoplast—a biotechnological tool. Oxford/IBH, New Delhi

Large-Scale Production of Lignocellulolytic Enzymes in Thermophilic Fungi

46

Manimaran Ayyachamy, Mary Shier,
and Maria G. Tuohy

Abstract

A semi-pilot scale process for the production of lignocellulolytic enzymes in the thermophilic fungus *Talaromyces emersonii* is presented. Each step involved in the scale-up process is described in detail.

Keywords

Lignocellulolytic enzymes • Thermophilic fungi • *T. emersonii* • Thermozymes • Fermenter • Submerged cultivation

Introduction

Talaromyces emersonii, a thermophilic, filamentous euascomycete fungus, has been reported as one of the potential candidates for lignocellulolytic enzymes production [1, 2]. *T. emersonii* produces highly specific thermozyme cocktails with efficient catalytic and long-term storage properties [3–6]. These thermostable fungal enzymes usually work 10–20 °C

M. Ayyachamy (✉) • M.G. Tuohy
Molecular Glycobiotechnology Research Group,
Department of Biochemistry, School of Natural Sciences,
National University of Ireland, University Road,
Galway, Ireland
e-mail: manimaran.ayyachamy@nuigalway.ie

M. Shier
Department of Biochemistry, National University
of Ireland, University Road, Galway, Ireland
e-mail: mary.shier@nuigalway.ie

higher than commercially available *Trichoderma* sp. enzymes [7].

T. emersonii has the ability to produce a wide range of extracellular polysaccharide degrading enzymes in higher amounts than bacteria or yeast [3]. Owing to the fact that enzymes are secreted into the culture medium, the downstream processing of enzymes is relatively simple. Enzyme systems from this fungus have been reported already and patents have been developed for significant applications [2, 5, 6]. Plant polymer hydrolysis studies using these crude thermozymes are usually carried out at high temperatures, which decrease reaction time, minimize microbial contamination and lower hydrolysate viscosity. The aforementioned catalytic properties of thermozymes are desirable characteristics for several biotechnological and industrial applications.

Thermozymes production by *T. emersonii* using different plant polymeric substrates and low-cost approaches has already been reported [2, 8]. Thermozymes have also been evaluated for food and biomass conversion applications

[7, 9–11]. These promising results undoubtedly indicate that lignocellulolytic thermozymes from *T. emersonii* have potential applications in food, biofuels and biorefineries sectors.

Materials

1. Fungal strain: *Talaromyces emersonii* IMI393751 (see Note 1).
2. Sarbouraud dextrose agar (SDA) from Oxoid (Basingstoke, UK).
3. TG medium: 2% (w/v) Glucose and 201 mL of 5× nutrients solution in 80 mL water.
4. Nutrient solution (5×) for TG (g/L): $(NH_4)_2SO_4$, 75; KH2PO4, 25; Corn steep liquor, 25 mL; Yeast extract, 25; $FeSO_4 7H_2O$, 0.312; ZnSO4 7H_2O, 0.0625; H_3BO_3, 0.0625; $MnSO_4 4H_2O$, Na_2MoO_4, 0.0625; $CoCl_3 6H_2O$, 0.0625; KI, 0.0625; $MgSO_4 7H_2O$, 2.5; $CaCl_2 2H_2O$, 2.5 g; Na_2SO_4, 5. Stir well and adjust to a final pH of 4.5 with 1 N NaOH.
5. EI medium: 2% (w/v) wheat bran in water.
6. Inexpensive carbon source: Wheat bran was collected from a local supplier.
7. Antifoam: Proflo Oil (Traders Protein, Tennessee, USA).
8. Buffers: pH 4.0, 7.0 and 11.0.
9. Electrolytes for pO_2 probe.
10. Steam generator (Fibrimatic/Camptel Dispositivo H2 48 kW minimum).
11. Air compressor (Bambi VT300D, oil-free 350 L/min, 8 bar, 3HP compressor, fitted with water condenser unit).
12. Cooling water supply pipeline.
13. Erlenmeyer flasks: 500 mL and 1 L.
14. Thermostatically controlled NewBrunswick Scientific Innova 44 platform, shaking incubator (New Brunswick Scientific, USA).
15. BIOSTAT® B-plus 10 L Fermenter (Sartorius BBI Systems, Germany).
16. BIOSTAT® D-plus 150 L Fermenter, equipped with control cabinet, touch screen controller and SCADA Fermenter Supervisory Control and Data Acquisition software (Sartorius BBI Systems, Germany).
17. Magnetic stirrer (Fisher Scientific, Dublin, Ireland).
18. Arium 61315 Reverse Osmosis (RO) water system with 360 L/day capacity, equipped with RO modules (Sartorius BBI Systems, Dublin, Ireland).
19. RO water 150-L capacity storage tank (Sartorius BBI Systems, Dublin, Ireland).
20. Trolley mounted Westfalia HSD® SD1 centrifuge with integrated control panel (Westfalia Separator AG, Germany).
21. Sartocon2 Sartoflow® membrane filtration unit (MF/UF crossflow cassette system), equipped with a Sartojet® pump and fitted with either 5 kDa or 10 kDa Hydrosart membranes (Sartorius BBI Systems, Germany).
22. BioTek Powerwave XS2 microplate reader with incubation facilities to 50 °C (BioTek, USA).
23. Filters: Sarstedt (0.2-μm and 0.4-μm filters).
24. Substrates: wheat arabino xylan (Megazyme International Ireland Ltd., Bray, Co. Wicklow), oat spelts xylan (Sigma-Aldrich, Dublin, Ireland), birch wood xylan (Sigma-Aldrich), β-glucan (from barley; Megazyme Intl. Ireland Ltd.), and carboxy methyl cellulose and 4-nitrophenyl α- or β-glycosides (Sigma-Aldrich). Unless otherwise stated, all chemicals and reagents were purchased from Sigma-Aldrich (Dublin, Ireland).
25. High-quality 20-L and 100-L plastic storage containers (Nalgene, UK, and Fisher Scientific, Ireland).
26. Laboratory plastics, i.e., disposable micropipette tips, microcentrifuge tubes and flat-bottomed, high-optical-quality, low-protein-absorbing microtitre plates (Sarstedt, Sinnottstown, Co. Wexford, Ireland).
27. Safety apparel (Caulfields Ltd., Galway, Ireland).

Methods

This section has been divided into seven subsections. The overall steps involved in the scale-up process are elaborated.

Pre-Inoculum Preparation

1. A ~1 cm^2 piece of *T. emersonii* IMI393751, either from a routinely sub-cultured SDA agar plate or a glycerol stock, is transferred aseptically onto a fresh SDA agar plate and incubated at 45 °C for 48–72 h (see Note 1).
2. To prepare a submerged (liquid) culture of the fungus, take 3–4 ~1 cm^2 pieces of *T. emersonii* IMI393751 mycelial mat (small section) from SDA agar plate transferred into 500-mL Erlenmeyer flasks containing 100 mL TG medium.
3. Incubate the culture flasks at 45 °C with shaking (200 rpm) for 48 h.

Inoculum Development in a 10-L Fermenter

1. Prepare and sterilize 6 L of TG medium in an autoclave (Sanyo Labo autoclave, Sanyo, USA) at 121 °C for 15 min (see Notes 2 and 3).
2. Prepare a seed culture of *T. emersonii* (36- to 48-h old) by inoculating this medium with *T. emersonii* mycelial suspension taken from the pre-inoculum shake flasks culture under aseptic conditions; a 5% (v/v) inoculum is used.
3. Agitation, aeration rate and temperature are maintained at 200 rpm, 1.5 vvm and 45 °C, respectively.
4. Dissolved oxygen should be maintained at 20% using the cascade mode (see Note 4).
5. Foaming is controlled by automatic addition of antifoaming agents.
6. At the end of 36–48 h of cultivation, the fungal cultures are harvested into a sterile container in a laminar air-flow chamber.

Fungal Cultivation in a 150-L Fermenter

1. Utilities (steam generator, air compressor and water pipe lines) were thoroughly checked prior to start of the fermenter run.
2. All utilities' valves should be opened. Make sure that the pressure bar values meet the technical specifications of the fermenter (see Notes 5–7).
3. Calibrate the pH and pO$_2$ probes according to the instructions given by the manufacturer.
4. Fill the fermenter with 100 L water from the storage tank (Sartorius Arium 61315), which is connected to a Sartorius RO purification system.
5. Add wheat bran at 2% (w/v) and mix it thoroughly with the water by setting the agitation rate at 400 rpm.
6. Before initiating the sterilization process, a pressure hold test for the vessel should be performed to ensure that there is no leakage in the vessel. This is achieved by closing the air inlet and gas vent valves manually and checked to see whether the pressure in the vessel is maintained over a holding period of 30–45 min (see Notes 8 and 9).
7. Set the cultivation and sterilization temperatures and cultivation period using the DCU (digital control unit) cocabinet, which is connected to an MS Windows-based MFCS (Multi Fermenter Control System) SCADA, and allows data acquisition and batch management.
8. Sterilize the contents of the fermenter (in situ) at 105 °C for 30 min.
9. At the end of sterilization process, allow the fermenter to cool down automatically until it reaches the cultivation temperature; the cultivation temperature has been set already during the sterilization process.
10. The *T. emersonii* cultures (36- to 48-h old), developed in a 10-L reactor, are used to inoculate the 150-L fermenter at 5% (v/v) under aseptic conditions (see Note 10).
11. Antifoaming agent (20 mL) is added soon after inoculating the cultures by setting the antifoam pump into manual mode. The antifoam pump is then set back to automatic mode.
12. Cultivation parameters are the same as described in the section Inoculum Development in a 10-L Fermenter.
13. Samples should be collected from two different sampling ports every 24 h and the desired enzyme activities and total sugar analysis determined (see Notes 11–13).

14. At the end of cultivation, the culture broth is mixed thoroughly at 300 rpm for 15–20 min before harvesting.

Harvesting and Concentrating the Enzyme Cocktails

1. Because the target enzymes produced during fermentation are secreted (i.e., extracellular), fungal broth cultures are harvested by separating the mycelia (fungal biomass) from the fermentation broth in a high-speed Westfalia SD1 discharge centrifuge, with integrated control panel. The contents of the 150-L fermenter are pumped to the centrifuge using a Sartojet (or other) pump, by a connecting line (Sartorius BBI Systems). The mycelial mass and left-over substrate are trapped in the centrifuge bowl and subsequently removed (see Note 14).

2. The supernatant from the HSD centrifuge is pumped into a Sartoflow microfiltration (MF) unit using a Sartojet pump, to remove residual mycelia and particulate material. The microfiltered crude enzyme is then concentrated using the Sartoflow system, this time operating the ultrafiltration mode. Fold concentrations vary from 20- to 50-fold, depending on the desired concentration factor and the viscosity of the enzyme product. The unconcentrated and concentrated enzyme samples are then assayed for protein content and enzyme activity.

3. All unconcentrated and concentrated enzyme samples should be stored at 4 °C (or at −20 °C for longer-term storage) until required.

Determination of Enzyme Activities

1. The crude enzyme samples are centrifuged prior to assay at $1,000 \times g$ for 10 min and the clarified supernatant is used as crude enzyme for enzyme assays, protein and reducing sugar analysis (see Note 8).

2. All enzyme assays are conducted at a working assay temperature of 50 °C in sodium acetate buffer (100 mM, pH 5.0), unless otherwise specified [2, 4–10]. Each test reaction and all controls are prepared in triplicate.

3. The hydrolysis of wheat arabino xylan (1% w/v, 15 min), oat spelts xylan (1% w/v, 15 min), birch wood xylan (1% w/v, 15 min), β-glucan (from barley; 1% w/v, 15 min) and carboxy methyl cellulose (1% or 3% w/v; 30 min) are determined as reducing sugars released by the dinitrosalicyclic acid (DNS) reagent method [2, 4–8, 12–14].

4. Absorbance values are measured at 550 nm in a 96-well microplate reader equipped with the appropriate filters (BioTek Powerwave XS2 microplate reader, BioTek, USA).

5. Exo-acting glycosyl hydrolase activities, including acetyl esterase (4-nitrophenyl acetate), β-glucosidase (4-nitrophenyl-β-D-glucopyranoside), α-glucosidase (4-nitrophenyl α-D-glucopyranoside), β-xylosidase (4-nitrophenyl β-D-xylopyranoside) and α-L-arabinofuranosidase (4-nitrophenyl-α-L-arabinofuranoside), were determined using the appropriate 1 mM 4-nitrophenyl α- or β-glycoside as substrate [2, 4, 5].

6. Suitably diluted enzyme is incubated with the appropriate substrate for 15–30 min at 50 °C and the reaction is stopped using 1 M Na_2CO_3. The increase in absorbance at 405 nm on release of the nitrophenolate anion is measured in a 96-well microplate reader equipped with the appropriate filters (BioTek, USA).

7. Enzyme activity is expressed as IU, i.e., μ mol of reducing equivalents (xylose, glucose or nitrophenol) released/mL of enzyme/min reaction time, under standard assay conditions.

Total Protein and Sugar Analysis

1. The protein content in the crude extracellular enzyme was determined preferably by the Bensadoun and Weinstein modification of the method of Lowry et al. [15] or by the Bradford method [16], using bovine serum albumin (BSA, fraction V) as standard.

2. The total sugars in the culture broth is quantified using the phenol-sulfuric acid method; [17] reducing sugars are quantified by DNS reagent method [14].

Monitoring the Fungal Cultivation

Temperature, pH and dissolved oxygen are monitored by the Biostat D⁺150, which is equipped with a DCU-control cabinet system. The DCU system is a local control system for the automation of the fermentation process. Thus, growth parameters for *T. emersonii* submerged cultures are automatically monitored in the fermenter.

Notes

1. *T. emersonii* is routinely sub-cultured on SDA plate at 45 °C.
2. All ports in the fermenter should be clamped properly. Make sure that one of the ports is plugged with non-absorbent cotton wool, if the fermenter is not autoclaved with gas cooling exit device.
3. The pH probe must be calibrated prior to sterilization. The pO_2 probe should be calibrated after sterilization. All probes and the motor must be wrapped with aluminium foil while sterilizing the fermenter.
4. Dissolved oxygen is one of the critical parameters for fungal cultivation. Make sure that the set agitation rate values during the cascade mode operation do not rupture the fungal mycelia.
5. Turn on the steam generator first and allow it to reach 5 bar. Open the steam valve and set the supply pressure value to 3–3.5 bar. Steam filters should be replaced after three fermenter runs.
6. The cooling water supply pressure should be maintained between 3 and 3.5 bar.
7. The water condensate from the air compressor should be removed completely before passing the air into the fermenter. All automatic valves in the pipeline are manufactured in such a way that they operate only in the presence of air. It is important that the pressure should be maintained between 5 and 6 bar throughout the fermenter run.
8. A leakage test is important to make sure that all of the ports are sufficiently tightened. If leakage occurs in the vessel, it will be difficult to attain the desired temperature. First, the fermenter gas vent should be closed manually and then air passed into the vessel until it reaches 1 bar in pressure. The pneumatic and air inlet valves are then closed manually.
9. After performing the leak test, all valves should be returned to the automatic mode setting. Before starting the sterilization, steam in the pipelines should be released through the drain valves to ensure that high-quality steam only is used for sterilization. Gas cooling exit pipes should be kept open so that liquid loss in the vessel can be prevented.
10. The inoculation port and nearby surface should be cleaned with 75% ethanol. A multitorch or spirit lamp can be used to create sterile conditions in and around the inoculation port. Minimize the exposure time when opening the inoculation port during inoculation.
11. Before and after collecting the samples from the fermenter, the sampling port should be sterilized with steam for 5–10 min. This will prevent microbial contamination.
12. All substrates used for enzyme assays should be freshly prepared and the appropriate standards curves should be conducted in parallel.
13. The enzyme cocktails used for the assay should be diluted appropriately with buffer prior to enzyme assay.
14. When the HSD centrifuge bowl becomes clogged, the resulting supernatant becomes dirty and unclear. To clear the bowl, a full discharge is required.

References

1. Stolk AC, Sampson RA (1972) The genus *Talaromyces*: studies in mycology, 2nd edn. Centraal bureau Voor Schimmelcultures Publishers, Baarn, The Netherlands
2. Tuohy MG, Murray PG, Gilleran CT, Collins CM, Reen FJ, McLoughlin L et al. (2007). *Talaromyces emersonii* enzyme systems. Patent WO/2007/091231.
3. Polizeli ML, Rizzatti AC, Monti R, Terenzi HF, Jorge JA, Amorim DS (2005) Xylanases from fungi: properties and industrial applications. Appl Microbiol Biotechnol 67:577–591

4. Murray PG, Collins C, Grassick A, Penttila M, Saloheimo M, Tuohy MG (2004) Expression in *Trichoderma reesei* and characterisation of a thermostable family 3 β-glucosidase from the moderately thermophilic fungus *Talaromyces emersonii*. Protein Expr Purif 38:248–257

5. Tuohy MG, Puls J, Claeyssens M, Vrsanska M, Coughlan MP (1993) The xylan-degrading enzyme system of *Talaromyces emersonii*: novel enzymes with activity against aryl β-D-xylosides and unsubstituted xylans. Biochem J 290:515–523

6. McCarthy T, Hanniffy O, Savage AV, Tuohy MG (2003) Catalytic properties and mode of action of three endo-betaglucanases from *Talaromyces emersonii* on soluble β-1,4- and β-1,3;1,4-linked glucans. Int J Biol Macromol 33:141–148

7. Waters DM, Murray PG, Ryan LA, Arendt EK, Tuohy MG (2010) *Talaromyces emersonii* thermostable enzyme systems and their applications in wheat baking systems. J Agric Food Chem 58:7415–7422

8. Gilleran CT, Hernon AT, Murray PG, Tuohy MG (2010) Induction of enzyme cocktails by low cost carbon sources for production of monosaccharide-rich syrups from plant materials. BioResources 5:634–649

9. McCarthy T, Hanniffy O, Lalor E, Savage AV, Tuohy MG (2005) Evaluation of three thermostable fungal endo-β-glucanases from *Talaromyces emersonii* for brewing and food applications. Proc Biochem 40: 1741–1748

10. Waters DM, Ryan LA, Murray PG, Arendt EK, Tuohy MG (2011) Characterization of a *Talaromyces emersonii* thermostable enzyme cocktail with applications in wheat dough rheology. Enzyme Microb Technol 49:229–236

11. Forbes C, O'Reilly C, McLaughlin L, Gilleran G, Tuohy M, Colleran E (2009) Application of high rate, high temperature anaerobic digestion to fungal thermozyme hydrolysates from carbohydrate wastes. Water Res 43:2531–2539

12. Bailey MJ, Biely P, Poutanen K (1992) Interlaboratory testing of methods for assay of xylanase activity. J Biotechnol 23:257–270

13. Ghose TK (1987) Measurement of cellulase activities. Pure Appl Chem 59(2):257–268

14. Miller GL (1959) Use of dinitrosalicylic acid reagent for determination of reducing sugar. Anal Chem 31:426–428

15. Bensadoun A, Weinstein D (1976) Assay of proteins in the presence of interfering materials. Anal Biochem 70:241–250

16. Bradford MM (1976) A rapid and sensitive method for the quantitation of microgram quantities of protein utilizing the principle of proteindye binding. Anal Biochem 72:248–254

17. Dubois M, Gilles KA, Hamilton JK, Rebers PA, Smith F (1956) Colorimetric method for determination of sugars and related substances. Anal Chem 28:350–356

Malik M. Ahmad, Pravej Alam, M.Z. Abdin,
and Saleem Javed

Abstract

Aflatoxigenic molds contaminate the processed or raw agricultural commodities during their storage and transportation. Traditional methods offer a laborious and tedious process; also, detection can be performed only after toxin production. Molecular methods provide a rapid, sensitive, and specific detection. The major advantage of molecular methods is that the detection can be observed prior to the toxin production. Molecular methods for fungal detection can be applied to distinguish the aflatoxin-secreting fungi from non-aflatoxigenic molds. The DNA-based PCR amplification of a consensus sequence can make it possible to monitor the range of aflatoxigenic fungi from contaminated products.

Keywords

Aflatoxigenic molds • DNA • Consensus sequence • Panfungal PCR • Molecular detection

Introduction

The past two decades have witnessed various scientific innovations in diagnostic microbiology owing to technological advances in molecular microbiology. For example, speedy detections of nucleic acid-based amplification process, with its mutual characterization through automated and user-friendly bio-computational tools, have significantly widened the area of diagnostic detections for clinical microbiologists [1]. The accessible traditional methods for aflatoxin detection are time-consuming, labor-intensive, difficult to standardize, have low sensitivity and non-specificity, and require days to weeks before the results are achieved. Serological methods work on the "one substance one assay" concept [2, 3]. With such performance, researchers have diverted their scientific thoughts toward molecular advancements [4]. Most molecular diagnostic assays use polymerase chain reaction (PCR).

M.M. Ahmad • P. Alam • M.Z. Abdin (✉)
Department of Biotechnology, Hamdard University,
New Delhi, 110062, India
e-mail: mzabdin@rediffmail.com

S. Javed
Department of Biochemistry, Hamdard University,
New Delhi, 110062, India

V.K. Gupta et al. (eds.), *Laboratory Protocols in Fungal Biology: Current Methods in Fungal Biology*,
Fungal Biology, DOI 10.1007/978-1-4614-2356-0_47, © Springer Science+Business Media, LLC 2013

It has become one of the easiest methods, as it rapidly amplifies the target DNA into millions of copies, for detection of microorganisms [4]. Variations in the technique increased the sensitivity of the detection method (nested PCR), while multiplex PCR detects a number of pathogens simultaneously. Nested PCR, however, employs two primer sets (outer and inner), whereas multiplex PCR uses a series of primer sets for identification. The basic problem with both these PCR variants is that the annealing temperature for each primer sets is first optimized to work correctly within a single reaction. This problem can be resolved by having a single consensus sequence of multiple microorganisms for primer design and its amplification by conventional PCR. In a similar way, phytopathogenic molds can also be detected by this method, especially those secreting mycotoxins such as aflatoxins. Aflatoxigenic molds can directly damage crops and indirectly impair human beings, as the latter ingest mycotoxin-contaminated agricultural products. These are aflatoxin-secreting fungi group, particularly from *Aspergillus* species [4]. *Aspergillus flavus* and *A. parasiticus* are the two prominent candidates that secrete the extrolites, although there are other aflatoxin-secreting species, such as *A. tamarii, A. pseudotamarii, A. bombycis, A. oryzae, A. nomius, A. parvisclerotigenus, A. minisclerotigenes, A. toxicarius, A. versicolor,* and so forth, with some other genus like *Emericella astellata* and *E. nidulans* [4]. The following protocol can be used to detect aflatoxigenic molds from agricultural commodities, including foods and feeds.

Materials

Chemicals and Glasswares

1. Nuclease-free water
2. Agarose (molecular biology grade, HiMedia)
3. TAE buffer: 50 mM Tris (HiMedia), 50 mM acetic acid (HiMedia), 1 mM EDTA (HiMedia). Dilute when needed from a 50× stock to 1× for use (see Note 1)
4. Ethidium bromide (EtBr; Sigma-Aldrich): 5.0 mg/mL stock (see Note 1)

5. Gel-loading dye: 60 % glycerol, 60 mM EDTA, 0.3 % bromophenol blue, 0.3 % xylene cyanol, 10 mM Tris–HCl (pH 7.6) (MBI Fermentas)
6. Molecular marker 100 bp: ready to use (MBI Fermentas)
7. Disposable polypropylene micropippet tips (Tarsons): 200 μL and 20 μL (see Note 1)
8. Disposable polypropylene microcentrifuge tubes (Tarsons): 1.5 mL, tight-fitting snap-capped (see Note 1)
9. PCR tubes (Tarsons) (see Note 1)

Equipment

1. Biofuge (Heraeus Pico)
2. Thermal cycler (GStorm)
3. Horizontal electrophoresis equipment (BioRad Wide Mini Sub Cell)
4. Gel documentation system (UVI Tech)

PCR Reagents

1. Template DNA of aflatoxigenic and non-aflatoxigenic fungi
2. Oligonucleotide primers (custom-made, Sigma-Aldrich, USA) re-suspended to a concentration of 100 mM using nuclease-free water and stored at −20 °C
3. dNTPs: a mixture of dATP, dCTP, dGTP, and dTTP (2.5 mM each, Merck), stored at −20 °C
4. Thermostable DNA polymerase (*Taq*) and reaction buffer supplied by manufacturer (Merck). We typically use 10× reaction buffer of 100 mM Tris pH 9.0 (25 °C), 500 mM KCl, 15 mM $MgCl_2$, and 0.1 % gelatin. Separate stocks of $MgCl_2$ (25 mM) are also supplied for adding to the reaction. All of these reagents are stored at −20 °C

Method

The method given as follows describes the general procedure for primer designing of a consensus sequence from aflatoxin-secreting fungi followed by its amplification and subsequent testing with non-aflatoxigenic molds. The volumes and number of tubes used per sample may vary, depending on the type of sample and the number of fungi being processed.

Multiple Sequence Alignment and Primer Design

The most commonly used target for molecular fungal detection is DNA encoding the ribosomal RNA genes including ITS regions, but other targets include housekeeping genes such as β-tubulin, calmodulin, actin, and cytochrome b [5–9] for consensus sequence selection and primer design. Thus, the primers amplifying these regions show conserved sequences in nature and that is why they are known as internal controls. However, to detect the aflatoxigenic molds at the molecular level, one has to look for a consensus region present in all the aflatoxin-secreting species. The following steps are involved in the selection of consensus sequence and its primer designing:

1. To identify a consensus region, all the genes of *A. flavus* and *A. parasiticus* involved in aflatoxin biosynthesis were collected in FASTA format from the NCBI Genbank. For example, collect the sequences of *afl*P gene from all strains of *A. flavus* and *A. parasiticus* (see Note 2).

2. Compare their similarities through multiple sequence alignment (ClustalW2, EBI tools) using default settings. Gene sequences of *afl*P of both fungal strains are matched for similarity.

3. Deduce a consensus sequence from the alignment that provides maximum identity. For example, *afl*P gives 96 % similarity. So, a sufficient consensus sequence can be obtained for primer design. Mark the mutations present in between the consensus sequence (see Note 3).

4. Design a primer set for the consensus sequence for PCR amplification considering all parameters (see Notes 4 and 5).

PCR Amplification

After DNA isolation, quantification, and primer design, PCR amplification of the aflatoxigenic molds can be performed. Detection of aflatoxin production before its secretion into food products is critical for maintaining the quality and economic value of those products. Thus, PCR method by amplification can forecast the prior production of aflatoxins and thus the quality of products can be taken care of. The following steps are involved in PCR amplification:

1. Keep all reagents and DNA samples on ice until used. Prepare adequate reaction mixture for the number of samples to be tested. To detect aflatoxigenic molds, for example, take four fungal species (two known for aflatoxin-producing *Aspergillus* species, *A. flavus*, and *A. parasiticus*; one non-aflatoxigenic *Aspergillus* species; and the other *Fusarium* species). Prepare PCR "master mix" of 92 μL containing 64 μL nuclease-free water, 8 μL 10× PCR buffer, 8 μL 10 mM dNTP mix, 4 μL each primer, and 4 μL *Taq* polymerase (5 U/μL). Mix reagents properly through micropipette and finally vortex and centrifuge briefly to remove the air bubbles before dispensing 23 μL aliquots into four PCR tubes and place on ice (see Note 6).

2. Add template DNA (100 ng) of 2 μL to each of the tubes, using microtip and mix the content briefly (see Note 7).

3. Set a PCR for positive control reaction with 18S ribosomal amplifying region as internal control and a separate multiplex PCR by mixing 18S ribosomal and *omt*A primers in another PCR master mix for all four fungal species with same cycling programme (see Note 8).

4. Put the PCR tubes in thermal cycler and run the appropriate cycling program: initial denaturation of 1 min at 95 °C followed by 30 amplification cycles of 1 min at 94 °C, 30 s at 56.7 °C and 1 min at 72 °C, with final extension of 2 min at 72 °C.

5. As the amplification completes, analyze 5 μL of each sample by agarose gel electrophoresis.

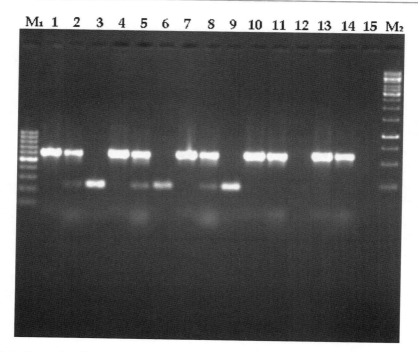

Fig. 47.1 Detection of aflatoxigenic and non-aflatoxigenic molds by PCR amplification. Lane M₁, 100 bp marker; Lanes 1, 4, 7, 10, 13, amplification of internal control (18S ribosomal gene) of *A. flavus*, *A. parasiticus*, plant-root isolated *A. flavus*, *A. versicolor*, *Fusarium* sp.; Lanes 2, 5, 8, multiplex PCR-positive control; Lanes 11, 14, multiplex PCR-negative control; Lanes 3, 6, 9, PCR of *omt*A gene from *A. flavus*, *A. parasiticus*, and plant-root isolated *A. flavus*-positive detection; Lanes 12, 15, PCR of *omt*A region in non-aflatoxigenic mold (*A. versicolor* and *Fusarium* sp.)—negative detection; Lane M₂, 1 Kb marker

6. Prepare a gel, add 1.2 % agarose to 1× TAE buffer, and heat to dissolve (e.g., using a microwave oven) (see Note 9).

7. Add EtBr of 3.5 μL per 100 mL of concentration 5 mg/mL, in agarose, and allow to cool at approximately 40 °C or till when your skin feels suitable (see Note 10).

8. Pour the agarose gel into a suitable gel casting tray with teflon comb and allow to set for approximately 20 min.

9. Remove the comb and place the gel in a horizontal electrophoresis tank containing 1× TAE, so that the wells are just covered with buffer.

10. Add 2 μL of gel-loading dye in 5 μL of each PCR samples, transfer the mixture into wells of agarose gel.

11. Pour 2 μL of each 100 bp and 1 kb molecular marker to the first and last wells and, finally, electrophores (see Note 11).

12. Visualize the agarose gel on gel documentation system and capture the photograph after proper separation of markers and amplicons (Fig. 47.1).

Notes

1. All the microtips, microcentrifuge tubes, PCR tubes, and necessary reagents should be autoclaved properly.

2. Select a gene whose product converts penultimate precursor or immediate precursor to aflatoxin. Thus, all the aflatoxins (B1, G1, B2, and G2) are covered for the detection purpose.

3. While selecting consensus sequence, it would be better to consider exonic region for primer design. Exons are the transcribed regions of the any genes and thus mostly conserved in nature. Therefore, these regions have higher similarity

and have large enough consensus region. Thus, all the aflatoxigenic molds that secret aflatoxin can be included for detection while those which are non–aflatoxin-secreting will not be amplified. These steps will finally enhance the specificity and sensitivity of the detection.

4. While designing the primer, do not select the primer region where the mutations are present in consensus region, particularly at 3' end of primer.

5. Design the primer with GC content of 50–60 %, T_m difference within 4–5 °C, no repetition of single bases and 3' ends with G, C, GC, or CG.

6. After the addition of all reagents, mix them properly with microtips. Mild vortexing and centrifugation are necessary to remove the air bubbles. If present, they can hinder the amplification of the target DNA.

7. DNA isolated should be pure, and prior to amplification, its quantity should be accessed. If the DNA quantity is high, it should be diluted in nuclease-free water.

8. A positive reaction of internal control reduces the possibility of false-positive PCR amplification and, thus, enhances the specificity of the PCR reaction.

9. Agarose should be melted until bubbles come from the bottom of the flask ends.

10. Gloves should be worn throughout these procedures, and particular care should be taken while handling EtBr (consult safety data sheets). It is a suspected mutagen.

11. Electrophoresis of agarose gel should usually be done at 8–10 V/cm. The gel should be run for 1.5 h for proper separation of marker and amplicons.

Acknowledgements M. M. Ahmad is thankful to receive RFSMS grant-aided support from UGC, Government of India, India.

References

1. Ahmad MM, Ahmad M, Ali A, Hamid R, Javed S and Abdin MZ (2010) Molecular methods for detecting pathogenic fungi. In: Gupta VK, Tuohy M and Gaur RK (eds) Fungal Biochemistry and Biotechnology, Germany: Lap Lambert Academic Publishing AG & Co. KG. pp. 154–169

2. Konietzny U, Greiner R (2003) The application of PCR in the detection of mycotoxigenic fungi in foods. Braz J Microbiol 34:283–300

3. Yong RK, Cousin MA (2001) Detection of moulds producing aflatoxins in maize and peanuts by an immunoassay. Int J Food Microbiol 65:27–38

4. Abdin MZ, Ahmad MM, Javed S (2010) Advances in molecular detection of *Aspergillus*: an update. Arch Microbiol 192:409–425

5. Mauchline T, Kerry BR, Hirsch PR (2002) Quantification in soil and the rhizosphere of the nematophagous fungus *Verticillium chlamydosporium* by competitive PCR and comparison with selective plating. Appl Environ Microbiol 68:1846–1853

6. Fraaije BA, Lovell DJ, Rohel EA, Hollomon DW (1999) Rapid detection and diagnosis of *Septoria tritici* epidemics in wheat using a polymerase chain reaction/Pico green assay. J Appl Microbiol 86: 701–708

7. Carbone I, Kohn LM (1999) A method for designing primer sets for speciation studies in filamentous ascomycetes. Mycologia 91:553–556

8. Foster SJ, Singh G, Fitt BDL, Ashby AM (1999) Development of PCR based diagnostic techniques for the two mating types of *Pyrenopeziza brassicae* (light leaf spot) on winter oilseed rape (*Brassica napus* ssp. oleifera). Physiol Mol Plant Pathol 55: 111–119

9. Fountaine JM, Shaw MW, Napier B, Ward E, Fraaije BA (2007) Application of real-time and multiplex PCR assays to study leaf blotch epidemics in barley. Phytopathology 97:297–303

Protocols for the Quantification of dsDNA and Its Fragmentation Status in Fungi

48

Ioannis Papapostolou, Konstantinos Grintzalis, and Christos Georgiou

Abstract

The presented protocols describe methodologies for the accurate quantification of dsDNA concentration and fragmentation status of fungal DNA (and that of any organism). The protocols can be used to quantify dsDNA concentration, even up to picogram level, using the PicoGreen fluorescent dye, and to discriminate the fragmentation status of a DNA sample as fragmented (0–23 kb) and intact (>23 kb) or as small size (0–1 kb) fragmented DNA, using DNA samples as low as 2.5 µg mL-1.

Keywords

dsDNA quantification • DNA fragmentation • Small size (0–1 kb) fragmented DNA • PEG precipitation • Fluorescence • PicoGreen • Fungi

Introduction

In many biological applications it is necessary to accurately quantify both dsDNA concentration as well as its fragmentation status. DNA damage originating from oxidative attack of reactive oxygen species to DNA [1] may be non-repairable (e.g., fragmentation) or repairable (e.g., nicks, sugar/base modifications). Traditionally, fragmented DNA damage was estimated qualitatively by the Comet assay [2] and by the DNA smearing observed in an electrophoresis agarose gel [3], visualized by ethidium bromide staining in both methods. Other approaches frequently used are the semi-qualitative micronuclei detection assay that estimates the % of micronuclei [4], and the DNA non-specific diphenylamine assay. Furthermore, other protocols use fluorescent dyes (e.g., ethidium bromide) to measure DNA concentration without taking into account the underestimation of DNA concentration due to the fragmentation degree of the DNA (i.e., the fluorescence of dsDNA is decreased even up to 70 % when it is fragmented ≤23 kb) [5]. An additional disadvantage of the traditionally used dsDNA fragmentation protocols is that they do not offer any specific discrimination according to the size of the fragmented segments. The present protocols resolve these problems by quantifying accurately any dsDNA sample and by discriminating

I. Papapostolou • K. Grintzalis • C. Georgiou(✉)
Department of Biology, University of Patras,
University Campus, Patras 25600, Greece
e-mail: c.georgiou@upatras.gr

V.K. Gupta et al. (eds.), *Laboratory Protocols in Fungal Biology: Current Methods in Fungal Biology*,
Fungal Biology, DOI 10.1007/978-1-4614-2356-0_48, © Springer Science+Business Media, LLC 2013

its fragmentation status as fragmented (0–23 kb) and intact (>23 kb) dsDNA fraction (replacing the DNA agarose electrophoresis gel method) and as the small size (0–1 kb) fragmented dsDNA (replacing the Comet assay) after its fractionation by polyethylene glycol (PEG), the later being an index for both necrotic and apoptotic dsDNA damage [6, 7].

Materials

1. Balance (Kern, 770/65/6 J)
2. Centrifugal vacuum concentrator (Savant, model SPD111V), connected to a vacuum pump (KNF, N 820.3 FT.18)
3. Centrifuge tubes, 15 mL (ISC BioExpress, catalog no. C-3394-1)
4. DNA genomic, from calf thymus, unsheared, (Sigma-Aldrich, catalog no. D-4764)
5. DNase I from bovine pancreas, 580 units/mg solid (Sigma-Aldrich, catalog no. DN-25)
6. Dimethyl sulfoxide, DMSO, anhydrous (Sigma-Aldrich, catalog no. 276855). *Caution*: Harmful
7. Ethylenediaminetetraacetic acid, Na_2EDTA (Merck, catalog no. 34033918). *Caution*: Irritant, dangerous for the environment
8. Glass tubes, 15 mL
9. Hydrochloric acid, HCl, ≥ 37 % (Fluka, catalog no. 84415). *Caution*: Corrosive
10. Manganese chloride (Merck, catalog no. 5833.0250)
11. Micropipettes (adjustable volume pipettes), 2.5 μL, 10 μL, 20 μL, 100 μL, 200 μL, 1 mL, and tips (Eppendorf Research)
12. Microcuvette for fluorescence measurements, quartz (SOG/Q) $45 \times 4 \times 4$ mm (0.5 mL) with its FCA4 adaptor (Starna, England)
13. NucleoSpin® Extract II kit (by Macherey-Nagel, Duren, Germany)
14. Sodium phosphate, Na_2HPO_4 (Merck, catalog no. 30412)
15. Sonicator, model Dr. Hielscher UP-50 H, equipped with a 2-mm-diameter MS2 microtip (Dr. Hielscher GmbH, Teltow, Germany)
16. pH meter (Metrohm, 827 pHlab)
17. Polyethylene glycol 6000, PEG-6000 or PEG (Serva, catalog no. 33137)
18. Refrigerated microcentrifuge (Eppendorf, model 5417R)
19. Sodium chloride, NaCl (Merck, catalog no. 567440)
20. Sodium hydroxide, NaOH (Merck, catalog no. 567530). *Caution*: Corrosive
21. Spectrofluorometer (Shimadzu, model RF-1501)
22. Quant-iT™ PicoGreen (Invitrogen Molecular Probes, catalog no. P7581). *Caution*: Very toxic
23. Tris–HCl (Merck, catalog no. 648313)
24. Water, ddH_2O, purified by a Milli-Q system (Millipore Corp)

Methods

Solutions and Standard Curves

For applying the presented protocols, the following stock solutions and a standard curve need to be performed.

Tris-EDTA (TE) Buffer
Prepare 100 mL TE buffer (10 mM Tris–HCl and 1 mM EDTA, pH 7.5), by dissolving 0.121 g Tris–HCl and 0.037 g EDTA in ddH_2O, and adjust pH to 8.

Fragmented DNA Stock Solution
Dissolve without stirring overnight at 4 °C 1 mg unsheared calf thymus genomic DNA in 4 mL TE buffer. Place 0.5 mL of the 0.25 mg mL^{-1} calf thymus DNA standard solution in a 1.5-mL Eppendorf tube, and sonicate it for 15 s at 350 W cm^{-2}.

PicoGreen Stock Solution
Prepare fresh by mixing 10 μL of the commercial reagent (made in 100 % anhydrous DMSO) with 0.99 mL TE buffer. This solution is sufficient for 15 tests.

PEG-NaCl Stock Solution
Prepare fresh 20 mL, by weighing accurately 2.822 g PEG and dissolving it in 12.344 mL ddH_2O.

Then, add to it 1.47 mL 0.2 M phosphate buffer, pH 7.0, and 4.041 mL 4 M NaCl made in 10 mM phosphate buffer pH 7.0. The resulting solution is somewhat viscous.

Construction of Totally Fragmented dsDNA Standard Curve

Prepare a series of final volume 0.225 mL TE solutions containing different concentrations of fragmented dsDNA (up to 200 pg mL^{-1}) and to each of them add 75 µL PicoGreen stock solution. Measure the fluorescence units (F.U.) of the above DNA solutions against a reagent blank (without DNA) at ex/em 480/530 nm (in a 0.5 mL quartz microcuvette) with a spectrofluorometer (e.g., a Shimadzu RF-1501 spectrofluorometer, set at 10 nm excitation/emission slit width and at high sensitivity).

Fungal dsDNA Isolation

The dsDNA samples to be analyzed by the presented protocols can be isolated from any type of fungi by the appropriate standard methods and commercial kits.

Note: Artificial fragmentation of the isolated dsDNA during its isolation should be avoided.

Note: Any buffers and equipment used for dsDNA isolation should be sterilized.

Protocol for dsDNA Concentration Quantification

The concentration of the isolated fungal dsDNA sample is estimated after its complete fragmentation (via sonication) from the fragmented dsDNA standard curve, and is performed as follows:

1. Dilute the previously isolated dsDNA sample and sonicate it (as in the section Fragmented dsDNA Stock Solution).
2. Measure the F.U. of the sonicated dsDNA sample (as in the section Construction of Totally Fragmented dsDNA Standard Curve) and convert its F.U. to dsDNA concentration using the totally fragmented dsDNA standard curve.

Protocol for the Discrimination of dsDNA as Fragmented and Intact

This protocol is used to quantify the fragmented (0–23 kb) and intact (>23 kb) dsDNA fractions in any sample. There is a precaution that should be taken into account, since the discrimination by fluorescence is limited to ≥23 kb, the analyzed DNA sample should be above 23 kb in its native condition (given the fact that DNA can also exist in some organisms in sizes <23 kb) [8, 9]. Regarding the sensitivity limit for discrimination of dsDNA fragmentation status, this protocol requires at least dsDNA concentration of 135 pg mL^{-1}. The protocol is performed as follows:

1. Initially, the dsDNA sample is diluted appropriately and divided in two 0.3-mL portions.
2. One 0.3-mL portion, designated as "fragmented," is sonicated (as stated in the section Fragmented dsDNA Stock Solution) and the F.U. (using 0.225 mL of it and 75 µL PicoGreen stock solution) is measured (F.U.$_{sonicated}$), as stated in the section Construction of Totally Fragmented DNA Standard Curve. From this F.U.$_{sonicated}$ the concentration of the DNA sample is estimated using the totally fragmented DNA standard curve.
3. The fluorescence of the other 0.3-mL portion (using 0.225 mL of it and 75 µL PicoGreen stock solution), designated as "unsonicated," is measured (F.U.$_{unsonicated}$, i.e., without subjecting it to sonication).
4. By taking into account any appropriate dilutions, the above F.U. measurements are compared and the percentage of fragmented/intact DNA is expressed as follows:
   a. If F.U.$_{sonicated}$ and F.U.$_{unsonicated}$ are equal (with S.E. ±3 %), the dsDNA sample is totally fragmented.
   b. If F.U.$_{sonicated}$ and F.U.$_{unsonicated}$ differ by 67–73 %, the dsDNA sample is composed only of intact DNA (as previously mentioned, the difference in F.U. of a DNA sample if it is totally intact and totally fragmented is 70 %).
   c. If F.U.$_{sonicated}$ and F.U.$_{unsonicated}$ differ below 67 %, the dsDNA sample is partially fragmented, and exists as a mixture of

fragmented and intact fractions, which are quantified by the following mathematical treatment of F.U.$_{sonicated}$ and F.U.$_{unsonicated}$: [5, 7].

   i. To estimate the F.U. the dsDNA sample would have if it were intact (F.U.$_{intact}$), multiply the F.U.$_{sonicated}$ value by 3.333 (=100/30 %) and set the resulting value as 100 %. Then use the resulting value in order to convert the F.U.$_{unsonicated}$ value to %, designated as y % (=100 % F.U.$_{unsonicated}$/F.U.$_{intact}$).

   ii. Enter this y % value in the equation x % = (100 % − y %)/0.7, which calculates the percentage of the fragmented DNA fraction expressed as %x of the total dsDNA concentration (which is determined using the F.U.$_{sonicated}$ in step 1). The concentration of the intact dsDNA fraction, expressed as 100-x %, can be determined accordingly.

Protocol for the Isolation and Quantification of Small Size (0–1 kb) Fragmented dsDNA

This protocol quantifies the small-size (0–1 kb) fragmented DNA after its isolation by PEG precipitation which is very important as an index of necrotic or apoptotic events. The protocol can be applied to any previously isolated DNA sample and is performed as follows:

1. Initially, the concentration of the isolated fungal dsDNA sample is determined as in "Protocol for dsDNA concentration quantification." For applying this protocol the limitation in dsDNA concentration of the sample for accurate PEG precipitation is 0.2 mL 2.5 µg mL^{-1} [6].

2. In a 1.5-mL Eppendorf tube add 0.16 mL of the isolated dsDNA sample and 0.34 mL PEG-NaCl stock solution. Mix gently and incubate the resulting DNA-PEG-NaCl mixture for minimum 12 h (or overnight) in ice-water bath. *Note:* Accurate pipetting is required due to the small volumes used in this step.

3. Centrifuge in a refrigerated microcentrifuge (4°C) at 15,000 g for 5 min and collect the supernatant in a new Eppendorf tube. The supernatant contains the 0–1 kb fragmented DNA fraction, and the pellet the intact plus the >1 kb fragmented dsDNA.

4. For measuring small-size fragmented dsDNA concentration in the PEG supernatant in step 3 by fluorescence, the interfering PEG is removed and the dilute small fragment dsDNA is concentrated as follows. Dilute the supernatant 2× with TE buffer and concentrate the 0–1 kb fragmented dsDNA with the Nucleospin Extract II kit (following manufacturer's instructions) in 0.05 mL of the kit's elution buffer.

5. Dilute the eluate appropriately and determine the dsDNA concentration of the isolated small-size (0–1 kb) fragmented DNA as in the section Protocol for dsDNA Concentration Quantification.

References

1. Halliwell B, Gutteridge CMJ (1999) Free radicals in biology and medicine, 3rd edn. Oxford University Press, Oxford, UK

2. Olive PL, Banath JP (2006) The comet assay: a method to measure DNA damage in individual cells. Nat Protoc 1:23–29

3. Sambrook J, Fritsch FE, Maniatis T (1989) Molecular cloning: a laboratory manual. Cold Spring Harbor Laboratory Press, Cold Spring Harbor, New York

4. Fenech M (2000) The in vitro micronucleus technique. Mutat Res 455:81–95

5. Georgiou DC, Papapostolou I (2006) Assay for the quantification of intact/fragmented genomic DNA. Anal Biochem 358:247–256

6. Georgiou DC, Patsoukis N, Papapostolou I (2005) Assay for the quantification of small-sized fragmented genomic DNA. Anal Biochem 339:223–230

7. Georgiou CD, Papapostolou I, Grintzalis K (2009) Protocol for the quantitative assessment of DNA concentration and damage (fragmentation and nicks). Nat Protoc 4:125–131

8. Brown TA (2001) Gene cloning and DNA analysis: an introduction. Blackwell Publishing, Malden, MA

9. Griffiths AJF, Miller JH, Suzuki DT, Lewontin RC, Gelbart WM (2000) An introduction to genetic analysis. W. H. Freeman, New York, NY

Rapid Identification and Detection of Pathogenic Fungi by Padlock Probes

49

Clement K.M. Tsui, Bin Wang, Cor D. Schoen, and Richard C. Hamelin

Abstract

Fungi are important pathogens of human diseases, as well as to agricultural crop and trees. Molecular diagnostics can detect diseases early, and improve identification accuracy and follow-up disease management. The use of padlock probe is effective to facilitate these detections and pathogen identification quickly and accurately. In this chapter we describe three diagnostic assays that utilize padlock probes in combination with various technologies for the detection of pathogenic fungi.

Keywords

Molecular diagnostics • Padlock probes • Pathogenic fungi • Single nucleotide polymorphisms • Rolling circle amplification

C.K.M. Tsui (✉)
Department of Forest Sciences, The University of British Columbia, 2424 Main Mall, Vancouver, BC, Canada V6T 1Z4
e-mail: clementsui@gmail.com

B. Wang
Centre of Virus Research, Westmead Millennium Institute, University of Sydney, Westmead Hospital, Darcy Road, Westmead, NSW 2145, Australia

C.D. Schoen
Department of Bio-Interactions and Plant Health, Plant Research International B. V, Droevendaalsesteeg 1, Wageningen 6708PB, The Netherlands

R.C. Hamelin
Department of Forest Sciences, The University of British Columbia, 2424 Main Mall, Vancouver, BC, Canada V6T 1Z4

Natural Resources Canada, Laurentian Forestry Centre, 1055 rue du P.E.P.S, Quebec, QC, Canada G1V 4C7

Introduction

Fungi represent the greatest eukaryotic diversity on earth and they are the primary decomposers in the ecosystems. Of the estimated 1.5 million species, more than 400 species have been associated with human diseases and about 100 species can cause infection to normal individuals [1]. Representatives of genera such as *Candida* (Saccharomycetes, Ascomycota), *Cryptococcus* (Tremellales, Basidiomycota) and *Aspergillus* (Eurotiales, Ascomycota) are prevalent human diseases that cause superficial, systematic and lung infections [1]. Fungi also represent one of the most important groups of pathogens to plants and they cause significant economic losses to agriculture and forestry business [2, 3]. In addition, some fungal pathogens can cause ecosystem-wide

V.K. Gupta et al. (eds.), *Laboratory Protocols in Fungal Biology: Current Methods in Fungal Biology*, Fungal Biology, DOI 10.1007/978-1-4614-2356-0_49, © Springer Science+Business Media, LLC 2013

disturbances that can endanger other plant and tree species. *Fusarium* (Hypocreales, Ascomycota), *Cronartium* (Pucciniales, Basidiomycota), *Grosmannia* (Ophiostomatales, Ascomycota), and *Phytophthora* (Peronosporales, Oomycota (fungus-like Stramenopiles)) are some examples of genera that comprise serious pathogens of crops and trees [2, 3].

Rapid and accurate detection and identification of fungal pathogens at species and subspecies level in the clinical setting and natural environment, or on plant materials are the keys to proper patient treatment and disease/pathogen surveillance, containment and eradication [1, 4]. However, many fungal pathogens exist as species-complexes or they have very low abundance in the clinical specimen and natural environment. Different molecular types/varieties also exist within species and they may have different pathogenic profiles and virulence level to the hosts. These issues pose increasing difficulties to accurate pathogen genotyping/identification [1, 4].

Culture-based morphological features are the most predominant approach used for fungal pathogen identification. But the major drawback of this approach is that many pathogenic species do not produce fruiting bodies readily in culture that can be useful for morphological identification [5]. In addition, identification procedures are time-consuming or laborious, or require taxonomical or specialized expertise in interpretation [5]. Other approaches commonly used in clinical pathogen identification are histopathological, biochemical or serological information and molecular analyses such as polymerase chain reaction (PCR) fingerprinting, amplified fragment length polymorphism (AFLP) and restriction fragment length polymorphism (RFLP) [1]. Similarly, these techniques may be time-consuming and sensitive to contamination, which can result in low accuracy and specificity.

Routine DNA sequencing of the internal transcribed spacer (ITS) and large subunit (LSU) regions of rRNA gene followed by comparative sequence analysis to resources in GenBank facilitate the design of novel molecular method(s) for rapid species identification. This approach

includes the selection of genomic regions with low sequence polymorphisms such as single nucleotide polymorphisms (SNPs) because many pathogenic molecular types or subspecies differ by only a few base pairs in various regions of the genome. Although detection and characterization of SNPs is becoming increasing popular for pathogen identification, it was considered as a major challenge for conventional real-time qPCR using regular oligos and probes targeting SNPs detected by fluorescent dyes (e.g., SYBR green or TaqMan probes). In order to recognize a single SNP among different genotypes, techniques other than conventional qPCR are required.

Padlock probes could be an alternative platform to the conventional approaches [6]. Padlock probes comprise two target-complementary sequence regions at both ends for hybridization to specific DNA sequence, as well as a non target-complementary segment. The sequence regions at both ends of the probe are designed that they are joined together after target hybridization and DNA ligation (Fig. 49.1). The helical nature of double-stranded DNA (dsDNA) enables the probe to topologically bind to the target strand. The advantage of padlock probes is that the sequences at both ends of the probe are sensitive to mismatches [7].

Padlock probes were initially introduced for *in situ* DNA localization and detection [6]. They were developed for the discrimination of centromeric sequence variation in human chromosomes [8]. However, the method has been applied to detection of genetically modified organisms [9]. In addition, this technology also has multiplexing potential as the interaction between padlock probes does not give rise to circular molecules, which can be easily removed from the detection system [10].

The sensitivity of a padlock probe is comparable to a conventional oligonucleotide probe, but padlock probes have greater specificity with less noise background [8]. To improve the detection sensitivity, various technologies have been introduced to detect and amplify the signals from the circularized probes targeting various fungal pathogens (Table 49.1) [11, 18].

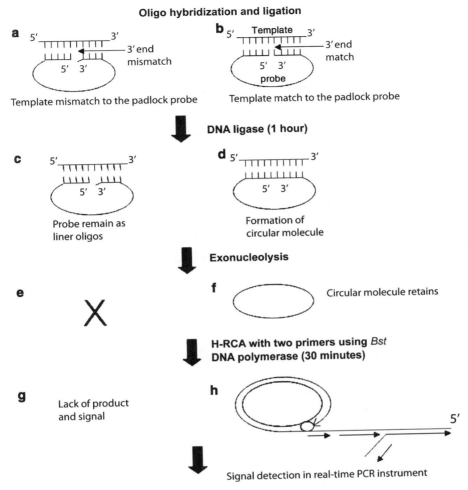

Fig. 49.1 Overview of the padlock probe technology coupled with hyperbranching rolling circle amplification (H-RCA) method for detection of the single nucleotide polymorphisms (SNPs). (**a**) Hybridization of padlock probe to a target template with a mismatch at 3′ end. (**b**) Padlock probe matches perfectly to target template. (**c**) High-fidelity DNA ligase is unable to ligate the mismatched template-probe complex. (**d**) In perfectly matched template-probe complex, the 5′ and 3′ ends of padlock probe can be joined, forming a circular molecule. (**e**) Templates and linear probes are removed by exonucleolysis reaction (Exonuclease I and III). (**f**) Only the circularized probe remains in the assay system. (**g**) H-RCA is performed using two primers and *Bst* DNA polymerase; however, no amplification will take place without a circular probe. (**h**) In the presence of a circular template, the two primers generate a self-propagating, ever-increasing pattern of alternating strand displacement, branching, and DNA fragment release events. (I) Exponential accumulation of product during isothermal H-RCA of DNA minicircles can be detected by real-time PCR instrument with the addition of SYBR green

Materials

1. Vortexer.
2. Water bath.
3. Microcentrifuge.
4. Disposable polypropylene microcentrifuge tubes (1.5-mL conical, 2-mL screw-capped).
5. PCR strips (flat cap, 0.2-mL thin-walled).
6. PCR Thermal cycler (e.g., Applied Biosystems GeneAmp® 9700).
7. Real-Time PCR Thermal cycler (e.g., Corbett RotorGene 3000).
8. Sterile distilled water (PCR grade)/Sigma water.

Table 49.1 Summary of padlock probe investigations in the detection of fungal pathogens

Type of samples	Gene target	Detection system	Genera or species identified[a]	References
DNA from cultures	ITS rRNA	Hyperbranched rolling circle amplification	Trichophyton rubrum, T. tonsurans, T. soudanense, T. violaceum, T. mentagrophytes, Epidermophyton floccosum, Microsporum canis, M. gypseum	Kong et al. [14]
DNA from cultures	ITS rRNA	Hyperbranched rolling circle amplification	Candida albicans, C. glabrata, C. krusei, C. tropicalis, C. dubliniensis, C. guilliermondii, Aspergillus fumigatus, A. flavatus, Scedosporium apiospermum, S. prolifercans	Zhou et al. [15]
DNA from cultures and clinical specimens (sputum, blood)	ITS rRNA	Hyperbranched Rolling circle amplification	Cryptococcus neoformans var. grubii, Cryptococcus neoformans var. neoformans, Cryptococcus gattii	Kaocharoan et al. [17]
DNA from cultures	Lanosterol 14α-demethylase (ERG11p)—enzyme in fungal ergosterol synethesis	Hyperbranched rolling circle amplification	Candida albicans	Wang et al. [16]
DNA from cultures and bark beetles	ITS, LSU rRNA	Hyperbranched rolling circle amplification	Grosmannia clavigera, Leptographium longiclavatum	Tsui et al. [3]
DNA from cultures	ITS rRNA	Real-time PCR using Taqman probe	Phytophthora nicotianae, P cactorum, P. infestans, P. sojae, Pythium ultimum (Oomycetes), Rhizotonia solani, Fusarium oxysporum f. sp. radicis-lycopersici, Myrothecium roridum, Myrothecium verrucaria, Verticillium dahlia, Verticillium alboatrum	Szemes et al. [13]
DNA from cultures	ITS rRNA	Real-time PCR on OpenArrays™ system	Phytophthora nicotianae, P cactorum, P. infestans, P. sojae, Pythium ultimum (Oomycetes), Rhizotonia solani, Fusarium oxysporum f. sp. radicis-lycopersici, Myrothecium roridum, Myrothecium verrucaria, Verticillium dahlia, Verticillium alboatrum	van Doorn et al. [25]
DNA from cultures, and the filters in which horticultural re-circulation water spiked with various pathogens has passed through	ITS rRNA	A desthiobiotin moiety connected to the linker sequence	Phytophthora nicotianae, P cactorum, P. infestans, P. sojae, Pythium ultimum (Oomycetes), Rhizotonia solani, Fusarium oxysporum f. sp. radicis-lycopersici, Myrothecium roridum, Myrothecium verrucaria, Verticillium dahlia, Verticillium alboatrum	van Doorn et al. [27]
DNA from cultures and clinical specimens (urine, charcoal swabs, sputum, vaginal swab culture broth)	ITS rRNA	Rolling circle amplification and Luminex™	Aspergillus flavus, A. fumigatus, A. nidulans, A. niger, Candida albicans, C. glabrata, C. tropicalis, C. parasilopsis, Cryptococcus neoformans, Pneumocystis jirovecii	Eriksson et al. [18]

[a]Including Oomycetes, which are fungus-like organisms classified under Stramenopiles. Traditionally, they have been studied by fungal biologists

9. dNTPs, a mixture of dATP, dCTP, dGTP and dTTP 400 mM each (Promega, Australia), stored at −20 °C.

10. DMSO (Dimethyl sulfoxide)—used to minimize the secondary structure formed in DNA and RNA (Sigma-Aldrich, Australia).

11. Thermostable DNA polymerase (e.g., *Taq*, and reaction buffer supplied by manufacturer). We typically use *Taq* from MBI Fermentas (York, UK), which is supplied with a 10× reaction buffer of 100 mM Tris–HCl pH 8.8 (25 °C), 500 mM KCl and 0.8 % Nonidet P40. Separate stocks of BSA (20 mg/mL) and MgCl$_2$ (25 mM) are also supplied for adding to the reaction. All of these reagents are stored at −20 °C.

12. TBE buffer—50 mM Tris, 50 mM boric acid, 1 mM EDTA. Dilute when needed from a 10× stock.

13. Padlock probe (custom-made, Sigma-Aldrich); page purified, re-suspended to a stock concentration of 100 μM and diluted to 1 μM working concentration.

14. *Pfu* DNA ligase with 10× reaction buffer (Stratagene) 20 mM Tris–HCl (pH 7.5), 20 mM KCl, 10 mM MgCl$_2$, 0.1 % Igepal, 0.01 mM rATP and 1 mM DTT. All of these reagents are stored at −20 °C.

15. Exonuclease I and III with 10× reaction buffer 1 (New England Biolabs, Ipswich, MA, USA), stored at −20 °C. (Note: Degraded nucleic acids and nonspecific products due to cross reactions from the 5′ and/or 3′ ends are degraded in order to remove any unreacted probes, while preserving reacted endless probes.)

16. *Bst* DNA polymerase with 10× ThermoPol Reaction Buffer (New England Biolabs, Ipswich, MA, USA). All of these reagents are stored at −20 °C.

17. Oligonucleotide primers targeting the linker sequence (custom-made, Sigma-Aldrich), re-suspended to a concentration of 10 μM.

18. SYBR Green I (Sigma-Aldrich, St Louis, MO, USA), stored at −20 °C

19. A confocal ScanArray 4000 laser scanning system (Packard GSI Lumonics) containing a GreNe 543-nm laser for PE and a HeNe 633-nm laser for Cy5 fluorescence measurement in microarrays analysis.

20. OpenArrays® Real-Time PCR Instrument (Applied Biosystems).

21. The ZipCodes were chosen from the GeneFlex TagArray set (Affymetrix). Potential for secondary structures and ZipCode specificity was examined with Visual OMP 6.0 software (DNA Software Inc.).

22. The cZipCode oligonucleotides carrying a C$_{12}$ linker and a 5′ NH$_2$ group were synthesized and spotted onto Nexterion MPX-E16 epoxy-coated slides by Isogen B.V. according to the manufacturer's instructions (Schott Nexterion).

23. Enzymes *EcoRI*, *HindIII* and *BamHI* (New England Biolabs Inc., Ipswich, USA).

24. MyOne™ Streptavidin C1 Dynabeads® (Dynal Biotech ASA, Oslo, Norway).

Design of Padlock Probes

1. To select a gene with enough resolution as target for the padlock probe design.

2. To improve the efficiency of hybridization, the padlock probes should be designed with minimum secondary structure, and the melting temperature (*Tm*) of 5′ end probe complementary sequence will be close to or above ligation temperature (60 °C).

3. To increase 3′ end complementary sequence specificity, *Tm* will be optimized with 10–15 °C below the ligation temperature.

4. The genetic linker region will be designed/created to minimize its similarity to any known species or sequences by performing BlastN search.

Differentiation of Fungal Pathogens by Padlock Probe and Rolling Circle Amplification

Rolling circle amplification (RCA) is based on the rolling replication of short single-stranded DNA (ssDNA) circular molecules [12, 19]. RCA involves a single forward primer complementary to the linker region of the padlock probe and a DNA polymerase with strand displacement activity in an isothermal condition [20]. As a result, the padlock probe signal can be amplified several thousand-fold because the polymerase extends the bound primer along the padlock probes for many cycles and displaces upstream sequences, producing a long ssDNA molecule comprising multiple repeats of the probe sequence. While two primers; a first forward primer that binds to the padlock probe and initializes RCA, and a second primer that targets the repeated ssDNA sequence of the primary RCA product, can generate large numbers of copies of the DNA fragments. This is called hyperbranching RCA (H-RCA) (see Fig. 49.1) [12].

Padlock probe coupled with H-RCA offers a significant advantage for the detection of SNPs. The formation of circular probes via ligation happens when both ends of the padlock probes perfectly hybridize to the target at juxtaposition. The subsequent H-RCA amplification of target probe can be carried out when circularized probes become available. These two strict conditions create an ideal detection platform for highly sensitive and specific SNPs detection. By increasing the hybridization temperature and shortening the 3′ complementary sequence (below the reaction temperature), the discrimination of SNP can be further improved [20, 21]. This method for SNPs detection has been developed for various groups of pathogenic microorganisms [3, 14–17, 22–24].

Below we describe the general procedures to perform assays using padlock probes coupled with H-RCA amplification of the probe signal. Modifications would be required to adjust for differences in probe annealing temperature. The size of tubes used for genotyping may be varied, depending on the type of real-time PCR machines.

1. Mix 10^{11} copies purified PCR amplicons, 1 pmol of padlock probe (Sigma-Aldrich), 2 U of *Pfu* DNA ligase (Stratagene) and 1 μL 10× reaction buffer with a total reaction volume of 10 μL in a 0.2 mL thin-walled PCR tube (see Note 1).
2. Centrifuge briefly and place the tube in a thermal cycler with a heated lid.

3. The ligation reaction included one cycle of 5 min at 94 °C to denature the target template followed by 5 cycles of 94 °C for 30 s and 4 min ligation at 65 °C.
4. After the ligation reaction, quickly transfer the tube to 4 °C.
5. Add 10 U each of Exonuclease I and III (New England Biolab, Ipswich, MA, USA), 2 μL of 10× reaction buffer 1 and sterile distilled water to PCR tube containing ligation mixture to make up the final volume to 20 μL.
6. Incubate the PCR tube at 37 °C for 30 min in thermal cycler with heated lid followed by 94 °C for 3 min to inactivate Exonuclease.
7. Prepare H-RCA master mixture by adding 8 U of *Bst* DNA polymerase (New England Biolabs, Ipswich, MA, USA), 5 μL reaction buffer, 400 μM dNTP mix, 10 pmol of each of two H-RCA primers [P1: 5′ ATGGGCACCG AAGAAGCA3′;P2:5′CGCGCAGACACGATA 3′], 5 % of DMSO (v/v) and 1× SYBR green I to a final volume of 30 μL.
8. Add 30 μL of H-RCA master mixture to the PCR tube; incubate at 65 °C for 30 min in Corbett RotorGene 3000 Real-Time thermal cycler. Collect florescence signal every minute for up to 30 min.

Quantitative Multiplex Detection by the Padlock Lock System Coupled with Universal, High-Throughput Real-Time PCR on OpenArray® Instrument

This assay enables specific, high-throughput, quantitative detection of multiple pathogens over a wide range of target concentrations. The ligation padlock probes (also known as PRI-lock probes) are long oligonucleotides with target complementary regions at their 5′ and 3′ ends [25]. Upon target hybridization, the PRI-lock probes are circularized via enzymatic ligation, subsequently serving as templates for individual amplification via unique probe-specific primers. Adaptation to the OpenArray, which can accommodate up to 3,072×33 nL PCR amplifications, allowed high-throughput real-time quantification (Fig. 49.2) [26].

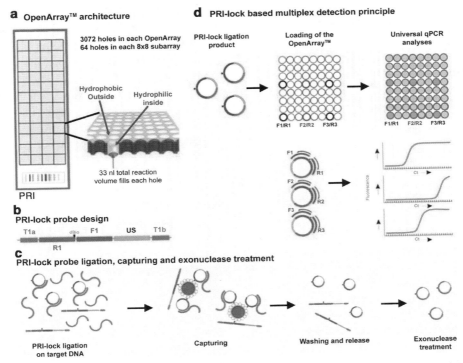

Fig. 49.2 (a) OpenArray™ architecture. The OpenArray™ has 48 subarrays and each subarray contains 64 microscopic through-holes of 33 nl volume. The primers are pre-loaded into the holes. The sample combined with the reaction mix is auto-loaded due to the surface tension, provided by the hydrophilic coating of the holes and the hydrophobic surface of the array. (b) PRI-lock probe design. T1a and T1b indicate target complementary regions. Unique primer sites ensure specific amplification (forward: F1 and reverse: R1) and each PRI-lock contains a universal sequence (US) and a desthiobiotin moiety (dBio). (c) Multiple target specific PRI-lock probes are ligated on fragmented DNA samples. T1a and T1b bind to adjacent sequences of the target and in case of a perfect match, the probe is circularized by a ligase. The probes are captured via the desthiobiotin moiety using magnetic streptavidin-coated beads. The PRI-lock probes are washed and quantitatively eluted from the beads. Unreacted probes are removed by exonuclease treatment. (d) Circularized probes are loaded and independently amplified on the Biotrove OpenArray™ platform using PRI-lock probe specific primers. The amplification is monitored using SYBR-Green and the ligated PRI-lock probes are quantified based on the threshold cycle number (CT)

The following protocol uses the PRI-lock detection system targeting plant pathogens at different taxonomic levels. The nucleic acid targets can be reliably quantified over 5 orders of magnitude with a dynamic detection range of more than 10^4 copies/μL. Pathogen quantification is equally robust in single target versus mixed target assays.

1. Genomic DNA (i.e., 500 pg) (Note 2) is fragmented by digestion using EcoRI, HindIII and BamHI (New England Biolabs Inc., Ipswich, MA, USA) for 15 min at 37 °C.

2. Add 250 pM of the individual PRI-lock probes (Seraing, Belgium) [Note 3], 20 mM Tris–HCl, pH 9.0, 25 mM KCH$_3$COO, 10 mM Mg(CH$_3$COO)$_2$, 1 mM NAD, 10 mM DTT, 0.1 % Triton X-100, 20 ng sonicated salm sperm DNA and 20 U Taq ligase (New England Biolabs Inc., Ipswich, MA, USA) to the fragmented DNA sample to an end volume of 10 μL (Note 4).

3. Place the tubes in the PCR machine and run the appropriate ligation cycle programme. Samples are denatured at 95 °C for 5 min and subsequently subjected to 20 cycles of 30 s at 95 °C and 5 min at 65 °C, followed by enzyme inactivation at 95 °C for 15 min.

4. When the ligation is complete 30 μL distilled water is added to each reaction.

5. The desthiobiotin moiety of the PRI-lock probes is captured through the addition of 40 μL solution containing 2 M NaCl, 10 mM Tris–HCl, pH 7.5, 1 mM EDTA, 0.2 M NaOH and 200 μg magnetic MyOne™ Streptavidin C1 Dynabeads (Dynal Biotech ASA, Oslo, Norway).

6. Rotate at 4 °C for 1 h.

7. Centrifuge the samples at 2,000×g for 10 s, and collect the Dynabeads via application of a magnetic field.

8. The Dynabeads are washed with 100 μL 100 mM Tris–HCl, pH 7.5 and 50 mM NaCl.

9. The Dynabeads are re-suspended in 10 μL distilled water and incubated at 95 °C for 10 min, allowing quantitative elution of the PRI-lock probes from the Dynabeads (Note 5).

10. Samples are transferred onto ice and the empty magnetic streptavidin beads removed via application of a magnetic field, leaving the washed PRI-lock probes in the solution.

11. 10 μL of Exonuclease mixture (10 mM Tris–HCl, pH 9.0, 4.4 mM MgCl$_2$, 0.1 mg/mL BSA, 0.5 U Exonuclease I (USB Europe GmbH, Staufen, Germany) and 0.5 U Exonuclease III (USB Europe GmbH, Staufen, Germany)) is added to each reaction, and incubated at 37 °C for 30 min.

12. When exonuclease step is complete the enzyme is inactivated at 95 °C for 2.5 h.

13. Amplification of ligated PRI-lock probes is followed in real-time using an OpenArray Real-Time PCR Instrument (Applied Biosystems, Foster City, USA) (Note 5). Each subarray is loaded with 5.0 μL master-mix containing 2.5 μL ligated padlock mixture and reagents in a final concentration of 1× LightCycler® FastStart DNA Master SYBR Green I mix (Roche Diagnostics GmbH, Mannheim, Germany), 0.2 % Pluronic F-68 (Gibco, Carlsbad, USA), 1 mg/mL BSA (Sigma-Aldrich, St Louis, USA), 1:4000 SYBR Green I (Sigma-Aldrich), 0.5 % (v/v) Glycerol (Sigma-Aldrich), 8 % (v/v) deionized formamide (Sigma-Aldrich) and 1.0 pg PCR control template (Note 6).

14. The PCR OpenArray thermal cycling protocol is fixed and consist a 90 °C step for 10 min, followed by 27 cycles of 28 s at 95 °C, 1 min at 55 °C and 70 s at 72 °C (imaging step).

15. The OpenArray Real-time PCR Software is used for data analyses. The positive amplification reactions are analyzed for amplicon specificity by studying the individual melting curves (Note 70).

Detection and Identification of Multiple Microorganisms in Environmental Samples by a Cleavable Padlock Probe-Based Ligation Detection Assay

The ligation detection (LD) system uses a single compound detector probe per target. The padlock probes contain asymmetric target complementary regions at both their 5′ and 3′ ends that confer specific target detection and have both a desthiobiotin moiety and an internal endonuclease IV cleavage site [27]. PCR amplification of universal phylogenetic target genes (e.g., 16S, 18S, and 23S rRNA genes), a number of microorganism-specific genetic markers [28–30], or random amplification of genomic DNA (gDNA) fragments [31] serve as potential targets for PRI-lock probe ligation. Upon target hybridization, the PRI-lock probes are circularized via enzymatic ligation, captured, and cleaved, allowing only the originally ligated padlock probes to be visualized on a universal microarray (Fig. 49.3) (Note 8). Unlike previous procedures, the probes themselves are not amplified, thereby allowing a simple padlock cleavage to yield a background-free assay.

The following LD protocol enables specific detection and identification of multiple pathogens over a wide range of target concentrations and is adaptable to a variety of applications.

1. Targets for ligation are generated by PCR preamplification of 10 ng extracted gDNA (Notes 2 and 9). Preamplification is performed in a 2× Master Mix (Applied Biosystems) containing AmpliTaq Gold

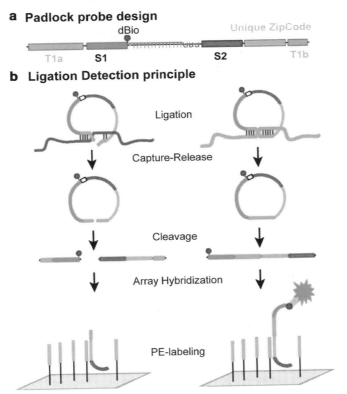

a Padlock probe design

b Ligation Detection principle

Ligation

Capture-Release

Cleavage

Array Hybridization

PE-labeling

Fig. 49.3 (a) PLP design. T1a and T1b are asymmetric target complementary regions. Each PLP contains a unique ZipCode sequence for universal array hybridization, two spacer sequences (S1 and S2), a desthiobiotin moiety (dBio) for probe capture, a polyoligo(dT) linker sequence, and a polydeoxyuracil sequence for probe cleavage. (b) Multiple target-specific PLPs are ligated to PCR-preamplified DNA samples. T1a and T1b bind to adjacent sequences of the target, and in the case of a perfect match, the probe is circularized by enzymatic ligation. The PLPs are reversibly captured and washed via the desthiobiotin moiety with magnetic streptavidin-coated beads. Next, the washed probes are cleaved at the polydeoxyuracil sequences with UNG and endonuclease enzymes. The sample containing the cleaved PLPs is hybridized on a universal microarray. Finally, only the hybridized PLPs that were originally ligated can be labeled and visualized with streptavidin R-PE by using the desthiobiotin moiety

DNA polymerase, deoxynucleoside triphosphates with dUTP, 0.12 μL UNG (Applied Biosystems), 10 ng gDNA extract and 300 nM of each primer (Note 11). The reaction mixture is incubated at 50 °C for 2 min, followed by 10 min denaturation at 95 °C and 40 cycles consisting of a 30 s incubation at 95 °C, an annealing step at 60 °C for 30 s (ITS), and an elongation step at 72 °C for 60 s (Note 9). After the last cycle, the reaction is immediately cooled to 4 °C.

2. The total PCR-amplified products are fragmented by digestion with EcoRI, HindIII, and BamHI (New England BioLabs, Inc.) for 15 min at 37 °C.

3. Add 5 μL of the fragmented DNA sample to 20 pM of the individual LD padlock probes, 20 mM Tris–HCl (pH 7.5), 20 mM KCl, 10 mM $MgCl_2$, 0.1 % Igepal, 0.01 mM rATP, 1 mM dithiothreitol, 20 ng sonicated salmon sperm DNA, and 4 U *Pfu* DNA ligase (Stratagene) with an end volume of 10 μL.

4. Place the tubes in the PCR machine and run the appropriate ligation cycle programme. Samples are denatured at 95 °C for 5 min and subsequently subjected to 20 cycles of 30 s at 95 °C and 5 min at 65 °C, followed by enzyme inactivation at 95 °C for 15 min.

5. When the ligation is complete 30 μL distilled water is added to each reaction.

6. The desthiobiotin moiety of the LD probes is captured through the addition of 40 μL containing 2 M NaCl, 10 mM Tris–HCl, pH 7.5, 1 mM EDTA, 0.2 M NaOH, and 200 μg magnetic MyOne™ Streptavidin C1 Dynabeads (Dynal Biotech ASA, Oslo, Norway).

7. Rotate at 4 °C for 1 h.

8. Centrifuge the samples at 2,000×g for 10 s, and collect the Dynabeads via application of a magnetic field.

9. Wash the Dynabeads with 100 μL 100 mM Tris–HCl, pH 7.5, and 50 mM NaCl.

10. Re-suspended the Dynabeads in 10 μL distilled water and incubated at 95 °C for 10 min, allowing quantitative elution of the LD probes from the Dynabeads [32].

11. Transfer the samples onto ice and remove the empty magnetic streptavidin beads via application of a magnetic field, leaving the washed LD probes in the solution.

12. 10 μL of cleavage mixture (10 U uracil-N-glycosylase (UNG; Applied Biosystems), 10 U endonuclease IV (New England BioLabs), 2× NEBuffer 3 (New England BioLabs), 2× bovine serum albumin) is added to each reaction mixture and incubated at 37 °C for 1.5 h.

13. 2 μl of 1.1 M NaOH is added and incubated at 95 °C for 10 min. Finally, 8 μL of 0.5 M Tris buffer (pH 6.8) is added to the samples for neutralizing the solution.

14. The cleaved ligation samples are heated for 10 min at 95 °C and then cooled rapidly on ice (Note 12).

15. 40 μL of the array hybridization mixture is added to each well (Note 10); the chambers are sealed and the arrays are hybridized at 55 °C overnight under high humidity.

16. The array is washed once at 55 °C for 5 min in preheated 1× SSC (0.15 M NaCl plus 0.015 M sodium citrate), 0.1 % sodium dodecyl sulfate and twice for an additional 1 min at room temperature in 0.1× SSC-0.1 % sodium dodecyl sulfate and in TNT (0.1 M Tris–HCl (pH 7.6), 0.15 M NaCl, 0.05 % Tween 20), respectively.

17. The array is incubated in blocking solution (0.1 M Tris–HCl (pH 7.5), 0.15 M NaCl, 0.5 % blocking reagent (Perkin-Elmer)) for 10 min at high humidity at room temperature and washed for 1 min in TNT.

18. 20 μl of staining solution (15 μg/mL streptavidin R-phycoerythrin (PE; Qiagen) in 20 μL of blocking solution) is added to each well.

19. The array is incubated in the dark at room temperature for 15 min. The silicon structures are removed from the slides, washed three times for 1 min in TNT and twice for 1 min in 0.1× SSC, respectively. Finally, the slides are dried by spinning at 250×g for 3 min.

20. Microarrays are analyzed with a confocal ScanArray 4000 laser scanning system (Packard GSI Lumonics) containing a GreNe 543-nm laser for PE and a HeNe 633-nm laser for Cy5 fluorescence measurement (Note 13).

Notes

1. Gloves should be worn throughout these procedures.

2. Genomic DNAs from all microorganisms are isolated using the Puregene Genomic DNA isolation kit (Gentra/Biozym, Landgraaf, the Netherlands) according to the manufacturer's instructions; other methods to prepare gDNA from organisms can be used as well.

3. The padlock probe target complementary regions are engineered according to previously described design criteria [13] and are connected by a 60 bp compound linker sequence. The linker sequence contains a 20 bp, generic sequence and two, unique primer binding sites for specific PCR amplification. All primer pairs have equal melting temperatures to allow universal SYBR-Green-based detection in real-time PCR. The primer pairs are chosen from GeneFlex™ TagArray set (Affymetrix) in a way to minimize padlock probe secondary structures and optimize both, primer T_M and primer specificity. Potential for secondary structures, primer T_M and primer specificity

are predicted using Visual OMP 6.0 software (DNA Software Inc., Michigan, USA). The prediction parameters are set to match ligation ([monovalent] = 0.025 M; [Mg^{2+}] = 0.01 M; T = 65 °C [probe] = 250 pM) and PCR conditions ([monovalent] = 0.075 M; [Mg^{2+}] = 0.005 M, T = 60 °C). When necessary, PRI-lock probe arm sequences are adjusted to avoid strong secondary structures that might interfere with efficient ligation as described previously [13].

Between the primer sites a thymine-linked desthiobiotin molecule is introduced for specific capturing and release with streptavidin-coated magnetic beads. The rationale of using desthiobiotin instead of biotin is the approximately 1,000 times lower affinity for streptavidin [33, 34], which permits the reversible release of the padlock probes.

4. Reaction mixtures were prepared on ice and rapidly transferred into a thermal cycler.

5. OpenArray® real-time PCR amplification of ligated PRI-lock probes can be followed in real-time using an OpenArray® Real-Time PCR Instrument (Applied Biosystems, Foster City, USA). OpenArray® subarrays are preloaded by Applied Biosystems with selected primer pairs. Each primer pair is spotted in duplicate to a final assay concentration of 128 nM.

6. A PCR control is developed to monitor differences in PCR efficiency within and between different OpenArrays®. To this end, a 99 bp PCR control template (5′-CTAACGAATCTGGGAC GTGCATCCGGTCTCATCGCTG AATCGCTCGTGAGGGCAGGGCCGGG AGGGGGGTCCGCAGGCGCAACACTGT AGTCGGTGCTA-3′) was amplified using the forward 5′-CTAACGAATCTGGGACGTGC-3′ and reverse 5′-TAGCACCGACTACAGTG TTG-3′ primer pair.

7. The OpenArray® Real-time PCR Software uses a proprietary calling algorithm that estimates the quality of each individual C_T value by calculating a C_T confidence value for the amplification reaction. In our assay, C_T values with C_T confidence values below 700 were regarded as background signals.

8. The cZipCode oligonucleotides carrying a C_{12} linker and a 5′ NH_2 group were synthesized and spotted onto Nexterion MPX-E16 epoxy-coated slides by Isogen B.V. according to the manufacturer's instructions (Schott Nexterion).

9. The internal transcribed spacer (ITS) regions of the fungal and oomycetal rRNA genes are amplified using the primers ITS1 and ITS4 [35].

10. Sixteen-well silicon structures (Schott Nexterion) were attached to the arrays to create 16 separate subarray chambers. Before hybridization, the arrays are washed and blocked according to the manufacturer's instructions.

11. Preamplification reactions are performed in the presence of 0.1 aM internal amplification control (IAC) oligonucleotide (5′TCCGT AGGTGAACCTGCGGCGGATCGTTA CAAGGGTCTCCAACTACGTCTAG CGCATAGACCACGTATCGAAGCTAGG T GCATATCAATAAGCGGAGGA 3′). It is often observed that DNA extracted from environmental samples contain PCR-inhibiting compounds [36], which may lead to false negatives. In order to monitor PCR inhibition during the preamplification reaction, an IAC was developed. This IAC consist a single-stranded oligonucleotide containing a random nonsense sequence flanked by ITS1-ITS4 primer regions, which is added to each preamplification reaction mixture. The IAC LD probe containing ligation arms targeting the reverse complement of the IAC oligonucleotide is added to each ligation mixture. In the case of PCR inhibition, the complementary strand of the IAC oligonucleotide will not be generated. Therefore, the target of the IAC padlock probes will not be present during the ligation reaction and, consequently, no IAC padlock probe signal will be observed on the array.

12. The array hybridization mixtures are made up of 15 μL of sample and 1 μL of 0.1 μM Cy5-labeled corner oligonucleotide in 1× tetramethylammonium chloride in a final volume of 48 μL.

13. Laser power is fixed at 70 % for both lasers, while the photomultiplier tube power ranged from 50 to 70 %, depending on the signal intensity. Fluorescence intensities is quantified by using QuantArray1 (Packard GSI Lumonics), and the mean signal minus the mean local background (mean PE-B) is used. The absolute signal intensity is defined as the mean PE-B minus the assay background. The assay background for each sub-array is defined as the mean PE−B+2 standard deviations (SDs) of the fluorescence of the unused cZipCodes. The cZipCode oligonucleotides are spotted in threefold triplicates (nine parallels) or twofold quadruplicates (eight parallels), depending on the array batch used. After exclusion of the outliers (mean PE−B±2 SDs), signals are averaged for the probes, and SDs calculated.

Acknowledgements We are grateful to Drs. N. Saksena, S. Chen, and F. Kong (Westmead Hospital, Australia) for assistance and advice. Part of the funding for this research has been provided by Genome Canada and Genome British Columbia in support of The Tria I and Tria II Projects http://www.thetriaproject.ca.

References

1. Richardson M, Warnock D (2003) Fungal infection: diagnosis and management, 3rd edn. Blackwell, Oxford
2. Agrios G (2005) Plant pathology, 5th edn. Elsevier Academic, London
3. Tsui CK, Wang B, Khadempour L, Alamouti SM, Bohlmann J, Murray BW et al (2010) Rapid identification and detection of pine pathogenic fungi associated with mountain pine beetles by padlock probes. J Microbiol Methods 83:26–33
4. Miller SA, Beed FD, Harmon CL (2009) Plant disease diagnostic capabilities and networks. Annu Rev Phytopathol 47:15–38
5. Lievens B, Thomma BP (2005) Recent developments in pathogen detection arrays: implications for fungal plant pathogens and use in practice. Phytopathology 95:1374–1380
6. Nilsson M, Malmgren H, Samiotaki M, Kwiatkowski M, Chowdhary BP, Landegren U (1994) Padlock probes: circularizing oligonucleotides for localized DNA detection. Science 265:2085–2088
7. Landegren U, Kaiser R, Sanders J, Hood L (1988) A ligase-mediated gene detection technique. Science 241:1077–1080
8. Nilsson M, Krejci K, Koch J, Kwiatkowski M, Gustavsson P, Landegren U (1997) Padlock probes reveal single-nucleotide differences, parent of origin and in situ distribution of centromeric sequences in human chromosomes 13 and 21. Nat Genet 16: 252–255
9. Prins TW, van Dijk JP, Beenen HG, Van Hoef AA, Voorhuijzen MM, Schoen CD et al (2008) Optimised padlock probe ligation and microarray detection of multiple (non-authorised) GMOs in a single reaction. BMC Genomics 9:584
10. Nilsson M, Dahl F, Larsson C, Gullberg M, Stenberg J (2006) Analyzing genes using closing and replicating circles. Trends Biotechnol 24:83–88
11. Baner J, Nilsson M, Mendel-Hartvig M, Landegren U (1998) Signal amplification of padlock probes by rolling circle replication. Nucleic Acids Res 26:5073–5078
12. Lizardi PM, Huang X, Zhu Z, Bray-Ward P, Thomas DC, Ward DC (1998) Mutation detection and single-molecule counting using isothermal rolling-circle amplification. Nat Genet 19:225–232
13. Szemes M, Bonants P, de Weerdt M, Baner J, Landegren U, Schoen CD (2005) Diagnostic application of padlock probes–multiplex detection of plant pathogens using universal microarrays. Nucleic Acids Res 33:e70
14. Kong F, Tong Z, Chen X, Sorrell T, Wang B, Wu Q et al (2008) Rapid identification and differentiation of Trichophyton species, based on sequence polymorphisms of the ribosomal internal transcribed spacer regions, by rolling-circle amplification. J Clin Microbiol 46:1192–1199
15. Zhou X, Kong F, Sorrell TC, Wang H, Duan Y, Chen SC (2008) Practical method for detection and identification of *Candida, Aspergillus,* and *Scedosporium* spp. by use of rolling-circle amplification. J Clin Microbiol 46:2423–2427
16. Wang H, Kong F, Sorrell TC, Wang B, McNicholas P, Pantarat N et al (2009) Rapid detection of ERG11 gene mutations in clinical *Candida albicans* isolates with reduced susceptibility to fluconazole by rolling circle amplification and DNA sequencing. BMC Microbiol 9:167
17. Kaocharoen S, Wang B, Tsui KM, Trilles L, Kong F, Meyer W (2008) Hyperbranched rolling circle amplification as a rapid and sensitive method for species identification within the *Cryptococcus* species complex. Electrophoresis 29:3183–3191
18. Eriksson R, Jobs M, Ekstrand C, Ullberg M, Herrmann B, Landegren U et al (2009) Multiplex and quantifiable detection of nucleic acid from pathogenic fungi using padlock probes, generic real time PCR and specific suspension array readout. J Microbiol Methods 78:195–202
19. Fire A, Xu SQ (1995) Rolling replication of short DNA circles. Proc Natl Acad Sci U S A 92:4641–4645
20. Pickering J, Bamford A, Godbole V, Briggs J, Scozzafava G, Roe P et al (2002) Integration of DNA

ligation and rolling circle amplification for the homogeneous, end-point detection of single nucleotide polymorphisms. Nucleic Acids Res 30:e60

21. Faruqi AF, Hosono S, Driscoll MD, Dean FB, Alsmadi O, Bandaru R et al (2001) High-throughput genotyping of single nucleotide polymorphisms with rolling circle amplification. BMC Genomics 2:4

22. Wang B, Dwyer DE, Chew CB, Kol C, He ZP, Joshi H et al (2009) Sensitive detection of the K103N nonnucleoside reverse transcriptase inhibitor resistance mutation in treatment-naive HIV-1 infected individuals by rolling circle amplification. J Virol Methods 161:128–135

23. Wang B, Dwyer DE, Blyth CC, Soedjono M, Shi H, Kesson A et al (2010) Detection of the rapid emergence of the H275Y mutation associated with oseltamivir resistance in severe pandemic influenza virus A/H1N1 09 infections. Antiviral Res 87:16–21

24. Tong Z, Kong F, Wang B, Zeng X, Gilbert GL (2007) A practical method for subtyping of Streptococcus agalactiae serotype III, of human origin, using rolling circle amplification. J Microbiol Methods 70:39–44

25. van Doorn R, Szemes M, Bonants P, Kowalchuk GA, Salles JF, Ortenberg E et al (2007) Quantitative multiplex detection of plant pathogens using a novel ligation probe-based system coupled with universal, high-throughput real-time PCR on OpenArrays®. BMC Genomics 8:276

26. Morrison T, Hurley J, Garcia J, Yoder K, Katz A, Roberts D et al (2006) Nanoliter high throughput quantitative PCR. Nucleic Acids Res 34:e123

27. van Doorn R, Slawiak M, Szemes M, Dullemans AM, Bonants P, Kowalchuk GA et al (2009) Robust detection and identification of multiple oomycetes and fungi in environmental samples by using a novel cleavable padlock probe-based ligation detection assay. Appl Environ Microbiol 75:4185–4193

28. Call DR, Brockman FJ, Chandler DP (2001) Detecting and genotyping Escherichia coli O157:H7 using multiplexed PCR and nucleic acid microarrays. Int J Food Microbiol 67:71–80

29. Gonzalez SF, Krug MJ, Nielsen ME, Santos Y, Call DR (2004) Simultaneous detection of marine fish pathogens by using multiplex PCR and a DNA microarray. J Clin Microbiol 42:1414–1419

30. Wilson WJ, Strout CL, DeSantis TZ, Stilwell JL, Carrano AV, Andersen GL (2002) Sequence-specific identification of 18 pathogenic microorganisms using microarray technology. Mol Cell Probes 16:119–127

31. Vora GJ, Meador CE, Stenger DA, Andreadis JD (2004) Nucleic acid amplification strategies for DNA microarray-based pathogen detection. Appl Environ Microbiol 70:3047–3054

32. Holmberg A, Blomstergren A, Nord O, Lukacs M, Lundeberg J, Uhlen M (2005) The biotin-streptavidin interaction can be reversibly broken using water at elevated temperatures. Electrophoresis 26:501–510

33. Hirsch JD, Eslamizar L, Filanoski BJ, Malekzadeh N, Haugland RP, Beechem JM (2002) Easily reversible desthiobiotin binding to streptavidin, avidin, and other biotin-binding proteins: uses for protein labeling, detection, and isolation. Anal Biochem 308:343–357

34. Gregory KJ, Bachas LG (2001) Use of a biomimetic peptide in the design of a competitive binding assay for biotin and biotin analogues. Anal Biochem 289:82–88

35. White T, Bruns T, Lee S, Taylor J (1990) Amplification and direct sequencing of fungal ribosomal RNA genes for phylogenetics. In: Innis MA, Gelfand DH, Sninski JJ, White TJ (eds) PCR-protocols a guide to methods and applications. Academic, San Diego, pp 315–320

36. Watson RJ, Blackwell B (2000) Purification and characterization of a common soil component which inhibits the polymerase chain reaction. Can J Microbiol 46:633–642

Maria D. Mayan, Alexandra McAleenan, and Priscilla Braglia

Abstract

One common problem for researchers working with yeast is the difficulty of efficiently treating whole cells with drugs or chemicals, as their uptake from the growth medium is very limited. Several methods have been described to increase drug penetration. However, most of them do not allow yeast growth under normal conditions. This chapter describes two protocols for permeabilizing whole yeast cells using either detergents at 4°C, to allow in vitro assays to be executed, or low amounts of the sesquiterpene dialdehyde polygodial, which permits experiments to be performed in vivo, without affecting yeast morphology or growth.

Keywords

Permeabilization • Sesquiterpene dialdehyde polygodial • Budding yeast • Detergent

M.D. Mayan (✉)
Biomedical Research Center-INIBIC,
Xubias de Arriba, 84, A Coruña, Galicia 15006, Spain
e-mail: MA.Dolores.Mayan.Santos@sergas.es

A. McAleenan
Imperial College London, Clinical Sciences Centre,
Hammersmith Hospital Campus, Du Cane Road,
London W12 0NN, UK

P. Braglia
Sir William Dunn School of Pathology, University
of Oxford, South Parks Road, Oxford OX1 3RE, UK

Introduction

Yeast cells are surrounded by a cell wall and are not normally permeable to many drugs, including the transcription inhibitors α-amanitin and DRB (5,6-Dichloro-1-β-D-ribofuranosyl-benzimidazole), which are commonly used to treat mammalian cells. Digestion of the cell wall can be achieved using cell lytic enzyme or zymolyase, but handling of the spheroplasts can be tricky and does not allow studies to be performed under physiological conditions. Different protocols have been described to increase drug uptake in yeast cells. For example, the *ERG6* gene has been mutated or deleted to make yeast permeable to inhibitors of the 26S proteasome. *ERG6* mutations affect ergosterol biosynthesis and hence alter the membrane lipid composition [1–3]. Another example is the use of the antibiotic amphotericin B which has been used to allow treatment of yeast cells with the transcription inhibitors rifampicin and actinomycin D. However, amphotericin B binds to ergosterol disrupting the fungal cell membrane [4–7]. Different organic solvents, including ethanol, have also been used for cell permeabilization [8]. In this chapter, we describe the use of detergents at 4°C to perform in vitro assays and polygodial under normal growth conditions to perform in vivo assays. Low concentrations of the sesqueterpene dialdehyde polygodial increase the penetration of components of the medium into the cells without affecting either the cell membrane or growth [9–12]. It is therefore of particular interest for the treatment of living cells under standard conditions.

Materials

Yeast Permeabilization Using Detergents

1. Yeast growth medium: prepare the medium to grow *S. cerevisiae* as required for the experiment.
2. Sterile H_2O, ice-cold.
3. 10% solution sodium *N*-lauroyl sarcosine (Sigma-Aldrich).
4. 2.5x Transcription Buffer: 50 mM Tris/Cl pH 7.7; 500 mM KCl; 80 mM $MgCl_2$.

Yeast Permeabilization Using Polygodial

1. Polygodial stock solution 3,200×: 1.25 mg of polygodial (Santa Cruz Biotechnology) dissolved in 20 mL of absolute ethanol. The stock solution can be stored for up to 3 months at −20°C, according to manufacturer's instructions.

Methods

Yeast Permeabilization Using Detergents

The use of detergents to permeabilize yeast cells is best exemplified by the protocol commonly used for transcriptional run-on [13]. TRO assays allow the density of elongating RNA polymerases over a desired target gene to be directly measured, providing a measure of nascent transcription at a set time-point [14, 15]. Here whole cells are used to perform *in vitro* transcription for a short period of time after providing radioactively labelled NTPs. The hot RNA is then extracted and hybridized to single-stranded DNA probes, immobilized on a filter. RNA labelling takes place after the yeast are permeabilized with sodium *N*-lauroyl-sarcosine; this also allows the effective treatment of cells with α-amanitin (AM), to which yeast cells are normally impermeable, to measure polymerase sensitivity to this inhibitor.

1. Grow 100 mL cultures to OD_{600} between 0.05 and 0.2. If, for example, a temperature shift is required, start it at $OD_{600} = 0.05$ and allow the cells to grow for two generations, to $OD_{600} = 0.2$ (see Note 1).

Working at 4°C:

2. Harvest the cells by centrifugation, 4 min at 3,000 rpm.
3. Discard the supernatant and wash the cell pellet with 5 mL ice-cold sterile H_2O. Resuspend by pipetting.
4. Spin again 4 min at 3,000 rpm, discard the supernatant and resuspend the cell pellet in 950 μL ice-cold sterile H_2O. Transfer to eppendorf tubes.
5. To the cell suspension add 50 μL of 10% sodium *N*-lauroyl sarcosine. Mix by inverting the tubes 5–6 times.
6. Incubate 20 min on ice.
7. Pellet the cells in microcentrifuge, 1 min at 6,000 rpm. Remove supernatant.
8. Resuspend the cell pellet in 60 μL 2.5× transcription buffer and proceed with run-on protocol [14–16].

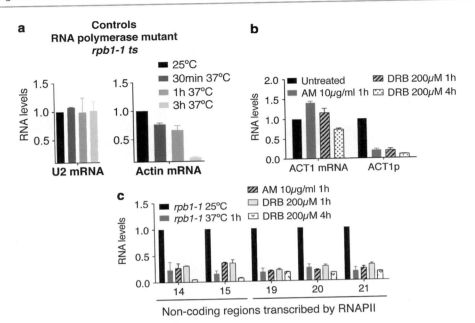

Fig. 50.1 AM and DRB efficiently inhibit transcription by RNAP-II in permeabilized yeast using low concentrations of polygodial. (**a**) Analysis of RNA levels by RT-PCR in *rpb1-1* cells at 25°C and following incubation at 37°C. The temperature shift inactivates RNAP-II in this mutant strain. Data were normalized to the mature form of the U2 small nuclear RNA. Note that mature snRNAs have been shown to be stable following RNAP-II inactivation [20]. (**b**) In cells permeabilized with polygodial, *ACT1* mRNA (*ACT1 mRNA*, exon1) is stable after 1 h of treatment with the transcription inhibitors AM or DRB. However the level of primary transcript (*ACT1p*, exon1/intron1 junction) decreases by ~90% after treatment with the inhibitors. (**c**) Non-coding transcripts at the ribosomal locus (labelled *14*, *15*, *19*, *20* and *21*) are transcribed by RNAP-II. The data relative to the *rpb1-1* strain, a temperature sensitive mutant in a key subunit of RNAPII, are shown in parallel with the data obtained and in the presence of the transcription inhibitors AM and DRB. Data are presented as mean ± SEM, $n = 2$

Yeast Permeabilization Using Polygodial

Adding polygodial to a yeast cell culture a few minutes before treatment with different drugs, immediately promotes uptake of the drug into the cell [11, 12]. Polygodial should not be used at concentrations higher than 1 µg/mL as higher levels of polygodial affect the integrity of the membrane and therefore the cell metabolism, interfering with the results obtained. A good example of the use of polygodial is provided by the treatment with the RNA polymerase II (RNAP-II) inhibitors AM or DRB. AM and DRB inhibit transcription elongation by different mechanisms [17–19]. After treating the cells with 0.39 µg/mL polygodial, the level of RNAP-II transcription drops to 10–20% of the initial level after only 1 h in the presence of 10 µg/mL AM or 45 min in the presence of 200 µM DRB (Fig. 50.1). Our experience indicates that adding polygodial to the yeast culture only a few minutes before the treatment is enough to promote the entry of different drugs.

1. Grow the yeast as desired (see Note 2).
2. Add 0.39 µg/mL polygodial resuspended in absolute ethanol. Mix the culture (see Note 3).
3. Split the culture into two flasks, untreated and treated culture.
4. Add the desired drug to the medium at an appropriate concentration.
5. Grow the cells for the desired treatment time.
6. Collect the cells as required for subsequent protocols.

Notes

1. When comparing different strains/conditions, ensure cultures are grown to similar OD_{600}. This might require dilution of the faster-growing strain [16]. Cultures grown to a higher OD_{600} have empirically resulted in lower TRO signal.

2. Temperatures from 23 to 39°C have been used in the presence of 0.39–1 µg/mL of polygodial without affecting cell membrane structure.

3. For 200 mL of culture add 62.4 µL of polygodial 1.25 mg/mL. Using this stock, the ethanol is diluted 3,200 times in the culture. No further incubation time with polygodial is needed: the 5–10 min spent dividing the culture into two different flasks, is enough to promote the entrance of the drugs after adding polygodial.

Acknowledgements The experimental work was performed in Professor Luis Aragon's lab. We thank R. Young for the Z118 strain (rpb1-1). This work was supported by the Medical Research Council of the United Kingdom. Maria Mayán is currently funded by the Xunta de Galicia.

References

1. Nitiss J, Wang JC (1988) DNA topoisomerase-targeting antitumor drugs can be studied in yeast. Proc Natl Acad Sci U S A 85:7501–7505
2. Lee DH, Goldberg AL (1996) Selective inhibitors of the proteasome-dependent and vacuolar pathways of protein degradation in Saccharomyces cerevisiae. J Biol Chem 271:27280–27284
3. Kaur R, Bachhawat AK (1999) The yeast multidrug resistance pump, Pdr5p, confers reduced drug resistance in erg mutants of Saccharomyces cerevisiae. Microbiology 145:809–818
4. Battaner E, Kumar BV (1974) Rifampin: inhibition of ribonucleic acid synthesis after potentiation by amphotericin B in Saccharomyces cerevisiae. Antimicrob Agents Chemother 5:371–376
5. Kwan CN, Medoff G, Kobayashi GS, Schlessinger D, Raskas HJ (1972) Potentiation of the antifungal effects of antibiotics by amphotericin B. Antimicrob Agents Chemother 2:61–65
6. Medoff G, Kobayashi GS, Kwan CN, Schlessinger D, Venkov P (1972) Potentiation of rifampicin and 5-fluorocytosine as antifungal antibiotics by amphotericin B (yeast-membrane permeability-ribosomal RNA-eukaryotic cell-synergism). Proc Natl Acad Sci U S A 69:196–199
7. Medoff G, Kwan CN, Schlessinger D, Kobayashi GS (1973) Potentiation of rifampicin, rifampicin analogs, and tetracycline against animal cells by amphotericin B and polymyxin B. Cancer Res 33:1146–1149
8. Panesar PS, Panesar R, Singh RS, Bera MB (2007) Permeabilization of yeast cells with organic solvents for β-galactosidase activity. Res J Microbiol 2:7
9. Andres MI, Forsby A, Walum E (1997) Polygodial-induced noradrenaline release in human neuroblastoma SH-SY5Y cells. Toxicol In Vitro 11:509–511
10. Taniguchi M, Yano Y, Motoba K, Oi S, Haraguchi H, Hashimoto K, Kubo I (1988) Polygodial-induced Sensitivity to Rifampicin and Actinomycin D of. Agric Biol Chem 52:1881–1883
11. Mayan MD (2010) Drug-induced permeabilization of S. cerevisiae. Curr Protoc Mol Biol Chapter 13:Unit 13.2B
12. Mayan M, Aragon L (2010) Cis-interactions between non-coding ribosomal spacers dependent on RNAP-II separate RNAP-I and RNAP-III transcription domains. Cell Cycle 9:4328–4337
13. Vennstrom B, Persson H, Pettersson U, Philipson L (1979) A DRB (5,6 dichloro-beta-D-ribofuranosylbenzimidazole)-resistant adenovirus mRNA. Nucleic Acids Res 7:1405–1418
14. Hirayoshi K, Lis JT (1999) Nuclear run-on assays: assessing transcription by measuring density of engaged RNA polymerases. Methods Enzymol 304:351–362
15. Elion EA, Warner JR (1986) An RNA polymerase I enhancer in Saccharomyces cerevisiae. Mol Cell Biol 6:2089–2097
16. Braglia P, Kawauchi J, Proudfoot NJ (2010) Co-transcriptional RNA cleavage provides a failsafe termination mechanism for yeast RNA polymerase I. Nucleic Acids Res 39:1439–1448
17. Mittleman B, Zandomeni R, Weinmann R (1983) Mechanism of action of 5,6-dichloro-1-beta-D-ribofuranosylbenzimidazole. II. A resistant human cell mutant with an altered transcriptional machinery. J Mol Biol 165:461–473
18. Bushnell DA, Cramer P, Kornberg RD (2002) Structural basis of transcription: alpha-amanitin-RNA polymerase II cocrystal at 2.8 A resolution. Proc Natl Acad Sci U S A 99:1218–1222
19. Bregman DB, Halaban R, van Gool AJ, Henning KA, Friedberg EC, Warren SL (1996) UV-induced ubiquitination of RNA polymerase II: a novel modification deficient in Cockayne syndrome cells. Proc Natl Acad Sci U S A 93:11586–11590
20. Allmang C, Kufel J, Chanfreau G, Mitchell P, Petfalski E, Tollervey D (1999) Functions of the exosome in rRNA, snoRNA and snRNA synthesis. EMBO J 18:5399–5410

Extraction and Characterization of Taxol: An Anticancer Drug from an Endophytic and Pathogenic Fungi

M. Pandi, P. Rajapriya, and P.T. Manoharan

Abstract

The basic characteristic feature of cancer is the transmissible abnormality of cells that is manifested by reduced control over growth and cell division, leading to serious adverse effects on the host through invasive growth and metastases. Abnormal development of cells leads to the growth of tumor; when the tumor is malignant in nature, it is termed as cancer. Cancer is one of the most common causes of premature death in the world. The most recent estimate of cancer indicates that 8.1 million new cases are diagnosed worldwide each year. Breast cancer is the second most prevalent cancer worldwide and its incidence is gradually increasing. Paclitaxel (taxol) is the most effective antitumor agent developed in the past three decades. Taxol has been hailed by many in the cancer community as a major breakthrough in the treatment of cancer. Taxol was originally isolated from the bark of the Pacific yew *Taxus brevifolia* in 1971. The increased demand for taxol, coupled with its limited availability from the protected Pacific yew, has had researchers scrambling for alternate sources. The endophytic and pathogenic fungi can produce taxol as a cheaper and more widely available product, eventually via industrial fermentation. This chapter deals with the isolation and identification of endophytic and pathogenic fungi and the extraction and characterization of taxol, an anticancer drug from endophytic and pathogenic fungi.

M. Pandi (✉)
Department of Molecular Microbiology,
School of Biotechnology, Madurai Kamaraj University,
Madurai, Tamil Nadu 625 021, India
e-mail: an_pandi@rediffmail.com

P. Rajapriya
Department of Microbiology, Srinivasan College
of Arts and Science, Perambalur,
Tamil Nadu 621 212, India

P.T. Manoharan
Department of Botany, Vivekananda College, Madurai,
Tamil Nadu 625 217, India

V.K. Gupta et al. (eds.), *Laboratory Protocols in Fungal Biology: Current Methods in Fungal Biology*,
Fungal Biology, DOI 10.1007/978-1-4614-2356-0_51, © Springer Science+Business Media, LLC 2013

Keywords

Cancer • Paclitaxel (taxol) • Antitumor • Endophytic fungi • Pathogenic fungi

Introduction

Taxol is a chemotherapic drug specifically effective against prostate, ovarian, breast, and lung cancer. Its primary mechanism of action is related to the ability to stabilize the microtubules and to disrupt their dynamic equilibrium [1–4]. Taxol inhibits cell proliferation by promoting the stabilization of microtubules at the G-M phase of the cell cycle, by which depolymerization of microtubules to soluble tubulin is blocked [5–7]. Taxol was originally isolated from the bark of the Pacific yew, *Taxus brevifolia* [8]. The limited availability of mature yew trees, slow growth rate of cultivated plants, and the low yield of the taxol has resulted in its high cost and also has raised concerns about environmental damage from excessive exploitation of wild trees. This makes taxol a financial burden for many patients. The search for an alternative source of taxol other than the bark of the yew trees (Taxus sp.) has been carried out by scientists all over the world to meet the demand in clinics. The most significant finding in the last decade might be the discovery of endophytic taxol-producing fungus in Gymnosperm, particularly in yew trees. It is remarkable that the taxol produced by the endophytes is identical to that produced by Taxus spp., chemically and biologically [9–16].

The production of taxol by using fungi has given rise to the possibility of reduced cost and wider availability, as taxol may eventually be available via large-scale industrial fermentation. The fungus can serve as a potential material for genetic engineering to improve taxol production. The purpose of this chapter is to focus on the isolation and identification of endophytic and pathogenic fungi and the extraction and characterization of taxol, an anticancer drug from endophytic and pathogenic fungi.

Materials

1. Sterile distilled water.
2. Leaves of selected plants.
3. Ethanol (75% V/V).
4. Sodium hypochloride (2.5% V/V).
5. Potato dextrose agar medium (PDA).
6. Ampicillin (200 µg/mL) and streptomycin (200 µg/mL).
7. Hand lens.
8. Clean glass slides.
9. Petri plates and blotting papers.
10. Stereomicroscope.
11. Lactophenol and DPX mountant.
12. Carl Zeiss Axiostar Plus—Photomicroscope.
13. 0.1% mercuric chloride solution.
14. 5 mg/g chloramphenicol.
15. MID medium supplemented with 1 g soy tone/L.
16. 0.25 g of Na_2Co_3 (0.025% W/V).
17. Dichloromethane.
18. Rotary evaporator.
19. 1% vanillin sulfuric acid (w/v).
20. 100% methanol.
21. Beckman DU-40 UV-Spectrophotometer.
22. IR grade potassium bromide (KBr).
23. Methanol/acetonitrile/water (25:35:40, by vol.).

Methods

Isolation and Identification of Endophytic Fungi

The leaves of selected plants were collected from different places. The plant materials were subjected to endophytic isolation within 3 h after harvest. The endophytic fungal cultures were separated from the healthy leaves according to the general mycological procedure [11–14, 16].

1. The leaves were washed with running tap water, sterilized with ethanol (75% V/V) for 1 min and sodium hypochloride (2.5% V/V) for 5 min, then rinsed in sterile water for three times and cut into 1 cm long segments.
2. Plant segments were then transferred to potato dextrose agar containing Petri plates amended with ampicillin (200 μg/mL) and streptomycin (200 μg/mL) to inhibit bacterial growth.
3. After 2 days of incubation, mycelia of fungi were observed growing from the inner leaf segments in the plates.
4. Individual hyphal tips of the various fungi were carefully removed from the agar plates with the help of an inoculation loop; then they were placed on new PDA medium and incubated at 25 °C for at least 7–10 days.
5. Each fungal culture was checked for purity and subcultured to another agar plate by the hyphal tip method.
6. Fungal identification was based on the morphology of the fungal culture, the mechanism of spore production, and the characteristics of the spores [17].

Isolation of Plant Pathogenic Fungi

1. The pathogenic infected leaves of various plants with symptoms showing Coelomycetes fungi were collected from different places in Tamilnadu, India.
2. Coelomycetes fungi are found on dead twigs, leaf litter, bark, and infected leaves.
3. The fungal infection was confirmed by the presence of conidiomata on the substrate using a hand lens in the field.
4. Then the specimen is collected. The data should be recorded on the envelope used to transport the material to the laboratory. It is essential that the identity of the substrate should be accurately known for the identification of many species is still based on a host basis.
5. Unless the material is examined immediately it should be dried thoroughly to prevent the growth of molds and unwanted saprophytes.

Examination of the Specimens

The following methods were used to study Coelomycetes fungi on the specimen.

Direct Examination of the Specimen

Once the selected conidioma has been removed from the substrate, it is transferred to a clean glass slide and mounted with water, for microscopic examination.

Moist Chamber Incubation Method

1. This technique was used to induce sporulation.
2. The specimen was incubated in a 15-cm diameter sterilized Petri plate lined with moist blotting paper.
3. The plates were kept moist by adding sterile distilled water periodically, but the blotting paper was never flooded with water.
4. The specimens were examined after a week under a stereomicroscope and the conidiomata on them were studied.
5. The fungi found in sporulating conditions were isolated, examined, and identified down to species level [17].

Hand Section

1. Cutting vertical section of conidiomata is essential to confirm the nature of the fructification.
2. Specimens with fruit bodies were sectioned with the help of razor blade.
3. Good sections were selected through stereomicroscope; then they were mounted in lactophenol and viewed under a light microscope.

Preparation of Permanent Slides

For the preparation of permanent slides, water and lactophenol were used. The slides were sealed by DPX mountant.

Illustration and Photomicrographs

Photomicrographs of conidia were taken with the help of Carl Zeiss Axiostar Plus—Photomicroscope (Phase contrast) with Nikon FM 10 camera and Nikon HF X Labophot (bright field) with Nicon Fx—35A by using Konica films.

Isolation of Single Pathogenic Fungi

1. The infected plant parts of approximately 4 mm² were sterilized by immersing for 1 min in 0.1% mercuric chloride solution and washed by successive transfer through sterile distilled water three times.
2. These were then incubated in sterile distilled water containing 5 mg/g chloramphenicol.
3. A few drops of spore suspensions were spread over Potato dextrose agar plates and incubated at 25 °C for 6–48 h.
4. Germinated spores were transferred to oatmeal agar or Potato dextrose agar plates. Subcultures were also made from spores extruding from pycnidia produced in culture.
5. Individual hyphal tips of the various fungi were removed from the agar plates, placed on new PDA medium and incubated at 25 °C for at least 2 weeks.
6. Each fungal culture was checked for purity and sub cultured to another agar plate by the hyphal tip method.
7. Fungal identification methods were based on the morphology of the fungal culture, the mechanism of spore production, and the characteristics of the spores [17].

Cultivation and Extraction of Taxol from Selected Fungal Isolates

1. The selected endophytic or pathogenic isolates were inoculated into a 2,000-mL Hopkins flask of MID medium supplemented with 1 g soy tone/L, incubated for 12 h under light and dark cycle at temperature between 22 and 25 °C for 21 days.
2. After 21 days, the cultures were passed through four layers of cheesecloth and 0.25 g of Na_2CO_3 (0.025% W/V) was added to the culture filtrate to avoid fatty acid contamination.
3. The culture filtrate was further extracted with twice the volume of dichloromethane and the organic phase was taken to dryness under reduced pressure at 50 °C using a rotary evaporator.
4. Then the dry solid residue was re-dissolved in methanol for the subsequent separation.
5. The presence of taxol was confirmed in the crude extracts, which were analyzed using different chromatographic and spectroscopic methods.

Thin-Layer Chromatography

1. TLC analysis was carried out on Merck 1-mm (20 × 20 cm) silica gel precoated plates.
2. The plates were developed by the solvent system reported by Strobel et al. [18, 19].
3. The taxol was detected with 1% vanillin sulfuric acid (w/v) and heating. It appears as a bluish spot that faded to dark grey after 24 h.
4. Then the area of the plate containing putative taxol was carefully removed by scraping off the silica at the appropriate R_f value and eluted with methanol. Then they were further analyzed by UV, IR, and HPLC to confirm the production of taxol.

Ultraviolet Spectroscopic Analysis

1. After chromatography, the area of the TLC plate containing putative taxol was carefully removed by scrapping off the silica at the appropriate R_f and exhaustively eluting it with methanol.
2. The purified sample of taxol was dissolved in 100% methanol and analyzed by Beckman DU-40 UV-Spectrophotometer and compared with authentic taxol.

Infrared Spectroscopic Analysis

1. The IR spectra of the compound were recorded on Shimadzu FTIR 8000 series instrument.
2. The purified taxol was ground with IR-grade potassium bromide (KBr) (1:10) pressed into discs under vacuum using spectra lab Pelletiser and compared with authentic Taxol.
3. The IR spectrum was recorded in the region between 4,000 and 5,000 cm.

High-Performance Liquid Chromatography Analysis

1. To confirm the presence of taxol, the fungal extract was subjected to high performance liquid chromatography (HPLC).
2. Taxol was analyzed by HPLC (Shimatzu 9A model) using a reverse phase C_{18} column with a UV detector.
3. Twenty microliter of the sample were injected each time and detected at 232 nm. The mobile phase was methanol/acetonitrile/water (25:35:40, by vol.) at a flow rate of 1.0 mL/min.
4. The sample and the mobile phase were filtered through 0.2 μm PVDF filter before injecting into the column.
5. Fungal taxol was confirmed by comparing the peak area of the samples with authentic taxol.

Acknowledgement I thank the University Grants Commission, New Delhi, India, for the financial support of the Research Grant.

References

1. Horwitz SB (1992) Mechanism of action of taxol. Trends Pharmacol Sci 13:131–6
2. Rao S, Orr GA, Chaudhary AG, Kingston DGY, Horwitz SB (1995) Characterization of the taxol binding site on the microtubule, 2-(m-Azidobenzoyl) Taxol photolabels a peptide (amino acids 217–231) of beta-tubulin. J Biol Chem 270:20235–8
3. Jordan MA, Wilson L (1998) Microtubules and actin filaments: dynamic targets for cancer chemotherapy. Curr Opin Cell Biol 10:123–30
4. Caplow M, Shanks J, Ruhlen R (1994) How taxol modulates microtubule disassembly. J Biol Chem 38:23399–402
5. Horwitz SB (1994) Taxol (paclitaxel): mechanisms of action. Ann Oncol 6:3–6
6. Nicolaou KC, Yang Z, Liu JJ, Ueno H, Nantermet PG, Guy RK (1994) Total synthesis of taxol. Nature 367:630–634
7. Jennewein S, Croteau R (2001) Taxol: biosynthesis, molecular genetics, and biotechnological applications. Appl Microbiol Biotechnol 57:13–19
8. Wani MC, Taylor HL, Wall ME, Coggon P, McPhail AT (1971) Plant antitumor agents VI. The isolation and structure of taxol, a novel antileukemic and antitumor agent from *Taxus brevifolia*. J Am Chem Soc 93:2325–2327
9. Stierle A, Strobel G, Stierle D (1993) Taxol and taxane production by *Taxomyces andreamae*, an endophytic fungus of Pacific yew. Science 260:214–6
10. Wang J, Li G, Lu H, Zheng Z, Huang Y, Su W (2000) Taxol from *Tubercularia* sp. strain TF5, an endophytic fungus of *Taxus mairei*. FEMS Microbiol Lett 193:249–53
11. Gangadevi V, Muthumary J (2007) Taxol, an anticancer drug produced by an endo-phytic fungus *Bartalinia robillardoides* Tassi, isolated from a medicinal plant *Aegle marmelos* Correa ex Roxb. World J Microbiol Biotechnol 24:717–24
12. Gangadevi V, Muthumary J (2009) Taxol production by *Pestalotiopsis terminaliae*, an endophytic fungus of *Terminalia arjuna*. Biotechnol Appl Biochem 52:9–15
13. Gangadevi V, Muthumary J (2008) Isolation of *Colletotrichum gleosporioides*, a novel endophytic Taxol-producing fungus from the leaves of a medicinal plant *Justicia gendarussa*. Mycol Balcanica 5:1–4
14. Senthil Kumaran R, Hur BK, Muthumary J (2008) Production of taxol from Phyllosticta spinarum, an endophytic fungus of Cupressus sp. Eng Life Sci 8(4):1–10
15. SenthilKumaran R, Muthumary J, Hur B-K (2008) Taxol from *Phyllosticta citricarpa*, a leaf spot fungus of the angiosperm *Citrus medica*. J Biosci Bioeng 106(1):103–6
16. Pandi M, Senthil KR, Choi Y-K, Kim HJ, Muthumary JP (2011) Isolation and detection of Taxol, an anticancer drug produced from *Lasiodiplodia theobromae*, an endophytic fungus of the medicinal plant *Morinda citrifolia*. Afr J Biotechnol 10(8):1428–1435
17. Sutton BC (1980) The coelomycetes. Fungi imperfecti with pycnidia, acervuli and stromata. Commonwealth Mycological Institute, Kew, Surrey
18. Strobel G, Yang X, Sears J, Kramer R, Sidhu RS, Hess WM (1996) Taxol from *Pestalotiopsis microspora*, an endophytic fungus of *Taxus wallichiana*. Microbiology 142:435–440
19. Strobel GA, Hess WM, Ford E, Sidhu RS, Yang X (1996) Taxol from fungal endophytes and the issue of biodiversity. J Ind Microbiol 17:417–423

Identification of Mycotoxigenic Fungi Using an Oligonucleotide Microarray

Eugenia Barros

Abstract

Mycotoxins are secondary metabolites produced by fungi; they can play a role as food contaminants and have the ability to negatively influence human and animal health. To improve food safety and to protect consumers from harmful contaminants, numerous detection tools have been developed for the detection and analysis of various mycotoxigenic fungi. These include PCR-based assays and microarrays targeting different areas of the fungal genome depending on its application. This chapter describes the development of an oligonucleotide microarray specific for eleven mycotoxigenic fungi isolated from different food commodities in South Africa. This array is suitable for the detection and identification of cultures of potential mycotoxigenic fungi in both laboratory samples and commodity-derived food samples.

Keywords

Mycotoxins • Fungi • Oligonucleotide microarray • Food contaminants

Introduction

Fungi can grow on many food commodities but when these fungi are able to produce health threatening substances, such as mycotoxins, they then become a threat to human and animal health. Although the presence of a mycotoxigenic fungus in a food commodity does not necessarily indicate the production of the respective mycotoxin it is however necessary to control the presence of fungal contaminants in the food production chain and develop quicker methods to identify food-borne fungi.

Many molecular biology techniques have been developed as alternative approaches to detect and quantify fungal growth so as to prevent contaminated commodities from entering the food chain. Most of the developed molecular techniques used restriction enzymes and the polymerase chain reaction (PCR) to assess the genetic variability within and among fungal species taking advantage of the polymorphisms that occur naturally in the DNA of a given species. Successful assays

E. Barros (✉)
Department of Biosciences, Council for Scientific and Industrial Research (CSIR), Meiring Naude Road, Brummeria, Pretoria 0001, South Africa
e-mail: ebarros@csir.co.za

V.K. Gupta et al. (eds.), *Laboratory Protocols in Fungal Biology: Current Methods in Fungal Biology*, Fungal Biology, DOI 10.1007/978-1-4614-2356-0_52, © Springer Science+Business Media, LLC 2013

have utilized the highly conserved ribsosomal RNA gene sequences to design species-specific PCR primers [1] and gene-specific PCR detection assays for genes involved in the biosynthesis of mycotoxins [2, 3].

The microarray technology provides a tool to potentially identify and quantify levels of gene expression for all genes in an organism. Small spots of DNA are fixed to a matrix; this can be either glass or a nylon membrane. Microarrays can be constructed using cDNAs, genomic sequences or oligonucleotides synthesized in silico. Typically, cDNAs used to construct microarrays are partial gene sequences that are derived from coding DNA and generally have a high degree of sequence conservation. Genomic sequences include sequences that target specific genes of interest like toxin producing genes in the case of fungal biology and, oligonucleotides include those synthesized from polymorphic sequences that occur naturally in the DNA of a given species.

In fungal biology microarrays can be used for a variety of applications. Some applications include identification of fungal species; identification of potential mycotoxigenic fungi; to study gene expression and thus identify differentially expressed genes; and identification of toxin genes. However the design of the microarray is dependent on its application and the most common arrays are cDNA microarrays [4] and oligonucleotide microarrays [5]. There are different types of oligonucleotide microarrays and they include oligonucleotides synthesized from (1) polymorphic fragments identified by molecular marker techniques, like for example, amplified fragment length polymorphisms (APFLs); (2) conserved regions like the internal transcribed spacer (ITS) regions and (3) gene sequences like fungal toxin genes, among others.

The present contribution focuses on the generation of probes that showed polymorphisms within the internal transcribed spacer (ITS) regions of rRNA of eleven mycotoxigenic fungi and the development of an oligonucleotide microarray that can detect and identify these fungi from different food commodities. The eleven mycotoxin producing fungi are amongst the most prevalent in South African food commodities [5].

Furthermore, the technique used to label the target DNA was random labelling and it does not involve amplification of target DNA prior to microarray hybridization. This technique avoids amplification bias, diminishes secondary structures and ensures a more efficient target [5, 6].

Materials

Fungal Strains

Isolates of eleven food-borne fungi known to produce mycotoxins were obtained from the Agricultural Research Council (ARC) culture collection in Pretoria, South Africa. These included *Aspergillus carbonarius*, *Aspergillus clavatus*, *Aspergillus niger*, *Alternaria alternata*, *Eurotium amstelodami*, *Penicillium corylophilum*, *Penicillium expansum*, *Penicillium fellutanum*, *Penicillium islandicum*, *Penicillium italicum*, and *Stenocarpella maydis*.

Culture Media

Fungal strains were grown on Malt Salt Agar (MSA) at 25 °C for 1–2 weeks (see Note 1). MSA was prepared by dissolving 90 g NaCl in 360 mL dH$_2$O in a Schott bottle (solution 1); 24 g of malt extract (Merck, South Africa) and 24 g of agar (Merck, South Africa) were added to 840 mL of dH$_2$O in a Schott bottle (solution 2). The solutions were autoclaved for 20 min at 121 °C. Solution 1 was then added aseptically to solution 2 and the medium poured into 90-mm petri dishes and allowed to settle.

DNA Extraction Buffer

Genomic DNA was extracted from the eleven fungal cultures using the method of Raeder and Broda [7]. The DNA extraction buffer contained 200 mM Tris–HCl (pH 8), 150 mM NaCl, 25 mM EDTA (pH 8), 0.5% SDS and 1% PVP. Just before use, 0.2% (v/v) 2-Mercapto-ethanol was added to the extraction buffer.

Methods

Fungal DNA Extraction

1. Fungal mycelium was gently scrapped off the cultures grown on MSA media and placed into 1.5-mL eppendorf tubes together with metal yellow tungsten carbide beads (3 mm) and in 500 μL DNA extraction buffer, as described in the section DNA Extraction Buffer, and according to the method of Raeder and Broda [7].

2. The tubes were span in a FastPrep® machine (MP Biomedicals, Cambridge, UK) at speed 4 for 5 s (see Note 2).

3. The metal beads were removed and the contents transferred to a 1.5-mL eppendorf tube to which 50 μL 1 M Tris–HCl (pH 8.0), 100 μL phenol and 170 μL chloroform were added and the tubes placed on ice for 5 min.

4. The suspension was then centrifuged at $20,817 \times g$ for 15 min and the aqueous phase transferred to a new eppendorf tube.

5. The aqueous phase was cleaned once again with 50 μL 1 M Tris–HCl (pH 8.0), 100 μL phenol and 170 μL chloroform as described in the previous step and the cleaning procedure repeated until the interface was clean.

6. Chloroform, 1× volume, was added to the clean aqueous phase and the samples centrifuged at $10,621 \times g$ for 15 min.

7. The aqueous phase was transferred to a new eppendorf tube to which 2.5× volume of ethanol and 0.5 M ammonium acetate were added and the solution allowed to precipitate overnight at −20 °C.

8. The suspension was centrifuged at $10,621 \times g$ for 10 min, the supernatant discarded and the pellet washed with 2.5× volume of 70% ethanol followed by centrifugation at $10,621 \times g$ for 10 min.

9. The supernatant was discarded, the pellet allowed to air dry and then resuspended with 50 μL of ddH$_2$O.

10. The DNA concentration and purity were assessed from the absorbance measurements with the Nanodrop 1000 instrument.

PCR Amplification of Fungal DNA with ITS1 and ITS4 Universal Primers

The ITS regions of the eleven fungi were amplified using universal fungal primers for ITS1 (5′-TCCGTAGGTGAACCTGCGG-3′) and for ITS4 (5′-TCCTCCGCTTATTGATATGC-3′) according to White et al. [8]. The PCR amplifications were carried out using the following reaction mixture in a total volume of 25 μL: 8 ng fungal template DNA, 1.5 mM MgCl$_2$, 0.2 mM of each dNTP, 0.5 U Taq polymerase (Bioline), 1× PCR reaction buffer (Bioline) and 0.4 μM of ITS1 primer and 0.4 μM of ITS4 primer. The PCR amplification consisted of an initial denaturation step of 94 °C for 5 min; followed by 35 cycles of denaturation at 94 °C for 30 s, primer annealing at 50 °C for 45 s and primer extension at 72 °C for 1 min; and a final extension at 72 °C for 5 min.

Amplicon Sequencing, Identification of Polymorphisms and Probe Design

Aliquots of the PCR products were separated on an agarose gel (1.4%) for quality control. The remainder of each PCR product was precipitated in a NaAc/EtOH solution made up of 90% ethanol and 0.9 mM NaAc (pH 5.2). The precipitate was collected by centrifugation at $3,600 \times g$ for 30 min. The pellets were washed in 70% ethanol, dried, and then resuspended in 50 μL dionized H$_2$O. Aliquots of the resuspended amplicons were sequenced by Inqaba Biotec (Pretoria, South Africa).

The sequenced fragments were aligned using ClustlX software as described by Thompson et al. [9] and polymorphisms were identified. These sequences were then used to design genus-specific and species-specific probes of various lengths, ranging from 14 to 25 bases, and within a narrow range of melting temperature, 56 ± 5 °C. The oligonucleotide probes were designed using the Primer Designer 4 Package, Version 4.2 (Scientific and Educational Software, Cary, NC). The probe sequences generated for each of the eleven fungi as well as the specific annealing temperatures are shown in Table 52.1.

Table 52.1 Fungal isolates, potential mycotoxins produced, and probe sequences generated to construct the oligonucleotide microarray

Probe name and reference	Probe sequence (5'→3')[a]	Fungal isolates	Mycotoxins produced	Annealing temperature (°C) for PCR amplification
AR1 [5]	ATCTGCTGCACAGTTGGCT	Aspergillus carbonarius	Ochratoxin A	56
ACIF [5]	ATTCGGAAACCUGCTCAGTACG	Aspergillus clavatus	Cytochalacin E; Patulin	58
ANIG [5]	ACGTTATCCAACCAT	Aspergillus niger	Ochratoxin A	55
Aaf AaR [5]	GACCGC7TTCG7GGTATGCA	Alternaria alternata	Tenuazonic acid	56
EurAF EurAR [5]	TGGCGGCACCATGTC TGGTTAAAAGATTGGTTGCGA	Eurotium amstelodami	Sterigmatocystin	58
PenCorF PenCorR [5]	GTCCAAACCCTCCCACCCA GTCAGACTTGCAATCTTCAGACTGT	Penicillium corylophilum	Cyclopiazonic acid	55
PenExF PenExR [5]	TTACCGAGTGAGGCCGT GCCAGCC7GACAGCTACG	Penicillium expansum	Patulin	58
PenFeF PenFeR [5]	CTGAGTGCGGGCCTCT CGCCGAAGCAACACTGTAAG	Penicillium fellutanum	Patulin	55
PenIsF PenIsR [5]	CGAGTGCGGGTTCGACA GGCAACGCGGTAACGGTAG	Penicillium islandicum	5,6, dihydro-4-meth-oxy-2 H-pyran-2-one	57
PenItF PenItR [5]	CTCCCACCCGTGTTTATTTATCA TCACTCAGACGACAATCTTCAGG	Penicillium italicum	Patulin	57
4 F 4R [11]	CAAACGTCGGGTCAGAAGAAGCGAC AGGAACCGTCCCCGCGACGTTTG	Stenocarpella maydis	Diplosporin	57

[a]Locked nucleic acids (LNAs) that were used to increase the specificity of a probe are in bold and italic

Ensuring Uniqueness and Specificity of Probes

The specificity of each oligonucleotide probe was further assessed by subjecting the sequence/s to similarity searches in public databases; the databases used included NCBI (http://www.ncbi.nlm.nih.gov) and EMBL (www.ebi.ac.uk/embl). BLAST searches were used and only unique oligonucleotide probes were selected for printing on the array. In cases where probes had similar sequences their specificity was enhanced by substituting an oligonucleotide with a high affinity DNA analogue known as locked nucleic acid (LNA) according to the method of Johnson, Haupt, and Griffiths [10]. This technique was also used in the design of some of the probes to ensure that the set of oligonucleotide probes that were printed on the array had similar hybridization efficiencies. This approach was used for the probes specific for *A. alternata* and *P. expansum* and the LAN is indicated in bold and italic in Table 52.1. All the probes, including the designed probes, were synthesized by Inqaba Biotech (Pretoria, South Africa).

Construction of the Array

The 18 uniquely designed species-specific oligonucleotide probes were used to construct the array together with three control probes consisting of ITS1, ITS3 and ITS4 fragments. Equal volumes (10 μL each) of 100 pmol/mL oligonucleotide and 100% DMSO were transferred into a 384-well plate (Amersham Pharmacia Biotech). Sixteen replicates of each of each oligonucleotide were arrayed onto Vapour Phase-coated Glass Slides (Amersham Pharmacia Biotech) using an Array Spotter Generation III (Molecular Dynamics, Sunnyvale, CA, USA) at the Microarray Facility of the African Centre for Gene Technologies (ACGT), University of Pretoria, South Africa (http://fabinet.up.ac.za/microarray). Following printing the slides were allowed to dry overnight at 45–50% relative humidity. Spotted DNA was bound to the slides by UV cross-linking at 250 mJ and then baked for 2 h at 80 °C. The control ITS fragments were spotted at concentrations of 50 ng/μL, 100 ng/μL, 150 ng/μL, and 200 ng/μL.

Preparation of Labelled Target DNA

DNA was extracted from the eleven fungal cultures following the DNA extraction protocol described in the section Fungal DNA Extraction. Two micrograms of DNA were labelled with red-fluorescent dye Cyanine 5 (Cy5) using the Cy™ Dye Postlabelling Reactive Dye Pack (GE Healthcare, UK). For each labelling reaction the DNA was diluted in 5 μL 0.2 M Na_2CO_3 (pH 9) and 2.5 μL Cy5 mono NHS ester 4,000 pmol dye resuspended in 12 μL DMSO. The reactions were incubated in the dark, at room temperature, for 90 min. After labelling the dye-coupling reaction was column-purified using the QIAquick PCR purification Kit (QIAGEN Germany).

Preparation of Labelled Control Probes

Each of the control probes ITS1, ITS3, and ITS4 were labelled by incorporation of green-fluorescent dye Cyanine 3 (Cy3-dUTP) (Amersham Biosciences, Buckinghamshire, UK) using the Klenow fragment DNA polymerase I (Roche Diagnostics). Each labelling reaction contained 5 μg of probe DNA, 1.8 mM dNTP mix (0.3 mM dATP, 0.3 mM dGTP, 0.3 mM dCTP and 0.8 mM dTTP), 0.1 mM Cy3-dUTP, 1× hexanucleotide mix (Roche Diagnostics) and 8 U Klenow enzyme (Roche Diagnostics). The reaction was incubated at 37 °C overnight. After labelling the dye-coupling reaction was column-purified using the QIAquick PCR purification Kit (QIAGEN Germany).

Hybridization of Array Slides

The Cy5-labelled target DNA and the Cy3-labelled control probes (0.3 pmol) were resuspended in 40 μL of hybridization mixture containing 50% formamide (Sigma-Aldrich),

25% 2× hybridization buffer (Amersham Pharmacia Biotech) and 25% dionized H_2O. The mixture was denatured at 95 °C for 5 min and stored on ice for hybridization. The hybridization mixture was then pipetted onto a glass slide (24×60 mm, No. 1, Marienfeld, Germany), covered with a cover slip and inserted into a custom-made hybridization chamber (N.B. Engineering Works, Pretoria, South Africa) and allowed to hybridize overnight at 53 °C. The slides were then washed twice in 2× SSC and 0.2% SDS at 37 °C for 6 min, once in 0.2× SSC and 0.2% SDS at room temperature for 5 min and then twice in 0.075× SSC at room temperature for 5 min. The slides were rinsed in de-ionized H_2O for 2 s and dried by centrifugation at $1,000 \times g$ for 5 min.

Scanning and Data Processing

The oligonucleotide arrays were scanned using a GenePix 4000B Scanner (Molecular Dynamics, USA) and the mean pixel intensity within each spot and the local background were determined using Array Vision, version 6.0 software (Imaging Research Inc., Molecular Dynamics, USA). All signal intensities were background corrected by subtracting the local background from the raw spot intensity value. The spot intensity data generated by the control samples were used as a reference for normalization of all spot intensity data. Irregular spots were manually flagged for removal and further data analysis was performed in the Microsoft Excel software (Microsoft, Richmond, WA).

Protocols describing the downstream data analysis confirming the reproducibility of the array, data processing and statistical analysis fall outside the scope of this chapter.

Notes

1. Growth of different fungal species on solid media is variable. However, 2 weeks incubation at 25 °C was found to be the optimum period to obtain sufficient fungal mycelium

for DNA extractions, but it can be extended if necessary.
2. If a FastPrep® machine is not available, mix thoroughly the tubes containing the mycelium, the metal beads, and the DNA extraction buffer and incubate for 15 min and at 60 °C in an ultrasonic bath (e.g., S10H from Elma, Singen, Germany).

References

1. Mishra PK, Fox RTV, Culham A (2003) Development of a PCR-based assay for rapid and reliable identification of pathogenic *Fusaria*. FEMS Microbiol Lett 218:329–332
2. Paterson RRM, Archer S, Kozakiewicz Z, Lea A, Locke T, O'Grady E (2000) A gene probe for the patulin metabolic pathway with potential for use in patulin and novel disease control. Biocontrol Sci Technol 10:509–512
3. Waalwijk C, van der Lee T, de Vries I, Hesselink T, Arts J, Kema GHJ (2004) Synteny in toxigenic *Fusarium* species: the fumonisin gene cluster and the mating type region as examples. Eur J Plant Pathol 110:533–544
4. Barros E, van Staden C, Lezar S (2009) A microarray-based method for the parallel analysis of genotypes and expression profiles of wood-forming tissues in *Eucalyptus grandis*. BMC Biotechnol 9:51
5. Lezar S, Barros E (2010) Oligonucleotide microarray for the identification of potential mycotoxigenic fungi. BMC Microbiol 10:87
6. Lane S, Everman J, Logea F, Call DR (2004) Amplicon structure prevents target hybridization to oligonucleotide microarrays. Biosensors Bioelectron 20:728–735
7. Raeder U, Broda P (1985) Rapid preparation of DNA from filamentous fungi. Lett Appl Microbiol 1:17–20
8. White TJ, Bruns T, Lee S, Taylor J (1990) Amplification and direct sequencing of fungal ribosomal RNA genes for phylogenetics. In: Gelfand DH, Sninsky JJ, White TJ, Innis MA (eds) PCR protocols. a guide to methods and applications. Academic, San Diego, CA, pp 315–322
9. Thompson JD, Gibson TJ, Plewniak F, Jeanmougin F, Higgins DG (1997) The CLUSTL_X windows interface: flexible strategies for multiple sequence alignment aided by quality analysis tool. Nucleic Acids Res 25:4876–4882
10. Johnson MP, Haupt LM, Griffiths LR (2004) Locked nucleic acids (LNA) single nucleotide polymorphism (SNP) genotype analysis and validation using real-time *PCR*. Nucleic Res 32:e55
11. Barros E, Crampton M, Marais G, Lezar S (2008) A DNA-based method to quantify *Stenocarpella maydis* in maize. Maydica 53:125–129

DNA Microarray-Based Detection and Identification of Fungal Specimens

Minna Mäki

Abstract

Novel DNA-based molecular methods can be used to detect the fungal species faster than with the conventional methods. Combination of PCR and microarray provides rapid, sensitive, and reliable detection of pathogenic fungi. The advantages over other DNA-based methods are that microarray technologies allow broader coverage of detectable targets and simultaneous detection of multiple targets in a single assay. Furthermore, microarray technologies have the potential to discriminate between closely related fungal species. Although the use of microarray technologies in clinical diagnostics is still rare, the microarray-based approaches are believed to have great clinical potential in the field of infectious diseases.

Keywords

DNA • Polymerase chain reaction • Microarray • Pathogen • Fungi • Identification

Introduction

Conventional microbiological diagnostics of a fungal infection mainly rely on microscopic and cultural techniques that are time-consuming, labor-intensive, and require expertise. These methods usually yield diagnostic results in days or in some cases up to weeks after sampling. Furthermore, cultivation of fungi is not always successful under laboratory conditions. Such failures may occur due to unsuitable culturing media and conditions for the fungal species in question. Molecular methods based on detection of nucleic acid (NA) from clinical samples aim to circumvent these problems. In addition, they aim to improve the diagnosis of fungal infections by shortening time to result and increasing sensitivity and accuracy. Polymerase chain reaction (PCR)-based assays can amplify and detect minute quantities of DNA isolated from a pathogenic fungus in few hours, having a limit of detection of only a few genome copies per reaction. Although the multiplex PCR is slowly gaining ground in fungal diagnostics, most of the tests are still amplifying only one or few fungal targets in

M. Mäki(✉)
Program Leader, NAT, Orion Diagnostica Oy,
P.O. Box 83, FI-02101 Espoo, Finland
e-mail: Minna.Maki@mobidiag.com;
Minna.Maki@welho.com

a single reaction [1, 2]. Multiplex or broad-range PCR in combination with microarray allows rapid detection of microbial DNA and species identification of multiple microbial targets in a single assay [3–7]. The simple, array-type technologies with broad target coverage have especially been believed to have great clinical potential in the field of infectious diseases [8–12]. Rapid clinical diagnostics reduces the use of antimicrobials in addition to allowing a faster switch to the most optimum treatment, which improves patient outcome and reduces both side-effects and treatment costs [13–17].

Breakspear and Momany have reviewed the use of fungal microarray in research settings, including studies of fungal metabolisms, development, pathogenesis, symbiosis, and industrial fermentations [18]. Recently, several publications have also demonstrated the applicability of fungal microarray in clinical diagnostic purposes [7, 19–22]. These publications described the use of a multiplex/broad-range PCR with oligonucleotide probe array, targeting highly conserved and variable species-specific regions of the internal transcribed spacers (ITS) of rRNA gene complex of clinically relevant fungal pathogens. More often the ITS regions have been chosen as targets in fungal microarrays due to their presence in numerous copies in the fungal genome, which enables highly sensitive amplification by PCR. The high level of sequence variability of the ITS regions also allows reliable differentiation of closely related fungal taxa and species. Moreover, the comprehensive rRNA gene complex database is rapidly expanding and, thus, supporting the in silico design of primers and oligonucleotide probes.

Three commercial PCR and microarray-based products for fungal diagnostics stand out from the rest: CLART® SeptiBac+ (Genomica, Madrid, Spain), MycArray™ (Myconostica, Manchester, UK), and Prove-it™ Sepsis, v2.0 (Mobidiag, Helsinki, Finland). All of these assays use similar methodologies for detection of pathogenic fungal species from clinical samples. The assays involve the use of PCR as an amplification method prior to microarray phase, where the actual identification of fungal species occurs. The ArrayTube™ or an analog microarray is used as a platform for the oligonucleotide probes. The ArrayTube™ has been demonstrated to detect and identify viral and bacterial pathogens or bacterial pathotypes with a high degree of sensitivity [23–27] and to be capable of detecting antimicrobial resistance genes [28–30] from an isolated DNA sample. Also, Monecke et al. have published a case report of peritonitis where the ArrayTube™ harboring the fungal content was used to detect the causative agent, *Rhizopus microspores* [31].

In contrast to the previously mentioned fungal microarray publications, in these three commercial platforms, the principle behind the visualization of a positive hybridization on the microarray is based on a colorimetric reaction instead of fluorescent-based methods. In the workflow, biotin labeled amplicons are first hybridized with the specific oligonucleotide probes pre-printed on the microarray surface and then streptavidin-horseradish peroxidase (HRP) conjugate is attached to the biotin label. Finally, the presence of the HRP is visualized in the precipitation reaction by which HRP catalyzes the conversion of 3,3′,5,5′-Tetramethyl Benzidine (TMB) substrate or an analog into a precipitate thus forming a colored spot on the specific microarray position. An image is then captured from the microarray by dedicated reader device. The image is analyzed, and the result of the analysis, typically the name of the causative agent and signal intensities of each oligonucleotide probes, are reported by the software.

The fungal panels of the CLART SeptiBac+, MycArray, and Prove-it Sepsis assays vary, but all of them target the clinically relevant *Candida* species, that is, *C. albicans*, *C. krusei*, and *C. glabrata*. The assays aim at identification of fungal species from the positive blood culture used in sepsis diagnostics. Sepsis necessitates rapid and accurate diagnostics to improve the chances of a positive outcome for the patient. The fungal sepsis is associated with significant mortality and morbidity, especially when *Candida* spp. is the causative agent. Fluconazole is the choice for first-line therapy in candidemia; therefore, rapid differentiation between fluconazole-sensitive and

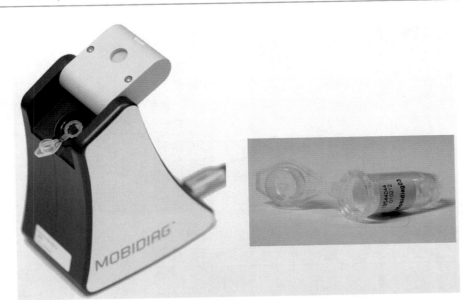

Fig. 53.1 The images of the Prove-it™ Sepsis TubeArray reader and Prove-it™ TubeArray, which is a plastic microreaction tube containing a microarray at the bottom

potentially fluconazole-resistant *Candida* species is of the essence. Recent studies have shown that appropriate and early antifungal therapy (treatment started within the 48 h after the onset of candidemia) is a major factor associated with a good prognosis in fungal infection [16, 17].

The performance of Prove-it Sepsis assay in routine clinical settings for sepsis diagnostics has been recently published [32]. In the multicenter study, the definitive identification of bacterial species with the Prove-it microarray platform and the corresponding assay protocol was considered highly sensitive (95%) and specific (99%). It was concluded that the assay was faster than the gold-standard culture-based methods and it could thus enable earlier evidence-based management for clinical sepsis. Furthermore, it was also stated that the microarray platform's robust nature, ease of implementation, software-controlled decision support for results, and portability has potential for successful strategic implementation in low resource settings (Fig. 53.1). The current generation of the Prove-it

Sepsis v2.0 assay consists of a pathogen panel that covers the majority of sepsis-causing pathogens, including over 60 g-negative and gram-positive bacterial species, the methicillin resistance marker together with 13 fungal species. The fungal detection is realized by broad-range PCR primers that originate from the conserved regions of ITS together with specific oligonucleotide probes located at hyper-variable regions flanked by the primers. Each probe on the array matches either a particular pathogen species or higher-level taxon. The turnaround time of the assay is three hours, excluding DNA extraction. The fungal pathogen panel of the assay covered the following clinically relevant species: *C. albicans, C. glabrata, C. parapsilosis, C. tropicalis, C. guilliermondii, C. lusitaniae, C. dubliniensis,* and *C. krusei* and pan-yeast identification covering *C. pelliculosa, C. kefyr, C. norvegensis, C. haemulonii,* and *Saccharomyces cerevisiae*. The protocol below is based on the procedure of Prove-it Sepsis StripArray (Figs. 53.2 and 53.3).

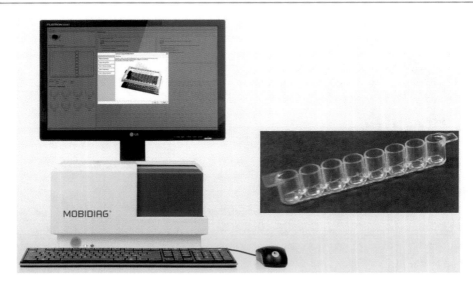

Fig. 53.2 The images of the Prove-it™ Sepsis StripArray system and Prove-it™ StripArray, which consists of eight plastic microreaction vials containing a microarray at the bottom of each well

Results	Candida glabrata, Listeria monocytogenes, Psuedomonas aeruginosa
Sample ID	-
Controls	Pass
Date and time	2010-10-19 17:27
Operator / Performer	
Assay ID	2649 / 834 / 70310
Comments	P.aer / L.mon / C.glab
Product	StripArray_Proveit_Bacteria_Fungi
Software version	1.1.0.0 (StripArray System)

Fig. 53.3 The image of the Prove-it™ Advisor result. The top section of the result view presents the end result of the assay including icons for the identified target(s) and the analyzed microarray image. The results view has the following tabs: Summary, Details, Graphs, Images, and Panel, from which the details can be viewed. Also, the detailed result of bacterial content and bacterial controls (pass/fail), the detailed result of fungal content and fungal controls (pass/fail), and other assay information (from Sample ID to Software version) that is common for all contents are shown

Materials

1. Prove-it Sepsis v2.0 kit.
2. Prove-it StripArray System.
3. Distilled water.
4. Nucleic acid and nuclease-free, aerosol-resistant pipette tips.
5. Sterile, nucleic acid-free 1.5-mL microfuge tubes.

6. Sterile, nucleic acid-free PCR tubes suitable for the PCR instrument.

7. Racks for tubes.

8. Disposable gloves and laboratory coats.

9. PCR Thermal Cycler. The performance of Prove-it Sepsis has been evaluated using Eppendorf Mastercycler® epGradient S. The selection of the PCR cycler instrument may affect the assay protocol duration and the assay sensitivity.

10. At least two thermal mixers capable of 25, 30, and 66 °C with microtiter plate adapter.

11. A vortex mixer.

12. A spin microfuge.

13. Adjustable micropipettes for pre-PCR and post-PCR areas.

14. A vacuum suction system.

Methods

Detection and Identification of Fungal Species Using PCR- and Microarray-Based Methods

The protocol of the commercial Prove-it Sepsis v2.0 assay is modified from the protocol published by Järvinen et al [3]. and Aittakorpi et al. [33].

Preparing the Fungal PCR

1. Take PCR reagents, except for polymerase, to room temperature.

2. Vortex and spin down all reagents.

3. Prepare the fungal master mixture to a clean laboratory tubes. Add 3.1 µL of PCR water (Mobidiag), 1.5 µL of 10× Buffer (Qiagen, Hilden, Germany), 1.1 µL of BSA (Mobidiag), 0.3 µL of $MgCl_2$ (Qiagen), 2.3 µL of dNTP-mix (Mobidiag), 0.8 µL of Prove-it Fungi Primer-F (Mobidiag), and 2.0 µL of Prove-it Fungi Primer-R (Mobidiag) to the master mixture. Make 10% more of the master mixture than needed.

4. Add polymerase to the master mixture, 0.4 µL of HotStarTaq® DNA Polymerase (Qiagen) per reaction. Always store polymerase at −20 °C.

5. Vortex and spin down the master mixture. Aliquot it to PCR tubes or strips (11.5 µL of master mixture per tube).

6. Add 2 µL of the internal PCR control to each tube. It is not recommended to store or handle the PCR controls in the same facilities where the PCR master mixes and primers are handled. The PCR control can be added into the mixture in the same facilities with the DNA template.

7. Add 1.5 µL of DNA sample.

8. Place all the tubes/strips to PCR machine and start PCR program: a denaturation step at 95 °C for 15 min, 36 cycles of 10 s at 96 °C, 35 s at 52 °C, 10 s at 72 °C, 5 cycles of 5 s at 96 °C, 30 s at 65 °C, 5 cycles of 5 s at 96 °C and finally 30 s at 68 °C.

Preparing Fungal Hybridization onto Prove-it Sepsis StripArrays

1. Pick up the number of microarrays and seals needed.

2. Take all hybridization reagents and distilled water to room temperature and make sure that they are equilibrated to RT.

3. Switch on the thermal blocks and make sure that they are at right temperatures of +30 and +66 °C. Also check correct agitation speed of 550 rpm.

4. Prepare a fresh hybridization buffer by mixing together the 2× hybridization buffer and Hybridization buffer diluents (1:2).

5. Prepare a fresh conjugate solution by mixing together conjugate stock and conjugate diluent (1:80).

6. Prewashing.
   (a) Add 200 µL of distilled water into microarray wells.
   (b) Incubate at +30 °C for 10 min with 550 rpm agitation.
   (c) After incubation, carefully remove all liquid from microarray wells.

7. Hybridization.
   (a) Add 97 µL of fresh hybridization buffer to the microarray wells.
   (b) Add into the same microarray well 3 µL of fungal PCR product.

(c) Incubate at +66 °C for 20 min with 550 rpm agitation.

(d) After incubation, carefully remove all liquid from microarray wells.

8. Washing.

(a) Add 200 μL of washing buffer into microarray wells.

(b) Incubate at +30 °C for 1 min with 550 rpm agitation.

(c) After incubation, carefully remove all liquid from microarray wells.

9. Conjugation

(a) Add 100 μL of freshly mixed conjugate solution into microarray wells.

(b) Incubate at +30 °C for 10 min with 550 rpm agitation.

(c) After incubation, carefully remove all liquid from microarray wells.

10. Washing

(a) Add 200 μL of washing buffer into microarray wells.

(b) Incubate at +30 °C for 5 min with 550 rpm agitation.

(c) After incubation, carefully remove all liquid from microarray wells.

11. Precipitation staining

(a) Add 100 μL of substrate into microarray wells.

(b) Incubate at +25 °C for 10 min. NO agitation for this incubation!

(c) After Incubation, carefully remove all liquid from microarray wells.

12. Analysis with the Prove-it StripArray reader and Prove-it Advisor software.

DNA Extraction

A prerequisite for a successful DNA-based analysis of fungal specimens is the efficiency of cell wall disruption step and subsequent recovery of fungal DNA without putative PCR inhibitors originated from the specimen. Hence, the most appropriate sample preparation and DNA extraction method for any particular application depends also on the specimen type and quantity used. Khot and Fredericks have reviewed both in-house and commercial DNA extraction

methods used with various clinical specimens in PCR-based fungal diagnostics [1]. A variety of in-house methods are available. Also, many manufacturers are providing DNA purification kits that are suitable for preparation of DNA from fungal specimens [34]. However, when adapting a protocol to be used in clinical diagnostics, it is of high importance that the used reagents and materials are free of fungal bioburden. Any risks for false-positive result reporting due to the fungal bioburden should be avoided. Since the contamination of DNA extraction reagents with fungal DNA is common [35], the upstream methods to be used in conjunction with PCR and microarray-based analysis should always be evaluated carefully. When blood culture samples is used as a specimen type, it should also be taken into account that blood culture media contains a common additive polyanetholesulfonate (SPS), which is a potent inhibitor of PCR and resistant to removal by some DNA purification methods [36].

No traces of fungal bioburden have been observed from the current production versions of three commercially available DNA extraction methods. These methods are also efficient regarding the disruption fungal cell wall and the removal of SPS from positive blood culture samples. The protocols for automated solution of NorDiag Arrow (Nordiag, Oslo, Norway) and Nuclisens®easyMAG® (bioMérieux, Marcy l'Etoile, France), and manual solution of MycXtra Fungal DNA Extraction Kit (Myconostica, Manchester, UK) are introduced below.

Extraction of DNA from Blood Culture Using NorDiag Arrow

The NorDiag Arrow pipetting instrument is recommended to be used according to the manufacturer's instructions and recommendations with the Arrow Viral NA kit and Viral NA v.1.0 program (www.nordiag.com). Shortly, NorDiag Arrow is an automated extraction instrument for NAs, using a magnetic bead-based method and running 1–12 samples simultaneously. Arrow provides cost-efficient purchasing and running costs. The following protocol is to be used in conjunction with Prove-it Sepsis assay:

1. Switch the NorDiag Arrow instrument on and on the select protocol menu, select the protocol dedicated for Arrow VIRAL NA kit to be run.
2. Load the instrument with the required consumables, i.e., pumps and pump-tips.
3. Place the cartridge containing the extraction reagents onto the Arrow rack and place the Arrow rack to the instrument. Note: The foil on the cartridges must be peeled off prior to starting a run.
4. For the DNA eluate, place a clean microcentrifuge tube to the appropriate place in the Arrow rack.
5. Mix 240 μL of blood culture and 10 μL of proteinase K in a microcentrifuge tube.
6. Place the sample solution to the appropriate place in the Arrow rack.
7. From the protocol touch screen, choose the sample input volume of 250 μL.
8. From the protocol touch screen, choose the sample elution of 100 μL.
9. Start the protocol. The run is carried out automatically.
10. Run is finished within ~50 min, after which DNA is ready to be used in PCR applications.

Extraction of DNA from Blood Culture Using NucliSENS easyMAG

NucliSENS® easyMAG® instrument is recommended to be used according to the manufacturer's instructions and recommendations (www.biomerieux.com). NucliSENSeasyMAG is an automated system for total nucleic acid extraction from a variety of sample types and volumes, capable of running 1–24 samples simultaneously. NA extraction method is based on the magnetic silica particles. The target NAs bind to the magnetic silica particles during the incubation of lysed sample. The magnetic device is then introduced to the silica particles, enabling the system to purify the NAs trough several washing steps. After washing, the heating step releases DNA

from the silica, after which it is ready to be used in PCR applications. The following protocol is to be used in conjunction with the Prove-it Sepsis assay:

1. Switch the NucliSENS easyMAG instrument on and select the protocol Generic 2.0.1 to be run.
2. Start the off-board lysis extraction protocol Generic 2.0.1 and set the sample material; i.e., blood culture media and the elution volume of 55 μL www.biomerieux.com.
3. Add 100 μL of blood culture to 2 mL of NucliSENS Lysis buffer. Vortex thoroughly. Incubate 10 min at room temperature.
4. Insert aspiration tip sets into the instrument.
5. Load the lysed sample into the 1 well of the 8-well sample vessel.
6. Mix 550 μL of distilled water and 550 μL magnetic silica together.
7. Add 100 μL of silica mixture to the well of the sample vessel containing the lysed sample and mix thoroughly.
8. Insert the vessel into the instrument.
9. Start the run which is carried out automatically. The instrument automatically verifies if there are sufficient amount of the on-board reagents.
10. After the run of 40 min, the extraction protocol is completed and the eluted DNA can be moved from the vessel to a clean laboratory tube for the use in PCR applications.

Extraction of DNA from Clinical Specimen Using MycXtra Fungal DNA Extraction Kit

MycXtra kit is recommended to be used according to the manufacturer's instructions and recommendations (www.myconostica.com). The principle of the kit is to lyse the fungal cells in the sample by combining the use of a detergent and a mechanical force against specialized beads. The cellular components are lysed by a mechanical action on a vortex. From the lysed cells, the released DNA is bound to a silica spin filter. The filter is washed and DNA is harvested in a buffered solution. The protocol is manual.

1. Centrifuge the sample for 20 min at $3,000 \times g$. Decant the supernatant and retain it.
2. Resuspend the pellet in 800 μL of the retained supernatant and transfer it to microcentrifuge tube.
3. Centrifuge for 2 min at $10,000 \times g$ and discard the supernatant. Resuspend the pellet in the solution remaining in the tube and transfer the entire amount to a 2-mL Bead Solution tube.
4. Gently invert to mix.
5. Add 60 μL of Solution S1 to the Bead Solution tube and invert several times.
6. Add 200 μL of Solution IRS to the Bead Solution tube.
7. Vortex at maximum speed for 10 min.
8. Centrifuge the Bead Soltuin tube at $10,000 \times g$ for 30 s.
9. Transfer 450 μL of supernatant to a clean microcentrifuge tube taking care not to disturb the beads. Discard the Bead Solution tube.
10. Add 250 μL of Solution S2 to the supernatant and vortex for 5 s. Incubate at 4–8 °C for 5 min.
11. Centrifuge tubes for 1 min at $10,000 \times g$.
12. Avoiding the pellet, transfer the entire volume of the supernatant to a clean microcentrifuge tube.
13. Add 1.1 mL of Solution S3 to the supernatant. Mix by inverting.
14. Load approximately 650 μL on to a spin filter and centrifuge at $10,000 \times g$ for 30 s. Discard the flow trough. Repeat this step until all supernatant has passed through the spin filter.
15. Add 300 μL of Solution S4 to the spin filter and centrifuge for 30 s at $10,000 \times g$.
16. Discard the flow through.
17. Centrifuge again for 1 min to remove the last traces of S4, which will inhibit the PCR reaction.
18. Carefully place spin filter in a new microcentrifuge tube and add 40 μL of Solution S5 to the center of the white spin filter membrane. Leave at room temperature for 2 min.
19. Centrifuge for 30 s at $10,000 \times g$.
20. Discard the spin filter.
21. DNA in the tube is now ready for use in a PCR application.

Notes

1. Specimens should always be considered as potentially infectious.
2. Store and extract DNA from a specimen separately from the reagents and the pre-PCR area.
3. The procedure should be performed in physically separated areas (pre-PCR area and post-PCR area) to avoid contamination with microbial organisms or nucleic acids or any other extraneous material or agents; e.g., amplicons from previous PCR runs. In the pre-PCR area, the preparation of the PCR mixture should be conducted in an area separate from where the addition of the DNA sample takes place.
4. Each pre/post-PCR area should have its own dedicated working materials assigned; e.g., pipettes, spin microfuge, and disposable gloves. Any material in the post-PCR area must never come into contact with that of the pre-PCR area.
5. Always follow the unidirectional workflow from the pre-PCR area to the post-PCR area. Never reverse the direction.
6. Always wear suitable protective clothing and gloves during the procedure.
7. Follow the recommendation of the manufacturer of thermal cycler regarding to the quality of PCR plastic ware.
8. Avoid scratching the microarrays in the bottom.
9. Avoid bubble formation on the microarray surface. Reverse pipetting is recommended to avoid bubble formation.
10. Keep the microarray bottom clean at all times to avoid any interference when detecting the assay result.
11. Be careful not to let the microarray wells dry for longer than necessary between the hybridization steps.
12. TMB-based substrate must be protected from light.
13. Proceed to the PCR step immediately after the DNA extraction step. Also, proceed to the microarray step after the PCR step. Storing the DNA extract or PCR product may affect the assay result.

Acknowledgments I thank my colleagues in Mobidiag, especially Anne Aittakorpi, for her constructive comments on the manuscript.

References

1. Khot PD, Fredricks DN (2009) PCR-based diagnosis of human fungal infections. Expert Rev Anti Infect Ther 7:1201–1221
2. Eggimann P, Garbino J, Pittet D (2003) Epidemiology of Candida species infections in critically ill non-immunosuppressed patients: review. Lancet Infect Dis 3:685–702
3. Järvinen AK, Laakso S, Piiparinen P, Aittakorpi A, Lindfors M, Huopaniemi L et al (2009) Rapid identification of bacterial pathogens using a PCR- and microarray-based assay. BMC Microbiol 9: 161–177
4. Jääskeläinen AJ, Piiparinen H, Lappalainen M, Koskiniemi M, Vaheri A (2006) Multiplex-PCR and oligonucleotide microarray for detection of eight different herpesviruses from clinical specimens. J Clin Virol 37:83–90
5. Jääskeläinen AJ, Piiparinen H, Lappalainen M, Vaheri A (2008) Improved multiplex-PCR and microarray for herpesvirus detection from CSF. J Clin Virol 42:172–175
6. Wiesinger-Mayr H, Vierlinger K, Pichler R, Kriegner A, Hirschl AM, Presterl E et al (2007) Identification of human pathogens isolated from blood using microarray hybridisation and signal pattern recognition. BMC Microbiol 7:78–95
7. Yoo SM, Choi JY, Yun JK, Choi JK, Shin SY, Lee K et al (2010) DNA microarray-based identification of bacterial and fungal pathogens in bloodstream infections. Mol Cell Probes 24:44–52
8. Mikhailovich V, Gryadunov D, Kolchinsky A, Makarov AA, Zasedatelev A (2008) DNA microarrays in the clinic: infectious diseases. Bioassays 30:673–682
9. Muldrew KL (2009) Molecular diagnostics of infectious diseases. Curr Opin Pediatr 21:102–111
10. Yoo SM, Choi JH, Lee SY, Yoo NC (2009) Applications of DNA microarray in disease diagnostics. J Microbiol Biotechnol 19:635–646
11. Leski TA, Malanoski AP, Stenger DA, Lin B (2010) Target amplification for broad spectrum microbial diagnostics and detection. Review. Future Microbiol 5:191–203
12. Miller MB, Tang YW (2009) Basic concepts of microarrays and potential applications in clinical microbiology: review. Clin Microbiol Rev 22:611–633
13. Howard D, Cordell R, McGowan JE Jr, Packard RM, Scott RD II, Solomon SL (2001) Workshop group. Measuring the economic costs of antimicrobial resistance in hospital settings: summary of the Centers for Disease Control and Prevention-Emory workshop. Clin Infect Dis 33:1573–1578
14. Barenfanger J, Drake C, Kacich G (1999) Clinical and financial benefits of rapid bacterial identification and antimicrobial susceptibility testing. J Clin Microbiol 37:1415–1418
15. Davey PG, Marwick C (2008) Appropriate vs inappropriate antimicrobial therapy. Clin Microbiol Infect 14:15–21
16. Guery BP, Arendrup MC, Auzinger G, Azoulay E, Borges Sá M, Johnson EM et al (2008) Management of invasive candidiasis and candidemia in adult non-neutropenic intensive care unit patients: Part I. Epidemiology and diagnosis. Review. Intensive Care Med 35:55–62
17. Wamola B, Mulholland R, Riordan A (2011) Management and outcome of candida blood stream infections within a regional paediatric hospital. Arch Dis Child 96:A52–A53
18. Breakspear A, Momany M (2007) The first fifty microarray studies in filamentous fungi. Review. Microbiology 153:7–15
19. Leinberger DM, Schumacher U, Autenrieth IB, Bachmann TT (2005) Development of a DNA microarray for detection and identification of fungal pathogens involved in invasive mycoses. J Clin Microbiol 43:4943–4953
20. Spiess B, Seifarth W, Hummel M, Frank O, Fabarius A, Zheng C et al (2007) DNA microarray-based detection and identification of fungal pathogens in clinical samples from neutropenic patients. J Clin Microbiol 45:3743–3753
21. Lu W, Gu D, Chen X, Xiong R, Liu P, Yang N et al (2010) Application of an oligonucleotide microarray-based nano-amplification technique for the detection of fungal pathogens. Clin Chem Lab Med 48:1507–1514
22. Campa D, Tavanti A, Gemignani F, Mogavero CS, Bellini I, Bottari F et al (2008) DNA microarray based on arrayed-primer extension technique for identification of pathogenic fungi responsible for invasive and superficial mycoses. J Clin Microbiol 46:909–915
23. Ehricht R, Slickers P, Goellner S, Hotzel H, Sachse K (2006) Optimized DNA microarray assay allows detection and genotyping of single PCR-amplifiable target copies. Mol Cell Probes 20:60–63
24. Sachse K, Hotzel H, Slickers P, Ellinger T, Ehricht R (2005) DNA microarray-based detection and identification of Chlamydia and Chlamydophila spp. Mol Cell Probes 19:41–50
25. Anjum MF, Mafura M, Slickers P, Ballmer K, Kuhnert P, Woodward MJ et al (2007) Pathotyping Escherichia coli by using miniaturized DNA microarrays. Appl Environ Microbiol 73:5692–5697
26. Borel N, Kempf E, Hotzel H, Schubert E, Torgerson P, Slickers P et al (2008) Direct identification of chlamydiae from clinical samples using a DNA microarray assay-A validation study. Mol Cell Probes 22:55–56
27. Cannon GA, Carr MJ, Yandle Z, Schaffer K, Kidney R, Hosny G et al (2009) A low density oligonucleotide microarray for the detection of viral and atypical bacterial respiratory pathogens. J Virol Methods 28: 3723–3734

28. Batchelor M, Hopkins KL, Liebana E, Slickers P, Ehricht R, Mafura M et al (2008) Development of a miniaturised microarray-based assay for the rapid identification of antimicrobial resistance genes in Gram-negative bacteria. Int J Antimicrob Agents 31:440–451

29. Shore AC, Deasy EC, Slickers P, Brennan G, O'Connell B, Monecke S et al (2011) Detection of staphylococcal cassette chromosome mec Type XI encoding highly divergent mecA, mecI, mecR1, blaZ and ccr genes in human clinical clonal complex 130 methicillin-resistant *Staphylococcus aureus*. Antimicrob Agents Chemother 55(8):3765–73 [Epub 2011 Jun 2]

30. Anjum MF, Choudhary S, Morrison V, Snow LC, Mafura M, Slickers P et al (2011) Identifying antimicrobial resistance genes of human clinical relevance within Salmonella isolated from food animals in Great Britain. J Antimicrob Chemother 66: 550–559

31. Monecke S, Hochauf K, Gottschlich B, Ehricht R (2006) A case of peritonitis caused by Rhizopus microsporus. Mycoses 49:139–141

32. Tissari P, ZumLa A, Tarkka E, Mero S, Savolainen L, Vaara M, Aittakorpi A et al (2010) Accurate and rapid identification of bacterial species from positive blood cultures with a DNA-based microarray platform: an observational study. Lancet 375:224–230

33. Aittakorpi A, Kuusela P, Koukila-Kähkölä P, Vaara M, Petrou M, Gant V et al. Accurate and Rapid Identification of *Candida* spp. Frequently Associated with Fungemia by Using PCR and the Microarray-Based Prove-it Sepsis Assay. J Clin Microbiol 50:11

34. Metwally L, Fairley DJ, Coyle PV, Hay RJ, Hedderwick S, McCloskey B et al (2008) Improving molecular detection of *Candida* DNA in whole blood: comparison of seven fungal DNA extraction protocols using real-time PCR. J Med Microbiol 57:296–303

35. Loeffler J, Hebart H, Bialek R, Hagmeyer L, Schmidt D, Serey FP et al (1999) Contaminations occurring in fungal PCR assays. J Clin Microbiol 37:1200–1202

36. Fredricks DN, Relman DA (1998) Improved amplification of microbial DNA from blood cultures by removal of the PCR inhibitor sodium polyanetholesulfonate. J Clin Microbiol 36:2810–2816

Bioinformatic Protocols and the Knowledge-Base for Secretomes in Fungi

54

Gengkon Lum and Xiang Jia Min

Abstract

Fungal secreted proteins play important roles in cell signaling, metabolism, and regulation of fungal growth and development. The secretome refers to all secreted proteins in a proteome that are identified from completely sequenced genomes. The majority of secreted proteins are classical, signal peptide-dependent proteins that can be predicted using bioinformatics tools. In this chapter, we describe some commonly used tools for secreted protein prediction in fungi and propose a relatively accurate bioinformatic protocol for fungal secretome identification. The protocol combines multiple signal peptide or subcellular location predictors, including SignalP, WoLF PSORT, and Phobius, with TMHMM for removing transmembrane proteins and PROSITE PS-Scan for removing endoplasmic reticulum (ER) proteins. Applying this protocol, we have built the fungal secretome knowledge-base (FunSecKB). The utility of FunSecKB is described in detail. FunSecKB serves the community as a central portal for search and deposition of fungal secretome information.

Keywords

Secreted proteins • Secretome • Signal peptide • Fungi • Prediction • Knowledge-base • Database

Introduction

Secreted proteins are proteins which are synthesized within cells and then secreted to extracellular space and matrix to play their roles. Secreted proteins play important roles in cell signaling, metabolism, and regulation in growth and development of all organisms. As the genomes have been completely sequenced in many organisms, the proteomes could be predicted using the information in

G. Lum
Department of Computer Science and Information Systems, Youngstown State University, Youngstown, OH 44555, USA

X.J. Min (✉)
Department of Biological Sciences, Center for Applied Chemical Biology, Youngstown State University, One University Plaza, Youngstown, OH 44555, USA
e-mail: xmin@ysu.edu

V.K. Gupta et al. (eds.), *Laboratory Protocols in Fungal Biology: Current Methods in Fungal Biology*,
Fungal Biology, DOI 10.1007/978-1-4614-2356-0_54, © Springer Science+Business Media, LLC 2013

genomes. The term "secretome" was first used to include all proteins secreted to extracellular space and matrix and proteins involved in the secretion pathway including endoplasmic reticulum, Golgi apparatus, and transportation vesicles [1–3]; however, more recently, the term was used to include secreted proteins only [4,5]. In this work, the secretome only refers the complete set of secreted proteins in an organism.

Secretomes are an important part of the fungal proteome. These secreted proteins include enzymes, growth factors, cell wall proteins, and other bioactive molecules which play important roles in host–pathogen interactions. Fungal secreted enzymes are used to break down potential food sources for transport into the cells. As there are many types of fungi producing a great variety of enzymes that are able to break down lignocelluloses and other biopolymers, fungi have an important function in the biosphere as decomposers. Since secreted proteins are useful in their ability to break down biopolymers, they have found a role in many applications including pharmaceutical and industrial [6]. Therefore, the ability to analyze a protein to determine if it is secreted and what functions it may have is useful as a tool in research to more easily focus on or to eliminate potential targets. Increased understanding of the secretome biology of fungi will further promote exploration of the potential applications of fungal secreted proteins in environmental remediation and industrial processing including bio-fuel production.

Most of secreted proteins in fungi are classical secreted proteins, which have a signal peptide on the N-terminus of protein sequences. A signal peptide is typically 15–30 amino acids long, located at the N-terminus of the protein and is cleaved off during translocation across the membrane. The presence of a signal peptide directs the protein to the rough endoplasmic reticulum (ER) and the Golgi complex in preparation for transport through the secretory vesicles. This is referred to as the classical secretory pathway. Although not all proteins excreted extracellularly contain a signal peptide, it is believed that that the majority of fungal proteins are secreted in this manner [6]. The presence of a transmembrane domain in the protein sequence, however, indicates that although the protein passes through the classical secretory pathway, it is not secreted extracellularly but instead becomes part of the cell membrane. By combining the results of one or more predictions for the presence of a signal peptide along with the absence of a transmembrane domain, the likelihood of the protein being secreted is very high. Our recent evaluation reveals that combining the results of multiple programs increases the accuracy by reducing the number of false positives and negatives [7].

Two fungal-specific secretome databases are currently available. The Fungal Secretome Database[1] developed by Choi et al. used nine bioinformatics tools and protein sequences from completely sequenced genomes including some work in progress draft genomes [8]. The Fungal Secretome Knowledge-Base (FunSecKB)[2] developed by us used all fungal protein sequences available in the NCBI RefSeq database and being linked and supplemented with protein sequences in the UniProt database. The detailed comparison of the two databases was described by Lum and Min [5]. In this work, we focus on how to utilize FunSecKB.

Materials (Data)

For individual secreted protein identification, the input is a fungal protein sequence in FASTA format. For a species-specific secretome prediction from a whole proteome, which often is obtained from a completely sequenced genome, a set of proteins in multiple FASTA format are used as input.[3] We will use a glucoamylase enzyme from *Aspergillus niger* (gi 145235763) and a *Schizosaccharomyces pombe* protein (gi 19115161) as examples to explain the input and output of the tools mentioned in Sect. "Methods."

---

[1]http://fsd.snu.ac.kr/.

[2]http://proteomics.ysu.edu/secretomes/fungi.php.

[3]A description of the FASTA format may be found at http://www.ncbi.nlm.nih.gov/BLAST/fasta.shtml.

Methods

The programs used for fungal secretome prediction include SignalP 3.0 [9], TargetP 1.1 [10], TMHMM 2.0 [11], Phobius [12], WoLF PSORT [13], PS-Scan for PROSITE [14], and FragAnchor [15]. SignalP and TargetP predict the presence and location of an N signal peptide and a potential cleavage site. TMHMM predicts the presence of a transmembrane domain. Phobius is a combined signal peptide and transmembrane topology predictor. WoLF PSORT (WolfPsort) predicts the subcellular location(s) of a protein. PS-Scan is a PROSITE scanning tool which predicts whether or not a protein contains an endoplasmic reticulum (ER) targeting sequence (Prosite: PS00014). FragAnchor is used to predict if there is a glycosylphosphatidylinositol (GPI) anchor in the protein, which may indicate if the secreted protein is a cell wall protein or attaches to the outside of the plasma membrane.

There are two methods of using these tools. The first one, most often used by a biologist to process an individual protein, uses the online Webserver tool. The second one, often used by bioinformaticians to process proteome-wide secretome identification, use a standalone package which may be downloaded and run on a UNIX (Linux) platform.

SignalP 3.0

SignalP 3.0 uses both neural network (NN) and hidden Markov model (HMM) algorithms in two different predictors to predict whether a protein has a signal peptide and where the most likely cleavage site would be if one is detected [9].[4] For each protein processed by SignalP 3.0, scores are calculated and returned in two sections: SignalP-NN result (Fig. 54.1a) and SignalP-HMM result (Fig. 54.1b).

In the Signal-NN results, two different neural networks are used for each prediction, one for predicting the presence of the signal peptide, the other for predicting the position of the cleavage site. For each position in the protein, a C, S, and Y score is calculated. The C score is the cleavage site score with values being high at potential cleavage sites. The S score is reported for every position submitted with high scores for amino acids which are part of the signal peptide and low score for those which are part of a mature protein. The Y score is a derivative of C and S with a likely cleavage point being when the slope of S is steep and there is a high C score resulting in a high Y score. The mean S score is the average of the S scores from the N-terminus to the highest Y score. The D score is average of the Y score and the mean S score. The D score is used to determine whether or not a protein is predicted to be secreted.

In the Signal-HMM results, the positions are evaluated to determine the likelihood of being a part of the n-region, h-region, or c-region. Signal peptides commonly have a hydrophobic central core (h-region) surrounded by the N- and C-terminal hydrophilic regions. The HMM makes a prediction of a signal peptide, a nonsecretory protein or a signal anchor. A protein with a signal anchor passes through the membrane but the uncleaved signal peptide remains anchored to the membrane resulting in a type II membrane protein [9]. The results also include a probability for both a signal peptide and a signal anchor.

Phobius

Phobius is a combined signal peptide and a transmembrane topology predictor [5] [12]. A known problem with signal peptide and transmembrane topology predictors is the high similarity of the hydrophobic regions of both the signal peptide h-region and transmembrane helix. Due to this similarity, pure signal peptide predictors and transmembrane topology predictors sometimes

---

[4]http://www.cbs.dtu.dk/services/SignalP/.

[5]http://phobius.sbc.su.se/index.html.

>gi|145235763| glucan 1,4-alpha-glucosidase glaA-Aspergillus niger

SignalP-NN result:

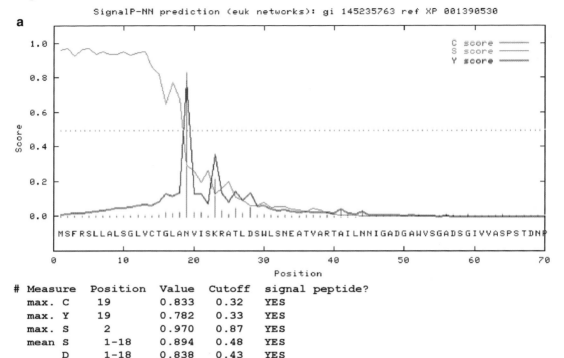

```
# Measure   Position   Value   Cutoff   signal peptide?
  max. C     19        0.833   0.32     YES
  max. Y     19        0.782   0.33     YES
  max. S      2        0.970   0.87     YES
  mean S    1-18       0.894   0.48     YES
       D    1-18       0.838   0.43     YES
# Most likely cleavage site between pos. 18 and 19: GLA-NV
```

SignalP-HMM result:

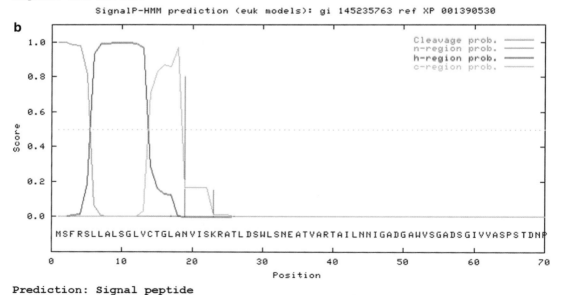

```
Prediction: Signal peptide
Signal peptide probability: 0.998
Signal anchor probability: 0.001
Max cleavage site probability: 0.806 between pos. 18 and 19
```

Fig. 54.1 (**a**) Neural network results output for SignalP 3.0 Server of *Aspergillus niger* glucoamylase protein. (**b**) Hidden Markov model results output for SignalP 3.0 Server of *Aspergillus niger* glucoamylase protein

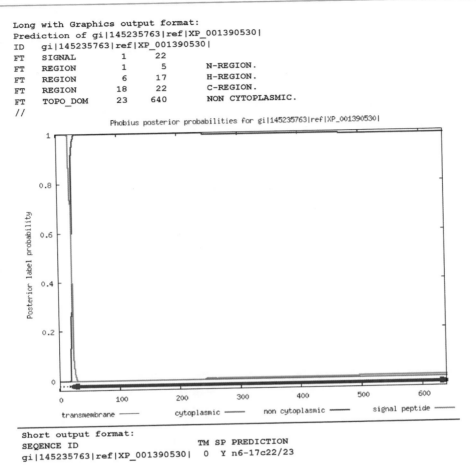

```
Long with Graphics output format:
Prediction of gi|145235763|ref|XP_001390530|
ID    gi|145235763|ref|XP_001390530|
FT    SIGNAL       1    22
FT    REGION       1     5        N-REGION.
FT    REGION       6    17        H-REGION.
FT    REGION      18    22        C-REGION.
FT    TOPO_DOM    23   640        NON CYTOPLASMIC.
//
```

Phobius posterior probabilities for gi|145235763|ref|XP_001390530|

transmembrane ———   cytoplasmic ———   non cytoplasmic ———   signal peptide ———

```
Short output format:
SEQENCE ID                    TM SP PREDICTION
gi|145235763|ref|XP_001390530|  0  Y  n6-17c22/23
```

Fig. 54.2 Results output for Phobius of *Aspergillus niger* glucoamylase protein

results in false classifications. To this end, Phobius was designed to do both signal peptide and transmembrane topology prediction and to distinguish between the two regions.

The output formats available are long with graphics, long without graphics, or short format (Fig. 54.2). The default output (long with graphics) shows the prediction of probable locations for sections of the protein. Some possible predictions are: SIGNAL for signal peptide, REGION for N-, H-, and C-regions, TOPO_DOM for topology (cytoplasmic or non-cytoplasmic) and TRANSMEM for positions predicted to be within the membrane. The range of positions is given for each predicted segment. If the entire sequence is labeled cytoplasmic or non-cytoplasmic though,

the prediction is that there are no membrane helices and is not an actual prediction of location, but the most probable location.

The short output format gives **TM** as the number of predicted transmembrane segments, **SP** as the prediction of whether or not there is a signal peptide, and **PREDICTION** as the predicted topology. The format of the predicted topology is given as a series of numbers and letters. If a signal peptide is detected, it is given in the format: n#-#c#/# where # represents a position in the sequence. The numbers between n and c is the range of the hydrophobic h-region and the #/# is the cleavage site. Following this is either an "i" if the loop is cytoplasmic or an "o" if it is on the non-cytoplasmic side and then numbers in the

```
k used for kNN is: 27
gi|145235763|ref|XP_001390530| details extr: 26.0
```

Fig. 54.3 Results output for WoLF PSORT of *Aspergillus niger* glucoamylase protein

```
### targetp v1.1 prediction results ##################################
Number of query sequences:  1
Cleavage site predictions included.
Using NON-PLANT networks.
```

Name	Len	mTP	SP	other	Loc	RC	TPlen
gi_145235763_ref_XP_	640	0.183	0.801	0.050	S	2	18
cutoff		0.000	0.000	0.000			

Fig. 54.4 Results output for TargetP 1.1 Server of *Aspergillus niger* glucoamylase protein

format #-# indicating the range of the transmembrane helix. This format is repeated until the end of the sequence.

WoLF PSORT

WoLF PSORT[6] is a program for predicting the subcellular location of proteins [13]. It takes the amino acid sequences and converts them into numerical vectors which are then classified using a weighted *k*-nearest neighbor classifier. The predictions are based on known sorting signal motifs and the content of the amino acids. It requires the selection of the organism type: animal, plant, or fungi. For our example protein (Fig. 54.3), the result was based on using *k*=27 nearest neighbors. Of these 27 closest, 26 were extracellular and the result is displayed as extr: 26.0. A list of the localization site definitions is available on the Website and include locations such as golg for the Golgi apparatus, mito for mitochondria, and nucl for nuclear.

TargetP 1.1

TargetP is designed to predict the subcellular locations of eukaryotic proteins[7] [10]. TargetP predicts in the N-terminus the presence of any of the N-terminal presequences such as signal peptide (SP), chloroplast transit peptide (cTP), or mitochondrial targeting peptide (mTP). The output is given in Fig. 54.4. **Name** is the sequence name truncated to 20 characters. **Len** is the length of the sequence. **cTP, mTP, SP, other** are the final neural network (NN) scores. **cTP** is only used if the organism group on the submission page is set to Plant since it is used to detect a cTP.

TMHMM 2.0

TMHMM 2.0 uses a HMM to predict the presence and topology of transmembrane helices and their orientation to the membrane (in/out)[8][11]. The output shows the results of the prediction (Fig. 54.5).

---

[6]http://wolfpsort.org/

[7]http://www.cbs.dtu.dk/services/TargetP/.
[8]http://www.cbs.dtu.dk/services/TMHMM/.

```
Long format (default):
# gi_19115161_ref_NP_594249_ Length: 324
# gi_19115161_ref_NP_594249_ Number of predicted TMHs:  7
# gi_19115161_ref_NP_594249_ Exp number of AAs in TMHs: 147.53314
# gi_19115161_ref_NP_594249_ Exp number, first 60 AAs:  40.80425
# gi_19115161_ref_NP_594249_ Total prob of N-in:        0.01841
# gi_19115161_ref_NP_594249_ POSSIBLE N-term signal sequence
gi_19115161_ref_NP_594249_   TMHMM2.0  outside      1     9
gi_19115161_ref_NP_594249_   TMHMM2.0  TMhelix     10    32
gi_19115161_ref_NP_594249_   TMHMM2.0  inside      33    38
gi_19115161_ref_NP_594249_   TMHMM2.0  TMhelix     39    56
gi_19115161_ref_NP_594249_   TMHMM2.0  outside     57    70
gi_19115161_ref_NP_594249_   TMHMM2.0  TMhelix     71    93
gi_19115161_ref_NP_594249_   TMHMM2.0  inside      94   112
gi_19115161_ref_NP_594249_   TMHMM2.0  TMhelix    113   135
gi_19115161_ref_NP_594249_   TMHMM2.0  outside    136   154
gi_19115161_ref_NP_594249_   TMHMM2.0  TMhelix    155   177
gi_19115161_ref_NP_594249_   TMHMM2.0  inside     178   205
gi_19115161_ref_NP_594249_   TMHMM2.0  TMhelix    206   223
gi_19115161_ref_NP_594249_   TMHMM2.0  outside    224   237
gi_19115161_ref_NP_594249_   TMHMM2.0  TMhelix    238   260
gi_19115161_ref_NP_594249_   TMHMM2.0  inside     261   324
```

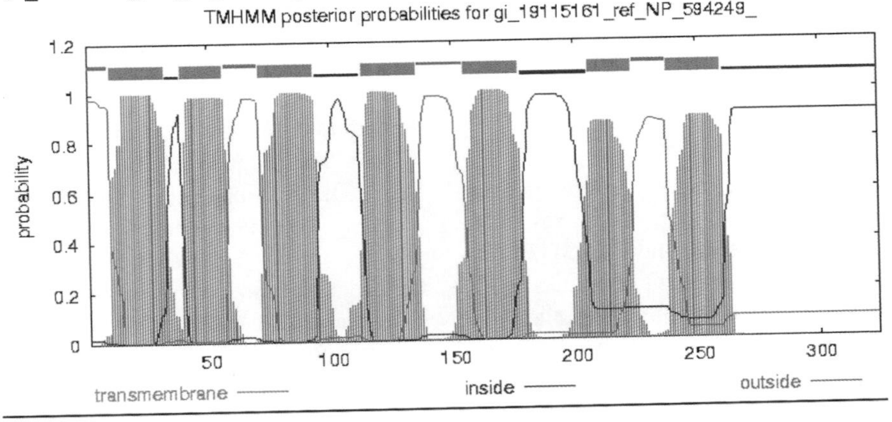

TMHMM posterior probabilities for gi_19115161_ref_NP_594249_

```
Short format:
gi_19115161  len=324  ExpAA=147.53  First60=40.80  PredHel=7  Topology=o10-
32i39-56o71-93i113-135o155-177i206-223o238-260i
```

Fig. 54.5 Results output for TMHMM Server 2.0 of a *Schizosaccharomyces pombe* protein

Using the default long format: **Length** is the length of the sequence submitted. **Number of predicted TMHs** is the number of predicted transmembrane helices. **Exp number of AAs in TMHs** is the expected number of amino acids in transmembrane helices. **Exp number, first 60 AAs** is the expected number of amino acids in transmembrane helices within the first 60 positions. **Total prob of N-in** is the total probability that the N-terminus is on the cytoplasmic side of the membrane. Following this section is the prediction of where specific parts of the protein are likely to be: inside, outside, or TM helix (part of the transmembrane helix). The structure is the identifier used, followed by the program name (TMHMM2.0), the predicted location, then the starting and ending position of the segment. In this example, the prediction is that the N-terminus is outside the membrane and the protein crosses the membrane seven times and the C-terminus ends on the inside of the cell. Using the short format: **len**, length of sequence, **ExpAA**, expected number of amino acids in transmembrane helices, **First60**, expected number of amino acids in transmembrane helices within the first 60 positions, **PredHel**, number of predicted transmembrane helices by N-best, and **Topology**, the topology predicted by N-best with "o" indicating sections outside and "i" indicating sections inside the cell.

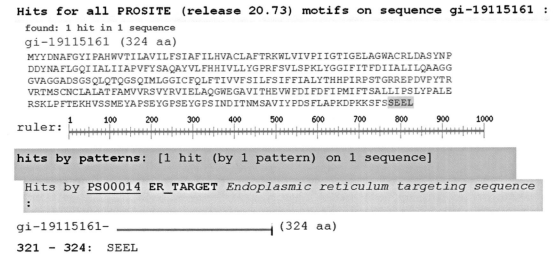

Fig. 54.6 Results output for ScanProsite of a *Schizosaccharomyces pombe* protein

PS-Scan for PROSITE

PROSITE is a database containing protein families, domains, and functional sites. The ScanProsite Website[9] scans the PROSITE database for motifs matching the input sequence [14]. The output from the Website lists any hits found in their database matching sections within our sequence. In our FunSecKB database, we used the standalone program PS-Scan to determine if there was an ER retention signal (Prosite: PS00014), which if found could rule out the possibility that the particular protein would be secreted. The output of a *Schizosaccharomyces pombe* protein (gi 19115161) shows an ER targeting sequence detected (Fig. 54.6) at positions 321–324.

FragAnchor

FragAnchor[10] is a tool to detect the presence of a glycosylphosphatidylinositol (GPI) anchor [15]. It uses a combination of a neural network to select

potential GPI-anchored sequences and a HMM to classify those sequences into categories of likelihood. The four categories are highly probable, probable, weakly probable, and potential false positive. Our example did not contain a potential GPI-anchored sequence and thus was rejected with the HMM classification never being run, thus the output is not shown here. However, the detailed information of the GPI-anchored secreted proteins and the correlations with proteome size and genome size can be found in Lum and Min [5]. This Webserver support a batch of sequences, but no standalone tool is available. There are some other tools available for GPI anchor prediction, including Big-PI predictor [11] and PredGPI.[12]

SecretomeP

SecretomeP[13] is a program that uses a sequence-based method for prediction of secreted proteins based on nonclassical secretory pathways. The original program was trained on bacteria and support for mammalian proteins was added

[9]http://expasy.org/tools/scanprosite/.

[10]http://navet.ics.hawaii.edu/~fraganchor/NNHMM/NNHMM.html.

[11]http://mendel.imp.ac.at/sat/gpi/gpi_server.html.

[12]http://gpcr.biocomp.unibo.it/predgpi/.

[13]http://www.cbs.dtu.dk/services/SecretomeP/.

Table 54.1 Linux
commandline summary
for standalone packages[a]

Tools	Commands
SignalP	signalp –t euk –f summary input_file>output_file
Phobius	phobius input_file –short>output_file
WoLFPsort	runWolfPsortSummaryOnly fungi<input_file>output_file
TargetP	targetp –c –N input_file>output_file
TMHMM	tmhmm input_file –short –noplot>output_file
PS-Scan	ps_scan.pl input_file –p PS00014 –o scan –d prosite.dat>output_file

[a]Input_file is the protein sequences in FASTA format. output_file is the file to save the
results of the program

afterward. The Webserver currently has support for gram-negative and gram-positive bacteria along with mammalian proteins but its use in prediction of secreted fungal proteins by nonclassical pathways has not been tested. Choi et al. used this tool to predict nonclassical, signal peptide independent secreted proteins in constructing the Fungal Secretome Database[14] [8]. However, as the accuracy of the tool in fungal secretome prediction was not reported, we did not use this tool in FunSecKB (see discussion in Sect. "TMHMM 2.0").

Commands of Standalone Tools

We described the online Webservers above. The online Webservers normally have a limit for the maximum number of sequences allowed to be submitted at once; therefore, to process a large number (i.e., a proteome of a whole species) the standalone tools are needed. For the standalone tools that need to be installed on a Linux system, the commands of how to run them are summarized in Table 54.1. Detailed explanations of how to run each tool often can be found in the "readme" page in each downloaded package.

Protocol Evaluation

The accuracy of a prediction tool can only be evaluated using a set of sequence data. Min reported the accuracy of some of the tools mentioned above in prediction of fungal secretomes [7]. The tools were evaluated individually and in combination with others. The dataset contained 241 secreted proteins and 5,992 nonsecreted proteins and the results were measured using sensitivity (Sn) (Equation 54.1), specificity (Sp) (Equation 54.2), and Mathews' Correlation Coefficient (MCC) (Equation 54.3) [16–18],

$$\text{Sn}(\%) = \text{TP} / (\text{TP} + \text{FN}) \times 100 \quad (54.1)$$

$$\text{Sp}(\%) = \text{TN} / (\text{TN} + \text{FP}) \times 100 \quad (54.2)$$

$$\text{MCC}(\%) = (\text{TP} \times \text{TN} - -\text{FP} \times \text{FN}) \times 100 / \left((\text{TP} + \text{FP})(\text{TP} + \text{FN})(\text{TN} + \text{FP})(\text{TN} + \text{FN}) \right)^{1/2} \quad (54.3)$$

where TP represents the number of true positives, FN is the number of false negatives, TN is the number of true negatives, and FP is the number of false positives. When tools were combined, a true positive was counted only when all the tools used predicted the protein as positive. The results were provided in Table 54.2, which was adopted from Ref. [7]. Based on the results, we used the combination of SignalP, Phobius, WoLF PSORT, TMHMM, and PS-Scan, which gave the highest MCC (83.4 %) result, as the prediction protocol for fungal secretome prediction in FunSecKB development [5]. The TargetP 1.1 can be used for

[14]http://fsd.snu.ac.kr/.

Table 54.2 Prediction accuracies of secreted proteins in fungi[a]

Methods	TP	FP	TN	FN	Sn (%)	Sp (%)	MCC (%)
SignalP	232	329	5663	9	96.3	94.5	61.2
Phobius	226	203	5789	15	93.8	96.6	68.8
TargetP	228	583	5409	13	94.6	90.3	48.6
WolfPsort	230	167	5825	11	95.4	97.2	73.1
SignalP/TMHMM	228	168	5824	13	94.6	97.2	72.6
Phobius/TMHMM	224	200	5792	17	92.9	96.7	68.6
TargetP/TMHMM	224	265	5727	17	92.9	95.6	63.5
WolfPsort/TMHMM	227	135	5857	14	94.2	97.7	75.8
SignalP/TMHMM/WolfPsort	226	86	5906	15	93.8	98.6	81.6
SignalP/TMHMM//WolfPsort/Phobius	222	69	5923	19	92.1	98.8	83.1
SignalP/TMHMM/WolfPsort/Phobius/PS-Scan	**222**	**67**	**5925**	**19**	**92.1**	**98.9**	**83.4**
SignalP/TMHMM/WolfPsort/Phobius/TargetP/PS-Scan	218	66	5926	23	90.5	98.9	82.6

TP true positives; *FP*, false positives; *TN*, true negatives; *FN*, false negatives; Sn, sensitivity; Sp, specificity; MCC, Mathews' correlation coefficient.
[a]The table is reproduced with permission from Min [7].

individual secreted protein prediction, however, adding it to the pipeline for secretome prediction slightly reduced the accuracy (see Table 54.2).

The Fungal Secretome Knowledge-Base

The Fungal Secretome Knowledge-Base (FunSecKB) is a database of fungal proteins collected from NCBI and UniProt on which we have performed various analyses for prediction of possible extracellular secretion [5].[15] From this site (Fig. 54.7), you can look up specific proteins using either NCBI's gi or RefSeq accession or UniProt's accession numbers. In addition you can enter a keyword to search for such as species, function, or cellular location. You may also search for secreted proteins of a specific species or perform BLAST (Basic Local Alignment Search Tool) search against our fungal database. When a keyword or species secretome search is performed, a list of results will be displayed with an identifier to the left followed by a description. The identifier is a link and clicking on it will display the details page for that protein. Similarly,

searching for a specific protein by gi or accession will display that particular protein. This page shows the results of the different tests performed on the protein along with the sequence in FASTA format and any available manually curated data.

The Web page is divided up into five main sections: Search individual proteins by ID or keyword(s), Search secretome information by species, BLAST search, and Community Annotation.

Search by ID or Key Words

This section allows searching for a specific protein by using NCBI's RefSeq accession or gi number or UniprotKB's accession number. A search by keyword(s) will return a list of proteins containing the keyword(s) based on the UniProt Protein name. Details of an individual protein's results may be found by clicking on the identifier. For each protein which has been tested in our database, the results of those tests are displayed. The first area includes the various identifiers from NCBI and UniProt along with a clickable direct link to those sites. Also listed are the species, RefSeq definition, UniProt name, and a UniProt annotation for subcellular location (if any). The second area is a summary of the test results consisting of a yes/no for prediction of a secreted protein for each test. Also listed is a

---
[15]It is an online resource available at http://proteomics.ysu.edu/secretomes/fungi.php.

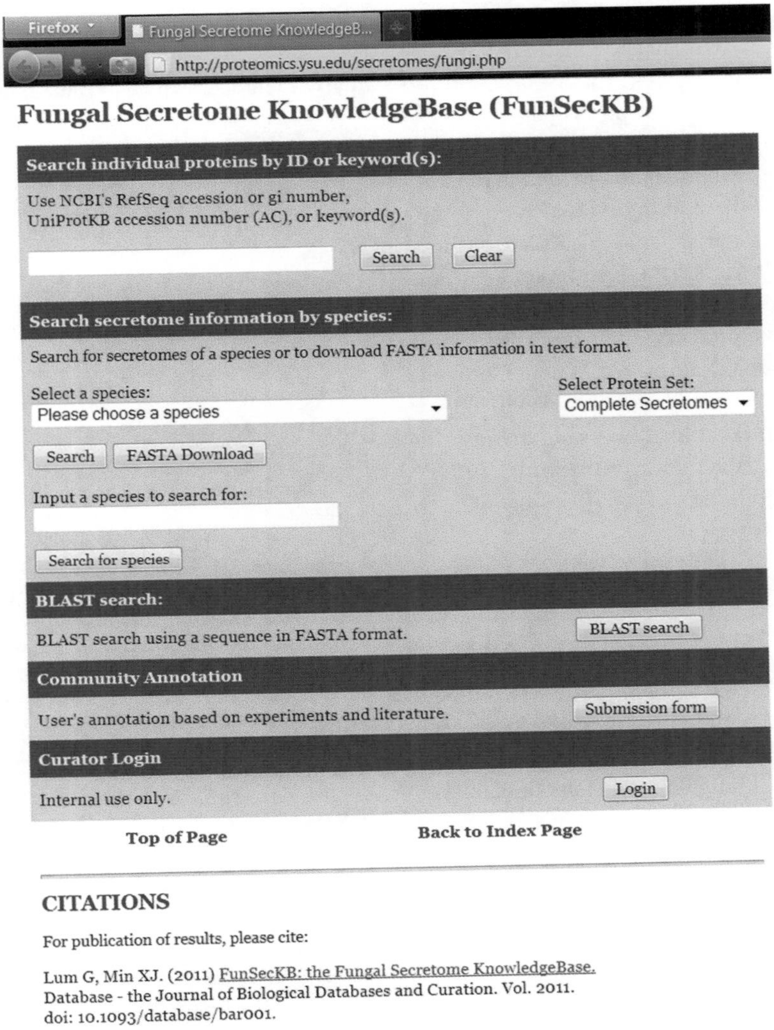

Fig. 54.7 Home page of Fungal Secretome Knowledge-Base

conclusion of whether or not this protein is belonged to a **Secretome**: based on our own combination prediction algorithm as mentioned above, that is, SignalP/Phobius/WoLF PSORT predicted to have a signal peptide, TMHMM predicted not have a transmembrane domain, and PS-Scan did not find an ER retention signal. The third area is the details for each of the tests along with a link to the original site's page on how to interpret the results (if available) or the Web site for the program. After the test results is listed the protein sequence used in FASTA format and if manual curation was done for the particular

protein, the experimental evidence and the PubMed reference to the paper is given.

Search or Download Secretomes by Species

This section allows searching by species of secreted proteins, which are either predicted or curated. You can either select from a drop-down menu one of 53 species or manually input a species to search for. When using the drop-down menu, you may also select a protein set, either

Complete Secretome or Curated Proteins. The complete secretome is all proteins in the species predicted or curated by UniProt or our curator to be secretomes. From these options you may either search or download the FASTA. Search gives a listing of proteins similar to searching for keyword as above where an individual protein may be clicked to view details. FASTA download allows you to download the FASTA for a particular species. When "Curated Proteins" is selected, a list of available proteins is given on a Web page which may be copied and pasted. When "Complete Secretomes" is selected a window will appear allowing you to download and save a ".fas" file for that species since usually the entire FASTA file would be too large to display on screen.

BLAST Search

This section allows either a BLASTP or a BLASTX search against either of our two fungal databases. One is the secretome database containing our predicted and confirmed proteins and the other is for all fungal proteins in our database. The input format is a sequence or file in FASTA format. The NCBI BLAST page provides more information about how to use BLAST.[16]

Community Annotation

This section is a Community Annotation submission page allowing the user community to submit a protein for manual curation and addition into our database. The required entries are email, RefSeq gi and accession numbers, subcellular location of the protein, evidence and reference for the submission. Entries will be curated and if confirmed, entered into our database. Currently, we have manually curated secreted proteins from *Aspergillus niger* based on Tsang et al. [19] and *A. oryzae* based on Oda et al. [20] We would like to request the fungal secretome research community to submit experimentally verified secreted

fungal proteins to FunSecKB using this utility. Once a protein has been curated, it will be permanently included as part of the database.

Acknowledgments We thank Dr. Gary Walker for his mentoring support and Jessica Orr for assistance in manual data curation. The work is supported by Youngstown State University (YSU) Research Council (Grants 2009-10 #04-10 and 2010-2011 #12-11), YSU Research Professorship (2009–2011), and the College of Science, Technology, Engineering, and Mathematics Dean's reassigned time for research to XJM.

References

1. Tjalsma H, Bolhuis A, Jongbloed JD, Bron S, van Dijl JM (2000) Signal peptide-dependent protein transport in Bacillus subtilis: a genome-based survey of the secretome. Microbiol Mol Biol Rev 64:515–547
2. Greenbaum D, Luscombe NM, Jansen R, Gerstein M (2001) Interrelating different types of genomic data, from proteome to secretome: 'oming in on function. Genome Res 11:1463–1468
3. Hathout Y (2007) Approaches to the study of the cell secretome. Expert Rev Proteomics 4:239–248
4. Simpson JC, Mateos A, Pepperkok R (2007) Maturation of the mammalian secretome. Genome Biol 8:211
5. Lum, G. and Min, X. J. (2011) FunSecKB: the Fungal Secretome KnowledgeBase. *Database* 2011: doi:10.1093/database/bar001.
6. O'Toole N, Min XJ, Storms R, Butler G, Tsang A (2006) Sequence-based analysis of fungal secretomes. Appl Mycol Biotechnol 6:277–296
7. Min XJ (2010) Evaluation of computational methods for secreted protein prediction in different eukaryotes. J Proteomics Bioinformatics 3:143–147
8. Choi J, Park J, Kim D et al (2010) Fungal secretome database: integrated platform for annotation of fungal secretomes. BMC Genomics 11:105
9. Bendtsen JD, von Nielsen H, Heijne G, Brunak S (2004) Improved prediction of signal peptides: SignalP 3.0. J Mol Biol 340:783–795
10. Emanuelsson O, Nielsen H, Brunak S, von Heijne G (2000) Predicting subcellular localization of proteins based on their N-terminal amino acid sequence. J Mol Biol 30:1005–1016
11. Krogh A, Larsson B, von Heijne G, Sonnhammer ELL (2001) Predicting transmembrane protein topology with a hidden Markov model: application to complete genomes. J Mol Biol 305:567–580
12. Käll L, Krogh A, Sonnhammer ELL (2004) A combined transmembrane topology and signal peptide prediction method. J Mol Biol 338:1027–1036
13. Horton P, Park KJ, Obayashi T, Fujita N, Harada H, Adams-Collier CJ et al (2007) WoLF PSORT: protein localization predictor. Nucleic Acids Res 35:W585–W587 (Web Server issue)

[16]http://blast.ncbi.nlm.nih.gov/Blast.cgi.

14. Sigrist CJA, Cerutti L, de Casro E, Langendijk-Genevaux PS, Bulliard V, Bairoch A et al (2010) PROSITE, a protein domain database for functional characterization and annotation. Nucleic Acids Res 38:161–166

15. Poisson G, Chauve C, Chen X, Bergeron A (2007) FragAnchor a large scale all Eukaryota predictor of glycosylphosphatidylinositol-anchor in protein sequences by qualitative scoring. Genomics Proteomics Bioinformatics 5:121–130

16. Matthews BW (1975) Comparison of the predicted and observed secondary structure of T4 phage lysozyme. Biochim Biophys Acta 405:442–451

17. Baldi P, Brunak S, Chauvin Y, Andersen CA, Nielsen H (2000) Assessing the accuracy of prediction algorithms for classification: an overview. Bioinformatics 16:412–424

18. Menne KM, Hermjakob H, Apweiler R (2000) A comparison of signal sequence prediction methods using a test set of signal peptides. Bioinformatics 16:741–742

19. Tsang A, Butler G, Powlowski J et al (2009) Analytical and computational approaches to define the *Aspergillus niger* secretome. Fungal Genet Biol 46:S153–S160

20. Oda K, Kakizono D, Yamada O, Iefuji H, Akita O, Iwashita K (2006) Proteomic analysis of extracellular proteins from *Aspergillus oryzae* grown under submerged and solid-state culture conditions. Appl Environ Microbiol 72:3448–3457

Gagan Garg and Shoba Ranganathan

Abstract

With the advent of next-generation sequencing approaches and mass spectrometry techniques, there is a huge explosion in nucleotide and protein sequence data. Despite this increase in sequence data, several proteins remain unannotated, such as hypothetical proteins. Annotation and extraction of secretory proteins from the proteome using labor-intensive wet-lab techniques is prohibitive. Computational tools can be used to provide putative functionality, prior to experimental validation. This chapter introduces a bioinformatics workflow system using the best currently available free computational tools for the annotation of hypothetical proteins and prediction and analysis of secreted proteins as therapeutic targets, applied to pathogenic fungi, *Cryptococcus gattii*, and *Cryptococcus neoformans var. grubii*.

Keywords

Annotation • Drug targets • Interproscan • Protein domains • BRITE • FASTA • KEGG • KAAS • SPAAN

Introduction

The proteome is the entire set of proteins expressed by an organism. It is a valuable tool for studying molecular function, development and progression of different life stages, and much more. With many fungal genomes sequenced or under sequencing led to the identification of whole genome and protein sequences. Usually genes are predicted by using gene prediction tools followed by prediction of coding regions. These putative proteins are annotated based on the similarity of other organisms' proteins in the same taxonomic class. Many proteins still remain unannotated using these practices. This chapter

G. Garg
Department of Chemistry and Biomolecular Sciences, Macquarie University, Sydney, NSW 2109, Australia

S. Ranganathan (✉)
Department of Chemistry and Biomolecular Sciences, Macquarie University, Sydney, NSW 2109, Australia

Department of Biochemistry, Yong Loo Lin School of Medicine, National University of Singapore, 8 Medical Drive, Singapore 117597, Singapore
e-mail: shoba.ranganathan@mq.edu.au

outlines an approach to functionally annotate putative proteins in terms of pathways, gene ontology, and protein domains from recently sequenced fungal genomes, followed by secretory peptides and therapeutic target prediction.

Materials

Data

1. Hypothetical protein or nucleotide fungal sequences to be annotated are obtained from NCBI[1] or from other fungal genomes sources in FASTA format. We downloaded a total of 6,210 proteins for *Cryptococcus gattii* (Serotype B) [1] and 6,967 proteins for *Cryptococcus neoformans var. grubii* (Serotype A) [2].
2. Mouse and human proteins[2] are obtained from NCBI proteins database.
3. Known drug targets dataset[3] are obtained from DrugBank [3, 4].

Software

Protein Level Annotation

1. NCBI's ORF finder[4] or EMBOSS [5] getorf tool (for local set up) for prediction of open reading frames from nucleotide sequences.
2. Fungal genomes Blast[5] to search for proteins homologous to known fungal proteins.
3. Blast2go[6] [6, 7] for gene ontology annotation of proteins.
4. KAAS[7] [8] for pathway mapping of proteins.
5. iPath2[8] [9] for graphical representation of pathway associations of proteins in the dataset.

6. Interproscan[9] [10] for protein domains mapping.
7. BLASTP[10] [11] for sequence similarity search of the query sequence against different datasets.

Secretory Proteins Prediction

1. SignalP[11] [12] for prediction of classically secreted proteins (CSP).
2. SecretomeP[12] [13] for prediction of nonclassical secreted proteins (NCSP).
3. TargetP[13] [14] for prediction of mitochondrial proteins.
4. TMHMM[14] [15] for prediction of transmembrane proteins.

Therapeutic Target Prediction

1. SPAAN [16] to predict the probability of a protein to act as adhesins. This tool is available under free academic license from the developers of the tool.

Methods

The protocol described here is a general approach to functionally annotate hypothetical proteins based on similarity searches. However, many parts of the workflow (shown in Fig. 55.1) are independent of others, so some parts can be deleted according to the requirements. We applied the protocol to download (refer to the Sect. Data) *C. gattii* and *C. grubii* proteins.

Translate Nucleotide Sequences into Protein Sequences

1. Submit nucleotide sequences (if used) in FASTA format at NCBI ORF finder web server for conceptual translation into putative

---

[1]http://www.ncbi.nlm.nih.gov.

[2]http://www.ncbi.nlm.nih.gov/protein.

[3]http://www.drugbank.ca/downloads.

[4]http://www.ncbi.nlm.nih.gov/projects/gorf/.

[5]http://www.ncbi.nlm.nih.gov/sutils/genom_table.cgi? organism=fungi.

[6] http://www.blast2go.org/.

[7] www.genome.jp/kaas.

[8] http://pathways.embl.de/.

[9]http://www.ebi.ac.uk/Tools/pfa/iprscan/.

[10]ftp://ftp.ncbi.nih.gov/blast/executables/blast+/LATEST/.

[11]http://www.cbs.dtu.dk/services/SignalP/.

[12]http://www.cbs.dtu.dk/services/SecretomeP/.

[13]http://www.cbs.dtu.dk/services/TargetP/.

[14]http://www.cbs.dtu.dk/services/TMHMM/.

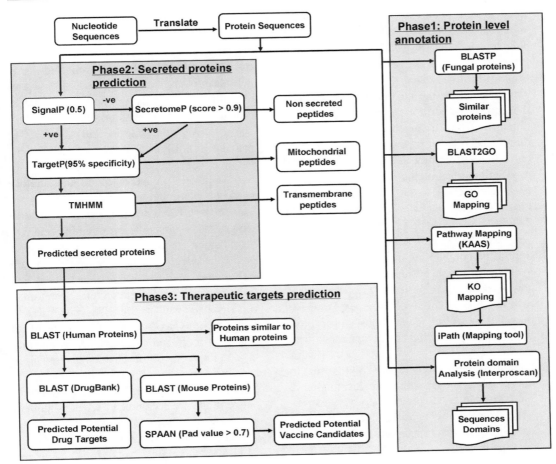

Fig. 55.1 The workflow protocol described here is a general approach to functionally annotate hypothetical proteins based on similarity searches. Many parts of the workflow are independent of others, so some parts can be deleted according to the requirements

proteins from predicted open reading frames. Use standard code (translation table 1) or the alternative yeast nuclear code (translation table 12), depending on the fungal organism. Consider the minimum length of 100 bases for nucleotide sequences translation to get reliable predictions in the following steps. For local setup, install EMBOSS getorf package on your local machine.

After all the nucleotide sequences get translated into proteins, run them through each phase of the work flow shown in Fig. 55.1.

Protein Level Annotation

For annotation of proteins, BlastP against known fungal proteins, Blast2go, KAAS, and Interproscan are used as shown in Phase 1 of Fig. 55.1.

1. Paste protein sequence on fungal genomes blast page. Choose query and database as protein and blast program as BlastP. We found 1,278 (20.6 %) *C. gattii* proteins dissimilar to *Saccharomyces cerevisiae* proteins. 508 (7.3 %) *C. grubii* proteins were found dissimilar to Aspergillus known proteins.

2. Load protein sequences into Blast2go software for gene ontology annotation.[15] We were able to annotate 135 (2.17 %) hypothetical proteins.

3. Submit protein sequences to KAAS web server. Uncheck the nucleotide box. Choose the same organism for proteins or the closest organism from organism list box using bi-directional best hit method. KAAS will map our KEGG orthology (KO) terms along with association of proteins with pathways and BRITE objects. All the files can be downloaded on local machine. We were able to map 2,288 (36.8 %) proteins with KO terms for *C. gattii* and 2,399 (34.4 %) for *C. grubii*.

4. Submit unique KO ids from KAAS mapping to iPath for graphical representation of pathway associations in proteins sample on global pathway maps. This tool provides extensive map customization capabilities. After plotting, the map can be downloaded on local machine.

5. Submit protein sequences in FASTA format to Interproscan for protein domains mapping. This program is computationally very expensive to run locally, especially for large datasets. We were able to annotate 6,001 (96.6 %) proteins for *C. gattii* and 6,614 (94.9 %) proteins for *C. grubii* with protein domains.

Secretory Proteins Prediction

For prediction of ES proteins, a combination of four tools, SecretomeP, SignalP, TargetP, and TMHMM is used as shown in Phase 2 of Fig. 55.1. All the methods described here have been implemented previously in fungal studies [17, 18].

1. Submit protein sequence in FASTA format at SignalP 3.0 server. Use eukaryotes as an organism group for fungal proteins. SignalP is based on neural networks (NN) and Hidden Markov models (HMM). We recommend the use of both methods for reliable prediction

results with standard output. Truncation field needs to be set according to the protein sequence. In the output, choose sequence having D score in SignalP-NN result and signal peptide probability in SignalP-HMM greater than 0.5 as classically secreted. These thresholds have been very well tested in other studies [19]. We were able to find 419 (6.7 %) proteins as CSP for *C. gattii* and 462 (6.6 %) for *C. grubii* using SignalP.

2. Submit protein sequences found nonsecretory in the previous step to SecretomeP 2.0 server. This server predicts nonclassical secretory proteins based on large number of amino acid features along with results of other feature prediction servers such as SignalP to obtain information on various post-translational and localization aspects of the protein. SecretomeP run SignalP as well but SignalP is run separately in the previous step because of the stringent cut offs of D score and signal peptide probability used in the protocol for reliable results. This is not the case for SignalP running inside SecretomeP. SecretomeP is based on neural network and we recommend selecting protein as nonclassical secreted where NN score is greater than equal to 0.9. We were able to find 94 (1.5 %) proteins as NCSP for *C. gattii* and 106 (1.5 %) for *C. grubii* using SecretomeP.

3. Combine protein sequences predicted as classically and nonclassically secreted and submit to TargetP 1.1 server. Select Non-plant as organism group and specificity greater than 0.95 in cut offs section. Consider a protein sequence as mitochondrial if Loc column is M in the output. Delete these proteins from the set of secretory proteins predicted in previous steps. A total of 19 (0.3 %) for *C. gattii* and 20 (0.3 %) for *C. grubii* were predicted as mitochondrial proteins.

4. Submit final proteins dataset after the deletion of mitochondrial proteins to TMHMM 2.0 server. Consider a protein sequence as non-transmembrane protein if number of predicted TMHs (transmembrane helices) is 0 or 1 in the final output. Such proteins are finally considered as secretory proteins. We consider proteins having one transmembrane helix in our final set

---

[15]A detailed documentation of how to run Blast2go is available at http://www.blast2go.org/start_blast2go.

of secretory proteins as these can be surface proteins having therapeutic value as potential vaccine candidates. A total of 384 (6.2 %) for *C. gattii* and 440 (6.3 %) for *C. grubii* were finally predicted as ES proteins.

In addition to the computational approach (shown in Fig. 55.1) for the prediction of ES proteins, sequence similarity search can be performed against known fungal ES proteins for the prediction of ES proteins, using BlastP.

Therapeutic Target Prediction

Lot of fungal species are pathogenic to humans. ES proteins of pathogens play a key role during pathogenic infections [20]. ES proteins predicted in Phase 2 can be checked computationally for therapeutic value.

This phase of the protocol is a tricky one. All the steps of this phase need to be performed on a local machine and command line operation is necessary. No specific web server like other parts of protocol is available. Different components are combined together for therapeutic targets prediction as shown in Phase 3 in Fig. 55.1.

1. To be a potential therapeutic target, a protein should not be present in human. To find out the secretory proteins similar to human proteins, search secretory proteins for sequence similarity against human proteins using BlastP. Sample command line for this operation is as follows:

blastall -i query -d human.fa –m 8 –e 1e-08 -o blast.out

Here query is the input file of protein sequences in fasta format, human.fa is the database file used for blast search, blast.out is the blast output file. Use –m 8 option to provide blast result in tabular format, which is easy to parse. E-value threshold is 1e-08. The command line shown here can be altered according to the datasets.[16]

Proteins found dissimilar to human proteins are searched for therapeutic value in terms of drug targets and potential vaccine candidates. We found 245 (3.9 %) secreted proteins for *C. grubii* and 261 (3.8 %) for *C. gattii*, similar to human proteins.

2. For drug target prediction, human dissimilar ES proteins are searched for sequence similarity against known drug targets from DrugBank using BlastP. Use the same command line mentioned above for this operation by changing the input, database, and output files. We found six potential drug targets for *C. gattii* and four for *C. grubii* in respective secreted proteins, mappable to known drug targets.

3. To test human dissimilar ES proteins for a potential vaccine candidate, a protein should not be present in mouse along with humans because most of the vaccine candidates are tested on mouse before they are tested on humans, so proteins that are found dissimilar to humans are tested for similarity against mouse proteins. Use the same command line mentioned above for this operation changing the input, database, and output files. We found further 86 (1.4 %) for *C. gattii* and 108 (1.5 %) for *C. grubii* secreted proteins similar to human dissimilar ES proteins.

Adhesins are cell surface proteins that are present during host pathogen invasion. These proteins play an important part in pathogenicity. Due to their important role in pathogenic infection, these proteins are good vaccine candidates.

4. To predict the probability of a protein to act as adhesins run SPAAN program according to program guidelines for proteins found dissimilar to human and mouse proteins. Consider a protein to be predicted as adhesion if Pad-value in the SPAAN output is greater than 0.7. This tool has been applied previously to fungal proteins for prediction of adhesins and adhesin-like molecules [21]. We predict 33 (0.5 %) for *C. gattii* and 35 (0.5 %) for *C. grubii* as potential vaccine candidates.

Detailed result files from our fungal protein annotation are available from http://estexplorer. biolinfo.org/fungal_annotation/

Notes

1. All the tools except ORF finder and iPath used in the protocol are available free to install locally under academic license. Although some of the tools available as web servers, it is recommended to install these tools locally on a Linux machine by following tool installation guidelines for big sequence datasets.

2. All the databases used for blast search locally needs to converted to blastable format before use by formatdb (provided in blast executables).

3. All the therapeutic targets predicted using this protocol are preliminary predictions which need to be further validated by additional computation analysis such as structural modeling and by experimental assays.

Acknowledgments We would like to thank Mr. Ben Herbert, for introducing us to the pathogenic Cryptococcal fungal genomes. GG acknowledges the award of Australian Postgraduate Award scholarship from Macquarie University.

References

1. *Cryptococcus gattii* Sequencing Project, Broad Institute of Harvard and MIT. http://www.broadinstitute.org/. Accessed 20 Jul 2011.

2. *Cryptococcus neoformans var. grubii* H99 Sequencing Project, Broad Institute of Harvard and MIT. http://www.broadinstitute.org/. Accessed 22 Jul 2011.

3. Wishart DS, Knox C, Guo AC, Cheng D, Shrivastava S, Tzur D et al (2008) DrugBank: a knowledgebase for drugs, drug actions and drug targets. Nucleic Acids Res 36:D901–D906

4. Wishart DS, Knox C, Guo AC, Shrivastava S, Hassanali M, Stothard P et al (2006) DrugBank: a comprehensive resource for in silico drug discovery and exploration. Nucleic Acids Res 34:D668–D672

5. Rice P, Longden I, Bleasby A (2000) EMBOSS: the European Molecular Biology Open Software Suite. Trends Genet 16:276–277

6. Gotz S, Garcia-Gomez JM, Terol J, Williams TD, Nagaraj SH, Nueda MJ et al (2008) High-throughput functional annotation and data mining with the Blast2GO suite. Nucleic Acids Res 36:3420–3435

7. Conesa A, Gotz S, Garcia-Gomez JM, Terol J, Talon M, Robles M (2005) Blast2GO: a universal tool for annotation, visualization and analysis in functional genomics research. Bioinformatics 21:3674–3676

8. Moriya Y, Itoh M, Okuda S, Yoshizawa AC, Kanehisa M (2007) KAAS: an automatic genome annotation and pathway reconstruction server. Nucleic Acids Res 35:W182–W185

9. Letunic I, Yamada T, Kanehisa M, Bork P (2008) iPath: interactive exploration of biochemical pathways and networks. Trends Biochem Sci 33:101–103

10. Zdobnov EM, Apweiler R (2001) InterProScan—an integration platform for the signature-recognition methods in InterPro. Bioinformatics 17:847–848

11. Altschul SF, Gish W, Miller W, Myers EW, Lipman DJ (1990) Basic local alignment search tool. J Mol Biol 215:403–410

12. Bendtsen JD, Nielsen H, von Heijne G, Brunak S (2004) Improved prediction of signal peptides: SignalP 3.0. J Mol Biol 340:783–795

13. Bendtsen JD, Jensen LJ, Blom N, Von Heijne G, Brunak S (2004) Feature-based prediction of non-classical and leaderless protein secretion. Protein Eng Des Sel 17:349–356

14. Emanuelsson O, Nielsen H, Brunak S, von Heijne G (2000) Predicting subcellular localization of proteins based on their N-terminal amino acid sequence. J Mol Biol 300:1005–1016

15. Krogh A, Larsson B, von Heijne G, Sonnhammer EL (2001) Predicting transmembrane protein topology with a hidden Markov model: application to complete genomes. J Mol Biol 305:567–580

16. Sachdeva G, Kumar K, Jain P, Ramachandran S (2005) SPAAN: a software program for prediction of adhesins and adhesin-like proteins using neural networks. Bioinformatics 21:483–491

17. Jain P, Podila G, Davis M (2008) Comparative analysis of non-classically secreted proteins in *Botrytis cinerea* and symbiotic fungus *Laccaria bicolor*. BMC Bioinformatics 9:O3

18. Choi J, Park J, Kim D, Jung K, Kang S, Lee YH (2010) Fungal secretome database: integrated platform for annotation of fungal secretomes. BMC Genomics 11:105

19. Nagaraj SH, Gasser RB, Ranganathan S (2008) Needles in the EST haystack: large-scale identification and analysis of excretory-secretory (ES) proteins in parasitic nematodes using expressed sequence tags (ESTs). PLoS Negl Trop Dis 2:e301

20. Ranganathan S, Garg G (2009) Secretome: clues into pathogen infection and clinical applications. Genome Med 1:113

21. Upadhyay SK, Mahajan L, Ramjee S, Singh Y, Basir SF, Madan T (2009) Identification and characterization of a laminin-binding protein of *Aspergillus fumigatus*: extracellular thaumatin domain protein (AfCalAp). J Med Microbiol 58:714–722

Application of Support Vector Machines in Fungal Genome and Proteome Annotation

56

Sonal Modak, Shimantika Sharma, Prashant Prabhakar, Akshay Yadav, and V.K. Jayaraman

Abstract

Support Vector Machines (SVM) is a statistical machine learning algorithm that has been used extensively in the past 3 years in computational biology. SVM has been widely used to detect, classify, and predict complex biological patterns. SVMs have been widely applied to many areas of bioinformatics, including protein function prediction, functional site recognition, transcription initiation site prediction, and gene expression data classification. This chapter gives a brief overview of SVM algorithm along with its specific applications in fungal genome and proteome annotations.

Keywords

Support vector machines (SVM) algorithm • Kernel functions • Hyperplane equation

Introduction

Some fungi are pathogens and cause diseases in plants and humans, whereas others are a rich source of therapeutic metabolites and useful enzymes, thus serving as basic models for molecular and cellular biology [1]. Genome sequencing and functional genomics provide an insight into the biological mechanisms in these fungi. Around 40 fungal genome sequences are now available and a further 50 genome sequencing projects are in progress. The advances in genomic technologies, such as microarrays and high-throughput sequencing, have enabled detailed analysis of fungal biochemistry. This analysis is automated using various bioinformatics tools in order to save time and effort.

S. Modak
Bioinformatics Centre, University of Pune, Ganeshkhind, Pune, Maharashtra 411007, India

S. Sharma • P. Prabhakar
Department of Biotechnology, Dr. D.Y. Patil University, Mumbai- Bangalore Highway, Tathawade, Pune, Maharashtra 411033, India

A. Yadav • V.K. Jayaraman (✉)
Centre for Development of Advanced Computing (C-DAC), Scientific and Engineering Computing Group (SECG), University of Pune, Ganeshkhind, Pune, Maharashtra 411007, India
e-mail: jayaramanv@cdac.in

V.K. Gupta et al. (eds.), *Laboratory Protocols in Fungal Biology: Current Methods in Fungal Biology*, Fungal Biology, DOI 10.1007/978-1-4614-2356-0_56, © Springer Science+Business Media, LLC 2013

Rapid developments in fungal genomics and proteomics have generated a large amount of biological data. Development and application of machine learning tools can help to solve challenging problems and pave the way for discovery of potent drugs. Analyzing fungal data sets requires understanding of the data by inferring structure or patterns from the data. Examples of this type of analysis include fungal protein structure prediction, fungal gene classification, and so forth.

Artificial Intelligence and machine learning-based methods can be employed for the purpose of the aforementioned annotations. Different classifiers based rigorously on principles of learning theories and machine learning formalisms can be employed for recognizing patterns in genomic and proteomic data. These classifiers capture the patterns provided by the domain information and group the data based on functionalities of the data. Classification is in general a supervisory learning method. In this method the algorithm needs domain features along with their functionalities for training data. As an example, we can illustrate the classification of fungal adhesins as follows: We start with a set of sequences which are experimentally annotated already as adhesins and a set of non-adhesins. Domain features can be extracted from these sequences and the classifier builds a model to group the data into adhesins and non-adhesins. This model, known popularly as the supervisory learning model, can readily be used for annotating new query sequences. Many different machine learning methods have already been applied for classification, like k-nearest neighbors [2], hierarchical clustering, self-organizing maps, [3, 4], and Support Vector Machines (SVM) or Bayesian networks. For solving classification problems, machine learning techniques first obtain information from a set of already labelled instances, called training data, and then use this information to classify unknown instances, called test data.

Material

Support Vector Machine

SVM is a classifier that has been formulated from statistical learning theory and structural risk minimization principle by Vapnik (1995) [5] and is widely employed in different fields of science and engineering. It is a parametric statistical linear classifier that performs a nonlinear mapping of the input space to a new feature space to which a linear machine can be applied. SVM constructs a hyperplane separating the positive examples from negative ones in the new space representation (Fig. 56.1). To avoid overfitting, SVM chooses the Optimal Separating Hyperplane that maximizes the margin in feature space [6]. The margin is defined as the minimal distance between the hyperplane and the training examples. The selected data points that support the hyperplane

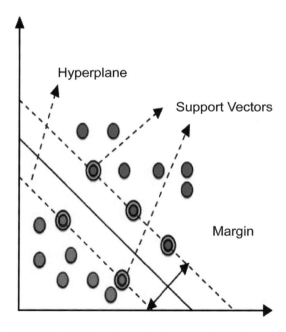

Fig. 56.1 SVM-based classification using hyperplane

are called support vectors. A smaller number of support vectors reflect a better generalization for linearly separable problems. SVM employs a maximum margin hyperplane for separating examples belonging to two different classes. For nonlinearly separable problems, the data are first transformed into a higher dimensional feature space and, subsequently, SVM employs a maximum margin linear hyperplane. Appropriate kernel functions are then employed for carrying out all the simulations in the input space itself.

During the past 3 years the popularity of SVMs as classification and prediction tools has increased drastically [7]. The main reason for the superior performance of SVMs is due to the fact that they minimize the structural risk as opposed to the empirical risk employed by some popular classifiers. This feature of SVM produces a more generalized model which performs well, both on the training data and unseen data.

The tremendous success of SVMs in prediction and classifications has enabled them to be used in fungal genome and proteome annotations. Fungi are eukaryotic saprophytic organisms that produce many commercially important products. Additionally, many fungi are important clinically because they cause many diseases in humans, plants, and other animals. During the postgenomic era, the amount of data available on fungal genomes and proteomes has grown enormously. Hence, there is an urgent need for development of new, rapid, high-throughput in silico techniques for fungal genome and proteome annotations, as the experimental techniques, though highly reliable, are very expensive and time-consuming.

For the purpose of identification of gene and protein functions employing SVM we need to provide domain features. As an example, for identification of defensin proteins the presence of certain amino acids or dipeptides in high-propensity levels may be necessary. Hydrophobicity can be an important domain feature required for prediction of membrane localized proteins. Similarly structural features, like accessible surface area and contact order, may correlate with ligand binding proteins. From the sequences these features can be extracted and subsequently employed by SVM for building appropriate models.

Some of the features that can be used for genome annotation using SVM are listed below:

- Nucleotide compositions
- Dinucleotide frequencies
- Position-specific scoring matrices (PSSM) profiles
- K-mer counts
- Motif conservation
- Promoter melting temperature
- DNA bending
- Transcript length
- Binding site degeneracy
- Binding site conservation

Some of the features that can be used for proteome annotation using SVM are listed below:

- Amino acid composition
- Frequencies of doublet, triplet, and multiplet of amino acids
- Charge composition
- Hydrophobicity composition
- Multiplet composition
- Position-specific scoring matrix (PSSM)
- Number of protein–protein interactions
- Gapped amino acid composition
- Signal sequences
- Isoelectric point
- Mean hydropathy
- Codon usage bias
- Remote Homology using PSSM
- Secondary Structure information
- Structural features like surface accessibility, coordinates of atoms, contact order

Methods

Feature Selection

In order to understand the molecular mechanisms in fungi, it is essential to identify the relevant biological features/attributes in large and complex datasets that regulate biological processes underlying the fungal data. Each feature pattern carries information relevant to describe a specific function in fungal data. However, fungal datasets may contain redundant features that do not contribute at all to the classification task. To overcome this

Fig. 56.2 Advantages of feature selection

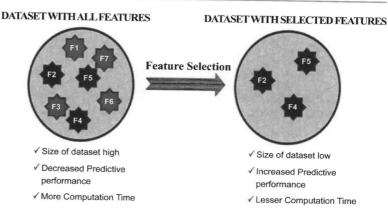

Fig. 56.3 Filter and Wrapper: the two approaches for feature selection

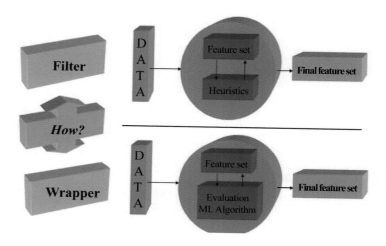

problem, one way is to select only a small subset of informative features from the fungal data (Fig. 56.2). This technique helps in getting rid of noisy features that do not correlate with the annotation problem at hand and thereby differentiates classes well. If such noisy features are not eliminated from the dataset, the classification performance of the learning algorithm may decrease dramatically, which will increase the computation time. Thus, there is a need to incorporate techniques that search for the best set of features with maximum classification performance. These dimensionality reduction techniques are often referred to as feature selection [8].

Feature selection algorithms mainly fall into two categories: wrappers and filters (Fig. 56.3). Wrappers make use of a learning algorithm to estimate the quality of features. Methods like Ant Colony Optimization and Genetic Algorithm in combination with a classifier like SVM fall into this category. On the other hand, filters evaluate the quality of features considering the inherent properties of the individual features without making use of a learning algorithm. Methods based on statistical tests and mutual information fall into this category. Since wrappers employ a learning algorithm for evaluating the quality of the genes, they give more accurate results than the filter methods. However, they need to train the learning algorithm, which makes the wrappers more time-consuming than filters. In addition, wrappers need to be re-executed when switching from one learning algorithm to another [9].

Most fungal classification algorithms involve three main steps: feature extraction, feature selection, and classification (Fig. 56.4). In the feature

Fig. 56.4 Application of
SVM in fungal genome
annotation

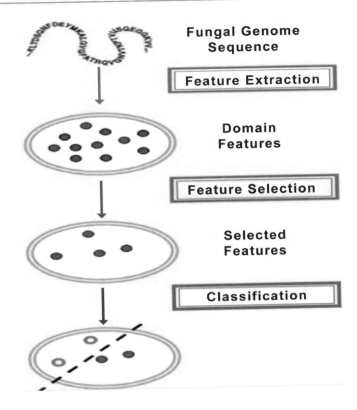

extraction step, a set of features is extracted that
are capable of describing a large fungal data
accurately. This step is followed by the feature
selection step, which is a dimensionality reduction step in which only the informative features
are selected from the previously extracted domain
features. Finally, the fungal data are divided into
different groups using a classifier such as SVM.

CASE STUDY: SVM-Based Prediction Method for Fungal Adhesins and Adhesin-Like Proteins

In this following case study, the authors have
tried to build a tool using SVM for prediction of
adhesin and adhesin-like proteins from fungal
domain (Fig. 56.5) [10]. Adhesins are cell surface proteins that confer on the microbes the ability to attach to cells, tissues, and/or abiotic
surfaces. Adhesins are the first line of pathogen's

strategy of host cell invasion and therefore
determine its virulence. Mostly fungal adhesins
have a definite modular structure consisting of
an N-terminal carbohydrate or peptide-binding
domain, central Ser and Thr-rich glycosylated
domains and C-terminal region that mediates
covalent cross-linking to the wall through
modified glycosylphosphatidylinositol (GPI)
anchors. But all fungal adhesins cannot be
identified on this basis. Experimental
identification of adhesins is very expensive and
time-consuming. Two computational algorithms
for the prediction of adhesins are currently available: Software Program for Prediction of Adhesins
and Adhesin-like proteins using Neural Networks
(SPAAN) and Malarial Adhesins and Adhesin-
like proteins Predictor (MAAP). Identification of
adhesion molecules would further our understanding of host-tissue adhesion in fungi, thereby
aiding the exploration of novel antifungal drug
targets and vaccine candidates.

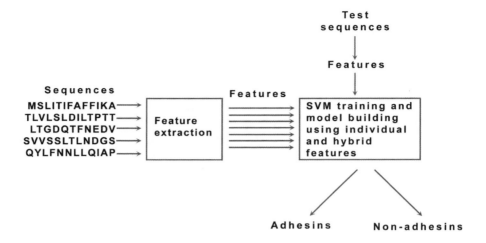

Fig. 56.5 SVM-based prediction system for Adhesins

Dataset Collection

For the compilation of the dataset, the authors used different keywords like "adhesin," "flocculin," and "agglutinin," while limiting the search to fungal domain to compile a raw pool of fungal adhesin sequences from sequence (Genbank and UniProt) databases. For a non-adhesin set they collected the proteins with known intracellular locations, such as nucleus, cytoplasm, mitochondria, endoplasmic reticulum, and so forth, and hypothetical proteins and protein fragments were removed.

Redundancy Removal

The CD-HIT program was used to reduce the redundancy to 50% in both positive and negative datasets. Thus, well-curated 75 adhesins and 341 non-adhesins from fungi were collected.

Feature Extraction

The authors used different types of models using different types of features. Hybrid models were also built combining different types of features. The lists of features used are as follows:

- Amino acid composition (AAC)
- Dipeptide composition (DPC)
- Charge composition (CC)
- Hydrophobicity composition (HC)
- Multiplet composition (MPC)

- Position-specific scoring matrix (PSSM) profile
- Hybrid features combining AAC, CC, HC, and MPC features (ACHM)

Results

Employing different types of features like AAC, DPC, and so on as, inputs to the SVM, different maximum margin hyperplane SVM models were developed. For maximizing performance they tried to use different kernels and tuned the kernel parameters for maximizing classification performance. The SVM models were trained on all the aforementioned features and their accuracy was checked. The best models were the hybrid model ACHM, PSSM-a, and PSSM-b, which had accuracies of 86.05%, 86.29%, and 81.97%, respectively.

Selection of Informative Features

The most appropriate domain features need to be input as features for SVM classification. Irrelevant features may intrude with the classification process and reduce algorithm performance. There is a need to incorporate techniques that search for an "informative" set of features with "increased" classification performance. These are referred to as feature selection. We have employed a filter-based feature selection methodology to further improve adhesin prediction performance.

Table 56.1 Filter-based feature selection methodology to improve adhesin prediction performance

	Selected number of features	Total number of features	Cross-validation accuracy	Kernel function
Selected Monopeptide	15	20	91%	RBF
Selected Dipeptide	192	400	88%	RBF
Selected Monopeptide and Dipeptide	207	420	89%	RBF

The Waikato Environment for Knowledge Analysis (WEKA) is a machine learning workbench that provides a general-purpose environment for automatic classification, regression, clustering, and feature selection for common data-mining problems in bioinformatics research. It contains an extensive collection of machine learning algorithms and data preprocessing methods complemented by graphical user interfaces for data exploration and the experimental comparison of different machine learning techniques on the same problem. WEKA can process data given in the form of a single relational table. Its main objectives are to (1) assist users in extracting useful information from data and (2) enable them to easily identify a suitable algorithm from which to generate an accurate predictive model.

In this work we have employed the information gain heuristics available in WEKA to rank the features. The top-ranking features that provided the best accuracy were subsequently employed in the final SVM model. It can be seen from Table 56.1 that the performance of adhesin prediction has improved by employing feature selection.

Review of Some Recent Fungal Bioinformatics Applications Using SVM

In this section, we outline a few important problems in bioinformatics where SVM and feature selection have been applied with interesting results on many real-world fungi case studies.

FaaPred: A SVM-based Prediction Method for Fungal Adhesins and Adhesion-like Proteins

This particular application has already been discussed previously. The server has been developed for SVM-based prediction of adhesins [10].

The prediction was based on two of the best-performing models, as discussed above. The prediction web server is freely accessible at http://bioinfo.icgeb.res.in/faap.

In silico Prediction of Yeast Deletion Phenotypes

Gene function can be predicted by studying gene duplications, gene deletions, and so forth. A modified simulated annealing algorithm for feature selection and weighting was developed to evaluate such phenotypic effects of gene deletions on yeast [11]. The high weight selected features comprised phylogenetic conservation scores for bacteria, fungi, *Ascomycota*, plants and mammals, degree of paralogy, and protein–protein interactions (PPIs) count. SVM along with weighted k-nearest neighbor were used for classification. The validation was done by prediction of essential genes that cause morphological changes in yeast.

Regulatory Analysis and Transcription Factor Target Prediction in Yeast

A number of regulatory binding sites within fungal genomes have been identified using high-throughput technologies such as array-based chromatin immunoprecipitation. However, in lower eukaryotes these sites are not identified completely. SVM was used to identify these binding sites, and the results were compared to those obtained using PSSMs for a set of 104 *Saccharomyces cerevisiae* regulators [12]. Results indicate that when specificity and positive predictive values are the same, SVM-based target classification is more sensitive (73% vs. 20%). SVM classifier was applied for each transcriptional regulator to all promoters in the yeast genome to obtain new targets. SVM predicted possible new roles for transcription factors like Gnc4 and Rap1. For instance, Rap1 was predicted to be involved in regulation of fermentative growth. Promoter melting temperature curves were also examined for the targets YJR060W, and the targets showed unique physical properties which distinguish them from other genes. High-quality predictions and biological functions can be studied, employing the feature reduction, and clustering strategies accuracy can be increased. Predictions for 104 transcription factors are available, and remaining factors can also be built.

FGsub: *Fusarium Graminearum* Protein Subcellular Localizations Predicted from Primary Structures

Fusarium graminearum is a fungal pathogen that causes several destructive crop diseases. The *F. graminearum* proteins should be assigned to different subcellular localizations in order to function properly. Therefore, these subcellular localizations can be used to gain insights into functions and pathogenic mechanism studies of the respective fungus. A new prediction method, Fgsub, was developed to predict the *F. gramineaurum* protein subcellular localizations from the primary structures [13]. A fungi data set with subcellular localization information is collected from UniProtKB database and used as training set. The subcellular locations were classified into ten groups. SVM was used for training of data and for prediction of the *F. graminearum* protein subcellular localizations. Ten-fold cross-validation accuracy showed efficient prediction results. It was also found that *F. graminearum* proteins bear significant sequence similarity to those present in the training set. For efficient annotations, BLAST was used, which increases the prediction coverage. Subcellular localizations of 12786 *F. graminearum* proteins were predicted, thus providing insights into protein functions and pathogenic mechanisms of this destructive pathogen fungus.

GPI-Anchored-Protein Identification System to Mine the Protein Databases of *Aspergillus fumigatus*, *Aspergillus nidulans*, and *Aspergillus oryzae*

Various computational approaches have been developed in order to identify GPI-anchored proteins in protein sequence databases. A new

sequence-based approach for identification using SVM algorithm was developed [14]. The algorithm recognizes COOH-terminal sequences and uses a classifier based on voting strategy to recognize appropriate NH2-terminal sequences. The classifier achieved a high accuracy of 96% in five-fold cross-validation testing. The voting-based classifier gave a higher accuracy of 98.88% when used on a test dataset of eukaryotic proteins. *S. cervisiae* protein sequences were used, which showed the classifier's ability to classify new unseen data. On using the predictor on three aspergilla species, 115 GPI-anchored proteins in *Aspergillu fumigatus*, 129 in *Aspergillus nidulans*, and 136 in *Aspergillus oryzae* were identified. Half of these proteins had conserved domains when the sequence-based conserved domain search was applied.

Screening Noncoding RNAs in Transcriptomes from Neglected Species Using PORTRAIT: Case Study of the Pathogenic Fungus *Paracoccidioides brasiliensis*

Structural genomic information can be obtained from transcriptome sequences. Such sequences can also be useful in characterizing neglected species that do not have the chance of undergoing whole-genome sequencing. But sequencing of these organisms is difficult owing to low quality of reads and incomplete coverage of transcripts. Another factor is lack of known protein homologs; also, noncoding RNAs may be caught during sequencing. These RNAs do not code for protein and instead perform the unique function of folding into other structural conformations. The analysis of such transcriptome sequences is limited. PORTRAIT is an algorithm that has been developed for analysis of such ncRNAs from poorly characterized species [15]. These sequences are translated by software that identifies sequencing errors and predicts putative proteins along with their transcripts. These are evaluated for coding potential by SVM. Two models have been developed. First, if a putative protein is found, a protein-dependent SVM model is used. If one is not found, a protein-independent SVM model is used. No homology information is used, as only ab-initio features are extracted. PORTRAIT was used for predicting ncRNAs from the transcriptome of pathogenic fungus *Paracoccidioides brasiliensis* and five related other fungi.

Determination of Protein Content of *Auricularia auricula* Using Near-Infrared Spectroscopy Combined with Linear and Nonlinear Calibrations

Near-infrared (NIR) spectroscopy has been used to determine the protein content of *Auricularia auricula*, also known as woody ear or tree ear, using partial least-squares (PLS), multiple regressions (MLR), and least-squares-support vector machine (LS-SVM) [16]. These were tested and compared against other different methods such as Savitzky-Golay (SG) smoothing, standard normal variate, multiplicative scatter correction (MSC), first derivative, second derivative, and direct orthogonal signal correction. A successive projections algorithm (SPA) was also used to find effective wavelength selection. Different combinations of pretreatment and calibration methods were compared based on the predictions. The optimal full-spectrum PLS model was obtained by raw spectra, whereas the optimal SPA-MLR, SPA-PLS, and SPA-LS-SVM model obtained MSC spectra. The best performance was achieved by SPA-LS-SVM model. NIR spectroscopy combined with SPA-LS-SVM is useful in determining the protein content of *A. auricula*.

ESLpred2: Improved Method for Predicting Subcellular Localization of Eukaryotic Proteins

ESLpred method uses new features as an input to SVM [17]. Three kingdom-specific protein sequence sets, 1198 fungi sequences, were included in the study for predictions. Evolutionary information in the form of profile composition along with whole and N-terminal sequence composition as an input feature vector of 440 dimensions was given. Overall accuracies of greater than 72% were achieved for fivefold cross-validation. Similarity search-based results, when used with whole and N-terminal sequence and

profile composition information, gave accuracies of 75.9%, 80.8%, and 76.6%, respectively. [1]

Vector-G: Multimodular SVM-Based Heterotrimeric G Protein Prediction

Heterotrimeric G proteins interact with GPCRs in response to any stimulus generated by hormones, neurotransmitters, chemokines, and sensory signals to intracellular signaling cascades. G protein subunits have been found to play an important role in different eukaryotic diseases including inflammation, cardiovascular diseases, and neurological disorders as well as in plant pathogens response, and differentiation and virulence of pathogenic fungi. All the methods available for finding new G proteins are based on homology search analyses, which are not robust. Vector-G is an SVM-based pattern recognition algorithm for finding new G proteins and their homologs [18]. Properties such as physicochemical and dipeptide, tripeptide, and hydrophobicity composition are used for generating SVM classifiers. This method gave 96.17%, 95.38%, and 97.6% sensitivity and 99.45%, 100%, and 100% specificity on tests sets for G protein alpha, beta, and gamma subunits, respectively. The algorithm correctly predicts known alpha, beta, and gamma subunits. New G protein subunits are predicted in 31 genome covering plant, fungi, and animal kingdom. The software is available at the website http://biomine.cs.uah.edu/bioinformatics/svm_prog/scripts/GProteins/vectorg.html. The supplementary files are available on http://www.bioinfo.de/isb/2008/08/0013/supplementary_material/

Physical PPIs Predicted from Microarrays

Microarray data reveal functionally associated proteins. Predicting physical interactions directly through microarrays are both important and challenging as most proteins that are associated are not actually in direct physical contact. Thus, an SVM-based method was developed to predict the pairs that are likely to interact [19]. This method was applied to predict interactions in yeast

---

[1]The ESLpred2 server is also developed and available with many models at http://www.imtech.res.in/raghava/eslpred2/.

(*S. cerevisiae*). Literature search revealed several new predictions that could be experimentally validated. This new method holds a promise to improve the annotation of interactions as one component of multisource integrated systems.

Kernel-Based Machine Learning Algorithm for the Efficient Prediction of Type III Polyketide Synthase Family of Proteins

Type III polyketide synthase is a protein family that has a significant role in biosynthesis of various polyketides in plants, fungi, and bacteria. These proteins have positive effects on human health; thus, developing a tool to identify the probability of sequence being a type III polyketide synthase is helpful. PKSIIIpred is a prediction server for type III PKS where SVM is used for classification [20]. The tool predicts efficiently the type III PKS superfamily of proteins with high sensitivity and specificity.[2]

Predicting PPIs from Protein Sequences using Meta Predictor

PPIs prediction based on meta approach has been developed, which predicts PPIs by using SVM that combines results by six independent predictors [21]. The method was used of *S. cerevisiae* and *Helicobacter pylori* datasets .The final predicted model trained on the PPIs dataset of *S. cerevisiae* was used to predict interactions in other species. The results obtained showed that this model is also capable in performing cross-species predictions.[3]

Conserved Codon Composition of Ribosomal Protein Coding Genes in *Escherichia coli, Mycobacterium tuberculosis*, and *S. cerevisiae*: Lessons from Supervised Machine Learning in Functional Genomics

Advances in genomics have resulted in creation of large sequence data. Annotation is currently relied on sequence comparison and homologs.

Codon composition, a fusion of codon usage and amino acid composition signals was used to accurately discriminate, in the absence of homology information, cytoplasmic ribosomal protein genes from other genes of known function in *S. cervisiae, E. coli*, and *M. tuberculosis* using SVM (light) [22]. Such analysis is helpful in determining features that provide individuality to ribosomal protein genes. Each of the sets of positively charged, negatively charged, hydrophobic residues, codon bias, contribute to their individual composition profile. SVM is used to detect, combine, and augment the representation of these signals in order to perform efficient classification. This method can have several alternatives by combining codon composition with other gene/protein classification attributes.

Softwares for SVM

A number of SVM softwares have been developed recently. Table 56.2 presents a list of some of these softwares.

Summary

Several fungi, form a rich source of therapeutic metabolites and the others act as pathogens causing diseases. It is thus essential to understand the biological mechanisms within fungi in order to gain insights into various biological problems. The amount of fungal genome data is thus increasing exponentially and in order to handle such a vast amount of data, machine learning techniques are applied in the field of fungal bioinformatics. SVM is a statistical learning tool that employs simple but very powerful machine learning algorithm for classification and regression. Feature selection is a method that helps in reducing the dimensionality of large fungal datasets by selecting only the relevant features. In this chapter, we have highlighted some of the recent fungal genome-proteome annotation problems in bioinformatics where SVM and feature selection have been applied successfully. A number of biological problems, such as identification of adhesion and

---

[2]The server is available at http://type3pks.in/prediction.
[3]The source code and the datasets are available at http://home.ustc.edu.cn/~jfxia/Meta_PPI.html.

Table 56.2 Software for support vector machines

Software	Salient features	Availability
LIBSVM	• Integrated software for support vector classification, regression, and distribution estimation. • Developed at National Taiwan University by Chang and Lin • Developed in C++ and Java. • Supports weighted SVMs for unbalanced data, multiclass classification, cross-validation, and automatic model selection. • Linear, polynomial, radial basis function, and neural (tanh) kernels available.	http://www.csie.ntu.edu.tw/~cjlin/libsvm/
SVM_light_	• Has a fast optimization algorithm. • Efficient implementation of the leave–one–out cross-validation and can be applied to very large datasets. • Developed in C++ • Polynomial, radial basis function, and neural (tanh) kernels available	http://svmlight.joachims.org/
SVMstruct	• Can model complex (multivariate) output data, such as trees, sequences, or sets. • Several implementations like SVMmulticlass, SVMcfg, SVMalign, SVMhmm,	http://svmlight.joachims.org/svm_struct.html
Weka	• Popular collection of machine learning algorithms developed in Java. • Supports a number of data-mining tasks such as data preprocessing, clustering, classification, regression, visualization, and feature selection. • Also contains an SVM implementation.	http://www.cs.waikato.ac.nz/ml/weka/
BioWeka	• Weka for data analysis tasks in biology, biochemistry and bioinformatics, and knowledge discovery. • Includes integration of the Weka LibSVM project.	http://sourceforge.net/projects/bioweka/
mySVM	• It is available as C++ source code and Windows binaries. • Kernels available include linear, polynomial, radial basis function, neural (tanh), and anova.	http://www-ai.cs.uni-dortmund.de/SOFTWARE/MYSVM/index.html
mySVM/db	• Efficient extension to mySVM. • Designed to run directly inside a relational database using an internal JAVA engine	http://www-ai.cs.uni-dortmund.de/SOFTWARE/MYSVMDB/index.html
Gist	• C implementation of SVM classification and kernel Principal Components Analysis. • Available as an interactive Web server at http://svm.sdsc.edu. • Kernels available include linear, polynomial, and radial.	http://www.chibi.ubc.ca/gist/
SmartLab	• Provides several SVM implementations, including cSVM, mcSVM, rSVM, and javaSVM1, and javaSVM2.	http://www.smartlab.dibe.unige.it/
MATLAB SVM Toolbox	• Developed by Gunn, implements SVM classification and regression. • Various kernels available include linear, polynomial, Gaussian radial basis function, exponential radial basis function, neural (tanh), Fourier series, spline, and B spline.	http://www.isis.ecs.soton.ac.uk/resources/svminfo/

(continued)

Table 56.2 (continued)

Software	Salient features	Availability
TinySVM	• C++ implementation of C-classification and C-regression that uses sparse vector representation. • Can handle thousands of training instances and feature dimensions.	http://chasen.org/~taku/software/TinySVM/
LS-SVMlab	• Developed by Suykens and is a MATLAB implementation of Least-Squares Support Vector Machines (LS–SVMs) • LS–SVMprimal–dual formulations, formulated for kernel PCA, kernel CCA, and kernel PLS	http://www.esat.kuleuven.ac.be/sista/lssvmlab/
LSVM	• Lagrangian Support Vector Machine is a very fast SVM implementation developed in MATLAB	http://www.cs.wisc.edu/dmi/lsvm/
ASVM	• Active Support Vector Machine is a very fast linear SVM script for MATLAB, by Mangasarian and Musicant, developed for large datasets.	http://www.cs.wisc.edu/dmi/asvm/
PSVM	• Proximal Support Vector Machine is a MATLAB script developed by Fung and Mangasarian.	http://www.cs.wisc.edu/dmi/svm/psvm/
GPDT	• Developed by Serafini, et al., is a C++ implementation for large-scale SVM classification in both scalar and distributed memory parallel environments	http://dm.unife.it/gpdt/
Spider	• Object-orientated environment for machine learning written in MATLAB • Implements SVM multiclass classification and regression	http://www.kyb.tuebingen.mpg.de/bs/people/spider/
LEARNSC	• Learning and Soft Computing by Kecman is available as MATLAB script.	http://www.support-vector.ws/html/downloads.html
SVM Toolbox	• MATLAB toolbox, contains many classification algorithms like linear and quadratic penalization, multiclass classification, e-regression, n-regression, wavelet kernel, and SVM feature selection.	http://asi.insa-rouen.fr/%7Earakotom/toolbox/index.html
SVMTorch	• Developed by Collobert and Bengio is a part of the Torch machine learning library • Distributed as C++ source code or binaries for Linux and Solaris	http://bengio.abracadoudou.com/SVMTorch.html

adhesion-like proteins, prediction of transcription factor target in yeast, prediction of subcellular localizations from primary structures, G-protein prediction, prediction of type III polyketide synthase family of proteins, PPIs predictions, and so on, have been solved using these machine learning approaches. These methods can be further enhanced by making use of high-performing computational systems, such as parallel computers, in order to solve complex fungal bioinformatics problems at a much faster rate.

References

1. Castrillo JI, Oliver SG (2006) Metabolomics and systems biology in *Saccharomyces cerevisiae*. Fungal Genomics 13(1):3–18
2. Li L, Weinberg CR, Darden TA, Pedersen LG (2001) Gene selection for sample classification based on gene expression data: study of sensitivity to choice of parameters of the ga/knn method. Bioinformatics 17:1131–1142
3. Baumgartner R, Windischberger C, Moser E (1998) Quantification in functional magnetic resonance

imaging: fuzzy clustering vs. correlation analysis. Magn Reson Imaging 16:115–125

4. Kohonen T (ed) (1997) Self-organizing maps, 3rd edn. Springer-Verlag, NewYork
5. Vapnik V (1995) The Nature of Statistical Learning Theory. Springer-Verlag, New York
6. Boser BE, Guyon IM, Vapnik VN (1992) A training algorithm for optimal margin classifiers. In: Haussler D (ed) 5th Annual ACM Workshop on COLT. ACM Press, Pittsburgh, PA, pp 144–152
7. Noble WS (2004) Support vector machine applications in computational biology. In: Schoelkopf B, Tsuda K, Vert J-P (eds) Kernel methods in computational biology. MIT, Cambridge, MA
8. Hall MA (2000) Correlation-based feature selection for discrete and numeric class machine learning. In: Proceedings of the 17th International Conference on Machine Learning, pp. 359–366
9. Hall MA (1999) Correlation-based feature selection for machine learning. PhD Thesis, Department of Computer Science, University of Waikato, Hamilton, NZ
10. Ramana J, Gupta D (2010) FaaPred: a SVM-based prediction method for fungal adhesins and adhesin-like proteins. PLoS One 5(3):9695
11. Saha S, Heber S (1998) In silico prediction of yeast deletion phenotypes. Genet Mol Res 5(1):224–232
12. Holloway DT, Kon M, Delisi C (2007) Machine learning for regulatory analysis and transcription factor target prediction in yeast. Syst Synth Biol 1(1):25–46
13. Sun C, Zhao XM, Tang W, Chen L (2010) FGsub: Fusarium graminearum protein subcellular localizations predicted from primary structures. BMC Syst Biol 4(2):S12
14. Cao W, Maruyama J, Kitamoto K, Sumikoshi K, Terada T, Nakamura S et al (2009) Using a new GPI-anchored-protein identification system to mine the protein databases of Aspergillus fumigatus, Aspergillus nidulans, and Aspergillus oryzae. J Gen Appl Microbiol 55(5):381–393
15. Arrial RT, Togawa RC, Brigido MM (2009) Screening noncoding RNAs in transcriptomes from neglected species using PORTRAIT: case study of the pathogenic fungus Paracoccidioides brasiliensis. BMC Bioinformatics 10:239
16. Liu F, He Y, Sun G (2009) Determination of protein content of Auricularia auricula using near infrared spectroscopy combined with linear and nonlinear calibrations. J Agric Food Chem 57(11):4520–4527
17. Garg A, Raghava GP (2008) ESLpred2: improved method for predicting subcellular localization of eukaryotic proteins. BMC Bioinformatics 9:503
18. Jain P, Wadhwa P, Aygun R, Podila G (2008) Vector-G: multi-modular SVM-based heterotrimeric G protein prediction. In Silico Biol 8(2):141–155
19. Soong TT, Wrzeszczynski KO, Rost B (2008) Physical protein-protein interactions predicted from microarrays. Bioinformatics 24:2608–2614
20. Mallika V, Sivakumar KC, Jaichand S, Soniya EV (2010) Kernel based machine learning algorithm for the efficient prediction of type III polyketide synthase family of proteins. J Integr Bioinform 7(1):143
21. Xia JF, Zhao XM, Huang DS (2010) Predicting protein-protein interactions from protein sequences using meta predictor. Amino Acids 39(5):1595–1599
22. Lin K, Kuang Y, Joseph JS, Kolatkar PR (2002) Conserved codon composition of ribosomal protein coding genes in Escherichia coli, Mycobacterium tuberculosis, and Saccharomyces cerevisiae: lessons from supervised machine learning in functional genomics. Nucleic Acids Res 30:2599

Bioinformatics Tools for the Multilocus Phylogenetic Analysis of Fungi

Devarajan Thangadurai and Jeyabalan Sangeetha

Abstract

Mycologists are generally identifying fungal communities by microscopic and macroscopic assessment. This conventional approach has several limitations due to the growth and environmental factors. Hence, molecular techniques and bioinformatics tools are essential in the field identification and characterization of fungi. Multilocus sequences are widely used in most of the bioinformatics tools and they can be used to recognize species boundaries. Nucleic acid and protein sequences-based analysis in fungal studies are revolutionizing the view on mycology. Numerous bioinformatics tools are available online to guide molecular biologists and biotechnologists. This chapter provides a guide to utilizing the available bioinformatics tools on the World Wide Web for sequence alignment, editing, and multilocus phylogenetic analysis.

Keywords

Bioinformatics • Tools • Softwares • Databases • Multilocus phylogenetic analysis • Fungi

Introduction

Conventional biochemical methods and phenotypic tests for fungal species differentiation are tedious and time-consuming and may require specialized tests. Recent developments in molecular biology and bioinformatics allow the consideration of other methods that are more universal and less time consuming [1]. Currently, scientific research requires parallel strategy to simultaneously gather, examine, integrate, and store the large volumes of data. In this scenario, researchers cannot attain their decisive goal without good

D. Thangadurai (✉)
Department of Botany, Karnataka University, Dharwad, Karnataka 580003, India
e-mail: drdthangadurai@gmail.com

J. Sangeetha
Department of Zoology, Karnataka University, 580003, Dharwad, Karnataka, India
e-mail: drsangeethajayabalan@gmail.com

data handling and analyzing skills. Bioinformatics is a new field of science that examines complex biological data on the basis of statistics and computer science [2, 3]. Computer programs in biology that help to reveal the principle mechanisms in biological problems related to the structure and function of macromolecules, biochemical pathways, disease processes, genetic analysis and evolution are discussed [4].

Studies of fungal diversity in natural environment have been detonated in the last 20 years, in large part due to the advent of the molecular techniques [5]. Regrettably, many of the molecular methods developed for prokaryotes are not appropriate for fungi. This is due to the fact that the ribosomal small subunit, which is the important key of prokaryotic diversity analysis, is insufficiently variable to precisely distinguish fungal taxa [6, 7]. A major challenge in mycology is to make sense of the enormous quantities of sequence data and structural data that are generated by genome sequencing projects, proteomics, and other large-scale molecular biology efforts [4]. Bioinformatics tools and techniques are playing an important

role in the study of fungal systems, especially the function and evolution of fungal genes and genomes. These methods are necessary to achieve a complete, quantitative, and reliable understanding of functions of fungi in molecular level.

The systematic comparison of genomic sequences and protein sequences from different fungal species represents a central focus of contemporary fungal genome analysis. Multilevel sequencing provides accurate species identifications as predicting evolutionary relationships among species. For the identification of most species a database of 26S 5'-end base sequences is adequate. Additional sequences are needed to predict more secluded relationships. Sequence alignment and editing can be made rapidly on computational methods [8, 9]. Various tools like CLUSTAL W/X, T-Coffee, DbClustal, Kalign, and MAFFT are available for multiple sequence alignments (MSAs) (Table 57.1). BLAST, FASTA, and FRACTURA are available for analyzing and interpreting algorithm, modeling and computer graphics of the databases on genomics and proteomics [10].

Table 57.1 List of bioinformatics tools, software, and programs used in multilocus phylogenetic analysis in fungi

Name	Description	Web link	References
A-Bruijn Alignment	Multiple alignment of sequences with repeated and shuffled elements	http://nbcr.sdsc.edu/euler/aba_v1.0_dl/	[11, 12]
Advanced PipMaker	Multiple sequence alignment program	http://pipmaker.bx.psu.edu/pipmaker/	[13]
ALIGN Query	Multiple sequence alignment program	http://xylian.igh.cnrs.fr/bin/align-guess.cgi	[14]
COBALT	Multiple protein sequence alignment using conserved domain and local sequence similarity information	http://www.ncbi.nlm.nih.gov/tools/cobalt/	[14]
AMAP	Multiple alignment tool for peptide sequences	http://packages.debian.org/sid/amap-align	[15]
Bali-Phy	Bayesian alignment and phylogeny estimation	http://www.biomath.ucla.edu/msuchard/bali-phy/	[14]
BLAST	Compare gene or protein sequences for similarity between the sequences	ftp://ftp.ncbi.nlm.nih.gov/blast	[16]
BOXSHADE	Pretty-printing of multiple sequence alignments	http://www.ch.embnet.org/software/BOX_doc.html	[14]
CINEMA	Color interactive editor for multiple alignments	http://www.bioinf.manchester.ac.uk/dbbrowser/CINEMA2.1/	[17]
CLUSTAL W	Multiple sequence alignment program for DNA and protein	http://clustal.org/clustal2	[18–20]
CodonCode aligner	Program for sequence assembly, editing and mutation detection	http://www.codoncode.com/aligner/	[21]

(continued)

Table 57.1 (continued)

Name	Description	Web link	References
ComAlign	Combining many multiple alignments in one improved alignment	http://www.daimi.au.dk/~ocaprani/ComAlign/ComAlign.html	[14]
Consensus	Calculates the consensus for the CLUSTAL or MSF multiple alignment	http://coot.embl.de/Alignment/consensus.html	[22]
CVTree	Infer phylogenetic relationships between microbial organisms by comparing their proteomes using a composition vector approach	http://cvtree.cbi.pku.edu.cn	[23]
DIALIGN-TX	Multiple sequence alignment program	http://dialign-tx.gobics.de/	[14]
DSC	Divide-and-Conquer Multiple Sequence Alignment	http://bibiserv.techfak.uni-bielefeld.de/dca/	[24–28]
eProbalign	Generation and manipulation of multiple sequence alignments using partition function posterior probabilities	http://probalign.njit.edu/probalign/login	[29]
ESPript 2.2	Multiple sequence alignments in PostScript	http://espript.ibcp.fr/ESPript/ESPript/	[30]
FFAS03	Profile-profile alignment and fold recognition algorithm	http://ffas.ljcrf.edu/ffas-cgi/cgi/document.pl?ses=#over	[31, 32]
FOLDALIGN	Semi-automated multiple global alignment and structure prediction	http://foldalign.ku.dk/	[33, 34]
GeneOrder 3.0	Ideal for comparing multiple GenBank genomes	http://binf.gmu.edu:8080/GeneOrder3.0/	[35, 36]
GOAnno	Multiple alignments of complete sequences	http://bips.u-strasbg.fr/GOAnno/GOAnnoHelp.html	[37]
JDotter	Multiple alignments of complete sequences	http://athena.bioc.uvic.ca/pbr/jdotter/	[38]
LALIGN	Finds multiple matching subsegments in multiple sequences	http://www.ch.embnet.org/software/LALIGN_form.html	[39]
LocARNA	Multiple alignment of RNA molecules	http://rna.informatik.uni-freiburg.de:8080/LocARNA.jsp	[22]
LVB	Parsimony and simulated annealing in the search for phylogenetic trees	http://biology.st-andrews.ac.uk/cegg/lvb.aspx	[40]
MAFFT	Offers a range of multiple alignment methods	http://mafft.cbrc.jp/alignment/software/	[41, 42]
MANVa 0.982b	Multilocus analysis of nucleotide variation	http://www.ub.edu/softevol/manva/	[22]
MASIA	Recognition of common patterns and properties in multiple aligned protein sequences	http://born.utmb.edu/masia/	[43]
Meme	Motif-based multiple sequence analysis	http://www.sdsc.edu/~tbailey/MEME-protocol-draft2/protocols.html	[22]
mlcoalsim 1.42	Multilocus coalescent simulations for evolution	http://www.ub.edu/softevol/mlcoalsim/	[44]
MOTIF	Sequence motif search	http://www.genome.jp/tools/motif/MOTIF3.html	[22]
MULALBLA	Multiple alignment with Blast	http://www-archbac.u-psud.fr/MULALBLA/mulalbla.html	[22]
Multilocus 1.3b	Facilitate analysis of multilocus population genetic data for genotypic diversity indices, linkage disequilibrium indices and population differentiation	http://www.agapow.net/software/multilocus/1.3b/view	[22]
Multiple Align Show	Multiple sequence alignment	http://www.bioinformatics.org/sms/multi_align.html	[45]

(continued)

Table 57.1 (continued)

Name	Description	Web link	References
Mumsa	Multiple sequence alignment	http://msa.sbc.su.se/cgi-bin/msa.cgi	[46]
MUSCLE	Progressive alignment of nucleic acids	http://www.drive5.com/muscle/docs.htm	[47]
MuSiC	Multiple sequence alignment with constraints	http://genome.life.nctu.edu.tw/MUSIC	[48]
NAST	Multiple sequence alignment server for comparative analysis of 16S rRNA genes	http://greengenes.lbl.gov/NAST	[49]
Opal	Multiple sequence alignment	http://opal.cs.arizona.edu/	[22]
ParAlign	Rapid and sensitive similarity search tool	http://www.paralign.org/	[50–52]
Pebble 1.0	Tool for analysis and simulation of maximum likelihood and least-squares methods	http://www.cebl.auckland.ac.nz/software2.php	[22]
Prof	Multiple alignment and structure prediction	http://www.aber.ac.uk/~phiwww/prof/	[22]
SCANMOT	Searching similarity using simultaneous scan of multiple sequence motifs	http://caps.ncbs.res.in/scanmot/scanmot.html	[22]
T-Coffee	Align DNA or RNA or protein sequences	http://www.tcoffee.org/	[53]
VerAlign	Multiple sequence alignment comparison	http://www.ibivu.cs.vu.nl/programs/veralignwww/	[22]
WebVar	Rapid estimation of relative site variability from multiple sequence alignments	http://www.pesolelab.it/Tools/WebVar.html	[54]
YASS	Sequence similarity search tool	http://bioinfo.lifl.fr/yass/	[55]

In general, bioinformatics tools are used to determine gene functional annotation, gene family evolution, and genome organization. These bioinformatics tools are developed with the help of internet tools and World Wide Web. Many universal programs and software are available, and they are providing with either free access or with charges to biotechnology databases [10]. Several programs, databases, and softwares can be used to perform MSA, multiple sequence editing, and phylogenetic analysis of individual DNA or protein sequences.

MSA method is a comparative method at the molecular level and it is a vital component of most of the bioinformatics techniques. The effectiveness of sequence analysis is based on the addition of more data to yield stronger analyses and also, if provided sequences which have <40 % residue identity the analyses will becomes unreliable [56]. Many methods are available for MSA of three or more biological sequences such as protein or nucleic acid of similar length. From the output of MSA, homology can be inferred and phylogenetic analysis can be carried out to study the evolutionary relationship between the sequences. Computational methods and algorithms are used to analyze the sequence alignments. Progressive alignment methods are more efficient to execute for many sequences and are commonly available on World Wide Web services [20, 53, 57, 58].

Materials

1. Computers with a lot of memory.
2. Macintosh, Windows, and Unix platforms.
3. Hi-speed internet connectivity.
4. Bioinformatics tools, softwares, and databases (freely available or on purchase).

Methods

Understanding the basic principles and methodologies used in bioinformatics is more important for mycologists to utilize available tools and to interpret results for the multilocus phylogenetic analysis. This chapter mainly focuses on several bioinformatics tools that are commonly used in the multilocus phylogenetic analysis of fungi with step-by-step guidance.

CLUSTAL W

Clustal programs text menu are provided with all of the options to do MSA and to create simple phylogenetic tree. With these simple menus, this program is highly convenient for the users and run on all computers. The output of MSA can be used to take printout or to manipulate and it can incorporate secondary structure information into the process [20, 57, 59, 60]. CLUSTAL X is a new graphical windows interface for the CLUSTAL W. CLUSTAL X includes new features like the ability to cut-and-paste sequences, selection of a subset of the sequences to be realigned, selection of a subrange of the alignment can be realigned and insert back to the original alignment and coloring option allows to highlight required features or exceptional residues in the alignment. CLUSTAL W is the most accurate and faster package [56].

CLUSTAL W can easily be installed by copying executable file to the system directory. Several parameters (named*.par) and an online help text file (clustalx.hlp for MS Windows, otherwise clustalx_help) are also required. These files should be copied to the directories specified by the PATH environment variable or user's current directory [18, 19].

1. Download and install the CLUSTAL W program.[1]

2. Open the CLUSTAL W program; in UNIX type ">CLUSTAL W" at the prompt within the appropriate directory.
3. In CLUSTAL W menu select option 1.
4. Enter the file name of the sequences; the screen will be back to the main menu.
5. Align the sequences by selecting option 2 and then choosing option 1 inside the multiple alignment menu.
6. Go to main menu of CLUSTAL W and choose option 4 for Phylogenetic trees.
7. Choose option 4 inside the phylogenetic tree menu and choose output file name.
8. Execute the draw tree command.
9. Go to main menu, select option X, and exit CLUSTAL W.
10. Choose the retree program from Phylip package; in UNIX type ">retree" prompt within the directory.
11. Type "Y" and as input give the tree file from CLUSTAL W; use "?" to find all the option.
12. To find the node number of the sequence use page up and down and choose a node as the out group.
13. Exit retree after writing the tree with the new root.
14. To draw the tree using diagram command in the UNIX type "drawgram" prompt.
15. The program will ask for the input file name; give the file name "outtree."
16. Enter the file name of the tree from CLUSTAL W.
17. Then the program will ask for font file name; the user should have the font file font 1 among the files.
18. Phylip will take to a series of dragram menus after giving the path and file name of font 1.
19. To see the postscript of the phylogenetic tree choose L; choose N to preview the tree.
20. Choose 1 from the main drawgram menu and choose phylogram.
21. Give option P, select 4, and provide an angle of 90 ° to get the standard format of phylogenetic trees.
22. The output of phylogenetic tree will be given in plot file in ps format.
23. View the phylogenetic tree using the command "ghostview."

[1]Available from http://www.clustal.org/clustal2.

T-Coffee

T-Coffee is a MSA program that provides improvement in accuracy and speed as compared to other alternatives [61]. T-Coffee is used to align nucleic acid and protein sequences, to compare alignments, reformat, and also allows to combine results obtained from several alignment methods and produces new MSAs. To install T-Coffee user need to have GCC, G77, CPAN, internet connection, and root password to install SOAP [62]. Gathering of the pairwise alignment is the main step for T-Coffee method. This collection is called as library. After computation of library, they can be pooled and used to compute MSA. Mocca is a special mode of T-Coffee that extracts a series of repetition from a sequence [61].

1. Download the T-Coffee program.[2]
2. Type "uncompress distribution.tar.Z" prompt to install the program.
3. Install CLUSTAL W [3] if not available.
4. Indicate the address and name of the CLUSTAL W on system.
5. Set the global variable CLUSTAL W_4_ TCOFFEE to "path/name_of_CLUSTAL W": Setenv CLUSTAL W_4_TCOFFEE "path/ name_of_CLUSTAL W."
6. Go to the main directory and type "./install."
7. Appearance of "Installation of t_coffee Successful" on the screen indicates successful completion of installation.
8. Add the bin folder to path "set path=($path.<address of the t_coffee bin folders>)."
9. Type the sequence (Swiss-prot, Fasta, or Pir) prompt "t_coffee sample_seq1.fasta" (see Note 1).
10. Type the sequence in same file "mocca<sequence>."
11. The output of this file will be "<sequence>. mocca_lib."

---

[2]Available at http://www.tcoffee.org/Projects_home_page/ t_coffee_home_page.html.
[3]from http://www.ebi.ac.uk/clustal.

Bayesian Estimation of Species Trees

Bayesian Estimation of Species Trees (BEST) (version 2.3) is a phylogenetic package of programs written by Liang Liu to estimate the posterior distribution of species trees using multilocus molecular data that accounts for deep coalescence of alleles [63, 64]. All the BEST parameters are usually defined in the popular Bayesian phylogenetic package, MrBayes using the preset command [65, 66]. The program estimates both the posterior joint distribution of gene trees and the posterior distribution of the species trees jointly in one Markov Chain Monte Carlo (MCMC) algorithm. BEST can also estimate gene trees and the species trees with divergence times and population sizes. The posterior distribution can be summarized in a program like MrBayes or PAUP [67–70].

1. Compile the program by simply typing "make" in the folder where the source code is available (see Note 2).
2. Run BEST 2.3 by simply typing "./best" which will prompt a working environment where one can type commands to manipulate and analyze aligned sequences.
3. Alternatively, use the command line "./best -idata.nex" in which commands in the MrBayes block will be executed by the program to analyze pre-prepared input file "data.nex."
4. Create the input file for BEST 2.3 in NEXUS format consisting data block and MrBayes block.
5. Concatenate the multilocus sequences across loci in the data block as in the regular MrBayes.
6. Replace the missing nucleotides/sequences with question marks.
7. Duplicate the haploid genes if any compatible to the diploid genes (see Note 3).
8. Otherwise, randomly choose one of the two sequences for the diploid genes compatible to the haploid genes.
9. Command substitution models and prior distributions at gene tree level.
10. Set prior distribution for the species tree, population sizes, and variable mutation rates across genes.

11. Define the location of each gene by the command CHARSET (see Note 4).

12. Divide sequences into genes specified by CHARSET by activating the "partition" command ("set partition=gene").

13. Set the sequence-species relationship by the command "TAXSET" which tells the program which sequences belong to which species.

14. Specify the substitution models for genes for each partition by the command "lset"

15. Set priors for the parameters in the substitution model.

16. Set priors for the species tree, mutation rates across genes and population sizes in the command "prset."

17. BEST 2.3 produces .p, .mcmc and .t files for which the description can be found in the MrBayes manual with species trees generated from the posterior distribution and saved in .sptree format.

18. For multiple runs, it also produces a .sptree file for each run.

19. Summarize the estimated posterior distribution of the species tree by the command "sumt."

MrBayes

MrBayes (version 3.1) is a program for the Bayesian inference of phylogeny that utilizes MCMC simulation [71] in combination with the chosen model and data to estimate the posterior probability distribution of trees, which eliminates much of the complex summation and integration and leaves comparatively simple calculations [72]. Typically, the posterior probability of phylogenies cannot be calculated analytically. However, it can be approximated by sampling trees from the posterior probability distribution. MCMC can be used to sample phylogenies according to their posterior probabilities and the Metropolis-Hastings-Green algorithm [71, 73] has now been used successfully to approximate the posterior probabilities of trees [74, 75]. This plain-vanilla program has a command-line interface to run on a variety of computer platforms including Macintosh, Windows, and Unix. Depending on the size of the data matrix, the

computer should be reasonably fast and should have a lot of memory. The program implements a wide variety of evolutionary models by generating phylogenetic trees for nucleotide, amino acid, restriction site (binary), and standard discrete data [65, 70, 76, 77].

1. Download and install the current version of MrBayes.[4]

2. Start by double-clicking the MrBayes application icon or type "./mb" to execute the program (see Note 5).

3. Change the size of MrBayes window to make it easier to read the output.

4. Type "execute fungi.nex" at the MrBayes>prompt to bring the data into the program (see Note 6).

5. Type "lset nst=6 rates=invgamma" at the MrBayes>prompt, to specify the evolutionary model that will be used in the analysis with gamma-distributed rate variation across sites and a proportion of invariable sites.

6. Set the priors of topology, branch lengths, four stationary frequencies of the nucleotides, six different nucleotide substitution rates, proportion of invariable sites and the shape parameter of the gamma distribution of rate variation for the model.

7. Type "showmodel" to check the model before starting the analysis which will give an overview of the model settings.

8. Review the run settings by typing "help mcmc" to start the analysis.

9. Type "mcmc" to run the analysis and print the state of the chains for every 100th generation after the initial log likelihoods.

10. Type "mcmc ngen=10,000 samplefreq=10" at the MrBayes>prompt, which will ensure to get at least 1,000 samples from the posterior probability distribution.

11. Stop the run by answering "no" when the program asks "Continue the analysis? (yes/no)" if the standard deviation of split frequencies is below 0.01 after 100,000 generations.

---

[4]Available from http://morphbank.ebc.uu.se/mrbayes/.

12. Type "sump burnin=250" (value corresponds to 25 % of samples) to summarize the parameter values as table including mean, mode, and 95 % credibility interval of each parameter.

13. Type "sumt burnin=250" (or value corresponds to 25 % of samples) to summarize the trees as cladogram with posterior probabilities for each split and a phylogram with mean branch lengths.

14. Print the trees as a file that can be read by tree drawing programs such as TreeView, MacClade, and Mesquite.

Phylogenetic Analysis Using Parsimony

Phylogenetic Analysis Using Parsimony (PAUP*) (version 4.0) is one of the most widely used gold standard optimization methods for phylogenetic tree reconstruction using algorithms such as UPGMA and neighbor joining through parsimony, distance-based and maximum likelihood methods [78]. It is a commercial program available in variety of platforms and licensed by Sinauer Associates, Sunderland, MA [5] for inferring and interpreting phylogenetic trees in both interactive and batch mode. The basic procedure uses the parsimony approach to infer phylogenetic tree and can be used for DNA, RNA, and protein sequences [79]. The process of maximum-likelihood tree search using PAUP* includes essential steps as getting a tree, selecting a model of DNA substitution, and searching for the optimal tree under the selected model. Moreover, this program can be used to infer phylogenetic trees using several maximum-likelihood models for DNA and RNA sequences only, but not provide maximum-likelihood models for amino acid substitution. Additionally, it also includes many useful tools for visualizing and examining trees. The initial data file with aligned DNA sequences in NEXUS, PHYLIP, or PIR format can be recog-

nized by PAUP* [80]. PAUP* default setting are chosen in general as they are compatible with all the data types which can be read by the program. In addition to reconstructing phylogenies, the program is also useful in diagnosing characters, inferring ancestral states, and testing the robustness of phylogenetic trees using several statistical techniques [81–87].

1. Purchase and download the most current version of PAUP.[6]

2. Double-click on the application file to start PAUP* in Macintosh and Windows or type "./paup," where as type "paupx.x" in the command line for Unix-like environments.

3. Set the initial mode to Execute and then select the batch file modelblock.

4. Select Import Data from the File Menu and execute NEXUS file with aligned sequences in the PAUP* interface (File>Execute).

5. Specify the sites to be included in the analysis by excluding noncoding regions (include coding/only;).

6. Specify the sequences (taxa) to be used.

7. Set an optimality criterion to parsimony for selecting a tree and define assumptions (set criterion=parsimony;).

8. Set character weighting (weights 2:2ndpos;) and character types (ctype 2_1:all;).

9. Check current character settings (cstatus;) before going on to the search for a tree.

10. Define search strategy as exact and heuristic (hsearch addseq=random;).

11. PAUP* will start processing the command modelblock first to estimate a neighbor-joining tree.

12. PAUP* will then calculate the parameter estimates corresponding to the best-fit model based on likelihood scores for 56 substitution models which will take few minutes to even days depending on the complexity of the data.

13. PAUP* will create two files during the run as sample.scores and sample.log, which is intended to check everything has proceeded normally.

[5]http://www.sinauer.com.

[6]Available from http://paup.csit.fsu.edu/index.html.

14. Define the outgroup sequence and display the tree found by the search (show trees all;) with the branching order of the sequences.
15. Save or print a high-resolution tree (savetrees file = parsTree.tre brlens = yes;) and exit the program (quit;).
16. Interpret the results to find out the best-fit model of nucleotide substitution.

MacClade

MacClade (version 4.08) is a compact, easy-to-use phylogenetic program written by David Maddison and Wayne Maddison for the purpose of analyzing character evolution, entering and editing of phylogenetic data for MSAs and for producing tree diagrams and charts. It has been widely used to map matrotrophy indices onto the mitochondrial DNA phylogeny and for assembling data matrices for downstream phylogenetic analyses. It allows to move branches of the tree around, build own trees, explore evolutionary patterns, and trace the evolution of characters on different trees [88]. The program provides an interactive environment to manipulate hypothesized phylogenetic trees and to visualize character evolution upon them. In order to manipulate the tree, many tools are provided to move branches, reroot clades, create polytomies, and search automatically for more parsimonious trees [89]. The summaries of changes in all characters are depicted on the tree in graphics and charts with statistics such as number of trees of each length, number of characters on the tree with different consistency indices and so on. There are also charts exclusively designed for DNA/RNA sequence data showing the number of changes on the tree, codon positions, and the relative frequencies of various transitions and transversions [90–93]. The MacClade's data editor has numerous features including rows, columns, and blocks of data for manipulating and recoding systematic and comparative data in tune with the abundance of molecular data. Several display features and tools were also available in the editor specifically for graphical manipulation and alignment of molecular sequences [94–101].

1. Download MacClade.[7]
2. Open the MacClade 4.08 folder and then double-click on the sample file.
3. Import the file by opening it in MacClade.
4. Data matrix appears something like an Excel file with taxa in the left column, characters along the top of each column (indicated by a number) and character states (T, A, C, or G) in each cell.
5. Check the entire data set using the scroll bar at the bottom.
6. Click the Display menu and select Go To Tree Window to see the tree.
7. Choose the file that represents the most parsimonious tree by selecting "Open tree file" in the dialog box.
8. Click the "Σ" menu to select "Tree Changes" and repeat with Consistency Index (see Note 7).
9. Click the "Trace" menu and select "Trace Character" which will highlight each branch on which the different character states have evolved in color.
10. Drag the branches around the screen to create new tree topologies.
11. Click through the characters until getting an informative character with two colors on the tree.
12. Examine the total number of changes that occur along each branch by going to the Trace menu and choosing "Trace all changes" which will be illustrated by colors on different branches.
13. Alternatively, view the number of changes directly by changing the "trace all changes option" under the Trace menu.
14. Click "graphics options" and change the setting to "label by amount of change."
15. Print out the tree with the amounts of change labeled.
16. Identify the two "longest" and two "shortest" branches in the printout.
17. Pull down the Trees menu and select Save Trees at the bottom to save the tree.
18. Choose Quit from the File menu to quit MacClade.

_____

[7]Freely available from http://macclade.org/download.html.

TREE-PUZZLE

TREE-PUZZLE (version 5.2) is a fast tree search algorithm that provides methods for reconstruction, comparison and testing of trees and models on DNA as well as protein sequences using quartets and parallel computing [102, 103]. The program computes pairwise maximum likelihood distances as well as branch lengths with and without the molecular-clock assumption for user specified trees [104, 105]. It also offers likelihood mapping to visualize the phylogenetic content of a sequence alignment by investigating the support of internal branches without computing an overall tree and conducts chi-square test for homogeneity of base composition, likelihood ratio to test clock hypothesis, one and two-sided Kishino-Hasegawa test, Shimodaira-Hasegawa test, and Expected Likelihood Weights on the data set [106–109]. In addition, the program computes the number and the percentage of completely unresolved maximum likelihood quartets which is an indicator of the suitability of the data for phylogenetic analysis [110]. In order to avoid overflow of internal integer variables, the program has a built-in limit to allow data sets only up to 257 sequences and TREE-PUZZLE terminates the program execution if problems are encountered during the data analysis [111–113].

1. Download and install the recent version of TREE-PUZZLE.[8]
2. Type "b" to switch between tree reconstruction by maximum likelihood and likelihood mapping.
3. Open the "data" directory to load sequence input file in CLUSTAL W or PHYLIP output format or tree input file in DRAWTREE or DRAWGRAM output format.
4. Type "y" at the input prompt to start the analysis.
5. TREE-PUZZLE computes pairwise maximum likelihood distances for all the sequences in the data file automatically.
6. TREE-PUZZLE displays quartet puzzling tree with its support values and maximum likelihood branch lengths in "INFILENAME. puzzle"/"outfile."
7. View the tree both with its branch lengths and with the support values for the internal branches using TreeView and TreeTool.
8. Print the tree topology along with branch lengths.
9. Type "q" to quit the analysis.

TreeView

TreeView (version 1.6) is a simple program useful for displaying and printing phylogenies. This software runs on almost all identical interfaces, reads many different tree file formats including NEXUS, PHYLIP, Hennig86, NONA, MEGA and CLUSTAL W and supports native graphics file format for copying pictures into other applications and for saving graphics files. The current version reads trees with up to 1,000 taxa and provides tree editor tools for moving branches, rerooting, polytomy formation, and rearranging the appearance of the tree. TreeView also displays a scale bar in the bottom left corner of the tree window if the tree being viewed as branch lengths, unrooted or as a phylogram [114].

1. Obtain the current version of TREEVIEW.[9]
2. Double-click on the data file icon to launch TREEVIEW.
3. Right click on TREEVIEW document to display context menu containing both an Open and an Edit command.
4. Load the file into TREEVIEW by choosing "Open."
5. TREEVIEW will display a single tree in the tree window with tree's name on the status bar.
6. Enable "Previous" and "Next" buttons to browse among the trees in the file, if it contains more than one tree.

---

[8]Available from http://www.tree-puzzle.de/.

[9]Available from http://taxonomy.zoology.gla.ac.uk/rod/treeview.html.

7. Choose "Edit" to open the file in the program Notepad to quickly see the tree file in a text editor.

8. Click tree editor to display the tools to move branches, reroot, form polytomy, and rearrange the appearance of the tree.

9. Click on the icon to select a tool which will change to the appropriate shape.

10. Click the cursor on a branch and drag it to the new position to manipulate the tree.

11. Specify the font used as Plain, Bold, Italic, Size, Font Type when drawing the tree.

12. Choose "Undo" from the "Edit" menu to undo a change and recover the old tree, if essential (see Note 8).

13. Save trees to different file formats.

14. Click Print preview to view how the tree will appear on the printed page.

15. Set the page orientation and choose a printer to print the tree.

16. Click Quit to exit TREEVIEW.

Notes

1. User will get two output files as a multiple alignment (sample_seq1.aln) and a dendrogram (sample_seq1.dnd).

2. Make sure that Architecture in the file Makefile is correctly set to the platform available.

3. If the dataset contains both diploid genes (nucleotide DNA) and haploid genes (mtDNA).

4. For example, CHARSET gene1 = 1–400 which indicates the first 400 nucleotides belong to the gene, gene1.

5. Type "help" or "help <command>" for information on commands that are available.

6. The data file (fungi.nex) must be in the same directory in the MrBayes program and must have aligned nucleotide or amino acid sequences, morphological ("standard") data, restriction site (binary) data or any mix of these data types in Nexus file format.

7. Consistency Index is a ratio of the number of characters in the data set to the length of the current tree displayed and measures how good the tree is at evolving those characters.

8. Undo is only available for move branch, collapse branch, collapse clade and reroot; cosmetic changes such as rotate branches cannot be undone.

References

1. Kolbert CP, Persing DH (1999) Ribosomal DNA sequencing as a tool for identification of bacterial pathogens. Curr Opin Microbiol 2:299–305
2. Abd-Elsalam KA (2003) Bioinformatics tools and guideline for PCR primer design. Afr J Biotechnol 2:91–95
3. Yan PV (2005) Bioinformatics: new research. Nova Science Publishers, New York
4. Pevsner J (2009) Bioinformatics and functional genomics. John Wiley & Sons, New York, pp 1–13
5. Giovannoni SJ, Britschgi TB, Moyer CL, Field KG (1990) Genetic diversity in Sargasso Sea bacterioplankton. Nature 345:60
6. Gardes M, Bruns T (1993) ITS primers with enhanced specificity for Basidiomycetes—application to the identification of mycorrhizae and rusts. Mol Ecol 2:113–118
7. Seifert KA (2009) Progress towards DNA barcoding of fungi. Mol Ecol Resour 9:83–89
8. Muller GM, Bills GF, Foster MS (2004) Biodiversity of fungi: inventory and monitoring methods. Elsevier Academic, San Diego, CA, 341
9. Thomas JW, Touchman JW, Blakesley RW, Bouffard GG, Beckstrom-Sternberg SM, Margulies EH et al (2003) Comparative analyses of multi-species sequences from targeted genomic regions. Nature 424:788–793
10. Tripathi KK (2000) Bioinformatics: the foundation of present and future biotechnology. Curr Sci 79:570–575
11. Jones NC, Zhi D, Raphael BJ (2006) AliWABA: alignment on the web through an A-Bruijn approach. Nucleic Acids Res 34:613–616
12. Raphael B, Zhi S, Tang H, Pevzner P (2004) A novel method for multiple alignment of sequences with repeated and shuffled elements. Genome Res 14:2336–2346
13. Schwartz S, Zhang Z, Frazer KA, Smit A, Riemer C, Bouk J et al (2000) PipMaker—a web server for aligning two genomic DNA sequences. Genome Res 10:577–586
14. Baxevanis AD, Ouellette BFF (2001) Bioinformatics: a practical guide to the analysis of genes and proteins. John Wiley & Sons, Inc., New York
15. Schwartz A, Pachter L (2006) Multiple alignment by sequencing annealing. Bioinformatics 23: 24–29
16. Madden T (2005) The BLAST sequence analysis tool. In: McEntyre J, Ostell J (eds) NCBI handbook. National Library of Medicine, Bethesda, M. D

17. Parry-Smith DJ, Payne AWR, Michie AD, Attwood TK (1997) CINEMA—a novel colour interactive editor for multiple alignments. Gene 211: GC45–GC56

18. Larkin MA, Blackshields G, Brown NP, Chenna R, McGettigan PA, McWilliam H et al (2007) Clustal W and Clustal X version 2.0. Bioinformatics 23:2947–2948

19. Chenna R, Sugawara H, Koike T, Lopez R, Gibson TJ, Higgins DG et al (2003) Multiple sequence alignment with the Clustal series of programs. Nucleic Acids Res 31:3497–3500

20. Thompson JD, Higgins DG, Gibson TJ (1994) CLUSTAL W: improving the sensitivity of progressive multiple sequence alignment through sequence weighting, position-specific gap penalties and weight matrix choice. Nucleic Acids Res 22:4673–4680

21. Crottini A, Dordel J, Köhler J, Glaw F, Schmitz A, Vences M (2009) A multilocus phylogeny of Malagasy scincid lizards elucidates the relationships of the fossorial genera Androngo and Cryptoscincus. Mol Phylogenet Evol 53:345–350

22. Ye SQ (2008) Bioinformatics: a practical approach. Chapman & Hall/CRC Press, Boca Raton, FL

23. Qi J, Luo H, Hao B (2004) CVTree: a phylogenetic tree reconstruction tool based on whole genomes. Nucleic Acids Res 32:W45–W47

24. Stoye J (1997) Divide-and-Conquer multiple sequence alignment. Dissertation thesis, Universitat Bielefeld, Bielefeld, Germany

25. Stoye J, Perrey SW, Dress AWM (1997) Improving the Divide-and-Conquer approach to sum-of-pairs multiple sequence alignment. Appl Math Lett 10:67–73

26. Brinkmann A, Dress AMW, Perrey SW, Stoye J (1997) Two applications of the divide & conquer principle in the molecular sciences. Math Program 79:71–97

27. Stoye J, Moulton V, Dress AWM (1997) DCA: an efficient implementation of the Divide-and-Conquer multiple sequence alignment algorithm. CABIOS 13:625–626

28. Stoye J (1998) Multiple sequence alignment with the Divide-and-Conquer method. Gene 211:GC45–GC56

29. Chikkagoudar S, Roshan U, Livesay DR (2007) eProbalign: generation and manipulation of multiple sequence alignments using partition function posterior probabilities. Nucleic Acids Res 35:W675–W677

30. Gouet P, Courcelle E, Stuart DI, Metoz F (1999) ESPript: multiple sequence alignments in PostScript. Bioinformatics 15:305–308

31. Jaroszewski L, Li Z, Cai XH, Weber C, Godzik A (2011) FFAS server: novel features and applications. Nucleic Acids Res 39:W38–W44

32. Jaroszewski L, Rychlewski L, Li Z, Li W, Godzik A (2005) FFAS03: a server for profile-profile sequence alignments. Nucleic Acids Res 33:W284–W288

33. Torarinsson E, Havgaard JH, Gorodkin J (2007) Multiple structural alignment and clustering of RNA sequences. Bioinformatics 23:926–932

34. Havgaard JH, Lyngsø RB, Gorodkin J (2005) The FOLDALIGN web server for pairwise structural RNA alignment and mutual motif search. Nucleic Acids Res 33:W650–W653

35. Nikhat Z, Mazumder R, Seto D (2001) Comparisons of gene co-linearity in genomes using GeneOrder2.0. Trends Biochem Sci 26:514–516

36. Mazumder R, Kolaskar A, Seto D (2001) GeneOrder compares the order of genes in small genomes. Bioinformatics 17:162–166

37. Chalmel F, Lardenois A, Thompson JD, Muller J, Sahel JA, Léveillard T et al (2005) GOAnno: GO annotation based on multiple alignment. Bioinformatics 21:2095–2096

38. Brodie R, Roper RL, Upton C (2004) JDotter: a Java interface to multiple dotplots generated by dotter. Bioinformatics 20:279–281

39. Huang X, Miller W (1991) The lalign program implements the algorithm of Huang and Miller. Adv Appl Math 12:337–357

40. Barker D (2004) LVB: parsimony and simulated annealing in the search for phylogenetic trees. Bioinformatics 20:274–275

41. Toh K (2008) Recent developments in the MAFFT multiple sequence alignment program. Brief Bioinform 9:286–298

42. Toh K (2010) Parallelization of the MAFFT multiple sequence alignment program. Bioinformatics 26: 1899–1900

43. Zhu H, Schein CH, Braun W (2000) MASIA: recognition of common patterns and properties in multiple aligned protein sequences. Bioinformatics 16:950–951

44. Sebastian W, Reiche K, Hofacker IL, Stadler PF, Backofen R (2007) Inferring non-coding RNA families and classes by means of genome-scale structure-based clustering. PLoS Comput Biol 3(4):e65

45. Stothard P (2000) The sequence manipulation suite: JavaScript programs for analyzing and formatting protein and DNA sequences. Biotechniques 28:1102–1104

46. Lassmann T, Sonnhammer ELL (2006) Kalign, Kalignvu and Mumsa: web servers for multiple sequence alignment. Nucleic Acids Res 34: W596–W599

47. Edgar RC (2004) MUSCLE: multiple sequence alignment with high accuracy and high throughput. Nucleic Acids Res 32:1792–1797

48. Tsai YT, Huang YP, Yu CT, Lu CL (2004) MuSiC: a tool for multiple sequence alignment with constraints. Bioinformatics 20:2309–2311

49. DeSantis TZ Jr, Hugenholtz P, Keller K, Brodie EL, Larsen N, Piceno YM et al (2006) NAST: a multiple sequence alignment server for comparative analysis of 16S rRNA genes. Nucleic Acids Res 34: W394–W399

50. Rognes T (2001) ParAlign: a parallel sequence alignment algorithm for rapid and sensitive database searches. Nucleic Acids Res 29:1647–1652

51. Sæbø PE, Andersen SM, Myrseth J, Laerdahl JK, Rognes T (2005) PARALIGN: rapid and sensitive

sequence similarity searches powered by parallel computing technology. Nucleic Acids Res 33:W535–W539

52. Rognes T, Andersen SM (2005) PARALIGN user's guide. Sencel Bioinformatics AS, Oslo, Norway

53. Notredame C, Higgins DG, Heringa J (2000) T-Coffee: a novel method for fast and accurate multiple sequence alignment. J Mol Biol 302:205–217

54. Mignone F, Horner DS, Pesole G (2004) WebVar: a resource for the rapid estimation of relative site variability from multiple sequence alignments. Bioinformatics 20:1331–1333

55. Noe L, Kucherov G (2005) YASS: enhancing the sensitivity of DNA similarity search. Nucleic Acids Res 33:W540–W543

56. Thompson JD, Gibson TJ, Plewniak F, Jeanmougin F, Higgins DG (1997) The CLUSTAL X windows interface: flexible strategies for multiple sequence alignment aided by quality analysis tools. Nucleic Acids Res 25:4876–4882

57. Higgins DG, Sharp PM (1988) CLUSTAL: a package for performing multiple sequence alignment on a microcomputer. Gene 73:237–244

58. Sze SH, Lu Y, Yang Q (2006) A polynomial time solvable formulation of multiple sequence alignment. J Comput Biol 13:309–319

59. Higgins DG, Bleasby AJ, Fuchs R (1992) CLUSTAL V: improved software for multiple sequence alignment. Comput Appl Biosci 8:189–191

60. Higgins DG, Thompson JD, Gibson TJ (1996) Using CLUSTAL for multiple sequence alignments. Methods Enzymol 266:383–402

61. Notredame C (2001) User documentation and F.A.Q. T-COFFEE (1.35) MOCCA. http://www.bioinformatics.nl/tools/t_coffee_doc.html. Accessed 10 Sept 2011

62. Poirot O, O'Toole E, Notredame C (2003) Tcoffee@igs: a web server for computing, evaluating and combing multiple sequence alignment. Nucleic Acids Res 31:3503–3506

63. Liu L, Pearl DK (2007) Species trees from gene trees: reconstructing Bayesian posterior distributions of a species phylogeny using estimated gene tree distributions. Syst Biol 56:504–514

64. Edwards SV, Liu L, Pearl DK (2007) High resolution species trees without concatenation. PNAS 104:5936–5941

65. Ronquist F, Huelsenbeck JP (2003) MrBayes version 3.0: Bayesian phylogenetic inference under mixed models. Bioinformatics 19:1572–1574

66. Liu L (2008) BEST: Bayesian estimation of species trees under the coalescent model. Bioinformatics 24:2542–2543

67. Liu L, Pearl DK, Brumfield RT, Edwards SV (2008) Estimating species trees using multiple-allele DNA sequence data. Evolution 62:2080–2091

68. Brito PH, Edwards SV (2009) Multilocus phylogeography and phylogenetics using sequence-based markers. Genetica 135:439–455

69. Nichols R (2001) Gene trees and species trees are not the same. Trends Ecol Evol 16:358–364

70. Rannala B, Yang Z (2003) Bayes estimation of species divergence times and ancestral population sizes using DNA sequences from multiple loci. Genetics 164:1645–1656

71. Metropolis N, Rosenbluth AW, Rosenbluth MN, Teller AH, Teller E (1953) Equations of state calculations by fast computing machines. J Chem Phys 21:1087–1091

72. Huelsenbeck JP (2002) Testing a covariotide model of DNA substitution. Mol Biol Evol 19:698–707

73. Hastings WK (1970) Monte Carlo sampling methods using Markov chains and their applications. Biometrika 57:97–109

74. Larget B, Simon D (1999) Markov chain Monte Carlo algorithms for the Bayesian analysis of phylogenetic trees. Mol Biol Evol 16:750–759

75. Yang Z, Rannala B (1997) Bayesian phylogenetic inference using DNA sequences: a Markov chain Monte carlo method. Mol Biol Evol 14:717–724

76. Romeralo M, Baldauf SL, Cavender JC (2009) A new species of cellular slime mold from southern Portugal based on morphology, ITS and SSU sequences. Mycologia 101:269–274

77. Huelsenbeck JP, Ronquist F (2001) MrBayes: Bayesian inference in phylogenetic trees. Bioinformatics 17:754–755

78. Swofford DL, Olsen GJ, Waddell PJ, Hillis DM (1996) Phylogenetic inference. In: Hillis DM, Moritz C, Mable BK (eds) Molecular systematics. Sinauer Associates, Sunderland, MA, pp 407–514

79. Wilgenbusch JC, Swofford D (2003) Inferring evolutionary trees with PAUP*. Curr Protoc Bioinformatics 6:21–28

80. Swofford DL (2002) PAUP*. Phylogenetic analysis using parsimony (*and other methods). Sinauer Associates, Sunderland, USA

81. Overton BE, Stewart EL, Geiser DM, Jaklitsch WM (2006) Systematics of *Hypocrea citrina* and related taxa. Stud Mycol 56:11–38

82. Huang T, Yeh Y, Tzeng DD (2010) Heteroduplex mobility assay for identification and phylogenetic analysis of anthracnose fungi. J Phytopathol 158:46–55

83. Grünig CR, Duò A, Sieber T, Holdenrieder O (2008) Assignment of species rank to six reproductively isolated cryptic species of the *Phialocephala fortinii* s.l.-*Acephala applanata* species complex. Mycologia 100:47–67

84. Sung G, Sung J, Hywel-Jones NL, Spatafora JW (2007) A multi-gene phylogeny of Clavicipitaceae (Ascomycota, Fungi): identification of localized incongruence using a combinational bootstrap approach. Mol Phylogenet Evol 44:1204–1223

85. Brenn N, Menkis A, Grünig CR, Sieber TN, Holdenrieder O (2008) Community structure of *Phialocephala fortiniis. lat.* in European tree nurseries, and assessment of the potential of the seedlings as dissemination vehicles. Mycol Res 112:650–662

86. Bomberg M, Timonen S (2009) Effect of tree species and mycorrhizal colonization on the archaeal

population of Boreal Forest rhizospheres. Appl Environ Microbiol 75:308–315

87. Grebenc T, Martín MP, Kraigher H (2009) Ribosomal ITS diversity among the European species of the genus *Hydnum* (Hydnaceae). Anales del Jardín Botánico de Madrid 66S1:121–132

88. Maddison DR, Maddison WP (2000) MacClade 4: analysis of phylogeny and character evolution. Version 4.0. Sinauer Associates, Sunderland, Massachusetts

89. Simmons MP, Bailey CD, Nixon KC (2000) Phylogeny reconstruction using duplicate genes. Mol Biol Evol 17:469–473

90. Wang Z, Nilsson RH, Lopez-Giraldez F, Zhuang W, Dai Y, Johnston PR et al (2011) Tasting soil fungal diversity with earth tongues: phylogenetic test of SATé alignments for environmental ITS data. PLoS One 6:e19039

91. James TY, Letcher PM, Longcore JE, Mozley-Standridge SE, Porter D, Powell MJ et al (2006) A molecular phylogeny of the flagellated fungi (Chytridiomycota) and description of a new phylum (Blastocladiomycota). Mycologia 98:860–871

92. Little AEF, Currie CR (2007) Symbiotic complexity: discovery of a fifth symbiont in the attine ant–microbe symbiosis. Biol Lett 3:501–504

93. Kim SY, Park SY, Ko KS, Jung HS (2003) Phylogenetic analysis of *Antrodia* and related taxa based on partial mitochondrial SSU rDNA sequences. Antonie Van Leeuwenhoek 83:81–88

94. Suh S, Noda H, Blackwell M (2001) Insect symbiosis: derivation of yeast-like endosymbionts within an entomopathogenic filamentous lineage. Mol Biol Evol 18:995–1000

95. Schultz TR, Brady SG (2008) Major evolutionary transitions in ant agriculture. PNAS 105: 5435–5440

96. Reeb V, Lutzoni F, Roux C (2004) Contribution of *RPB 2* to multilocus phylogenetic studies of the euascomycetes (Pezizomycotina, Fungi) with special emphasis on the lichen-forming Acarosporaceae and evolution of polyspory. Mol Phylogenet Evol 32:1036–1060

97. Lutzoni F, Kauff F, Cox CJ, McLaughlin D, Celio G, Dentinger B et al (2004) Assembling the fungal tree of life: progress, classification, and evolution of subcellular traits. Am J Bot 91:1446–1480

98. Bidochka MJ, St Leger RJ, Stuart A, Gowanlock K (1999) Nuclear rDNA phylogeny in the fungal genus *Verticillium* and its relationship to insect and plant virulence, extracellular proteases and carbohydrases. Microbiology 145:955–963

99. U'ren JM, Dalling JW, Gallery RE, Maddison DR, Davis EC, Gibson CM et al (2009) Diversity and evolutionary origins of fungi associated with seeds of a neotropical pioneer tree: a case study for analysing fungal environmental samples. Mycol Res 113:432–449

100. Berbee ML (2001) The phylogeny of plant and animal pathogens in the Ascomycota. Physiol Mol Plant Pathol 59:165–187

101. Gerardo NM, Mueller UG, Price SL, Currie CR (2004) Exploiting a mutualism: parasite specialization on cultivars within the fungus-growing ant symbiosis. Proc Biol Sci 271:1791–1798

102. Schmidt HA (2009) Testing tree topologies. In: Lemey P, Salemi M, Vandamme AM (eds) The phylogenetic handbook: a practical approach to phylogenetic analysis and hypothesis testing. Cambridge University Press, Cambridge, U.K., pp 381–404

103. Schmidt HA, von Haeseler A (2009) Phylogenetic inference using maximum likelihood methods. In: Lemey P, Salemi M, Vandamme AM (eds) The phylogenetic handbook: a practical approach to phylogenetic analysis and hypothesis testing. Cambridge University Press, Cambridge, U.K., pp 181–209

104. Felsenstein J (1981) Evolutionary trees from DNA sequences: a maximum likelihood approach. J Mol Evol 17:368–376

105. Felsenstein J (1988) Phylogenies from molecular sequences: inference and reliability. Annu Rev Genet 22:521–565

106. Kishino H, Hasegawa M (1989) Evaluation of the maximum likelihood estimate of the evolutionary tree topologies from DNA sequence data, and the branching order in Hominoidea. J Mol Evol 29:170–179

107. Strimmer K, von Haeseler A (1996) Quartet puzzling: a quartet maximum likelihood method for reconstructing tree topologies. Mol Biol Evol 13:964–969

108. Strimmer K, Goldman N, von Haeseler A (1997) Bayesian probabilities and quartet puzzling. Mol Biol Evol 14:210–213

109. Strimmer K, von Haeseler A (1997) Likelihood-mapping: a simple method to visualize phylogenetic content of a sequence alignment. Proc Natl Acad Sci U S A 94:6815–6819

110. Trelles O (2001) On the parallelisation of bioinformatics applications. Brief Bioinform 2:181–194

111. Schmidt HA, Petzold E, Vingron M, von Haeseler A (2003) Molecular phylogenetics: parallelized parameter estimation and quartet puzzling. J Parallel Distrib Comput 63:719–727

112. Schmidt HA, Strimmer K, Vingron M, von Haeseler A (2002) TREE-PUZZLE: maximum likelihood phylogenetic analysis using quartets and parallel computing. Bioinformatics 18:502–504

113. Petzold E, Merkle D, Middendorf M, von Haeseler A, Schmidt HA (2006) Phylogenetic parameter estimation on COWs. In: Zomaya AY (ed) Parallel computing for bioinformatics and computational biology. John Wiley & Sons, New York, pp 347–368

114. Page RDM (1996) TREEVIEW: an application to display phylogenetic trees on personal computers. Comput Appl Biosci 12:357–358